Lecture Notes in Artificial Intelligence   8444

Subseries of Lecture Notes in Computer Science

LNAI Series Editors

Randy Goebel
  University of Alberta, Edmonton, Canada
Yuzuru Tanaka
  Hokkaido University, Sapporo, Japan
Wolfgang Wahlster
  DFKI and Saarland University, Saarbrücken, Germany

LNAI Founding Series Editor

Joerg Siekmann
  DFKI and Saarland University, Saarbrücken, Germany

T0139962

Vincent S. Tseng   Tu Bao Ho
Zhi-Hua Zhou   Arbee L.P. Chen
Hung-Yu Kao (Eds.)

Advances in Knowledge Discovery and Data Mining

18th Pacific-Asia Conference, PAKDD 2014
Tainan, Taiwan, May 13-16, 2014
Proceedings, Part II

 Springer

Volume Editors

Vincent S. Tseng
National Cheng Kung University, Tainan, Taiwan, R.O.C.
E-mail: tsengsm@mail.ncku.edu.tw

Tu Bao Ho
Japan Advanced Institute of Science and Technology, Nomi, Ishikawa, Japan
E-mail: bao@jaist.ac.jp

Zhi-Hua Zhou
Nanjing University, China
E-mail: zhouzh@nju.edu.cn

Arbee L.P. Chen
National Chengchi University, Taipei, Taiwan, R.O.C.
E-mail: alpchen@cs.nccu.edu.tw

Hung-Yu Kao
National Cheng Kung University, Tainan, Taiwan, R.O.C.
E-mail: hykao@mail.ncku.edu.tw

ISSN 0302-9743   e-ISSN 1611-3349
ISBN 978-3-319-06604-2   e-ISBN 978-3-319-06605-9
DOI 10.1007/978-3-319-06605-9
Springer Cham Heidelberg New York Dordrecht London

Library of Congress Control Number: 2014936624

LNCS Sublibrary: SL 7 – Artificial Intelligence

Typesetting: Camera-ready by author, data conversion by Scientific Publishing Services, Chennai, India

Printed on acid-free paper

Springer is part of Springer Science+Business Media (www.springer.com)

Preface

PAKDD 2014 was the 18th conference of the Pacific Asia Conference series on Knowledge Discovery and Data Mining. The conference was held in Tainan, Taiwan, during May 13–16, 2014. Since its inception in 1997, the PAKDD conference series has been a leading international conference in the areas of data mining and knowledge discovery. It provides an inviting and inspiring forum for researchers and practitioners, from both academia and industry, to share new ideas, original research results, and practical experience. The 18th edition continues the great tradition with three world-class keynote speeches, a wonderful technical program, a handful of high quality tutorials and workshops, and a data mining competition.

The PAKDD 2014 conference received 371 submissions to the technical program, involving more than 980 authors in total. Each submitted paper underwent a rigorous double-blind review process and was reviewed by at least three Program Committee (PC) members as well as one senior PC member. Based on the extensive and thorough discussions by the reviewers, the senior PC members made recommendations. The Program Co-chairs went through each of the senior PC members' recommendations, as well as the submitted papers and reviews, to come up with the final selection. Overall, 100 papers were accepted in the technical program among 371 submissions, yielding a 27% acceptance rate. 40 of which (10.8%) had full presentations and 60 of which (16.2%) had short presentations. The technical program consisted of 21 sessions, covering the general fields of data mining and KDD extensively. We thank all reviewers (Senior PC, PC and external invitees) for their great efforts in reviewing the papers in a timely fashion. Without their hard work, we would not have been able to see such a high-quality program.

The conference program this year included three keynote talks by world-renowned data mining experts, namely, Professor Vipin Kumar from the University of Minnesota (*Understanding Climate Change: Opportunities and Challenges for Data Driven Research*); Professor Ming-Syan Chen from the National Taiwan University (*On Information Extraction for Social Networks*); Professor Jian Pei from the Simon Fraser University (*Being a Happy Dwarf in the Age of Big Data*). The program also included 12 workshops, which covered a number of exciting and fast growing hot topics. We also had 7 very timely and educational tutorials, covering the hot topics of social networks and media, pattern mining, big data, biomedical and health informatics mining and crowdsourcing. PAKDD 2014 also organized a data mining competition for those who wanted to lay their hands on mining interesting real-world datasets.

Putting together a conference on a scale like PAKDD 2014 requires tremendous efforts from the organizing team as well as financial support from the sponsors. We would like to express our special thanks to our honorary chairs,

Hiroshi Motoda and Philip S. Yu, for providing valuable advice and kind support. We thank Wen-Chih Peng, Haixun Wang, and James Bailey for organizing the workshop program. We also thank Mi-Yen Yeh, Guandong Xu and Seung-Won Hwang for organizing the tutorial program. As well, we thank Shou-De Lin, Nitesh Chawla and Hung-Yi Lo for organizing the data mining competition. We also thank Hung-Yu Kao for preparing the conference proceedings. Finally, we owe a big thank you to the great team of publicity co-chairs, local arrangement co-chairs, sponsorship chair and helpers. They ensured the conference attracted many local and international participants, and the conference program proceeded smoothly.

We would like to express our gratitude to all sponsors for their generous sponsorship and support. Special thanks are given to AFOSR/AOARD (Air Force Office of Scientific Research/Asian Office of Aerospace Research and Development) for their support to the success of the conference. We also wish to thank the PAKDD Steering Committee for offering the student travel support grant.

Finally, we hope you found the conference a fruitful experience and trust you had an enjoyable stay in Tainan, Taiwan.

May 2014   Vincent S. Tseng
                                                                Tu Bao Ho
                                                                Zhi-Hua Zhou
                                                                Arbee L.P. Chen
                                                                Hung-Yu Kao

Organization

Honorary Co-chairs

Hiroshi Motoda   Osaka University, Japan
Philip S. Yu   University of Illinois at Chicago, USA

General Co-chairs

Zhi-Hua Zhou   Nanjing University, China
Arbee L.P. Chen   National Chengchi University, Taiwan

Program Committee Co-chairs

Vincent S. Tseng   National Cheng Kung University, Taiwan
Tu Bao Ho   JAIST, Japan

Workshop Co-chairs

Wen-Chih Peng   National Chiao Tung University, Taiwan
Haixun Wang   Google Inc., USA
James Bailey   University of Melbourne, Australia

Tutorial Co-chairs

Mi-Yen Yeh   Academia Sinica, Taiwan
Guandong Xu   University of Technology Sydney, Australia
Seung-Won Hwang   POSTECH, Korea

Publicity Co-chairs

Takashi Washio   Osaka University, Japan
Tzung-Pei Hong   National University of Kaohsiung,
     Taiwan
Yu Zheng   Microsoft Research Asia, China
George Karypis   University of Minnesota, USA

Proceedings Chair

Hung-Yu Kao   National Cheng Kung University, Taiwan

Contest Co-chairs

Shou-De Lin	National Taiwan University, Taiwan
Nitesh Chawla	University of Notre Dame, USA
Hung-Yi Lo	Shih-Chien University, Taiwan

Local Arrangements Co-chairs

Jen-Wei Huang	National Cheng Kung University, Taiwan
Kun-Ta Chuang	National Cheng Kung University, Taiwan
Chuang-Kang Ting	National Chung Cheng University, Taiwan
Ja-Hwung Su	Kainan University, Taiwan

Sponsorship Chair

Yue-Shi Lee	Ming Chuan University, Taiwan

Registration Co-chairs

Hsuan-Tien Lin	National Taiwan University, Taiwan
Chien-Feng Huang	National University of Kaohsiung, Taiwan

Steering Committee

Chairs

Graham Williams	Australian Taxation Office, Australia
Tu Bao Ho (Co-Chair)	Japan Advanced Institute of Science and Technology, Japan

Life Members

Hiroshi Motoda	AFOSR/AOARD and Osaka University, Japan (Since 1997)
Rao Kotagiri	University of Melbourne, Australia (Since 1997)
Ning Zhong	Maebashi Institute of Technology, Japan (Since 1999)
Masaru Kitsuregawa	Tokyo University, Japan (Since 2000)
David Cheung	University of Hong Kong, China (Since 2001)
Graham Williams (Treasurer)	Australian National University, Australia (Since 2001)
Ming-Syan Chen	National Taiwan University, Taiwan (Since 2002)
Kyu-Young Whang	Korea Advanced Institute of Science & Technology, Korea (Since 2003)

Members

Huan Liu	Arizona State University, USA (Since 1998)
Chengqi Zhang	University of Technology Sydney, Australia (Since 2004)
Tu Bao Ho	Japan Advanced Institute of Science and Technology, Japan (Since 2005)
Ee-Peng Lim	Singapore Management University, Singapore (Since 2006)
Jaideep Srivastava	University of Minnesota, USA (Since 2006)
Zhi-Hua Zhou	Nanjing University, China (Since 2007)
Takashi Washio	Institute of Scientific and Industrial Research, Osaka University, Japan (Since 2008)
Thanaruk Theeramunkong	Thammasat University, Thailand (Since 2009)
P. Krishna Reddy	International Institute of Information Technology, Hyderabad (IIIT-H), India (Since 2010)
Joshua Z. Huang	Shenzhen Institutes of Advanced Technology, Chinese Academy of Sciences, China (Since 2011)
Longbing Cao	Advanced Analytics Institute, University of Technology Sydney, Australia (Since 2013)
Jian Pei	School of Computing Science, Simon Fraser University, Canada (Since 2013)
Myra Spiliopoulou	Information Systems, Otto-von-Guericke-University Magdeburg, Germany (Since 2013)

Senior Program Committee Members

James Bailey	University of Melbourne, Australia
Michael Berthold	University of Konstanz, Germany
Longbing Cao	University of Technology Sydney, Australia
Sanjay Chawla	University of Sydney, Australia
Lei Chen	Hong Kong University of Science and Technology, Hong Kong
Ming-Syan Chen	National Taiwan University, Taiwan
Peter Christen	The Australian National University, Australia
Ian Davidson	UC Davis, USA
Wei Fan	IBM T.J. Watson Research Center, USA
Bart Goethals	University of Antwerp, Belgium
Xiaohua Hu	Drexel University, USA
Ming Hua	Facebook, USA
Joshua Huang	Shenzhen Institutes of Advanced Technology, Chinese Academy of Sciences, China

Program Committee Members

Shafiq Alam	University of Auckland, New Zealand
Aijun An	York University, Canada
Hideo Bannai	Kyushu University, Japan
Gustavo Batista	University of Sao Paulo, Brazil
Bettina Berendt	Katholieke Universiteit Leuven, The Netherlands
Chiranjib Bhattachar	Indian Institute of Science, India
Jiang Bian	Microsoft Research, China
Marut Buranarach	National Electronics and Computer Technology Center, Thailand
Krisztian Buza	University of Warsaw, Poland
Mary Elaine Califf	Illinois State University, USA
Rui Camacho	Universidade do Porto, Portugal
K. Selcuk Candan	Arizona State University, USA
Tru Cao	Ho Chi Minh City University of Technology, Vietnam
James Caverlee	Texas A&M University, USA
Keith Chan	The Hong Kong Polytechnic University, Hong Kong
Chia-Hui Chang	National Central University, Taiwan
Muhammad Cheema	Monash University, Australia
Chun-Hao Chen	Tamkang University, Taiwan
Enhong Chen	University of Science and Technology of China, China
Jake Chen	Indiana University-Purdue University Indianapolis, USA
Ling Chen	University of Technology Sydney, Australia
Meng Chang Chen	Academia Sinica, Taiwan
Shu-Ching Chen	Florida International University, USA
Songcan Chen	Nanjing University of Aeronautics and Astronautics, China
Yi-Ping Phoebe Chen	La Trobe University, Australia
Zheng Chen	Microsoft Research Asia, China
Zhiyuan Chen	University of Maryland Baltimore County, USA
Yiu-ming Cheung	Hong Kong Baptist University, Hong Kong
Silvia Chiusano	Politecnico di Torino, Italy
Kun-Ta Chuang	National Cheng Kung University, Taiwan
Bruno Cremilleux	Universite de Caen, France
Bin Cui	Peking University, China
Alfredo Cuzzocrea	ICAR-CNR and University of Calabria, Italy
Bing Tian Dai	Singapore Management University, Singapore
Dao-Qing Dai	Sun Yat-Sen University, China

Irena Koprinska	University of Sydney, Australia
Walter Kosters	Universiteit Leiden, The Netherlands
Marzena Kryszkiewicz	Warsaw University of Technology, Poland
James Kwok	Hong Kong University of Science and Technology, China
Wai Lam	The Chinese University of Hong Kong, Hong Kong
Wang-Chien Lee	Pennsylvania State University, USA
Yue-Shi Lee	Ming Chuan University, Taiwan
Yuh-Jye Lee	University of Science and Technology, Taiwan
Philippe Lenca	Telecom Bretagne, France
Carson K. Leung	University of Manitoba, Canada
Chengkai Li	The University of Texas at Arlington, USA
Chun-hung Li	Hong Kong Baptist University, Hong Kong
Gang Li	Deakin University, Australia
Jinyan Li	University of Technology Sydney, Australia
Ming Li	Nanjing University, China
Tao Li	Florida International University, USA
Xiaoli Li	Institute for Infocomm Research, Singapore
Xue Li	The University of Queensland, Australia
Xuelong Li	Chinese Academy of Sciences, China
Yidong Li	Beijing Jiaotong Univeristy, China
Zhenhui Li	Pennsylvania State University, USA
Grace Lin	Institute of Information Industry, Taiwan
Hsuan-Tien Lin	National Taiwan University, Taiwan
Shou-De Lin	National Taiwan University, Taiwan
Fei Liu	Bosch Research, USA
Qingshan Liu	NLPR Institute of Automation Chinese Academy of Science, China
David Lo	Singapore Management University, Singapore
Woong-Kee Loh	Sungkyul University, South Korea
Chang-Tien Lu	Virginia Polytechnic Institute and State University, USA
Hua Lu	Aalborg University, Denmark
Jun Luo	Hua Wei Noah's Ark Lab, Hong Kong
Ping Luo	Institute of Computing Technology, Chinese Academy of Sciences, China
Shuai Ma	Beihang University, China
Marco Maggini	Università degli Studi di Siena, Italy
Luong Chi Mai	Inst. of Information Technology, Vietnam Academy of Science and Technology, Vietnam
Bradley Malin	Vanderbilt University, USA
Hiroshi Mamitsuka	Kyoto University, Japan
Giuseppe Manco	Università della Calabria, Italy
David Martens	University of Antwerp, Belgium

Florent Masseglia	Inria, France
Tao Mei	Microsoft Research Asia, China
Xiaofeng Meng	Renmin University of China, China
Nguyen Le Minh	JAIST, Japan
Pabitra Mitra	Indian Institute of Technology Kharagpur, India
Mohamed Mokbel	University of Minnesota, USA
Yang-Sae Moon	Kangwon National University, Korea
Yasuhiko Morimoto	Hiroshima University, Japan
J. Nath	Indian Insitute of Technology, India
Richi Nayak	Queensland University of Technologies, Australia
See-Kiong Ng	Institute for Infocomm Research, A*STAR, Singapore
Wilfred Ng	Hong Kong University of Science and Technology, Hong Kong
Ngoc Thanh Nguyen	Wroclaw University of Technology, Poland
Xuan Vinh Nguyen	University of Melbourne, Australia
Tadashi Nomoto	National Institute of Japanese Literature, Japan
Masayuki Numao	Osaka University, Japan
Manabu Okumura	Japan Advanced Institute of Science and Technology, Japan
Salvatore Orlando	University of Venice, Italy
Jia-Yu Pan	Google, USA
Dhaval Patel	Indian Institute of Technology, Roorkee, India
Yonghong Peng	University of Bradford, UK
Jean-Marc Petit	INSA Lyon, France
Clifton Phua	SAS Institute, Singapore
Dinh Phung	Deakin University, Australia
Vincenzo Piuri	Università degli Studi di Milano, Italy
Oriol Pujol	University of Barcelona, Spain
Weining Qian	East China Normal University, China
Chedy Raissi	Inria, France
Chandan Reddy	Wayne State University, USA
Patricia Riddle	University of Auckland, New Zealand
Hong Shen	Adelaide University, Australia
Jialie Shen	Singapore Management University, Singapore
Yi-Dong Shen	Chinese Academy of Sciences, China
Masashi Shimbo	Nara Institute of Science and Technology, Japan
Lisa Singh	Georgetown University, USA
Andrzej Skowron	University of Warsaw, Poland
Min Song	New Jersey Institute of Technology, USA
Mingli Song	Zhejiang University, China

Aixin Sun	Nanyang Technological University, Singapore
Yizhou Sun	Northeastern University, USA
Thepchai Supnithi	National Electronics and Computer Technology Center, Thailand
David Taniar	Monash University, Australia
Xiaohui (Daniel) Tao	The University of Southern Queensland, Australia
Tamir Tassa	The Open University, Israel
Srikanta Tirthapura	Iowa State University, USA
Ivor Tsang	Nanyang Technological University, Singapore
Jeffrey Ullman	Stanford University, USA
Sasiporn Usanavasin	SIIT, Thammasat University, Thailand
Marian Vajtersic	University of Salzburg, Austria
Kitsana Waiyamai	Kasetsart University, Thailand
Hui Wang	University of Ulster, UK
Jason Wang	New Jersey Science and Technology University, USA
Lipo Wang	Nanyang Technological University, Singapore
Xiang Wang	IBM TJ Watson, USA
Xin Wang	University of Calgary, Canada
Chih-Ping Wei	National Taiwan University, Taiwan
Raymond Chi-Wing Wong	Hong Kong University of Science and Technology, Hong Kong
Jian Wu	Zhejiang University, China
Junjie Wu	Beihang University, China
Xintao Wu	University of North Carolina at Charlotte, USA
Guandong Xu	University of Technology Sydney, Australia
Takehisa Yairi	University of Tokyo, Japan
Seiji Yamada	National Institute of Informatics, Japan
Christopher Yang	Drexel University, USA
De-Nian Yang	Academia Sinica, Taiwan
Min Yao	Zhejiang University, China
Mi-Yen Yeh	Academia Sinica, Taiwan
Dit-Yan Yeung	Hong Kong University of Science and Technology, China
Jian Yin	Hong Kong University of Science and Technology, China
Xiaowei Ying	Bank of America, USA
Jin Soung Yoo	IUPU, USA
Tetsuya Yoshida	Hokkaido University, Japan
Clement Yu	University of Illinois at Chicago, USA
Aidong Zhang	State University of New York at Buffalo, USA
Bo Zhang	Tsinghua University, China
Daoqiang Zhang	Nanjing University of Aeronautics and Astronautics, China

Table of Contents – Part II

Recommendation

Feature Selection and Reduction

Machine Learning

Temporal and Spatial Data

Novel Algorithms

Clustering

Biomedical Data Mining

Stream Mining

Outlier and Anomaly Detection

Multi-Sources Mining

Unstructured Data and Text Mining

Table of Contents – Part I

Pattern Mining

Social Network and Social Media

Classification

Graph and Network Mining

Applications

Privacy Preserving

CTROF: A Collaborative Tweet Ranking Framework for Online Personalized Recommendation

Kaisong Song[1], Daling Wang[1,2], Shi Feng[1,2], Yifei Zhang[1,2], Wen Qu[1], and Ge Yu[1,2]

[1] School of Information Science and Engineering, Northeastern University
[2] Key Laboratory of Medical Image Computing (Northeastern University),
Ministry of Education, Shenyang 110819, P.R. China
songkaisongabc@126.com,
{wangdaling,fengshi,zhangyifei,yuge}@ise.neu.edu.cn

Abstract. Current social media services like Twitter and Sina Weibo have become an indispensable platform, and provide a large number of real-time messages. However, users are often overwhelmed with large amounts of information delivered via their followees, and may miss out on much enjoyable or useful content. An information overload problem has troubled many users, especially those with many followees and thousands of tweets arriving every day. In this case, real-time personalized recommendation plays an extreme important role in microblog, which needs analyzing users' preference and recommending most relevant and newest content. Both of them pose serious challenges. In this paper, we focus on personal online tweet recommendation and propose a Collaborative Tweet Ranking Online Framework (CTROF) for the recommendation, which has integrated the Optimized Collaborative Tweet Ranking model CTR+ and Reservoir Sampling algorithm together. The experiment conducted on a real dataset from Sina microblog shows good performance and our algorithm outperforms the other baseline methods.

Keywords: Bayesian Personalized Ranking, Latent Factor Model, Online Recommendation, Reservoir Sampling.

1 Introduction

In recent years, microblog such as Weibo and Twitter becomes very popular, because it allows users to post a short message named tweet or status for sharing viewpoints and acquiring knowledge in real time. According to statistics, more than 400 million tweets are generated per day in Twitter. As a result, the rich information in microblog not only expands our horizon, but also has wide applications in public opinions supervision, natural disaster prediction and political upheaval detection.

Microblog facilitates our life, but the information overload problem prevents it from developing further. A user usually follows many interested users such as friends, stars and organizations, and receives a lot of tweets at all times because of frequent updates. So the users are hard to consume so much content instantly in an effective way. In some cases, as for those with limited time to read, it's necessary to filter out those irrelevant and boring tweets by online recommending expected content.

V.S. Tseng et al. (Eds.): PAKDD 2014, Part II, LNAI 8444, pp. 1–12, 2014.

There are several challenges to be tackled for online personalized tweet recommendation. Firstly, although some latent factor models such as CTR [1] (Collaborative Tweet Ranking) have been proposed, yet they show poor performance as the dataset grows larger. Secondly, online algorithms can shorten the processing time, but they also reduce prediction quality. Thirdly, most online algorithms are generally based on recent data, and don't consider the history records which are relevant with users' customs and preferences. All the problems above have posed severe challenges for online tweets recommendation.

In this paper, we propose a Collaborative Tweet Ranking Online Framework (CTROF). Figure 1 below shows the algorithm process. Suppose we have some tweet history data in advance, we build an initial model called Optimized Collaborative Tweet Ranking model CTR+. For recommending the interested tweets in real time, we sample the tweet stream to update Double Reservoir for capturing the "sketch" of the stream. Then the training set is sampled and the initial CTR+ base model can be updated incrementally. Eventually, we get an online model for recommending interested tweets to users.

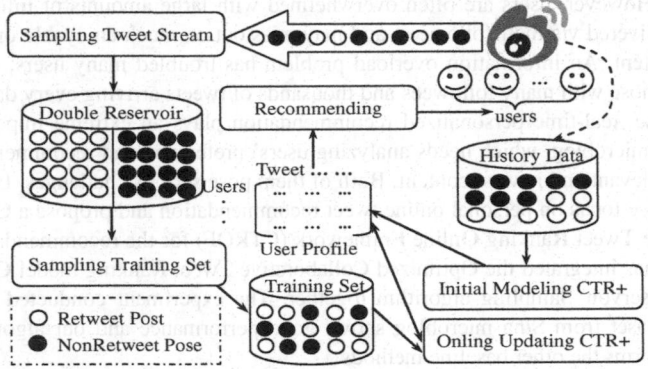

Fig. 1. Overview of CTROF in sampling and modeling online tweet stream

To the best of our knowledge, this work is the first experimental study to integrate tweet content (past and online), social structure and personal profile into collaborative filter model, and demonstrate a practical online personalized tweet recommendation framework. To summarize, the main contributions of our work are as follows.

(1) We propose a novel CTROF for online personal tweet recommendation. The novelty lies in a complete stream processing framework for real-time tweet ranking.

(2) We improve the performance of state-of-the-art tweet recommendation model CTR [1] by introducing personal hashtags to optimize BPR (Bayesian Personalized Ranking) [2], and creatively apply the model into online scenario.

(3) We use Reservoir Sampling algorithm for acquiring the "sketch" of incoming tweet stream dataset, which considers both the historical and the changing preferences.

(4) Finally, our algorithm considers the performance of time and space, and is able to balance the recommendation quality and the complexity of time and space well.

The structure of the rest of the paper is as follows. Related work is discussed briefly in Section 2. In Section 3, we briefly review CTR and propose a novel CTR+ recommendation model. In Section 4, we introduce the collaborative tweet ranking online framework CTROF. Section 5 introduces the classification of explicit features in detail. Section 6 compares CTROF with the other baseline models. In Section 7, we make a conclusion and point out the directions for our future work.

2   Related Work

CF (collaborative filtering) technique behind RS (recommender system) has been developed for years and kept to be a hotspot in academic and industrial field. Real applications include goods at Amazon, news at Google, movies at Yahoo and CDs at Netflix. In recent years, latent factor model proposed by Simon Funk has been widely applied in CF. Koren [3,4] and Xiang & Yang [5] improved it by considering neighborhood or time. Besides explicit feedback (rating), abundant implicit feedback [6] was also used. In high-order setting, Tensor Decomposition models [7,8] were studied. In addition, some learning methods were studied, such as Stochastic Gradient Descent (SGD), Alternating Least Squares [9] and Markov Chain Monte Carlo [10].

RS in microblog generally contains six categories factors in content: followee, follower, hashtag, tweet, retweet and URL. In recommendation pattern, it includes offline and online RS. In offline RS, [11] ranked incoming tweets by using author profile, syntactic feature, content and followee feature. Hong et al. [12] proposed a co-factorization machine to model interest and recommend relevant tweets. Chen et al. [1] proposed CTR model with high precision, however extra user preference signs, i.e. labels, were not considered. In addition, offline models are not suitable in online scenario with large continuous incoming data. In online RS, Diaz-Aviles et al. [13] proposed a RMFX online framework, however it had complex sampling algorithm and just considered hashtags of tweet. Work for ranking tweets also includes [14,15].

In this paper, we focus on recommending tweets in real time by integrating offline model and stream sampling algorithm together. We propose a novel CTROF framework by improving the drawbacks of Chen and Diaz-Aviles's work and absorbing their advantages. For this purpose, we first review CTR model briefly and propose an innovative CTR+ model by considering content, social relation and hashtags in Section 3. And then we utilize Reservoir Sampling [16] algorithm and propose a CTROF algorithm framework for ranking tweet stream in Section 4.

3   Optimized Collaborative Tweet Ranking Model

3.1   Notations Definition

We firstly define some notations being frequently used later. Let $U=\{u_1, u_2, ..., u_n\}$ be a user set, and $I=\{i_1, i_2, ..., i_m\}$ be a tweet set. Suppose there is interaction between any two entities, which shows the degree of interest. Then we get an interactive matrix X: $U \times I$, and each element $x_{ui} \in X$ represents an observation value. Predicting $\hat{x}_{ui}$ value can be seen as the task of estimating $\hat{X}$. As we aim to get a personalized total ranking $>_u \subset I^2(?)$ of all tweets for a specific user u, we use Bayesian Personalized

Ranking for estimating $\hat{\mathbf{X}}$, instead of Root Mean Square Error (RMSE). As for any two entries $\hat{x}_{ui}$ and $\hat{x}_{uj}$ $(i{\neq}j)$ in $\hat{\mathbf{X}}$, if $\hat{x}_{ui} > \hat{x}_{uj}$, then $i{>}_u j$.

After above modeling process, a basic offline model is built. Based on the model, we give out an optimized collaborative tweet ranking model CTR+ in offline scenario.

3.2   Optimizing Offline CTR to CTR+ Model

CTR [1] is an excellent offline RS, considers content, social relation, and explicit features simultaneously. In view of data sparsity and expandability, each tweet is decomposed into several words at topic level. All words from tweet set I constitute bag of words. Therefore, let $u{\in}U$ and $i{\in}I$, CTR model is described as follows:

$$\hat{x}_{u,i} = bias + p_u^T (\frac{1}{Z} \sum_{w \in W_i} q_w) \tag{1}$$

where W_i is word set of tweet i, q_w is vector of word $w{\in}W_i$, Σq_w is vector combination of tweet i, p_u is vector of user u, $bias{=}b_u{+}\Sigma b_w$ is bias term, b_u and b_w are user bias and word bias, and Z is normalization term which equals $|W_i|^{1/2}$ in general.

In addition, social relations are also important because users are more likely to retweet favorite publisher's tweets. As for any incoming tweet i, it can be mapped into corresponding publisher $p(i)$. So the formula can be further rewritten as:

$$\hat{x}_{u,i} = bias' + p_u^T (\frac{1}{Z} \sum_{w \in W_i} q_w + \kappa d_{p(i)}) \tag{2}$$

where $d_{p(i)}$ is publisher vector of i, $bias' {=}b_{p(i)}{+}bias$ is bias term, $b_{p(i)}$ is publisher bias and κ is an adjustable weighting parameter indicating publisher's importance relative to content.

In this paper, we introduce personal hashtags, a profile of personal interests and hobbies, into CTR model. Suppose users with similar interests are more likely to retweet each other. As for any tweet i, it can be mapped into its publisher's personal hashtag set $H_{p(i)}$. So we rewrite Formula (2) and represent CTR+ as:

$$\hat{x}_{u,i} = bias'' + p_u^T (\frac{1}{Z} \sum_{w \in W_i} q_w + \kappa d_{p(i)} + \frac{\beta}{Z'} \sum_{h \in H_{p(i)}} g_h) \tag{3}$$

where g_h is hashtag vector of any tag h in $H_{p(i)}$, $bias''{=}bias' {+}\Sigma b_h$, b_h is tag bias, β is an adjustable weighting parameter indicating hashtags' importance relative to the content and $Z' {=}|H_{p(i)}|^{1/2}$ is normalization term.

Besides the above latent features, information such as tweet quality can also be incorporated into CTR+ as explicit features. Then $bias''$ term is replaced by Σbr, a weighted linear combination of $bias''$ and explicit feature biases. In the final CTR+ offline model, we get Formula (4) shown below:

$$\hat{x}_{u,i} = \sum_j b_j r_j + p_u^T (\frac{1}{Z} \sum_{w \in W_i} q_w + \kappa d_{p(i)} + \frac{\beta}{Z'} \sum_{h \in H_{p(i)}} g_h) \tag{4}$$

where b_j is any latent or explicit feature bias and r_j is weighting parameter represented by explicit feature value. For simplified formula, the weight r of b_u, b_w, $b_{p(i)}$ and b_h is set 1 by default. Details about explicit feature classification are discussed in Section 5.

Different from rating prediction, users just need to be recommended a list of sorted tweets. Similar to [1,12], retweet represents users' preference. Slightly different from Root Mean Square Mean in rating prediction, a BPR method is used instead.

Given a tweet set I, we should transform I into training set D in the form of tuples at first. For convenience, we define retweet set $R_u \subset I$ for any user u. Let $((u, i), (u, j)) \in D$ denotes a training instance, where $i \in R_u$ has been retweeted and $j \notin R_u$ not. Thus, D is formally defined as the tuple set from I, and we can describe it as: $D=\{((u, i), (u, j))|i \in R_u \wedge j \notin R_u \wedge u \in U\}$. According to BPR Optimization Criterion, probability $p(\Theta)$ follows normal distribution $N(0, \Sigma_\Theta)$, in which diagonal matrix $\Sigma_\Theta = \lambda_\Theta E$, E is a unit diagonal matrix and λ_Θ is a constant, we aim to maximize the formula below:

$$\prod_{((u,i),(u,j)) \in D} \delta(\hat{x}_{uij}(\Theta)) \times p(\Theta) \tag{5}$$

where δ is sigmoid function. For convenience, Formula (5) is transferred as equivalent Formula (6) below by maximizing logarithm of posterior probability:

$$BPR-Opt := \max_{\Theta} \sum_{((u,i),(u,j)) \in D} \ln(1/1+e^{-(\hat{x}_{u,i}-\hat{x}_{u,j})}) - \lambda_\Theta \|\Theta\|^2 \tag{6}$$

In general, SGD is used for estimating parameter space Θ, and $\lambda_\Theta \|\Theta\|^2$ is a L2 regularization term. The training process of CTR+ model is shown below.

Algorithm. Training Offline Model CTR+ based on SGD for Θ estimation;

Input: Tweet training set D; Parameter space Θ; Relative weighting β and κ; Explicit feature weighting vector r; Latent factor number f; Learning rate η; Regularization parameters λ_Θ; Number of iterations T_Θ;
Output: Θ;
Description:
   1) procedure CTR+Model(D, Θ, λ_Θ, f, η, T_Θ, β, κ, r);
   2) initialize Θ;
   3) for t=1 to T_Θ
   4)   for each $p=((u,i), (u,j)) \in D$
   5) $\Theta \leftarrow \Theta + \eta((e^{-\hat{x}_{uij}}/1+e^{-\hat{x}_{uij}}) \cdot (\partial \hat{x}_{uij} / \partial \Theta) - \lambda_\Theta \Theta)$;
   6) return Θ;

---

4   Collaborative Tweet Ranking Online Framework

4.1   Building Online CTROF Model

In Section 3, we discussed CTR+ with tweet training set D. CTR+ is an offline model, because D is a static training set. As for new incoming tweet $i+$, we calculate $\hat{x}_{ui+}$ by decomposing $i+$ into words, publisher and hashtag vectors. The larger $\hat{x}_{ui+}$ is, the higher $i+$ is ranked. Based on offline CTR+, we introduce CTROF in real-time scenario, which update model dynamically every time new tweets arrive.

In social network (such as Facebook) or microblogging service (like Twitter and Sina Weibo), messages are updated rapidly. A flow of messages constitutes data stream, called tweet stream in Twitter or Weibo. Diaz-Aviles [13] proved that Reservoir Sampling outperformed Single Pass, User Buffer, and captured the "sketch" of history under the constraint of fixed space quite well. CTROF uses it and achieves online model by training CTR+ incrementally without retraining model completely.

Under the background of tweet stream, we use S to represent incoming tweet stream $i_1, i_2...$ that arrives sequentially. As for tweet stream S, it is divided into retweet stream S_{ret} and non-retweet stream S_{nret}. Our algorithm maintains two fixed size Reservoirs R^+ and R^-, which contains random samples from S_{ret} and S_{nret}. So the key is to define reservoir $R^+=\{s_1, s_2,...,s_{|R^+|}\}$ for S_{ret}, and reservoir R^- for S_{nret} as well. Similarly, let notation t^+ and t^- be tweet index for S_{ret} and S_{nret} respectively, reflecting the order of arrival of data in the stream. At the beginning, all incoming tweets will be pushed into reservoir R^+ and R^- continually and indiscriminately until $t^+=|R^+|$ and $t^-=|R^-|$. For subsequent t, we will decide whether a new incoming tweet will be put in reservoirs or not, and in which the old record will be replaced instead. The process of Collaborative Tweet Recommendation Online framework CTROF is shown below.

Algorithm. CTROF Framework;

Input: Tweet stream S; Reservoirs R^+ and R^-; Offline model parameters Θ'; Relative weighting β and κ; Explicit feature weighting vector r; Latent factor number f; Regularization parameters λ_Θ; Learning rate η; Number of iterations T_Θ for updateCTR+Model; Parameters c_r and c_{nr} control updates frequency of model;

Output: Θ;

Description:

   1) procedure CTROF(S, Θ', R^+, R^-,λ_Θ, f, η, T_Θ, β, κ, r, c_r, c_{nr});
   2) initialize $\Theta=\Theta'$; $count_r \leftarrow 0$; $count_{nr} \leftarrow 0$;
   3) for $t=1$ to $|S|$ do
   4)   if t is retweet
   5) $R^+ \leftarrow$ ReservoirSampling(R^+, i_t);
   6) $count_r \leftarrow count_r+1$;
   7)   else if t is non-retweet
   8) $R^- \leftarrow$ ReservoirSampling (R^-, i_t);
   9) $count_{nr} \leftarrow count_{nr}+1$;
  10)   if $count_r=c_r$ and $count_{nr}=c_{nr}$
  11) $\Theta \leftarrow$ updateCTR+Model (Θ,R^+, R^-, λ_Θ, f, η, T_Θ, β, κ, r);
  12) $count_r \leftarrow 0$, $count_{nr} \leftarrow 0$;
  13) Return Θ;

In above process, we selectively update two fixed-size reservoirs R^+ and R^- by Reservoir Sampling algorithm every time a new tweet arrives. For convenience, let R^* denotes R^+ or R^-. During R^* initialization, a new incoming tweet i_t is saved in the corresponding R^* directly until $t=|R^*|$. For subsequent t, random index μ is selected randomly within the scope of $|t|$. If $\mu \leq |R^*|$, we replace t-th tweet in R^* with tweet i_t. Above Reservoir Sampling ensures that each tweet is selected with equal probability.

4.2   Updating Online CTROF Model

In Algorithm CTROF Framework, updateCTR+Model (Line 11) updates the model incrementally by sampling training instances from R^*. We design a simple but effective sampling strategy by computing time distance between retweet and nonretweet. For formulization, as for any particular user u in each iteration, let pair (u, i) be retweet selected randomly from reservoir R^+, and closest pair (u, j) from R^-, where distance $\delta=\min|Time_i-Time_j|$ and $1 \le j \le |R^-|$. Then we select randomly m training instance pairs as *TrainSet*, and perform model update based on it.

Algorithm. Online Updating CTR+ based on Reservoir for Θ estimation;

Input: Tweet stream reservoirs R^+ and R^-; Relative weighting β and κ; Explicit feature weighting vector r; Latent factor number f; base model Θ'; Regularization parameters λ_Θ; Learning rate η; Number of iterations T_Θ;

Output: Θ;

Description:

   1) procedure updateCTR+Model($\Theta', R^+, R^-, \lambda_\Theta, f, \eta, T_\Theta, \beta, \kappa, r$);

   2) initialize $\Theta=\Theta'$; *TrainSet* ={};

   3) for t=1 to SampleNum

   4)   Draw pair (u, i) from R^+ randomly and closest negative (u, j) from R^-,
          save triple (u, i, j) into training set *TrainSet*;

   5) for t=1 to T_Θ

   6) for each triple (u, i, j) from *TrainSet*

   7) $p_u \leftarrow p_u + \eta(\hat{e}(\frac{1}{Z^+}\sum_{w \in W_i} q_w^+ + \frac{1}{Z'^+}\sum_{h \in H_{p(i)}} g_h^+ - \frac{1}{Z^-}\sum_{w \in W_j} q_w^- - \frac{1}{Z'^-}\sum_{h \in H_{p(j)}} g_h^+ +$

         $\kappa d_{p(i)}^+ - \kappa d_{p(j)}^-) - \lambda_u p_u)$;

   8) $d_{p(i)}^+ \leftarrow d_{p(i)}^+ + \eta(\kappa \hat{e} p_u - \lambda_{p(i)} d_{p(i)}^+)$;

   9) $d_{p(j)}^- \leftarrow d_{p(j)}^- - \eta(\kappa \hat{e} p_u + \lambda_{p(j)} d_{p(j)}^-)$;

  10)   for each $w \in W_i$ // W_i is the word set of tweet i

  11) $q_w^+ \leftarrow q_w^+ + \eta(\hat{e} p_u / Z^+ - \lambda_w q_w^+)$;

  12)   for each $w \in W_j$ // W_j is the word set of tweet j

  13) $q_w^- \leftarrow q_w^- - \eta(\hat{e} p_u / Z^- + \lambda_w q_w^-)$;

  14)   for each $h \in H_{p(i)}$ // $H_{p(i)}$ is the hashtag set of followee $p(i)$

  15) $g_h^+ \leftarrow g_h^+ + \eta(\beta \hat{e} p_u / Z'^+ - \lambda_h g_h^+)$;

  16)   for each $h \in H_{p(j)}$ // $H_{p(j)}$ is the hashtag set of followee $p(j)$

  17) $g_h^- \leftarrow g_h^- - \eta(\beta \hat{e} p_u / Z'^- + \lambda_h g_h^-)$;

  18)   for each explicit or latent feature bias k

  19) $b_k \leftarrow b_k + \eta(\hat{e}(r_k^+ - r_k^-) - \lambda_k b_k)$;

  20) return $\Theta=(p^*, d^*, q^*, g^*, b^*)$; // o^* represents any estimated parameter

---

Here notation o^+ denotes parameter vector of pair (u, i), while o^- for (u, j). We use $\hat{e}=(e^{-\hat{x}_{uij}}/1+e^{-\hat{x}_{uij}})$ for convenience. Given new reservoirs R^+ and R^-, we update model incrementally. Therefore, CTROF captures the history "sketch" and the current interest, and can overcome the problem of short-memory and avoids retraining model.

5   Relevant Features

In Section 3, we introduce CTR+ integrating linear combination of explicit features with *bias"* by bias term Σbr. In this section, we will further classify explicit features for capturing users' interests. Although [1,12] have defined different categories respectively, yet we will propose a more complete solution including four categories.

1) User Relationship Features: User relationship feature refers to the relationship between target user u and his/her friend v. It makes an assumption that: The more familiar with each other, the more likely to retweet his/her messages.

- **Co-Friends Score:** The similarity between u's followee set and v's.
- **Co-Follow Score:** The similarity between u's follower set and v's.
- **Mention Score:** The number of times u mentions v.
- **Retweet Score:** The number of times u retweets v.
- **Reply Score:** The number of times v replys to u.
- **Mutual Friend Score:** If u and v follow each other, it is 1, else 0.

2) Content Features: The features are the relevance between new incoming tweet i and the profiles of a target user u. Let $\mathbf{w}(i)$ as term set of i, $SP(u)$ as word set of u's status data, $RP(u)$ as word set of u's retweet data, $LP(u)$ as hashtag set of u's profile, $NP(u) = SP(u) \cup RP(u) \cup CP(u)$, and $\Re(\mathbf{w}_1, \mathbf{w}_2)$ as similarity between two term sets.

- **Relevance to Status:** $\Re(\mathbf{w}(i), SP(u))$ is the similarity between $\mathbf{w}(i)$ and $SP(u)$.
- **Relevance to Retweets:** $\Re(\mathbf{w}(i), RP(u))$ is similarity between $\mathbf{w}(i)$ and $RP(u)$.
- **Relevance to Hash Tags:** $\Re(\mathbf{w}(i), LP(u))$ is similarity between $\mathbf{w}(i)$ and $LP(u)$.
- **Relevance to Neighborhood:** $\Re(\mathbf{w}(i), NP(u))$ is similarity of $\mathbf{w}(i)$ and $NP(u)$.

3) Tweet Features: Tweet features refer to the attributes of tweet, including general tweet length, hash tag count, URL count, reply count, retweet count. In addition, we add another two new features, that is, thumb up score and view score.

- **Thumb Up Score:** The number of times that tweet i is favorable or agreed.
- **View Score:** The number of times that tweet i is viewed.

4) Publisher Features: Publisher features represent the influence power of corresponding publisher of i, including not only mention, followee, follower and status count in [1], but also activity degree and loyalty degree. Let $time_{ui}$ be the time u publish tweet i, $\tau_u = \max\{time_{ui} - time_{uj}\}(i \neq j)$ as the period from first status to the last.

- **Loyal activity:** The feature measures how long the publisher u is active in RS. In general, we use τ_u to show the degree of loyalty.
- **Activity Degree:** The feature shows the activity of publisher and we may use $NT(u)/\tau_u$ to measure it, $NT(u)$ is the number of tweets that u has published.

6   Experiments

6.1   Experiment Setup

Our experiments are based on Sina Weibo platform and utilize the API tool [17]. Our work focuses on real-time personal tweet recommendation in Chinese microblog scenario and we use ICTCLAS [18] to handle word segmentation. For getting dataset, we randomly select a user and adopt user-based breadth-first traversal method by following followers and followees' links. Different from [1], our dataset includes tweet content, retweet action, personal hashtags and social relation. Retweeted and non-retweeted tweets are named positive and negative samples respectively. The dataset includes 46385 users' profile (user id, tags, followees) and their publishing historical data (tweet id, content, time, repost number). We select 675 users with more than 20 retweets, and others as their followees. Then tweets flood continuously from followees into corresponding followers in chronological order. Three fifths of dataset is as

training set and the others as testing set. Finally, training set for offline training contains 171,937 positive samples and 1,124,840 negative ones. Testing set contains
113,941 positive samples and 458,104 negative ones for offline testing. For testing
stream dataset, we set parameters c_r and c_{nr} for controlling update frequency, and testing set can be further divided into tweet stream set $S=\{s_1, s_2 \ldots s_{11}\}$ by parameters, of
which s_n ($1 \leq n \leq 10$) is for incremental online training, and s_{n+1} for online testing. The
experiment shows that the crawled dataset coincides with our proposed model.

FM (Factorization Machines) [19] and SVD Feature [20] are generic factorization
models tools. Considering coding workload and algorithm efficiency, we use the later.

Different from rating prediction, we focus on ranking tweets. So we measure recommendation precision by $P@N$ and recommendation quality by MAP metric. Let
$MAP=\Sigma AP_u/|U|$ and $P@N=\Sigma p_u@n/|U|$. Given $u \in U$, $p_u@n$ is retweet proportion of top
n in list, averaging precision (AP_u) is the average precision of each user:

$$AP_u = \frac{\sum_{n=1}^{N} p_u @ n \times \delta(n)}{|R_u|} \tag{8}$$

where $\delta(n)$ is an indicator function, which returns 1 if n-th tweet in the list is ret-
weeted, and 0 otherwise. $|R_u|$ is total number of retweeted tweets in top N list, and
$|R_u| \leq N$. And $p_u@n$ measures the precision of top n tweets.

6.2   Experiment Result

In Section 3, we have proposed CTR+ model for offline tweet recommendation system modeling. CTR+ includes necessary components (explicit factor, term factor,
social factor and hash tag factor). For studying components' influence, we make a
comparison by $s_1 \in S$ in Figure 2. We compare MAP by $N=15$ and $P@N$ by setting N
to 5, 10 and 15. The number of iterations and factors is set to 40 and 64 respectively.
Relative weight parameters β and κ are set 0.8 uniformly. CTR performs well
(MAP=0.8074~0.8114) when β is around 0.8, so we choose best parameter 0.8. Given
fixed β, we randomly select $\kappa=0.8$ because MAP remains stable when κ ranges from
0.7 to 1. As large training dataset rarely encounter the over-fitting problem, the regularization parameter λ is set 0.005. For approaching optimal value, the learning rate η
is set small value 0.004, despite a certain loss in convergence rate.

Figure 2 shows the precision of CTR+ is always higher than CTR, and reflects the
importance of single component and their combination. Chronological method's precision is shown for reference. For simplicity, we just choose explicit features (Text
Length, Retweet Score, and Relevance to Hash Tags) as global features. We find that
explicit features, term, hash tag, and social component improve MAP by 54%, 70%,
86% and 92% respectively relative to chronological method, which indicates all components are necessary and effective. CTR contains all components except hash tag,
and outperforms any single component. However, CTR+, compared with CTR, improves precision by 12.3%, which indicates that our model is better.

Figure 3 shows that runtime convergence of different models. All models have different convergence rate and converge to steady values after 30 rounds. So our offline
base model is reasonably set to 40 rounds. In addition, we calculate $P@5$ value by

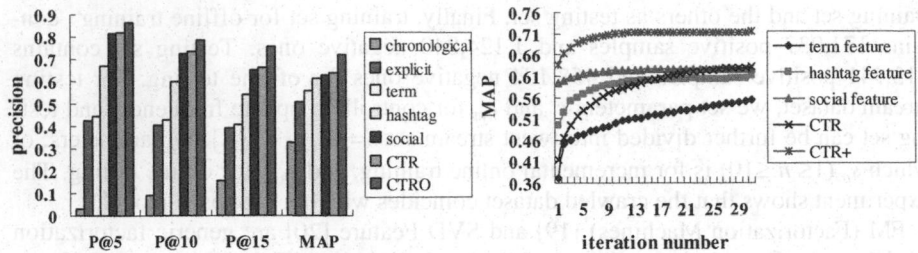

Fig. 2. Evaluation of Compared Methods   Fig.3. Runtime Convergence of CTR+

setting factor number to 32, 64, 80, 96, and 112. Term Feature (0.4438~0.4439), Tag Feature (0.6644~0.6752), Social Feature (0.8114~0.8124), CTR (0.8219~0.8229) and CTR+ (0.8637~0.8641) remains stable. Therefore, we set 64 factors reasonably.

We imitate tweets flow into our framework in chronological sequence continually. In addition, we let each one' tweet stream arrive at the same opportunity. Retweetes in testing set are far less than non-retweets, and the ratio is about 1/4. So we set the size of reservoirs R^+ and R^- to 10,000 and 40,000 respectively for reflecting real data distribution. Therefore, we won't update our base model until 10,000 retweets and 40,000 non-retweets arrive. In order to verify prediction precision of top-N items in recommendation list, N is set to 5, 10 and 15 respectively. We compare our stream framework CTROF with two offline models CTR and CTR+. The MAP for each method, in different list size, is shown in Figure 4 below.

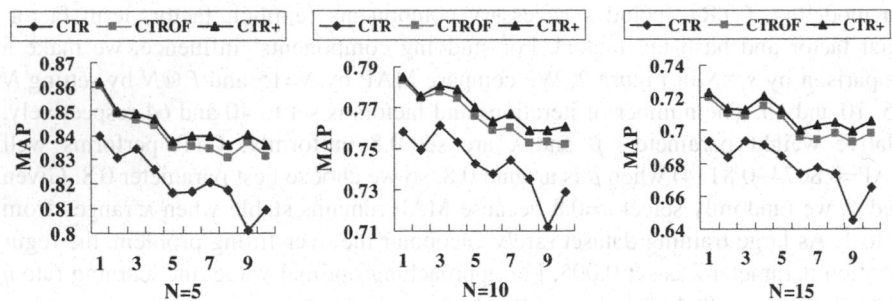

Fig. 4. Recommendation precision of different sized list. The number of factor is 64 for CTR, CTR+, and CTROF. S_i $(1 \leq i \leq 10)$ denotes the incoming retweet and nonretweet tweet stream by setting c_r=10,000 and c_{nr}=40,000, which is also as big as the size of reservoir R^+ and R^-.

Figure 4 shows that online model CTROF achieves better performance over offline CTR, and slightly below than offline CTR+ model. Compared with CTR+, CTROF just capture information sketch by sampling, so precision is slightly lower. In addition, additional hash tag factor represents personal preference and makes CTROF outperform CTR model. As online recommendations focus more on comprehensive performance of runtime, space and precision, our method saves lots of runtime and space and precision is close to best offline model CTR+.

Next, we further discuss time, space and recommendation precision comparison of CTROF and CTR+, which are implemented by SVDFeature tool in C++. We ran

CTR+ and CTROF on an Intel Core i7-2600 3.4GHz CPU and 2G memory virtual machine with Linux 32bit operating system. As none of the methods is parallel, we run the program on a single CPU. As the platform and implementation technique influence the performance greatly, so the setting can be used as a reference indicator.

The performance of CTROF is not only related to the number of factors and iterations, but also related to the training set size of the initial base model and the reservoir size. If we train base model with a very big training set about 1.3 million tweets, then the size of reservoir has little impact on the recommendation quality. That's because long-term accumulated dataset may include almost all possible situations in terms, retweet relation and tags, so a small amount of tweet stream will not change the model greatly. So we just choose about 0.4 million training instances, and set different sized reservoirs. The experiment comparison result is shown below.

Table 1. Comparison among time, space, and recommendation quality with different sized reservoir R^+ and R^-. The number of iteration and factors is set to 40 and 64. The base model CTR+ is trained by 0.4 million tweets. List length N is 40. The testing set has 0.1 million tweets. Time is the runtime sum of six tests, and Space is the disk space of training text.

Method (64 factors, 40 r) Reservoir size of R^+, R^-	Time (second)	Space	MAP	Recommendation Quality of CTROF
CTR+ [Baseline]	464s	100%	0.802	100%
CTROF R^+=1,000 R^-=4,000	40s	1.11%	0.752	93.82%
CTROF R^+=5,000 R^-=20,000	66s	5.25%	0.771	96.13%
CTROF R^+=10,000 R^-=40,000	92s	10.43%	0.780	97.23%

Table 1 shows that the reservoir size can influence the final recommendation quality obviously. Offline model CTR+ is used as a reference. When reservoir is big enough, the data distribution is much more appropriate and close to real data distribution. In this paper, we set reservoir size to $T_1=(R^+=1,000: R^-=4,000)$, $T_2=(R^+=5,000: R^-=20,000)$, and $T_3=(R^+=10,000: R^-=40,000)$ respectively. By comparison, we find T_2 is close to T_3 in recommendation quality, but faster than T_3 by 28.2%. In addition, T_2 outperforms T_1 by 2.5% within the scope of tolerable time. So we can draw a conclusion that recommendation quality is good enough when $5,000 \leq |R^+| \leq 10,000$ and $R^-=4|R^+|$.

7   Conclusions and Future Work

In this paper, we propose an offline ranking model CTR+ which considers explicit feature, content, social relation and personal hashtags. Moreover, we propose a novel tweet ranking online framework CTROF for real-time personalized recommendation. CTROF integrates Reservoir Sampling algorithm and CTR+ together, which captures "sketch" of tweet historical data, and absorbs new preference change from incoming tweet stream in the meantime. By experiments, we show that CTR+ outperforms CTR offline model and CTROF can capture real data distribution and achieve quite good precision, which demonstrates that our proposed method is effective and efficient.

Future work includes further analyzing semantics of content and studying more accurate and efficient sampling methods for improving the recommendation quality. We will consider more media factors in tweet such as images and videos. In addition, the tensor factorization for tweet recommendation is also our future direction.

Acknowledgements. This work is supported by the State Key Development Program for Basic Research of China (Grant No. 2011CB302200-G), State Key Program of National Natural Science of China (Grant No. 61033007), National Natural Science Foundation of China (Grant No. 61100026, 61370074), and Fundamental Research Funds for the Central Universities (N100704001, N120404007, N100304004).

References

1. Chen, K., Chen, T., Zheng, G., Jin, O., Yao, E., Yu, Y.: Collaborative personalized tweet recommendation. SIGIR, 661–670 (2012)
2. Rendle, S., Freudenthaler, C., Gantner, Z., Schmidt-Thieme, L.: BPR: Bayesian Personalized Ranking from Implicit Feedback. CoRR abs/1205.2618 (2012)
3. Koren, Y.: Factorization meets the neighborhood: A multifaceted collaborative filtering model. In: KDD 2008, pp. 426–434 (2008)
4. Koren, Y.: Collaborative filtering with temporal dynamics. In: KDD 2009, pp. 447–456 (2009)
5. Xiang, L., Yang, Q.: Time-Dependent Models in Collaborative Filtering Based Recommender System. In: Web Intelligence, pp. 450–457 (2009)
6. Oard, D.W., Kim, J.: Implicit Feedback for Recommender Systems. In: Proc. 5th DELOS Workshop on Filtering and Collaborative Filtering, pp. 31–36 (1998)
7. Rendle, S., Marinho, L.B., Nanopoulos, A., Schmidt-Thieme, L.: Learning optimal ranking with tensor factorization for tag recommendation. In: KDD 2009, pp. 727–736 (2009)
8. Symeonidis, P., Nanopoulos, A., Manolopoulos, Y.: Tag recommendations based on tensor dimensionality reduction. In: RecSys 2008, pp. 43–50 (2008)
9. Pilászy, I., Zibriczky, D., Tikk, D.: Fast als-based matrix factorization for explicit and implicit feedback datasets. In: RecSys 2010, pp. 71–78 (2010)
10. Salakhutdinov, R., Mnih, A.: Bayesian probabilistic matrix factorization using Markov chain Monte Carlo. In: ICML 2008, pp. 880–887 (2008)
11. Uysal, I., Croft, W.B.: User oriented tweet ranking: A filtering approach to microblogs. In: CIKM 2011, pp. 2261–2264 (2011)
12. Hong, L., Doumith, A.S., Davison, B.D.: Co-factorization machines: Modeling user interests and predicting individual decisions in Twitter. In: WSDM 2013, pp. 557–566 (2013)
13. Diaz-Aviles, E., Drumond, L., Schmidt-Thieme, L., Nejdl, W.: Real-time top-n recommendation in social streams. In: RecSys 2012, pp. 59–66 (2012)
14. Feng, W., Wang, J.: Retweet or not?: Personalized tweet re-ranking. In: WSDM 2013, pp. 577–586 (2013)
15. Hong, L., Bekkerman, R., Adler, J., Davison, B.: Learning to rank social update streams. In: SIGIR 2012, pp. 651–660 (2012)
16. Vitter, J.S.: Random Sampling with a Reservoir. ACM TOMS 11(1), 37–57 (1985)
17. http://open.weibo.com/
18. http://ictclas.nlpir.org/
19. Rendle, S.: Factorization Machines with libFM. ACM TIST 3(3), 57 (2012)
20. Chen, T., Zhang, W., Lu, Q., Chen, K., Zheng, Z., Yu, Y.: SVDFeature: A Toolkit for Feature-based Collaborative Filtering. JMLR 13(Dec), 3619–3622

Two-Phase Layered Learning Recommendation via Category Structure

Ke Ji[1], Hong Shen[2,3], Hui Tian[4], Yanbo Wu[1], and Jun Wu[1]

[1] School of Computer and Information Tech., Beijing Jiaotong University, China
[2] School of Information Science and Technology, Sun Yat-sen University, China
[3] School of Computer Science, University of Adelaide, Australia
[4] School of Electronics and Info. Engineering, Beijing Jiaotong University, China
{12120425,htian,ybwu,wuj}@bjtu.edu.cn, hongsh01@gmail.com,

Abstract. Context and social network information have been introduced to improve recommendation systems. However, most existing work still models users' rating for every item directly. This approach has two disadvantages: high cost for handling large amount of items and unable to handle the dynamic update of items. Generally, items are classified into many categories. Items in the same category have similar/relevant content, and hence may attract users of the same interest. These characteristics determine that we can utilize the item's content similarity to overcome the difficultiess of large amount and dynamic update of items. In this paper, aiming at fusing the category structure, we propose a novel two-phase layered learning recommendation framework, which is matrix factorization approach and can be seen as a greedy layer-wise training: first learn user's average rating to every category, and then, based on this, learn more accurate estimates of user's rating for individual item with content and social relation ensembled. Based on two kinds of classifications, we design two layered gradient algorithms in our framework. Systematic experiments on real data demonstrate that our algorithms outperform other state-of-the-art methods, especially for recommending new items.

Keywords: Collaborative filtering, Matrix Factorization, Recommender Systems, Layered Learning.

1 Introduction

With the rapid development of the Internet, information growth has gone beyond the capacity of our social infrustucture. Recommendation systems that can suggest users with useful information become a powerful way to solve the information overload. A successful technique in recommendation systems is *collaborative filtering* (CF) [1]. It has been applied in many areas, such ase ecommerce (e.g., Amazon) and social networks (e.g., Twitter). Two primary approaches to CF are memory based [2] and model based [3,4] algorithms. The basic difference is that memory based algorithms predict the missing rating based on similar users or items which can be found from the whole user-item rating matrix

V.S. Tseng et al. (Eds.): PAKDD 2014, Part II, LNAI 8444, pp. 13–24, 2014.
© Springer International Publishing Switzerland 2014

(Figure 1(a)) using the similarity measurement (PCC, VSS [5]), whereas model based algorithms explore the training data to train a model, which can make fast prediction using only a few parameters of the model instead of manipulating the whole matrix.

Traditional CF algorithms have several challenges. Due to the sparsity, they cannot make reliable recommendation for lazy users who have rated few items or cold start users who have never rated any items because of insufficient data to capture their tastes accurately. Mining purely the rating matrix may give unrealistic recommendation. In order to solve these problems, lots of studies have been done. Matrix factorization can solve the sparsity problem [4]. Context-aware algorithms [6] that incorporate contextual information have improved the accuracy. With the popularity of online social networks, social recommendation models [7,8,9,10,11] that incorporate social networks information (Figure 1(b)) not only improve the recommendation quality, but also solve the cold start problem.

Even so, there are still some drawbacks. They typically model users' rating for every item. As the number of items increases, the rating matrix becomes very large so that matrix operations in all CF algorithms become exceedingly expensive which may even go beyond the physical computation/storage power. Beside that, attention to an individual item does not reveal users' tastes explicitly, and provides no ability to deal with new items to arrive in the future. There is therefore an urgent need to establish a general system that can provide scalable solutions for both the large amount and dynamic update of data. As nowadays we can easily get a greater variety of data than ever before, information extraction methods that can extract keyword from item content (Figure 1(c)) are widely adopted. There are classification methods that can accurately classify the items into many categories (Figure 1(d)). Intuitively, for items under the same category, their content is relevant, and hence the user's tastes to them may well be similar. This means that we can explore the category structure to find user's similar tastes. Since this information is comparatively static, we can use it to improve the scalability of a recommendation system. However, the current models cannot be adopted to incorporate this information. Therefore, a more flexible recommendation mechanism that can efficiently integrate this information is needed.

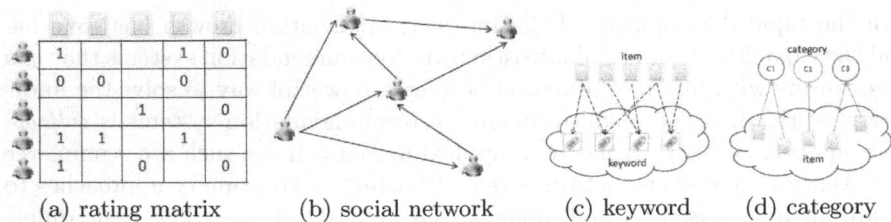

| (a) rating matrix | (b) social network | (c) keyword | (d) category |

Fig. 1. A Toy Example

To address the above problems, we apply a new strategy of *layered learning* toconsider separately different factors in different layers. Motivated by this idea, we propose a two-phase layered learning recommendation framework integrating various information. The main process is defined as: we first learn user's average tastes to every category of items in phase one, then we regard them as baseline estimates and learn more accurate estimates of user's rating for each item with content and social relation ensembled in phase two. We employ matrix factorization to factorize different user preference matrixes: user-category preference matrix and user-keyword preference matrix. According to the two kinds of classification, we design two layered gradient algorithms in our framework, and conduct experiments on real dataset. The experimental result and analysis demonstrate that our framework not only increases the classification accuracy, but also has good performance for dynamic updates of items.

The rest of this paper is organized as follows. In Section 2, we introduce the related work. Our recommendation framework is formulated in Section 3, and experimental results are reported in Section 4. Section 5 is the conclusion.

2   Related Work

2.1   Matrix Factorization(MF)

Matrix factorization is one of the most popular approaches for low-dimensional matrix decomposition. Here, we review the basic MF method [4]. The rating matrix $R \in R^{M \times N}$ (M is the number of users and N is the number of items) can be predicted by UV^T with the user latent factor matrix $U \in R^{D \times M}$ and item latent factor matrix $V \in R^{D \times N}$, where D is the dimension of the vectors. In order to learn the two matrices, the sum-of-squared-error function $\mathcal{L}$ is defined (with Frobenius regularization $\| . \|_F$).

$$\mathcal{L} = \sum_{i=1}^{M} \sum_{j=1}^{N} I_{ij} \left(R_{ij} - U_i^T V_j \right)^2 + \lambda_1 \|U\|_F^2 + \lambda_2 \|V\|_F^2 \tag{1}$$

where λ_1 or λ_2 is the extent of regularization and I_{ij} is the indicator function that is equal to 1 if user i rated item j and equal to 0 otherwise. The optimization problem $\arg \min_{U,V} \mathcal{L}$ can be solved using gradient descent method.

2.2   Classification Based on Flat Approache and Top-Down Approache

Classification is an important data analysis method. It can help us better understand data. Classification can be artificial, also can be automatic based on machine learning. According to the division structure, there are two main classification methods: flat approach and top-down approach [12] (Figure 2). Flat approach divides the data into multi-category directly, not considering the hierarchical relation between categories. Top-down approach uses the divide and conquer technique: classify the current category into some small-scale subcategories, perform the step iteratively until a reasonable classification. In this paper, we introduce the category of items to find the similarity among items.

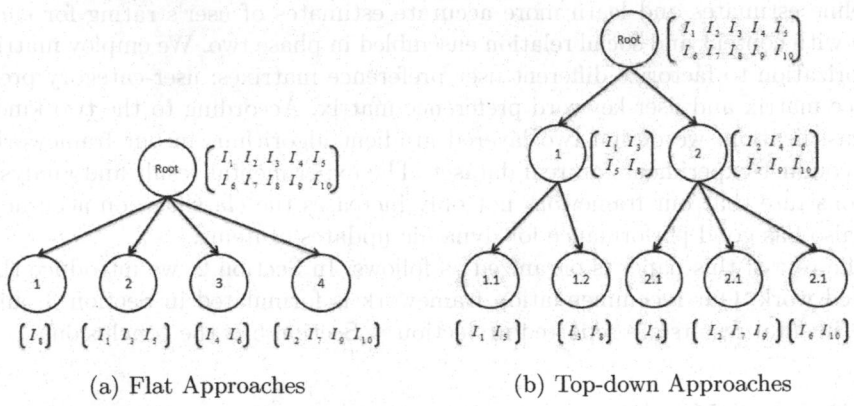

(a) Flat Approaches   (b) Top-down Approaches

Fig. 2. Two kinds of classifications

2.3   Social Recommendation

Traditional recommendation systems assume users are *i.i.d* (independent and identically distributed). In real life, people's decision is often affected by friends' action or recommendation. How to utilize social information has been extensively studied. Trust-aware models [7,9,13] fusing users' social network graph with the rating matrix move an important step forward for recommendation systems. Recently, social-based models make some further improvements. [10] proposed two better methods to leverage the social relation. [8] revealed two important factors: individual preference and interpersonal influence for better utilization of social information. CircleCon [11] used the domain-specific "Trust Circles" to extend the SocialMF [7]. However, all of them give no consideration to item content and the similarity among items. In this paper, we incorporate this information to elaborate recommendation.

3   Layered Learning Frameworks for Recommendation

We introduce the problem description, basic idea and define notations in Section 3.1, and present two layered gradient algorithms in Section 3.2 and 3.3.

3.1   Preliminaries

Because of the weakness of directly modeling every rating mentioned in Section 1, we take advantage of user's tastes to the information of items and indirectly model the rating. Choosing the appropriate category and keyword from the information, we can present the rating matrix as the combination of the user-category preference matrix and user-keyword preference matrix. The first problem is how to fuse the two matrices into the CF model. We apply the *two-phase layered learning* strategy: **Find user's average tastes to every category** and **Ensemble it with content and social relation**. The second problem is how to deal with different classifications. For the flat approach, we directly learn user's tastes to every category , whereas for the top-down approach, we apply the same layered learning strategy: after learning user's average taste to current category, we learn their tastes to the subcategories.

Suppose that we have M users, N items and K keywords. Every item belongs to one category. For the flat approach, we assume that the values of a category are discrete variables in the range $c = \{1, 2, \ldots, n\}$. For the top-down approach, we assume that the category is expressed hierarchically as a string $c_1.c_2.c_3.c_4$, where the categories are delimited by the character '.', ordered in top-down fashion (i.e., category 'c_1' is a parent category of 'c_2', and category 'c_3' is a parent category of 'c_4', and so on). $c_i = \{1, 2, \ldots, n_i\}$ is the set of discrete values of a category in the i-th layer. The rating matrix is denoted by $R \in R^{M \times N}$. We also have a directed social follow graph $G = (\nu, \varepsilon)$ where ν represents the users and the edge set ε represents the following relationships between users.

3.2 Layered Learning Framework on Flat Approach

For the flat approach, we directly learn user's average tastes to every category.

Phase One: Find User's Average Tastes to Every Category. We associate user i with factor vector $U_i \in R^D$ and category k with factor vector $C_k \in R^D$. R_{ij} can be computed by $\hat{R}_{ij} = U_i^T C_{Ca(j)}$, where $Ca(j)$ is the category that item j belongs to. The sum-of-squared-error function $\mathcal{L}_1$ is defined:

$$\mathcal{L}_1 = \sum_{i=1}^{M} \sum_{j=1}^{N} I_{ij} \left(R_{ij} - \hat{R}_{ij} \right)^2 + \lambda_u \|U\|_F^2 + \lambda_c \|C\|_F^2 \qquad (2)$$

We perform gradient descent in U_i and C_k (Eq.3 and 4) to minimize $\mathcal{L}_1$.

$$\frac{\partial \mathcal{L}_1}{\partial U_i} = \sum_{j=1}^{N} I_{ij} C_{Ca(j)} \left(\hat{R}_{ij} - R_{ij} \right) + \lambda_u U_i, \quad \frac{\partial \mathcal{L}_1}{\partial C_k} = \sum_{j \in \phi(k)} \sum_{i \in \varphi(j)} \left(\hat{R}_{ij} - R_{ij} \right) + \lambda_c C_k$$

$$(3)$$

where $\phi(k)$ is the set of the items belong to category k, $\varphi(j)$ is the set of users who have rated item j and λ_u or λ_c is the extent of regularization. After the optimization, we can get the user-category preference matrix $R^c = U^T C$. The matrix *base*, where $base_{ij} = R^c_{iCa(j)} = U_i^T C_{Ca(j)}$ means user i's average taste to item j's category is taken as the initial prediction to R.

Phase Two: Ensemble User's Rating with Content and Social Relation. Although we have user's tastes to every category, user's preference for individual

---

Algorithm 1. Layered gradient algorithm for flat approach

Require: $0 < \alpha_u, \alpha_c, \alpha_k < 1$, $t = 0$.

Ensure: $\mathcal{L}_1^{(0)}(U_i^{(0)}, C_k^{(0)}) \geq 0$, $\mathcal{L}_2^{(0)}\left(U_i^{(0)}, k_z^{(0)}\right) \geq 0$, $\mathcal{L}_1^{(t+1)} < \mathcal{L}_1^{(t)}$, $\mathcal{L}_2^{(t+1)} < \mathcal{L}_2^{(t)}$.

Phase one:

 $Initialization{:} U_i^{(0)}, C_k^{(0)}$

 for $t = 1, 2, \cdots$ do

  Calculate $\frac{\partial \mathcal{L}_1^{(t-1)}}{\partial U_i}$, $\frac{\partial \mathcal{L}_1^{(t-1)}}{\partial C_k}$

  $U_i^{(t)} = U_i^{(t-1)} - \alpha_u \frac{\partial \mathcal{L}_1^{(t-1)}}{\partial U_i}$, $C_k^{(t)} = C_k^{(t-1)} - \alpha_c \frac{\partial \mathcal{L}_1^{(t-1)}}{\partial C_k}$

 end for

Generate the baseline estimate matrix $base$ whose elements are $base_{ij} = U_i^T C_{Ca(j)}$

Phase two:

 $Initialization$: $K_z^{(0)}$. Take current value U_i as the initial value: $U_i^{(0)} \leftarrow U_i$

 for $t = 1, 2, \cdots$ do

  Calculate $\frac{\partial \mathcal{L}_2^{(t-1)}}{\partial U_i}$, $\frac{\partial \mathcal{L}_2^{(t-1)}}{\partial K_z}$

  $U_i^{(t)} = U_i^{(t-1)} - \alpha_u \frac{\partial \mathcal{L}_2^{(t-1)}}{\partial U_i}$, $K_z^{(t)} = K_z^{(t-1)} - \alpha_k \frac{\partial \mathcal{L}_2^{(t-1)}}{\partial K_z}$

 end for

---

item is around the average estimate. For example, a user's taste to one category is 3, but the user's rating for individual item may be some higher 3.3 or some lower 2.9. We introduce user's preference for item's keywords to help optimize the initial estimates. We associate keyword t with factor vector $K_t \in R^D$. The user-keyword preference matrix is denoted by $R^K = U^T K$. $I(j)$ is the set of the keywords extracted from item j. User i's preference for item j's keywords is denoted by $\tilde{R}_{ij} = \sum_{t \in I(j)} R_{it}^K = \sum_{t \in I(j)} U_i^T K_t$. Given the $base_{ij}$, we define the new prediction:$\hat{R}_{ij} = base_{ij} + \tilde{R}_{ij}$. The error function is redefined:

$$\mathcal{L} = \sum_{i=1}^{M} \sum_{j=1}^{N} I_{ij} \left(R_{ij} - base_{ij} - \tilde{R}_{ij} \right)^2 + \lambda_u \|U\|_F^2 + \lambda_k \|K\|_F^2 \qquad (4)$$

Beside item content, we have social network information. Inspired by SoReg [10], with the same assumption that if user i has a friend f, there is a similarity between their tastes, the regularization term to impose constraints between one user and their friends is formulated as:

$$\lambda_f \sum_{i=1}^{M} \sum_{f \in \mathcal{F}^+(i)} Sim\,(i, f)\, \|U_i - U_f\|_F^2 \qquad (5)$$

where $\mathcal{F}^+(i)$ is the set of outlink friends of user i and $Sim\,(i, f) \in [0, 1]$ is the similarity function. We use PCC to compute this value. We change Eq.4 to $\mathcal{L}_2$:

$$\mathcal{L}_2 = \mathcal{L} + \frac{\lambda_f}{2} \sum_{i=1}^{M} \sum_{f \in \mathcal{F}^+(i)} Sim\,(i, f)\, \|U_i - U_f\|_F^2 \qquad (6)$$

We perform gradient descent in U_i and K_z (Eq.7 and 8) to minimize $\mathcal{L}_2$.

$$\frac{\partial \mathcal{L}_2}{\partial U_i} = \sum_{j=1}^{N} I_{ij} \left(\sum_{t \in I(j)} K_t \right) \left(base_{ij} + \tilde{R}_{ij} - R_{ij} \right) + \lambda_f \sum_{f \in \mathcal{F}^+(i)}^{|\mathcal{F}^+(i)|} Sim\,(i,f)\,(U_i - U_f)$$

$$+ \lambda_f \sum_{g \in \mathcal{F}^-(i)}^{|\mathcal{F}^-(i)|} Sim\,(i,g)\,(U_i - U_g) + \lambda_u U_i \tag{7}$$

$$\frac{\partial \mathcal{L}_2}{\partial K_z} = \sum_{j \in \psi(z)} \sum_{i \in \varphi(j)} U_i \left(base_{ij} + \tilde{R}_{ij} - R_{ij} \right) + \eta K_z \tag{8}$$

where $\psi(z)$ is the set of the items that contain the keyword z, $\mathcal{F}^-(i)$ is the set of inlink friends of user i and $|\,\mathcal{F}^+(i)\,|/|\,\mathcal{F}^-(i)\,|$ denote the number of friends in the set $\mathcal{F}^+(i)/\mathcal{F}^-(i)$. The whole algorithm is presented in Algorithm 1.

3.3 Layered Learning Framework on Top-Down Approach

In order to adapt our framework to the multi-layer category, we do some adjustments to Algorithm 1. The improved algorithm is shown in Algorithm 2.

Phase One: Find User's Average Tastes to Every Category. Suppose that the category has L layers. $Ca^l(j)$ is the category that item j belongs to in the l-th layer. We associate user i with latent factor $U_i^l \in R^D$ and category k with latent factor $C_k^l \in R^D$ in the l-th layer. In the $1st$ layer, the method is consistent with phase one of Algorithm 1. In the l-th layer, user's taste to item's category is denoted by $\tilde{R}_{ij}^l = (U_i^l)^T C_{Ca^l(j)}$. Given user's average taste to the parent category $base_{ij}^{l-1}$ in the $l-1$-th layer, R_{ij} can be predicted by $base_{ij}^{l-1} + \tilde{R}_{ij}^l$. The sum-of-squared-error function $\mathcal{L}_1^l$ given by:

$$\mathcal{L}_1^l = \sum_{i=1}^{M} \sum_{j=1}^{N} I_{ij} \left(R_{ij} - base_{ij}^{l-1} - \tilde{R}_{ij}^l \right)^2 + \lambda_u \|U^l\|_F^2 + \lambda_c \|C^l\|_F^2 \tag{9}$$

We perform gradient descent in U_i^l and C_k^l (Eq.10 and 11) to minimize $\mathcal{L}_1^l$ in the l-th layer given by Eq.9.

$$\frac{\partial \mathcal{L}_1^l}{\partial U_i^l} = \sum_{j=1}^{N} I_{ij} C_{Ca^l(j)} \left(base_{ij}^{l-1} + \tilde{R}_{ij}^l - R_{ij} \right) + \lambda_u U_i^l \tag{10}$$

$$\frac{\partial \mathcal{L}_1^l}{\partial C_k^l} = \sum_{j = \in \phi^l(k)} \sum_{i = \in \varphi(t)} U_i^l \left(base_{ij}^{l-1} + \tilde{R}_{ij}^l - R_{ij} \right) + \lambda_c C_k^l \tag{11}$$

where $\phi^l(k)$ is the set of the items belonging to category j in the l-th layer. Then basic estimate $base_{ij}^l$ for the category in the l-th layer is given by: $base_{ij}^l = base_{ij}^{l-1} + (U_i^l)^T C_{Ca^l(j)}$. Repeat the operation down the categories until the

---

Algorithm 2. Layered gradient algorithm for top-down approach

---

Require: $0 < \alpha_u, \alpha_c, \alpha_k < 1$, $t = 0$, $l = 1$.

Ensure: $\mathcal{L}_1^{(0)}(U_i^{(0)}, C_k^{(0)}) \geq 0$, $\mathcal{L}_2^{(0)}\left(U_i^{(0)}, k_z^{(0)}\right) \geq 0$, $\mathcal{L}_1^{(t+1)} < \mathcal{L}_1^{(t)}$, $\mathcal{L}_2^{(t+1)} < \mathcal{L}_2^{(t)}$.

Phase one:

for $l = 1, 2, \cdots, L$ do

   Initialization:$U_i^{l\,(0)} \leftarrow U_i^{l-1}$, $C_k^{l\,(0)}$

   for $t = 1, 2, \cdots$ do

      Calculate $\frac{\partial \mathcal{L}_1^{l\,(t-1)}}{\partial U_i^l}$, $\frac{\partial \mathcal{L}_1^{l\,(t-1)}}{\partial C_k^l}$

      $U_i^{l\,(t)} = U_i^{l\,(t-1)} - \alpha_u \frac{\partial \mathcal{L}_1^{l\,(t-1)}}{\partial U_i^l}$, $C_k^{l\,(t)} = C_k^{l\,(t-1)} - \alpha_k \frac{\partial \mathcal{L}_1^{l\,(t-1)}}{\partial C_k^l}$

   end for

   Generate the baseline estimate matrix $base^l$ in the l-th layer.

   $base_{ij}^l = base_{ij}^{l-1} + \left(U_i^l\right)^T C_{Ca^l(j)}$

end for

Generate the baseline estimate matrix $base$ whose elements are $base_{ij} = base_{ij}^L$

Phase two:

   Initialization:$K_z^{(0)}$. Take current value U_i as the initial value:$U_i^{(0)} \leftarrow U_i^L$

   The following process is the same as phase two in Algorithm 1

---

lowest layer. Finally, we can get the more accurate baseline estimate $base_{ij} = base_{ij}^L$ in the L-th layer.

Phase Two: Ensemble User's Rating with Content and Social Relation. Given the baseline estimate $base$, the phase is the same as the phase two in Algorithm 1.

4   Experimental Results

4.1   Datasets and Metrics

In our experiments, we use the real Tencent Weibo[1] data published by KDD Cup 2012[2]. Beside the social network information, it contains much context information such as keyword, category and timestamp. The items have been organized using four-layer categories, such as "1.2.5.8"; each category belongs to another category, and all categories together form a hierarchy. This structure is suitable for our framework. We predict user's action to items, where "1" represents that the user accepts the item, and "0" otherwise.

We extract a small dataset over a period of time randomly. It is much bigger and richer than other datasets used by [7,11]. The statistics of the dataset are summarized in Table 1.

For the flat approach, we only use the categories in the $4th$ layer. The density of the rating matrix is $\frac{379598}{12518 \times 3610} = 0.84\%$. We divide the dataset into three parts: the training set $R_{train \cdot}$, test set R_{test}, and set R_{new} containing all items not in R_{train}.

---

[1] http://t.qq.com/

[2] http://www.kddcup2012.org/

Table 1. Statistics of dataset extracted

(a) The basic statistics

Description	Number	Description	Number
user	12518	user-item rating	375989
item	3610	item-keyword pair	85107
keyword	1102	Min.Num.of Rating per user	1
Social link	3898	Max.Num.of rating per user	325

(b) Category

category	Number
1st layer	6
2nd layer	23
3rd layer	83
4th layer	258

The evaluation metrics we use in our experiments are two popular error metrics: Mean Absolute Error (MAE) and Root Mean Square Error (RMSE). A smaller MAE or RMSE value means higher accuracy.

4.2 Implementation and Comparisons

We compare our algorithms with four state-of-art CF algorithms.

- PMF [4]: It is a Low-rank matrix factorization based on minimizing the sum-of-squared-error. It does not take into account the social information.
- SocialMF [7] is a trust-based model incorporating the mechanism of trust propagation. It can reduce recommendation error for cold start users.
- SoReg [10]: It is a matrix factorization model with social regularization, which treats dissimilar tastes of friends with different social regularization.
- CircleCon [11]: It incorporates the concept of circle-based recommendation, which only considers the trust circle specific to one category.

We call our Algorithm 1/Algorithm 2 proposed in section 3 LLR1/LLR2. In all the experiments, the tradeoff parameter settings are $\lambda_u = \lambda_c = \lambda_k = \lambda_f = 0.001$.

JAMA[3] is an open matrix package for Java, developed at NIST and the University of Maryland. It provides the fundamental operations of numerical linear algebra. All algorithms are implemented using this library.

4.3 Impacts of Different Factors

The Number of Layers of Category: The difference between LLR1 and LLR2 is the number of layers of category. The results of LLR1/LLR2 (Figure 3) show in phase one, LLR1 achieves 0.1906/0.2910 on MAE/RMSE, but for LLR2, the training of each layer decreases the values: the training of the 1st layer and 2nd layer reduce the values greatly, the training of the 3rd layer and 4th layer have made only minor changes to the results of the 2nd layer, and after the training of the 4th layer, the values can be reduced to 0.1503/0.2739. Contrasts looked, LLR2 has smaller prediction error than LLR1 in phase one. So our framework benefits more from the top-down approaches than the flat approach. We believe classification based on hierarchy can better model the similarity among items.

---

[3] http://math.nist.gov/javanumerics/jama/

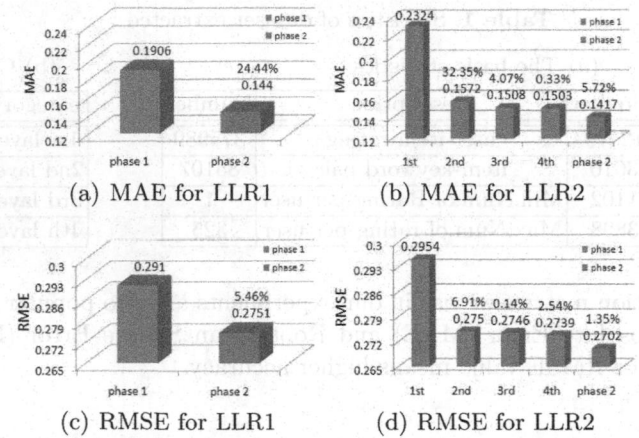

(a) MAE for LLR1   (b) MAE for LLR2

(c) RMSE for LLR1   (d) RMSE for LLR2

Fig. 3. The results of different phases of LLR1 and LLR2 (Dimensionality = 5)

The Item Content and Social Networks Information: After we get the baseline estimate, we discuss how the content and social information may contribute to improving the values. The results of phase two (Figure 3) show we get more accurate estimate of user's rating for individual item: LLR1 and LLR2 achieve 0.1440/0.2751 and 0.1417/0.2702 on MAE/RMSE respectively. For LLR1, this information improves the accuracy as high as 24.44%/5.46% in contrast to phase one. For LLR2, this information improves as high as 5.72%/1.35%. The improvement demonstrates that the content and social information are helpful to boost the performance, especially for LLR1, although classification on the flat approach improves much less than LLR2 based on the top-down approach in phase one, the information significantly enhance more accuracy than LLR2 in phase two. Overall, final results show LLR2 achieves better performance than LLR1.

4.4   Analysis of Recommendation Performance

Figure 4 shows the results of the algorithms on different amounts of training data (25%, 50%, 75%). We observe PMF has the worst MAE/RMSE. SocialMF and SoReg have almost the same accuracy, both superior to PMF, but SoReg is a bit lower because it uses better social regularization terms. CircleCon viewed as an extension of SocialMF is better than the three algorithms. This demonstrates only considering the trust circle belong to one category is useful for learning user's tastes. Our algorithms have the minimum of MAE/RMSE: when the training is 25%, LLR2 gets the decrease by 5.57%/2.99% over CircleCon/SoReg , when the training is 50%, the decrease is 13.68%/10.18% over CircleCon, when the training is 75%, the decrease is 20.88%/6.37% over CircleCon. Experiments demonstrate that our algorithms have higher accuracy than purely using the user-item rating matrix, purely utilizing social networks information or purely considering category-specific circles.

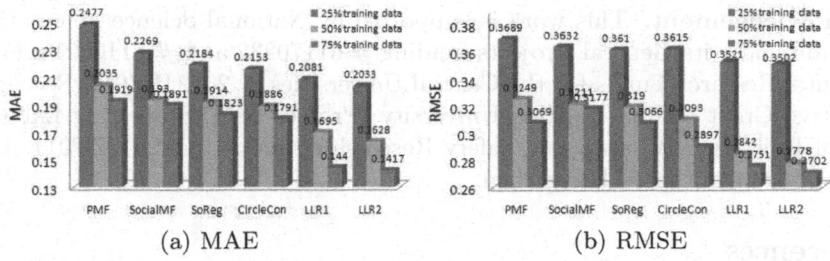

Fig. 4. Performance comparison with other algorithms (Dimensionality = 5)

4.5 Performance on Dynamic Update of Items

We analyze the performance of our algorithms on dynamic update of items, i.e., addition of new items. Except some very special items, usually we can know the keywords and category of new items before their addition. The predicted value of $R_{ij_{new}}$ in R_{new} can be computed by $base_{ij_{new}} + \tilde{R}_{ij_{new}}$. Figure 5 shows the results of our algorithms on training data (25%, 50%, 75%). We observe although the new items are not in the rating matrix, our algorithms still make very good prediction using the new item's category and keywords.

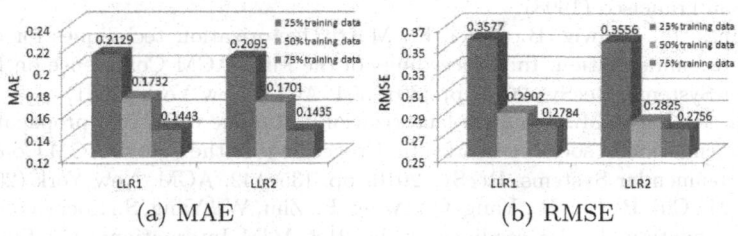

Fig. 5. The results on addition of new items (Dimensionality = 5)

5   Conclusion

In this paper, based on the similarity in the classification, we proposed a novel two-phase layered learning framework, which incorporates the category, item content and social networks information. For two kind of classifications, we designed two layered gradient algorithms in our framework. We conducted extensive experiments on real data. Comparison results of different phases of LLR1 and LLR2 show that the top-down approaches are more helpful to find user's similar tastes than the flat approach, and item content and social networks information contribute to improve the classification accuracy. The analysis results show that our algorithms outperform other state-of-the-art methods. The results also show that our algorithms has good scalability for the dynamic update of items to cope with addition of new items.

Acknowledgement. This work is supported by National Science Foundation of China under its General Projects funding # 61170232 and # 61100218, Fundamental Research Funds for the Central Universities # 2012JBZ017, Research Initiative Grant of Sun Yat-Sen University (Project 985), State Key Laboratory of Rail Traffic Control and Safety Research Grant # RCS2012ZT011. The corresponding author is Hong Shen.

References

1. Su, X., Khoshgoftaar, T.M.: A survey of collaborative filtering techniques. Adv. in Artif. Intell. 2009, 4:2–4:2 (2009)
2. Wang, J., de Vries, A.P., Reinders, M.J.T.: Unifying user-based and item-based collaborative filtering approaches by similarity fusion. In: Proceedings of the 29th Annual International ACM SIGIR Conference on Research and Development in Information Retrieval, SIGIR 2006, pp. 501–508. ACM, New York (2006)
3. Hofmann, T.: Latent semantic models for collaborative filtering. ACM Trans. Inf. Syst. 22(1), 89–115 (2004)
4. Salakhutdinov, R., Mnih, A.: Probabilistic matrix factorization. In: Platt, J., Koller, D., Singer, Y., Roweis, S. (eds.) Advances in Neural Information Processing Systems 20, pp. 1257–1264. MIT Press, Cambridge (2008)
5. Breese, J.S., Heckerman, D., Kadie, C.: Empirical analysis of predictive algorithms for collaborative filtering. In: Proceedings of the Fourteenth Conference on Uncertainty in Artificial intelligence, UAI 1998, pp. 43–52. Morgan Kaufmann Publishers Inc., San Francisco (1998)
6. Baltrunas, L., Ludwig, B., Ricci, F.: Matrix factorization techniques for context aware recommendation. In: Proceedings of the Fifth ACM Conference on Recommender Systems, RecSys 2011, pp. 301–304. ACM, New York (2011)
7. Jamali, M., Ester, M.: A matrix factorization technique with trust propagation for recommendation in social networks. In: Proceedings of the Fourth ACM Conference on Recommender Systems, RecSys 2010, pp. 135–142. ACM, New York (2010)
8. Jiang, M., Cui, P., Liu, R., Yang, Q., Wang, F., Zhu, W., Yang, S.: Social contextual recommendation. In: Proceedings of the 21st ACM International Conference on Information and Knowledge Management, CIKM 2012, pp. 45–54. ACM, New York (2012)
9. Ma, H., King, I., Lyu, M.R.: Learning to recommend with social trust ensemble. In: Proceedings of the 32nd International ACM SIGIR Conference on Research and Development in Information Retrieval, SIGIR 2009, pp. 203–210. ACM, New York (2009)
10. Ma, H., Zhou, D., Liu, C., Lyu, M.R., King, I.: Recommender systems with social regularization. In: Proceedings of the Fourth ACM International Conference on Web Search and Data Mining, WSDM 2011, pp. 287–296. ACM, New York (2011)
11. Yang, X., Steck, H., Liu, Y.: Circle-based recommendation in online social networks. In: Proceedings of the 18th ACM SIGKDD International Conference on Knowledge Discovery and Data Mining, KDD 2012, pp. 1267–1275. ACM, New York (2012)
12. Silla Jr., C.N., Freitas, A.A.: A survey of hierarchical classification across different application domains. Data Min. Knowl. Discov. 22(1-2), 31–72 (2011)
13. Bedi, P., Kaur, H., Marwaha, S.: Trust based recommender system for semantic web. In: IJCAI, pp. 2677–2682 (2007)

Dynamic Circle Recommendation:
A Probabilistic Model

Fan-Kai Chou[1], Meng-Fen Chiang[1], Yi-Cheng Chen[2], and Wen-Chih Peng[1]

[1] Department of Computer Science, National Chiao Tung University, Hsinchu,
Taiwan
[2] Department of Computer Science and Information Engineering,
Tamkang University, Taiwan
{plapla.cs00g,mfchiang.cs95g}@nctu.edu.tw, ycchen@mail.tku.edu.tw,
wcpeng@cs.nctu.edu.tw

Abstract. This paper presents a novel framework for dynamic circle
recommendation for a query user at a given time point from
historical communication logs. We identify the fundamental factors that
govern interactions and aim to automatically form dynamic circle for
scenarios, such as, *who should I dial to in the early morning? whose
mail would I reply first at midnight?* We develop a time-sensitive
probabilistic model (TCircleRank) that not only captures temporal
tendencies between the query user and candidate friends but also
blends frequency and recency into group formation. We also utilize the
model to support two types of dynamic circle recommendation: **Seedset
Generation**: single-interaction suggestion and **Circle Suggestion**:
multiple interactions suggestion. We further present approaches to infer
relevant time interval in determining circles for a query user at a
given time. Experimental results on Enron dataset, Call Detail Records
and Reality Mining Data prove the effectiveness of dynamic circle
recommendation using TCircleRank.

1 Introduction

As the emergence of on-line social media, users can easily share information to
their friends via Mobile Social Media Apps such as Gmail, WhatsApp, Facebook
using their mobile devices. Social media gather and syndicate these information
to target users. Users can browse through the information shared by their friends.
Most existing social media generally render information based on recency, that
is, latest information always appear on top of personal feed walls. Some may
provide manual tools for users to explicitly adjust friend circles so that users can
control how information are rendered on their walls or which friend circles to
share information with. Such great efforts motivate us to wonder: Is it possible to
design a dynamic circle recommendation system which can automatically suggest
a ranked list of friend candidates driven by both historical interaction statistics
and contextual information such as time point?

Most studies on formation of groups mainly focus on static group formation,
where a group is a fixed set of friedns manually pre-defined by a user. We argue

V.S. Tseng et al. (Eds.): PAKDD 2014, Part II, LNAI 8444, pp. 25–37, 2014.

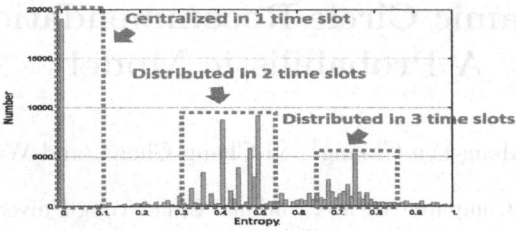

Fig. 1. Distribution of Time Centrality

that the notion of group should dynamically adapt to context information such as location, time, etc. That is, some users may have the tendency to share information to different groups of friends at certain time points while some users may share information to the same group of friends at all time. For example, a user may have the tendency to share information to his/her family during daytime and share information to his/her close colleagues in the evening. To discover time-dependency for a target user, we need to identify his/her tendency at different time point. Following this, we need to provide a ranked list of friends that a user has the highest probability to interact with at each time point.

The general problem of recommendation system has been widely studied [6]. Several prior studies attempt to consider temporal factor in designing recommendation systems [9][3][2][11]. For example, [9] leveraged user's long-term and short-term preferences for temporal recommendation. Nonetheless, non of them addresses the fact that user interactions are not always correlate with time as users present diverse variation of temporal dependency. For example, some users have higher temporal dependency in sharing information. Moreover, a user may only be sensitive to certain time points during a day. In this paper, we argue that temporal tendency should be analyzed individually for each pair of query user and friend candidate at each time point. As an evidence, Figure 1 illustrates a distribution of time centrality for all pairs of users. If a pair of users' interactions only fall into a few time slots during a day, they have lower entropy and thus indicating higher time centrality and vice versa. We observe that over 60% pairs of users' have higher time centrality in interactions (entropy ≤ 0.5), meaning the rest 40% user interactions are driven or dominated by other factors.

In this paper, we propose a framework to discover personalized dynamic circle for a given time point. Given a query user, a time point, and historical communication logs, our recommendation system returns a ranked list of friends (referred to as Circle) for the query user at given time point. To achieve this, we propose a temporal probabilistic model (*TCircleRank*) to capture user behaviors in terms of three factors: frequency, recency and time-dependency. After this, we utilize TCircleRank to derive two types of dynamic circle recommendation: **Seedset Generation**: single interaction suggestion and **Circle Suggestion**: multiple interactions suggestion. TCircleRank considers the dynamic importance of each candidate user for a query user to incorporate the factor, *different users show different temporal dependency with related to a target user at different time.*

Seedset Generation aims to generate a candidate user by TCircleRank for two purposes: shifting the burden for query users (especially mobile users) to provide a list of users who intent to interact with at the very beginning, and the query user merely interact with a single user at given time.

Recommending dynamic circle is useful in many applications. For example, dynamic circle can be utilized to enhance the ranking results for content-based on-line social media (e.g., Gmail, WhatsApp, Facebook), where the information for each user can be adjusted based on the dynamic circle. Moreover, it can be used in location sharing services (e.g., Foursquare), where the ranking of locations can be adjusted based on a user's dynamic circle at particular time point. To summarize, our contributions are as follows.

- We propose a framework to discover personalized dynamic circle for a query user at given time point.
- We propose a temporal probabilistic model (TCircleRank) to capture user's interaction tendency at different time point.
- We consider three fundamental factors in user interactions and propose approaches to support: single interaction suggestion and multiple interactions suggestion.
- We proposes two methods to find the most appropriate time interval for our probabilistic model.
- We conduct experiments on real datasets to demonstrate the effectiveness of our framework and report empirical insights.

This paper is organized as follows. Section 2 presents the related work for this paper. Section 3 introduces TCircleRank and then discusses the two types of dynamic circle recommendation. Section 4 presents two methods to infer time interval for TCircleRank. Section 5 shows the experimental results using the three real datasets. Section 6 concludes this paper.

2   Related Work

2.1   Relationship Link Prediction

Friends suggestion can be modeled as relationship link prediction, if we predict the occurrence of an interaction at a given time. Liben-Nowell and Kleinberg [4] formalized the link prediction problem and employed random walk methods to address this problem. Yang et al. [10] proposed FIP model bridges between Collaborative Filtering (CF) and link prediction to provide a unified treatment for interest targeting and friendship prediction. Sun et al. [7] built a relationship building time prediction model, which uses learning algorithms to fit different distributions and then gets a probability for building relationships between two nodes. However, the edges are only constructed once, so we cannot use it for communication networks which change over time.

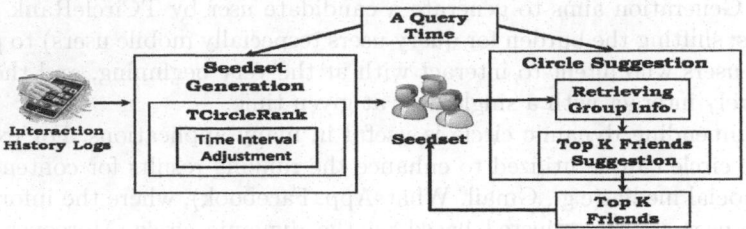

Fig. 2. Framework Overview

2.2 Friends Suggestion System

Our main idea is based on Roth *et al.* [5], who proposed a friends recommendation system for *Gmail* using group information and three criteria. *Gmail* is a well-known mail system constructed by *Google*, which may have many history records to retrieve for friends suggestion. However, the algorithm in [5] could not work effectively for sparse data, insufficient interaction history resulting in some recommendation lists to be empty. Moreover, *Time-Dependency* of user interactions is not addressed in their work. Bartel and Dewan [1] enhanced [5] with a hierarchical structure, which re-orders the recommendation list by ranking past communication group and hierarchically predicts next group. Wu *et al.* [8] proposed a interactive learning framework to formulate the problem of recommending patent partners into a factor graph model. Similarly, no attention has been paid to address the problem of *Time-Dependency* of user interactions.

3   Dynamic Circle Recommendation

We propose a framework for dynamic circle recommendation without requiring query users to provide any information as a prior. The system framework overview is illustrated in Fig. 2. Our system consists of two phases: Seedset Generation and Circle Suggestion. Seedset Generation automatically derives a set of core users (*referred to as seedset*) with the highest probability to be contacted with the query user. Seedset Generation is achieved by mining frequent and time-dependent communication patterns from historical interaction logs. Circle Suggestion phase aims to provide a group of friends whenever the query user intends to interact with multiple users at the same time (*referred to as circle*) based on the derived seedset. Once the query user chooses partial members from the list, our system updates the circle suggestion list by adding selected users to current seedset and then launching Circle Suggestion again to update the ranked list of friends. This process continues until no more friends can be suggested or the query user drops this session. Notice that our recommendation system provides a generic framework where the Circle Suggestion component can be replaced by other state-of-the-art algorithms to serve different requirements.

3.1   TCircleRank

When a query user attempts to share information (e.g., photos), the query user forms a list of friends in his/her mind. Without any assistance, query users has to manually and sequentially select the list of friends by scanning through their friends pool. This brings lots of unnecessary efforts. To solve this problem, we first propose Seedset Generation that uses TCircleRank to predict a set of users as seeds for Circle Suggestion.

We claim that *if a query user interacts with a user in a particular time interval, the query user has a higher probability to interact with the user in the time interval as well.* Fig. 1 verifies this by showing that, over 60% of interactions are strongly temporal-correlated with entropy is no greater than 0.5.

To address this, we propose a framework, TCircleRank, to predict a ranked list of friends who are most likely to be interacted with the query user a given time point. There are three factors considered in TCircleRank:

1. Frequency: Receivers who have more interactions with the query user are more important than those who interact less with the query user.
2. Recency: More recent interactions should have more importance whereas older interactions decay over time.
3. Time-Dependency: If receivers always interact with the query user in a similar time interval, they should have more importance in that time interval.

Frequency is a straightforward yet effective measurement. Inspired from Interaction Rank [5], we unify *Frequency* with *Recency* into a single measurement as shown in Equation (1). [5] introduced a decaying parameter λ, to control the importance of every interaction according to its time. Namely, every interaction decays exponentiation over time with a half life λ. To fit TCircleRank, we form the two factors into a probability, which can be expressed as:

$$P(R_n) = \frac{\sum_{i \in I(R_n)}(\frac{1}{2})^d}{\sum_{i \in I}(\frac{1}{2})^d} \tag{1}$$

where $P(R_n)$ is the probability of the query user interacting with R_n in the past, I is a set of all the query user's interactions, and $I(R_n)$ is a set of all interactions between query user and R_n. d is a decay function which is expressed as $\frac{t_{now}-t_i}{\lambda}$, where t_{now} is the current time, t_i is the time of interaction $i \in I$, and a half-life parameter λ that assigns score 1 to an interaction at current time and decays the importance of an interaction to $\frac{1}{2}$ with the half-life λ.

To incorporate the third factor, *Time-Dependency*, we formulate a conditional probability as:

$$P(R_n|t) = \frac{P(R_n \cap t)}{P(t)}. \tag{2}$$

Equation (2) shows the probability of the query user interacting with R_n in a time interval t, where $P(R_n \cap t)$ and $P(t)$ can be derived like Equation 1 if we change $I(R_n)$ to $I(R_n \cap t)$ and $I(t)$.

To take into the following three factors into consideration, *Frequency*, *Recency* and *Time-Dependency*. Intuitively, we combine $P(R_n)$ and $P(R_n|t)$ by a linear combination with a tunable parameter α, which can be formulated as follows:

$$Score(R_n) = (1 - \alpha)P(R_n) + \alpha \cdot P(R_n|t) \qquad (3)$$

where α is the weight of *Time-Dependency* and the range of α is between 0 and 1. In general, Equation (3) does not make sense, because when a candidate receiver R_n has higher $P(R_n)$ and also has higher $P(R_n|t)$, it should be chosen with more chances. When both probabilities are not relative to each other, we should think about other methods to merge them. Calculating the mean between $P(R_n)$ and $P(R_n|t)$ is a good idea to balance Equation (3), because it considers the influence from not only specific time intervals but also all time intervals. We adjust Equation (3) by using geometric mean, and thus the equations can be expressed as follows:

$$Score_{geo}(R_n) = (1 - \alpha)P(R_n) + \alpha \cdot \sqrt[1+\omega]{P(R_n)(P(R_n|t))^\omega} \qquad (4)$$

where ω represents the weight of a specific time interval. We find that geometric mean makes sense for our assumption: *if one of $P(R_n)$ and $P(R_n|t)$ is much lower than the other, their mean should be closer to the lower one.*

To refine Equation (4), we need to define the best α. According to our observations, we find that not all receivers have high time-dependency, as some shows similar behaviors regardless of any time points. In other words, receivers have different time-dependencies in different time intervals. Thus, time-dependencies will vary from person to person. To achieve this, we change α to another conditional probability, $P(t|R_n)$, which is indicates the probability of R_n interacting with the query user in time interval t. If $P(t|R_n)$ is higher, R_n has a higher time-dependency with the query user and vice versa. We then utilize Z-score to normalize importance of time-dependency. Because Z-score may be negative, we normalize Z-score by considering the central point from the range [-3, 3] to [0, 1]. Therefore, we can reformulate Equation (4) as follows:

$$Score_{final}(R_n) = (1 - NZ(R_n)) \cdot P(R_n)$$
$$+ NZ(R_n) \cdot \sqrt[1+\omega]{P(R_n)(P(R_n|t))^\omega}$$

$$(5)$$

where $NZ(R_n)$ is the normalized Z-score and the range is from 0 to 1.

3.2   Seedset Generation

Seedset Generation phase derives a set of core friends who are most likely to be the receivers with related to the query user at given time. In a sense, Seedset Generation can serve as a Circle Suggestion in a special case when query users intend to communicate with a single user instead of a group of users. In that case, Seedset Generation phase returns the potential receivers as a top-k list of users.

Without specific groups information, Seedset Generation adopts TCircleRank mentioned before predicting which friends in the past are most likely to be the receivers, merely based on specified query time. The algorithm of Seedset Generation is summarized in Algorithm 1.

Algorithm 1. Seedset Generation Algorithm

Input: query user's history interactions I and current time interval t
Output: a set of core friends S

1 $S = \phi$;
2 **foreach** $i \in I$ **do**
3   Sum scores of i for TCircleRank;
4      $C = GetFriend(i)$;
5   **foreach** $c \in C$ **do**
6   **if** $c \notin S$ **then**
7   Put c into S;
8 **foreach** $c \in S$ **do**
9   Calculate all probabilities $P(c)$, $P(t)$, $P(c|t)$ and $P(t|c)$;
10    $S[c] = Score_{final}(c)$;

3.3 Circle Suggestion

Circle Suggestion can be applied to any seed-based suggestion approach. In this subsection, we propose an enhanced approach, Circle Suggestion, by incorporating the state-of-the-art ranking model [5] with TCircleRank.

TCircleRank can be combined with *Interaction Rank* [5]. *Interaction Rank* only considered three factors, *Frequency*, *Recency* and *Direction*, and we consider one additional factor, *Time-Dependency*. *Interaction Rank* is formally defined as follows:

$$\mathcal{IR}(g) = \theta_{out} \sum_{i \in I_{out}(g)} (\frac{1}{2})^d + \sum_{i \in I_{in}(g)} (\frac{1}{2})^d \qquad (6)$$

where $I_{out}(g)$ is the set of outgoing interactions between a query user and a group, $I_{in}(g)$ is the set of incoming interactions and θ_{out} is the weight of outgoing interactions to represent *Direction*. To form a circle of friends, we adopt *Intersection Weighed Score*, which considers the intersection of group and seedset to weight the score of the group. As reported in [5], *Interaction Weighted Score* achieves the best performance among their proposals.

4 Time Interval Adjustment

Considering the following scenario: *A user A has a regular behavior to call user B after user A finishes his works during 5:00pm and 6:00pm. One day, user A has finished his works early at 3:30pm and he calls user B immediately. Should the interaction at 3:30pm be considered as reference interactions in suggesting* friends? To answer this question, we propose two approaches to identify the time intervals as references in ranking friends. The main idea is to analyze the time distribution of interactions in one day and then determine an optimal time interval to describe the interaction behaviors.

4.1 Entropy Examination

We utilize entropy as a measurement to determine the optimal time interval. A narrow time interval indicates regular behavior and a broad time interval indicates relatively irregular behavior. To measure the regularity of user behaviors, we start with 24 time slots and calculate the entropy for user interactions across each time slots. If the entropy is lower than a threshold, which means the level of regularity is higher enough, we choose $(h-1)/2$ as the optimal time interval, where h is the length of each time slot. Otherwise, we continue to split 24 hours into 16, 12, 8, 6 or 4 time slots until the entropy is lower than a threshold.

4.2 Close Peak Detection

To detect the close peak, we only need to know the trends between each time slot. The goal is to find the cluster that contains the current time slot and then we can choose this cluster as optimal time interval. First, we consider the trend between two adjacent time slots. Larger number of interactions time slot should be less or equal τ times than smaller number of interactions time slot, where τ is a threshold for clustering time slots. Otherwise, the detection would be terminated and the final cluster has been determined. Algorithm 2 describes Close Peak Detection in detail.

Algorithm 2. Close Peak Detection

Input: Time Distribution in 24 hours D, Current Time h and Threshold τ
Output: Time Interval Start T_s and Time Interval End T_e

1 $T_s = h$;
2 $T_e = h$;
3 **foreach** *Clockwise Time Slots:* $T_e, x_2 \in D$ **do**
4   **if** $p(x_2) > p(T_e) \& p(x_2) \leq \tau * p(T_e)$ **then**
5          $T_e = x_2$;
6   **else if** $p(x_2) < p(T_e) \& p(T_e) \leq \tau * p(x_2)$ **then**
7          $T_e = x_2$;
8 **foreach** *Counterclockwise Time Slots:* $T_s, x_2 \in D$ **do**
9   **if** $p(x_2) > p(T_s) \& p(x_2) \leq \tau * p(T_s)$ **then**
10         $T_s = x_2$;
11   **else if** $p(x_2) < p(T_s) \& p(T_s) \leq \tau * p(x_2)$ **then**
12         $T_s = x_2$;

---

5   Experiment

5.1   Datasets

Social interactions present in calling and mailing behaviors. Therefore, we use calling behavior and mailing behavior datasets to simulate general social behavior dataset. In our experiment, we use three real datasets, Enron Mail[1], call detail records (cdr) from Chunghwa Telecom (CHT)[2] and Reality Mining Dataset (RMD) from MIT[3]. The basic information of each dataset is shown in Table 1, where Enron Mail contains multiple interaction data and the others only contains single interaction data. Therefore, we adopt Enron Mail to evaluate Seedset Generation and Circle Suggestion, and the others two dataset to evalute Seedset Generation.

Table 1. Basic Information on the Enron/CHT/RMD Datasets

Element	Enron	CHT	RMD
No. of user	65,182	76,263	92
No. of interactions	236,505	2,443,667	78,110
No. of group interactions	67,631	-	-
time	1998/01/04 - 2002/12/21	2010/08	2004/01/19 - 2005/07/15

5.2   Time Centrality Analysis

In time centrality experiment, we constrained the number of interaction between the user and the test query user exceeds four times, because we split the time of one day into four time slots of six hours.

We observe the difference of time centrality distribution between Enron and other datasets on Fig. 3 and find that Enron Mail has higher time centrality because its entropy is relatively lower than those of CHT and RMD. This indicates that mailing behavior is relatively regular for the same receiver, i.e., most user tend to send their mail to the same receiver at particular time points. Unlike mailing behavior, calling behavior does not show strong time centrality. The calling behaviors in CHT and RMD are similar and they distribute around entropy 0.5. This explains that when the entropy is 0.5, the users call callees not only at the same time slot but also at the adjacent time slots. In other words, the regular calling behavior may shift to the temporally close time points occasionally.

5.3   Experimental Setup

For Enron Mail, we chose 21,262 mails from Enron Mail to be the testing data and extracted 30 days before testing data to be the training data, where the

---

[1] The Enron Mail data can be downloaded from
   http://www.cs.cmu.edu/~enron/
[2] The CHT data is not in public, and Chunghwa Telecom's website is
   http://www.cht.com.tw/
[3] The Reality Mining Dataset can be downloaded from
   http://realitycommons.media.mit.edu/realitymining4.html

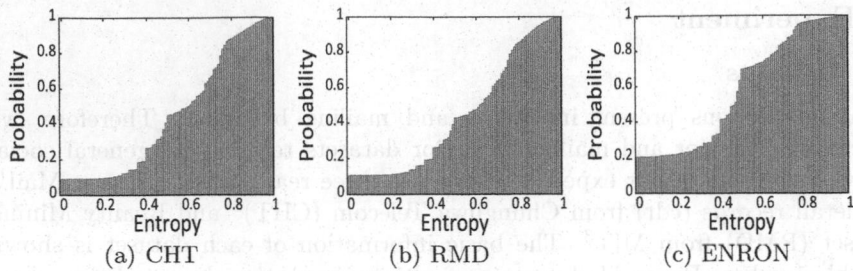

Fig. 3. CDF of Time Centrality with 4 time slots in (a) CHT (b) RMD (c) Enron

Table 2. Parameter Settings

Parameter	Meaning	Enron	CHT	RMD
λ	time decay parameter	7 days	3 days	3 days
θ_{out}	outlink weight parameter	5	-	-
ω	time dependency parameter	1	1	1
Time Interval	additional hours next to the current hour	1 hour	1 hour	1 hour

rule in selecting testing data is as follows: (1) the mail should be sent to at least two receivers, or a *group*, and (2) the sender of the mail had sent no less than four mails before. CHT, which is a single interaction data, do not have group information, because CHT consists of cell phone call records and we only need to predict the most likely callee. We chose 30,295 records from CHT to be the testing data and extracted 30 days before testing data to be the training data. The testing data is all in the last day in CHT. We chose 44,166 records from RMD to be the testing data and extracted 30 days before testing data to be the training data. Parameter settings are shown in Table 2.

To evaluate the recommendation quality, we adopt normalized discounted cumulative gain (nDCG) as the measurements. DCG measures the *gain* of a hit result based on its rank in the list, where the top rank has more gain and the lower rank has less gain.

5.4 Circle Recommendation Quality

Evaluation on Seedset Generation: Figure 4(a)(b)(c) shows the impact of each fundamental factor: Frequency (F), Recency (R) and Time-dpendency (T) on Seedset Generation quality. We also compare with RecentLog which directly generates the recommendation list in order by the recent contacts.

In Fig. 4(a), the pink line (All) is our proposal which considers all factors and outperforms other models with at most 4.2% increase in accuracy compare to the baseline. In Fig. 4(b), it is worth mentioning that the lines assemble the log-likelihood, because CHT only has one receiver for recommendation in each record. Our proposal outperforms other models with 26% increase in accuracy compare to the baseline when k is 5. The similar results could be found on

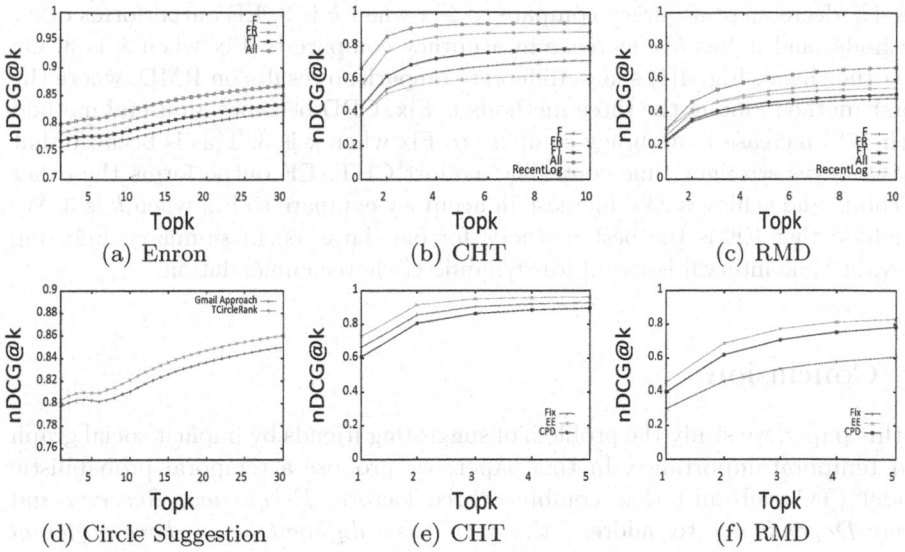

Fig. 4. nDCG Comparison for Seedset Generation, Circle Suggestion and Time Interval Adjustment

RMD. Fig. 4(c) shows the nDCG comparison for four models and our proposal outperforms other models with 16% increase in accuracy to the baseline when k is 5. Based on above results, we conclude that TCircleRank presents consistent improvement than straightforward suggestion such as frequency or recency.

Evaluation on Circle Suggestion: Fig. 4(d) shows the performance comparison of Gmail Approach [5] and TCircleRank. Because Gmail Approach is a seed-based suggestion approach, we use Seedset Generation to generate a seedset with $k = 3$ and pass the input to Gmail and TCircleRank respectively. The final recommendation list contains the seeds which are different from the original Gmail Approach, but it will not affect the recommendation result because the seeds appears at the top of the list and they are also uncertain receivers for the query user. We use the same test data from Enron Mail as in Fig. 4(a). In Fig. 4(d), the x-axis is top-k ($1 \leq k \leq 30$) and the y-axis is the nDCG value, where the red line is Gmail Approach and the green line is TCircleRank. We can see that no matter in what situation, TCircleRank always has higher nDCG than Gmail Approach.

5.5   Time Interval Adjustment Quality

We compare fixed time interval and our proposed methods. We set δ to 0.5 for Entropy Examination (EE) and τ to 2 for Close Peak Detection (CPD). In Fix Time Interval (Fix), we fixed the time interval to 1 hour because it results in highest nDCG among all fixed time intervals.

Fig. 4(e) shows three methods comparison on CHT. The x-axis is top-k and the y-axis is the nDCG. CPD has the lowest nDCG among three methods, and it has 4% decrease in accuracy compare to Fix when k is 3. EE outperforms other methods, and it has 5% increase in accuracy compare to Fix when k is 3. On the other hand, Fig. 4(f) shows different comparison results on RMD, where the worst method among the three methods is Fix. CPD becomes an useful method with 17% increase in accuracy compare to Fix when k is 3. This is because that RMD shows stronger time centrality against CHT. EE outperforms the other methods and achieves 24% increase in accuracy compare to Fix when k is 3. We conclude that EE is the best methods for our datasets. In summary, inferring relevant time interval is useful for dynamic circle recommendation.

6   Conclusion

In this paper, we study the problem of suggesting friends by implicit social graph and temporal importance. In this paper, we propose a temporal probabilistic model (TCircleRank) that combine three factors, *Frequency*, *Recency* and *Time-Dependency* to address the fact that *different users have different importance of time for a query user*. Based on TCircleRank, Seedset Generation generates a set of seeds automatically. To recommend circles, we utilize the seedset generated by TCircleRank and considers an additional feature, *Direction* of interactions in our Circle Suggestion approach. We enhance the probabilistic model by further dynamically determine the time interval, which is a parameter to identify time-dependent interactions in derived time intervals. Our experiment results show that TCircleRank and dynamic circle recommendation system are effective on three real datasets, Enron Mail, CHT call detail records and Reality Mining Dataset. We also show that inferring optimal time interval is useful for dynamic circle recommendation. We will extend TCircleRank by automatically deciding the number of seeds and using user clusters. We will further apply our approach in other applications such as content-based sharing and temporal community detection.

References

1. Bartel, J., Dewan, P.: Towards hierarchical email recipient prediction
2. Koren, Y.: Collaborative filtering with temporal dynamics. Communications of the ACM 53(4), 89–97 (2010)
3. Lathia, N., Hailes, S., Capra, L., Amatriain, X.: Temporal diversity in recommender systems. In: Proceedings of the 33rd Annual International ACM SIGIR Conference on Research and Development in Information Retrieval (SIGIR 2010), pp. 210–217 (2010)
4. Liben-Nowell, D., Kleinberg, J.: The link-prediction problem for social networks. Journal of the American Society for Information Science and Technology 58(7), 1019–1031 (2007)
5. Roth, M., Ben-David, A., Deutscher, D., Flysher, G., Horn, I., Leichtberg, A., Leiser, N., Matias, Y., Merom, R.: Suggesting friends using the implicit social graph. In: Proceedings of the 16th ACM SIGKDD International Conference on Knowledge Discovery and Data Mining, pp. 233–242. ACM (2010)

6. Su, X., Khoshgoftaar, T.M.: A survey of collaborative filtering techniques. Advances in Artificial Intelligence 2009, 4 (2009)
7. Sun, Y., Han, J., Aggarwal, C.C., Chawla, N.V.: When will it happen?: Relationship prediction in heterogeneous information networks. In: Proceedings of the Fifth ACM International Conference on Web Search and Data Mining, pp. 663–672. ACM (2012)
8. Wu, S., Sun, J., Tang, J.: Patent partner recommendation in enterprise social networks. In: Proceedings of the Sixth ACM International Conference on Web Search and Data Mining, pp. 43–52. ACM (2013)
9. Xiang, L., Yuan, Q., Zhao, S., Chen, L., Zhang, X., Yang, Q., Sun, J.: Temporal recommendation on graphs via long-and short-term preference fusion. In: Proceedings of the 16th ACM SIGKDD International Conference on Knowledge Discovery and Data Mining, pp. 723–732. ACM (2010)
10. Yang, S.-H., Long, B., Smola, A., Sadagopan, N., Zheng, Z., Zha, H.: Like like alike: Joint friendship and interest propagation in social networks. In: Proceedings of the 20th International Conference on World Wide Web, pp. 537–546. ACM (2011)
11. Zheng, N., Li, Q.: A recommender system based on tag and time information for social tagging systems. Expert Systems with Applications 38(4), 4575–4587 (2011)

HOSLIM: Higher-Order Sparse LInear Method for Top-N Recommender Systems

Evangelia Christakopoulou and George Karypis

Computer Science & Engineering
University of Minnesota, Minneapolis, MN
{evangel,karypis}@cs.umn.edu

Abstract. Current top-N recommendation methods compute the recommendations by taking into account only relations between pairs of items, thus leading to potential unused information when higher-order relations between the items exist. Past attempts to incorporate the higher-order information were done in the context of neighborhood-based methods. However, in many datasets, they did not lead to significant improvements in the recommendation quality. We developed a top-N recommendation method that revisits the issue of higher-order relations, in the context of the model-based Sparse LInear Method (SLIM). The approach followed (Higher-Order Sparse LInear Method, or HOSLIM) learns two sparse aggregation coefficient matrices S and S' that capture the item-item and itemset-item similarities, respectively. Matrix S' allows HOSLIM to capture higher-order relations, whose complexity is determined by the length of the itemset. Following the spirit of SLIM, matrices S and S' are estimated using an elastic net formulation, which promotes model sparsity. We conducted extensive experiments which show that higher-order interactions exist in real datasets and when incorporated in the HOSLIM framework, the recommendations made are improved. The experimental results show that the greater the presence of higher-order relations, the more substantial the improvement in recommendation quality is, over the best existing methods. In addition, our experiments show that the performance of HOSLIM remains good when we select S' such that its number of nonzeros is comparable to S, which reduces the time required to compute the recommendations.

1 Introduction

In many widely-used recommender systems [1], users are provided with a ranked list of items in which they will likely be interested in. In these systems, which are referred to as top-N recommendation systems, the main goal is to identify the most suitable items for a user, so as to encourage possible purchases. In the last decade, several algorithms for top-N recommendation tasks have been developed [12], the most popular of which are the neighborhood-based (which focus either on users or items) and the matrix-factorization methods. The neighborhood-based algorithms [6] focus on identifying similar users/items based on a user-item purchase/rating matrix. The matrix-factorization algorithms [5] factorize

V.S. Tseng et al. (Eds.): PAKDD 2014, Part II, LNAI 8444, pp. 38–49, 2014.

the user-item matrix into lower rank user factor and item factor matrices, which represent both the users and the items in a common latent space.

Though matrix factorization methods have been shown to be superior for solving the problem of rating prediction, item-based neighborhood methods are shown to be superior for the top-N recommendation problem [3,6,9,10]. In fact the winning method in the recent million song dataset challenge [3] was a rather straightforward item-based neighborhood top-N recommendation approach.

The traditional approaches for developing item-based top-N recommendation methods (k-Nearest Neighbors, or k-NN) [6] use various vector-space similarity measures (e.g., cosine, extended Jaccard, Pearson correlation coefficient, etc.) to identify for each item the k most similar other items based on the sets of users that co-purchased these items. Then, given a set of items that have already been purchased by a user, they derive their recommendations by combining the most similar unpurchased items to those already purchased. In recent years, the performance of these item-based neighborhood schemes has been significantly improved by using supervised learning methods to learn a model that both captures the similarities (or aggregation coefficients) and also identifies the sets of neighbors that lead to the best overall performance [9,10]. One of these methods is SLIM [10], which learns a sparse aggregation coefficient matrix from the user-purchase matrix, by solving an optimization problem. It was shown that SLIM outperforms other top-N recommender methods [10].

However, there is an inherent limitation to both the old and the new top-N recommendation methods as they capture only pairwise relations between items and they are not capable of capturing higher-order relations. For example, in a grocery store, users tend to often buy items that form the ingredients in recipes. Similarly, the purchase of a phone is often combined with the purchase of a screen protector and a case. In both of these examples, purchasing a subset of items in the set significantly increases the likelihood of purchasing the rest. Ignoring these types of relations, when present, can lead to suboptimal recommendations.

The potential of improving the performance of top-N recommendation methods was recognized by Mukund et al. [6], who incorporated combinations of items (i.e., itemsets) in their method. In that work, the most similar items were identified not only for each individual item, but also for all sufficiently frequent itemsets that are present in the active user's basket. This method referred to as HOKNN (Higher-Order k-NN) computes the recommendations by combining itemsets of different size. However, in most datasets this method did not lead to significant improvements. We believe that the reason for this is that the recommendation score of an item is computed simply by an item-item or itemset-item similarity measure, which does not take into account the subtle relations that exist when these individual predictors are combined.

In this paper, we revisit the issue of utilizing higher-order information, in the context of model-based methods. The research question answered is whether the incorporation of higher-order information in the recently developed model-based top-N recommendation methods will improve the recommendation quality further. The contribution of this paper is two-fold: First, we verify the existence of higher-order information in real-world datasets, which suggests that higher-order relations do exist and thus if properly taken into account, they can lead to performance improvements. Second, we develop an approach referred to as Higher-Order Sparse Linear Method, (HOSLIM) in which the itemsets capturing the higher-order information are treated as additional items and their contribution to the overall recommendation score is estimated using the model-based framework introduced by SLIM. We conduct a comprehensive set of experiments on different datasets from various applications. The results show that this combination improves the recommendation quality beyond the current best results of top-N recommendation. In addition, we show the effect of the support threshold chosen on the quality of the method. Finally, we present the requirements that need to be satisfied in order to ensure that HOSLIM computes the predictions in an efficient way.

The rest of the paper is organized as follows. Section 2 introduces the notations used in this paper. Section 3 presents the related work. Section 4 explains the method proposed. Section 5 provides the evaluation methodology and the dataset characteristics. In Section 6, we provide the results of the experimental evaluation. Finally, Section 7 contains some concluding remarks.

2 Notations

In this paper, all vectors are represented by bold lower case letters and they are column vectors (e.g., $\mathbf{p}$, $\mathbf{q}$). Row vectors are represented by having the transpose superscript ^{T}, (e.g., $\mathbf{p}^T$). All matrices are represented by upper case letters (e.g., R, W). The ith row of a matrix A is represented by $\mathbf{a}_i^T$. A predicted value is denoted by having a $\sim$ over it (e.g., $\tilde{r}$).

The number of users will be denoted by n and the number of items will be denoted by m. Matrix R will be used to represent the *user-item implicit feedback* matrix of size $n \times m$, containing the items that the users have purchased/viewed. Symbols u and i will be used to denote individual users and items, respectively. An entry (u, i) in R, r_{ui}, will be used to represent the feedback information for user u on item i. R is a binary matrix. If the user has provided feedback for a particular item, then the corresponding entry in R is 1, otherwise it is 0. We will refer to the items that the user has bought/viewed as purchased items and to the rest as unpurchased items.

Let $\mathcal{I}$ be the set of sets of items that are co-purchased by at least σ users in R, where σ denotes the minimum support threshold. We will refer to these sets as *itemsets* and we will use p to denote the cardinality of $\mathcal{I}$ (i.e., $p = |\mathcal{I}|$). Let R' be a matrix whose columns correspond to the different itemsets in $\mathcal{I}$ (the size of this matrix is $n \times p$). In this matrix r'_{uj} will be one, if user u has purchased *all* the items corresponding to the itemset of the jth column of R' and zero otherwise. We refer to R' as the *user-itemset implicit feedback* matrix. We will use $\mathcal{I}_j$ to denote the set of items that constitute the itemset of the jth column of R'. In the rest of the paper, every itemset will be of size two (unless stated otherwise) and considered to be frequent, even if it is not explicitly stated.

3   Related Work

In this paper, we combine the idea of higher-order models introduced by HOKNN with SLIM. The overview of these two methods is presented in the following subsections.

3.1   Higher-Order k-Nearest Neighbors Top-N Recommendation Algorithm (HOKNN)

Mukund et al. [6] had pointed out that the recommendations could potentially be improved, by taking into account higher-order relations, beyond relations between pairs of items. They did that by incorporating combinations of items (itemsets) in the following way: The most similar items are found not for each individual item, as it is typically done in the neighborhood-based models, but for all possible itemsets up to a particular size l.

3.2   Sparse LInear Method for top-N Recommendation (SLIM)

SLIM computes the recommendation score on an unpurchased item i of a user u as a sparse aggregation of all the user's purchased items:

$$\tilde{r}_{ui} = \mathbf{r}_u^T \mathbf{s}_i, \tag{1}$$

where $\mathbf{r}_u^T$ is the row-vector of R corresponding to user u and $\mathbf{s}_i$ is a sparse size-m column vector which is learned by solving the following optimization problem:

$$\begin{aligned}
\underset{\mathbf{s}_i}{\text{minimize}} \ & \tfrac{1}{2}||\mathbf{r}_i - R\mathbf{s}_i||_2^2 + \tfrac{\beta}{2}||\mathbf{s}_i||_2^2 + \lambda||\mathbf{s}_i||_1, \\
\text{subject to } & \mathbf{s}_i \geq 0 \\
& s_{ii} = 0,
\end{aligned} \tag{2}$$

where $||\mathbf{s}_i||_2^2$ is the l_2 norm of $\mathbf{s}_i$ and $||\mathbf{s}_i||_1$ is the entry-wise l_1 norm of $\mathbf{s}_i$. The l_1 regularization gets used so that sparse solutions are found [13]. The l_2 regularization prevents overfitting. The constants β and λ are regularization parameters. The non-negativity constraint is applied so that the matrix learned will be a positive aggregation of coefficients. The $s_{ii} = 0$ constraint makes sure that when computing the weights of an item, that item itself is not used as this would lead to trivial solutions. All the $\mathbf{s}_i$ vectors can be put together into a matrix S, which can be thought of as an item-item similarity matrix that is learned from the data. So, the model introduced by SLIM can be presented as $\tilde{R} = RS$.

4   HOSLIM: Higher-Order Sparse LInear Method for Top-N Recommendation

The ideas of the higher-order models can be combined with the SLIM learning framework in order to estimate the various item-item and itemset-item similarities. In this approach, the likelihood that a user will purchase a particular item is computed as a sparse aggregation of both the items purchased and the itemsets that it supports. The predicted score for user u on item i is given by

$$\tilde{r}_{ui} = \mathbf{r}_u^T \mathbf{s}_i + \mathbf{r'}_u^T \mathbf{s'}_i, \tag{3}$$

where $\mathbf{s}_i$ is a sparse vector of size m of aggregation coefficients for items and $\mathbf{s'}_i$ is sparse vector of size p of aggregation coefficients for itemsets.

Thus, the model can be presented as:

$$\tilde{R} = RS + R'S', \tag{4}$$

where R is the user-item implicit feedback matrix, R' is the user-itemset implicit feedback matrix, S is the sparse coefficient matrix learned corresponding to items (size $m \times m$) and S' is the sparse coefficient matrix learned corresponding to item-sets (size $p \times m$). The ith columns of S and S' are the $\mathbf{s}_i$ and $\mathbf{s'}_i$ of Equation 3.

Top-N recommendation gets done for the uth user by computing the scores for all the unpurchased items, sorting them and then taking the top-N values.

The sparse matrices S and S' encode the similarities (or aggregation coefficients) between the items/itemsets and the items. The ith columns of S and S' can be estimated by solving the following optimization problem:

$$\begin{aligned}
\underset{s_i, s'_i}{\text{minimize}} \quad & \tfrac{1}{2}||\mathbf{r}_i - R\mathbf{s}_i - R'\mathbf{s'}_i||_2^2 \quad + \tfrac{\beta}{2}||\mathbf{s}_i||_2^2 + \tfrac{\beta}{2}||\mathbf{s'}_i||_2^2 \\
& \hspace{8em} + \lambda||\mathbf{s}_i||_1 + \lambda||\mathbf{s'}_i||_1 \\
\text{subject to} \quad & \mathbf{s}_i \geq 0 \\
& \mathbf{s'}_i \geq 0 \\
& s_{ii} = 0, \text{ and} \\
& s'_{ji} = 0, \text{ where } \{i \in \mathcal{I}_j\}.
\end{aligned} \tag{5}$$

The constraint $s_{ii} = 0$ makes sure that when computing r_{ui}, the element r_{ui} is not used. If this constraint was not enforced, then an item would recommend itself. Following the same logic, the constraint $s'_{ji} = 0$ ensures that the itemsets j for which $i \in \mathcal{I}_j$ will not contribute to the computation of r_{ui}.

The optimization problem of Equation 5 can be solved using coordinate descent and soft thresholding [7].

5   Experimental Evaluation

5.1   Datasets

We evaluated the performance of HOSLIM on a wide variety of datasets, both synthetic and real. The datasets we used include point-of-sales, transactions, movie ratings and social bookmarking. Their characteristics are shown in Table 1.

The *groceries* dataset corresponds to transactions of a local grocery store. Each user corresponds to a customer and the items correspond to the distinct products purchased over a period of one year. The *synthetic* dataset was generated by using the IBM synthetic dataset generator [2], which simulates the

Table 1. Dataset Characteristics

Name	#Users	#Items	#Transactions	Density	Average Basket Size
groceries	63,035	15,846	1,997,686	0.2%	31.69
synthetic	5000	1000	68,597	1.37%	13.72
delicious	2,989	2,000	243,441	4.07%	81.44
ml	943	1,681	99,057	6.24%	105.04
retail	85146	16470	820,414	0.06%	9.64
bms-pos	435,319	1,657	2,851,423	0.39%	6.55
bms1	26,667	496	90,037	0.68%	3.37
ctlg3	56,593	39,079	394,654	0.017%	6.97

Columns corresponding to #users, #items and #transactions show the number of users, number of items and number of transactions, respectively, in each dataset. The column corresponding to density shows the density of each dataset (i.e., density=#transactions/(#users×#items)). The average basket size is the average number of transactions for each user.

behavior of customers in a retail environment. The parameters we used for generating the dataset were: average size of itemset= 4 and total number of itemsets existent= 1, 200. The *delicious* dataset [11] was obtained from the eponymous social bookmarking site. The items on this dataset correspond to tags. A non-zero entry indicates that the corresponding user wrote a post using the corresponding tag. The *ml* dataset corresponds to MovieLens 100K dataset, [8] which represents movie ratings. All the ratings were converted to one, showing whether a user rated a movie or not. The *retail* dataset [4] contains the retail market basket data from a Belgian retail store. The *bms-pos* dataset [14] contains several years worth of point-of-sales data from a large electronics retailer. The *bms1* dataset [14] contains several months worth of clickstream data from an e-commerce website. The *ctlg3* dataset corresponds to the catalog purchasing transactions of a major mail-order catalog retailer.

5.2   Evaluation Methodology

We employed a 10-fold leave-one-out cross-validation to evaluate the performance of the proposed model. For each fold, one item was selected randomly for each user and this was placed in the test set. The rest of the data comprised the training set. We used only the data in the training set for both the itemset discovery and model learning.

We measured the quality of the recommendations by comparing the size-N recommendation list of each user and the item of that user in the test set. The quality measure used was the hit-rate (HR). HR is defined as follows,

$$HR = \frac{\#hits}{\#users}, \tag{6}$$

where "#users" is the total number of users (n) and "#hits" is the number of users whose item in the test set is present in the size-N recommendation list.

5.3 Model Selection

We performed an extensive search over the parameter space of the various methods, in order to find the set of parameters that gives us the best performance for all the methods. We only report the performance corresponding to the parameters that lead to the best results. The l_1 regularization λ was chosen from the set of values: $\{0.0001, 0.001, 0.01, 0.1, 1, 2, 5\}$. The l_F regularization parameter β ranged in the set: $\{0.01, 0.1, 1, 3, 5, 7, 10\}$. The larger β and λ were, the stronger the regularizations were. The number of neighbors examined lied in the interval $[1 - 50, 60, 70, 80, 90, 100, 200, 300, 400, 500, 600, 700, 800, 900, 1000]$. The support threshold σ took on values $\{10, 15, 20, 25, 30, 35, 40, 45, 50, 55, 60, 65,$ $70, 75, 80, 85, 90, 95, 100, 150, 200, 250, 300, 350, 400, 450, 500, 550, 600, 650,$ $700, 750, 800, 850, 900, 950, 1000, 1500, 2000, 2500, 3000\}$.

6   Experimental Results

The experimental evaluation consists of two parts. First, we analyze the various datasets in order to assess the extent to which higher-order relations exist in them. Second, we present the performance of HOSLIM and compare it to SLIM as well as HOKNN.

6.1   Verifying the Existence of Higher-Order Relations

We verified the existence of higher-order relations in the datasets, by measuring how prevalent are the itemsets with strong association between the items that comprise it (beyond pairwise associations). In order to identify such itemsets, (which will be referred to as "good"), we conducted the following experiment. We found all frequent itemsets of size 3 with σ equal to 10. For each of these itemsets we computed two quality metrics. The first is

$$dependency\_max = \frac{P(ABC)}{max(P(AB), P(AC), P(BC))},$$

(7)

which measures how much greater the probability of a purchase of all the items of an itemset is than the maximum probability of the purchase of an induced pair. The second is

$$dependency\_min = \frac{P(ABC)}{min(P(AB), P(AC), P(BC))},$$

(8)

which measures how much greater the probability of the purchase of all the items of an itemset is than the minimum probability of the purchase of an induced pair. These metrics are suited for identifying the "good" itemsets, as

Table 2. Coverage by Affected Users/Non-zeros

Name	Percentage (%) with at least one "good" itemset of dependency:							
	max$\geq$ 2		max$\geq$ 5		min$\geq$ 2		min$\geq$ 5	
	users	non-zeros	users	non-zeros	users	non-zeros	users	non-zeros
groceries	95.17	68.30	88.11	47.91	97.53	84.69	96.36	73.09
synthetic	98.04	76.50	98.00	75.83	98.06	76.80	98.06	76.79
delicious	81.33	59.02	55.34	22.88	81.80	59.97	72.57	44.14
ml	99.47	69.77	28.42	3.75	99.89	77.94	63.63	37.62
retail	23.54	13.69	8.85	4.10	49.70	40.66	38.48	25.63
bms-pos	59.66	81.51	32.61	44.77	66.71	91.92	51.53	80.09
bms1	31.52	63.18	29.47	60.82	31.55	63.22	31.54	63.21
ctlg3	34.95	24.85	34.94	24.81	34.95	24.85	34.95	24.85

The percentage of users/non-zeros with at least one "good" itemset. The itemsets considered have a support threshold of 10, except in the case of *delicious* and *ml*, where the support threshold is 50, (as *delicious* and *ml* are dense datasets and thus a large number of itemsets is induced).

they discard the itemsets that are frequent just because their induced pairs are frequent. Instead, the above-mentioned metrics discover the frequent itemsets that have all or some infrequent induced pairs, meaning that these itemsets contain higher-order information.

Given these metrics, we then selected the itemsets of size three that have quality metrics greater than 2 and 5. The higher the quality cut-off, the more certain we are that a specific itemset is "good".

For these sets of high quality itemsets, we analyzed how well they cover the original datasets. We used two metrics of coverage. The first is the percentage of users that have at least one "good" itemset, while the second is the percentage of the non-zeros in the user-item matrix R covered by at least one "good" itemset (shown in Table 2). A non-zero in R is considered to be covered, when the corresponding item of the non-zero value participates in at least one "good" itemset supported by the associated user.

We can see from Table 2 that not all datasets have uniform coverage with respect to high quality itemsets. The *groceries* and *synthetic* datasets contain a large number of "good" itemsets that cover a large fraction of non-zeros in R and nearly all the users. On the other hand, the *ml*, *retail* and *ctlg3* datasets contain "good" itemsets that have significantly lower coverage with respect to both coverage metrics. The coverage characteristics of the good itemsets that exist in the remaining datasets is somewhere in between these two extremes. These results suggest that the potential gains that HOSLIM can achieve will vary across the different datasets and should perform better for *groceries* and *synthetic* datasets.

Table 3. Comparison of 1st order with 2nd order models

| | | SLIM models | | | | | | | k-NN models | | | |
| | | SLIM | | HOSLIM | | | Improved | k-NN | | HOKNN | | Improved |
Dataset	β	λ	HR	σ	β	λ	HR	%	nnbrs	HR	nnbrs	σ	HR	%
groceries	5	0.001	0.259	10	10	0.0001	**0.338**	32.03	1000	0.174	800	10	0.240	37.93
synthetic	0.1	0.1	0.733	10	3	1	**0.860**	17.33	41	0.697	47	10	0.769	10.33
delicious	10	0.01	0.148	50	10	0.01	**0.156**	5.41	80	0.134	80	10	0.134	0
ml	1	5	0.338	180	5	0.0001	**0.349**	3.25	15	0.267	15	10	0.267	0
retail	10	0.0001	0.310	10	10	0.1	**0.317**	2.26	1000	0.281	1,000	10	0.282	0.36
bms-pos	7	2	0.502	20	10	5	**0.509**	1.39	700	0.478	600	10	0.480	0.42
bms1	15	0.01	0.588	10	10	0.001	**0.594**	1.02	200	0.571	200	10	0.571	0
ctlg3	5	0.1	0.581	15	5	0.1	**0.582**	0.17	700	0.559	700	11	0.559	0

For each method, columns corresponding to the best HR and the set of parameters with which it is achieved are shown. For SLIM (1st order), the set of parameters consists of the $l2$ regularization parameter β and the $l1$ regularization parameter λ. For HOSLIM (2nd order), the parameters are β, λ and the support threshold σ. For k-NN (1st order), the parameter used is the number of nearest neighbors (nnbrs). For HOKNN (2nd order), the parameters are the number of nearest neighbors (nnbrs) and the support threshold σ. The columns "Improved" show the percentage of improvement of the 2nd order models above the 1st order models. More specifically, the 1st column "Improved" shows the percentage of improvement of HOSLIM beyond SLIM. The 2nd column "Improved" shows the percentage of improvement of HOKNN beyond k-NN.

6.2   Performance Comparison

Table 3 shows the performance achieved by HOSLIM, SLIM, k-NN and HOKNN. The results show that HOSLIM produces recommendations that are better than the other methods in nearly all the datasets. We can also see that the incorporation of higher-order information improves the recommendation quality, especially in the HOSLIM framework.

Moreover, we can observe that the greater the existence of higher-order relations in the dataset, the more significant the improvement in recommendation quality is. For example, the most significant improvement happens in the *groceries* and the *synthetic* datasets, in which the higher-order relations are the greatest (as seen from Table 2). On the other hand, the *ctlg3* dataset does not benefit from higher-order models, since there are not enough higher-order relations. These results are to a large extent in agreement with our expectations based on the analysis presented in the previous section. The datasets for which HOSLIM achieves the highest improvement are those that contain the largest number of users and non-zeros that are covered by high-quality itemsets.

Figure 1 demonstrates the performance of the methods for different values of N (i.e., 5, 10, 15 and 20). HOSLIM outperforms the other methods for all different values of N as well. We choose N to be quite small, as a user will not see an item that exists in the bottom of a top-100 or top-200 list.

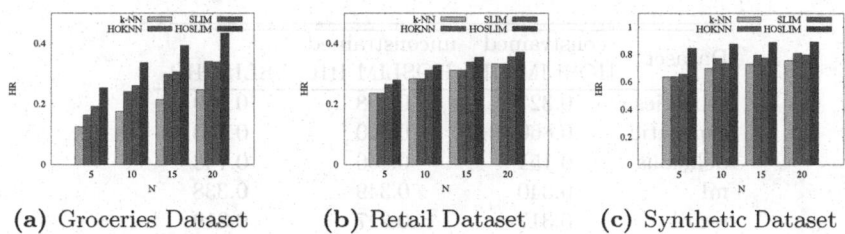

(a) Groceries Dataset   (b) Retail Dataset   (c) Synthetic Dataset

Fig. 1. HR for different values of N

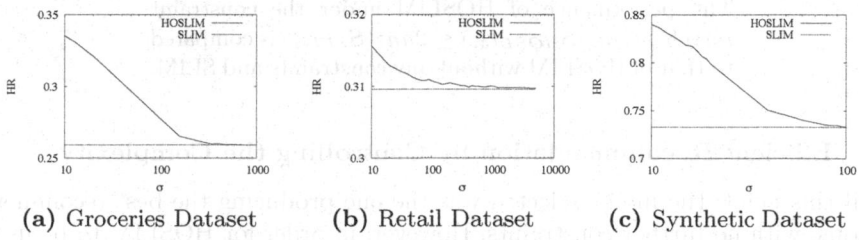

(a) Groceries Dataset   (b) Retail Dataset   (c) Synthetic Dataset

Fig. 2. Effect of the support threshold on HR

6.3 Performance only on the Users Covered by "Good" Itemsets

In order to better understand how the existence of "good" itemsets affects the performance of HOSLIM, we computed the correlation coefficient of the percentage improvement of HOSLIM beyond SLIM (presented in Table 3) with the product of the affected users coverage and the number of non-zeros coverage (presented in Table 2). The correlation coefficient is 0.712, indicating a strong positive correlation between the coverage (in terms of users and non-zeros) of higher-order itemsets in the dataset and the performance gains achieved by HOSLIM.

6.4 Sensitivity on the Support of the Itemsets

As there are lots of possible choices for support threshold, we analyzed the performance of HOSLIM, with varying support threshold σ. The reason behind this is that we wanted to see the trend of the performance of HOSLIM with respect to σ. Ideally, we would like HOSLIM to perform better than SLIM, for as many values of σ, as possible; not for just a few of them.

Figure 2 shows the sensitivity of HOSLIM to the support threshold σ. We can see that there is a wide range of support thresholds for which HOSLIM outperforms SLIM. Also, a low support threshold means that HOSLIM benefits more from the itemsets, leading to a better HR.

Table 4. Comparison of unconstrained HOSLIM with constrained HOSLIM and SLIM

Dataset	constrained HOSLIM HR	unconstrained HOSLIM HR	SLIM HR
groceries	0.327	0.338	0.259
synthetic	0.860	0.860	0.733
delicious	0.154	0.156	0.148
ml	0.340	0.349	0.338
retail	0.317	0.317	0.310
bms-pos	0.509	0.509	0.502
bms1	0.594	0.594	0.588
ctlg3	0.582	0.582	0.581

The performance of HOSLIM under the constraint $nnz(S')+nnz(S_{HOSLIM}) \leq 2nnz(S_{SLIM})$ is compared to that of HOSLIM without any constraints and SLIM.

6.5   Efficient Recommendation by Controlling the Complexity

Until this point, the model selected was the one producing the best recommendations, with no further constraints. However, in order for HOSLIM to be used in real-life scenarios, it also needs to be applied fast. In other words, the model should compute the recommendations fast and this means that it should have non-prohibitive complexity.

The question that normally arises is the following: If we find a way to control the complexity, how much will the performance of HOSLIM be affected? In order to answer this question, we did the following experiment: As the cost of computing the top-N recommendation list depends on the number of non-zeros in the model, we selected from all learned models the ones that satisfied the constraint:

$$nnz(S') + nnz(S_{HOSLIM}) \leq 2nnz(S_{SLIM}). \tag{9}$$

With this constraint, we increased the complexity of HOSLIM little beyond the original SLIM (since the original number of non-zeros is now at most doubled).

Table 4 shows the HRs of SLIM and constrained and unconstrained HOSLIM. It can be observed that the HR of the constrained HOSLIM model is close to the optimal one. This shows that a simple model can incorporate the itemset information and improve the recommendation quality in an efficient way, making the approach proposed in this paper usable, in real-world scenarios.

7   Conclusion

In this paper, we revisited the research question of the existence of higher-order information in real-world datasets and whether its incorporation could help the recommendation quality. This was done in the light of recent advances in the top-N recommendation methods. By coupling the incorporation of higher-order associations (beyond pairwise) with state-of-the-art top-N recommendation methods like SLIM, the quality of the recommendations made was improved beyond the current best results.

References

1. Adomavicius, G., Tuzhilin, A.: Toward the next generation of recommender systems: A survey of the state-of-the-art and possible extensions. IEEE Transactions on Knowledge and Data Engineering 17(6), 734–749 (2005)
2. Agrawal, R., Srikant, R., et al.: Fast algorithms for mining association rules. In: Proc. 20th Int. Conf. Very Large Data Bases, VLDB, vol. 1215, pp. 487–499 (1994)
3. Aiolli, F.: A preliminary study on a recommender system for the million songs dataset challenge. Preference Learning: Problems and Applications in AI, 1 (2012)
4. Brijs, T., Swinnen, G., Vanhoof, K., Wets, G.: Using association rules for product assortment decisions: A case study. In: Proceedings of the Fifth ACM SIGKDD International Conference on Knowledge Discovery and Data Mining, pp. 254–260. ACM (1999)
5. Cremonesi, P., Koren, Y., Turrin, R.: Performance of recommender algorithms on top-n recommendation tasks. In: Proceedings of the Fourth ACM Conference on Recommender Systems, pp. 39–46. ACM (2010)
6. Deshpande, M., Karypis, G.: Item-based top-n recommendation algorithms. ACM Transactions on Information Systems (TOIS) 22(1), 143–177 (2004)
7. Friedman, J., Hastie, T., Tibshirani, R.: Regularization paths for generalized linear models via coordinate descent. Journal of Statistical Software 33(1), 1 (2010)
8. Herlocker, J.L., Konstan, J.A., Borchers, A., Riedl, J.: An algorithmic framework for performing collaborative filtering. In: Proceedings of the 22nd Annual International ACM SIGIR Conference on Research and Development in Information Retrieval, pp. 230–237. ACM (1999)
9. Kabbur, S., Ning, X., Karypis, G.: Fism: Factored item similarity models for top-n recommender systems (2013)
10. Ning, X., Karypis, G.: Slim: Sparse linear methods for top-n recommender systems. In: 2011 IEEE 11th International Conference on Data Mining (ICDM), pp. 497–506. IEEE (2011)
11. Pan, R., Zhou, Y., Cao, B., Liu, N.N., Lukose, R., Scholz, M., Yang, Q.: One-class collaborative filtering. In: Eighth IEEE International Conference on Data Mining, ICDM 2008, pp. 502–511. IEEE (2008)
12. Ricci, F., Shapira, B.: Recommender systems handbook. Springer (2011)
13. Tibshirani, R.: Regression shrinkage and selection via the lasso. Journal of the Royal Statistical Society. Series B (Methodological), 267–288 (1996)
14. Zheng, Z., Kohavi, R., Mason, L.: Real world performance of association rule algorithms. In: Proceedings of the Seventh ACM SIGKDD International Conference on Knowledge Discovery and Data Mining, pp. 401–406. ACM (2001)

Mining GPS Data for Trajectory Recommendation

Peifeng Yin[1], Mao Ye[2], Wang-Chien Lee[1], and Zhenhui Li[3]

[1] Department of Computer Science and Engineering, Pennsylvania State University
[2] Pinterest, San Francisco Bay Area, CA
[3] College of Information Science and Technology, Pennsylvania State University
{pzy102,wlee}@cse.psu.edu, mao@pinterest.com, jessieli@ist.psu.edu

Abstract. The wide use of GPS sensors in smart phones encourages people to record their personal trajectories and share them with others in the Internet. A recommendation service is needed to help people process the large quantity of trajectories and select potentially interesting ones. The GPS trace data is a new format of information and few works focus on building user preference profiles on it. In this work we proposed a trajectory recommendation framework and developed three recommendation methods, namely, Activity-Based Recommendation (ABR), GPS-Based Recommendation (GBR) and Hybrid Recommendation. The ABR recommends trajectories purely relying on activity tags. For GBR, we proposed a generative model to construct user profiles based on GPS traces. The Hybrid recommendation combines the ABR and GBR. We finally conducted extensive experiments to evaluate these proposed solutions and it turned out the hybrid solution displays the best performance.

1 Introduction

With the rapid development of mobile devices, wireless networks and Web 2.0 technology, a number of location-based sharing services, e.g., Foursquare[1], Facebook Place[2], Everytrail[3] and GPSXchange[4], have emerged in recent years. Among them, Everytrail and GPSXchange are particularly unique because they allow users to share their outdoor experiences by uploading GPS *trajectory data* of various outdoor activities, e.g., hiking and biking. By sharing trajectory information, these Web 2.0 sites provide excellent resources for their users to plan or explore outdoor activities of interests.

The rich amount of trajectories available in those web sites brings significant challenges for users to find what they search for. Also, different from conventional items with enrich texts, it is difficult to judge whether the trajectory is interesting or not based on the activity tag or GPS raw data. Therefore, in order to automatically discover interesting trajectories, a trajectory recommendation service is highly desirable.

---

[1] http://www.foursquare.com
[2] http://www.facebook.com/places/
[3] http://www.everytrail.com
[4] http://www.gpsxchange.com/

V.S. Tseng et al. (Eds.): PAKDD 2014, Part II, LNAI 8444, pp. 50–61, 2014.
© Springer International Publishing Switzerland 2014

Conventional collaborative filtering (CF) techniques do not fit the problem trajectory recommendation. The CF requires people to access the same items to compute user interest similarity. However, in trajectory sharing website, there are no two people who generate exactly the same trajectory and the user similarity can not be calculated by "accessing the same item".

In this work, we explore the ideas of content-based recommendation techniques [1,8,13]. We consider two types of trajectory "content", activity tags and GPS points. The activitiy tags, such as hiking or biking, are annotated by the users themselves. The trajectory is represented as a sequence of GPS points with corresponding time stamps.

Recommendation based on tags is named as *activity-based recommendation (ABR)*, which utilizes the tag content (if available) to make trajectory recommendation. Since the tags are manually labeled by the creator, they can be treated as a good feature for a trajectory. Unfortunately, activity tags are not always available for a GPS trajectory. In the Everytrail data we collected, about 12.61% of the trajectories do not have tags. Additionally, ABR may not be able to make recommendation if there are too many candidates with the same tag. For example, in our collected data, 14% of all tagged trajectories, are tagged with "hiking". One intuitive solution would be using geographical region as a filtering to eliminate infeasible candidates. However, it does not really solve the problem. For example, after constraining the search result into "San Fran", we still found 96 hiking trajectories in the collected Everytrail dataset. Finally, trajectories with the same tag may have different moving patterns, which the ABR is unable to capture. Let's consider two hiking fans. The first one likes to take a gentle walk so she can take a lot of photographs but the other one treats hiking as a physical exercise. Naturally, the two trajectories, although both labeled as "hiking", may contain very different features, which ABR fails to capture.

Considering these weak points of ABR discussed above, we also exploit the sampled points in GPS trajectories for recommendation and call the proposed technique *GPS based recommendation (GBR)*. The raw GPS data contains plentiful movement information (e.g., speed, change of speed, etc.), which captures the user's outdoor experiences implicitly. For example, techniques for using raw GPS data to infer the transportation modes (e.g., taking bus, taking subway, biking and walking) of trajectories have been studied [22,21,17,18,7,6]. However, these techniques are not applicable to our trajectory recommendation service since we aim to capture users' moving habits and use them to differentiate the trajectories of the same activity type. Take the example of hiking fans mentioned earlier, existing techniques can only classify them as "hiking". However, what a recommender system needs are more personalized moving habits, e.g., gentle walking or intense trotting. We argue that such information is embedded in GPS data and we aim to mine them out to facilitate trajectory recommendation.

The rest of the paper is organized as follows. Section 2 formally defines the problem, introduces ABR and reviews the related work. Section 3 and 4 respectively detail the GPS feature extraction and the generative model in GBR. Section 5 presents the evaluation of our proposed solutions. Finally, Section 6 concludes the paper.

2   Preliminaries

In this section, we first formally introduce the trajectory recommendation problem and discuss the sub-tasks to tackle the proposed problem. Then we provide a comprehensive literature review on recommendation and trajectory related research work.

2.1   Problem Formulation

A trajectory consists of two parts, i.e., an activity tag (could be absent) and a raw GPS trace. Formally, a trajectory is represented as $T = \langle a, T^G \rangle$, where $a \in \{"hiking", "biking", \cdots, "null"\}$ denotes the activity tag and T^G stands for the raw GPS trace.

The GPS trace is obtained via GPS sensor which sampled the moving object's current location together with the sampled time stamp. Thus the original format is a series of triple tuples defined below.

Definition 1 (Raw GPS Trace). *A GPS trace $T^G = \{pt_1, \cdots, pt_n\}$ is defined as a series of sample points, $pt_i = \langle x_i, y_i, t_i \rangle$ where x_i, y_i represent the latitude and longitude of the i_{th} point and t_i stands for the time stamp.*

The recommendation problem is to find a subset of candidate trajectories that could be of interest to an active user. More formally, given a collection of trajectories $S = \{T_1, \cdots, T_n\}$ and a person u, recommendation needs to find k trajectories $S' = \{T_{r_1}, \cdots, T_{r_k}\}$ that u is most interested in. Suppose we have a ranking function $\text{Score}(T, u)$ that can compute the "interest degree" of a trajectory to a user, the recommendation can be formulated as follows.

Definition 2 (Top-k Trajectory Recommendation). *Given a trajectory set $S = \{T_1, \cdots, T_n\}$, the recommendation service for user u needs to find a subset of k trajectories $S' = \{T_{r_1}, \cdots, T_{r_k}\}$ so that $\forall T_i \in S - S'$, we have*

$$\text{Score}(T_i, u) \leq \min_{T_j \in S'} \text{Score}(T_j, u) \tag{1}$$

The above definition reveals three problems for trajectory recommendation. The first two problems are how to represent the trajectory (Feature Extraction) and the user (User Profile Modeling) in a proper way to facilitate the computation of a ranking score. And the final one is how to design an effective ranking function $\text{Score}(T, u)$ to measure the "interest degree".

2.2   Activity-Based Recommendation

The ABR tries capturing a person's activity preferences based on her previously shared trajectories. This preference to different activities is represented as a series of probabilities, whose values are obtained by maximizing the joint probability of observed data.

Let $A = \{a_1, \cdots, a_n\}$ denote the collection of all activity tags and $p_i, 1 \leq i \leq n$ denote the probability that the user u is interested in activity a_i. Obviously $\sum_{i=1}^{n} p_i = 1$. For the user's previously published trajectories, the activity tags are $X = \{x_1, \cdots, x_m\}$ where $x_j \in A, 1 \le j \le m$. X is the observed data for the user and the solution is to guess the user's preference, or exactly the value of p_i based on these experiences. We assume that the instance $x_j \in X$ is independent of each other and the probability of observing X is given in Equation (2).

$$P(X|p_1, \cdots, p_m) = \prod_{j=1}^{m} P(x_j|p_1, \cdots, p_m) = \prod_{j=1}^{m} \sum_{i=1}^{n} p_i \cdot 1_{x_j=a_i} = \prod_{i=1}^{n} p_i^{n_i} \quad (2)$$

where n_i represents the number of trajectories that is tagged with a_i in X.

To learn the value of p_i, we need to maximize the Equation (2) under the constraint that the sum of all probabilities is equal to 1, i.e., the objective function as shown in Equation (3).

$$\mathcal{L}(p_1, \cdots, p_n) = \log P(X|p_1, \cdots, p_n) + \lambda(1 - \sum_{i=1}^{n} p_i) = \sum_{i=1}^{n} n_i \log p_i + \lambda(1 - \sum_{i=1}^{n} p_i)$$
$$(3)$$

where λ is a Lagrange multiplier.

The objective function is solved by setting each partial differential $\frac{\partial \mathcal{L}}{\partial p_i}$ to 0. For ABR, the ranking function is thus defined as:

$$\texttt{Score}_{abr}(T, u) = \log \sum_{i=1}^{n} p_i \cdot 1_{T.activity=p_i} \qquad p_i = \frac{n_i}{\sum_{j=1}^{n} n_j} \quad (4)$$

2.3   Related Work

Due to the wide use of GPS-equipped smart phones, much attention is focused on the use of the trajectory data to improve people's life, among which *transportation mode detection* is most related to our work.

Zheng et. al. [21,22] collected 47 people's GPS data and compared different machine learning techniques to classify transportation modes. The methods however can not be used for recommendation. Trajectory recommendation requires to give a ranking score to each candidate trajectory while classification algorithms, e.g., decision tree, can only output binary values. In [17,18], Reddy et. al. compared and even ranked different types of trajectory features. One of the most important features in their work is the instant acceleration recorded by accelerometer. This information is usually unavailable for common trajectory information since most of the smart phones are not equipped with accelerometer. In [6,7], different trajectories of moving objects, including eye-tracking, are collected for transportation mode classification.

Trajectories contain plenty of valuable information. Previous classification works explored different types of features that can well capture the trajectory modes. However they did not pay attention to user's moving habit that is also contained in trajectory data. Li et. al. [10,11] tries to mine moving patterns from GPS data of animals. GPS data in our case are records of a person's trips that happen at different places and few of them overlap with each other. Therefore no periodic patterns can be mined out of such "scattered" data. In [9,14,15], Discrete Fourier Transformation is also used to extract features from trajectory data. However, their goal is for clustering, which is quite straightforward with the extracted data. Our work is to develop generative model based on these features to learn user moving habits for recommendation.

Other works related to recommendation are based on semantic information of trajectory [3,4,20]. These works treats trajectory as a sequence of "meaningful places" and use the semantic information of the locations, e.g., restaurant, shopping centers.In our case, trajectories do not have semantic tags. Furthermore, not all trajectories contain meaningful locations. For example, a hiking trajectory is unlikely to pass places such as restaurant, shopping center.

3   GPS Feature Extraction

In this section we focus on extracting features from GPS data. Specifically, we introduce two types of features, i.e., partial-view feature (PVF) and entire-view feature (EVF). The PVF mainly consists of physic values such as speed, velocity, etc., and is easy to understand.

Specifically, given a trajectory's raw GPS data, average velocity, average acceleration and other physical measurements can be easily computed and they represent some characteristics of that trajectory. In this work, the PVF contains the total length of trajectory Len, the total time of the trajectory $Time$ and top-pf_1 maximum velocity $\hat{V}_1, \cdots, \hat{V}_{pf_1}$ and top-pf_2 acceleration $\hat{A}_1, \cdots, \hat{A}_{pf_2}$.

The EVF tries to capture the global features and is harder to understand semantically. We adopt Discrete Fourier Transform (DFT) to transform the GPS data and a discussion is provided in Section 3.2.

3.1   Entire-View Feature

Before applying DFT on GPS, there are two issues need to be addressed. Firstly, different trajectories may have different lengths, i.e., different number of sampling points. If we take the whole GPS trace as input, DFT will generate features that have different dimensions. This situation makes it difficult to compare two trajectories as they might be in different frequency spectrums. Secondly, there are three kinds of signals that can be obtained from GPS traces, i.e., distance signal, velocity signal and acceleration signal. We need to decide which one should be used as DFT input.

For the first problem, a sliding window of fixed size is used to split the GPS trace into several segments. DFT coefficients of these segments are then refined to form a GPS feature of the same size. This processing method is similar to music compression and classification [12,16]. As for the second problem, we choose speed signal because i) it suffers less impact of sampling rates than the distance signal and ii) it is more accurate in reflecting the moving status than acceleration signal. Given two trajectories which have the same sampled data points (i.e., latitude, longitude and the number of points) except for the time stamp, the DFT features will be same. However, the moving status for the two trajectories could be quite different if the sampling rates are not the same. The speed series can avoid this weakness. Also, note that the acceleration signal is converted from the velocity signal under the assumption that the object is moving at a constant acceleration between two sampled points. Each manipulation of the GPS data, e.g., linear interpolation, converting to distance signal etc., may introduce some errors. Therefore the velocity signal is better than the acceleration signal in terms of accuracy.

A final feature of size $ws \times r$ is obtained by keeping the top-r max value for each dimension. The process is both shown in Figure 1. The use of overlapping window is to mitigate the "loss" at the edges of the window[5]. The refined value r is to mitigate the impact of path condition. because extreme cases are usually free of road limitation. For example, if it is found that a user drove 35 miles per hour for some segment while 25 miles per hour for other. The it is more reasonable to believe that the speed 35 mph is more likely to represent the user's habit instead of 25 mph.

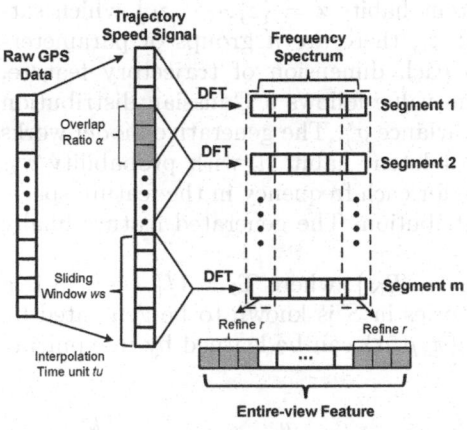

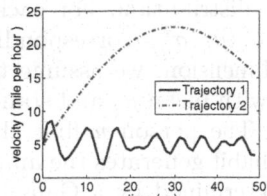

(a) Original Speed Signal

Fig. 1. Illustration of EVF extraction. Four parameters, interpolation time unit tu, window size ws, overlap ratio α and refine value r, are involved.

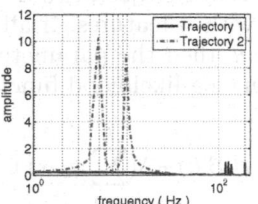

(b) Frequency Spectrum

Fig. 2. Illustration of feature meaning

3.2 Discussion of the Entire-View Feature

The basic waves have two components, i.e., frequency and amplitude. The frequency indicates the intensity of speed change. That is, lower frequency suggests that the moving is smoother while higher one means the moving object frequently changes its speed. The amplitude of the wave reflects the strength of a signal, or exactly, the absolute value of the speed. We use the example shown in Figure 2 to illustrate this point.

As shown in Figure 2(a), the speed signal of two trajectories are drawn. It is easily seen that the trajectories are quite different. Firstly, trajectory 1 has a smaller speed than trajectory 2. Also, the signal of trajectory 2 is smooth while the speed of trajectory 1 suffers frequent change. Figure 2(b) displays the result of DFT. Trajectory 2's amplitudes are bigger than that of Trajectory 1, corresponding to its original bigger speed. Furthermore, the frequency of trajectory

---

[5] http://en.wikipedia.org/wiki/Window_function#Overlapping_windows

1 lies mainly in higher spectrum while that of trajectory 2 in lower part. This is consistent with the change of their original speed signal where low frequency reflects smooth signal and high frequency corresponds to intense change.

4   User Profile Modeling and Trajectory Ranking

In this section we formally define the proposed generative model and introduce the designed ranking functions.

4.1   Generative Model

Formally, suppose a person u has k latent habits $\mathbf{z} = \{z_1, \cdots, z_k\}$ which satisfies some distribution. For each habit z_i, there are n groups of parameters $\langle \mu_1^i, \sigma_1^i \rangle, \cdots, \langle \mu_n^i, \sigma_n^i \rangle$ corresponding to each dimension of trajectory feature. For each dimension, we assume that the value follows a Gaussian distribution $\mathcal{N}(\mu_j^i, \sigma_j^i)$ with mean μ_j^i and standard variance σ_j^i. The generative model works as follows. The person u first chooses a latent habit z_k with probability π_k. Then this habit generates the amplitude for each frequency in the feature space, which is determined by a Gaussian distribution. The generated feature finally constructs a trajectory.

Given a set of trajectories $\mathbf{S} = \{\mathbf{T_1}, \cdots, \mathbf{T_N}\}$, where $\mathbf{T_i} = \langle f_1^i, \cdots, f_n^i \rangle$ is a trajectory with n features. If all trajectories in S is known to be generated by the person u, then the parameters $\theta = \{\pi, \mu, \sigma\}$ can be learned by maximizing the following log-likelihood function:

$$L(\theta) = \log P(\mathbf{S}|\theta) + \lambda(\sum_{i=1}^{k} \pi_i - 1) = \sum_{i=1}^{N} \sum_{j=1}^{k} \tau_{ij}(\log \pi_j + \sum_{m=1}^{n} \log \mathcal{G}(\cdot)) + \lambda(\sum_{i=1}^{k} \pi_i - 1)$$

(5)

where $\mathcal{G}(f_m^i, \mu_m^j, \sigma_m^j) = \frac{1}{\sqrt{2\pi}\sigma_m^j} e^{-\frac{(f_m^i - \mu_m^j)^2}{2(\sigma_m^j)^2}}$ is the probability of value f_m^i for Gaussian distribution $\mathcal{N}(\mu_m^j, \sigma_m^j)$, λ is the Lagrange multiplier and τ_{ij} is an indicator function whose value is 1 if and only if the trajectory T_i is generated by habit z_j.

We use *EM algorithm* [5] to solve this problem. In the following, let $\theta(t)$ denote the values of parameters at t^{th} iteration.

E-Step

$$E[\tau_{ij}] = \frac{P(T_i|z_j)P(z_j)}{\sum_{m=1}^{k} P(T_i|z_m)P(z_m)} = \frac{\prod_{d=1}^{n} \mathcal{G}(f_d^i, \mu_d^j(t-1), \sigma_d^j(t-1))\pi_j^{(t-1)}}{\sum_{m=1}^{k} \prod_{d=1}^{n} \mathcal{G}(f_d^i, \mu_d^m(t-1), \sigma_d^m(t-1))\pi_m^{(t-1)}}$$

(6)

M-Step

$$\pi_j^{(t)} = \frac{\sum_{i=1}^{N} E[\tau_{ij}]}{\sum_{i=1}^{N} \sum_{m=1}^{k} E[\tau_{im}]} \qquad \mu_m^j(t) = \frac{\sum_{i=1}^{N} E[\tau_{ij}]f_m^i}{\sum_{i=1}^{N} E[\tau_{ij}]}$$

$$\sigma_m^j(t) = \sqrt{\frac{\sum_{i=1}^{N} E[\tau_{ij}](\mu_m^j(t) - f_m^i)^2}{\sum_{i=1}^{N} E[\tau_{ij}]}}$$

4.2   Ranking Function

We provide two ranking functions, of which one is developed from the generative model and the other is hybrid of ABR and GBR.

The generative model discussed above captures the user's moving habit and can then be used for recommendation. Each user has a profile expressing her latent habits $\langle z_1, \cdots, z_k \rangle$ with probability $\langle \pi_1, \cdots, \pi_k \rangle$. For each habit z_i, there is a n-dimension Gaussian distribution $\{\langle \mu_1, \sigma_1 \rangle, \cdots, \langle \mu_n, \sigma_n \rangle\}$. The ranking of a trajectory to a given user's profile is to compute the probability that the trajectory is generated by that user. More specifically, for a trajectory $T = \langle f_1, \cdots, f_n \rangle$, its ranking score is computed below.

$$\text{Score}_{gm}(T, u) = \log P(T|u) = \log \sum_{i=1}^{k} P(T|z_i)P(z_i) = \log \sum_{i=1}^{k} \pi_i \prod_{j=1}^{n} \mathcal{G}(f_j, \mu_j, \sigma_j)$$

$$(7)$$

The probability $P(u)$ is omitted since it is a constant to all candidate trajectories.

The ABR introduced in Section 2.2 is aware of a person's historical activities but can not distinguish two trajectories with the same activity. On the other hand, the GBR described in Section 4 ranks trajectories purely based on its GPS data and the user profile. In other words, it gives different ranking scores to trajectories disregarding their activity tags. Therefore it indicates a potential improvement when the two methods are combined. Equation (8) shows the hybrid ranking function that merges the two recommendation methods.

$$\text{Score}_{hybrid}(T, u) = (1 - \lambda)\text{Score}_{gm}(T, u) + \lambda\text{Score}_{abr}(T, u) \qquad (8)$$

where $0 \leq \lambda \leq 1$ is a balance parameter adjusting the weight of the two separate ranking scores. Note that when $\lambda = 0$ the hybrid ranking becomes pure generative model and $\lambda = 1$ reduces the method to pure ABR.

5   Evaluation

All the experiments are based on a real data set collected from Everytrail. Everytrail is a trajectory sharing website encouraging people to publish their trip trajectories recorded by the smart phones. Each trajectory may also be associated with an activity tag, e.g., hiking, road biking, driving and so on. We crawled the website from June 05, 2010 to August 07, 2010 and obtains 8,444 users and 63,760 trajectories in total.

5.1   Data Preparation

Because the trajectory data is uploaded by different people and there is no strict examination, the raw data contains much noise for mining. We take the

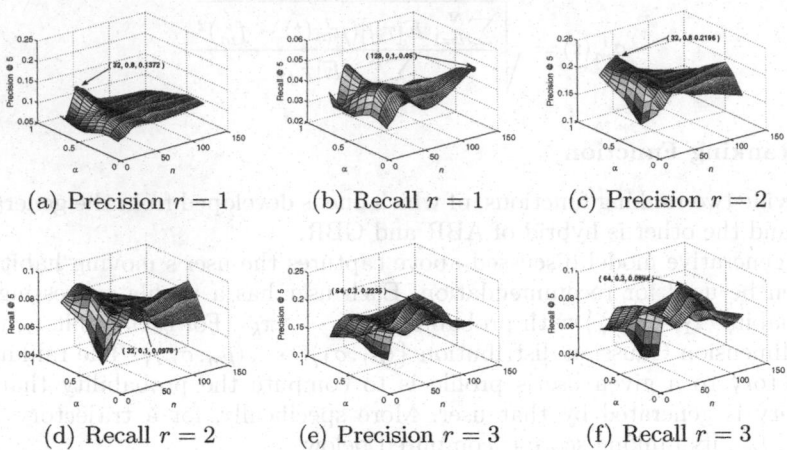

(a) Precision $r = 1$ (b) Recall $r = 1$ (c) Precision $r = 2$

(d) Recall $r = 2$ (e) Precision $r = 3$ (f) Recall $r = 3$

Fig. 3. Parameter Tuning for the EVF

following steps to clean the data. Firstly, all untagged trajectories are removed to guarantee a fair comparison of ABR and GBR. Also, we removed trajectories which have "illegal" GPS traces, defined as those i) whose sampled time stamp are not monotone increasing ii) whose total time length is less than 300 seconds. Finally, users whose shared trajectories are less than 20 are removed since small sample may hurt the model accuracy and thus the recommendation performance. After this preprocessing, there are 252 users and 9,120 trajectories remaining.

For each user in the preprocessed data, we randomly masked 20% of her trajectories. These masked data serve as the test data to evaluate our proposed recommendation methods and the remaining part is used as training data to build the user's profile. Furthermore precision and recall are metrics used to evaluate the recommendation performance.

5.2 Parameter Tuning for Entire-View Feature

Figure 3 shows the tuning process for the remaining parameters, where ws ranges from 8 to 128, α from 0.1 to 0.8 and r from 1 to 3. Also, the optimal configuration for different r is highlighted in the figure. As is shown, best performance is achieved when the feature dimension is 64, the overlapping window percentage is 30% and the refined window size is 3.

Figure 4 shows experiments on time unit tu which varies from 3 seconds to 20 seconds. It can be seen that the time unit should be neither too small nor too large. For a small time unit, many pseudo points have to be added via linear interpolation and it may introduce too many errors. On the other hand, a big time unit may discard some true sample points, which may miss important information. Judging from the result, tu is properly set to 5 seconds.

Finally Figure 5 shows the tuning of λ for hybrid recommendation. Note that when $\lambda = 0$ the method is reduced to the ABR and when $\lambda = 1$ it is equivalent to the GBR. Based on the result we set λ to 0.9 for the rest experiments.

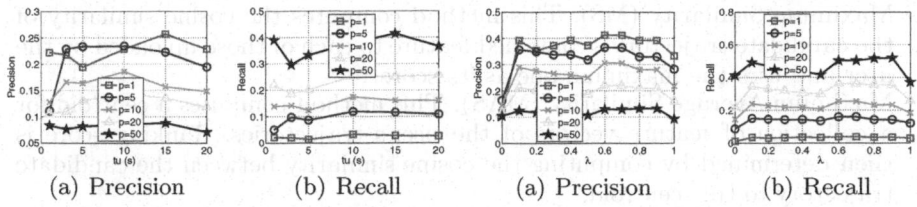

(a) Precision   (b) Recall   (a) Precision   (b) Recall

Fig. 4. Parameter Tuning for tu **Fig. 5.** Parameter Turing for λ

5.3   Comparison of Different Features

After determining the optimal configuration of EVF, we compare its recommendation performance with two other features, namely, partial-view feature (PVF) and combined feature. The PVF, as introduced in Section 3, consists of top-pf_1 maximum velocity, top-pf_2 acceleration, the length and time of the trajectory. In the empirical study, the best performance of PVF is obtained when both pf_1 and pf_2 are set to 3. Combined feature is the combination of the two types of features. As is seen in Figure 6, the combined feature outperforms the other two in terms of both precision and recall, especially for top-1 recommendation. This scenario proves our earlier argument that these two types of features represent different aspects of a trajectory. The PVF aims to capture the locally extreme characteristics while the EVF places more emphasis on the global picture of a trajectory. They are complimentary to each other and can achieve the best performance when combined together.

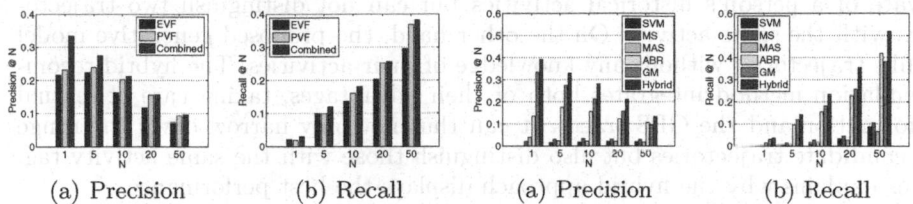

(a) Precision   (b) Recall   (a) Precision   (b) Recall

Fig. 6. Feature Comparison   **Fig. 7.** Method Comparison

5.4   Comparison of Different Recommendation Approaches

In this set of experiment, we evaluate the performance of the proposed recommendation methods, i.e., ABR in Section 2.2, GBR in Section 4 and Hybrid Recommendation where the λ is set to 0.9. We also include several baseline methods for comparison. The details of them are listed below.

– Support Vector Machine (SVM)[6]. The SVM treats each user as a class and all the trajectories' combined features as training data. Then each candidate trajectory will be assigned a series of probabilities indicating how likely it belongs to each class (user). These values are thus treated as ranking scores.

---

[6] In this experiment we used LIBSVM [2].

- Maximum Similarity (MS). This method computes the cosine similarity of the candidate trajectory's combined feature to each of those uploaded by the user and uses the maximum one as its score.
- Maximum Average Similarity (MAS). This method computes a centroid for a collection of feature vectors of the user's trajectories. Ranking score is then determined by computing the cosine similarity between the candidate trajectory to this centroid.

The MS and MAS can also be treated as an variant of item-based recommendation [19]. Note that the above baselines share the same features (i.e., GPS combined features) with our proposed GBR but adopt different approaches for user-profile modeling and trajectory ranking. To distinguish the difference, we use generative model (GM) to denote our proposed GBR.

The experiment results are shown in Figure 7. It is easily seen that the generative model outperforms other methods that use GPS feature. These recommendation methods, SVM, MS, MAS, are solely based on user's previous trajectories and thus may be too biased to history data. The generative model, instead of limiting to the data, tries to learn the users' hidden moving pattern and thus can achieve higher precision. ABR is the only method that relies on the tag of the trajectory. In Figure 7, its performance, in terms of both precision and recall, is worse than GM except for $N = 50$. As mentioned in Section 1, this method can not distinguish trajectories of the same activity. This explains its low precision and recall when N is small. The performance improvement for bigger N, however, suggests a user's concentration on the number of different trajectory's activities. Finally, the hybrid recommendation, which combines the ranking functions of ABR and GM, shows the best performance. The ABR is aware of a person's historical activities but can not distinguish two trajectories with the same activity. On the other hand, the proposed generative model ranks trajectories without any knowledge of their activities. The hybrid recommendation method integrates both of their advantages, taking care of textual information and the GPS traces. It can thus not only narrow down the range of candidate trajectories but also distinguish those with the same activity tag. This explains why the hybrid approach displays the best performance.

6   Conclusion

In this paper we studied the problem of trajectory recommendation. Each trajectory usually consists a GPS trace and may contain an activity tag. A recommendation service is supposed to find potentially interesting trajectories and push them to particular people. We proposed a recommendation framework and divide the task into three subproblems, i.e., feature extraction, user profile modeling and trajectory ranking. Under this framework, we developed three recommendation methods, namely, Activity-Based Recommendation (ABR), GPS-Based Recommendation (GBR) and Hybrid recommendation. We conduct extensive experiments to evaluate our solutions. For GPS feature, it is shown that the combination of partial-view feature and entire-view feature achieves best performance. As for the recommendation approach, the hybrid one that combines ABR and GBR obtains highest precision and recall. In future work, we will focus on ways of integrating travelogues, if available, into recommendation methods.

References

1. Balabanovic, M., Shoham, Y.: Content-based collaborative recommendation. CACM 40(3), 66–72 (1997)
2. Chang, C.-C., Lin, C.-J.: LIBSVM: A library for support vector machines. ACM Transactions on Intelligent Systems and Technology 2, 27:1–27:27 (2011), http://www.csie.ntu.edu.tw/~cjlin/libsvm
3. Chen, Z., Shen, H.T., Zhou, X.: Discovering popular routes from trajectories. In: ICDE (2011)
4. Chen, Z., Shen, H.T., Zhou, X., Zheng, Y., Xie, X.: Searching trajectories by locations: An efficiency study. In: SIGMOD, pp. 255–266 (2010)
5. Dempster, A., Laird, N., Rubin, D.: Maximum likelihood from incomplete data via the em algorithm. Journal of the Royal Statistical Society 39(1), 1–38 (1977)
6. Dodge, S., Weibel, R., Forootan, E.: Revealing the physics of movement: Comparing the similarity of movement characteristics of different types of moving objects. Computers, Environment and Urban Systems 33(6), 419–434 (2009)
7. Dodge, S., Weibel, R., Laube, P.: Exploring movement-similarity analysis of moving objects. SIGSPATIAL Special 1, 11–16 (2009)
8. Ferman, A.M., Errico, J.H., van Beek, P., Sezan, M.I.: Content-based filtering and personalization using structured metadata. In: JCDL, p. 393 (2002)
9. Li, X., Hu, W., Hu, W.: A coarse-to-fine strategy for vehicle motion trajectory clustering. In: Pattern Recognition (2006)
10. Li, Z., Ding, B., Han, J., Kays, R., Nye, P.: Mining periodic behaviors for moving objects. In: KDD, pp. 1099–1108 (2010)
11. Li, Z., Han, J., Ji, M., Tang, L.-A., Yu, Y., Ding, B., Lee, J.-G., Kays, R.: Movemine: Mining moving object data for discovery of animal movement patterns. TIST (2010)
12. Logan, B.: Mel frequency cepstral coefficients for music modeling. In: ISMIR (2000)
13. Melville, P., Mooney, R.J., Nagarajan, R.: Content-Boosted Collaborative Filtering for Improved Recommendations. In: AAAI/IAAI, pp. 187–192 (2002)
14. Naftel, A., Khalid, S.: Classifying spatiotemporal object trajectories using unsupervised learning in the coefficient feature space. Multimedia Systems 12(3), 227–238 (2006)
15. Naftel, A., Khalid, S.: Motion trajectory learning in the dft-coefficient feature space. In: IEEE Conf. Comput. Vision Syst. (2006)
16. Pfeiffer, S., Vincent, T.: Formalisation of mpeg-1 compressed-domain audio features. Technical report, CSIRO mathematical and information, sciences, Australia (2001)
17. Reddy, S., Burke, J., Estrin, D., Hansen, M., Srivastava, M.: Determining transportation mode on mobile phones. In: ISWC (2008)
18. Reddy, S., Mun, M., Burke, J., Estrin, D., Hansen, M., Srivastava, M.: Using mobile phones to determine transportation modes. ACM Trans. Sen. Netw. 6, 13:1–13:27 (2010)
19. Sarwar, B., Karypis, G., Konstan, J., Riedl, J.: Item-based collaborative filtering recommendation algorithms. In: WWW, pp. 285–295 (2001)
20. Zheng, V.W., Zheng, Y., Xie, X., Yang, Q.: Collaborative location and activity recommendations with gps history data. In: WWW, pp. 1029–1038 (2010)
21. Zheng, Y., Chen, Y., Li, Q., Xie, X., Ma, W.-Y.: Understanding transportation modes based on gps data for web applications. ACM Trans. Web 4, 1:1–1:36 (2010)
22. Zheng, Y., Liu, L., Wang, L., Xie, X.: Learning transportation mode from raw gps data for geographic applications on the web. In: WWW, pp. 247–256 (2008)

Gaussian Processes Autoencoder
for Dimensionality Reduction

Xinwei Jiang[1], Junbin Gao[2], Xia Hong[3], and Zhihua Cai[1]

[1] School of Computer Science,
China University of Geosciences, Wuhan, 430074, China
ysjxw@hotmail.com, zhcai@cug.edu.cn
[2] School of Computing and Mathematics,
Charles Sturt University, Bathurst, NSW 2795, Australia
jbgao@csu.edu.au
[3] School of Systems Engineering,
University of Reading, Reading, RG6 6AY, UK
x.hong@reading.ac.uk

Abstract. Learning low dimensional manifold from highly nonlinear data of high dimensionality has become increasingly important for discovering intrinsic representation that can be utilized for data visualization and preprocessing. The autoencoder is a powerful dimensionality reduction technique based on minimizing reconstruction error, and it has regained popularity because it has been efficiently used for greedy pretraining of deep neural networks. Compared to Neural Network (NN), the superiority of Gaussian Process (GP) has been shown in model inference, optimization and performance. GP has been successfully applied in nonlinear Dimensionality Reduction (DR) algorithms, such as Gaussian Process Latent Variable Model (GPLVM). In this paper we propose the Gaussian Processes Autoencoder Model (GPAM) for dimensionality reduction by extending the classic NN based autoencoder to GP based autoencoder. More interestingly, the novel model can also be viewed as back constrained GPLVM (BC-GPLVM) where the back constraint smooth function is represented by a GP. Experiments verify the performance of the newly proposed model.

Keywords: Dimensionality Reduction, Autoencoder, Gaussian Process, Latent Variable Model, Neural Networks.

1 Introduction

Dimensionality Reduction (DR) aims to find the corresponding low dimensional representation of data in a high-dimensional space without incurring significant information loss and has been widely utilized as one of the most crucial preprocessing steps in data analysis such as applications in computer vision [15]. Theoretically the commonly-faced tasks in data analysis such as regression, classification and clustering can be viewed as DR. For example, in regression, one

V.S. Tseng et al. (Eds.): PAKDD 2014, Part II, LNAI 8444, pp. 62–73, 2014.

tries to estimate a mapping function from an input (normally with high dimensions) to an output space (normally with low dimensions).

Motivated by the ample applications, DR techniques have been extensively studied in the last two decades. Linear DR models such as Principal Component Analysis (PCA) and Linear Discriminant Analysis (LDA) may be the most well-known DR techniques used in the settings of unsupervised and supervised learning [2]. These methods aim to learn a linear mapping from high dimensional observation to the lower dimension space (also called latent space). However, in practical applications the high dimensional observed data often contain highly nonlinear structures which violates the basic assumption of the linear DR models, hence various non-linear DR models have been developed, such as Multidimensional Scaling (MDS) [11], Isometric Mapping (ISOMAP) [19], Locally Linear Embedding (LLE) [16], Kernel PCA (KPCA) [17], Gaussian Process Latent Variable Model (GPLVM) [9], Relevance Units Latent Variable Model (RULVM) [6], and Thin Plate Spline Latent Variable Model (TPSLVM) [7].

Among the above mentioned nonlinear DR approaches, the Latent Variable Model (LVM) based DR models attract considerable attention due to their intuitive explanation. LVM explicitly models the relationship between the high-dimensional observation space and the low-dimensional latent space, thus it is able to overcome the out-of-sample problems (projecting a new high-dimensional sample into its low-dimensional representation) or pre-image problems (projecting back from the low-dimensional space to the observed data space). The linear Probabilistic PCA (PPCA) [20] and GPLVM [9] may be the most well-known LVM based DR techniques, where the mapping from the low dimensional latent space (latent variables) to the high dimensional observation space (observed variables) is represented by a linear model and a nonlinear Gaussian Process (GP), respectively. Since the nonlinear DR technique GPLVM performs very well in many real-world data sets, this model has become popular in many applications, such as movement modelling and generating [9]. Meanwhile, many GPLVM extensions have been developed to further improve performance. For instance, the Gaussian Process Dynamical Model (GPDM) [22] allows modelling dynamics in the latent space. The back constraint GPLVM (BC-GPLVM) was proposed in [10] to maintain a smooth function from observed data points to the corresponding latent points thus enforcing the close observed data to be close in latent space. Other extensions, such as Bayesian GPLVM, shared GPLVM and supervised GPLVM which further extend the classical GPLVM to unsupervised and supervised settings, can be referred to [4,21,8].

The autoencoder [3] can be regarded as an interesting DR model, although originally it is a neural network (NN) architecture used to determine latent representation for observed data. The idea of autoencoder is to resolve the latent embedding within the hidden layer by training the NN to reproduce the input observed data as its output. Intuitively this model consists of two parts: the encoder which maps the input observed data to a latent representation, and the decoder which reconstructs the input through a map from the latent representation to the observed input (also called output). Basically, the two mappings in the encoder and decoder are modelled by neural network (NN). Recently, autoencoder has regained popularity because it has been efficiently used for greedy pre-training of deep neural network (DNN) [1].

The relationship between GP and NN was established by Neal [12], who demonstrated that NN could become GP in the limit of infinite hidden units and the model inference may be simpler. With specific covariance function (NN covariance) [14], the back constraint GPLVM (BC-GPLVM) can be seen as autoencoders[18], where the encoder is NN and the decoder is GP. The superiority of GP over NN lies in small-scale model parameters, easy model inference and training [12,14], and in many real-world applications GP outperforms NN. Motivated by the comparison, we propose the Gaussian Processes Autoencoder Model (GPAM), which can be viewed as BC-GPLVM where GP represents the smooth mapping from latent space to observation space, and also as an autoencoder where both the encoder and decoder are GPs. It is expected that the proposed GPAM will outperform typical GPLVM, BC-GPLVM and autoencoder models.

The rest of the paper is organized as follows. In Section 2 we briefly review the GP, GPLVM, and autoencoder models. The proposed Gaussian Processes Autoencoder Model (GPAM) will be introduced in Section 3. Then, real-world data sets are used to verify and evaluate the performance of the newly proposed algorithm in Section 4. Finally, the concluding remarks and comments are given in Section 5.

2   Related Works

In this and following section, we use the following notations: $X = [x_1, ..., x_N]^T$ are observed (inputs) data in a high dimensional space $\mathcal{R}^D$, i.e., $x_n \in \mathcal{R}^D$; $Y = [y_1, ..., y_N]^T$ are observed (outputs or labels) data with each $y_n \in \mathcal{R}^q$; and $Z = [z_1, ..., z_N]^T$ are the so-called latent variables in a low dimensional space $\mathcal{R}^p$ with $p \ll D$ where each z_n is associated with x_n. For the sake of convenience, we also consider X as an $N \times D$ matrix, Y an $N \times q$ and Z an $N \times p$ matrix. We call $\mathcal{D} = \{(x_n, y_n)\}_{n=1}^N$ (or $\mathcal{D} = \{x_n\}_{n=1}^N$ if no labels (outputs) given) the observed dataset. Data items are assumed to be i.i.d. Let $\mathcal{N}(\mu, \Sigma)$ denote the Gaussian distribution with mean μ and covariance Σ.

2.1   Gaussian Process

Given a dataset $\mathcal{D} = \{(x_1, y_1), \cdots, (x_N, y_N)\}$ as defined above, the classical Gaussian Process Regression (GPR) is concerned with the case when $q = 1$. It aims to estimate the predictive distribution $p(y|x^*)$ for any test data x^*. In the classical GPR model, each sample y_n is generated from the corresponding latent functional variable g with independent Gaussian noise

$$y = g(x) + \epsilon$$

where g is drawn from a (zero-mean) GP which only depends on the covariance/kernel function $k(\cdot, \cdot)$ defined on input space and ϵ is the additive Gaussian noise with zero mean and covariance σ^2.

Given a new test observation $\boldsymbol{x}^*$, it is easy to prove that the predictive distribution conditioned on the given observation is

$$g^*|\boldsymbol{x}^*, X, Y \sim \mathcal{N}(K_{\boldsymbol{x}^* X}(K_{XX} + \sigma^2 \mathbf{I})^{-1} Y,$$

$$K_{\boldsymbol{x}^* \boldsymbol{x}^*} - K_{\boldsymbol{x}^* X}(K_{XX} + \sigma^2 \mathbf{I})^{-1} K_{X \boldsymbol{x}^*}) \tag{2.1}$$

where Ks are the matrices of the covariance/kernel function values at the corresponding points X and/or $\boldsymbol{x}^*$.

2.2 Gaussian Process Latent Variable Model (GPLVM)

Lawrence introduced GPLVM in [9], including the motivation of proposing the GPLVM and the relationship between PPCA and GPLVM. Here we just review GPLVM from the view of GP straightforwardly.

Given a high dimensional dataset $\mathcal{D} = \{\boldsymbol{x}_1, ..., \boldsymbol{x}_N\} \subset \mathcal{R}^D$ without any given labels or output data. We aim to obtain the latent/unknown variables $\boldsymbol{z}_n \in \mathcal{R}^p$ corresponding to each data item $\boldsymbol{x}_n$ ($n = 1, 2, ..., N$). GPLVM [9] defines a generative mapping from the latent variables $\boldsymbol{z}_n$ to its corresponding observed variables $\boldsymbol{x}_n$ which is governed by a group of GPs $\boldsymbol{x}_n = \boldsymbol{g}(\boldsymbol{z}_n) + \epsilon$ where $\boldsymbol{g} = [g_1, ..., g_D]^T$ is assumed to be a group of D GPs, and ϵ is an independent Gaussian noise with zero mean and covariance $\sigma^2 \boldsymbol{I}$, which means the likelihood of the observations is Gaussian

$$P(X|\boldsymbol{g}, Z) = \prod_{n=1}^{N} \mathcal{N}(\boldsymbol{x}_n|\boldsymbol{g}(\boldsymbol{z}_n), \sigma^2 \boldsymbol{I}) \tag{2.2}$$

Suppose that each GP g_i ($i = 1, ..., D$) has the same covariance function $k(\cdot, \cdot)$, then the data likelihood defined by equation (2.2) can be marginalized with respect to the given GP priors over all g_ds, giving rise to the following overall marginalized likelihood of the observations X

$$P(X|Z) = \frac{1}{(2\pi)^{DN/2}|K|^{D/2}} \exp\left(-\frac{1}{2}\mathrm{tr}(K^{-1} X X^T)\right) \tag{2.3}$$

where $K = K_{ZZ} + \sigma^2 \boldsymbol{I}$ is the kernel matrix over latent variables Z.

The model learning is implemented by maximizing the above marginalized data likelihood with respect to the latent variables Z and the parameters of the kernel function k.

Although GPLVM provides a smooth mapping from latent space to the observation space, it does not ensure smoothness in the inverse mapping. This can be undesirable because it only guarantees that samples close in latent space will be close in data space, while points close in data space may be not close in latent space. Besides, due to the lack of direct mapping from observation space to latent space the out-of-sample problem becomes complicated, meaning that the latent representations of testing data must be optimized conditioned on the latent embedding of the training examples [9]. In order to address this issue, the back constraint GPLVM (BC-GPLVM) was proposed in [10]. The idea behind this model is to constrain latent points to be a smooth function of the corresponding data points, which forces points which are close in data space to be close in latent space. The back constraint smooth function can be written by

$$z_{mn} = f_m(x_n, \alpha) \qquad (2.4)$$

where α are parameters of the smooth function. Typically, we can use a linear model, a kernel based regression (KBR) model or a multi-layer perception (MLP) model to represent this function. As the function f_m is fully parameterized in terms of a set of new parameters α, the learning process becomes an optimization process aiming at maximizing the likelihood (2.3) w.r.t. the latent variables X and parameters α.

2.3   Autoencoder

The autoencoder[3] is based on NN, which will be termed NNAM (NN Autoencoder Model) for short throughout the paper. Basically it is a three-layer NN with one hidden layer where the input and output layers are the observation data. Our goal is to find the latent representation over the hidden layer of the model through minimizing reconstruction errors. The autoencoder model can be separated into two parts: an encoder (mapping the input into latent representation) and a decoder (reproducing the input through a map from the latent representation to input).

With the above notations, let's define the encoder as a function $z = f(x, \theta,$ and the decoder as a function $x = g(z, \gamma)$. Given a high dimensional dataset $\mathcal{D} = \{x_1, ..., x_N\} \subset \mathcal{R}^D$, we jointly optimize the parameters of the encoder θ and decoder γ by minimizing the least-squares reconstruction cost:

$$\{\theta, \gamma\} = \underset{\{\theta, \gamma\}}{\operatorname{argmax}} \sum_{n=1}^{N} \sum_{d=1}^{D} \left\{ x_n^d - g^d(f(x_n, \theta), \gamma) \right\}^2 \qquad (2.5)$$

where $g^d(\cdot)$ is the dth output dimension of $g(\cdot)$. When f and g are linear transformations, this model is equivalent to PCA. However, nonlinear projections show a more powerful performance. This function is also called the active function in NN framework. In this paper we use the sigmoidal function $f(x, \theta) = (1 + \exp(-x^T \theta))^{-1}$ as the active function. The model can be optimized by gradient-based algorithms, such as scaled conjugate gradient (SCG).

3   Gaussian Processes Autoencoder Model

Based on the relationship between GP and NN, we introduce the detailed model inference of Gaussian Processes Autoencoder Model (GPAM). The fundamental idea of this novel model is to use Gaussian Process (GP) to replace Neural Networks (NN) that was originally used in autoencoder.

Given a high dimensional dataset $\mathcal{D} = \{x_1, ..., x_N\} \subset \mathcal{R}^D$ without any labels or output data, where each sample x_n is assumed to be associated with the latent/unknown variables $z_n \in \mathcal{R}^p$ $(n = 1, 2, ..., N)$. Our goal is to find these latent variables which should clearly show the intrinsic structures of the observation data for data visualization or preprocessing.

The idea behind GPAM is to define a mapping from the observation variables x_n to the corresponding latent variables z_n (encoder) and a mapping from the latent variables z_n to the corresponding observation variables x_n (decoder) by using Gaussian Processes Regressions (GPRs) defined as follows

$$z = f(x, \theta) + \epsilon_1; \quad x = g(z, \gamma) + \epsilon_2 \tag{3.1}$$

where $f = [f^1, ..., f^p]^T$ and $g = [g^1, ..., g^D]^T$ are assumed to be two groups of p and D GPs with hyperparameters θ and γ, respectively, and both ϵ_1 and ϵ_2 are the independent Gaussian noises with zero mean and covariance $\sigma^2 I$. Thus it is easy to see that the likelihood of the observations is Gaussian,

$$P(Z|f, X, \theta) = \prod_{n=1}^{N} \mathcal{N}(z_n|f(x_n), \sigma_1^2 I); \quad P(X|g, Z, \gamma) = \prod_{n=1}^{N} \mathcal{N}(x_n|g(z_n), \sigma_2^2 I)$$

Let's further assume that both functions f and g are nonlinearly modelled by GPs

$$P(f|X, \theta) = \mathcal{N}(f|0, K_{X,X}); \quad P(g|Z, \gamma) = \mathcal{N}(g|0, K_{Z,Z}) \tag{3.2}$$

By marginalizing over the unknown functions f and g, we have

$$P(Z|X, \theta) = \frac{1}{(2\pi)^{pN/2}|K_X|^{p/2}} \exp\left\{ -\frac{1}{2}\mathrm{tr}(K_X^{-1} Z Z^T) \right\} \tag{3.3}$$

$$P(X|Z, \gamma) = \frac{1}{(2\pi)^{DN/2}|K_Z|^{D/2}} \exp\left\{ -\frac{1}{2}\mathrm{tr}(K_Z^{-1} X X^T) \right\}$$

with $K_X = K_{X,X} + \sigma_1^2 I$ and $K_Z = K_{Z,Z} + \sigma_2^2 I$ where $K_{X,X}$ and $K_{Z,Z}$ are the covariance matrices defined over the input data X, and the latent variables Z, respectively.

Furthermore, in order to do model inference let's assume that the input X of encoder function f is different from the output X of decoder function g, which is rewritten by X_c. Thus the notation of marginal likelihood $P(X|Z, \gamma)$ can be changed to $P(X_c|Z, \gamma)$. Based on the conditional independence property of graphical model the posterior distribution over latent variables Z given observation (X, X_c) can be derived as follows

$$P(Z|X, X_c, \theta, \gamma) = P(Z|X, \theta)P(X_c|Z, \gamma)/P(X_c|X, \gamma, \theta) \tag{3.4}$$

In order to learn the unknown variables (Z, θ, γ), we maximize the log posterior distribution $P(Z|X, X_c, \theta, \gamma)$ (3.4) w.r.t. (Z, θ, γ)

$$\max_{Z, \theta, \gamma} \left\{ \log P(Z|X, \theta) + \log P(X_c|Z, \gamma) - \log P(X_c|X, \theta, \gamma) \right\} \quad (3.5)$$

For the sake of convenience, we simply denote the negative log posterior distribution $P(Z|Y, X, \theta, \gamma)$ by

$$L = L^r + L^l = -\log P(Z|X, \theta) - \log P(X_c|Z, \gamma) \quad (3.6)$$

$$= \frac{1}{2} \left\{ pN \log 2\pi + p \log|K_X| + \mathrm{tr}(K_X^{-1} Z Z^T) \right\}$$

$$+ \frac{1}{2} \left\{ DN \log 2\pi + D \log|K_Z| + \mathrm{tr}(K_Z^{-1} X_c X_c^T) \right\}$$

where $P(X_c|X, \theta, \gamma)$ has been omitted because it is irrelevant to Z.

The process of model training is equal to simultaneously optimizing a GPR (corresponding to the encoder distribution $P(Z|X, \theta)$) and a GPLVM (corresponding to the decoder distribution $P(X_c|Z, \gamma)$). To apply a gradient based optimization algorithm like SCG algorithm to learn the parameters of the model, we need to find out the gradient of L w.r.t. the latent variables Z, and the kernel parameter (θ, γ).

Firstly for the part of GPR corresponding to $P(Z|X, \theta)$ we can simply obtain the following gradients

$$\frac{\partial L^r}{\partial Z} = K_X^{-1} Z \quad (3.7)$$

As for the parameter θ in kernel K_X, since we consider the output of the mapping $z = f(x, \theta)$ as the known quantity in the GPR model, the optimization process is identical to the procedure of determining parameters for a typical GPR model from training data. Thus we can derive the partial derivative of the hyperparameter θ by chain rule (refer to Chapter 5 in [14]) $\frac{\partial L^r}{\partial \theta} = \frac{\partial L^r}{\partial K_X} \frac{\partial K_X}{\partial \theta}$.

Subsequently for the second part of GPLVM corresponding to $P(X_c|Z, \gamma)$ it is easy to evaluate the gradients of L^l w.r.t. the latent variables Z

$$\frac{\partial L^l}{\partial Z} = \frac{\partial L^l}{\partial K_Z} \frac{\partial K_Z}{\partial Z} \quad (3.8)$$

where the gradients of log likelihood w.r.t. kernel matrix K_Z is evaluated by

$$\frac{\partial L^l}{\partial K_Z} = K_Z^{-1} - K_Z^{-1} Y Y^T K_Z^{-1}. \quad (3.9)$$

Similarly the gradient of L^l w.r.t. the hyperparameter γ can be calculated by

$$\frac{\partial L^l}{\partial \gamma} = \frac{\partial L^l}{\partial K_Z} \frac{\partial K_Z}{\partial \gamma} \quad (3.10)$$

and the computation of the derivative of the kernel matrix w.r.t. the latent variable Z and hyperparameter depend on a specific kernel function.

By combining equations (3.7) with equation (3.8) and (3.9), it is quite simple to get the complete gradients of L w.r.t. the latent variables Z ($\partial L/\partial Z$). Once we get all the derivative ready, the derivative based algorithms like SCG can be utilized to iteratively optimize these parameters. However, when we perform experiments, we find that the value of L^r (corresponding to the encoder distribution $P(Z|X, \theta)$) is much smaller than that of L^l (corresponding to the decoder distribution $P(X_c|Z, \gamma)$), leading to very little performance improvement compared to GPLVM. Thus we propose a novel algorithm to train the model based on two-stage optimization; this is to say, we try to asynchronously optimize the model consisting of GPR and GPLVM rather than simultaneously learn it. The algorithm is detailed in Algorithm 1.

Algorithm 1. Train and Test GPAM

Input: High dimensional training inputs $X \subset \mathcal{R}^{D \times N}$, pre-fixed latent dimensionality p, number of training iterations T and testing inputs $X^* \subset \mathcal{R}^{D \times M}$.
Output: $s = \{Z, Z^*, \theta, \gamma\}$.
1. Initialize $Z = \text{PPCA}(X, p)$, θ and γ (depending on specific kernel function);
2. For i = 1:T{
3. Optimize $\{Z^t, \gamma\} = \text{argmax}_{Z,\gamma} \log P(X|Z^{t-1}, \gamma)$;
4. Optimize $\{Z^{t+1}, \theta\} = \text{argmax}_{Z,\theta} \log P(Z^t|X, \theta)$;
5. Check converges: break if $\text{Error}(Z) = ||Z^{t+1}(:) - Z^t(:)||^2 \leq \eta\}$; //end loop
6. Compute latent variables $Z^* = K_{x^* X} K_X^{-1} Z$ with learnt hyperparameters θ for testing data X^*;
7. **return** s

To sum up, there are two ways to view the proposed GPAM. Firstly, it can be seen as the generalization of classic NNAM. While GPAM makes use of GPR model to encode and decode the data, NN is utilized to do encoding and decoding in classic NNAM. Based on the superiority of GP over NN, we believe that the proposed GPAM will outperform typical NNAM. Secondly, the proposed GPAM can also be considered as the BC-GPLVM where the back constrain function is modelled by GPR. Compared to classic BC-GPLVM, such as the KBR or MLP based models, the smooth mapping from the observation space to the latent space in the proposed GPAM is modelled by GPR, which results in better performance than typical KBR and MLP based BC-GPLVM.

4   Experiments

In this section, we compare the proposed GPAM with original GPLVM [9], BC-GPLVM [10] and NNAM [3], in two real-world tasks to show the better performance that GPAM provides. In order to assess the performance of these models in visualizing high dimensional data sets, we perform dimensionality reduction by using a 2D latent space for visualization. Moreover, the nearest neighbour classification error is tested in the low dimensional latent space to objectively evaluate the quality of visualization for training data. After the DR models are learnt, we further use them as feature extraction, followed by a k-Nearest Neighbour (kNN) classifier for testing data. Of course we can use other classifier such as GP Classifier (GPC) rather than a simple kNN to classify the testing data, but the final goal is to reduce the dimensionality of the observation, and the learnt low-dimensional data would be utilized for other proposes, such as data visualization and compression, so the simple kNN classifier is better to evaluate the quality of DR models. By comparing the classification accuracies in low dimensional latent space for testing data, we demonstrate the improvement of the proposed model again. The experimental results verify that the proposed GPAM is an efficient DR model and outperforms GPLVM, BC-GPLVM and NNAM.

For a fair comparison, we ran 500 iterations for all the models, and the covariance used for GPLVM, BC-GPLVM and GPAM was optimally selected from RBF(ARD), POLY(ARD), and MLP(ARD) in Neil D. Lawrence's MATLAB packages Kern. The back constraint function of BC-GPLVM is manually picked from KBR and MLP. The code GPLVM/BC-GPLVM and NNAM are based on Neil D. Lawrence's MATLAB packages FGPLVM[1], and R. B. Palm's Deep Learning Toolbox[2] [13], respectively. Since the learning rate of NNAM needs to be selected manually, we varied it between 0.1 and 10 optimally with sigmoidal active function.

4.1 Oil Flow Data

The oil flow data set [17] consists of 12 dimensional measurements of oil flow within a pipeline. There are 3 phases of flow associated with the data and 1000 samples in the data set. For all four models, we use 600 samples (200 points from each class) to learn the corresponding 2D latent data for the purpose of data visualization, and the remaining 400 samples are the testing data. RBF covariance function is used for GPLVM/BC-GPLVM (MLP back-constraint) and GPAM (RBF covariance for both GPR and GPLVM in the model). As can be seen from Figure 1, the proposed GPAM is superior to GPLVM/BC-GPLVM and NNAM remarkably because the novel model makes the points in the latent space which belong to the same class in the original feature space much closer than the rest of three models.

Furthermore, in order to objectively evaluate the new DR technique we compare the nearest neighbour errors and the classification accuracies based on kNN classifier in the learnt 2D latent space provided by the four models on this data set, respectively. All the four DR models are firstly learnt from training data with the 2D latent space corresponding to the training data where the nearest neighbour errors are evaluated, and then based on the trained four DR models the testing data will be projected to the low dimensional latent/feature space

---

[1] http://ml.sheffield.ac.uk/~neil/fgplvm
[2] https://github.com/areslp/matlab/tree/master/DeepLearnToolbox-master

(2D in our experiments) where kNN is performed to compare the testing accuracies ($K = 10$ in kNN). Table I tells us that the proposed GPAM outperforms GPLVM/BC-GPLVM and NNAM in terms of nearest neighbour errors and classification accuracies for training and testing data respectively, which verifies that the novel DR model is better than the other three techniques.

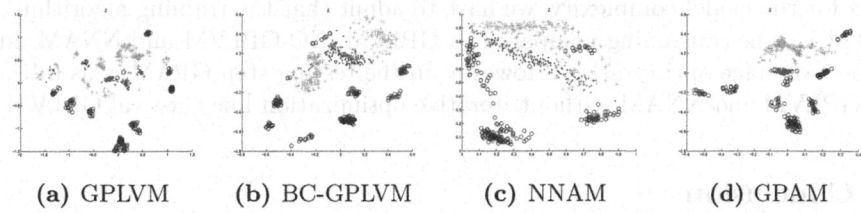

(a) GPLVM   (b) BC-GPLVM   (c) NNAM   (d) GPAM

Fig. 1. Oil data set is visualized by GPLVM, BC-GPLVM, NNAM and GPAM

Table 1. The comparisons of nearest neighbor classification errors for training data and kNN classification accuracies for testing data in oil flow data

	GPLVM	BC-GPLVM	NNAM	GPAM
NN Error	8	16	108	5
kNN Accuracy	96.75%	97.00%	90.50%	99.25%

4.2 Iris Data

The Iris data set [5] contains three classes of 50 instances each, where each class refers to a type of iris plant. There are four features for each instance. All 150 data points are utilized to learn the 2D latent space. POLY covariance achieves the best results than the other two covariance functions (RBF and MLP) for GPLVM/BC-GPLVM (MLP back-constraint) and GPAM (POLYARD and POLY covariances for GPR and GPLVM respectively). Figure 2 and Table II show the same conclusion as stated for the oil flow data set. Since there is no more testing data for this data set, the classification comparison for testing is not given.

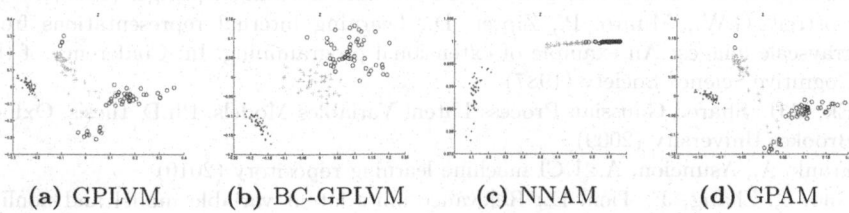

(a) GPLVM   (b) BC-GPLVM   (c) NNAM   (d) GPAM

Fig. 2. Iris data set is visualized by GPLVM, BC-GPLVM, NNAM and GPAM

Table 2. The comparisons of nearest neighbour classification errors for training data in iris data

	GPLVM	BC-GPLVM	NNAM	GPAM
NN Error	4	5	17	3

As for the model complexity, we have to admit that the training algorithm of GPAM is time-consuming compared to GPLVM, BC-GPLVM and NNAM due to the two-stage optimization. However, in the testing step GPAM is as fast as BC-GPLVM and NNAM without iterative optimization like classical GPLVM.

5 Conclusion

In this paper a novel LVM-based DR technique, termed Gaussian Processes Autoencoder Model (GPAM), has been introduced. It can be seen as the generalization of classic Neural Network Autoencoder Model (NNAM) model by replacing the NNs with GPs, leading to simpler model inference and better performance. Also, we can view the new model as the back constraint GPLVM where the smooth back constraint function is represented by GP, and the model is trained by minimizing the reconstruction error. The experimental results have demonstrated the performance of the newly developed model.

For the future work, inspired by recent works in deep learning [1] we will extend the GP Autoencoder to sparse and denoising GP Autoencoder models, and then we also want to study the deep GP model by stacking the GP Autoencoder.

Acknowledgments. Xinwei Jiang' work is supported by the Fundamental Research Funds for the Central Universities, China University of Geosciences (Wuhan). Junbin Gao and Xia Hong's work is supported by the Australian Research Council (ARC) through the grant DP130100364.

References

1. Bengio, Y., Lamblin, P., Popovici, D., Larochelle, H.: Greedy layer-wise training of deep networks. In: Advances in Neural Information Processing Systems (2007)
2. Bishop, C.M.: Pattern Recognition and Machine Learning. Springer (2006)
3. Cottrell, G.W., Munro, P., Zipser, D.: Learning internal representations from grayscale images: An example of extensional programming. In: Conference of the Cognitive Science Society (1987)
4. Ek, C.H.: Shared Gaussian Process Latent Variables Models. Ph.D. thesis, Oxford Brookes University (2009)
5. Frank, A., Asuncion, A.: UCI machine learning repository (2010)
6. Gao, J., Zhang, J., Tien, D.: Relevance units latent variable model and nonlinear dimensionality reduction. IEEE Transactions on Neural Networks 21, 123–135 (2010)

7. Jiang, X., Gao, J., Wang, T., Shi, D.: Tpslvm: A dimensionality reduction algorithm based on thin plate splines. IEEE Transactions on Cybernetics (to appear, 2014)
8. Jiang, X., Gao, J., Wang, T., Zheng, L.: Supervised latent linear gaussian process latent variable model for dimensionality reduction. IEEE Transactions on Systems, Man, and Cybernetics - Part B: Cybernetics 42(6), 1620–1632 (2012)
9. Lawrence, N.: Probabilistic non-linear principal component analysis with gaussian process latent variable models. Journal of Machine Learning Research 6, 1783–1816 (2005)
10. Lawrence, N.D., Quinonero-Candela, J.: Local distance preservation in the gp-lvm through back constraints. In: International Conference on Machine Learning (ICML), pp. 513–520. ACM Press (2006)
11. Mardia, K.V., Kent, J.T., Bibby, J.M.: Multivariate analysis. Academic Press, London (1979)
12. Neal, R.: Bayesian learning for neural networks. Lecture Notes in Statistics 118 (1996)
13. Palm, R.B.: Prediction as a candidate for learning deep hierarchical models of data. Master's thesis, Technical University of Denmark (2012)
14. Rasmussen, C.E., Williams, C.K.I.: Gaussian Processes for Machine Learning. The MIT Press (2006)
15. Rosman, G., Bronstein, M.M., Bronstein, A.M., Kimmel, R.: Nonlinear dimensionality reduction by topologically constrained isometric embedding. International Journal of Computer Vision 89, 56–68 (2010)
16. Roweis, S.T., Saul, L.K.: Nonlinear dimensionality reduction by locally linear embedding. Science 290, 2323–2326 (2000)
17. Scholkopf, B., Smola, A., Muller, K.R.: Nonlinear component analysis as a kernel eigenvalue problem. Neural Computation 10(5), 1299–1319 (1998)
18. Snoek, J., Adams, R.P., Larochelle, H.: Nonparametric guidance of autoencoder representations using label information. Journal of Machine Learning Research 13, 2567–2588 (2012)
19. Tenenbaum, J.B., de Silva, V., Langford, J.C.: A global geometric framework for nonlinear dimensionality reduction. Science 290, 2319–2323 (2000)
20. Tipping, M.E., Bishop, C.M.: Probabilistic principal component analysis. Journal of the Royal Statistical Society, Series B 61, 611–622 (1999)
21. Titsias, M.K., Lawrence, N.D.: Bayesian gaussian process latent variable model. In: International Conference on Artificial Intelligence and Statistics (2010)
22. Wang, J.M., Fleet, D.J., Hertzmann, A.: Gaussian process dynamical models. In: Advances in Neural Information Processing Systems (NIPS), pp. 1441–1448 (2005)

Semi-supervised Feature Analysis for Multimedia Annotation by Mining Label Correlation

Xiaojun Chang[1], Haoquan Shen[2], Sen Wang[1], Jiajun Liu[3], and Xue Li[1]

[1] The University of Queensland, QLD 4072, Australia
[2] Zhejiang University, Zhejiang, China
[3] CSIRO Brisbane

Abstract. In multimedia annotation, labeling a large amount of training data by human is both time-consuming and tedious. Therefore, to automate this process, a number of methods that leverage unlabeled training data have been proposed. Normally, a given multimedia sample is associated with multiple labels, which may have inherent correlations in real world. Classical multimedia annotation algorithms address this problem by decomposing the multi-label learning into multiple independent single-label problems, which ignores the correlations between different labels. In this paper, we combine label correlation mining and semi-supervised feature selection into a single framework. We evaluate performance of the proposed algorithm of multimedia annotation using MIML, MIRFLICKR and NUS-WIDE datasets. Mean average precision (MAP), MicroAUC and MacroAUC are used as evaluation metrics. Experimental results on the multimedia annotation task demonstrate that our method outperforms the state-of-the-art algorithms for its capability of mining label correlations and exploiting both labeled and unlabeled training data.

Keywords: Semi-supervised Learning, Multi-label Feature Selection, Multimedia Annotation.

1 Introduction

With the booming of social networks, such as Facebook and Flickr, we have witnessed a dramatical growth of multimedia data, *i.e.* image, text and video. Consequently, there are increasing demands to effectively organize and access these resources. Normally, feature vectors, which are used to represent aforementioned resources, are usually very large. However, it has been pointed out in [1] that only a subset of features carry the most discriminating information. Hence, selecting the most representative features plays an essential role in a multi-media annotation framework. Previous works [2,3,4,5] have indicated that feature selection is able to remove redundant and irrelevant information in the feature representation, thus improves subsequent analysis tasks.

Existing feature selection algorithms are designed in various ways. For example, conventional feature selection algorithms, such as Fisher Score [6], compute

V.S. Tseng et al. (Eds.): PAKDD 2014, Part II, LNAI 8444, pp. 74–85, 2014.

weights of all features, rank them accordingly and select the most discriminating features one by one. While dealing with multi-label problems, the conventional algorithms generally transform the problem into a couple of binary classification problems for each concept respectively. Hence, feature correlations and label correlations are ignored [7], which will deteriorate the subsequent annotation performance.

Another limitation is that they only use labeled training data for feature selection. Considering that there are a large number of unlabeled training data available, it is beneficial to leverage unlabeled training data for multimedia annotation. Over recent years, semi-supervised learning has been widely studied as an effective tool for saving labeling cost by using both labeled and unlabeled training data [8,9,10]. Inspired by this motivation, feature learning algorithms based on semi-supervised framework, have been also proposed to overcome the insufficiency of labeled training samples. For example, Zhao *et al.* propose an algorithm based on spectral analysis in [5]. However, similarly to Fisher Score [6], their method selects the most discriminating features one by one. Besides, correlations between labels are ignored.

Our semi-supervised feature selection algorithm integrates multi-label feature selection and semi-supervised learning into a single framework. Both labeled and unlabeled data are utilized to select features while label correlations and feature correlations are simultaneously mined.

The main contributions of this work can be summarized as follows:

1. We combine joint feature selection with sparsity and semi-supervised learning into a single framework, which can select the most discriminating features with an insufficient amount of labeled training data.
2. The correlations between different labels are taken into account to facilitate the feature selection.
3. Since the objective function is non-smooth and difficult to solve, we propose a fast iteration algorithm to obtain the global optima. Experimental results on convergence validates that the proposed algorithm converges within very few iterations.

The rest of this paper is organized as follows. In Section 2, we introduce details of the proposed algorithm. Experimental results are reported in Section 3. Finally, we conclude this paper in Section 4.

2 Proposed Framework

To mine correlations between different labels for feature selection, our algorithm is built upon a reasonable assumption that different class labels have some inherent common structures. In this section, our framework is described in details, followed by an iterative algorithm with guaranteed convergence to optimize the objective function.

2.1 Formulation of Proposed Framework

Let us define $X = \{x_1, x_2, \cdots, x_n\}$ as the training data matrix, where $x_i \in \mathbb{R}^d (1 \leq i \leq n)$ is the i-th data point and n is the total number of training data. $Y = [y_1, y_2, \cdots, y_m, y_{m+1}, \cdots, y_n]^T \in \{0, 1\}^{n \times c}$ denotes the label matrix and c is the class number. $y_i \in \mathbb{R}^c$ $(1 \leq i \leq n)$ is the label vector with c classes. Y_{ij} indicates the j-th element of y_i and $Y_{ij} := 1$ if x_i is in the j-th class, and $Y_{ij} := 0$ otherwise. If x_i is not labeled, y_i is set to a vector with all zeros. Inspired by [11], we assume that there is a low-dimensional subspace shared by different labels. We aim to learn c prediction functions $\{f_t\}_{t=1}^c$. The prediction function f_t can be generalized as follows:

$$f_t(x) = v_t^T x + p_t^T Q^T x = w_t^T x, \tag{1}$$

where $w_t = v_t + Q p_t$. v and p are the weights, Q is a transformation matrix which projects features in the original space into a shared low-dimensional subspace.

Suppose there are m_t training data $\{x_i\}_{i=1}^{m_t}$ belonging to the t-th class labeled as $\{y_i\}_{i=1}^{m_t}$. A typical way to obtain the prediction function f_t is to minimize the following objective function:

$$\arg \min_{f_t, Q^T Q = I} \sum_{t=1}^c (\frac{1}{m_t} \sum_{i=1}^{m_t} loss(f_t(x_i), y_i) + \beta \Omega(f_t)) \tag{2}$$

Note that to make the problem tractable we impose the constraint $Q^T Q = I$. Following the methodology in [2], we incorporate (1) into (2) and obtain the objective function as follows:

$$\min_{\{v_t, p_t\}, Q^T Q = I} \sum_{t=1}^c (\frac{1}{m_t} \sum_{i=1}^{m_t} loss((v_t + Q p_t)^T x_i, y_i) + \beta \Omega(\{v_t, p_t\})) \tag{3}$$

By defining $W = V + QP$, where $V = [v_1, v_2, \cdots, v_c] \in \mathbb{R}^{d \times c}$ and $P = [p_1, p_2, \cdots, p_c] \in \mathbb{R}^{sd \times c}$ where sd is the dimension of shared lower dimensional subspace, we can rewrite the objective function as follows:

$$\min_{W, V, P, Q^T Q = I} loss(W^T X, Y) + \beta \Omega(V, P) \tag{4}$$

Note that we can implement the shared feature subspace uncovering in different ways by adopting different loss functions and regularizations. Least square loss is the most widely used in research for its stable performance and simplicity. By applying the least square loss function, the objective function arrives at:

$$\arg \min_{W, P, Q^T Q = I} \|X^T W - Y\|_F^2 + \alpha \|W\|_F^2 + \beta \|W - QP\|_F^2 \tag{5}$$

As indicated in [12], however, there are two issues worthy of further consideration. First, the least square loss function is very sensitive to outliers. Second, it is beneficial to utilize sparse feature selection models on the regularization term for effective feature selection. Following [12,13,14], we employ $l_{2,1}$-norm to handle the two issues. We can rewrite the objective function as follows:

$$\arg\min_{W,P,Q^TQ=I} \|X^TW - Y\|_F^2 + \alpha\|W\|_{2,1} + \beta\|W - QP\|_F^2 \qquad (6)$$

Meanwhile, we define a graph Laplacian as follows: First, we define an affinity matrix $A \in \mathbb{R}^{n \times n}$ whose element A_{ij} measures the similarity between x_i and x_j as:

$$A_{ij} = \begin{cases} 1, & x_i \text{ and } x_j \text{ are } k \text{ nearest neighbours}; \\ 0, & otherwise. \end{cases} \qquad (7)$$

Euclidean distance is utilized to measure whether two data points x_i and x_j are k nearest neighbours in the original feature space. Then, the graph Laplacian, L, is computed according to $L = S - A$, where S is a diagonal matrix with $S_{ii} = \sum_{j=1}^{n} A_{ij}$.

Note that multimedia data have been normally shown to process a manifold structure, we adopt manifold regularization to explore it. By applying manifold regularization to the aforementioned loss function, our objective function arrives at:

$$\arg\min_{W,P,Q^TQ=I} Tr(W^TXLX^TW) + \gamma[\alpha\|W\|_{2,1} \\ +\beta\|W - QP\|_F^2 + \|X^TW - F\|_F^2] \qquad (8)$$

We define a selecting diagonal matrix U whose diagonal element $U_{ii} = \infty$, if x_i is a labeled data, and $U_{ii} = 0$ otherwise. To exploit both labeled and unlabeled training data, a label prediction matrix $F = [f_1, \cdots, f_n]^T \in \mathbb{R}^{n \times c}$ is introduced for all the training data. The label prediction of $x_i \in X$ is $f_i \in \mathbb{R}^c$. Following [5], we assume that F holds smoothness on both the ground truth of training data and on the manifold structure. Therefore, F can be obtained as follows:

$$\arg\min_{F} tr(F^TLF) + tr((F - Y)^TU(F - Y)). \qquad (9)$$

By incorporating (9) into (6), the objective function finally becomes:

$$\arg\min_{F,W,P,Q^TQ=I} tr(F^TLF) + tr(F - Y)^TU(F - Y) \\ +\gamma[\alpha\|W\|_{2,1} + \beta\|W - QP\|_F^2 + \|X^TW - F\|_F^2] \qquad (10)$$

As indicated in [13], W is guaranteed to be sparse to perform feature selection across all data points by $\|W\|_{2,1}$ in our regularization term.

2.2 Optimization

The proposed function involves the $l_{2,1}$-norm, which is difficult to solve in a closed form. We propose to solve this problem in the following steps. By setting the derivative of (10) w.r.t. P equal to zero, we have

$$P = Q^TW \qquad (11)$$

By denoting $W = [w^1, \ldots, w^d]^T$, the objective function becomes

$$\arg\min_{F,W,P,Q^TQ=I} tr(F^TLF) + tr(F - Y)^TU(F - Y)$$
$$+\gamma[\alpha tr(W^TDW) + \beta\|W - QP\|_F^2 + \|X^TW - F\|_F^2], \tag{12}$$

where D is a matrix with its diagonal elements $D_{ii} = \frac{1}{2\|w^i\|_2}$.

Note that for any given matrix A, we have $\|A\|_F^2 = tr(A^TA)$. Substituting P in (10) by (11), we can rewrite the objective function as follows:

$$\arg\min_{F,W,Q^TQ=I} tr(F^TLF) + tr((F - Y)^TU(F - Y))$$
$$+\gamma[\alpha tr(W^TDW) + \beta tr((W - QQ^TW)^T(W - QQ^TW)) \tag{13}$$
$$+\|X^TW - F\|_F^2],$$

According to the equation $(I - QQ^T)(I - QQ^T) = (I - QQ^T)$, we have:

$$\arg\min_{F,W,Q^TQ=I} tr(F^TLF) + tr((F - Y)^TU(F - Y))$$
$$+\gamma(\|X^TW - F\|_F^2 + tr(W^T(\alpha D + \beta I - \beta QQ^T)W)) \tag{14}$$

By the setting the derivative $w.r.t.$ W to zero, we have:

$$W = (M - \beta QQ^T)^{-1}XF \tag{15}$$

where $M = XX^T + \alpha D + \beta I$.

Then the objective function becomes:

$$\arg\min_{F,Q^TQ=I} tr(F^TLF) + tr((F - Y)^TU(F - Y))$$
$$+\gamma[tr(F^TF) - tr(F^TX^T(M - \beta QQ^T)^{-1}XF] \tag{16}$$

By setting the derivative $w.r.t.$ F to zero, we have:

$$LF + U(F - Y) + \gamma F - \gamma X^T(M - \beta QQ^T)^{-1}XF = 0 \tag{17}$$

Thus, we have

$$F = (B - \gamma X^TR^{-1}X)^{-1}UY, \tag{18}$$

where

$$B = L + U + \gamma I \tag{19}$$

$$R = M - \beta QQ^T. \tag{20}$$

Then, the objective function can be written as

$$\max_{Q^TQ=I} tr[Y^TU(B - \mu X^TR^{-1}X)^{-1}UY]. \tag{21}$$

According to the Sherman-Woodbury-Morrison matrix identity,

$$(B - \gamma X^TR^{-1}X)^{-1} = B^{-1} + \gamma B^{-1}X^T(R - \gamma XB^{-1}X^T)^{-1}XB^{-1}. \tag{22}$$

Thus, the objective function arrives at

$$\max_{Q^T Q=I} tr[Y^T U B^{-1} X^T J^{-1} X B^{-1} U Y], \tag{23}$$

where

$$J = (R - \mu X B^{-1} X^T) = (M - \beta Q Q^T - \gamma X B^{-1} X^T). \tag{24}$$

Theorem 1. *The global optimization Q^* can be obtained by solving the following ratio trace maximization problem:*

$$max_{Q^T Q=I} tr[(Q^T C Q)^{-1} Q^T D Q], \tag{25}$$

where

$$C = I - \beta (X X^T + \alpha D + \beta I - \gamma X B^{-1} X^T)^{-1} \tag{26}$$

$$D = N^{-1} X B^{-1} U Y Y^T U B^{-1} X^T N^{-1}. \tag{27}$$

Proof. See Appendix.

To obtain Q, we need to conduct eigen-decomposition of $C^{-1}D$, which is $O(d^3)$ in complexity. However, as the solution of Q requires the input of D which is obtained according to W, it is still not straightforward to obtain Q and W. So as shown in Algorithm 1, we propose an iterative approach to solve this problem.

The proposed iterative approach in Algorithm 1 can be verified to converge to the optimal W by the following theorem. Following the work in [12], we can prove the convergence of Algorithm 1.

3   Experiments

In this section, experiments are conducted on three datasets, *i.e.* MIML [16], MIRFLICKR [17] and NUS-WIDE [18] to validate performance of the proposed algorithm.

3.1   Compared Methods

To evaluate performances of the proposed method, we compare it with the following algorithms:

1. All features [All-Fea]: We directly use the original data for annotation without feature selection as a baseline.
2. Fisher Score [F-score] [6]: This is a classical method, which selects the most discriminative features by evaluating the importance of features one by one.
3. Feature Selection via Joint $l_{2,1}$-Norms Minimization [FSNM] [3]: This algorithm utilizes joint $l_{2,1}$-norm minimization on both loss function and regularization for joint feature selection.
4. Spectral Feature Selection [SPEC] [15]: It employs spectral regression to select features one by one.

Algorithm 1. The algorithm for solving the objective function

---

Data: The training data $X \in \mathbb{R}^{d \times n}$
      The training data labels $Y \in \mathbb{R}^{n \times c}$
      Parameters γ, α and β

Result:
      Optimized $W \in \mathbb{R}^{d \times c}$

1 Compute the graph Laplacian matrix $L \in \mathbb{R}^{n \times n}$;
2 Compute the selection matrix $U \in \mathbb{R}^{n \times n}$;
3 Set $t = 0$ and initialize $W_0 \in \mathbb{R}^{d \times c}$ randomly;
4 **repeat**
5   Compute the diagonal matrix D_t as:

6     $$D_t = \begin{bmatrix} \frac{1}{2\|w_t^1\|_2} & & \\ & \ddots & \\ & & \frac{1}{2\|w_t^d\|_2} \end{bmatrix}$$

7   Compute C according to $C = I - \beta(XX^T + \alpha D + \beta I - \gamma XB^{-1}X^T)^{-1}$.
8   Compute D according to $D = N^{-1}XB^{-1}UYY^TUB^{-1}X^TN^{-1}$.
9   Compute the optimal Q^* according to Theorem 1.
10   Compute W according to $W = (M - \beta QQ^T)^{-1}XF$.
11 **until** *Convergence*;
12 Return W^*.

---

5. Sub-Feature Uncovering with Sparsity [SFUS] [12]: It incorporates the latest advances in a joint, sparse feature selection with multi-label learning to uncover a feature subspace which is shared among different classes.
6. Semi-supervised Feature Selection via Spectral Analysis [sSelect] [5]: It is semi-supervised feature selection approach based on spectral analysis.

3.2 Dataset Description

Three datasets, *i.e.*, MIML [16] Mflickr [17] and NUS-WIDE [18] are used in the experiments. A brief description of the three datasets is given as follows.

MIML: This image dataset consists of $2,000$ natural scene images. Each image in this dataset is artificially marked with a set of labels. Over 22% of the dataset belong to more than one class. On average, each image has 1.24 class labels.

MIRFLICKR: The MIRFLICKR image dataset consists of 25 000 images collected from Flickr.com. Each image is associated with 8.94 tags. We choose 33 annotated tags in the dataset as the ground truth.

NUS-WIDE: The NUS-WIDE image dataset has $269,000$ real-world images which are collected from Flickr by Lab for Media Search in the National University of Singapore. All the images have been downloaded from the website, among which $59,563$ images are unlabeled. By removing unlabeled images, we use the remaining $209,347$ images, along with ground-truth labels in the experiments.

Table 1. Settings of the Training Sets

Dataset	Size(n)	Labeled Training Data (m)	Number of Selected Features
MIML	1,000	$5 \times c, 10 \times c, 15 \times c$	$\{200, 240, 280, 320, 360, 400\}$
NUS-WIDE	10,000	$5 \times c, 10 \times c, 15 \times c$	$\{240, 280, 320, 360, 400, 440, 480\}$
Mflickr	10,000	$5 \times c, 10 \times c, 15 \times c$	$\{200, 240, 280, 320, 360, 400\}$

3.3 Experimental Setup

In the experiment, we randomly generate a training set for each dataset consisting of n samples, among which m samples are labeled. The detailed settings are shown in Table 1. The remaining data are used as testing data. Similar to the pipeline in [2], we randomly split the training and testing data 5 times and report average results. The libSVM [19] with RBF kernel is applied in the experiment. The optimal parameters of the SVM are determined by grid search on a tenfold crossvalidation. Following [2], the graph Laplacian, k is set as 15. Except for the SVM parameters, the regularization parameters, γ, α and β, in the objective funciton (10), are tuned in the range of $\{10^{-4}, 10^{-2}, 10^0, 10^2, 10^4\}$. The number of selected features can be found in Table 1.

3.4 Performance Evaluation

Tables 2, 3 and 4 present experimental results measured by MAP, MicroAUC and MacroAUC when using different numbers of labeled training data ($5 \times c$, $10 \times c$ and $15 \times c$) respectively.

Taking MAP as an example, it is observed that: 1) The proposed method is better than All-Fea which does not apply feature selection. Specifically, the proposed algorithm outperforms All-Fea by about 5.5% using $10 \times c$ labeled training data in the MIML dataset, which indicates that feature selection can contribute to annotation performance. 2) Our method has consistently better performances than the other supervised feature selection algorithms. When using $5 \times c$ labeled training data in the MIML dataset, the proposed algorithm is better than the second best supervised feature selection algorithm by 3.8%. 3) The proposed algorithm gets better performances than the compared semi-supervised feature selection algorithm, which demonstrates that mining label correlations is beneficial to multimedia annotation.

Table 2. Performance Comparison(±Standard Deviation(%)) when $5 \times c$ data are labeled

Dataset	Criteria	All-Fea	F-Score	SPEC	FSNM	SFUS	sSelect	Ours
MIML	MAP	26.1 ± 0.1	26.9 ± 0.2	26.1 ± 0.2	26.1 ± 0.3	26.2 ± 0.2	28.9 ± 0.3	$\mathbf{31.4 \pm 0.1}$
	MicroAUC	54.6 ± 0.1	54.4 ± 0.2	54.6 ± 0.2	54.6 ± 0.2	54.7 ± 0.1	55.1 ± 0.2	$\mathbf{55.8 \pm 0.2}$
	MacroAUC	52.4 ± 0.3	52.6 ± 0.4	52.4 ± 0.2	52.4 ± 0.2	52.6 ± 0.3	53.1 ± 0.4	$\mathbf{54.4 \pm 0.2}$
NUS	MAP	5.8 ± 0.2	5.4 ± 0.1	5.9 ± 0.2	5.8 ± 0.3	6.0 ± 0.2	6.4 ± 0.3	$\mathbf{7.1 \pm 0.2}$
	MicroAUC	86.4 ± 0.4	86.1 ± 0.1	86.5 ± 0.3	87.2 ± 0.2	87.4 ± 0.4	87.9 ± 0.3	$\mathbf{89.1 \pm 0.2}$
	MacroAUC	64.0 ± 0.4	63.7 ± 0.2	64.2 ± 0.4	64.4 ± 0.3	64.9 ± 0.2	65.5 ± 0.3	$\mathbf{66.3 \pm 0.2}$
Mflickr	MAP	12.2 ± 0.2	12.2 ± 0.3	12.3 ± 0.2	12.3 ± 0.2	12.4 ± 0.3	13.6 ± 0.2	$\mathbf{15.8 \pm 0.1}$
	MicroAUC	75.2 ± 0.2	75.1 ± 0.3	75.4 ± 0.3	75.3 ± 0.4	75.5 ± 0.2	76.1 ± 0.3	$\mathbf{77.3 \pm 0.1}$
	MacroAUC	50.3 ± 0.3	50.3 ± 0.4	50.4 ± 0.3	50.5 ± 0.2	50.7 ± 0.4	51.3 ± 0.3	$\mathbf{52.6 \pm 0.2}$

Table 3. Performance Comparison($\pm$Standard Deviation(%)) when $10 \times c$ data are labeled

Dataset	Criteria	All-Fea	F-Score	SPEC	FSNM	SFUS	sSelect	Ours
MIML	MAP	31.6 ± 0.3	33.0 ± 0.2	31.6 ± 0.2	31.6 ± 0.3	33.0 ± 0.1	35.2 ± 0.2	**37.1 ± 0.1**
	MicroAUC	59.3 ± 0.4	58.9 ± 0.3	59.3 ± 0.2	59.4 ± 0.3	59.8 ± 0.2	60.4 ± 0.2	**61.7 ± 0.2**
	MacroAUC	62.0 ± 0.3	61.0 ± 0.2	62.0 ± 0.2	62.0 ± 0.1	62.0 ± 0.2	62.6 ± 0.2	**63.7 ± 0.2**
NUS	MAP	6.6 ± 0.2	6.0 ± 0.1	6.5 ± 0.2	6.4 ± 0.3	7.0 ± 0.2	6.9 ± 0.3	**8.0 ± 0.2**
	MicroAUC	87.3 ± 0.3	87.2 ± 0.2	87.4 ± 0.5	87.3 ± 0.2	87.6 ± 0.3	88.2 ± 0.3	**89.5 ± 0.3**
	MacroAUC	67.5 ± 0.4	67.4 ± 0.3	67.7 ± 0.4	67.6 ± 0.2	67.9 ± 0.3	68.2 ± 0.4	**69.4 ± 0.3**
Mflickr	MAP	12.8 ± 0.3	12.6 ± 0.2	12.3 ± 0.2	12.4 ± 0.3	12.9 ± 0.2	14.2 ± 0.3	**16.1 ± 0.2**
	MicroAUC	78.1 ± 0.2	78.2 ± 0.3	78.1 ± 0.2	78.4 ± 0.3	78.4 ± 0.2	78.8 ± 0.3	**80.0 ± 0.1**
	MacroAUC	55.1 ± 0.3	55.3 ± 0.2	55.2 ± 0.4	55.4 ± 0.3	55.6 ± 0.2	56.4 ± 0.4	**57.3 ± 0.2**

Table 4. Performance Comparison($\pm$Standard Deviation(%)) when $15 \times c$ data are labeled

Dataset	Criteria	All-Fea	F-Score	SPEC	FSNM	SFUS	sSelect	Ours
MIML	MAP	33.0 ± 0.2	34.7 ± 0.1	33.0 ± 0.2	33.5 ± 0.3	34.1 ± 0.1	35.8 ± 0.2	**37.9 ± 0.1**
	MicroAUC	63.4 ± 0.4	63.3 ± 0.3	63.5 ± 0.1	63.4 ± 0.3	63.7 ± 0.2	64.2 ± 0.3	**65.1 ± 0.2**
	MacroAUC	62.3 ± 0.3	62.5 ± 0.2	62.3 ± 0.2	62.3 ± 0.4	62.5 ± 0.2	63.1 ± 0.3	**64.2 ± 0.1**
NUS	MAP	6.9 ± 0.1	6.5 ± 0.3	6.8 ± 0.2	6.9 ± 0.2	7.3 ± 0.3	7.4 ± 0.3	**8.5 ± 0.2**
	MicroAUC	89.4 ± 0.2	89.1 ± 0.3	89.5 ± 0.2	89.8 ± 0.4	90.1 ± 0.3	90.7 ± 0.4	**91.9 ± 0.4**
	MacroAUC	69.2 ± 0.3	69.1 ± 0.2	69.3 ± 0.1	69.5 ± 0.3	69.7 ± 0.5	70.2 ± 0.3	**71.5 ± 0.5**
Mflickr	MAP	13.0 ± 0.2	12.9 ± 0.1	12.9 ± 0.1	12.8 ± 0.2	13.1 ± 0.3	14.8 ± 0.2	**16.7 ± 0.4**
	MicroAUC	79.2 ± 0.3	79.1 ± 0.3	79.2 ± 0.2	79.2 ± 0.4	79.5 ± 0.2	80.2 ± 0.4	**81.6 ± 0.2**
	MacroAUC	58.7 ± 0.4	58.5 ± 0.3	58.8 ± 0.2	59.1 ± 0.3	58.6 ± 0.3	59.9 ± 0.2	**60.4 ± 0.3**

3.5   Convergence Study

In this section, an experiment is conducted to validate that our proposed iterative algorithm monotonically decreases the objective function until convergence. $10 \times c$ labeled training data in MIML dataset are tested in this experiment. γ, α and β are fixed at 1 which is the median value of the tuned range of the parameters.

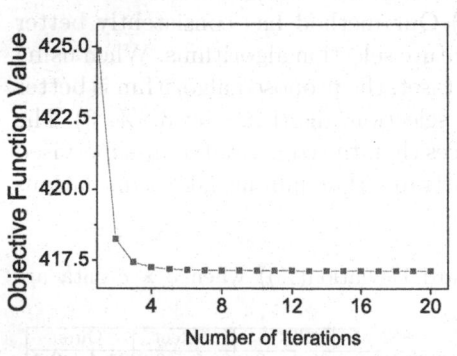

Fig. 1. Convergence

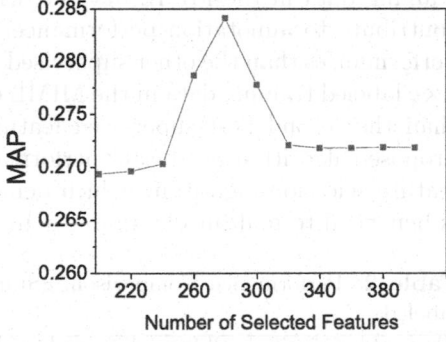

Fig. 2. Influence of selected feature number

Figure 1 shows the convergence curve of the proposed algorithm *w.r.t.* the objective function value in (10) on the MIML dataset. It is observed that the objective function values converge within 4 iterations.

3.6   Influence of Selected Features

In this section, an experiment is conducted to study how the number of selected features affect the performance of the proposed algorithm. Following the above experiment, we still use the same setting.

Figure 2 shows MAP varies *w.r.t.* the number of selected features. We can observe that: 1) When the number of selected features is relatively small, MAP of annotation is quite small. 2) When the number of selected features rises to 280, MAP increases from 0.269 to 0.284. 3) When we select 280 features, MAP arrives at the peak level. 4) MAP keeps stable when we increase the number of selected features from 320 to full features. From this figure, feature selection benefits to the annotation performance.

3.7   Parameter Sensitivity Study

Another experiment is conducted to test the sensitivity of parameters in (10). Among different parameter combinations, the proposed algorithm gains the best performance when $\gamma = 10^1$, $\alpha = 10^4$ and $\beta = 10^2$. We show the MAP variations *w.r.t.* γ, α and β. From Figure 3, we notice that the performance of the proposed algorithm changes corresponding to different parameters. In summary, better results are obtained when α, β and γ are in the range of $[10^{-2}, \cdots, 10^2]$.

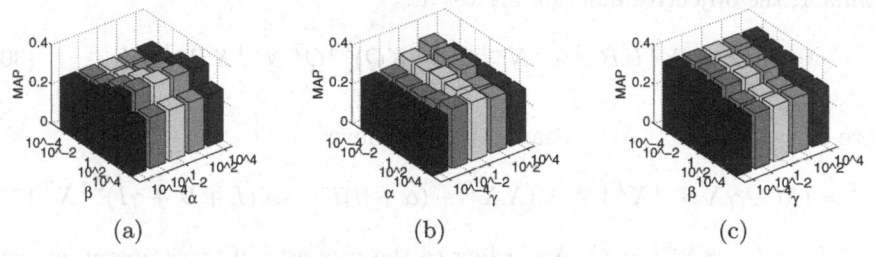

(a)   (b)   (c)

Fig. 3. The MAP variations of different parameter settings using the MIML dataset

4   Conclusion

In this paper, we have proposed a novel framework for semi-supervised feature analysis by mining label correlation. First, our method simultaneously discovers correlations between labels in a shared low-dimensional subspace to improve the annotation performance. Second, to make the classifier robust for outliers, $l_{2,1}$-norm is applied to the objective function. Third, this framework is extended into a semi-supervised scenario which exploits both labeled and unlabeled data. We evaluate the performance of a multimedia annotation task on three different datasets. Experimental results have demonstrated that the proposed algorithm consistently outperforms the other compared algorithms on all the three datasets.

Appendix

In this appendix, we prove Theorem 1.

To prove Theorem 1, we first give the following lemma and prove it.

Lemma 1. *With the same notations in the paper, we have the following equation:*

$$(R - \gamma X B^{-1} X^T)^{-1} = N^{-1} + \beta N^{-1} Q (Q^T (I - \beta N^{-1}) Q)^{-1} Q^T N^{-1}, \quad (28)$$

where

$$N = M - \gamma X B^{-1} X^T. \quad (29)$$

Proof.

$$(R - \gamma X B^{-1} X^T)^{-1}$$
$$= (M - \beta Q Q^T - \gamma X B^{-1} X^T)^{-1}$$
$$= N^{-1} + \beta N^{-1} Q (I - \beta Q^T N^{-1} Q)^{-1} Q^T N^{-1}$$
$$= N^{-1} + \beta N^{-1} Q (Q^T (I - \beta N^{-1}) Q)^{-1} Q^T N^{-1}$$

Proof of Theorem 1

Proof. From Eq. (29), we can tell that N is independent from Q. By employing Lemma 1, the objective function arrives at:

$$\max_{Q^T Q = I} tr[Y^T U B^{-1} X^T N^{-1} Q (Q^T K Q)^{-1} Q^T N^{-1} X B^{-1} U Y], \quad (30)$$

where $K = I - \beta N^{-1}$. At the same time, we have:

$$N^{-1} = (M - \gamma X B^{-1} X^T)^{-1} = (X X^T + (\alpha + \beta) I - \gamma X (L + U + \gamma I)^{-1} X^T)^{-1}.$$

Thus, $K = I - \beta N^{-1} = C$. According to the property of trace operation that $tr(UV) = tr(VU)$ for any arbitrary matrices U and V, the objective function can be rewritten as:

$$\max_{Q^T Q = I} tr[Q^T N^{-1} X B^{-1} U Y Y^T U B^{-1} X^T N^{-1} Q (Q^T K Q)^{-1}].$$

The objective function is equivalent to:

$$\max_{Q^T Q = I} tr[(Q^T C Q)^{-1} Q^T D Q].$$

Acknowledgement. This work was partially supported by the Australian Research Council the Discovery Project DP No. 130104614 and DP No. 140100104. Any opinions, findings and conclusions or recommendations expressed in this material are those of the author(s) and do not necessarily reflect the views of the Australian Research Council.

References

1. Yang, S.H., Hu, B.G.: Feature selection by nonparametric bayes error minimization. In: Proc. PAKDD, pp. 417–428 (2008)
2. Ma, Z., Nie, F., Yang, Y., Uijlings, J.R.R., Sebe, N., Hauptmann, A.G.: Discriminating joint feature analysis for multimedia data understanding. IEEE Trans. Multimedia 14(6), 1662–1672 (2012)
3. Nie, F., Henghuang, C.X., Ding, C.: Efficient and robust feature selection via joint l21-norms minimization. In: Proc. NIPS, pp. 759–768 (2007)
4. Wang, D., Yang, L., Fu, Z., Xia, J.: Prediction of thermophilic protein with pseudo amino acid composition: An approach from combined feature selection and reduction. Protein and Peptide Letters 18(7), 684–689 (2011)
5. Zhou, Z.H., Zhang, M.L.: Semi-supervised feature selection via spectral analysis. In: Proc. SIAM Int. Conf. Data Mining (2007)
6. Richard, D., Hart, P.E., Stork, D.G.: Pattern Classification. Wiley-Interscience, New York (2001)
7. Yang, Y., Wu, F., Nie, F., Shen, H.T., Zhuang, Y., Hauptmann, A.G.: Web and personal image annotation by mining label correlation with relaxed visual graph embedding. IEEE Trans. Image Process. 21(3), 1339–1351 (2012)
8. Cohen, I., Cozman, F.G., Sebe, N., Cirelo, M.C., Huang, T.S.: Semisupervised learning of classifiers: Theory, algorithms, and their application to human-computer interaction. In: IEEE Trans. PAMI, pp. 1553–1566 (2004)
9. Zhao, X., Li, X., Pang, C., Wang, S.: Human action recognition based on semi-supervised discriminant analysis with global constraint. Neurocomputing 105, 45–50 (2013)
10. Wang, S., Ma, Z., Yang, Y., Li, X., Pang, C., Hauptmann, A.: Semi-supervised multiple feature analysis for action recognition. IEEE Trans. Multimedia (2013)
11. Ji, S., Tang, L., Yu, S., Ye, J.: A shared-subspace learning framework for multi-label classification. ACM Trans. Knowle. Disco. Data 2(1), 8(1)–8(29) (2010)
12. Ma, Z., Nie, F., Yang, Y., Uijlings, J.R.R., Sebe, N.: Web image annotation via subspace-sparsity collaborated feature selection. IEEE Trans. Multimedia 14(4), 1021–1030 (2012)
13. Nie, F., Huang, H., Cai, X., Ding, C.: Efficient and robust feature selection via joint l21-norms minimization. In: Proc. NIPS, pp. 1813–1821 (2010)
14. Yang, Y., Shen, H., Ma, Z., Huang, Z., Zhou, X.: L21-norm regularization discriminative feature selection for unsupervised learning. In: Proc. IJCAI (July 2011)
15. Zhao, Z., Liu, H.: Spectral feature selection for supervised and unsupervised learning. In: Proc. ICML, pp. 1151–1157 (2007)
16. Zhou, Z.H., Zhang, M.L.: Multi-instance multi-label learning with application to scene classification. In: Proc. NIPS, pp. 1609–1616 (2006)
17. Huiskes, M.J., Lew, M.S.: The mir flickr retrieval evaluation. In: Proc. MIR, pp. 39–43 (2008)
18. Chua, T.S., Tang, J., Hong, R., Li, H., Luo, Z., Zheng, Y.: Nus-wide: A real-world web image database from national university of singapore. In: Proc. CIVR (2009)
19. Chang, C.C., Lin, C.J.: LIBSVM: A library for support vector machines. ACM Transactions on Intelligent Systems and Technology 2, 27:1–27:27 (2011), http://www.csie.ntu.edu.tw/~cjlin/libsvm

Highly Scalable Attribute Selection
for Averaged One-Dependence Estimators

Shenglei Chen[1,2], Ana M. Martinez[2], and Geoffrey I. Webb[2]

[1] College of Information Science,
Nanjing Audit University, Nanjing, China
tristan_chen@126.com
[2] Faculty of Information Technology,
Monash University, VIC 3800, Australia
anam.martinezf@gmail.com, geoff.webb@monash.edu

Abstract. Averaged One-Dependence Estimators (AODE) is a popular and effective approach to Bayesian learning. In this paper, a new
attribute selection approach is proposed for AODE. It can search in a
large model space, while it requires only a single extra pass through the
training data, resulting in a computationally efficient two-pass learning
algorithm. The experimental results indicate that the new technique significantly reduces AODE's bias at the cost of a modest increase in training time. Its low bias and computational efficiency make it an attractive
algorithm for learning from big data.

Keywords: Classification, Naive Bayes, AODE, Semi-naive Bayes,
Attribute Selection.

1 Introduction

Naive Bayes (NB) [1] is a simple, computationally efficient probabilistic approach
to classification learning. It assumes that all attributes are conditionally independent of each other given the class. As an improvement to NB, Averaged
One-Dependence Estimators (AODE) [2] relaxes the attribute independence assumption by averaging all models that assume all attributes are conditionally
dependent on the class and one common attribute, known as the super-parent.
This often improves the classification performance significantly. An extensive
comparative study [3] shows that AODE obtains significant lower error rates
than most alternative semi-naive Bayes algorithms with similar computational
complexity. One of the attractive features of AODE is that it has complexity
linear with respect to data quantity, making it a useful approach for big data.

Attribute selection has been demonstrated to be effective at improving the
accuracy of AODE [4,5]. However, the most effective conventional attribute selection techniques have high computational complexity and hence are not feasible
in the context of big data. In this paper we develop an efficient attribute selection algorithm for AODE that is linear with respect to data quantity, and of
low polynomial complexity in the number of attributes and hence well suited to

V.S. Tseng et al. (Eds.): PAKDD 2014, Part II, LNAI 8444, pp. 86–97, 2014.

big data. The empirical results show that this technique obtains lower bias than AODE, and thus usually achieves lower error on larger data sets, at the cost of only a modest increase in training time.

2 Background

The classification task can be described as follows, given a training sample $\mathcal{T}$ of t classified objects, we are required to predict the probability $P(y \mid \mathbf{x})$ that a new example $\mathbf{x} = \langle x_1, \ldots, x_a \rangle$ belongs to some class y, where x_i is the value of the attribute $\mathbf{x}_i$ and $y \in \{c_1, \ldots, c_k\}$.

In the following sections, we describe AODE for this classification task and a number of its key variants.

2.1 AODE

From the definition of conditional probability, we have $P(y \mid \mathbf{x}) = P(y, \mathbf{x})/P(\mathbf{x})$. As $P(\mathbf{x}) = \sum_{i=1}^{k} P(c_i, \mathbf{x})$ and $y \in \{c_1, \ldots, c_k\}$, it is reasonable to consider $P(\mathbf{x})$ as the normalizing constant and estimate only the joint probability $P(y, \mathbf{x})$ in the remainder of this paper.

Since the example $\mathbf{x}$ does not appear frequently enough in the training data, we cannot directly derive an accurate estimate of $P(y, \mathbf{x})$ and must extrapolate this estimate from observations of lower-dimensional probabilities in the data [6]. Applying the definition of conditional probabilities again, we have $P(y, \mathbf{x}) = P(y)P(\mathbf{x} \mid y)$. The first term $P(y)$ on the right side can be sufficiently accurately estimated from the sample frequencies, if the number of classes, k, is not too large. For the second term $P(\mathbf{x} \mid y)$, AODE assumes every attribute depends on the same parent attribute, the super-parent, thus obtains an one-dependence estimator (ODE), and then averages all eligible ODEs [2]. The joint probability $P(y, \mathbf{x})$ is estimated as follows,

$$\hat{P}(y, \mathbf{x}) = \frac{\sum_{i:1 \leq i \leq a \wedge F(x_i) \geq m} \hat{P}(y, x_i) \prod_{j=1}^{a} \hat{P}(x_j \mid y, x_i)}{|\{i : 1 \leq i \leq a \wedge F(x_i) \geq m\}|}, \tag{1}$$

where $|\cdot|$ denotes the cardinality of a set, $\hat{P}(\cdot)$ represents an estimate of $P(\cdot)$, $F(x_i)$ is the frequency of x_i and m is the minimum frequency to accept x_i as a super parent. The current research uses $m = 1$ [7].

2.2 Weightily AODE

In the classification of AODE, each ODE is treated equally, that is, all eligible models are averaged and contribute uniformly to the classification rule. However, in many real world applications, attributes do not play the same role in classification. This observation inspires the weightily AODE [8], in which the joint probability is estimated as,

$$\hat{P}(y, \mathbf{x}) = \frac{\sum_{i:1 \leq i \leq a \wedge F(x_i) \geq m} W_i \hat{P}(y, x_i) \prod_{j=1}^{a} \hat{P}(x_j \mid y, x_i)}{\sum_{i:1 \leq i \leq a \wedge F(x_i) \geq m} W_i}. \tag{2}$$

In practice, mutual information between the super-parent and the class is often used as the weight W_i.

2.3   AODE with Subsumption Resolution

One extreme type of inter-dependence between attributes results in a value of one being a generalization of a value of the other. For example, consider *Gender* and *Pregnant* as two attributes, then $Pregnant = yes$ implies that $Gender = female$. Therefore, $Gender = female$ is a generalization of $Pregnant = yes$. Likewise, $Pregnant = no$ is a generalization of $Gender = male$. Where one value x_i is a generalization of another, x_j, $P(y|x_i, x_j) = P(y|x_j)$. In consequence dropping the more general value from any calculations should not harm any posterior probability estimates, whereas assuming independence between them may.

Motivated by this observation, Subsumption Resolution (SR) [9] identifies pairs of attribute values such that one appears to subsume the other and deletes the generalization. Suppose that the set of indices of the resulting attribute subset is denoted by R, the joint probability is estimated as,

$$\hat{P}(y, \mathbf{x}) = \frac{\sum_{i:i \in R \wedge F(x_i) \geq m} \hat{P}(y, x_i) \prod_{j \in R} \hat{P}(x_j \mid y, x_i)}{|\{i : i \in R \wedge F(x_i) \geq m\}|} . \tag{3}$$

2.4   Forward and Backward Attribute Selection in AODE

In order to repair harmful inter-dependencies among highly correlated attributes, Zheng et al [5] proposed to select an appropriate attribute subset by hill climbing search. Two different search strategies can be used: FSS begins with the empty attribute set and successively adds attributes [10], while BSE starts with the complete attribute set and successively removes attributes [11]. Both strategies greedily select the attribute whose addition or elimination best reduces the leave-one-out cross validation error on the training set. The process is terminated if there is no error improvement.

To differentiate the selection of parent or child, they introduce the use of a *parent* (p) and a *child* (c) set, each of which contains the set of indices of attributes that can be employed in, respectively, a parent or a child role in AODE. The joint probability is estimated as,

$$\hat{P}(y, \mathbf{x}) = \frac{\sum_{i:i \in p \wedge F(x_i) \geq m} \hat{P}(y, x_i) \prod_{j \in c} \hat{P}(x_j \mid y, x_i)}{|\{i : i \in p \wedge F(x_i) \geq m\}|} . \tag{4}$$

As indicated in [5], the performance of BSE is better than FSS, so we focus on BSE in this paper. Four types of attribute elimination are considered, *parent elimination* (PE), *child elimination* (CE), *parent and child elimination* (P∧CE), *parent or child elimination* (P∨CE) which performs the former three types of attribute eliminations in each iteration, selecting the option that best reduces the error.

The last strategy allows flexible selection of parents and children, but comes at a high cost, since it needs to scan the training data $2a$ times in the worst case.

2.5   AnDE

The last extension to AODE we review here is AnDE [6], which allows children to depend on not just one super-parent, but a combination of n parents. The joint probability $P(y, \mathbf{x})$ is estimated as follows,

$$\hat{P}(y, \mathbf{x}) = \frac{\sum_{s:s\in\binom{A}{n}\wedge F(x_s)\geq m} \hat{P}(y, x_s) \prod_{j=1}^{a} \hat{P}(x_j \mid y, x_s)}{|\{s : s \in \binom{A}{n} \wedge F(x_s) \geq m\}|}, \tag{5}$$

where $\binom{A}{n}$ indicates the set of all size-n subsets of $\{1, \cdots, a\}$ and x_s means the set of attribute values indexed by the element in s.

Note that AnDE is in fact a superclass of AODE and NB. That is, AODE is AnDE with $n = 1$ (A1DE) and NB is AnDE with $n = 0$ (A0DE).

3   Our Proposal: Attribute Selective AODE

Previous work on attribute selection for AODE through BSE and FSS [4,5] has demonstrated attribute selection did succeed in reducing the harmful influence of inter-dependencies among attributes. This success may be attributed to their ability to search in a large model space. For P∨CE, the search space is of size 2^{a+1}, as it includes all subsets of attributes in parent role coupled with all subsets of attributes in child role.[1]

Nevertheless, this is achieved at a high computational overhead. The strategy of P∨CE needs to scan the training data $2a$ times, as each time either one child or one parent can be deleted. This is impractical for data sets with a large number of attributes.

In order to explore a large space of models in a single additional pass through the data, we propose a new attribute selection approach for AODE. Our proposal is based on the observation that it is possible to nest a large space of alternative models such that each is a trivial extension to another. Let p and c be the set of indices of parent and child attributes, respectively. For every attribute $\mathbf{x}_i$, the AODE models that use attributes in p as parents and attributes in $c \cup \{i\}$ as children are minor extensions of a model that uses attributes in p as parents and attributes in c as children. The same is true of models that use attributes in $p \cup \{i\}$ as parents and attributes in c as children. Importantly, multiple models that build upon one another in this way can be efficiently evaluated in a single set of computations. Using this observation, we create a space of models that are nested together, and then select the best model using leave-one-out cross validation in single extra pass through the training data.

Step by step information of the algorithm is provided in the following sections.

---

[1] Note that although the search space is of size 2^{a+1}, the actual number of models evaluated is $\mathcal{O}(a^2)$, which is much smaller.

3.1   Ranking the Attributes

Our method for nesting models depends on a ranking of the attributes. Models containing lower ranked attributes will be built upon models containing higher ranked attributes. The mutual information between an attribute and the class measures how informative this attribute is about the class [12], and thus it is a suitable metric to rank the attributes.

The advantage of using mutual information is that it can be computed very efficiently after one pass through the training data. Although the mutual information between an attribute and the class can help to identify the attributes that are individually most discriminative, it is important to note that it does not directly assess the discriminative power of an attribute in combination with other attributes. Nevertheless, the ranking of attributes based on mutual information with the class will permit the search over a large space of possible models and the deficiencies of this discriminative approach will be mitigated by the richness of the search space that is evaluated in a discriminative fashion.

3.2   Building the Model Space

Without loss of generality, in the following we assume that the attributes are ordered by mutual information. That is, $\mathbf{x}_i$ represents the attribute with the ith greatest mutual information with the class. As the attributes have been ranked, we can create, in total, a^2 nested submodels of attribute subsets. To be more specific, suppose we select top r attributes as parents and top s attributes as children, where $1 \leq r, s \leq a$, the candidate AODE model would be,

$$\hat{P}(y, \mathbf{x})_{r,s} = \frac{\sum_{i:1 \leq i \leq r \wedge F(x_i) \geq m} \hat{P}(y, x_i) \prod_{j=1}^{s} \hat{P}(x_j \mid y, x_i)}{|\{i : 1 \leq i \leq r \wedge F(x_i) \geq m\}|} \ . \tag{6}$$

Figure 1 gives an example of the model space with 3 attributes. For instance, model m_{21} considers the two attributes $\{\mathbf{x}_1, \mathbf{x}_2\}$ as parents and a single attribute $\{\mathbf{x}_1\}$ as a child. Then, when the attribute $\mathbf{x}_2$ is considered to be added as a child, we obtain a new model m_{22}. When instead the attribute $\mathbf{x}_3$ is considered to be added as a parent, we obtain a new model m_{31}. Both of these models are minor extensions to the existing model m_{21} and all three (and all their extensions) can be applied to a test instance in a single nested computation. Consequently all models can be efficiently evaluated in a single set of nested computations.

	children		
parents	$\{\mathbf{x}_1\}$	$\{\mathbf{x}_1, \mathbf{x}_2\}$	$\{\mathbf{x}_1, \mathbf{x}_2, \mathbf{x}_3\}$
$\{\mathbf{x}_1\}$	m_{11}	m_{12}	m_{13}
$\{\mathbf{x}_1, \mathbf{x}_2\}$	m_{21}	m_{22}	m_{23}
$\{\mathbf{x}_1, \mathbf{x}_2, \mathbf{x}_3\}$	m_{31}	m_{32}	m_{33}

Fig. 1. An example of the model space with 3 attributes

3.3   Selecting the Best Model

Once we have built the model space, we can perform model selection within this space. To evaluate the goodness of an alternative model, an evaluation function is required, which commonly measures the discriminative ability of the model among classes.

We use leave-one-out cross validation error to measure the performance of each model. Rather than building a new model for every fold, we use incremental cross validation [13], in which the contribution of the training example being left out in each fold is simply subtracted from the count table, thus producing a model without that training example. This method allows the model to be evaluated quickly, whilst obtaining a good estimate of the generalization error.

There are several loss functions to measure model performance for leave-one-out cross validation, zero-one loss and root mean squared error (RMSE) are among the most common and effective. Zero-one loss simply assigns a loss of '0' to correct classification, and '1' to incorrect classification, treating all misclassifications as equally undesirable. RMSE, however, accumulates for each example the squared error, which is the probability of incorrectly classifying the example, and then computes the root mean of the sum. As RMSE gives a finer grained measure of the calibration of the probability estimates compared to zero-one loss, with the error depending not just on which class is predicted, but also on the probabilities estimated for each class, we use RMSE to evaluate the candidate models in this research.

3.4   Algorithm and Analysis

Based on the methodology presented above, we develop the training algorithm for attribute selective AODE shown in Algorithm 1.

Algorithm 1. Training algorithm for attribute selective AODE

1: Form the table of joint frequencies of pairwise attribute-values and class
2: Compute the mutual information
3: Rank the attributes
4: **for all** example in $\mathcal{T}$ **do**
5:     Build all a^2 models while leaving the current example out
6:     Predict the current example using a^2 models
7:     Accumulate the squared error for each model
8: **end for**
9: Compute the root mean squared error for each model
10: Select the model with the lowest RMSE

As in AODE, we need to form the table of joint frequencies of pairs of attribute-values and class from which the probability estimates $\hat{P}(y, x_i)$, $\hat{P}(x_j \mid y, x_i)$ and the mutual information between the attributes and class are derived.

This is done in one pass through the training data (line 1). Note that this provides all of the information needed to create any selective AODE model with any sets of parent and child attributes.

In the second pass through the training data (line 4-8), the squared error is accumulated for each model. After this pass, the RMSE will be computed and used to select the best model.

At training time, the space complexity of the table of joint frequencies of attribute-values and class is $\mathcal{O}(k(av)^2)$ as in AODE, where v is the average number of values per attribute. Attribute selection will not require more memory. Derivation of the frequencies required to populate this table is of time complexity $\mathcal{O}(ta^2)$. Attribute selection needs one more pass through the training data, the time complexity of which is $\mathcal{O}(tka^2)$, since for each example we need to compute the joint probability in (1) for each class. So the overall time complexity is $\mathcal{O}(t(k+1)a^2)$.

Classification requires the table of probability estimates formed at training time of space complexity $\mathcal{O}(k(av)^2)$. The time complexity of classifying a single example is $\mathcal{O}(ka^2)$ in the worst-case scenario, because some attributes may be omitted after attribute selection.

4   Empirical Comparisons

In this section, we compare the newly proposed attribute selective AODE (AS-AODE) with AODE, weightily AODE (WAODE), AODE with subsumption resolution (AODESR), BSE selective AODE (BSEAODE) and A2DE.

Zheng et al [9] discussed three different subsumption resolution techniques, Lazy SR, Eager SR and Near SR. Lazy SR is used in this paper, as it can improve AODE with low training time and modest test time overheads. The minimum frequency for identifying generalizations is set to 100. The results in [5] show that BSE performs better than FSS, and the elimination of a child is more effective than the elimination of a parent. So we select only the children in BSEAODE. However, we do not perform statistical tests in BSEAODE, as we do not do this in ASAODE, either. We also include A2DE in the set of experiments so as to provide a comprehensive comparison.

The experimental system is implemented in C++. In order to deal with numerical data, Minimum Description Length (MDL) discretization [14] is implemented. More specifically, the cut points are computed on 100,000 examples randomly selected from training data or on all training examples if the training data is less than 100,000. These cut points are then used to discretize the training and test data. The base probabilities are estimated using m-estimation ($m = 1$) [15]. Missing values have been considered as a distinct value.

We run the above algorithms on 71 data sets from the UCI repository [16]. Table 1 presents the detailed characteristics of data sets in ascending order on the number of instances. We run the experiments on a single CPU single core virtual Linux machine running on a Sun grid node with dual 6 core Intel Xeon L5640 processors running at 2.27 GHz with 96 GB RAM.

Table 1. Data sets

No. Name	Inst	Att	Class	No. Name	Inst	Att	Class
1 contact-lenses	24	4	3	37 vowel	990	13	11
2 lung-cancer	32	56	3	38 german	1000	20	2
3 labor-negotiations	57	16	2	39 led	1000	7	10
4 post-operative	90	8	3	40 contraceptive-mc	1473	9	3
5 zoo	101	16	7	41 yeast	1484	8	10
6 promoters	106	57	2	42 volcanoes	1520	3	4
7 echocardiogram	131	6	2	43 car	1728	6	4
8 lymphography	148	18	4	44 segment	2310	19	7
9 iris	150	4	3	45 hypothyroid	3163	25	2
10 teaching-ae	151	5	3	46 splice-c4.5	3177	60	3
11 hepatitis	155	19	2	47 kr-vs-kp	3196	36	2
12 wine	178	13	3	48 abalone	4177	8	3
13 autos	205	25	7	49 spambase	4601	57	2
14 sonar	208	60	2	50 phoneme	5438	7	50
15 glass-id	214	9	3	51 wall-following	5456	24	4
16 new-thyroid	215	5	3	52 page-blocks	5473	10	5
17 audio	226	69	24	53 optdigits	5620	64	10
18 hungarian	294	13	2	54 satellite	6435	36	6
19 heart-disease-c	303	13	2	55 musk2	6598	166	2
20 haberman	306	3	2	56 mushrooms	8124	22	2
21 primary-tumor	339	17	22	57 thyroid	9169	29	20
22 ionosphere	351	34	2	58 pendigits	10992	16	10
23 dermatology	366	34	6	59 sign	12546	8	3
24 horse-colic	368	21	2	60 nursery	12960	8	5
25 house-votes-84	435	16	2	61 magic	19020	10	2
26 cylinder-bands	540	39	2	62 letter-recog	20000	16	26
27 chess	551	39	2	63 adult	48842	14	2
28 syncon	600	60	6	64 shuttle	58000	9	7
29 balance-scale	625	4	3	65 connect-4	67557	42	3
30 soybean	683	35	19	66 ipums.la.99	88443	60	19
31 credit-a	690	15	2	67 waveform	100000	21	3
32 breast-cancer-w	699	9	2	68 localization	164860	5	11
33 pima-ind-diabetes	768	8	2	69 census-income	299285	41	2
34 vehicle	846	18	4	70 poker-hand	1025010	10	10
35 anneal	898	38	6	71 record-linkage	5749132	11	2
36 tic-tac-toe	958	9	2				

4.1 Bias, Variance and RMSE

Because ASAODE explores a larger space of models than AODE and BSEAODE explores a larger space of models than ASAODE, we expect BSEAODE to have the lowest bias, followed by ASAODE then AODE and this order to be reversed for their relative variance. Hence we expect AODE to deliver the lowest error on smaller datasets, ASAODE to dominate at some intermediate data size, and for BSEAODE to deliver the lowest error on very large data. The bias and variance of ASAODE relative to WAODE, AODESR and A2DE can be expected to vary from dataset to dataset as these all embody different learning biases and none of their spaces of models subsumes the other.

In order to assess these expectations, we first perform bias variance decomposition using the experimental method proposed by Kohavi and Wolpert [17]. As this study is more meaningful with more data, we run these experiments only on the largest 28 data sets which have at least 2000 examples. For each data set, 1000 training examples and 1000 test examples are randomly selected. The bias variance decomposition is calculated from the error on the test examples. This process is repeated 10 times to obtain the mean bias and variance.

A summary of pairwise win/draw/loss records, which indicate the number of data sets on which one algorithm has lower, equal or higher outcome relative to the other, is presented in Table 2. Each entry in cell $[i, j]$ compares the algorithm in row i against the algorithm in column j. The p value following each win/draw/loss record is the outcome of a binomial sign test and represents the probability of observing the given number of wins and losses if each were equally likely. The reported p value is the result of a two-tailed test. We consider a difference to be significant if $p \leq 0.05$. All such p values have been changed to boldface in the table.

Table 2 shows that all five variants to AODE achieve significant reductions in bias relative to AODE. While ASAODE achieves lower bias than WAODE and AODESR more often than not, the reverse is true for BSEAODE and A2DE; although these differences are not significant.

Next, we conduct 10-fold cross validation experiments to obtain the error of the alternative algorithms. As attribute selection is based on the RMSE metric, we are inclined to evaluate the error by RMSE. The win/draw/loss records of alternative algorithms for RMSE on 71 data sets are also presented in Table 2.

We can see that all five improvements to AODE have achieved significant reductions in RMSE relative to AODE. ASAODE has also achieved significant reductions in RMSE relative to WAODE and AODESR. The p value (0.807)indicates that ASAODE and BSEAODE have achieved almost the same performance. But the advantages of BSEAODE over WAODE and AODESR are not as significant as those of ASAODE over WAODE and AODESR. While A2DE achieves significant reductions in RMSE relative to AODE, WAODE, AODESR and BSEAODE, its advantage over ASAODE is not significant.

The fact that ASAODE obtains, in general, lower bias and higher variance compared with WAODE and AODESR, indicates that it will perform better on larger datasets, since it will be able to capture more complex relationships from large amount of data [18]. In order to demonstrate this hypothesis, we also compile the win/draw/loss results in terms of RMSE on the 43 smallest data sets and the 28 largest data sets in Table 2. We can see that the performance of ASAODE is better on large data sets than on small data sets. While for even larger data sets BSEAODE and A2DE might outperform ASAODE for the same reason, both have high computational complexity that can be prohibitive for large data, since BSEAODE requires $2a$ pases on the whole training set and A2DE's memory requirements and classification time are very high (see the following Section 4.2).

4.2 Computation Time

The logarithmic means of training and classification time on the 71 data sets for all algorithms are shown in Fig. 2. We have added 1 to each mean before computing the logarithm to avoid negative bars. ASAODE requires more training time than such one pass algorithms as AODE, WAODE and AODESR. This is because ASAODE involves two passes through the training data. As BSEAODE needs at most $2a$ passes, it requires significantly more training time than ASAODE.

Table 2. Win/draw/loss records of bias, variance and RMSE with binomial sign test

		AODE		WAODE		AODESR		BSEAODE		ASAODE	
		W/D/L	p	W/D/L	p	W/D/L	p	W/D/L	p	W/D/L	p
Bias[1]	WAODE	21/2/5	**0.002**								
	AODESR	15/8/5	**0.041**	12/1/15	0.701						
	BSEAODE	19/5/4	**0.003**	17/4/7	0.064	17/2/9	0.169				
	ASAODE	21/3/4	**<0.001**	18/1/9	0.122	16/1/11	0.442	11/2/15	0.557		
	A2DE	23/2/3	**<0.001**	21/1/6	**0.006**	20/3/5	**0.004**	14/1/13	1	17/1/10	0.248
Variance[1]	WAODE	13/1/14	1								
	AODESR	7/8/13	0.263	13/0/15	0.851						
	BSEAODE	11/5/12	1	10/0/18	0.185	12/3/13	1				
	ASAODE	9/1/18	0.122	11/1/16	0.442	13/0/15	0.851	13/2/13	1		
	A2DE	14/1/13	1	14/0/14	1	14/3/11	0.69	14/1/13	1	14/0/14	1
RMSE[2]	WAODE	45/5/21	**0.004**								
	AODESR	32/27/12	**0.004**	28/6/37	0.321						
	BSEAODE	40/20/11	**<0.001**	40/4/27	0.142	35/14/22	0.111				
	ASAODE	43/6/22	**0.013**	42/4/25	**0.05**	42/5/24	**0.036**	35/4/32	0.807		
	A2DE	52/4/15	**<0.001**	47/2/22	**0.004**	48/3/20	**<0.001**	42/4/25	**0.05**	43/2/26	0.053
RMSE_S[3]	WAODE	26/3/14	0.081								
	AODESR	20/19/4	**0.002**	18/3/22	0.636						
	BSEAODE	19/14/10	0.136	23/2/18	0.533	16/10/17	1				
	ASAODE	19/5/19	1	19/4/20	1	18/5/20	0.871	20/4/19	1		
	A2DE	27/3/13	**0.038**	24/1/18	0.441	23/2/18	0.533	22/3/18	0.636	25/2/16	0.211
RMSE_L[4]	WAODE	19/2/7	**0.029**								
	AODESR	12/8/8	0.503	10/3/15	0.424						
	BSEAODE	21/6/1	**<0.001**	17/2/9	0.169	19/4/5	**0.007**				
	ASAODE	24/1/3	**<0.001**	23/0/5	**<0.001**	24/0/4	**<0.001**	15/0/13	0.851		
	A2DE	25/1/2	**<0.001**	23/1/4	**<0.001**	25/1/2	**<0.001**	20/1/7	**0.019**	18/0/10	0.185

[1] Bias and variance results on the 28 largest data sets.
[2] RMSE results on all the 71 data sets.
[3] RMSE_S: RMSE results on the 43 smallest data sets.
[4] RMSE_L: RMSE results on the 28 largest data sets.

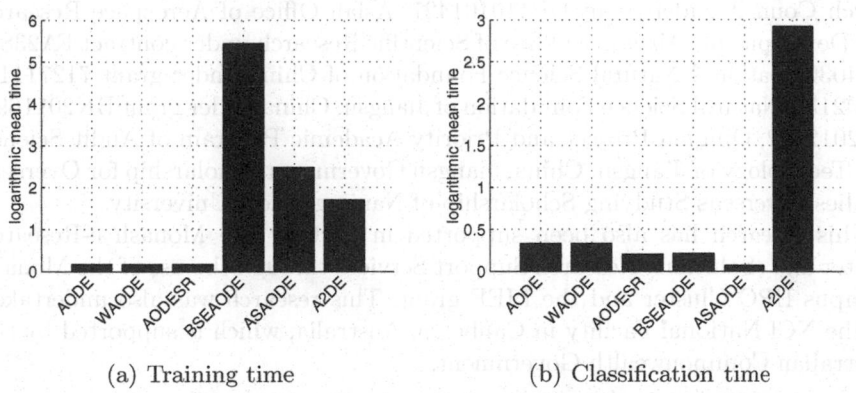

(a) Training time   (b) Classification time

Fig. 2. Computation time comparison of different algorithms (seconds)

As for the classification time, ASAODE, AODESR and BSEAODE require, in general, less time than AODE and WAODE because they might eliminate some attributes. Fig. 2 also shows that ASAODE requires even less classification time than AODESR and BSEAODE.

A2DE requires more training and classification time than AODE, as it needs to compile a more complicated table at training time and requires more computation at classification time.

5   Conclusion

In this paper, a new attribute selection algorithm is proposed for AODE. It is a two-pass algorithm, so compared to AODE, it just requires one more pass through the training data. The alternative attribute selection methods, such as FSA and BSE, need a number of passes that is linear to the number of attributes to obtain similar results.

The empirical results show that the new algorithm is significantly more accurate than AODE, WAODE and AODESR, has comparable error to BSEAODE, and as we expected, worse than A2DE. It requires significantly less training time than BSEAODE, and less classification time than AODE and all other variants, especially than A2DE.

It is worthwhile to note that the technique proposed in this paper is of squared complexity in the number of attributes, so it is not scalable to high dimensional data. On the other hand, it is compatible with weighting, subsumption resolution and higher orders of AnDE. Consequently, it might be possible to further improve the accuracy by combining it with weighting, subsumption resolution and A2DE. This is a promising direction for future research.

Acknowledgments. This research has been supported by the Australian Research Council under grant DP110101427, Asian Office of Aerospace Research and Development, Air Force Office of Scientific Research under contract FA2386-1214030, National Natural Science Foundation of China under grant 71271117, 61202135, Natural Science Foundation of Jiangsu, China under grant BK2011692, BK2012472, Qinglan Project and Priority Academic Program of Audit Science and Technology of Jiangsu, China, Jiangsu Government Scholarship for Overseas Studies, Overseas Studying Scholarship of Nanjing Audit University.

This research has also been supported in part by the Monash e-Research Center and eSolutions-Research Support Services through the use of the Monash Campus HPC Cluster and the LIEF grant. This research was also undertaken on the NCI National Facility in Canberra, Australia, which is supported by the Australian Commonwealth Government.

References

1. Duda, R.O., Hart, P.E.: Pattern Classification and Scene Analysis, 1st edn. John Wiley & Sons Inc. (1973)
2. Webb, G.I., Boughton, J.R., Wang, Z.: Not so naive Bayes: Aggregating one-dependence estimators. Machine Learning 58(1), 5–24 (2005)
3. Zheng, F., Webb, G.I.: A comparative study of semi-naive Bayes methods in classification learning. In: AusDM, pp. 141–156 (2005)

4. Yang, Y., Webb, G.I., Cerquides, J., Korb, K.B., Boughton, J., Ting, K.M.: To select or to weigh: A comparative study of linear combination schemes for superparent-one-dependence estimators. IEEE Transactions on Knowledge and Data Engineering 19(12), 1652–1665 (2007)
5. Zheng, F., Webb, G.I.: Finding the right family: Parent and child selection for averaged one-dependence estimators. In: Kok, J.N., Koronacki, J., de Lopez Mantaras, R., Matwin, S., Mladenič, D., Skowron, A. (eds.) ECML 2007. LNCS (LNAI), vol. 4701, pp. 490–501. Springer, Heidelberg (2007)
6. Webb, G.I., Boughton, J.R., Zheng, F., Ting, K.M., Salem, H.: Learning by extrapolation from marginal to full-multivariate probability distributions: Decreasingly naive Bayesian classification. Machine Learning 86(2), 233–272 (2012)
7. Cerquides, J., de Mántaras, R.L.: Robust Bayesian linear classifier ensembles. In: Gama, J., Camacho, R., Brazdil, P.B., Jorge, A.M., Torgo, L. (eds.) ECML 2005. LNCS (LNAI), vol. 3720, pp. 72–83. Springer, Heidelberg (2005)
8. Jiang, L., Zhang, H.: Weightily averaged one-dependence estimators. In: Yang, Q., Webb, G. (eds.) PRICAI 2006. LNCS (LNAI), vol. 4099, pp. 970–974. Springer, Heidelberg (2006)
9. Zheng, F., Webb, G.I., Suraweera, P., Zhu, L.: Subsumption resolution: An efficient and effective technique for semi-naive Bayesian learning. Machine Learning 87(1), 93–125 (2012)
10. Langley, P., Sage, S.: Induction of selective Bayesian classifiers. In: Proceedings of the Tenth International Conference on Uncertainty in Artificial Intelligence, pp. 399–406. Morgan Kaufmann Publishers Inc. (1994)
11. Kittler, J.: Feature selection and extraction. In: Handbook of Pattern Recognition and Image Processing, pp. 59–83 (1986)
12. MacKay, D.J.: Information theory, inference and learning algorithms. Cambridge university press (2003)
13. Kohavi, R.: The power of decision tables. In: Lavrač, N., Wrobel, S. (eds.) ECML 1995. LNCS, vol. 912, pp. 174–189. Springer, Heidelberg (1995)
14. Fayyad, U.M., Irani, K.B.: Multi-interval discretization of continuous-valued attributes for classification learning. In: IJCAI, pp. 1022–1027 (1993)
15. Cestnik, B.: Estimating probabilities: A crucial task in machine learning. In: ECAI, vol. 90, pp. 147–149 (1990)
16. Bache, K., Lichman, M.: UCI machine learning repository (2013)
17. Kohavi, R., Wolpert, D.H.: Bias plus variance decomposition for zero-one loss functions. In: ICML, pp. 275–283 (1996)
18. Brain, D., Webb, G.I.: The need for low bias algorithms in classification learning from large data sets. In: Elomaa, T., Mannila, H., Toivonen, H. (eds.) PKDD 2002. LNCS (LNAI), vol. 2431, pp. 62–73. Springer, Heidelberg (2002)

A New Evaluation Function for Entropy-Based Feature Selection from Incomplete Data

Wenhao Shu[1], Hong Shen[2,3], Yingpeng Sang[1], Yidong Li[1], and Jun Wu[1]

[1] School of Computer and Information Technology, Beijing Jiaotong University,
Beijing, China
[2] School of Information Science and Technology, Sun Yat-sen University, China
[3] School of Computer Science, University of Adelaide, Australia
`11112084@bjtu.edu.cn, hongsh01@gmail.com`

Abstract. In data mining and knowledge discovery, evaluation functions for evaluating the quality of features have great influence on the outputs of feature selection algorithms. However, in the existing entropy-based feature selection algorithms from incomplete data, evaluation functions are often inadequately computed as a result of two drawbacks. One is that the existing evaluation functions have not taken into consideration the differences of discernibility abilities of features. The other is that in the feature selection algorithms of forward greedy search, if the feature with the same entropy value is not only one, the arbitrary selection may affect the classification performance. This paper introduces a new evaluation function to overcome the drawbacks. A main advantage of the proposed evaluation function is that the granularity of classification is considered in the evaluation computations for candidate features. Based on the new evaluation function, an entropy-based feature selection algorithm from incomplete data is developed. Experimental results show that the proposed evaluation function is more effective than the existing evaluation functions in terms of classification accuracy.

Keywords: Evaluation function, Conditional entropy, Feature selection, Rough sets, Incomplete data.

1 Introduction

Feature reduction has been shown effective in dealing with high-dimensional data for efficient data mining, which refers to the study of methods for reducing the number of dimensions describing data [4, 10]. Its general purpose is to select relevant features to represent data and reduce computational cost, without deteriorating discriminative capability. It can bring many potential benefits: alleviating the curse of dimensionality, speeding up the learning process, and improving the generalization capability of a learning model. Many feature reduction algorithms have been developed at present. In general, they can be broadly classified into two categories: feature extraction and feature selection [5]. Feature extraction constructs new features with a linear or nonlinear transformation by projecting the original feature space to a lower dimensional one. Unlike feature

V.S. Tseng et al. (Eds.): PAKDD 2014, Part II, LNAI 8444, pp. 98–109, 2014.

extraction methods, feature selection methods preserve the original meaning of the features after reduction, which can be broadly categorized into wrapper [1] and filter [7, 9] methods. The wrapper method uses the predictive accuracy of a predetermined learning algorithm to determine the quality of selected features. One drawback of the wrapper method, however, is that it is very expensive to run for data with numbers of features. The filter method separates feature selection from classifier learning so that the bias of a learning algorithm does not interact with a feature selection algorithm. It relies on many feature measures such as distance [3], consistency [11], correlation [2] and so on. Much attention has been paid to filter feature selection.

Generally speaking, filter feature selection methods work under the framework consisting of four components [4]: subset generation, evaluation, stopping criterion and result validation. The main difference among various feature selection algorithms lies in how to evaluate the candidate features. Obviously, evaluation functions have great influence on outputs of feature selection algorithms. Rough set theory offers a formal methodology for filter feature selection. The main advantage of rough set theory is that no additional information about the data is required for data analysis such as thresholds or expert knowledge on a particular domain. It provides a mathematical tool to handle uncertainty in many data analysis tasks [6, 13]. The feature subset obtained by rough set-based feature selection is called a reduct. The features in the reduct are not only strongly relevant to the classification task, but also no redundant with each other, which keep consistency with the objective of feature selection.

It is clear that the feature selection work in classical rough set theory is based on complete data. However, in many real-world applications, it may happen that some feature values are missing because of many factors such as noise in data, prediction capability [12, 13, 15]. Here we briefly review the state of the art about feature selection algorithms from incomplete data. Sun et al. [12] introduced rough entropy to evaluate the roughness of knowledge in incomplete data, and developed a rough entropy-based feature selection algorithm. Slezak [14] proposed an algorithm based on information entropy to compute a reduct. As the uncertainty measure, conditional entropy, is one key issue in rough set theory, Dai et al. [15] proposed conditional entropy for incomplete data, and studied the application of feature selection based on conditional entropy. Evaluation functions, used to evaluate the quality of features, have great influence on the outputs of feature selection algorithms. However, there are some drawbacks in the existing evaluation functions. On the one hand, the existing evaluation functions only consider the differences of entropy values' variation, but there exists the differences of discernibility abilities for candidate features. As much as we know, the existing research work has not considered this aspect. Even if there are multiple features leading to the same entropy values, we can still compare the discernibility power of the features according to the granularity measure. On the other hand, for the forward greedy search, if the feature with the same entropy values is not only one, we often arbitrarily choose one of them, but the arbitrariness may affect the classification performance. Therefore, the main contribution of this paper is to present a new evaluation function to overcome the above stated drawbacks.

This paper is organized as follows. In Section 2, we review some basic concepts from the theory of rough sets. In Section 3, a simple example is firstly given to illustrate the drawbacks of existing evaluation functions, and then a new evaluation function together with an entropy-based feature selection algorithm are presented. In Section 4, comparison experiments are made to show the validity of the proposed evaluation function. Finally, the conclusions are presented in Section 5.

2   Preliminaries

Data sets are usually given as the form of tables, we call a data table as an information system, formulated as $IS =< U, A, V, f >$, where U is a set of nonempty and finite objects, called the universe; A is the set of features characterizing the objects; V is the union of feature domains, i.e., $V = \cup_{a \in A} V_a$, where V_a is the value set of feature a, called the domain of a; and $f : U \times A \to V$ is an information function, which assigns feature values to objects such as $\forall a \in A$, $x \in U$, and $f(x, a) \in V_a$, where $f(x, a)$ denotes the value of feature a for object x. If the feature set is divided into condition feature set C and decision feature set D, the information system is called a decision system. If there exist $x \in U$ and $a \in A$ such that $f(x, a)$ is equal to a missing value (a null or unknown value, denoted as "*"), i.e., $* \in V_a$, then the information system is an incomplete information system (IIS). If $* \notin V_D$ but $* \in V_C$, then the decision system is an incomplete decision system (IDS).

Given a complete information system $CIS =< U, A, V, f >$, for $\forall B \subseteq A$, the equivalence relation generated by B is defined by $IND(B) = \{(x, y) | \forall a \in B, f(x, a) = f(y, a)\}$. The family of all equivalence classes of $IND(B)$ is denoted as $U/IND(B)$. An equivalence class of $IND(B)$ containing x is denoted by $[x]_B$. Since there are missing values for some objects, the equivalence relation $IND(B)$ is not suitable for incomplete information systems.

Given an incomplete information system $IIS =< U, A, V, f >$, for $\forall B \subseteq A$, a tolerance relation between objects that are possibly indiscernible in terms of B is defined by $TR(B) = \{(x, y) | \forall a \in B, f(x, a) = f(y, a) \lor f(x, a) = * \lor f(y, a) = *\}$. It can be easily shown that $TR(B) = \cap_{a \in B} TR(\{a\})$. The tolerance class of object x with reference to a feature set B is denoted as $T_B(x) = \{y | (x, y) \in TR(B)\}$. Let $U/TR(B)$ denote the family set $\{T_B(x) | x \in U\}$, which is the classification induced by B. For $X \subseteq U$, the lower and upper approximation of X with respect to B can be defined as $\underline{B}(X) = \{x \in U | T_B(x) \subseteq X\}$ and $\overline{B}(X) = \{x \in U | T_B(x) \cap X \neq \ddot{A}\ddot{Y}\}$. The lower approximation is called the positive region, that is $POS_B(X) = \underline{B}(X)$. X is called $B-$definable iff $\overline{B}(X) = \underline{B}(X)$. Otherwise, $\overline{B}(X) \neq \underline{B}(X)$ and X is rough.

Given an incomplete decision system $IDS =< U, C \cup D, V, f >$, for $\forall B \subseteq C$, the objects are partitioned into n mutually exclusion crisp subsets $U/IND(D) = \{D_1, D_2, \cdots, D_n\}$ by the decision features D. The lower and upper approximations with respect to B of D are defined as $\underline{B}(D) = \{\underline{B}(D_1), \underline{B}(D_2), \cdots, \underline{B}(D_n)\}$

and $\overline{B}(D) = \{\overline{B}(D_1), \overline{B}(D_2), \cdots, \overline{B}(D_n)\}$. Denoted by $POS_B(D) = \bigcup_{i=1}^{n} \underline{B}(D_i)$, which is called the positive region of D with respect to B in the IDS. The lower approximation is a description of the domain objects which are known with absolute certainty to belong to the decision classes.

3 An Evaluation Function for Entropy-Based Feature Selection

In this section, a simple example is firstly given to illustrate the drawbacks of existing evaluation functions, and then a new evaluation function together with a entropy-based feature selection algorithm are presented.

The conditional entropy of Definition 1 can be used as a reasonable information measure in incomplete decision tables[15], and it is quite representative among other entropies. Correspondingly, the evaluation function in terms of conditional entropy is also defined.

Definition 1. Let $IDS =< U, C \cup D, V, f >$be an incomplete decision table, $U = \{x_1, x_2, \ldots, x_n\}$, for $B \subseteq C$, the classification induced by B is $U/TR(B) = \{T_B(x_1), T_B(x_2), \ldots, T_B(x_n)\}$, and $U/IND(D) = \{D_1, D_2, \cdots, D_m\}$ is a partition on decision attribute set D. The conditional entropy of D with respect to B is defined by

$$H(D|B) = -\sum_{i=1}^{n} \sum_{j=1}^{m} \frac{|T_B(x_i) \cap D_j|}{|U|} log \frac{|T_B(x_i) \cap D_j|}{|T_B(x_i)|}.$$

Definition 2. Given an incomplete decision table $IDS =< U, C \cup D, V, f >$, suppose $B \subseteq C$ is the selected feature subset, and $a \in C - B$ is a candidate feature. Then the evaluation function of candidate feature a is defined as $e(a) = H(D|B) - H(D|B \cup \{a\})$.

From Definition 2, the existing evaluation function can be used to evaluate the importance of features. The smaller the evaluation value is, the more important the feature will be. However, the drawbacks of above evaluation function can be explained with reference to the following example.

Example. Suppose there is an incomplete decision table $IDS =< U, C \cup D, V, f >$, where $U = \{x_1, x_2, x_3, x_4, x_5, x_6, x_7, x_8\}$ and $C = \{c_1, c_2, c_3, c_4\}$. In the feature selection process, Definition 2 is applied to compute the evaluation values of features. By computing, the descending sequence of four candidate features is listed as follows: $e(c_1) > e(c_2) > e(c_3) = e(c_4)$. Obviously, the features with the minimum evaluation value are c_3 and c_4. By direct computation the classifications induced by two features, $U/TR(c_3) = \{\{x_1, x_2, x_5, x_6\}, \{x_3, x_4, x_7, x_8\}\}$ and $U/TR(c_4) = \{\{x_1, x_2\}, \{x_3, x_4\}, \{x_5, x_6\}, \{x_7, x_8\}\}$, obviously, the discernibility abilities of them are different, feature c_3 can describe the stronger discernibility power than c_4. However, Definition 2 does not take into consideration this difference. Thus the evaluation function given by Definition 2 is inadequately computed as a result of this aspect.

On the other hand, in the feature selection algorithm of forward greedy search, due to $e(c_3) = e(c_4)$, we can select one feature arbitrarily. Consequently, feature c_3 or c_4 are chosen to the selected feature subset. The arbitrariness can surely not guarantee a selected feature subset is a reduct. Suppose that the selected feature subset containing feature c_3 and c_2 exhibit the best performance, but we obtain the final feature subset is $\{c_4, c_2\}$ due to the arbitrary selection. Obviously, this result may affect the classification performance. Therefore, we give a new evaluation function from a reasonable perspective to improve the above mentioned problems.

Definition 3. Given an incomplete decision table $IDS =< U, C \cup D, V, f >$, suppose $B \subseteq C$ is the selected feature subset, and $a \in C - B$ is a candidate feature, the classification induced by a consists of tolerance class $A_i (1 \leq i \leq k)$. Then a new evaluation function of candidate feature a is defined as $f(a) = e(a) + g(a)$, where $g(a) = \frac{1}{|U|^2} \sum_{i=1}^{k} |A_i|^2$, which is the granularity measure of feature a.

Theorem 1. *Given an incomplete decision table $IDS =< U, C \cup D, V, f >$, suppose $B \subseteq C$ is the selected feature subset, for $\forall a, b \in C - B$, there is $f(a \cup b) < f(a)$ or $f(a \cup b) < f(b)$.*

Proof. Suppose the classification induced by a consists of tolerance classes $A_i (1 \leq i \leq k)$, and the classification induced by $a \cup b$ consists of tolerance classes $B_j (1 \leq j \leq l)$, by Definition 2 and the definition of conditional entropy, it is obvious that $e(a \cup b) < e(a)$. Since $a \subseteq a \cup b$, according to the definition of tolerance class, there is $|B_j| < |A_i|$, obviously, it holds that $\sum_{j=1}^{l} |B_j|^2 < \sum_{i=1}^{k} |A_i|^2$, thus $g(a \cup b) < g(a)$. Therefore, $f(a \cup b) < f(a)$. In the same way, it can proof that $f(a \cup b) < f(b)$. ☐

Theorem 1 shows the rationality of the new evaluation function, which states the uncertainty decreases when the available knowledge increases. Obviously, the granularity measure can represent discernibility ability of candidate feature a, the smaller $g(a)$ is, the stronger its discernibility ability. Through comparison, the selection of survival features can be achieved. From above example, there is $g(c_3) > g(c_4)$, thus it also holds that $f(c_3) > f(c_4)$, the discernibility ability of candidate feature c_4 is stronger than that of feature c_3. Therefore, the survival feature is c_4. It is obvious that the new evaluation function is more reasonable. Combine the new evaluation function into feature selection, a selected feature subset (called reduct) can be characterized by the following statement.

Definition 4. Given an incomplete decision table $IDS =< U, C \cup D, V, f >$, a selected feature subset $B \subseteq C$ is called a reduct of the IDS if and only if $H(D|B) = H(D|C)$, and for $\forall B' \subset B$, $H(D|B') \neq H(D|C)$.

In this definition, the first one indicates that the selected feature subset preserves the same information measure as the whole set of features; the second one guarantees that all of the features are indispensable, i.e., there is not any redundant feature in the reduct.

In the following, we combine the proposed evaluation function with forward greedy search to construct the feature selection algorithm.

Algorithm 1. Entropy-based Feature Selection Algorithm from Incomplete Data

Input: An incomplete decision table $IDS = < U, C \cup D, V, f >$;
Output: A feature subset Red.
Begin

1. Initialize $Red = \emptyset$;
2. **For** each $c \in C$ **do**
3.   compute $H(D|C - \{c\}) - H(D|C)$;
4.   if $H(D|C - \{c\}) - H(D|C) > 0$, then $Red = Red \cup \{c\}$;
5. **End for**
6. **While** $H(D|Red) \neq H(D|C)$ **do**
7.   compute $f(c)$ for all $c \in C - Red$;
8.   choose the feature c_k that minimizes $f(c)$, and let $Red = Red \cup \{c_k\}$, $C = C - \{c_k\}$;
9. **End while**
10. **For** each $c \in Red$ **do**
11.   compute $H(D|Red) - H(D|Red - \{c\})$;
12.   if $H(D|Red) - H(D|Red - \{c\}) = 0$, then $Red = Red - \{c\}$;
13. **End for**
14. Return Red.

End

The algorithm begins with an empty subset Red, and adds some indispensable features to Red gradually. Then select the features with the minimal value by the new evaluation function into Red each loop until satisfying the stopping condition. Finally, a redundancy-removing step is carried out to avoid the redundancy in the selection result. The feature subset selected by this algorithm obtains the same information as the original feature set from incomplete data.

4   Experimental Analysis

In order to test the validity of the new proposed evaluation function, we conduct some experiments on a PC with Windows 7, Intel (R) Core(TM) Duo CPU 2.93 GHz and 4GB memory. Algorithms are coded in C++ and the software being used is Microsoft Visual 2008. The objective of the following experiments is to show the effectiveness of feature selection algorithm based on the new evaluation function. We perform the experiments on six real UCI data sets, which are downloaded from UCI Repository of machine learning databases in [16]. The characteristics of six data sets are described in Table 1. For the complete data sets, we randomly change 5% of the known features values from each original data set into missing values to create incomplete data sets. For the numerical features, we use the data tool Rosetta (http://www.lcb.uu.se/tools/rosetta/index.php) to discretize them.

Table 1. A description of six data sets

Data sets	Objects	Features	Classes
Hepatitis	155	19	2
Soybean-large	307	35	19
Synthetic	600	60	6
Cardiotocography	2126	21	3
Ticdate 2000	5822	85	2
Mushroom	8124	22	2

In what follows, we first make a comparative study on the feature selection algorithms in terms of feature subset size. The results are shown in Table 2 in which PFS represents the proposed feature selection algorithm, EFS represents the feature selection algorithm constructed in [15] and LFS denotes the lower approximation-based feature selection algorithm in [13]. Note that PFS selects candidate features by Definition 4, while EFS finds candidate features by Definition 2. The main difference between PFS and EFS is the evaluation function.

Table 2. Comparison of feature subset size by Algorithms PFS, EFS and LFS

Data sets	Original feature set size	Feature subset size		
		PFS	EFS	LFS
Hepatitis	19	12	14	14
Soybean-large	35	9	11	10
Synthetic	60	13	13	16
Cardiotocography	21	12	13	12
Ticdate 2000	85	24	24	24
Mushroom	22	4	5	5

As shown in Table 2, we can observe that Algorithm PFS selects fewer features comparing with EFS and LFS in most data sets. For example, as data set Hepatitis, PFS selects 12 features, while both of EFS and LFS select 14 features. The reason can be attributed to that the total number of objects in the data sets keep invariant, the more objects can be discerned with the selected features by proposed evaluation function in PFS than that of EFS at certain iterations, such that fewer features needed to discern all the objects in the data sets by PFS.

And it does shows that there is a decrease in feature subset size between PFS and LFS, demonstrating that there is other information contained in the entropy other than that in the lower approximation. This phenomenon indicates that the proposed feature selection algorithm can reduce data dimensions effectively, thus it verifies the validity of new evaluation function.

We employ two classifiers NaiveBayes and J48 to evaluate the classification performance of the selected feature subset. Each data set is divided into two parts: one for training and the other for test. On the basis of the training data, we employ feature selection algorithms to reduce the data sets. By NaiveBayes and J48, the rules are extracted from the training set. Using the rules the test set is classified and the classification results are obtained. The average classification accuracies and standard deviation are acquired based on tenfold cross-validation shown in Tables 3 and 4, where Raw depicts the classification performance on data sets with the original features, and the average classification accuracies are expressed in percentage. The "Average(ACC)" row records the average classification accuracy of the three algorithms on six data sets.

Table 3. Comparison of classification accuracy for NaiveBayes Classifier

Data sets	NaiveBayes Classifier			
	Raw	PFS	EFS	LFS
Hepatitis	84.07±0.99	86.12±0.75	85.30±0.61	85.28±0.73
Soybean-large	91.43±1.07	92.50±1.12	90.89±1.20	90.11±1.54
Synthetic	95.58±2.20	94.97±1.93	94.97±1.93	92.06±1.87
Cardiotocography	89.79±0.61	91.85±0.40	88.56±0.76	89.23±0.31
Ticdate 2000	76.04±1.14	78.07±0.86	77.58±1.39	76.90±2.15
Mushroom	95.52±0.76	98.19±0.58	96.72±0.60	98.95±0.71
Average(ACC)	88.73	90.28	89.00	88.76

Table 4. Comparison of classification accuracy for J48 Classifier

Data sets	J48 Classifier			
	Raw	PFS	EFS	LFS
Hepatitis	79.35±1.16	84.60±1.09	82.32±1.25	80.81±1.74
Soybean-large	88.01±0.63	87.92±0.54	87.09±0.41	87.75±0.90
Synthetic	84.51±1.02	89.40±0.66	89.40±0.66	86.03±0.82
Cardiotocography	95.07±0.84	97.26±0.59	94.01±1.20	95.92±1.03
Ticdate 2000	79.55±0.91	81.70±1.37	79.34±1.44	82.15±1.58
Mushroom	100.00±0.0	100.00±0.0	100.00±0.0	100.00±0.0
Average (ACC)	87.74	90.15	88.69	88.78

The results shown in Tables 3 and 4 indicate that PFS produces the better classification performances after feature selection based on the new evaluation function than those of EFS and LFS as to NaiveBayes and J48. Regarding Naive-Bayes, PFS is better than EFS on all the data sets other than data set Synthetic, and PFS also shows increases in classification accuracies comparing with LFS. As to J48, PFS outperforms EFS on four of six data sets; PFS outperforms LFS on most of the data sets. Considering the results between PFS and EFS, it can demonstrate the effectiveness of new evaluation function in feature selection. In addition, the three approaches improve the classification capability by selecting a small portion of the original features. From the experimental results, we can confirm that the proposed evaluation function leads to promising improvement on classification performance.

To further explain the reason why the classification performances are improved using the new evaluation function, we conduct the experiments on four large data sets using NaiveBayes classifier with Algorithms PFS, EFS and LFS. Fig.1 displays more detailed change trend of the three algorithms in classification accuracy with the number of selected features.

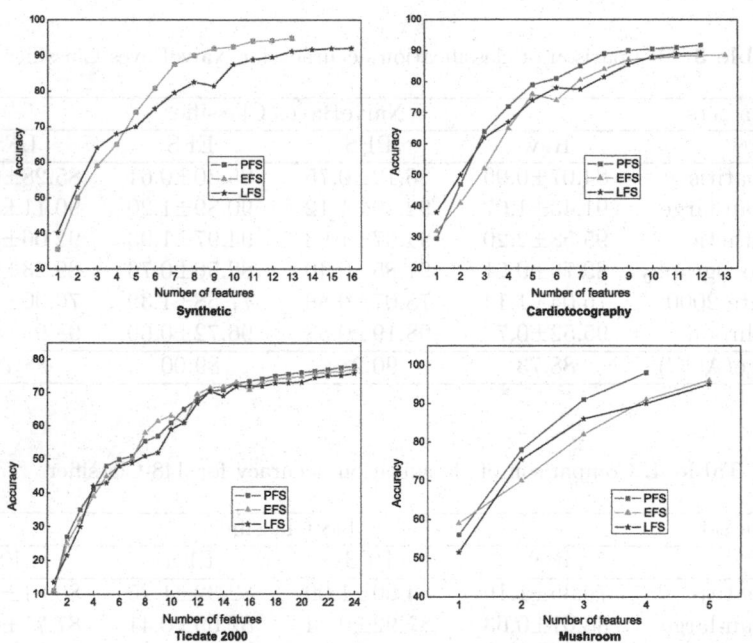

Fig.1. Trends of accuracies by NaiveBayes with number of features

From Fig.1, the curves between PFS and EFS in the data set Synthetic are overlapping. The reason is that PFS and EFS select the same features, thus the classification accuracies are the same for selecting the same number of features. However, most points in the curves of PFS are higher than those of EFS and LFS in the data sets. Take data set Cardiotocography as an illustration, the classification accuracies of PFS are higher than those of EFS and LFS since the beginning of selecting three features. The underlying reason perhaps is that though the number of features is the same by PFS and EFS, the selected features are different, PFS employed the new evaluation function always find the candidate features that can discern more objects for classification learning, such that the classification performance is better than that of EFS. The similar situations can be found in two other data sets. Observing the curves, we can find that PFS can keep a steady increase in accuracy value, whereas EFS and LFS incur a fluctuant increase, even a decrease. This phenomenon may result from one possible reason that PFS has a redundancy-removing step, while EFS and LFS does not consider the redundant information between the selected features. It shows some dispensable features in the selected feature subset are superfluous, which deteriorate the classification performance.

Furthermore, we conduct the experiments on the four larger data sets using J48 classifier with the three algorithms. Fig.2 displays more detailed change trend of the three algorithms in classification accuracy with the number of selected features.

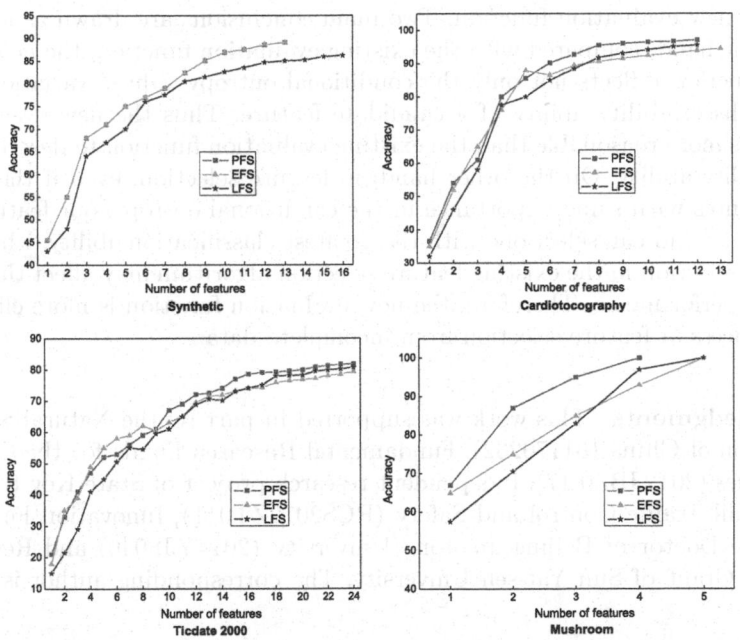

Fig.2. Trends of accuracies by J48 with number of features

As shown in Fig.2, the curves between PFS and EFS in the data set Synthetic are overlapping. However, one may observe that there are many points in the curves where the classification performance of PFS clearly surpasses those of EFS and LFS. We can see that, as data set Mushroom, when the selected feature number is two, the classification accuracy of PFS is higher than those of EFS and LFS. Though the same number of selected features, PFS can select the feature that discerns more objects for classification learning, correspondingly, the selected features are different, and the classification accuracy is higher than that of EFS. And comparing with LFS, PFS can find some other useful information contained in the entropy other than lower approximation, which would result in better classification performance. For the other three data sets, one may observe that the similar situations.

Based on the aforementioned experimental results, we can conclude that the new evaluation function gives an effective way to select satisfactory feature subset in the process of feature selection from incomplete data.

5　Conclusions

In this paper, we introduce a new evaluation function to overcome the drawbacks of existing evaluation functions. Based on the new evaluation function, we construct a conditional entropy-based feature selection algorithm with forward greedy search from incomplete data. The numerical experiments show the validity of the new evaluation function. Two main conclusions are drawn as follows. On the one hand, compared with the existing evaluation function, the new evaluation function reflects not only the conditional entropy values' variation, but also the discernibility ability of a candidate feature. Thus the new evaluation function is more reasonable than the existing evaluation function to describe the discernibility ability. On the other hand, in feature selection, even if there are more features with same importance in the conditional entropy, our feature selection algorithm can select one with the greatest classification ability, while the arbitrary selection in the existing feature selection algorithm may affect the classification performance. Therefore, the new evaluation function is more effective in the process of feature selection from incomplete data.

Acknowledgments. This work was supported in part by the Natural Science Foundation of China (61170232), Fundamental Research Funds for the Central Universities (2012JBZ0 17), Independent research project of State Key Laboratory of Rail Traffic Control and Safety (RCS2012ZT011), Innovation Funds of Excellence Doctor of Beijing Jiaotong University (2014YJS040) and Research Initiative Grant of Sun Yat-sen University. The corresponding author is Hong Shen.

References

1. Kohavi, R., John, G.H.: Wrappers for feature subset selection. Artificial Intelligence 97, 273–324 (1997)
2. Qu, G., Hariri, S., Yousif, M.: A new dependency and correlation analysis for features. IEEE Transactions on Knowledge and Data Engineering 17(9), 1199–1207 (2005)

3. Liang, J., Yang, S., Winstanley, A.: Invariant optimal feature selection: A distance discriminant and feature ranking based solution. Pattern Recognition 41(5), 1429–1439 (2008)
4. Dash, M., Liu, H.: Feature selection for classification. Intelligent Data Analysis 1(3), 131–156 (1997)
5. Steppe, J.M., Bauer, K.W., Rogers, S.K.: Integrated feature and architecture selection. IEEE Transactions on Neural Networks 7(4), 1007–1014 (1996)
6. Pawlak, Z., Skowron, A.: Rough sets and Boolean reasoning. Information Sciences 177(1), 41–73 (2007)
7. Xue, B., Cervante, L., et al.: A multi-objective particle swarm optimisation for filter-based feature selection in classification problems. Connection Science 24(2-3), 91–116 (2012)
8. Cervante, L., Xue, B., Shang, L., Zhang, M.J.: Binary particle swarm optimisation and rough set theory for dimension reduction in classification. In: IEEE Congress on Evolutionary Computation (CEC), pp. 2428–2435 (2013)
9. Sebban, M., Nock, R.: A hybrid filter / wrapper approach of feature selection using information theory. Pattern Recognition 35(4), 835–846 (2002)
10. Farahat, A.K., Ghodsi, A., Kamel, M.S.: An efficient greedy method for unsupervised feature selection. In: The 11th IEEE International Conference on Data Mining (ICDM), pp. 161–170 (2011)
11. Hu, Q.-H., Zhao, H., Xie, Z.-X., Yu, D.-R.: Consistency based attribute reduction. In: Zhou, Z.-H., Li, H., Yang, Q. (eds.) PAKDD 2007. LNCS (LNAI), vol. 4426, pp. 96–107. Springer, Heidelberg (2007)
12. Sun, L., Xu, J.C., Tian, Y.: Feature selection using rough entropy-based uncertainty measures in incomplete decision systems. Knowledge-Based Systems 36, 206–216 (2012)
13. Qian, Y.H., Liang, J.Y., Pedrycz, W., Dang, C.Y.: An efficient accelerator for attribute reduction from incomplete data in rough set framework. Pattern Recognition 44, 1658–1670 (2011)
14. Slezak, D.: Approximate entropy reducts. Fundamenta Informaticae 53, 365–390 (2002)
15. Dai, J.H., Wang, W.T., Xu, Q.: An uncertainty measure for incomplete decision tables and its applications. IEEE Transactions on Cybernetics 43(4), 1277–1289 (2013)
16. UCI Machine Learning Repository, http://www.ics.uci.edu/mlearn/MLRepository.html

Constrained Least Squares Regression
for Semi-Supervised Learning

Bo Liu[1,2], Liping Jing[1,*], Jian Yu[1], and Jia Li[1]

[1] Beijing Key Lab of Traffic Data Analysis and Mining,
Beijing Jiaotong University, Beijing, 100044, China
[2] College of Information Science and Technology,
Agricultural University of Hebei, Hebei, 071000, China
liubohbu@126.com, {lpjing,jianyu}@bjtu.edu.cn, jiali.gm@gmail.com

Abstract. The core tasks of graph based semi-supervised learning (GSSL) are constructing a proper graph and selecting suitable supervisory information. The ideal graph is able to outline the intrinsic data structure, and the ideal supervisory information could represent the whole data. In this paper, we propose a new graph learning method, called constrained least squares regression (CLSR), which integrates the supervisory information into graph learning process. To learn a more adaptive graph, regression coefficients and neighbor relations are combined in CLSR to capture the global and local data structures respectively. Moreover, as byproduct of CLSR, a new strategy is presented to select the high-quality data points as labeled samples, which is practical in real applications. Experimental results on different real world datasets demonstrate the effectiveness of CLSR and the sample selection strategy.

Keywords: graph based semi-supervised learning, graph construction, constrained least squares regression, labeled sample selection.

1 Introduction

Lack of sufficiently labeled data is a big problem when building supervised learner in real applications. Semi-supervised learning (SSL) can bridge the gap between labeled and unlabeled data, as it combines limited labeled samples with rich unlabeled samples to enhance the learner's ability [20]. As an important branch of SSL, graph based semi-supervised learning (GSSL) propagates the supervisory information (class labels) on a pre-defined graph and aims to make the similar samples share the common labels [12]. Under the cluster assumption [4] or the manifold assumption [8], there are many GSSL methods have been proposed, including Gaussian fields and harmonic functions (GFHF) [21], local and global consistency (LGC) [19], manifold regularization [2], and etc. For GSSL, there are several key issues to be solved including graph construction, labeled sample selection, learning model formulation, parameter adjustment and etc. In this paper, we will limit to highlight the former two issues.

---

* Corresponding Author.

V.S. Tseng et al. (Eds.): PAKDD 2014, Part II, LNAI 8444, pp. 110–121, 2014.
© Springer International Publishing Switzerland 2014

An adaptive graph construction is a main challenge of GSSL. Neighborhood-driven methods (e.g., k-nearest-neighbors (k-NN) [16], ϵ-ball neighborhood [1] and b-matching graph [7]) are unable to reflect the overall views of data and sensitive to noise. Recently, some researchers formulate the graph building process into a subspace learning problem. Under the subspace assumption, each sample can be represented as a linear combination of other samples, and intuitively, the representation coefficients could be accepted as a proper surrogate of similarity metric. In the literature, this measurement is referred to as self-expressive similarity [6]. There are several methods such as sparse representation (SR) [6], low-rank representation (LRR) [10], least squares regression (LSR) [13] to obtain the representation coefficients.

Although these approaches have gained great effects in some domains, there are still some drawbacks. First, labeled samples only work at propagating stage, so the supervisory information cannot directly influence the affinity learning process. Second, regardless of noise and outliers, data points may not strictly lie in a union of subspaces, which indicates that the graph's adaptability is restricted owing to the utilization of a single metric. Third, in the context of the subspace assumption, when we have to select some samples as a labeled set, however, the existing method, random sampling, does not leverage the structural characteristic of the original dataset.

Inspired by the work [13], we propose an effective graph construction framework, called constrained least squares regression (CLSR), and try to improve GSSL from three perspectives:

- The labeled samples are effectively integrated into the graph learning process of GSSL by representing them as additional pairwise constraints.
- Both local and global data structures are considered to build a more flexible graph via self-expressive similarity metric and k-NN.
- A greedy-like strategy is designed to pick out more representative samples as the labeled set.

2   Preliminaries and Related Works

Given a data set $X = [x_1, x_2, \ldots, x_l, x_{l+1}, \ldots, x_n] \in \mathbb{R}^{m \times n}$, the subset $X_l = \{x_i\}_{i=1}^{l}$ with cardinality $|X_l| = l$ contains labeled points and $X_u = \{x_i\}_{i=l+1}^{n}$ with cardinality $|X_u| = n - l$ contains unlabeled points. The target of graph learning is to generate a proper graph or weight matrix $W \in \mathbb{R}^{n \times n}$ and its element W_{ij} denotes the similarity between the ith point and the jth point under some measurement. SR [6] and LRR [10] are two popular affinity representation techniques. SR aims to construct a sparse graph or $\ell 1$-graph [17], where each point could be reconstructed by a combination of other limited points, and thus the sparse coefficients correspond to a kind of similarity. Basic SR is formulated as the following optimization problem:

$$\min_{Z} \|Z\|_1 \quad \text{s.t.} \ X = XZ, diag(Z) = 0 \tag{1}$$

where $Z = [z_1, z_2, \ldots, z_n] \in \mathbb{R}^{n \times n}$ denotes the coefficient matrix, $\|Z\|_1$ is the $\ell 1$-norm of Z which can promote sparse solution, $\|Z\|_1 = \sum_{i=1}^{n} \sum_{j=1}^{n} |z_{ij}|$. Then, the graph weight matrix W could be easily obtained by $W = (|Z| + |Z|^T)/2$

Compared with the k-NN graph, the $\ell 1$-graph avoids evaluating the hyperparameter k and therefore it outputs more robust result. Nevertheless, both $\ell 1$-graph and k-NN graph are lack of the global views of data, so their performance would be degenerated when there is no "clean" data available [22]. In order to capture the global data structure, Liu et al. [10] proposed LRR method which enforces a rank minimization constraint on the coefficient matrix. The basic LRR problem can be formulated as:

$$\min_{Z} \|Z\|_* \quad \text{s.t. } X = XZ \tag{2}$$

where $\|Z\|_*$ denotes the nuclear norm of Z, which is a usual surrogate of rank function, i.e., the sum of the singular values. Since the sparseness and lowrankness are merits of a graph, Zhuang et al. [22] presented a non-negative low-rank and sparse graph (NNLRS) learning method. Recently, Lu et al. [13] pointed out that, besides $\ell 1$-norm and nuclear norm, Frobenius norm is also an appropriate constraint for the coefficient matrix Z, and presented the LSR model with noise as follows:

$$\min_{Z} \lambda \|Z\|_F^2 + \|X - XZ\|_F^2 \tag{3}$$

where $\lambda > 0$ is the regularization parameter. Note there are little differences in (1), (2) and (3), but (3) has a close-form solution

$$Z^* = (X^T X + \lambda I)^{-1} X^T X \tag{4}$$

In this case, LSR can be solved efficiently.

Even though all the above approaches could output suitable graphs for GSSL, the graph learning itself is still unsupervised. In recent work [15], Shang et al. presented an enhanced spectral kernel (ESK) model, which makes use of pairwise constraints to favor graph learning, and is solved as a low-rank matrix approximation problem [9]. The main difference between ESK and our approach is as follows. ESK uses the Gaussian kernel to initialize the weight matrix, and encodes the known labels as the pairwise constraints. While in CLSR, we adopt the regression coefficients to measure the correlations among data points, and consider additional local constraints to promote the model's flexibility.

Additionally, the quality of labeled points play an important role in GSSL, thus, it is necessary to select the samples with high representability and discriminability as labeled set. In [5] and [11], k-means algorithm has been verified as an effective method for sample selection. But for GSSL, an extra step is needed to estimate the labels of clustering centers. Recently, some researchers pointed out that collaborative representation is a promising method for sample selection [18], [14]. In this paper, we propose a simple and effective method which applies minimal reconstruction error criterion to labeled sample selection.

3 Constrained Least Squares Regression for Graph Learning

In this section, we first introduce the label consistent penalty for encoding known labels, then integrate it with original LSR, and finally design its optimization algorithm.

3.1 Label Consistent Penalty

Given two sets for labeled points, $ML = \{(x_i, x_j)\}$ includes must-link constraints, where x_i and x_j have the same label, and $CL = \{(x_i, x_j)\}$ covers cannot-link constraints, where x_i and x_j have different labels. Let Ω be a set of indices which correspond to all pairwise constraints. The label consistent penalty is defined as:

$$f(Z) = \|S \circ Z - L\|_F^2 \tag{5}$$

where $\circ$ denotes the element-wise product. The sampling matrix $S \in \mathbb{R}^{n \times n}$ is defined as:

$$S_{ij} = \begin{cases} 1 & (i,j) \in \Omega \\ 0 & otherwise \end{cases} \tag{6}$$

The constraint matrix $L \in \mathbb{R}^{n \times n}$ is defined as:

$$L_{ij} = \begin{cases} 1 & (i,j) \in ML \\ 0 & (i,j) \in CL \\ 0 & otherwise \end{cases} \tag{7}$$

Equation (5) is a squared lose function to measure the consistency between the predicted affinity matrix induced by Z and the given pairwise constraints. Here, the pairwise constraints are expected to reflect the data structure. However, the number of labeled samples is usually few so that it is hard to sufficiently capture the essential structure of data with them. Thus, it is necessary to bring in more local pairwise constraints which are encoded as $L' \in \mathbb{R}^{n \times n}$:

$$L'_{ij} = \begin{cases} 1 & i \in N_j \text{ and } j \in N_i \\ 0 & otherwise \end{cases} \tag{8}$$

where N_i stands for the set of k-nearest neighbor of x_i. Actually, L' employs a k-NN graph to roughly recover the local relations among data points by $0/1$ assignments, and thus it will result in some wrong assignments. One way to fix these incorrect assignments is to utilize the original L with correct assignments from labeled samples, and the fixed $L^f \in \mathbb{R}^{n \times n}$ is defined as:

$$L_{ij}^f = \begin{cases} L_{ij} & (i,j) \in ML \text{ or } (i,j) \in CL \\ L'_{ij} & (i,j) \notin ML \text{ and } (i,j) \notin CL \end{cases} \tag{9}$$

From the perspective of matrix approximation, these wrong assignments in (5) can be taken as one kind of sparse noise. Therefore, the $\ell 1$-norm is used here instead of the Frobenius norm and we have

$$f(Z) = \|S \circ Z - L\|_1 \tag{10}$$

3.2   Objection Function

After adding the label consistent penalty to the LSR model, the objective function of CLSR is written as:

$$\min_{Z,E} \|Z\|_F^2 + \frac{\lambda_e}{2}\|XZ - X\|_F^2 + \lambda_s\|E\|_1 \quad \text{s.t. } E = S \circ Z - L \tag{11}$$

where $E \in \mathbb{R}^{n \times n}$ denotes the sparse error, λ_e and λ_s are parameters to trade off other terms. In (11), the first two items are used to hold the global structure of data by Z and the third item introduces the pairwise constraints by L which is defined in (9).

3.3   Optimization

Equation (11) could be solved by the alternating direction method of multipliers (ADMM) [3] method. To start, we introduce an auxiliary matrix $A \in \mathbb{R}^{n \times n}$ for variables separation, then obtain

$$\min_{Z,E} \|Z\|_F^2 + \frac{\lambda_e}{2}\|XA - X\|_F^2 + \lambda_s\|E\|_1 \quad \text{s.t. } E = S \circ Z - L, Z = A \tag{12}$$

The augmented Lagrangian function of (12) can be written as:

$$\mathcal{L} = \min_{Z,E} \|Z\|_F^2 + \frac{\lambda_e}{2}\|XA - X\|_F^2 + \lambda_s\|E\|_1 + <Y_1, Z - A> \\ + <Y_2, E - S \circ Z + L> + \frac{\mu}{2}(\|Z - A\|_F^2 + \|E - S \circ Z + L\|_F^2) \tag{13}$$

where $Y_1 \in \mathbb{R}^{n \times n}$ and $Y_2 \in \mathbb{R}^{n \times n}$ are two Lagrange multipliers. ADMM approach updates the variables Z, A and E alternately with other variables fixed, and we can get the updating rules as:

$$Z_{k+1} = \arg\min_Z \|Z_k\|_F^2 + \frac{\mu_k}{2}\|Z_k - A_k + \frac{Y_1}{\mu_k}\|_F^2 + \frac{\mu_k}{2}\|S \circ Z_K - E_k - L - \frac{Y_2}{\mu_k}\|_F^2 \\ = (1/(\frac{2}{\mu_k} + 1 + S)) \circ (A_k - \frac{Y_1}{\mu_k} + S \circ (\frac{Y_2}{\mu_k}) + S \circ E_k + S \circ L) \tag{14}$$

$$A_{k+1} = \arg\min_A \frac{\lambda_e}{2}\|XA_k - X\|_F^2 + \frac{\mu_k}{2}\|Z_{k+1} - A_k + \frac{Y_1}{\mu_k}\|_F^2 \\ = (\lambda_e X^T X + \mu_k I)^{-1}(\lambda_e X^T X + \mu_k Z_{k+1} + Y_1) \tag{15}$$

$$E_{k+1} = \arg\min_E \lambda_s\|E_k\|_1 + \frac{\mu_k}{2}\|E_k - (S \circ Z_{k+1} - L - \frac{Y_2}{\mu_k})\|_F^2 \\ = S_{\frac{\lambda_s}{\mu_k}}(S \circ Z_{k+1} - L - \frac{Y_2}{\mu_k}) \tag{16}$$

where $1 \in \mathbb{R}^{n \times n}$ stands for an all-one matrix, and $\mathcal{S}_\mu(\cdot)$ is the shrinkage-thresholding operator [9] which is defined as:

$$\mathcal{S}_\mu(\nu) = sign(\nu)(|\nu| - \mu)_+ \tag{17}$$

The complete algorithm is summarized in Algorithm 1.

Algorithm 1. Solving Problem (11) via ADMM

Input: data matrix X, sampling matrix S, constraint matrix L, and parameters λ_e, λ_s.
 1. Initialize $Z_0 = A_0 = E_0 = Y_1 = Y_2 = 0$, $\mu_0 = 0.1$, $\mu_{max} = 10^4$, $\rho = 1.1$, $\epsilon = 10^{-2}$
 2. **while** not converged **do**
 3.   Update Z, A and E by (14-16).
 4.   Update the multipliers Y_1, Y_2 as:
        $Y_1 = Y_1 + \mu(Z - A)$,
        $Y_2 = Y_2 + \mu(E - S \circ Z + L)$.
 5.   Update $\mu : \mu = \min(\rho\mu, \mu_{max})$.
 6.   Check the convergence conditions:
        $\|E - S \circ Z + L\|_\infty < \epsilon$ and $\|Z - A\|_\infty < \epsilon$.
 7. **end while**
Output: Z_k, E_k

---

4   Labeled Sample Selection via CLSR

In many real applications, we need select a small part of data set as a labeled set. Usually, a natural and simple method, random sampling is adopted. However, this method cannot guarantee the quality of labeled samples. Based on the subspace assumption, we could select a more representative data subset to upgrade graph's performance in GSSL.

In the CLSR framework, it is convenient to use the basic LSR model for labeled sample selection. We randomly select c subsets $\{X_i\}_{i=1}^c$ from $X, X_i \in \mathbb{R}^{m \times p}$ and each subset contains p samples, $p \ll n$. We consider each subset as a tiny dictionary and use it to reconstruct the whole data set, consequently, the representative ability of each subset could be ranked by the corresponding reconstruction error, therefore, the smaller reconstruction error it has, the more representative it is. The reconstruction error can be solved by

$$\min_{Z_i, E_i} \lambda \|Z_i\|_F^2 + \|E_i\|_F^2 \quad \text{s.t.} \quad E_i = X - X_i Z_i \tag{18}$$

where $X_i \in \mathbb{R}^{m \times p}$ denotes the selected subset, $E_i \in \mathbb{R}^{m \times n}$ is the reconstruction error, and $Z_i \in \mathbb{R}^{p \times n}$ is the coefficient matrix of X_i. Note problem (18) has a close-form solution

$$Z_i = (X_i^T X_i + \lambda I)^{-1} X_i^T X \tag{19}$$

The labeled sample selection method is summarized in Algorithm 2.

Algorithm 2. Labeled Sample Selection via Minimal Reconstruction Error

---

Input: data matrix X, selected subset $\{X_i\}_{i=1}^c$, parameter λ.
1. Initialize $\lambda = 10$
2. **for** $i = 1, \ldots, c$ **do**
3.   Get Z_i by (19).
4.   Computer the reconstruction error $r_i(X_i) = \|X - X_i Z_i\|_F^2$.
5. **end for**
6. Find X_i^* with minimal reconstruction error $\arg\min_i r_i(X_i)$.
Output: X_i^*

---

5   CLSR for Semi-Supervised Classification

In this section, we integrate CLSR with a popular label propagation approach, LGC [19], for semi-supervised classification. Define a label set $F = \{1, \ldots, k\}$, and an initial label matrix $Y \in \mathbb{R}^{m \times k}$ with $Y_{ij} = 1$ for x_i is labeled as j and $Y_{ij} = 0$ otherwise. The iterative scheme for propagation is

$$Y_{k+1} = \alpha \overline{W} Y_k + (1 - \alpha) Y_0 \qquad (20)$$

where $\overline{W}$ is a normalized affinity matrix with $\overline{W} = D^{-1/2} W D^{-1/2}$ and D is a diagonal matrix whose diagonal entries are equal to the sum of corresponding rows. We fix the parameter α to 0.01 in following experiments. The detail of the algorithm is summarized in Algorithm 3.

Algorithm 3. CLSR for Semi-Supervised Classification

Input: data matrix X, initial label matrix Y, parameters $\lambda_s, \lambda_e, \lambda$.
1. Initialize $\lambda = 10$
2. Get the labeled subset X_i by Algorithm 2 or random sampling.
3. Generate the sampling matrix S by (6) and the constraint matrix L by (9).
4. Get the coefficient matrix Z by Algorithm 1.
5. Normalize all column vectors of Z to unit-norm, $z_i = z_i / \|z_i\|_2$.
6. Get the weight matrix W by $W = (|Z| + |Z|^T)/2$.
7. Compute the label matrix Y by (20).
Output: Y

---

6   Experimental Results and Analysis

In this section, we evaluate the performance of CLSR and other popular graph construction methods on six public databases.

6.1   Datasets and Settings

We use two categories of public datasets in the experiments, including UCI data and image data (see Table 1).

1. **UCI data**[1]. We perform experiments on three UCI datasets including WDBC, Sonar and Parkinsons.
2. **Extended YaleB database**[2]. This face database contains 38 individuals under 9 poses and 64 illumination conditions. We choose the cropped images of first 10 individuals, and resize them to 48×42 pixels.
3. **ORL database**[3]. There are 40 distinct subjects and each of them has 10 different images. For some subjects, the images were taken at different times, varying the lighting, facial expressions and facial details. We resize them to 32×32 pixels.
4. **COIL20 database**[4]. This database consists of a set of gray-scale images with 20 objects. For each object, there are 72 images of size 32×32 pixels.

Table 1. Descriptions of datasets

Dataset	label Size	♯ of Features	♯ of Classes
WDBC	569	30	2
Sonar	208	60	2
Parkinsons	195	21	2
YaleB	640	2016	10
ORL	400	1024	40
COIL20	1440	1024	20

We compare following six graph construction algorithms. There are some parameters in each algorithm, and we tune the parameters on each dataset for every algorithm and record the best results.

1. k-**NN:** the Euclidean distance is used as similarity metric, and the Gaussian kernel is used to reweight the edges. The number of nearest neighbors is set to 5 for k-**NN5**, and 15 for k-**NN15**, respectively. The scale parameter of Gaussian kernel is set as [22]
2. **ESK:** Following the lines of [15], a low-rank kernel is learned as the affinity matrix. ESK model also use the Gaussian kernel to initialize the weight matrix.
3. **LSR:** Compared with CLSR, LSR [13] does not consider the pairwise constraints in graph leaning process.
4. **LRR:** Following [10], we construct the low-rank graph and adopt $\ell_{2,1}$-norm to model "sample-specific" corruptions.
5. **NNLRS:** Following [22], we construct the non-negative low-rank and sparse graph.
6. **CLSR:** In CLSR, the neighbor relations are encoded as the additional pairwise constraints for reflecting the local data structure. In the experiments, the sizes of nearest neighbors are set to 0, 5 and 15, respectively.

---

[1] http://archive.ics.uci.edu/ml/

[2] http://vision.ucsd.edu/~leekc/ExtYaleDatabase/ExtYaleB.html

[3] http://www.cl.cam.ac.uk/research/dtg/attarchive/facedatabase.html

[4] http://www.cs.columbia.edu/CAVE/software/softlib/coil-20.php

6.2   Results and Discussions

All experiments are repeated 20 times, for each dataset, the label rate varies from 10% to 40%. Table 2 lists the average accuracies.

From Table 2 we can get following observations.

1. LSR, LRR and NNLRS generally outperform k-NN and ESK on YaleB and ORL datasets, as these datasets have roughly subspace structures. Correspondingly, datasets WDBC, Parkinsons cater to Euclidean distance-based measurement, so k-NN, ESK can work well on these datasets.
2. NNLRS usually achieves better performance than LSR and LRR, owing to it considers both sparseness and low-rankness of the graph.
3. ESK generally outperforms k-NN with the increasing of the sampling percentage, which testifies the effectiveness of integrating pairwise constraints into the graph learning process.
4. In most cases, CLSR outperforms other algorithms, since it takes advantage of both the self-expressive similarity and local constraints to enhance the model's flexibility and performance.

Table 2. Average accuracies (mean and standard deviation) of different graphs integrated with LGC label propagation strategy (The best results are highlighted in bold)

Dataset	k-NN5	k-NN15	ESK	LSR	LRR	NNLRS	CLSR
WDBC(10%)	93.56±1.26	93.54±0.80	92.41±1.50	89.01±1.70	91.14±1.48	91.11±1.43	**94.27±1.23**
WDBC(20%)	94.02±0.56	94.08±0.67	94.20±1.15	91.86±1.34	93.27±1.21	92.41±0.87	**95.09±0.39**
WDBC(30%)	94.84±0.55	94.87±0.63	94.90±0.74	93.51±1.10	94.15±1.03	93.59±0.76	**96.18±0.27**
WDBC(40%)	95.52±0.51	95.10±0.59	95.60±0.43	94.75±0.89	95.34±0.94	94.64±0.60	**96.85±0.26**
Sonar(10%)	73.44±4.21	73.41±4.48	67.18±4.34	67.90±4.32	68.37±7.55	71.06±3.98	**74.40±3.19**
Sonar(20%)	75.60±3.38	76.87±3.08	76.54±3.56	74.25±3.06	75.10±3.20	76.39±3.11	**81.38±2.99**
Sonar(30%)	79.71±2.03	78.32±3.25	81.82±3.13	79.46±2.42	80.94±2.92	81.39±2.05	**85.60±1.66**
Sonar(40%)	83.55±1.39	85.20±2.76	85.37±2.24	83.54±1.85	83.31±2.01	85.38±1.13	**88.75±0.94**
Parkinsons(10%)	75.82±6.23	72.66±3.93	75.27±6.18	67.12±3.54	74.11±3.33	76.10±3.48	**77.95±3.26**
Parkinsons(20%)	79.24±5.49	72.00±3.17	82.44±5.10	74.43±3.29	77.58±2.34	80.72±1.88	**83.59±2.19**
Parkinsons(30%)	80.57±4.66	72.29±2.93	85.91±4.78	79.24±2.80	81.66±1.86	82.87±1.55	**87.44±1.06**
Parkinsons(40%)	81.19±3.98	72.10±2.44	88.34±3.81	82.37±2.10	84.68±2.12	86.26±1.19	**89.05±0.85**
YaleB(10%)	69.06±2.25	66.98±6.48	60.23±2.51	87.80±1.74	87.88±2.11	**88.86±2.35**	88.85±1.65
YaleB(20%)	75.91±2.17	74.72±1.53	71.96±2.12	94.21±1.05	94.08±0.96	93.70±1.02	**94.37±0.88**
YaleB(30%)	79.58±1.89	78.64±2.11	77.82±1.23	96.28±0.94	96.20±0.79	95.39±0.58	**96.65±0.43**
YaleB(40%)	79.92±1.72	79.96±1.76	81.69±0.83	97.39±0.73	97.00±0.57	96.45±0.41	**97.69±0.35**
ORL(10%)	40.47±3.39	44.88±3.03	53.50±3.71	49.95±3.78	50.49±3.85	52.58±4.23	**54.52±3.07**
ORL(20%)	63.88±2.10	66.05±3.36	72.60±2.33	72.15±2.79	71.95±3.24	75.10±2.92	**76.65±2.39**
ORL(30%)	78.63±5.88	78.95±5.55	82.98±2.35	83.75±2.19	83.98±2.73	84.33±2.73	**85.50±2.14**
ORL(40%)	86.17±3.19	84.52±3.93	88.20±2.01	91.53±1.65	90.19±2.24	90.95±2.38	**92.50±1.91**
COIL20(10%)	86.12±0.81	85.80±1.01	86.24±1.21	80.10±1.52	79.38±2.54	81.39±1.55	**88.12±1.07**
COIL20(20%)	88.24±0.76	86.23±0.91	88.50±1.18	87.85±1.16	87.39±1.18	87.26±1.06	**90.84±0.73**
COIL20(30%)	89.88±0.85	87.82±1.02	90.69±0.77	91.35±0.92	90.57±1.11	89.96±0.81	**92.93±0.69**
COIL20(40%)	90.17±0.80	88.61±0.84	92.10±0.60	93.28±0.71	92.98±0.92	92.47±0.74	**94.58±0.55**

Next, we study the effectiveness of sample selection strategy based on minimal reconstruction error. We first randomly select 50 labeled subsets from each dataset, and then sort them in ascending order to form a subset-residual array according to the representative residual of each subset. Secondly, these labeled subsets are used as the supervisory information for classification and the average accuracies are recorded. Furthermore, another two results are listed for comparison, one is the average accuracy of the top 10% of the array (denoted as AT-10%), the other is the average accuracy of the lowest 10% of the array (denoted as AL-10%). The percentage of labeled samples is 5% on WDBC, Parkinsons, YaleB, COIL20 and Sonar, because the selection strategy could be useful in case that there are only limited labeled samples available, especially, we select 20% of samples from ORL, since there are only 10 samples in each class of ORL.

The results are plotted in Fig. 1(a-f). It shows that our method is almost effective for all graph construction approaches on each dataset, except Parkinsons. The result on Parkinsons is unstable. The reason is that there are two classes in Parkinsons, but its imbalance ratio is nearly 3. In this case, our method tends to select more samples from the majority class to minimize the total reconstruction error, which leads to that the selected samples are incapable of capturing the true geometric structure of the dataset. We balance the sizes of two classes by randomly selecting some samples from the majority class, and the result shown in Fig. 1(g) is consistent with the other datasets'.

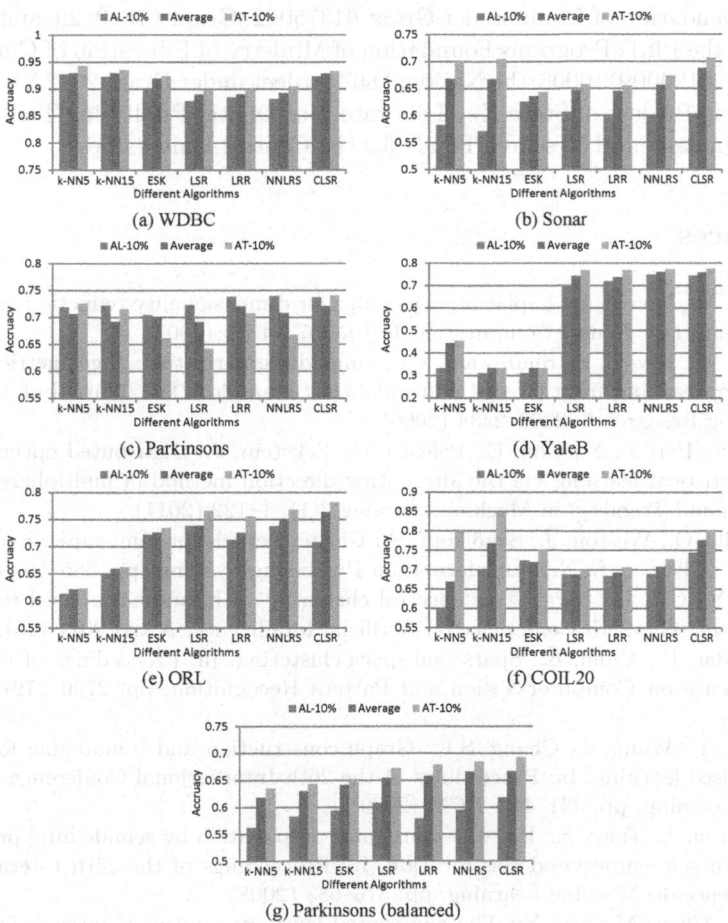

Fig. 1. Classification results of all graph construction algorithms on each dataset after applying sample selection strategy

7 Conclusion

We propose a new graph based semi-supervised learning approach called CLSR, which utilizes the pairwise constraints to guide the graph learning process. Beside the labeled information, there constraints also bring in local neighbor relations to enhance the graph's flexibility. In addition, based on CLSR, we design a labeled sample selection strategy which is used to select more representative points as a labeled set. Experimental results on real world datasets demonstrate the effectiveness of our method. Furthermore, given a small size of labeled set (e.g., 5% of total samples), our sample selection strategy could generally improve the performance of several state-of-the-art methods on most of the datasets used in the experiments.

Acknowledgments. This work was supported in part by the National Natural Science Foundation of China under Grant 61375062, Grant 61370129, and Grant 61033013, the Ph.D Programs Foundation of Ministry of Education of China under Grant 20120009110006, the National 863 project under Grant 2012AA040912, the Opening Project of State Key Laboratory of Digital Publishing Technology, and the Fundamental Research Funds for the Central Universities.

References

1. Belkin, M., Niyogi, P.: Laplacian eigenmaps for dimensionality reduction and data representation. Neural Computation 15(6), 1373–1396 (2003)
2. Belkin, M., Niyogi, P., Sindhwani, V.: Manifold regularization: A geometric framework for learning from labeled and unlabeled examples. The Journal of Machine Learning Research 7, 2399–2434 (2006)
3. Boyd, S., Parikh, N., Chu, E., Peleato, B., Eckstein, J.: Distributed optimization and statistical learning via the alternating direction method of multipliers. Foundations and Trends® in Machine Learning 3(1), 1–122 (2011)
4. Chapelle, O., Weston, J., Schölkopf, B.: Cluster kernels for semi-supervised learning. In: Advances in Neural Information Processing Systems, pp. 585–592 (2002)
5. Chen, X., Cai, D.: Large scale spectral clustering with landmark-based representation. In: The 25th Conference on Artificial Intelligence, AAAI 2011 (2011)
6. Elhamifar, E., Vidal, R.: Sparse subspace clustering. In: Proceedings of the 22th Conference on Computer Vision and Pattern Recognition, pp. 2790–2797. IEEE (2009)
7. Jebara, T., Wang, J., Chang, S.F.: Graph construction and b-matching for semi-supervised learning. In: Proceedings of the 26th International Conference on Machine Learning, pp. 441–448. ACM (2009)
8. Li, Z., Liu, J., Tang, X.: Pairwise constraint propagation by semidefinite programming for semi-supervised classification. In: Proceedings of the 25th International Conference on Machine Learning, pp. 576–583 (2008)
9. Lin, Z., Chen, M., Ma, Y.: The augmented lagrange multiplier method for exact recovery of corrupted low-rank matrices. UIUC Technical report UILU-ENG-09-2215 (2010)

10. Liu, G., Lin, Z., Yan, S., Sun, J., Yu, Y., Ma, Y.: Robust recovery of subspace structures by low-rank representation. IEEE Transactions on Pattern Analysis and Machine Intelligence, 171–184 (2013)
11. Liu, W., He, J., Chang, S.F.: Large graph construction for scalable semi-supervised learning. In: Proceedings of the 27th International Conference on Machine Learning, pp. 679–686 (2010)
12. Liu, W., Wang, J., Chang, S.F.: Robust and scalable graph-based semisupervised learning. Proceedings of the IEEE 100(9), 2624–2638 (2012)
13. Lu, C.-Y., Min, H., Zhao, Z.-Q., Zhu, L., Huang, D.-S., Yan, S.: Robust and efficient subspace segmentation via least squares regression. In: Fitzgibbon, A., Lazebnik, S., Perona, P., Sato, Y., Schmid, C. (eds.) ECCV 2012, Part VII. LNCS, vol. 7578, pp. 347–360. Springer, Heidelberg (2012)
14. Peng, X., Zhang, L., Yi, Z.: Scalable sparse subspace clustering. In: IEEE Proceedings of the 26th Conference on Computer Vision and Pattern Recognition (2013)
15. Shang, F., Jiao, L., Liu, Y., Tong, H.: Semi-supervised learning with nuclear norm regularization. Pattern Recognition 46(8), 2323–2336 (2013)
16. Tenenbaum, J.B., De Silva, V., Langford, J.C.: A global geometric framework for nonlinear dimensionality reduction. Science 290(5500), 2319–2323 (2000)
17. Yan, S., Wang, H.: Semi-supervised learning by sparse representation. In: SDM, pp. 792–801 (2009)
18. Zhang, L., Yang, M., Feng, X.: Sparse representation or collaborative representation: Which helps face recognition? In: Proceedings of the 12th International Conference on Computer Vision, pp. 471–478. IEEE (2011)
19. Zhou, D., Bousquet, O., Lal, T.N., Weston, J., Schölkopf, B.: Learning with local and global consistency. Advances in Neural Information Processing Systems 16(16), 321–328 (2004)
20. Zhu, X.: Semi-supervised learning literature survey. Technical report, Department of Computer Science, University of Wisconsin-Madison (2006)
21. Zhu, X., Ghahramani, Z., Lafferty, J., et al.: Semi-supervised learning using gaussian fields and harmonic functions. In: Proceedings of the 20th International Conference on Machine Learning, vol. 3, pp. 912–919 (2003)
22. Zhuang, L., Gao, H., Lin, Z., Ma, Y., Zhang, X., Yu, N.: Non-negative low rank and sparse graph for semi-supervised learning. In: Proceedings of the 25th Conference on Computer Vision and Pattern Recognition, pp. 2328–2335. IEEE (2012)

Machine Learning Approaches
for Interactive Verification

Yu-Cheng Chou and Hsuan-Tien Lin

Department of Computer Science, National Taiwan University, Taipei 106, Taiwan

Abstract. Interactive verification is a new problem, which is closely related to active learning, but aims to query as many positive instances as possible within some limited query budget. We point out the similarity between interactive verification and another machine learning problem called contextual bandit. The similarity allows us to design interactive verification approaches from existing contextual bandit approaches. We compare the performance of those approaches on interactive verification. In particular, we propose to adopt the upper confidence bound (UCB) algorithm, which has been widely used for the contextual bandit, to solve the interactive verification problem. Experiment results demonstrate that UCB reaches superior performance for interactive verification on many real-world datasets.

Keywords: active learning, contextual bandit, upper confidence bound.

1 Introduction

Breast cancer is the most frequently diagnosed cancer in woman (Rangayyan et al., 2007). Breast cancer screening is a strategy to achieve an earlier diagnosis in asymptomatic women for breast cancer. A common technique for screening is mammography. Somehow interpreting mammogram images is difficult and requires radiology experts, while hiring radiology experts is usually expensive. In breast cancer screening, most of the efforts are spent on interpreting mammogram images from healthy individuals. But actually only the mammogram images from the patients with breast cancer require the diagnosis from radiology experts. If we can select a subset of patients that are asymptomatic, we can save radiology experts a lot of efforts. One possible way to do so is to let computers select the subset automatically in a computer-aided diagnosis (CAD) system.

CAD systems are designed to assist radiology experts in interpreting mammogram images (Rangayyan et al., 2007; Li and Zhou, 2007). A CAD system can prompt potential unhealthy region of interests (ROIs) for radiology experts to verify. A typical CAD session can be decomposed into three stages: labeling stage, where radiology experts perform the reading of some mammogram images and record the label (malignant or benign) for each ROI; learning stage, where a learning algorithm within the CAD system builds a classifier to predict the labels of ROIs for future mammogram images based on the labels obtained from labeling stage; verification stage, where radiology experts analyze the prompts given by the CAD system to verify whether the ROIs are malignant or

V.S. Tseng et al. (Eds.): PAKDD 2014, Part II, LNAI 8444, pp. 122–133, 2014.

benign. A CAD system can reduce the efforts spent in breast cancer screening by select-ing worthy-verified ROIs for radiology experts. Such a problem, which allows human experts to verify something (malignant ROIs) selected by computers (CAD system), is named the "verification problem" in this work.

In a verification problem, there are two stages that require the efforts of human ex-perts: the labeling stage and the verification stage. These two stages are different from the point of view of the system. In the labeling stage, the system requests label of an ROI for learning; in the verification stage, the system prompts an ROI that is considered to be positive (malignant) for verification. Nevertheless, these two stages are similar from human experts' point of view. Both of them require radiology experts to diagnose on an ROI and return the diagnosis. We call the request of diagnosis as a "query" in the verification problem. Given the similarity between the labeling stage and verification stage, we propose to combine these two stages together: a human expert can do the verification while doing the labeling; and the feedback of the verification can be treat as the labeling result. By combining the learning and verification, the system can get the flexibility to decide how to distribute limited human resources on these two stages to achieve better performance. Given limited query budget, how could we most efficiently distribute and utilize the queries to verify as many malignant ROIs as possible? This is the main question of this work.

In this paper, we formalize the question above by defining a new problem called interactive verification. The problem describes a procedure that performs verification through the interaction between the system and the human experts. By interacting with humans, the system aims to verify as many positive instances as possible within limited query budget, and the query result can be immediately used to learn a better classifier. An effective approach for the problem can then help reduce the overall human efforts.

In our work, we first point out the similarity of interactive verification to the popular contextual bandit problem (Langford and Zhang, 2007). We also discuss the similarity of interactive verification to the active learning problem. Then, we design four possible interactive verification approaches based on the similarities. In particular, one of the four is called the upper confidence bound (UCB), which is adopted from a state-of-the-art family of contextual bandit algorithms. We conduct experiments on real world datasets to study the performance of these approaches. The results demonstrate that UCB leads to superior performance.

The rest of this paper is organized as follows. In Section 2, we define the interactive verification problem and compare it to other problems. We describe our design of the four approaches to solve the problem in Section 3. Finally, we present the experiment results in Section 4 and conclude our work in Section 5.

2 Problem Setting

Given a set of instances $X = \{x_1, ..., x_m\}$, where each instance x_i is associated with a label $Y(x_i) \in \{-1, 1\}$. We define the set of positive instances $P = \{x_i \in X | Y(x_i) = 1\}$, which is the set of the instances that require verification. Interactive verification is an iterative process. In the first iteration, we assume that an interactive verification learner knows the labels of one positive instance and one negative instance as initial instances and do not know the labels of other instances. On the t-th iteration, the learner is asked to select an instance s_t from unlabeled (un-verified) dataset U, where $U = \{x_i \in X | x_i \neq s_\tau, \forall \tau < t\}$. The learner then receives the label $Y(s_t)$ to update its internal model. The goal is to verify as many positive instances as possible within T iterations. That is, we want to maximize

$$\sum_{t=1}^{T} \frac{[Y(s_t) = 1]}{|P|}. \tag{1}$$

Sabato et al. (2013) also proposed an equivalent problem called "auditing", which aims to minimize the number of labeled negative instances needed to classify all of the instances accurately. The work compares the similarity and differences between auditing and active learning, and only studies one baseline auditing algorithm. In this work, we consider designing and comparing different approaches for the interactive verification problem.

As pointed out by Sabato et al. (2013), immediate tools for interactive verification can be easily found in active learning. Active learning is a form of supervised learning in which the learner can interactively ask for information (Settles, 2009). The spirit of active learning is to believe that the information amount carried by each instance is different. By choosing informative instances to query, the learner can obtain an accurate model with only few labeled instances, thereby reducing human efforts.

Pool-based active learning is a widely used setting for active learning, which assumes that the learner can only query the instances chosen from a given dataset pool (Lewis and Gale, 1994). The setting of pool-based active learning is almost the same as interactive verification: both of them allow the learner to query an instance to obtain its label in each iteration. The difference between them is the different goals. Active learning focuses on getting an accurate model; on the other hand, interactive verification aims to maximize the number of verified positive instances. Although the goals are different, the similar setting allows tools of active learning to be possibly used for interactive verification.

In this work, we will connect interactive verification to another problem called contextual bandit. The contextual bandit problem is a form of multi-armed bandit problem, where a player faces some slot machines and wants to decide in which order to play them (Auer et al., 2000). In every iteration, the player can select one slot machine (action) from some action set A. Then, the player will receive a randomize reward decided by the distribution under the corresponding slot machine (action). The goal is to maximize the rewards received by the player after a given number of iterations. One key property of the multi-armed bandit problem is that we could only get partial information from environment: only the reward of the selected action will be revealed. If an action has never been selected, the player will not have information about it. Thus, it is necessary to spend some iterations to explore the actions that the player is not familiar with. Somehow only doing the exploration cannot maximize the total rewards, and the player also needs to spend some iterations to exploit the action with high expected rewards. The key to solve the multi-armed bandit problem is to find the balance between the exploration and the exploitation. In addition to the setting above, the contextual bandit problem allows the learner (player) to receive some context information about the environment prior to making selections in every iteration (Langford and Zhang, 2007).

The context information makes it possible for contextual bandit algorithms to exercise a more strategic choice according to the context.

In a first glance, the setting of contextual bandit appears very different from interactive verification. A closer look at the two problems, however, reveal that the trade-off between the exploration and the exploitation in contextual bandit is similar to the trade-off between the learning stage and the verification stage in the interactive verification. In particular, if we define a special contextual bandit problem as follows: The action set A consists of the choices to query each unlabeled instance; the context represent the features of each unlabeled instance; the reward is 1 if the selected action (queried instance) is a positive one, and 0 otherwise. Then, we see that maximizing the cumulative rewards in such a contextual bandit problem is exactly the same as maximizing (1). The connection leads to new possibilities in designing interactive verification approaches, which will be discussed in the next section.

Although we find the similarity between contextual bandit and interactive verification, there is still a big difference. In a contextual bandit problem, each action is usually allowed to be selected several times. Then, the actions that are more likely to produce high rewards could be selected more often. In interactive verification, however, each instance is supposed to be queried at most once. That is, in the corresponding contextual bandit problem, each action can be selected at most once. The difference make it non-trivial to apply existing contextual bandit algorithms for interactive verification.

3   Approaches

For the convenience of discussion, we first outline a general framework for interactive verification approaches. In every iteration, we use a base learner to train the model from labeled instances, and then the learner chooses the next instance to be queried according to a scoring function computed from the model. The general framework is shown in Algorithm 1. By defining the scoring function, we define the behavior of an approach to interactive verification.

Algorithm 1. General approach to interactive verification

Require: Base learner, B; Unlabeled instances, U; Labeled instances, L; Number of
      iterations, T;
  1. **for** $t = 1$ to T **do**
  2.   model $M = B(L)$
  3.   **for all** $u \in U$ **do**
  4.   Compute scoring function: $S(u, M)$
  5.   **end for**
  6. $s_t = \arg\max_u S(u, M)$
  7. $L = L \cup \{(s_t, Y(s_t))\}$
  8. $U = U \setminus \{s_t\}$
  9. **end for**

In this work, we use support vector machine (SVM) with linear kernel as our base learner, and denote w_t to be the liner weights we get from the base learner in the beginning of every iteration.

3.1 Greedy Approach

The goal of our problem is to verify as many positive instances as possible. The most intuitive solution is querying the instance which be considered most likely to be positive by current model in every iteration, i.e. the instance with highest $p(y = 1|x_i)$. When using SVM as base learner, the instance to be queried comes with the largest decision value. That is, the scoring function of the greedy approach is simply

$$S(x_i, w_t) = x_i^\mathsf{T} w_t.$$

Greedy approach only considers how possible an instance to be positive in each iteration. It ignores the information amount carried by each instance. If we start from a biased model, the greedy approach may perform poorly. Here, we give an example that the greedy approach will fail. Consider the case shown in Figure 1. There are two clusters of red positive instances and one big cluster of blue negative instances in the figure. Without loss of generality, we assume the initial positive instance is in the top red positive cluster. The model we start with will be the dashed line. The optimal model is the solid line, which is very different from the dashed line. By running greedy approach on this dataset, we can easily verify the positive instances in top cluster. But after all the instances in top positive cluster is queried, greedy approach will prefer to query the instances in the negative cluster than query the instance in bottom positive cluster. To solve this issue, we may need to do some explorations to help us find the instances in the bottom positive cluster.

3.2 Random Then Greedy

In the previous subsection we discuss the risk of not doing exploration. Here we propose an approach using the random as exploration method to solve interactive verification problems: random then greedy (RTG). Randomly selecting an instance to query is a naive yet reasonable strategy to do the exploration. It can provide some unbiased information. Then, we use greedy approach described in he previous section for exploitation (verification). In this approach we do an one-time switching from exploration to exploitation. We use the parameter ϵ to decide the ratio between exploration and exploitation. That is, the scoring function of RTG is

$$S(x_i, w_t) = \begin{cases} random(), & \text{if } t \leq \epsilon T \\ x_i^\mathsf{T} w_t, & \text{otherwise} \end{cases}.$$

3.3 Uncertainty Sampling Then Greedy

As the discussion in Section 2, the setting of the interactive verification is pretty similar to the active learning problem. It is natural to attempt to use tools of active learning

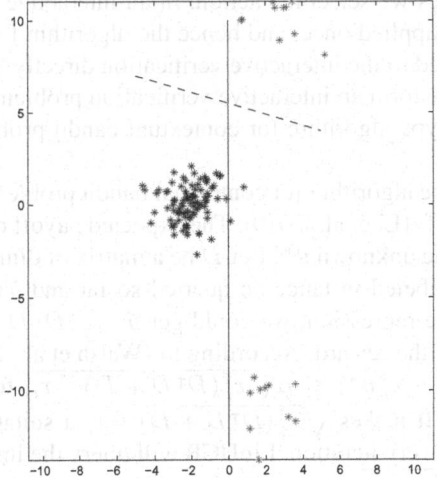

Fig. 1. Artificial dataset

for interactive verification. Uncertainty sampling is one of the most commonly used algorithm for active learning (Settles, 2009). The idea is to query the instances that the current model is least certain on how to label it. For probabilistic learning models, uncertainty sampling queries the instances with probability to be positive close to 50%. Uncertainty sampling can also be employed with non-probabilistic learning model. When using SVM as the base learning model, uncertainty sampling queries the instance closest to the linear decision boundary (Tong and Koller, 2001).

To apply the uncertainty sampling on the interactive verification, we can borrow the framework from RTG as described in previous section. We use greedy as exploitation method and use uncertainty sampling as our new exploration method to replace random sampling. We call this approach uncertainty sampling then greedy (USTG). The scoring function of USTG is

$$S(x_i, w_t) = \begin{cases} \frac{1}{|x_i^\mathsf{T} w_t| + 1}, & \text{if } t \leq \epsilon T \\ x_i^\mathsf{T} w_t, & \text{otherwise} \end{cases}.$$

Uncertainty sampling may suffer from a biased model like the greedy approach. When starting with a model of bad quality, the instances that are selected by uncertainty sampling may not be very informative. Thus, using the uncertainty sampling as exploration method cannot totally solve the issue of biased model in the greedy approach.

3.4   Upper Confidence Bound

Upper confidence bound (UCB) is an algorithm to solve the multi-armed bandit problem (Auer et al., 2000). The idea of UCB is to keep the upper bound of plausible rewards of the actions and select the action according this value. In the traditional multi-armed bandit problem, there is no contextual features. The prediction of confidence bound is based on how many times we select the action. In an interactive verification problem, each action can be only applied once, and hence the algorithm for multi-armed bandit problem cannot be applied to the interactive verification directly. But as our discussion in Section 2, we can transform an interactive verification problem to a contextual bandit problem. The UCB-type algorithm for contextual bandit problem may suit for the interactive verification.

LinUCB is a UCB-type algorithm for contextual bandit problem, which assumes the problem has linear payoffs (Li et al., 2010). The expected payoff of an action with context x_i is $x_i^\mathsf{T} w^*$ with some unknown w^*. Let D be a matrix of dimension $m \times d$, whose rows correspond to m labeled instance be queried so far and b as the corresponding labels. By applying ridge regression, we could get $\hat{w} = (D^\mathsf{T} D + I)^{-1} D^\mathsf{T} b$, so $x_i^\mathsf{T} \hat{w}$ will be the estimation of the reward. According to (Walsh et al., 2009), with probability at least $1 - \delta$, $|x_i^\mathsf{T} \hat{w} - x_i^\mathsf{T} w^*| \leq \hat{\alpha} \sqrt{x_i^\mathsf{T} (D^\mathsf{T} D + I_d)^{-1} x_i}$, for any $\delta > 0$, where $\hat{\alpha} = 1 + \sqrt{\ln(2/\delta)/2}$. It makes $\sqrt{x_i^\mathsf{T} (D^\mathsf{T} D + I_d)^{-1} x_i}$ a suitable upper confidence bound measurement. In every iteration, LinUCB will query the instance x_i with largest $x_i^\mathsf{T} \hat{w} + \hat{\alpha} \sqrt{x_i^\mathsf{T} (D^\mathsf{T} D + I_d)^{-1} x_i}$.

Since the interactive verification does not have the assumption of linear payoff, we use our original base learner SVM instead of ridge regression. We treat confidence term in LinUCB as a term to measure the uncertainty of each instance in unsupervised learning view. If the learner is not certain on the instance, the confidence term will be large; otherwise, it will be small. By using confidence term from LinUCB, we can find the instances that worthy to be explored. The value of confidence term can also help to decide the switching timing between exploration and exploitation. We add the confidence term to the decision value that is produced from SVM and connect these two terms with a parameter α. The scoring function of the UCB approach to interactive verification is

$$S(x_i, w_t) = x_i^\mathsf{T} w_t + \alpha \sqrt{x_i^\mathsf{T} (D^\mathsf{T} D + I_d)^{-1} x_i}.$$

3.5 Discussions

We have now discussed four different approaches to solve interactive verification problems. Among them, the greedy approach could be seen as a special case of the other three approaches. All four approaches all apply greedy approach during exploitation. But these four approaches have different philosophy for exploration. The greedy approach spend all the iterations for exploitation; the exploration method used by RTG is random sampling, which can get unbiased information; the exploration method used by USTG is uncertainty sampling, which is widely used for active learning; UCB uses the confidence term from LinUCB to decide which instances are worthy of being explored and when the learner should do the exploration.

Now we compare the strategies on switching between exploration and exploitation. Greedy approach does not do the switching at all; RTG and USTG share a similar framework by only doing a one-time switching from exploration to exploitation; UCB uses the confidence term to decide the switching between exploration and exploitation automatically. That is, it is possible for UCB to switch between exploration and exploitation several times.

4   Experiment

4.1   Datasets and Experiment Setting

We conduct experiments on eight real-world datasets to compare the performance of the four approaches proposed in Section 3. Table 1 shows the datasets that we use. Among them, the KDD Cup 2008 dataset is a breast cancer screening dataset as discussed in Section 1. As the table shows, the percentages of positive instances, which may greatly affect the performance of interactive verification algorithm, are very different from different datasets. To do a fair comparison, we do the re-sampling on all the datasets to control the percentages of positive instances in each dataset. We separate the positive instances from negative instances in original dataset, and sample P positive instances and N negative instances from corresponding set. For convenience, we set $N = 1000$ all the time and only adjust the value of P in our experiments. We repeat each experiment 1000 times with different initial instances, which include one positive instance and one negative instance. We used (1) as the evaluation metric. The results and the discussions can be seen in following sections. The KDD Cup 2008 dataset will be studied further in Section 4.4.

Table 1. Dataset characteristics

Dataset	Number of instances	Number of positive instances	Positive rate
KDD Cup 2008	102294	623	0.6%
spambase	4601	1813	39.4%
a1a	1605	395	24.6%
cod-rna	59535	19845	33.3%
mushrooms	8124	3916	48.2%
w2a	3470	107	3%
covtype.binary	581012	297711	51.2%
ijcnn1	49990	4853	9.7%

4.2   Effect of ϵ

In this section we demonstrate the effect of different ϵ in RTG and USTG. We conduct experiments on the KDD Cup 2008 dataset with $P = 50, 100$ and $T = 100$. We change the value of ϵ from 0 to 1. The results are shown in Figure 2. The performance decreases when ϵ increase both for RTG and USTG, and $\epsilon = 0$ is one of the best choice. The rest of the datasets show the same trend. RTG and USTG with $\epsilon = 0$ are actually the greedy approach. As our discussion before, the greedy approach spent all the iterations in exploitation. The results that greedy approach has best performance seem to suggest that spending queries on improving model quality is not important for interactive verification. Nevertheless, if we take a closer look on greedy approach, we will find out that instances selected by greedy approach could benefit on both verification and model quality.

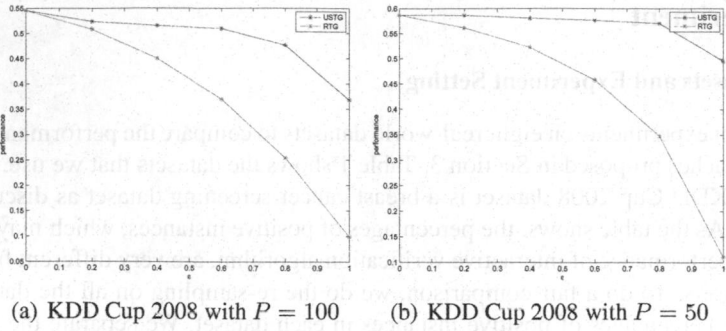

(a) KDD Cup 2008 with $P = 100$ (b) KDD Cup 2008 with $P = 50$

Fig. 2. The effect of ϵ

The story is that, the instance selected by greedy approach the instance with highest possibility to be positive among all the unlabeled instances. It will have the highest probability to be a positive instance, and hence the query is likely to be a successful verification; on the other hand, even if greedy approach queries a negative instance, it may not totally be a bad news. The instance selected by greedy approach is the instance that considered most possible to be positive by current model. The truth that the instance is actually a negative instance is very informative. The query result may greatly improve the model quality. So no matter what result we get from querying the instance selected by greedy approach, we either successfully verify a positive instance or label an informative negative instance. In other word, greedy approach often either does a successful exploitation or does an efficient exploration.

Although greedy approach has such good property in the interactive verification, it still will have poor performance on the dataset shown in Figure 1. The reason that the good property of greedy approach does not work is that the instance selected by greedy approach may actually have low possibility to be positive. It may happen when there is no better choice for greedy approach to select. Consider the biased model shown as dashed line in Figure 1, the instances in negative cluster are considered to be negative instances by the model. But since the instances in bottom positive cluster are misclassified as extremely negative ones, the greedy approach will still select the instance in negative cluster to query. To solve this issue, we should do the exploration when the instance selected by greedy approach does not have high enough possibility to be positive, and do the exploitation when the instance selected by greedy has high enough possibility to be positive. It is actually what UCB does: when the first term in UCB is large, it will do the exploitation; when the first term is small, it will do the exploration. So UCB may be a better choice to solve interactive verification problems than the greedy approach.

4.3 Comparison of All Approaches

In this section, we conduct experiments for comparing four approaches on all eight datasets. We set $P = 50, 100$ and $T = 100$. For RTG and USTG, we set ϵ to be 0.2,

Table 2. Experiment results

Dataset	Algorithm	$P = 50$	$P = 100$
KDD Cup 2008	greedy	0.5868 ± 0.0040 (3)	**0.5454 ± 0.0022 (2)**
	RTG($\epsilon = 0.2$)	0.5615 ± 0.0035 (5)	0.5080 ± 0.0018 (5)
	USTG($\epsilon = 0.2$)	0.5863 ± 0.0032 (4)	0.5235 ± 0.0023 (4)
	UCB($\alpha = 0.2$)	0.5968 ± 0.0031 (2)	0.5434 ± 0.0018 (3)
	UCB($\alpha = 0.4$)	**0.6055 ± 0.0027 (1)**	**0.5467 ± 0.0015 (1)**
spambase	greedy	**0.7467 ± 0.0024 (1)**	**0.6055 ± 0.0012 (1)**
	RTG($\epsilon = 0.2$)	0.7042 ± 0.0020 (4)	0.5422 ± 0.0012 (5)
	USTG($\epsilon = 0.2$)	0.7429 ± 0.0023 (2)	0.5905 ± 0.0012 (2)
	UCB($\alpha = 0.2$)	0.7306 ± 0.0020 (3)	0.5856 ± 0.0013 (3)
	UCB($\alpha = 0.4$)	0.6965 ± 0.0022 (5)	0.5559 ± 0.0013 (4)
a1a	greedy	**0.3883 ± 0.0034 (4)**	0.3754 ± 0.0020 (2)
	RTG($\epsilon = 0.2$)	0.3535 ± 0.0035 (5)	0.3413 ± 0.0018 (5)
	USTG($\epsilon = 0.2$)	**0.3898 ± 0.0035 (3)**	0.3585 ± 0.0018 (4)
	UCB($\alpha = 0.2$)	**0.3915 ± 0.0034 (1)**	**0.3775 ± 0.0019 (1)**
	UCB($\alpha = 0.4$)	**0.3909 ± 0.0031 (2)**	0.3711 ± 0.0019 (3)
cod-rna	greedy	0.7249 ± 0.0027 (3)	0.6251 ± 0.0012 (2)
	RTG($\epsilon = 0.2$)	0.6763 ± 0.0024 (5)	0.5610 ± 0.0012 (5)
	USTG($\epsilon = 0.2$)	0.7155 ± 0.0025 (4)	0.6074 ± 0.0012 (4)
	UCB($\alpha = 0.2$)	**0.7333 ± 0.0024 (1)**	**0.6265 ± 0.0012 (1)**
	UCB($\alpha = 0.4$)	0.7297 ± 0.0025 (2)	0.6236 ± 0.0012 (3)
mushrooms	greedy	0.9710 ± 0.0014 (4)	**0.9125 ± 0.0008 (1)**
	RTG($\epsilon = 0.2$)	0.9715 ± 0.0012 (3)	0.8112 ± 0.0006 (5)
	USTG($\epsilon = 0.2$)	0.9600 ± 0.0008 (5)	0.8776 ± 0.0005 (4)
	UCB($\alpha = 0.2$)	0.9776 ± 0.0007 (2)	0.9109 ± 0.0006 (2)
	UCB($\alpha = 0.4$)	**0.9837 ± 0.0006 (1)**	0.9031 ± 0.0005 (3)
w2a	greedy	0.5944 ± 0.0030 (3)	0.5498 ± 0.0016 (2)
	RTG($\epsilon = 0.2$)	0.5371 ± 0.0032 (5)	0.4933 ± 0.0016 (5)
	USTG($\epsilon = 0.2$)	0.5931 ± 0.0028 (4)	0.5393 ± 0.0015 (3)
	UCB($\alpha = 0.2$)	**0.6160 ± 0.0024 (1)**	**0.5601 ± 0.0013 (1)**
	UCB($\alpha = 0.4$)	0.6064 ± 0.0023 (2)	0.5314 ± 0.3883 (4)
covtype.binary	greedy	0.2202 ± 0.0026 (5)	0.2306 ± 0.0021 (5)
	RTG($\epsilon = 0.2$)	0.2342 ± 0.0027 (3)	0.2388 ± 0.0017 (4)
	USTG($\epsilon = 0.2$)	0.2294 ± 0.0026 (4)	0.2491 ± 0.0021 (3)
	UCB($\alpha = 0.2$)	0.2536 ± 0.0024 (2)	0.2554 ± 0.0021 (2)
	UCB($\alpha = 0.4$)	**0.2798 ± 0.0024 (1)**	**0.2649 ± 0.0021 (1)**
ijcnn1	greedy	0.5220 ± 0.0027 (3)	0.4705 ± 0.0023 (3)
	RTG($\epsilon = 0.2$)	0.4668 ± 0.0034 (5)	0.4247 ± 0.0015 (5)
	USTG($\epsilon = 0.2$)	0.5184 ± 0.0028 (4)	0.4607 ± 0.0019 (4)
	UCB($\alpha = 0.2$)	0.5402 ± 0.0029 (2)	0.4750 ± 0.0021 (2)
	UCB($\alpha = 0.4$)	**0.5598 ± 0.0025 (1)**	**0.4849 ± 0.0018 (1)**
Average Rank	greedy	3.25	2.25
	RTG($\epsilon = 0.2$)	4.38	4.88
	USTG($\epsilon = 0.2$)	3.75	3.5
	UCB($\alpha = 0.2$)	1.75	1.88
	UCB($\alpha = 0.4$)	1.88	2.5

the best observed choice among $\epsilon > 0$. For the parameter α in UCB, we consider 0.2 and 0.4. Table 2 shows the result of our experiments. We treat datasets with different P as different datasets. The results show that greedy outperform RTG and USTG. It is consistent to our finding in the previous subsection. The table also shows that the best α for UCB is dataset dependent, so parameter tuning may be necessary for UCB. Generally, $\alpha = 0.2$ is a good choice. UCB with $\alpha = 0.2$ has best performance both for $P = 50$ and $P = 100$ cases. When $P = 50$, UCB totally outperform greedy. But when $P = 100$, although UCB with $\alpha = 0.2$ still has the best performance, the gap between it and greedy is smaller. The reason behind that is when P increase from 50 to 100 while T is still fix to 100, there may be not much iterations left after greedy finish querying the instances with high probability to be positive, so the ability to dynamically switch

Table 3. KDD Cup 2008

Dataset	Algorithm	$T = 623$	$T = 1243$
KDD Cup 2008	greedy	**0.3649** ± **0.0037**	0.4831 ± 0.0059
	RTG($\epsilon = 0.2$)	0.3062 ± 0.0022	0.4482 ± 0.0023
	USTG($\epsilon = 0.2$)	**0.3659** ± **0.0013**	0.4802 ± 0.0058
	UCB($\alpha = 0.2$)	**0.3660** ± **0.0016**	**0.4917** ± **0.0029**
	UCB($\alpha = 0.4$)	**0.3655** ± **0.0013**	**0.4897** ± **0.0048**

to the exploration stage will be less significant. The results also show that UCB, which does dynamic switching from the exploration stage to the exploitation stage approach, has better performance than RTG and USTG, which does an one-time switching.

4.4 Real-World Task

In this subsection, we conduct experiments on the KDD Cup 2008 dataset without re-sampling. The KDD Cup 2008 challenge focuses on the problem of early detection of breast cancer from X-ray images of the breast. In this dataset, only 623 out of 102294 ROIs are malignant mass lesions. The percentage of positive instance is only around 0.6%. The P is given by the dataset, which equals to 623. We set T to be 623 and 1243 separately, which are the value of P and twice the P. We do each experiment 20 times. The result is shown in Table 3. Although the difference is small when $T = 623$, UCB apparently has best performance when $T = 1243$. The result is consistent with our experiments on the re-sampled datasets.

5　Conclusion

Interactive verification is a new problem. We pointed out that the trade-off between the learning stage and the verification stage is similar to the trade-off between exploration and exploitation in the contextual bandit problem, and transformed interactive verification to a special contextual bandit problem. We discussed the pros and cons of three basic approaches: greedy, RTG, and USTG, and showed that applying greedy on the interactive verification leads to better results. We also showed the potential risk of the greedy approach for interactive verification, and proposed to adopt UCB, which has been widely used for contextual bandit, to solve interactive verification. UCB avoids the risk that the greedy approach may encounter. The experimental results on re-sampled datasets and a real-world task show that greedy is quite competitive and UCB performs the best among four approaches.

Acknowledgment. We thank Profs. Yuh-Jye Lee, Shou-De Lin, the anonymous reviewers, and the members of the NTU Computational Learning Lab for valuable suggestions. This work is mainly supported by National Science Council (NSC 101-2628-E-002-029-MY2) of Taiwan.

References

Auer, P., Cesa-Bianchi, N., Fischer, P., Informatik, L.: Finite-time analysis of the multi-armed bandit problem. Machine Learning 2-3, 235–256 (2000)

Langford, J., Zhang, T.: The epoch-greedy algorithm for contextual multi-armed bandits. In: Proceedings of the Conference on Neural Information Processing Systems (2007)

Lewis, D.D., Gale, W.A.: A sequential algorithm for training text classifiers. In: Proceedings of the 17th Annual International ACM SIGIR Conference on Research and Development in Information Retrieval, pp. 3–12 (1994)

Li, L., Chu, W., Langford, J., Schapire, R.E.: A contextual-bandit approach to personalized news article recommendation. In: Proceedings of the International Conference on World Wide Web, pp. 661–670 (2010)

Li, M., Zhou, Z.H.: Improve computer-aided diagnosis with machine learning techniques using undiagnosed samples. IEEE Transactions on Systems, Man, and Cybernetics, Part A 37(6), 1088–1098 (2007)

Rangayyan, R.M., Fabio, J.A., Desautels, J.L.: A review of computer-aided diagnosis of breast cancer: Toward the detection of subtle signs. Journal of the Franklin Institute 344(3-4), 312–348 (2007)

Sabato, S., Sarwate, A.D., Srebro, N.: Auditing: Active learning with outcome-dependent query costs. In: Proceedings of the Conference on Neural Information Processing Systems (2013)

Settles, B.: Active learning literature survey. Tech. rep., University of Wisconsin–Madison (2009)

Tong, S., Koller, D.: Support vector machine active learning with applications to text classification. Journal of Machine Learning Research 2, 45–66 (2001)

Walsh, T.J., Szita, I., Diuk, C., Littman, M.L.: Exploring compact reinforcement-learning representations with linear regression. In: Proceedings of the Conference on Uncertainty in Artificial Intelligence, pp. 591–598 (2009)

Positional Translation Language Model for Ad-Hoc Information Retrieval

Xinhui Tu[1,2], Jing Luo[1,2], Bo Li[3], Tingting He[3], and Jinguang Gu[1,4]

[1] College of Computer Science and Technology,
Wuhan University of Science and Technology, Wuhan, China
[2] Hubei Province Key Laboratory of Intelligent Information Processing and
Real-time Industrial System, Wuhan, China
[3] Department of Computer Science, Central China Normal University, Wuhan, China
[4] State Key Lab of Software Engineering, Wuhan University, Wuhan, China
{tuxinhui,luoluocat,simongu}@gmail.com,
liboccnu@126.com, tthe@mail.ccnu.edu.cn

Abstract. Most existing language modeling approaches are based on the term independence hypothesis. To go beyond this assumption, two main directions were investigated. The first one considers the use of the proximity features that capture the degree to which search terms appear close to each other in a document. Another one considers the use of semantic relationships between words. Previous studies have proven that these two types of information, including term proximity features and semantic relationships between words, are both useful to improve retrieval performance. Intuitionally, we can use them in combination to further improve retrieval performance. Based on this idea, this paper propose a positional translation language model to explicitly incorporate both of these two types of information under language modeling framework in a unified way. In the first step, we present a proximity-based method to estimate word-word translation probabilities. Then, we define a translation document model for each position of a document and use these document models to score the document. Experimental results on standard TREC collections show that the proposed model achieves significant improvements over the state-of-the-art models, including positional language model, and translation language models.

Keywords: Positional Language Model, Translation Language Model, Information Retrieval.

1   Introduction

Language modeling (LM) for Information Retrieval (IR) has been a promising area of research over the past decade and a half. It provides an elegant mathematical model for ad-hoc text retrieval with excellent empirical results reported in the literature [12][20]. However, language models suffer from one problem: term independence assumption which is common for all retrieval models.

To address this problem, two main directions were investigated. The first one is based on the use of the proximity features. These features capture the degree to which

V.S. Tseng et al. (Eds.): PAKDD 2014, Part II, LNAI 8444, pp. 134–145, 2014.
© Springer International Publishing Switzerland 2014

search terms appear close to each other in a document. To incorporate the cues of term position and term proximity under language model framework, Lv and Zhai [10] proposed a positional language model (PLM). In PLM, a language model for each term position in a document is defined, and document is scored based on the scores of its PLMs.

The second one considers the use of semantic relationships between words. In order to reduce the semantic gap between documents and queries, statistical translation models (TLM) have been proposed for information retrieval to capture semantic word relations [2]. The basic idea of translation language models is to estimate the probabilities of translating a word in a document to query words. Since a word in a document could be translated into different words in the query, translation language models can avoid exact matching of words between documents and queries.

The previous studies have proven that term proximity features and semantic relationships between words are both useful information to improve retrieval performance (e.g., [2][7][10][11]). Intuitively, we can use them in combination to further improve retrieval performance. Based on this idea, this paper proposes a positional translation language model to explicitly incorporate these two types of information in a united way. In the first step, we present a proximity-based method to estimate word-word translation probabilities. Then, we define a translation document model for each position of a document and use these document models to score the document.

The main contribution of this paper is as follows: First, we propose a new proximity-based method, in which the proximity of co-occurrences is taking into account, to estimate word-word translation probabilities. Second, we propose a positional translation language model (PTLM) to explicitly incorporate term proximity features and semantic relationships between words in a unified way. Finally, extensive experiments on standard TREC collections have been conducted to evaluate the proposed model. Experimental results on standard TREC collections show that PTLM achieves significant improvements over the state-of-the-art models, including positional language model, and translation language models.

2   Background: PLM and TLM

2.1   Basic Language Modelling Approach

The basic idea of language models is to view each document to have its own language model and model querying as a generative process. Documents are ranked based on the probability of their language model generating the given query. Different implementations were proposed [20]. The general ranking formula is defined as follows:

$$logp(D|Q) \overset{rank}{=} \sum_{w \in V} c(w, Q) \, logp(w|D) \qquad (1)$$

where $\overset{rank}{=}$ means equivalence for the purpose of ranking documents, $c(w, Q)$ is the count of word w in query Q, and V is the vocabulary set. The challenging part is to estimate a document model $p(w|D)$. The simplest way to estimate $p(w|D)$ is the maximum likelihood estimator. However, this method is suffering from the data sparseness problem. To address this problem, some effective smoothing approaches, which combine the document model with the background collection model, have been proposed. One commonly used method is Dirichlet Prior smoothing methods [18], which is defined as follows:

$$p(w|Q) = \frac{|D|}{|D| + \mu} p_{ml}(w|D) + \frac{\mu}{|D| + \mu} p_{ml}(w|C) \tag{2}$$

2.2   Positional Language Model

To incorporate the cues of term position and term proximity under language model framework, Lv and Zhai [10] proposed a positional language model (PLM). In PLM model, for each document $D(w_1; \dots; w_i; \dots; w_j; \dots; w_N)$, where $1, i, j$, and N are absolute positions of the corresponding terms in the document, and N is the length of the document, a virtual D_i document is estimated at each position. This model is represented as a term frequency vector $D\langle c'(w_1; i); \dots; c'(w_N; i)\rangle$, where $c'(w; i)$ is the total propagated count of term w at position i from the occurrences of w in all the positions. That is $c'(w, i) = \sum_{j=1}^{N} c(w, j) k(i, j)$ where $c(w, j)$ is the count of term w at position i in document D. If w occurs at position i, it is 1, otherwise 0. $k(i, j)$ is the propagated count to position i from a term at position j. Several proximity-based density functions are used to estimate this factor: (Gaussian kernel, Triangle kernel, Circle kernel, Cosine kernel). Once the virtual document D_i is estimated, the language model of this virtual document can be estimated as follow

$$p(w|D, i) = \frac{c'(w, i)}{\sum_{w' \in V} c'(w', i)} \tag{3}$$

where V is the vocabulary, $p(w|D, i)$ is noted as a positional language model at position i. To compute the final score of document D, they used the position-specific scores. Different strategies were used: Best Position Strategy, Multi-Position Strategy, Multi-σ Strategy.

2.3   Statistical Translation Language Model

To incorporate the semantic relationship between terms under language model framework, Berger and Lafferty proposed translation language modelling approach to estimate $p(w|D)$ based on statistical machine translation [2]. In this approach, the document model $p(w|D)$ can be calculated by using the following "translation document model":

$$p_t(w|D) = \sum_{u \in D} p_t(w|u) p(u|D) \tag{4}$$

where $p(u|D)$ is the probability of seeing word u in document d, and $p_t(w|u)$ is the probability of "translating" word u into word w. In this way, a word can be translated into its semantically related words with non-zero probability, which allows us to score a document by counting the matches between a query word and semantically related words in document.

The key part for translation language model is estimating translation probabilities. Berger and Lafferty [2] proposed a method to estimate translation probabilities by generating synthetic query. This method is inefficient and does not have good coverage of query words. In order to overcome these limitations, Karimzadehgan and Zhai [7] proposed an effective estimation method based on mutual information. Recently, Karimzadehgan and Zhai [8] defined four constraints that a reasonable translation language model should satisfy, and proposed a new estimation method which is shown to be able to better satisfy the constraints. This new estimation method, namely conditional context analysis, is described in formula 5.

3 Positional Translation Language Model

In this section, we will describe the PTLM in detail. In the first part, a proximity-based method is presented to estimate word-word translation probabilities. Then, we will introduce how to estimate the translation document model for each position within a document. Finally, these positional document models are used to score the document.

3.1 Estimating Translation Probability

In the conditional context analysis method proposed in [8], the probability of translating word u into word w can be estimated as follows:

$$p(w|u) = \frac{c(w,u) + 1}{\sum_{w'} c(w',u) + |V|} \qquad (5)$$

where $c(w,u)$ is the co-occurrences of word u with word w, and $|V|$ is the size of the vocabulary.

In this method, any co-occurrence within the document is treated in the same way, no matter how far they are from each other. This strategy is not optimal as a document may cover several different topics and thus contain much irrelevant information. Intuitionally, closer words usually have stronger relationships, thus should be more relevant. Therefore, we introduce a new concept, namely proximity-based word co-occurrence frequency (pcf) to model the proximity feature of co-occurrences.

Recently, density functions based on proximity are proven to be effective to characterize term influence propagation. A number of term propagation functions (e.g. Gaussian, Triangle, Cosine and Circle) have been proposed [10][11]. In this section, we adopted Gaussian functions because it has been shown to be effective in most cases. The Gaussian-based pcf can be calculated as follows:

$$pcf(w,u) = \sum_{D \in Col(w,u)} exp\left[\frac{-(dist(w,u,D))^2}{2\sigma^2}\right] \tag{6}$$

where, σ is a parameter in Gaussian distribution, $Col(w,u)$ is the set of documents which contain both w and u, and $dist(w,u,D)$ is the distance score of word w and u in document D.

In this paper, three commonly used distance measures are adopted to late $dist(w,u,D)$. We will use the following short document D as an example to explain how to calculate distance score in the three distance measures.

$$\begin{array}{ccccccccccc} & 1 & 2 & 3 & 4 & 5 & 6 & 7 & 8 & 9 & 10 \\ D = \{ & w & u & c & k & w & u & k & e & w & g & \} \end{array}$$

Minimum pair distance: It is defined as the minimum distance between any occurrences of w and u in document D. In the example, $dist(w,u,D)$ is 1 and can be calculated from the position vectors.

Average pair distance: It is defined as the average distance between w and u for all position combinations in D. In the example, the distances from the first occurrence of w (in position 1) to all occurrences of u are: {1 and 5}. This is computed for the next occurrence of w (in position 5) and so on. $dist(w,u,D)$ for the example is (((2-1) + (6-1)) + ((5-2) + (6-5)) + ((9-2) + (9-6)))/(2 · 3) = 20/6 = 3.33.

Average minimum pair distance: It is defined as the average of the shortest distance between each occurrence of the least frequently occurring word and any occurrence of the other word. In the example, u is the least frequently occurring word so $dist(w,u,D)$ = ((2−1)+(6−5))/2 =1.

Then, the probability of translating word u into word w can be estimated as follows:

$$p(w|u) = \frac{pcf(w,u) + \epsilon}{\sum_{w'} pcf(w',u) + |V| * \epsilon} \tag{7}$$

where ϵ is a smoothing parameter in order to account for unseen words in the context of u. Here ϵ is set equals to the smallest of all pcf values in collection.

In order to satisfy the constraints defined in [8], we adjust self-translation probabilities as follows:

$$p_t(u|u) = s \ (s \geq 0.5) \tag{8}$$

$$p_t(w|u) = (1 - s) * \frac{p(w|u)}{\sum_{v \neq u} p(v|u)} \tag{9}$$

where parameter s is a constant value that could be set to $0.5 <= s <= 1$. Note that when $s = 1$, the query likelihood model are gained.

3.2   Estimating Translation Document Model

The state-of-art translation language models use an entire document as a unit to estimate the generative probability of the query [7][8]. This strategy is not optimal as a document may cover several different topics. Intuitively, the words referring to the same topic may occur close to each other. Positional language model has been proven to be an effective way to incorporate the cues of term position and term proximity under language model framework [10]. In this section, we will introduce a positional translation language model to naturally incorporate two types of information, including term proximity features and semantic relationships between terms, under language model framework in a united way.

The key idea of our method is to extend the translation language model from document level to positional level via the positional language model. The proposed model can capture the topic of the document at the position by giving more weight on words close to the position and less weight on words far away. The translation language model at each position can be estimated based on all the propagated counts of all the words to the position as if all the words had appeared actually at the position with discounted counts.

Previous studies have shown that translation language model works better with Dirichlet prior smoothing [7][8]. Therefore, in the rest of the paper, we further focus on PTLM with Dirichlet prior smoothing only. The final positional translation language model for position i in document D can be defined as follows:

$$p_t(w|D,i) = \frac{|D|}{|D|+\mu}\left[\sum_{u \in D} p_t(w|u)\, p(u|D,i)\right] + \frac{\mu}{|D|+\mu}p(w|C) \qquad (10)$$

where $p_t(w|u)$ is the translation probability from word u to word w, and can be estimated by formula 8 and 9; $p(u|D,i)$ is the positional document model at position i of document D, and can be estimated as follows:

$$p(u|D,i) = \frac{c'(u,i)}{\sum_{u' \in V} c'(u',i)} \qquad (11)$$

where $c'(u,i)$ is the total propagated count of term u at position i from the occurrences of u in all the positions. $c'(u,i)$ can be estimated using the Gaussian kernel function:

$$c'(u,i) = \sum_{j=1}^{|D|} c(u,j)\exp\left[\frac{-(i-j)^2}{2\sigma^2}\right] \qquad (12)$$

where i and j are absolute positions of the corresponding terms in document, and $|D|$ is the length of the document, $c(u,j)$ is the real count of term u at position j.

3.3   Ranking Document

In the section 3.2, we have obtained a translation language model for each position in a document. Intuitively, we can imagine that the PTLMs give us multiple representations of D. Thus given a query Q, we can adopt the KL-divergence retrieval model [19] to score each PTLM as follows:

$$S(Q, D, i) = -\sum_{w \in V} p(w|Q) log \frac{p(w|Q)}{p_t(w|D, i)} \qquad (13)$$

Then, the position-specific scores can be used to compute the final score of document D. In this paper, we compute the final score of document D using the best position strategy [10], which simply scores a document based on its best match position and can be defined as follows:

$$S(Q, D) = max_{i \in [1, N]}\{S(Q, D, i)\} \qquad (14)$$

4   Experiments

4.1   Data Set

We used six standard TREC data sets in our study. They represent different sizes and genre of text collections. Table 1 shows some basic statistics about these data sets. Each document is processed in a standard way for indexing. Words are stemmed (using porter-stemmer), and stop words are removed. In the experiments, we only use title of the queries because semantic word matching is necessary for such short queries.

Table 1. Document set characteristic

	TREC7	DOE	WSJ	TREC8	AP88-89	FR
queries	351-400	51-100	51-100	401-450	51-100	51-100
#doc	528,155	226,087	74,520	528,155	164,597	45,820

In each experiment, we use the KL-divergence model using Dirichlet prior smoothing (with prior parameter μ=1000) to retrieve 2000 documents for each query, and then use the PTLM to re-rank them. The top-ranked 1000 documents are used for comparison with other models. In order to evaluate our model and compare it to other models we use the MAP measure, which is widely accepted measure for evaluating effectiveness of ranked retrieval systems.

In the section 3.1, three different proximity measures are adapted to measure the distance score of two words in a document. The corresponding models based on the three different proximity measures are evaluated on standard TREC collections. The methods used for the experiments are:

- **QL:** baseline, query likelihood model with Dirichlet prior smoothing [18].
- **KL:** baseline, KL-divergence model with Dirichlet prior smoothing [19].
- **TM-MI:** translation language model with mutual information [7].
- **TM-CCON:** translation language model with conditional context analysis [8].
- **PLM:** positional language model with the best position strategy [10].
- **PTLM-1:** PTLM with minimum pair distance.

- **PTLM-2:** PTLM with average pair distance.
- **PTLM-3:** PTLM with average minimum pair distance.

4.2   Comparing with Existing Retrieval Models

As we can see from all the PTLM models used in our experiments, there are several controlling parameters to tune. In order to make the comparison fair, we evaluate PTLMs and PLM by a 5-fold cross-validation on each collection. For the two baselines (QL and KL), parameter μ in the Dirichlet smoothing is set to the optimal value for each collection. The results of TM-MI, TM-CCON are directly from [8].

Table 2 shows the results for these models with Dirichlet prior smoothing. Comparing the rows in the table indicates that the PTLM models achieve significant improvements over the state-of-the-art models, including positional language model and translation language models. In addition, the results confirm our hypothesis that the two types of information can be used in combination to improve retrieval performance. Comparing the three variants of PTLM, PTLM-3 is more effective and robust than PTLM-2 and PTLM-1. It also indicates that average-minimum-pair distance measure can capture the proximity feature of co-occurrences better than the other two measures. The significance test results using Wilcoxon signed-rank test indicate that the differences between the PTLM models and the start-of-art models are statistically significant.

Table 2. The comparison of experiment results
(* and + mean improvements over TM-CCON and PLM are statistically significant with Wilcoxon signed-rank test, respectively)

	TREC7	DOE	WSJ	TREC8	AP88-89	FR
QL	0.1852	0.1740	0.2600	0.2518	0.2154	0.2817
KL	0.1847	0.1742	0.2584	0.2509	0.2196	0.2697
TM-MI	0.1854	0.1750	0.2658	-	-	-
TM-CON	0.1920	0.1844	0.2780	-	-	-
PLM	0.1893	0.1795	0.2641	0.2548	0.2196	0.2842
PTLM-1	0.2003^{*+}	0.1952^{*+}	0.2896^{*+}	0.2672$^+$	0.2246$^+$	0.2885$^+$
PTLM-2	0.2021^{*+}	0.1967^{*+}	0.2913^{*+}	0.2685$^+$	0.2259$^+$	0.2891$^+$
PTLM-3	**0.2030^{*+}**	**0.1975^{*+}**	**0.2924^{*+}**	**0.2692$^+$**	**0.2276$^+$**	**0.2920$^+$**

4.3   Parameter Sensitivity Study

An important issue that may affect the robustness of the PTLM models is the sensitivity of their parameters s (in Equation 8, 9) and σ (in Equation 6, 12). The parameter s controls the amount of self-translation probabilities. The kernel parameter σ in

Equation 6 determines the distance in which words are considered to be related. Another kernel parameter σ in Equation 12 restricts the propagation scope of a virtual document. In this section, we study how sensitive these parameters are to MAP measure.

We investigate a large range of σ (in Equation 6) from 10 to 1000. Generally, the value of σ affects the performance of all PTLM models extensively. The experimental results show that the influence of σ is collection-based. For the three PTLM models, their curves fluctuate similarly on the same collection. However, the best σ values for these PTLM models are not the same. For example, on the TREC7 collection, optimal σ value for PTLM-1 is 80, and the corresponding value for PTLM-2 is 150. Thus, the optimal values of σ depend on the proximity measures and the collections. Figure 1 plots the evaluation metrics MAP obtained by the three PTLM models with σ values ranging from 10 to 1000 on TREC7.

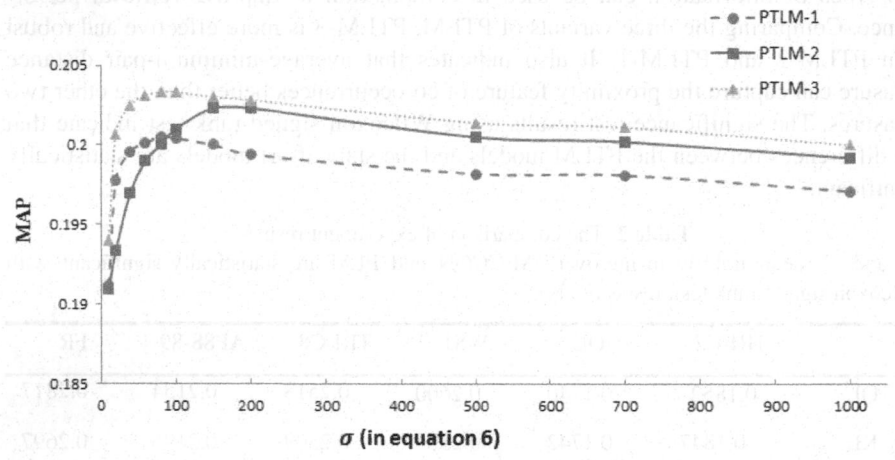

Fig. 1. PTLM-1, PTLM-2, PTLM-3 over TREC7 with σ (in Equation 6) values ranging from10 to 1000

The experimental results also show that the influence of s is collection-based. For one collection, the best s values for all PTLM models are the same. Specifically, the best s values are 0.7, 0.8, and 0.5 for TREC7, DOE, and WSJ, respectively.

In order to see how the propagation scope parameter σ (in Equation 12) affects the performance of the PTLM models, we test a set of values from 25 to 275 in increments of 25. Overall, we see that a relatively large often brings the best performance. It also seems that the performance of the PTLM models stabilizes after σ reach 175.

To investigate how the Dirichlet prior parameter μ (in Equation 10) affects the performance of the PTLM models, we also change the settings of the smoothing parameters for them. The results indicate that the optimal smoothing parameters are the same (equals to 500) for all the three PTLM models on all collections.

5   Related Work

Most existing information retrieval model including probabilistic and vector space models are based on the term independence hypothesis. Given common knowledge about language, such an assumption might seem unrealistic. To go beyond the term independency assumption in information retrieval, two main directions were investigated.

The first one considers the use of the proximity features that capture the degree to which search terms appear close to each other in a document. For example, it looks at the minimum span of the query terms appearing in the document. Term proximity, as an effective retrieval heuristic, has been studied extensively in the past few years. In these papers, various methods have been proposed to integrate proximity information into different retrieval models. Keen [9] firstly attempted to import term proximity in the Boolean retrieval model by introducing a "NEAR" operator. Buttcher et al. [3] proposed an integration of term proximity scoring into Okapi BM25 and obtain improvements on several collections. Tao et al. [14] systematically studied five proximity measures and compared their performance in various retrieval models. Zhao et al. [16] used a query term's proximity centrality as a hyper parameter in Dirichlet language model under the language modelling framework. Lv and Zhai [10] integrated the position and proximity information into the language model by defining a language model for each position within a document. Zhao et al. [17] introduce a pseudo term, namely cross term, to model term proximity for boosting retrieval model. Miao et al. [11] has attempted to incorporate proximity information into the Rocchio's model.

The second one considers the use of semantic relationships between words. Under this way, relevant words are used to enrich document or query representation.   Many studies have tried to bridge the vocabulary gap between documents and queries both based on term co-occurrences [1, 6, 13] and hand-crafted thesaurus [15]. Some other works have considered to combine both approaches [4]. Berger and Lafferty [2] firstly proposed a translation language model to corporate semantic relationship between words under the language modeling framework. To train translation models, they used synthetically generated query-document pairs. An alternation way of estimating the translation model is based on document titles [5]. Recent works have relied on document-based word co-occurrences to estimate the translation model [7][8].

6   Conclusion

Term proximity features and semantic relationships between words have proven to be two kinds of useful information to improve retrieval performance. In this paper, we proposed a positional translation language model to incorporate both of them in a unified way. In the first step, a new proximity-based method is presented to estimate the translation model. Three proximity measures are then adopted for calculating the distance score of two words within a document. The corresponding models based on these measures, PTLM-1, PTLM-2 and PTLM-3, are evaluated on six standard TREC collections. Our experiment results indicate that the PTLM models are more effective than the state-of-art models, including positional language model and translation language models. Comparing the three variants of PTLM, PTLM-3 is more effective than the other two.

Since the number of positions is much larger than the number of documents, the cost of estimating PTLMs can be extremely high. For the sake of efficiency, we use PTLM to re-rank the top 2000 documents from initial search results. However, such a strategy does not fully take advantage of the capacity of PTLM to potentially retrieve relevant documents that do not match any query word. In the future, we will try to study how to reduce the computational complexity of PTLM and to further improve retrieval performance.

Acknowledgments. This work was partially supported by the National Science Foundation of China under grants No. 60803160 and No. 61300144, the Key Projects of National Social Science Foundation of China under grant number 11ZD&189. It was partially supported by NSF of Hubei Prov. under grant number 2013CFB334, and the State Key Lab of Software Engineering Open Foundation of Wuhan University under grant number SKLSE2012-09-07.

References

1. Bai, J., Song, D., Bruza, P., Nie, J.Y., Cao, G.: Query expansion using term relationships in language models for information retrieval. In: Proceedings of the 14th ACM International Conference on Information and Knowledge Management, pp. 688–695. ACM, New York (2005)
2. Berger, A., Lafferty, J.: Information retrieval as statistical translation. In: Proceedings of the 22nd Annual International ACM SIGIR Conference on Research and Development in Information Retrieval, pp. 222–229. ACM, New York (1999)
3. Büttcher, S., Clarke, C.L.A., Lushman, B.: Term proximity scoring for ad-hoc retrieval on very large text collections. In: Proceedings of the 29th Annual International ACM SIGIR Conference on Research and Development in Information Retrieval, pp. 621–622. ACM, New York (2006)
4. Cao, G., Nie, J.Y., Bai, J.: Integrating word relationships into language models. In: Proceedings of the 28th Annual International ACM SIGIR Conference on Research and Development in Information Retrieval, pp. 298–305. ACM, New York (2005)
5. Jin, R., Hauptmann, A.G., Zhai, C.X.: Title language model for information retrieval. In: Proceedings of the 25th Annual International ACM SIGIR Conference on Research and Development in Information Retrieval, pp. 42–48. ACM, New York (2002)
6. Jing, Y., Croft, W.B.: An association thesaurus for information retrieval. In: Proceedings of RIAO, pp. 146–160 (1994)
7. Karimzadehgan, M., Zhai, C.X.: Estimation of statistical translation models based on mutual information for ad hoc information retrieval. In: Proceedings of the 33rd International ACM SIGIR Conference on Research and Development in Information Retrieval, pp. 323–330. ACM, New York (2010)
8. Karimzadehgan, M., Zhai, C.: Axiomatic analysis of translation language model for information retrieval. In: Baeza-Yates, R., de Vries, A.P., Zaragoza, H., Cambazoglu, B.B., Murdock, V., Lempel, R., Silvestri, F. (eds.) ECIR 2012. LNCS, vol. 7224, pp. 268–280. Springer, Heidelberg (2012)

9. Keen, E.M.: Some aspects of proximity searching in text retrieval systems. The Journal of Information Science 18(2), 89–98 (1992)
10. Lv, Y., Zhai, C.X.: Positional language models for information retrieval. In: Proceedings of the 32nd International ACM SIGIR Conference on Research and Development in Information Retrieval, pp. 299–306. ACM, New York (2009)
11. Miao, J., Huang, X.J., Ye, Z.: Proximity-based Rocchio's model for pseudo relevance. In: Proceedings of the 35th International ACM SIGIR Conference on Research and Development in Information Retrieval, pp. 535–544. ACM, New York (2012)
12. Ponte, J.M., Croft, W.B.: A language modeling approach to information retrieval. In: Proceedings of the 21st Annual International ACM SIGIR Conference on Research and Development in Information Retrieval, pp. 275–281. ACM, New York (1998)
13. Schütze, H., Pedersen, J.O.: A co-occurrence based thesaurus and two applications to information retrieval. Information Processing and Management 33(3), 307–318 (1997)
14. Tao, T., Zhai, C.X.: An exploration of proximity measures in information retrieval. In: Proceedings of the 30th Annual International ACM SIGIR Conference on Research and Development in Information Retrieval, pp. 295–302. ACM, New York (2007)
15. Voorhees, E.M.: Query expansion using lexical-semantic relations. In: Proceedings of the 17th Annual International ACM SIGIR Conference on Research and Development in Information Retrieval, pp. 61–69. ACM, New York (1994)
16. Zhao, J., Yun, Y.: A proximity language model for information retrieval. In: Proceedings of the 32nd International ACM SIGIR Conference on Research and Development in Information Retrieval, pp. 291–298. ACM, New York (2009)
17. Zhao, J., Huang, X., He, B.: CRTER: Using cross terms to enhance probabilistic information retrieval. In: Proceedings of the 34th International ACM SIGIR Conference on Research and Development in Information Retrieval, pp. 155–164. ACM, New York (2011)
18. Zhai, C.X., Lafferty, J.: A study of smoothing methods for language models applied to Ad Hoc information retrieval. In: Proceedings of the 24th Annual International ACM SIGIR Conference on Research and Development in Information Retrieval, pp. 334–342. ACM, New York (2001)
19. Zhai, C.X., Lafferty, J.: Model-based feedback in the language modeling approach to information retrieval. In: Proceedings of the Tenth International Conference on Information and Knowledge Management, pp. 403–410. ACM, New York (2001)
20. Zhai, C.X.: Statistical Language Models for Information Retrieval A Critical Review. Foundations and Trends in Information Retrieval 2(3), 137–213 (2008)

Topic Modeling Using Collapsed Typed Dependency Relations

Elnaz Delpisheh and Aijun An

Department of Electrical Engineering and Computer Science,
York University
Toronto, ON, Canada M3J 1P3
{elnaz,aan}@cse.yorku.ca

Abstract. Topic modeling is a powerful tool to uncover hidden thematic structures of documents. Many conventional topic models represent documents as a bag-of-words, where the important linguistic structures of documents are neglected. In this paper, we propose a novel topic model that enriches text documents with collapsed typed dependency relations to effectively acquire syntactic and semantic dependencies between consecutive and nonconsecutive words of text documents. In addition, we propose to enforce coherent topic assignments for conceptually similar words by generalizing words with their synonyms. Our experimental studies show that the proposed model and strategy outperform the original LDA model and the Bigram Topic Model in terms of perplexity; and our performance is comparable to other models in terms of stability, coherence, and accuracy.

1 Introduction

A large amount of text corpora and discrete data demands more on improving people's ability to interpret and comprehend them. Previously, texts were collected and stored in large text repositories and retrieved by a set of keywords. Documents were seldom analyzed using their themes, because there were very few technologies to extract their thematic structures. During the past decade, *topic modeling* has emerged to remedy the situation. Topic modeling is a powerful statistical tool to uncover hidden thematic structures of documents, to facilitate document summarization and organization in a variety of applications in natural language processing, vision, social network analysis, and text mining [1–3]. Most topic models consider documents to be a weighted mixture of topics, where each topic is a multinomial distribution over words. An inferred topic model of a corpus assigns high probability to members of the corpus as well as to other similar documents [1, 2]. Text documents are the only observed data in most conventional topic models. However, more recent topic models extend previous models by incorporating extra information [4]. Extra information is obtained by enriching text representation to include information, such as authors of the documents [5], images associated with the text [6], style of writing and reviewers of the documents [7]. The aforementioned topic models represent documents as

V.S. Tseng et al. (Eds.): PAKDD 2014, Part II, LNAI 8444, pp. 146–161, 2014.
© Springer International Publishing Switzerland 2014

a bag-of-words, where the order of words, thus important linguistic structures of documents are neglected [1, 2].

In order to include richer linguistic structures of text documents, many methods were proposed to incorporate local word dependencies into topic models [8–12]. Local word dependencies are either dependencies between a set of consecutive words, or a set of nonconsecutive words with arbitrary distances. For example, the term[1] *"data mining"* contains two words *"data"* and *"mining"* that are consecutively related. In addition, in sentence *"There are countries that deny human basic civil rights."*, the term *"human rights"* contains two nonconsecutive words *"human"* and *"rights"* that are syntactically related. In order to capture sequential consecutive dependencies between words, the Bigram Topic Model [8] and Topical n-gram Model [9] extend word generation by conditioning on n previous words. However, the n-gram topic models do not capture relations between nonconsecutive words.

To remedy this problem, some recent methods integrate grammatical regularities of text documents into topic models. HMM-LDA [11] uses the states of a Hidden Markov Model to represent syntactic and semantic words. Then, the model assumes that words are either sampled from topics randomly drawn from the topic mixture of the documents or from a syntactic class sampled from a distribution of associated syntactic classes [12]. Their model only considers local dependencies between variables of the syntactic states and fails to obtain syntactic or semantic dependencies between words. The Syntactic Topic Model (STM) [10] was proposed to integrate grammatical regularities in the text to detect syntactically relevant topics. In STM, documents are collections of dependency parse trees, in which words in the sentence are the nodes in the graph and grammatical regularities are the edge labels [13]. The root in the dependency parse tree is used as a governor. Topic assignment of the root node affects topic assignments of all its children. Moreover, STM does not draw words from just the document distribution over topics. Rather, it draws a word from a distribution formed by the document distribution over topics weighted by the parse tree distributions. Thus, topic assignment of a word depends on both the document's theme as well as the parents of the word in the parse tree. Although, STM improves topic modeling by combining syntactic and thematic structures of documents, it does not fully distinguish topic assignment of the words that share the same parent in the tree, i.e., children of a node. This problem specifically occurs when a root node has many children [10].

Moreover, text documents consist of words with possible conceptual similarities, called *synonyms*, defined in lexical resources like WordNet [14]. It is reasonable to expect the distribution of topics over synonymous words to be similar.

In this paper, a novel topic model is proposed to consider syntactic and semantic structures of text documents in probabilistic topic models. In essence, we enrich text documents with the *collapsed typed dependency relations* to circumvent obstacles in acquiring consecutive and nonconsecutive dependencies between words.

---

[1] A *term* consists of one or more words forming a unit of a sentence.

In addition, we investigate the influence of enforcing similar topic distribution over conceptually similar words by generalizing words with their synonyms.

The structure of this paper is as follows: In Section 2, we discuss our proposed topic model incorporated with collapsed typed dependency relations. We also explain our method for generalizing words using synonyms. Section 3 introduces some criteria to evaluate topic models. Then, it demonstrates the effectiveness of our approach through experiments. Finally, Section 4 concludes the paper with some remarks on our future work.

2 Main Contributions

In this section, we first explain the collapsed typed dependency relations and how to find them from the HPSG parse trees. These relations are used in capturing consecutive and nonconsecutive dependencies between words of text documents. We then describe our topic model and how it embodies collapsed typed dependency relations. In addition, we propose a method to enforce similar topic distribution over synonymous words of text documents. Lastly, we explain the relationship between our contributions and other related work.

2.1 Collapsed Typed Dependency Relations and HPSG Parse Trees

The bag-of-words representation of text documents is of particular interest in most topic models. However, this representation does not contain information about the relations between words. Relations could hold over a consecutive or nonconsecutive neighborhood of a word [15].

In this work, we use the collapsed typed dependency relations to acquire syntactic and semantic structures of text documents. This acquisition enables us to further capture consecutive and nonconsecutive relations between words of text documents. The collapsed typed dependency relations are extracted from typed dependency parse trees. The typed dependency parse tree of a sentence provides a tree representation of detailed grammatical relations between words in the sentence [16]. Words in the sentence are nodes of the tree and grammatical relations are the edge labels. The total number of grammatical relations that can be assigned by typed dependency parse trees is 48 [16]. Table 1 shows most common grammatical relations used in typed dependency parse trees. For more information on this set of relations, please see [13].

Typed dependency parse trees are constructed according to the *Head-Driven Phrase Structure Grammar* (HPSG). HPSG, developed by Pollard *et al.* [17], is a highly structured grammatical representation of text documents that effectively analyzes syntactic relations concerning multi-word constituents [15, 16]. The HPSG-based parse tree of a sentence starts from a root and ends in leaf nodes which represent words. Internal nodes of the tree represent syntactic roles of the connected leaf nodes. For example, Figure 1[2] represents the HPSG-based parse

---

[2] Enju is used to extract the HPSG parse tree. This parser is available at
http://www.nactem.ac.uk/enju

Table 1. Most common grammatical relations used in typed dependency parse trees, defined in de Marneffe *et al.* [13, 16]

Grammatical Relation	Definition	Example
root	It points to the root of the sentence; and acts as the root of the tree.	"I love French fries." root(ROOT, love)
amod	Adjective Modifier: An adjective that changes the meaning of the noun.	"Sam eats red meat." amod(meat, red)
rcmod	Relative Clause Modifier: A relative clause that changes the meaning of the noun.	"I saw the man you love." rcmod(man, love)
nsubj	Nominal Subject: A word that is the subject of the clause.	"Clinton defeated Dole." nsubj (defeated, Clinton)
dobj	Direct Object: A word that is the direct object of the verb.	"They win the lottery." dobj (win, lottery)
expl	Expletive: This relation captures the existential there.	"There is a ghost in the room." expl(is, There)

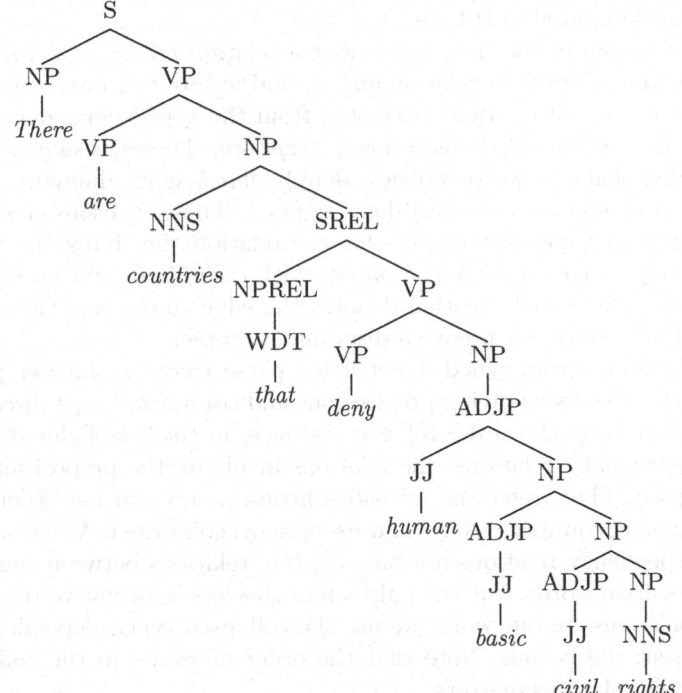

Fig. 1. The HPSG-based parse tree for the sentence *"There are countries that deny human basic civil rights."*. Abbreviations that are used in this tree are as follows: *S*: sentence; *VP*: verb phrase; *NNS*: plural noun; *SREL*: sentence relation; *NPREL*: noun phrase relation; *WDT*: wh-determiner; *ADJP*: adjective phrase; *JJ*: adjective.

tree of the sentence *"There are countries that deny human basic civil rights."*. In this tree, the leftmost branch, node *NP* represents the role of "noun phrase" for the leaf node *"There"*.

HPSG provides a high level syntactic representation of sentences in text documents [16]. However, we need to capture specific relations between every individual related pair of words. Thus, we need to elaborate HPSG to include additional labeled grammatical relations between words. This is achieved by constructing typed dependency parse trees from HPSG-based parse trees, using an algorithm described in [16]. This algorithm has two phases: dependency extraction and dependency typing. In the first phase, a sentence is parsed with a phrase structure grammar parser (HPSG). The output of this phase is arranged hierarchically and rooted with the most generic relation. In the second phase, when the relation between an internal node and its connected leaf node can be identified more precisely, more specific grammatical relations further down in the hierarchy is used. Figure 2 shows the typed dependency parse tree constructed from Figure 1 for the sentence *"There are countries that deny human basic civil rights."*. As illustrated in this figure, nonconsecutive relations between words with gaps, i.e. *"human rights"* is captured under the *amod* relation. Typed dependency parse trees are constructed using the Stanford parser toolkit that has phrase structured grammars integrated in [13, 16][3].

For each edge in the tree, we extract a relation $rel(w_i, w_j)$, where rel is the edge label representing a relation and w_i and w_j are two nodes of the edge. For example, the set of relations extracted from the typed dependency parse tree, illustrated in Figure 2, is as follows: {*expl*(are, There), *nsubj*(are, countries), *nsubj*(deny, that), *rcmod*(countries, deny), *amod*(rights, human), *amod*(rights, basic), *amod*(rights, civil), dobj(deny, rights)}. These relations enable us to better distinguish topic assignments for the relations involving the same parent. For instance, a tree including a parent with c children, will be represented by c relations, where each relation denotes the edge connecting the child and the parent. Each relation can have a discriminate topic.

The relations from typed dependency parse trees are further processed by collapsing relations involving prepositions and conjuncts to get direct dependencies between content words [16]. For instance, in the set of the aforementioned typed dependency relations, the relations involving the preposition *"that"* will be collapsed. Thus, relations *rcmod*(countries, deny) and *nsubj*(deny, that) will become *rcmod*(countries, deny) and *nsubj*(deny, countries). As a result, collapsed typed dependency relations not only capture relations between consecutive and nonconsecutive words, but they also eliminate less informative relations involving prepositions. In our work, we use the collapsed typed dependency relations to represent the corpus. Note that the order of words in the collapsed typed dependency relations matters.

2.2 Probabilistic Topic Model Using Collapsed Dependency Relations

We assume that corpus $\mathcal{D}$ consists of M documents denoted by $\mathcal{D} = \{d_1, d_2, \cdots, d_M\}$. Each document d_l contains n words denoted by $d_l = \{w_1, w_2, \cdots, w_n\}$.

---

[3] http://nlp.stanford.edu/software/lex-parser.shtml

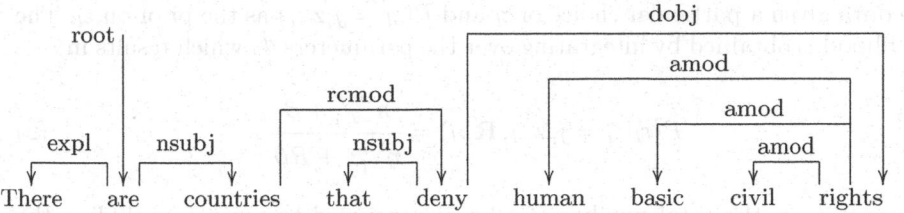

Fig. 2. The typed dependency parse tree of the sentence *"There are countries that deny human basic civil rights."*. See Table 1 for the explanation of each relation. As illustrated in this figure, the typed dependency parse tree effectively captures relations between nonconsecutive words, i.e., *dobj* relation between words *deny* and *right*.

Each document is represented by R collapsed dependency relations between words of the document, denoted by $\mathbf{R} = \{r_1, r_2, \cdots, r_R\}$. These relations are instances of the 48 grammatical relations described in Section 2.1, each of which consists of two words.

Our topic model assumes that each document d_l has a multinomial distribution over K topics with parameters $\Theta^{(d_l)}$. Thus, for a relation in document d_l, $P(z_l = j | \mathcal{D} = d_l) = \Theta_j^{(d_l)}$, where z_l denotes topic assignment to relation l. In our proposed model, the jth topic is represented by a multinomial distribution over R relations with parameters $\Phi^{(j)}$, thus $P(r_l | z_l = j) = \Phi_{r_l}^{(j)}$. Inspired from LDA [1, 2, 18], we provide a procedure to generate documents. In this procedure, each document d_l is generated by first drawing a distribution over topics ($\Theta^{(d_l)}$), generated from a Dirichlet distribution with parameter α. The relations in the document are then generated by drawing a topic j from this distribution and then drawing a relation from that topic according to a multinomial distribution with parameters ($\Phi_{r_l}^{(j)}$), generated from a Dirichlet distribution with parameter β.

Note that the only observed variables are the relations in the collection of documents. Document distribution over topics and topic distribution over relations are latent variables generated from Dirichlet distributions with parameters α and β, respectively. We use Gibbs sampling to obtain approximate estimates for the latent variables. Gibbs sampling is a simple Markov chain Monte Carlo algorithm that sequentially replaces the value of one of the latent variables by a value drawn from the distribution of that variable conditioned on the values of the remaining variables [19].

We adopt Gibbs sampling algorithm proposed by Griffiths *et al.* [2, 18] to draw a topic from the conditional distribution iteratively. For each topic j the distribution is given by

$$P(z_l = j | \mathbf{z}_{-l}, \mathbf{R}) \propto P(r_l | z_l = j, \mathbf{z}_{-l}, \mathbf{R}_{-l}) P(z_l = j | \mathbf{z}_{-l}), \qquad (1)$$

where $\mathbf{z}_{-l}$ and $\mathbf{R}_{-l}$ denote the $\mathbf{z}$ and $\mathbf{R}$ for all relations other than r_l. This expression is an instance of Bayes' rule with $P(r_l | z_l = j, \mathbf{z}_{-l}, \mathbf{R}_{-l})$ as the likelihood of the data given a particular choice of z_l and $P(z_l = j|\mathbf{z}_{-l})$ as the prior on z_l. The likelihood is obtained by integrating over the parameters Φ, which results in

$$P(r_l|z_l = j, \mathbf{z}_{-l}, \mathbf{R}_{-l}) = \frac{n_{-l,j}^{(r_l)} + \beta}{n_{-l,j}^{(\cdot)} + R\beta}, \tag{2}$$

where $n_{-l,j}^{(\cdot)}$ is the total number of relations assigned to topic j, excluding the current one, and $n_{-l,j}^{(r_l)}$ is the total number of relation r_l assigned to topic j, excluding the current one.

Similarly, the prior is calculated by integrating over the parameters Θ:

$$P(z_l = j|\mathbf{z}_{-l}) = \frac{n_{-l,j}^{(d_l)} + \alpha}{n_{-l,\cdot}^{(d_l)} + K\alpha}, \tag{3}$$

where $n_{-l,j}^{(d_l)}$ is the total number of relations from document d_l assigned to topic j, excluding the current one, and $n_{-l,\cdot}^{(d_l)}$ is the total number of relations in document d_l, excluding the current one.

Then, the conditional distribution for the topic assignments is given by

$$P(z_l = j|\mathbf{z}_{-l}, \mathbf{R}) \propto \frac{n_{-l,j}^{(r_l)} + \beta}{n_{-l,j}^{(\cdot)} + R\beta} \frac{n_{-l,j}^{(d_l)} + \alpha}{n_{-l,\cdot}^{(d_l)} + K\alpha}. \tag{4}$$

2.3   Generalizing Words Using Synonyms

Text documents often contain words that are synonyms. Sets of synonyms can be obtained from lexical resources like WordNet [14]. In this work, we investigate the influence of generalizing words using a synonym on topic modeling.

Similar to LDA [1], we assume that a document is a multinomial distribution over K topics, where each topic is a multinomial distribution over N words. We also assume that documents are represented by a sequence of words, denoted by $\mathbf{W} = \{w_1, w_2, \cdots, w_N\}$, where $w_n \in \mathbf{W}$ is the nth word in the sequence. Given the fact that a set of synonyms shares a similar concept, it is reasonable to expect them to have similar probabilities under topics. For example, if a text document is about happiness, the inferred topic should assign higher probabilities to words such as *delighted, blessed*, and *prosperity*; and lower probabilities to words such as *sad, bitter*, and *sorrow*. In order to ensure that topics are similarly distributed over synonyms, we propose the following algorithm to replace all synonyms of a word with an equivalent synonym with the highest frequency in WordNet:

1. Group the words from WordNet, based on their conceptual similarities. Each group will contain a set of synonyms.
2. For each group, find the frequency of the words in the group. The frequency of a word is the number of occurrences of the word in WordNet.
3. Select the most frequent word in the group as the *group representative*.

4. For each $w_i \in \mathbf{W}$:
   - Look for a group where w_i belongs to.
   - If a group is found, replace w_i with the group representative, found in Step 3;
   - else, leave the word as is.

For example, consider a text document that contains the word *prosperous*. This word belongs to the following group of synonyms { *delighted, blessed, prosperous, happy, fortunate* }. Our algorithm finds the frequency of each synonym in WordNet. It selects happy as the group representative because it is the most frequent word in the group. Finally, our algorithm replaces the word *prosperous* with the word *happy*.

2.4 Relationships to Other Work

In this work, we go beyond the bag-of-words representation of documents to incorporate syntax and semantics of text documents into topic models. This section reviews the theoretical relationships of our contributions with previous topic models that used syntactic and semantic structures of texts.

Our proposed topic model is similar to STM [10] due to using typed dependency trees to represent syntactic structures of sentences. However, our topic model has following major differences with STM. Firstly, STM draws a word from a single distribution formed by the document distribution over topics weighted by the parse tree distributions. Thus, topic assignment of a word depends on both the document's theme as well as the parent of the word in the parse tree. However, in our model we use two distributions: document distribution over topics and topic distribution over the collapsed dependency relations. We first draw a distribution over topics; then, we select a topic from this distribution and then draw a relation from that topic distribution over the collapsed dependency relations. Secondly, STM does not fully distinguish topic assignments of the words that share the same parent in the dependency parse tree, i.e., children of a node, as stated by Boyd-Graber *et al.* [10]. However, in our model each pair of related nodes in the parse tree introduces a discriminate relation. Thus, topic assignment to the relations involving the same parent is better distinguished. Thirdly, STM does not use labeled dependency relations and lexicalization. However, our model uses the labels of dependency relations to distinguish and further collapse relations involving prepositions and conjuncts to get direct dependencies between content words. Finally, STM computes the posterior topic distributions by Bayesian variational methods. Our model uses Gibbs sampling to infer posterior topic distributions. This final difference is complementary rather than competitive.

In addition, our proposed topic model differs from the n-gram topic models [8] in capturing dependencies between words of a sentence. Our topic model considers dependencies between nonconsecutive words with a distance; while the n-gram topic model is limited to capturing dependencies between consecutive words.

Moreover, our proposed model, uses WordNet to enforce topic similarity for words with conceptual similarities, by generalizing similar words with their synonyms. Lexical resources, i.e. WordNet, were previously used in topic models. Musat *et al.* [20] employs WordNet to improve topic models by removing unrelated words from the simplified topic descriptions. Mei *et al.* [21] used WordNet to label each topic in a multinomial topic model. Newman *et al.* [22] uses WordNet to evaluate topic coherence. None of them uses synonyms to generalize words prior to building topic models.

3   Experiments

We conducted experiments on two text corpora to compare the performance of four following topic models: LDA [1], LDA on generalized words using synonyms, explained in Section 2.3, the Bigram Topic Model [8], and the HPSG Topic Model, explained in Section 2.2[4]. The first three topic models were trained with 1000 iterations of Gibbs sampling [2, 18] used in the MALLET [23]. However, the HPSG Topic Model was trained with 1000 iterations of Gibbs sampling. Initial values for the hyperparameters (α, β) applied to all our experiments were $\alpha = 50.0$ and $\beta = 0.01$. Note that these parameters are default parameters of the MALLET [23].

In our experiments we used Associated Press corpus[5] that consists of 2246 Associated Press articles, 33872 words, and 454370 collapsed typed dependency relations. In addition, we used Reuters-21578 Distribution 1.0[6] that includes 22 files. Each of the first 21 files contain 1000 documents, while the last file contains 578 documents. This corpus contains a total number of 43012 words and 793345 collapsed typed dependency relations.

Table 2 illustrates top 10 terms of the most probable topics generated by aforementioned topic models on the Reuters corpus. The first column shows the words generated by LDA. Some words in this topic are ambiguous and can have multiple meanings. To identify the correct meaning of each word, one needs to consider other words in the topic. For example, the word *"share"* has many meanings. Observing other words in the topic, such as *"bank"* and *"profit"*, helps to identify the correct meaning of the word *"share"* that is *"assets belonging to an individual"*. The second column shows the results of LDA on generalized words using synonyms. These words are similar to the words in the first column and still suffer from ambiguity. The terms generated by the Bigram Topic Model and the HPSG Topic Model are shown in columns three and four, respectively. These topic models have less ambiguity, given the fact that they generate terms that include pairs of words that are more descriptive than single words. In addition,

---

[4] Section 2.4 provides a theoretical comparison between our proposed probabilistic topic model and STM [10]. Given the fact that the source codes of STM was not available prior to the submission of this paper, experimental comparisons with this method will be done in our future work.

[5] http://www.cs.princeton.edu/~blei/lda-c

[6] http://www.research.att.com/~lewis

Table 2. Top 10 terms of the most probable topic, generated by four topic models: LDA, LDA on generalized words using synonyms, the Bigram Topic Model, and the HPSG Topic Model from Reuters corpus

LDA	LDA on generalized words using synonyms	the Bigram Topic Model	the HPSG Topic Model
bank	financial	reconstruction plans	money funds
profit	international	debt repayment	overseas investments
foreign	net	private institute	raising stake
share	government	traders reported	foreign deposits
federal	billion	existing research	commercial banks
japanese	withdraw	payments improve	buyout transaction
policy	currency	banking office	lack assets
rates	rise	borrowing occurred	stock exchange
money	sale	federal supervisory	account balance
shares	february	bank consultancies	bank regulation

as opposed to the Bigram Topic Model, terms generated by the HPSG Topic Model are not only limited to consecutive pairs of words of a sentence, but they also contain pairs of related words with gaps.

Given the text corpora, we compare our work with other topic models based on the following criteria:

- High likelihood on a held-out test set (perplexity) [1].
- Stable distribution of topics over words across samples [5].
- Coherent distribution of words learned by individual topics [22].
- Accurate distribution of topics over words.

These criteria and experimental results are discussed in the subsequent sections.

3.1   Perplexity

Perplexity is the most common criterion to evaluate the quality of topic models [24]. Perplexity measures the cross-entropy between the word distribution learned by the topic model and the distribution of words in an unseen test document. Thus, lower perplexity score indicates that the model is better in predicting distribution of the test document [1, 25]. We evaluate perplexity as a function of number of topics for both Associated Press and Reuters corpora. We trained the topic models on 90% of the corpus to estimate the held out probability of previously unseen 10% of the corpus. We compute the perplexity of the held-out test set with respect to the HPSG Topic Model by

$$perplexity(\mathbf{D}_{test}) = exp\left(-\frac{\sum_{d=1}^{M} logP(\mathbf{R}_d)}{\sum_{d=1}^{M} |\mathbf{R}_d|}\right), \qquad (5)$$

where $\mathbf{D}_{test}$ is the test corpus with M documents, $\mathbf{R}_d$ denotes the set of collapsed typed dependency relations in document d, $|\mathbf{R}_d|$ is the total number of collapsed

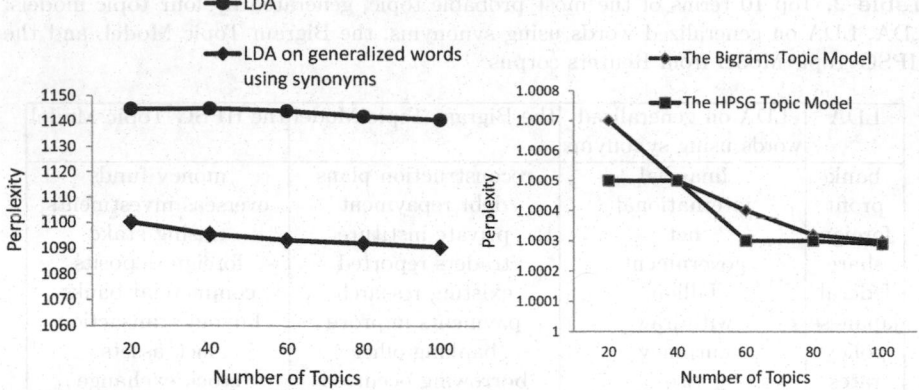

Fig. 3. Perplexity as a function of number of topics, using LDA, and LDA on generalized words using synonyms on Association Press corpus

Fig. 4. Perplexity as a function of number of topics, using the Bigram Topic Model, and the HPSG Topic Model on Association Press corpus

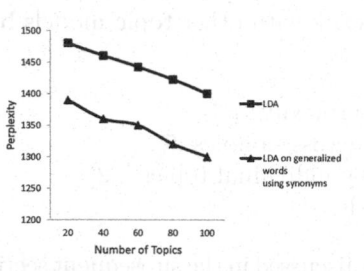

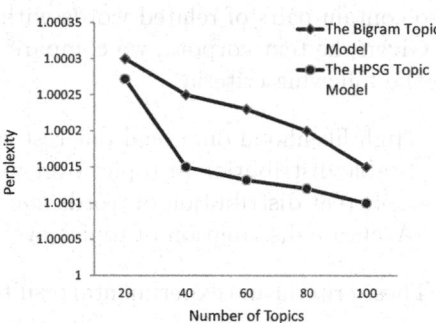

Fig. 5. Perplexity as a function of number of topics, using LDA, and LDA on generalized words using synonyms on Reuters corpus

Fig. 6. Perplexity as a function of number of topics, using the Bigram Topic Model, and the HPSG Topic Model on Reuters corpus

typed dependency relations in document d, and $P(\mathbf{R}_d)$ is the probability estimate assigned to $\mathbf{R}_d$ by the HPSG topic model. The perplexity of $\mathbf{D}_{test}$ by other topic models, such as LDA, is defined similarly, except that $\mathbf{R}_d$ is replaced by $\mathbf{W}_d$, the set of words in the corpus.

The results are illustrated in Figures 3, 4, 5, 6. The x-axis shows the number of topics (K) used in each model; the y-axis shows the perplexity. These figures clearly indicate that the perplexity of our proposed topic model drastically decreases the perplexity of LDA and LDA on generalized words using synonyms. Moreover, the perplexity of our proposed topic model is slightly better than the perplexity of the Bigrams Topic Model.

Table 3. Topic stability across two different runs of the HPSG Topic Model on Reuters corpus

Table 4. Topic stability across two different runs of LDA on Reuters corpus

Topics from sample 1	Best aligned topics from sample 2	Best KL	Topics from sample 1	Best aligned topics from sample 2	Best KL
Topic 1	Topic 14	0.834	Topic 1	Topic 5	0.821
Topic 2	Topic 20	1.630	Topic 2	Topic 12	1.073
Topic 3	Topic 13	0.835	Topic 3	Topic 8	0.533
Topic 4	Topic 3	0.730	Topic 4	Topic 19	0.721
Topic 5	Topic 11	0.454	Topic 5	Topic 3	1.031
Topic 6	Topic 18	0.951	Topic 6	Topic 18	1.050
Topic 7	Topic 19	0.450	Topic 7	Topic 7	0.836
Topic 8	Topic 18	0.760	Topic 8	Topic 8	0.754
Topic 9	Topic 15	0.420	Topic 9	Topic 15	0.428
Topic 10	Topic 13	0.939	Topic 10	Topic 13	0.765
Topic 11	Topic 5	0.526	Topic 11	Topic 7	0.818
Topic 12	Topic 17	0.439	Topic 12	Topic 8	0.798
Topic 13	Topic 12	0.953	Topic 13	Topic 6	0.961
Topic 14	Topic 7	1.053	Topic 14	Topic 5	0.764
Topic 15	Topic 6	1.013	Topic 15	Topic 12	1.161
Topic 16	Topic 14	1.139	Topic 16	Topic 8	0.867
Topic 17	Topic 5	1.041	Topic 17	Topic 6	0.791
Topic 18	Topic 9	1.172	Topic 18	Topic 4	0.921
Topic 19	Topic 10	1.026	Topic 19	Topic 18	1.064
Topic 20	Topic 17	1.226	Topic 20	Topic 8	1.091
Average		0.87955	Average		0.8624

3.2 Stability

Stability is the similarity of topic distributions over words across different samples [5]. We follow the algorithm proposed by Rosen-Zvi et al. [5] to find the best one-to-one topic alignment across samples. The algorithm finds the best aligned topic pair by calculating $min_{j=1,\cdots,K} d(S_1, S_2)$, where $d(S_1, S_2)$ denotes symmetrized Kullback Leibler (KL) divergences between the K topic distributions over relations from samples S_1 and S_2. KL divergence is calculated by $d(S_1, S_2) = \sum_{x \in X} S_1(x) log(S_1(x)/S_2(x))$, where X represents the set of relations in the samples [26]. We compare the stability of topic distributions over relations across samples, generated by the HPSG Topic Model and LDA on the Reuters corpus. The results, illustrated in Tables 3 and 4, show that our proposed topic model is comparably as stable as LDA in producing similar topic distributions over words across multiple samples. Similar results were obtained using the Bigram Topic Model.

3.3 Topic Coherence

Topic coherence measures the integrity or coherence of a set of words generated by a topic model. Words generated by topic T, denoted by $T = \{w_1, w_2, \cdots, w_n\}$,

Table 5. The average topic coherence of top 50 words of 20 topics generated from Reuters corpus

Topic model	Coherence
LDA	41.35
LDA on generalized words using synonyms	41.68
the Bigram Topic Model	39.18
the HPSG Topic Model	39.79

Table 6. The average accuracy of topic distribution over words from a subset of topic-labeled Reuters corpus

Topic model	Accuracy
LDA	0.225
LDA on generalized words using synonyms	0.220
the Bigram Topic Model	0.221
the HPSG Topic Model	0.223

are coherent if they are semantically similar. In order to calculate the topic coherence score, we adopted the method proposed by Newman *et al.* [22]. We calculate the semantic similarity scores between every pair of words in a topic using the Lesk algorithm [27][7]. Then, we compute their arithmetic means. We compared the topic coherence of top 50 words from 20 topics generated by LDA, LDA on generalized words using synonyms, the Bigram Topic Model, and the HPSG Topic Model on Reuters corpus. The results are shown in Table 5. The HPSG Topic Model generates slightly more coherent topic distributions over words than The Bigram Topic Model. The HPSG Topic Model performs comparable to LDA in topic coherence. However, LDA on generalized words using synonyms results in more coherent topic distribution over words. This coherence is due to the fact that we replaced conceptually related words with one general word, prior to modeling the topic assignments.

3.4 Accuracy

The accuracy of a topic model is the degree of closeness of the topic distribution over words of a test corpus to actual topic distribution over words of a topic-labeled corpus. Note that calculating accuracy depends on the availability of the topic-labeled corpus.

We assume that the test corpus $\mathcal{T}$ consists of M documents $\mathcal{T} = \{d_1, d_2, \cdots, d_M\}$. Each document consists of H actual topic labels, denoted by $L = \{l_1, l_2, \cdots, l_H\}$, where each $l_i \in L$ represents an actual topic-label for the document. As mentioned earlier, a topic model generates K topics, where each topic is a distribution over n words, denoted by $T = \{w_1, w_2, \cdots, w_n\}$. The accuracy score of the topic model is calculated by computing $Accuracy = \frac{\sum_{i=1}^{M} min_{j=1,\cdots,K} d(T_j, L)}{M}$, where $d(T_j, L)$ denotes the semantic similarity between two sets of T_j and L. This semantic similarity is measured using the Lesk algorithm, explained in Section 3.3.

We compared the accuracy of LDA, LDA on generalized words using synonyms, the Bigram Topic Model, and the HPSG Topic Model on a subset of

---

[7] The Lesk algorithm uses dictionary definitions of two words in a pair and counts the number of words that are shared between two definitions. The more overlapping the definitions are, the more related the words are. The Lesk toolkit is available at http://text-similarity.sourceforge.net

Reuters corpus that contains topic-labeled documents. As illustrated in Table 6, these algorithms are comparable in terms of accuracy. However, LDA is slightly better.

4   Conclusions

We proposed a novel method that incorporates syntactic and semantic structures of text documents into probabilistic topic models. This representation has several benefits. It captures relations between consecutive and nonconsecutive words of text documents. In addition, the labels of the collapsed typed dependency relations help to eliminate less important relations, i.e., relations involving prepositions. Also, words of text documents, regardless of their parents in the collapsed typed dependency parse trees, are distinguished in topic assignment. Furthermore, our experimental studies show that the proposed topic model significantly outperforms LDA and is also better than the Bigram Topic Model in terms of perplexity. We also show that our model achieves comparable results with other models in terms of stability, coherence, and accuracy. Besides, the results from our topic model have less ambiguity, given the fact the generated terms include pairs of words that are more descriptive than single words.

Moreover, we introduced a method to enforce topic similarity to conceptually similar words. As a result, this algorithm led to more coherent topic distribution over words.

In the future, we will extend our topic model to effectively capture more dependencies between words in sentence or document levels. In addition, we will investigate the influence of the order of words in the collapsed typed dependency relations.

Acknowledgement. This project is funded in part by the Center for Information Visualization and Data Driven Design (CIV/DDD) established by the Ontario research fund.

References

1. Blei, D.M., Ng, A.Y., Jordan, M.I.: Latent dirichlet allocation. J. Mach. Learn. Res. 3, 993–1022 (2003)
2. Griffiths, T.L., Steyvers, M.: Finding scientific topics. Proceedings of the National Academy of Sciences of the United States of America 101, 5228–5235 (2004)
3. Su, H., Tang, J., Hong, W.: Learning to diversify expert finding with subtopics. In: Tan, P.-N., Chawla, S., Ho, C.K., Bailey, J. (eds.) PAKDD 2012, Part I. LNCS, vol. 7301, pp. 330–341. Springer, Heidelberg (2012)
4. Andrzejewski, D.M.: Incorporating Domain Knowledge in Latent Topic Models. PhD thesis, University of Wisconsin-Madison, USA (2010)
5. Rosen-Zvi, M., Chemudugunta, C., Griffiths, T., Smyth, P., Steyvers, M.: Learning author-topic models from text corpora. ACM Trans. Inf. Syst. 28, 4:1–4:38 (2010)

6. Blei, D.M., Jordan, M.I.: Modeling annotated data. In: Proceedings of the 26th Annual International ACM SIGIR Conference on Research and Development in Informaion Retrieval, SIGIR 2003, pp. 127–134. ACM, New York (2003)
7. Mimno, D., McCallum, A.: Expertise modeling for matching papers with reviewers. In: Proceedings of the 13th ACM SIGKDD International Conference on Knowledge Discovery and Data Mining, KDD 2007, pp. 500–509. ACM, New York (2007)
8. Wallach, H.M.: Topic modeling: Beyond bag-of-words. In: NIPS 2005 Workshop on Bayesian Methods for Natural Language Processing (2005)
9. Wang, X., McCallum, A., Wei, X.: Topical n-grams: Phrase and topic discovery, with an application to information retrieval. In: Proceedings of the 2007 Seventh IEEE International Conference on Data Mining, ICDM 2007, pp. 697–702. IEEE Computer Society, Washington, DC (2007)
10. Boyd-Graber, J.L., Blei, D.M.: Syntactic topic models. CoRR abs/1002.4665 (2010)
11. Griffiths, T.L., Steyvers, M., Blei, D.M., Tenenbaum, J.B.: Integrating topics and syntax. In: Advances in Neural Information Processing Systems 17, pp. 537–544. MIT Press (2005)
12. Gruber, A., Rosen-zvi, M., Weiss, Y.: Hidden topic markov models. In: Proceedings of Artificial Intelligence and Statistics (2007)
13. de Marnee, M.C., Manning, C.D.: Stanford typed dependencies manual (2012)
14. Miller, G.A.: Wordnet: a lexical database for english. Commun. ACM 38, 39–41 (1995)
15. Levine, R.D., Meurers, W.D.: Head-driven phrase structure grammar: Linguistic approach, formal foundations, and computational realization. Elsevier, Oxford (2006)
16. de Marneffe, M.C., MacCartney, B., Manning, C.D.: Generating typed dependency parses from phrase structure parses. In: Proc. Intl. Conf. on Language Resources and Evaluation LREC, pp. 449–454 (2006)
17. Pollard, C., Sag, I.A.: Information-based syntax and semantics: Vol. 1: Fundamentals, Stanford, CA, USA. Center for the Study of Language and Information (1988)
18. Griffiths, T.: Gibbs sampling in the generative model of latent dirichlet allocation. Standford University 518, 1–3 (2002)
19. Bishop, C.M.: Pattern Recognition and Machine Learning (Information Science and Statistics). Springer-Verlag New York, Inc., Secaucus (2006)
20. Musat, C., Velcin, J., Rizoiu, M.-A., Trausan-Matu, S.: Concept-based topic model improvement. In: Ryżko, D., Rybiński, H., Gawrysiak, P., Kryszkiewicz, M. (eds.) Emerging Intelligent Technologies in Industry. SCI, vol. 369, pp. 133–142. Springer, Heidelberg (2011)
21. Mei, Q., Shen, X., Zhai, C.: Automatic labeling of multinomial topic models. In: Proceedings of the 13th ACM SIGKDD International Conference on Knowledge Discovery and Data Mining, KDD 2007, pp. 490–499. ACM, New York (2007)
22. Newman, D., Lau, J.H., Grieser, K., Baldwin, T.: Automatic evaluation of topic coherence. In: Human Language Technologies: The 2010 Annual Conference of the North American Chapter of the Association for Computational Linguistics, HLT 2010, pp. 100–108. Association for Computational Linguistics, Stroudsburg (2010)
23. McCallum, A.K.: Mallet: A machine learning for language toolkit (2002), http://mallet.cs.umass.edu
24. Jurafsky, D., Martin, J.H.: Speech and Language Processing: An Introduction to Natural Language Processing, Computational Linguistics, and Speech Recognition, 1st edn. Prentice Hall PTR, Upper Saddle River (2000)

25. Chemudugunta, C., Holloway, A., Smyth, P., Steyvers, M.: Modeling documents by combining semantic concepts with unsupervised statistical learning. In: Sheth, A.P., Staab, S., Dean, M., Paolucci, M., Maynard, D., Finin, T., Thirunarayan, K. (eds.) ISWC 2008. LNCS, vol. 5318, pp. 229–244. Springer, Heidelberg (2008)
26. Bigi, B.: Using kullback-leibler distance for text categorization. In: Sebastiani, F. (ed.) ECIR 2003. LNCS, vol. 2633, pp. 305–319. Springer, Heidelberg (2003)
27. Banerjee, S., Pedersen, T.: An adapted lesk algorithm for word sense disambiguation using wordnet. In: Gelbukh, A. (ed.) CICLing 2002. LNCS, vol. 2276, pp. 136–145. Springer, Heidelberg (2002)

An Iterative Fusion Approach to Graph-Based Semi-Supervised Learning from Multiple Views

Yang Wang[1,3], Jian Pei[2], Xuemin Lin[1], Qing Zhang[3,1], and Wenjie Zhang[1]

[1] The University of New South Wales, Sydney, Australia
{wangy,lxue,zhangw}@cse.unsw.edu.au
[2] Simon Fraser University, Canada
[3] Australian E-Health Research Center
jpei@cs.sfu.ca, qing.zhang@csiro.au

Abstract. Often, a data object described by many features can be naturally decomposed into multiple "views", where each view consists of a subset of features. For example, a video clip may have a video view and an audio view. Given a set of training data objects with multiple views, where some objects are labeled and the others are not, *semi-supervised learning with graphs from multi-views* tries to learn a classifier by treating each view as a similarity graph on all objects, where edges are defined by the similarity on object pairs based on the view attributes. Labels and label relevance ranking scores of labeled objects can be propagated from labeled objects to unlabeled objects on the similarity graphs so that similar objects receive similar labels. The state-of-the-art, one-combo-fits-all methods linearly and independently combine either the metrics or the label propagation results from multi-views and then build a model based on the combined results. However, the similarities between various objects may be manifested differently by different views. In such situations, the one-combo-fits-all methods may not perform well. To tackle the problem, we develop an iterative Semi-Supervised Metric Fusion (SSMF) approach in this paper. SSMF fuses metrics and label propagation results from multi-views iteratively until the fused metric and label propagation results converge simultaneously. Views are weighted dynamically during the fusion process so that the adversary effect of irrelevant views, identified at each iteration of fusion process, can be reduced effectively. To evaluate the effectiveness of SSMF, we apply it on multi-view based and content based image retrieval and multi-view based multi-label image classification on real world data set, which demonstrates that our method outperforms the state-of-the-art methods.

Keywords: Graph-based semi-supervised learning, multiple views.

1 Introduction

Semi-supervised learning with graphs [10] is an important and effective approach, which propagates limited label information to unlabeled data objects on a similarity graph. A similarity graph uses the set of objects as vertices, and links edges based on the similarity between objects. Edges in a similarity graph may

V.S. Tseng et al. (Eds.): PAKDD 2014, Part II, LNAI 8444, pp. 162–173, 2014.

take similarity scores as weights. After *label propagation* [10] or *manifold ranking* [9] in a similarity graph, the more similar two objects, the more likely they have similar labels or the similar label relevance ranking scores. This property is called *local smoothness* [8]. The labeled objects iteratively propagate the label information or label relevance ranking scores to unlabeled ones via graph edges until convergence, and the final labeling result based on the label relevance scores should be consistent to the initial label information, which is called *global consistency* [8].

Often, a data object described by many features can be naturally decomposed into multiple "views", where each view consists of a subset of features. For example, an image may have a color view and a shape view. Given a set of training data objects with multi-views, where some objects are labeled and the others are not, *semi-supervised learning with graphs from multi-views* tries to learn a classifier by incorporating the complementary information from multi-views. More often than not, the similarities between various objects may be manifested differently by different views. In such situations, the one-combo-fits-all methods [3,5,6] may not perform well, since they use the same linear fusion from multi-views for all objects. Moreover, different views in such methods don't collaborate with each other to achieve consistency when performing fusion process.

To tackle the problem, in this paper, we develop an iterative fusion approach, called SSMF (for semi-supervised metric fusion and cross-view label propagation). SSMF fuses metrics and label propagation results from multi-views iteratively until the fused metric and label propagation results converge simultaneously. Views are weighted dynamically during the fusion process so that the adversary effect of irrelevant views can be reduced effectively. Here, the similarity in an irrelevant view contributes negatively to the similarity measurement matching the ground truth. Specifically, in each iteration, there are two steps. In the *semi-supervised metric fusion* step, for each view we form a fused metric by combining the current metric of the view and the label propagation results from other views. Unlike the methods in [2,4] that obtain a fused metric from multi-views without label information, the metric fusion step in our method fully utilizes the label information from all views. In the *label propagation* step, in each view we conduct label propagation using the fused metric. This step incorporates the complementary information from other views rather than from a single view only. Our SSMF method iteratively conducts the two steps until convergence.

The critical idea here is that the metric fusion and cross-view label propagation processes are complementary to each other. Moreover, we fuse the similarity matrix from one view and label the relevance matrix from other views to yield a cross-view based query (label) driven similarity matrix.

Contributions. Our major contributions can be summarized as follows.

1. We develop an iterative fusion approach SSMF in this paper. SSMF fuses metrics and label propagation results from multi-views iteratively until the fused metric and label propagation results converge simultaneously. We prove the convergence in SSMF theoretically.

2. To further improve the performance of SSMF, we extend it to WSSMF, a novel strategy that automatically generates different weight parameters to views in the fusion process. WSSMF effectively addresses the problem of irrelevant views that are undesirable to fuse in the fusion process for each iteration.

3. Our comprehensive experiments on real image data sets show that our techniques significantly outperform the state-of-the-art methods in terms of accuracy evaluated by varied metrics.

2   Related Work

To our best knowledge, our proposed technique is the first co-training based method for multi-view and graph-based semi-supervised learning problem. Existing one-combo-fits-all methods linearly and independently combine either the metric (kernel) or the labeling propagation result from multiple views to yield a better performance than single view paradigms, as introduced in section 1. Wang *et al.* [4] proposed a related Unsupervised based Metric Fusion (UMF for short) method. However, it fuses equal weight as suggested by UMF. Unlike the adapted UMF that fuses the pair-wise similarity metric information, which cannot utilize the graph structure to evaluate the similarity between pair-wise objects. SSMF fuses label propagation and similarity metric information interactively for each view and at each iteration, the label propagation can be regarded as a variant of graph random walk. Wang *et al.* [7] proposed another metric fusion technique against multi-view data via a cross-view based graph random walk approach, however, they studied the unsupervised case rather than semi-supervised learning studied in this paper.

3   SSMF

In this section, we present SSMF and describe its two nice properties, namely *global consistency and local smoothness* [8]. We first review the preliminaries. Then, we discuss SSMF using two views. Last, we present the general iterative form of SSMF with multi-views.

3.1   Preliminaries

Let $\mathbf{X} = \{x_1, x_2, \cdots, x_n\}$ be a set of data points from $\mathbf{M}$ views, we construct $\mathbf{M}$ graphs each using a different feature. $\mathbf{G}^g$ denotes a k-NN graph constructed on $\mathbf{X}$ using g-th feature. Specifically, $\mathbf{G}^g$ is constructed by connecting every two vertices x_i and x_j if one is among the k nearest neighbors of the other. Here, the nearest neighbors are computed using Euclidean distance between the g-th feature vectors of the images. The Euclidean distance between the g-th feature vectors of x_i and x_j is denoted as $||x_i, x_j||_g$. $\mathbf{W}_g$ denotes the edge affinity matrix of $\mathbf{G}^g$. Each entry $\mathbf{W}_g(i,j)$ in $\mathbf{W}_g$ represents the similarity between x_i and x_j

according to the g-th feature vector. $\mathbf{W}_g(i,j)$ is defined by a Gaussian kernel and is set to

$$\mathbf{W}_g(i,j) = exp(-||x_i, x_j||_g^2/2\sigma^2) \qquad (1)$$

if there is an edge in $\mathbf{G}^g$ between x_i and x_j. Otherwise, $\mathbf{W}_g(i,j)$ is zero. $\mathbf{D}_g$ is the diagonal matrix of $\mathbf{G}^g$ where each element $\mathbf{D}_g(i,i)$ is defined as $\mathbf{D}_g(i,i) = \sum_{j=1}^{n} \mathbf{W}_g(i,j)$.

Without loss of generality, assume the first m points x_i ($i = 1, 2, \ldots, m$) are labeled points and the remaining points are unlabeled. Let the number of labels be c, and $\mathbf{L} \in \mathbb{R}^{n \times c}$ be the relevance labeling matrix with $\mathbf{L}(i,j) = 1$, if x_i is labeled by label j, denoted by $\mathbf{L}(x_i) = j$ ($1 \le j \le c$), and 0 otherwise. Here, we assume each point is associated with a single class label from the label set. Similarly, let $\mathbf{R}_g \in \mathbb{R}^{n \times c}$ be the relevance score of unlabeled point x_u belonging to label j regarding the g-th view. The closed form of optimal $\mathbf{R}_g$ is yielded by minimizing the objective function

$$\mathbf{F}(\mathbf{R}_g) = \frac{1}{2}(\sum_{i,j=1}^{n} \mathbf{W}_g(i,j)(\frac{1}{\sqrt{\mathbf{D}_g(i,i)}}(\mathbf{R}_g(i,\cdot)) \qquad (2)$$

$$-\frac{1}{\sqrt{\mathbf{D}_g(j,j)}}(\mathbf{R}_g(j,\cdot))^2 + \alpha_g \sum_{i=1}^{n}(\mathbf{R}_g(i,\cdot) - \mathbf{L}(i,\cdot))^2)$$

where $\mathbf{R}_g(i,\cdot)$ and $\mathbf{L}(i,\cdot)$ are the i-th row of $\mathbf{R}_g$ and $\mathbf{L}$, respectively. The first term in the right hand side of Eq. (2) represents the *local smoothness*, which means that $\mathbf{R}_g(i,\cdot)$ is similar to $\mathbf{R}_g(j,\cdot)$ if x_i and x_j are proximate to each other. The second term in Eq. (2) represents the *global consistency*, which means that the final labeling matrix $\mathbf{R}_g$ should be consistent to the initial labeling matrix $\mathbf{L}$. We minimize $\mathbf{F}(\mathbf{R}_g)$ by setting $\frac{\partial \mathbf{F}(\mathbf{R}_g)}{\partial \mathbf{R}_g} = 0$, and have

$$\mathbf{R}_g^\star = (I - \alpha_g \mathbf{S}_g)^{-1}\mathbf{L} \qquad (3)$$

where $\mathbf{S}_g = \mathbf{D}_g^{-\frac{1}{2}}\mathbf{W}_g\mathbf{D}_g^{-\frac{1}{2}}$, $\mathbf{D}_g$ is the diagonal matrix with the i-th diagonal element $\mathbf{D}_g(i,i) = \sum_{j=1}^{n} \mathbf{W}_g(i,j)$, and α_g is a real value such that $0 < \alpha_g < 1$. $\mathbf{R}_g^\star$ can also be regarded as the label propagation result on $\mathbf{G}_g$.

3.2   SSMF for Two Views

Instead of directly computing the similarity metric between any pair-wise points under unsupervised scenario [4], we achieve the similarity, under semi-supervised scenario, by indirectly measuring relevance between each point and all labels, formulated as labeling relevance matrix. As such, one can imagine that if both data points have large relevance regarding all labels, their similarity is large, otherwise, it is small. In order to learn semi-supervised metric regarding two views, we need to consider the following two challenges. That is, (1) the learned similarity metric should encode the relevance between data points and all labels. (2) the learned metric should well incorporate the complementary information from two views to achieve the consistency. Assume $\mathbf{W}_g^{[t+1]}$ ($g = 1, 2$) denote the metric similarity matrix for g-th view in $t + 1$ iterations, then we define the following semi-supervised fusion strategy:

$$\mathbf{W}_1^{[t+1]} = \mathbf{Q}_1^{[t]}\mathbf{Rn}_2^{[t]}(\mathbf{Rn}_2^{[t]})^T(\mathbf{Q}_1^{[t]})^T + \lambda I \tag{4}$$

$$\mathbf{W}_2^{[t+1]} = \mathbf{Q}_2^{[t]}\mathbf{Rn}_1^{[t]}(\mathbf{Rn}_1^{[t]})^T(\mathbf{Q}_2^{[t]})^T + \lambda I \tag{5}$$

where $\mathbf{Q}_g^{[t]}$ ($g = 1, 2$) is the normalized affinity matrices such that $\mathbf{Q}_g^{[t]}(i, j) = \frac{\mathbf{W}_g^{[t]}(i,j)}{\sum_{j=1}^n \mathbf{W}_g^{[t]}(i,j)}$, $\mathbf{Rn}_g^{[t]}(i, j) = \frac{\mathbf{R}_g^{[t]}(i,j)}{\sum_{j=1}^c \mathbf{R}_g^{[t]}(i,j)}$, the goal of using normalized form is to avoid the huge difference in scale of the label relevance matrices in different views. I is identity matrix, and λI is incorporated to make SSMF robust to the noise. To better explain the above fusion strategy, we take Eq. (4) as an example of refining the metric for the first view by applying SSMF.

Intuition. We divide the right-hand-side of Eq. (4) into two parts, as $\mathbf{Q}_1^{[t]}\mathbf{Rn}_2^{[t]}$, and its transpose $(\mathbf{Rn}_2^{[t]})^T(\mathbf{Q}_1^{[t]})^T$, we study each entry of $\mathbf{Q}_1^{[t]}\mathbf{Rn}_2^{[t]}$ for any iteration t

$$(\mathbf{Q}_1^{[t]}\mathbf{Rn}_2^{[t]})(i, y) = \sum_{m=1}^n \mathbf{Q}_1^{[t]}(i, m)\mathbf{Rn}_2^{[t]}(m, y) \tag{6}$$

$(\mathbf{Q}_1^{[t]}\mathbf{Rn}_2^{[t]})(i, y)$ represents the fused relevance scores between the y-th label and x_i in the first view, which can be seen as the summation of propagation of label relevance score between x_m and y-th label formulated as $\mathbf{Rn}_2^{[t]}(m, y)$, through the edge weight equivalent to similarity between x_i and x_m ($m \neq i$), formulated as $\mathbf{Q}_1^{[t]}(i, m)$ to x_i. Such $(\mathbf{Q}_1^{[t]}\mathbf{Rn}_2^{[t]})(i, y)$ is obtained by incorporating the metric, $\mathbf{Q}_1^{[t]}(i, m)$, $m \neq i$, from the first view, and label relevance matrix, $\mathbf{Rn}_2^{[t]}(m, y)$, $m \neq i$, from the second view to make the incorporation of the complementary information from two views. Following this principle, the refined $\mathbf{W}_1^{[t+1]}(i, j)$ in next iteration $t + 1$, for the first view, is yielded by considering relevance score between all labels and both two points (x_i and x_j, respectively), while effectively incorporates the complementary information from two views. Eq. (5) may be conducted similarly. ■

One natural question is how to calculate $\mathbf{R}_g^{[t]}$, and its normalized form $\mathbf{Rn}_g^{[t]}$ for each iteration t, we propose to adopt the general iterative form in the next section.

3.3   The General Iterative Form of SSMF

We can get $\mathbf{R}_g^{[t]}$ ($g = 1, 2, \ldots, M$) iteratively by

$$\mathbf{R}_g^{[t]} = \alpha_g \mathbf{P}_g^{[t]}\mathbf{R}_g^{[t-1]} + (1 - \alpha_g)\mathbf{L} \tag{7}$$

where $\mathbf{P}_g^{[t]}(i, j) = \frac{\mathbf{W}_g^{[t]}(i,j)}{\mathbf{D}_g^{[t]}(i,i)+\mathbf{D}_g^{[t]}(j,j)}$ ($g = 1, 2, \ldots, M$), and $0 \leq \alpha_g \leq 1$. $\mathbf{P}_g^{[t]}$ is a symmetric matrix. $\mathbf{L}$ is the initial labeling matrix mentioned in Section 3.1.

Generalizing Eqs. (4) and (5) regarding two views, $\mathbf{W}_g^{[t]}$ may be calculated as follows for multi-views.

$$\mathbf{W}_g^{[t+1]} = \mathbf{Q}_g^{[t]}(\frac{\sum_{j \neq g} \mathbf{Rn}_j^{[t]}}{M-1})(\frac{\sum_{j \neq g} (\mathbf{Rn}_j^{[t]})^T}{M-1})(\mathbf{Q}_g^{[t]})^T + \lambda I \qquad (8)$$

The iterative form of SSMF with multi-view by iteratively applying Eqs. (7) and (8) represents a novel label propagation process. Specifically, each weighted graph $\mathbf{G}_g^{[t]}$ associated with the matrix $\mathbf{W}_g^{[t]}$ or $\mathbf{P}_g^{[t]}$, incorporates the label propagation results inherent in $\mathbf{Rn}_j^{[t]}$ ($j \neq g$) from other views, as shown in Eq. (8), and hence we call the label propagation formulated as Eq. (7) as *cross view label propagation*.

Now, we are ready to prove the convergence of SSMF.

Theorem 1. *The iterative form of SSMF formulated in Eq. (7) converges.*

It suffices to prove the convergence on one view. Following Eq. (7), we have

$$\mathbf{R}_g^{[t]} = \alpha_g^t \mathbf{L} \prod_{i=1}^{t} \mathbf{P}_g^{[i]} + (1 - \alpha_g)\mathbf{L} \sum_{i=1}^{t-1} \alpha_g^i \prod_{j=1}^{i} \mathbf{P}_g^{[j]} \qquad (9)$$

where $\mathbf{R}_g^{[0]} = \mathbf{L}$. Apparently, since $0 < \alpha_g < 1$,

$$\lim_{t \to \infty} \alpha_g^t \mathbf{L}[\prod_{i=1}^{t} \mathbf{P}_g^{[i]}](i,j) = 0$$

Moreover, the largest eigenvalue of $\mathbf{P}_g^{[i]}$ ($i = 1, 2, \ldots, t$) is no more than 1 according to the Gershgorin circle theorem. For the second term in Eq. (9), $(1 - \alpha_g)\mathbf{L}$ is a constant matrix for all $\alpha_g^i[\prod_{j=1}^{i} \mathbf{P}_g^{[j]}]$ at any step i, thus, we only need to consider the series $\sum_{i=1}^{t-1} \alpha_1^i[\prod_{j=1}^{i} \mathbf{P}_g^{[j]}]$. We denote the i-th term by $\mathcal{H}_g[i]$ and study the convergence of entry $\mathcal{H}_g[i](l,m)$. We only need to prove the convergence of series $\sum_{i=1}^{t-1} \mathcal{H}_g[i](l,m)$, where $\mathcal{H}_g[i](l,m) = \alpha_g^i[\prod_{j=1}^{i} \mathbf{P}_g^{[j]}](l,m) < \alpha_g^i$, since $[\prod_{j=1}^{i} \mathbf{P}_g^{[j]}](l,m) < 1$, which can be easily verified by simple arithmetic operations. We construct the series $\sum_{i=1}^{t-1} \alpha_g^i$ ($0 < \alpha_g < 1$). Obviously, the series converges, since $[\mathcal{H}_g[i]](l,m) \leq \alpha_g^i$ and each item $[\mathcal{H}_g[i]](l,m)$ is positive. ∎

Let $(\mathbf{R}_g)^\star_{SSMF}$ be the convergent label relevance matrix regarding the g-th view by interactively applying Eq. (7) (cross-view label propagation) and Eq. (8) (semi-supervised metric fusion). The final label relevance matrix regarding multi-views is $\mathcal{R}^\star_{SSMF} = \sum_{g=1}^{M} \frac{(\mathbf{R}_g^\star)_{SSMF}}{M}$. We summarize the algorithm of SSMF in Algorithm 1.

Algorithm 1. The algorithm of SSMF

Input: Initial affinity matrix $W_g^{[1]}(g = 1, 2, \cdots, \cdot, M)$, $R_g^{[0]}$, α_g, λ , initial label
relevance matrix L in Eq. (7), the convergence threshold ϵ.

Output: The final label relevance matrix $R_{SSMF}^{\star}$

1 **for** $g = 1, \cdots, M$ **do**
2 | $t = 0$.
3 | Obtaining the label propagation $R_g^{[1]}$ by Eq. (7)
4 $t = 1$.
5 **repeat**
6 | **for** $g = 1, \cdots, M$ **do**
7 | | $Z_g^{[t]} = Q_g^{[t]}(\frac{\sum_{j \neq g} \mathbf{R} \mathbf{n}_j^{[t]}}{M-1})$
8 | | $\mathbf{W}_g^{[t+1]} = \mathbf{Z}_g^{[t]}(\mathbf{Z}_g^{[t]})^T + \lambda I$
9 | | $\mathbf{R}_g^{[t+1]} = \alpha_g \mathbf{P}_g^{[t+1]} \mathbf{R}_g^{[t]} + (1 - \alpha_g)\mathbf{L}$
10 | | $t = t + 1$;
11 **until** *change is smaller than* ϵ;
12 $R_{SSMF}^{\star} = \sum_{g=1}^{M} \frac{(\mathbf{R}_g)_{SSMF}^{\star}}{M}$;
13 // $(\mathbf{R}_g)_{SSMF}^{\star}$ is the converged relevance label matrix in the g-th view.

One important issue that SSMF does not consider is that there may be some *irrelevant views*, and simply fusing all views using the same weight in Eq. (8) may not achieve the best overall performance if there are irrelevant views during the fusion process. To address this issue, we devise an effective learning method to assign a weight to each view in each fusion iteration. Consequently, we extend SSMF to WSSMF, which will be described in next section.

4   WSSMF: Learning Weights for SSMF

The basic idea is to consider the labeling result of cross-view label propagation for unlabeled points in the set $\mathbf{U}$ in each iteration. Two views are regarded consistent if their labeling results are similar. Specifically, we denote by $\mathbf{V}_i^{[t]}$ the i-th view in iteration t. The more consistent $\mathbf{V}_i^{[t]}$ and $\mathbf{V}_j^{[t]}$ $(1 \leq j \neq i \leq M)$ are, the larger the weight parameter $\theta_{ij}^{[t]}$ is for $\mathbf{V}_j^{[t]}$. Note that the labeling result of cross-view label propagation may be different at various iterations. Therefore, we calculate the weight parameter in different iterations. We define a function $\mathbf{D}(\mathbf{V}_i^{[t]}, \mathbf{V}_j^{[t]})$ in Eq. (10) to measure the mismatch between i-th and j-th view in terms of cross-view labeling propagation result.

$$\mathbf{D}(\mathbf{V}_i^{[t]}, \mathbf{V}_j^{[t]}) = \sum_{x_u^{[0]} \in \mathbf{U}, \mathrm{L}(x_u^{[t]}) \neq 0} \mathbf{B}(\mathrm{L}(x_u^{[t]}[i]), \mathrm{L}(x_u^{[t]}[j])) \qquad (10)$$

where $\mathbf{B}(\mathrm{L}(x_u^{[t]}[i]), \mathrm{L}(x_u^{[t]}[j])) = ||\mathrm{L}(x_u^{[t]}[i]) - \mathrm{L}(x_u^{[t]}[j])||$, $||\cdot||$ is the absolute value operator, and $\mathrm{L}(x_u^{[t]})$ is the largest label relevance score of $x_u^{[t]}$ regarding all labels. We have $\mathrm{L}(x_u^{[t]}[i]) = \max_l\{\mathbf{Rn}_i^{[t]}(u, l)\}$.

Initially, we set the label relevance score of all unlabeled points to be 0, and $\mathbf{D}(\mathbf{V}_i^{[t]}, \mathbf{V}_j^{[t]})$ describes the inconsistency degree between $\mathbf{V}_i^{[t]}$ and $\mathbf{V}_j^{[t]}$ at iteration t. The larger $\mathbf{D}(\mathbf{V}_i^{[t]}, \mathbf{V}_j^{[t]})$ is, the more inconsistent $\mathbf{V}_i^{[t]}$ and $\mathbf{V}_j^{[t]}$ are. For $\mathbf{V}_i^{[t]}$, the weight parameter $\theta_{ij}^{[t]}$ $(i \neq j)$ for $\mathbf{V}_j^{[t]}$ is defined as

$$\theta_{ij}^{[t]} = 1 - \frac{\mathbf{D}(\mathbf{V}_i^{[t]}, \mathbf{V}_j^{[t]})}{\sum_{h \neq i} \mathbf{D}(\mathbf{V}_i^{[t]}, \mathbf{V}_h^{[t]})} \tag{11}$$

Immediately, we have $\theta_{ij}^{[t]} = \theta_{ji}^{[t]}$ and $0 \leq \theta_{ij}^{[t]} \leq 1$. They are the entries in the coefficient symmetric matrix in iteration t, denoted by $\Theta^{[t]}$. In iteration t, the j-th view $(1 \leq j \neq i \leq M)$ is said to be ***irrelevant*** with respect to the i-th view if $\theta_{ij}^{[t]} < \frac{\sum_{g \neq i} \theta_{ig}^{[t]}}{M-1}$, otherwise, the j-th view is said to be ***relevant***. For the i-th view, we denote ***the set of relevant views*** at iteration t by $\mathbf{Re}_i^{[t]}$.

Instead of computing global irrelevant views explicitly, for the i-th view, we only fuse the views from $\mathbf{Re}_i^{[t]}$ in iteration t, and set the correlation strength weight to be 0 for irrelevant views. Combining Eq. 11 and Eq. 8, we have the Weighted SSMF (WSSMF for short) for multi-views, which iteratively applies Eq. 7 and Eq. 12 until convergence.

$$\mathbf{W}_g^{[t+1]} = \mathbf{Q}_g^{[t]}(\frac{\sum_{j \in \mathbf{Re}_g^{[t]}} \theta_{gj}^{[t]} \mathbf{Rn}_j^{[t]}}{|\mathbf{Re}_g^{[t]}|}) \times$$
$$(\frac{\sum_{j \in \mathbf{Re}_g^{[t]}} (\theta_{gj}^{[t]} \mathbf{Rn}_j^{[t]})^T}{|\mathbf{Re}_g^{[t]}|})(\mathbf{Q}_g^{[t]})^T + \lambda I \tag{12}$$

Like SSMF, WSSMF also converges, which can be immediately proved in the same manner as Theorem 1. Therefore, the final optimal label relevance matrix can be obtained as $\mathcal{R}_{WSSMF}^\star = \sum_{g=1}^{M} \frac{(\mathbf{R}_g)_{WSSMF}^\star}{M}$, where M is the number of views, $(\mathbf{R}_g)_{WSSMF}^\star$ is the convergent label relevance matrix in the g-th view obtained using WSSMF. Based on Algorithm 1, we generate the algorithm of WSSMF by replacing $\mathbf{Z}_g^{[t]}$ in line **7** with $\mathbf{Z}_g^{[t]} = \mathbf{Q}_g^{[t]}(\frac{\sum_{j \in \mathbf{Re}_g^{[t]}} \theta_{gj}^{[t]} \mathbf{Rn}_j^{[t]}}{|\mathbf{Re}_g^{[t]}|})$, and $\mathbf{W}_g^{[t+1]}$ in line **8** with Eq. 12.

4.1   Complexity Analysis

Now, we analyze the time complexity of each iteration in SSMF and WSSMF.

The cost of SSMF mainly comes from two parts: cross-view label propagation and semi-supervised metric fusion. The iterative cross-view label propagation in Line 9 of Algorithm 1 takes $\mathcal{O}(Mn^2c)$ time, and the same time complexity holds for semi-supervised metric fusion in Lines 7-8. We remark that all the above cost is from the matrix multiplication rather than matrix inverse computation. It is well known that matrix multiplication implementation without inverse computation is efficient. Similar to SSMF, WSSMF also needs $\mathcal{O}(Mn^2c)$ time for both metric fusion and cross-view label propagation. In addition, $\mathcal{O}(M^2n)$ time is needed to obtain the view correlation matrix Θ in each iteration regarding M views. Therefore, the overall time complexity for WSSMF is $\mathcal{O}(Mn^2c)+\mathcal{O}(M^2n)$. As observed in our experiments(refer to Fig 2), both SSMF and WSSMF converge within quite limited iterations for most cases (less than 65 times).

5   Experiments

We evaluate both SSMF and WSSMF using multi-view content based image retrieval (CBIR) and multi-label image classification on real data sets. We set the convergence threshold ϵ to 10^{-4} for all methods.

In our experiments, we compare with the following state-of-the-art multi-view graph based methods for both multi-view CBIR and multi-label image classification.

- *The multi-modality graph (MMG) method* [3], which uses multiple graph models under different views. The final ranking score vector is obtained by combining the independent label propagation (manifold ranking) results carried by each image in each view with different weights.
- *The averaged distance of multiple feature based metric (ADF) method* [2], which constructs a single relevance graph using the metric of average distance from multiple views.
- *The unsupervised metric fusion (UMF) method* [4], which conducts metric fusion without considering label propagation result. It is adapted to tackle multi-view graph-based semi-supervised learning as follows. We first obtain the convergent affinity matrix $\mathbf{W}_g$ ($g = 1, 2, \ldots, M$) for the g-th view by applying UMF, and then obtain the ranking score vector by optimizing Eq. (2), where the affinity matrix $\mathbf{W}_g$ is the fused affinity matrix using UMF on multi-views.

5.1   Multi-view Content Based Image Retrieval (CBIR)

Multi-view CBIR is a typical problem where graph based multi-view semi-supervised learning is extensively applied. Specifically, a query image is a labeled data object in our model, and the label relevance matrix $R_g \in \mathbb{R}^{n \times c}$ in Eq. (2) is reduced to a ranking score vector $r_g \in \mathbb{R}^n$, and $R_g(i, \cdot) \in \mathbb{R}^n$ is reduced to $r_g(i) \in \mathbb{R}$, which represents the relevance score between x_i and the query image (labeled image). $\mathbf{L} \in \mathbb{R}^{n \times c}$ in Eq. (2) is reduced to an n dimensional vector $\mathbf{Y} \in \mathbb{R}^n$ with the i-th entry to be 1 if x_i is the query image, and 0 otherwise.

We set the number of nearest neighbors k to 20 to calculate the metric distance in Eq. (1) for all views, which is consistent with the UMF method [4].

Similar to [9], we set α_g to 0.99 in Eq. (7) for all views, set λ to 1 in both Eq. (8) and Eq. (12). All methods are tested on the COREL5K data set [1], which consists of 5000 images in 50 categories. Each category contains 100 images. Due to the same number of images in each category, we use the *precision-scope* [3] as the evaluation metric. We use HOG, color histogram, RGB-SIFT and Pyramid wavelet texture feature to construct different views, most of them are utilized by MMG. For each method, we select every sample of 5000 images as the query image (labeled objects) each time, and obtain the average precision value and its statistical distribution regarding all 5000 samples, shown using 3 points (mean, +1 standard deviation, and -1 standard deviation) in Fig. 1(a).

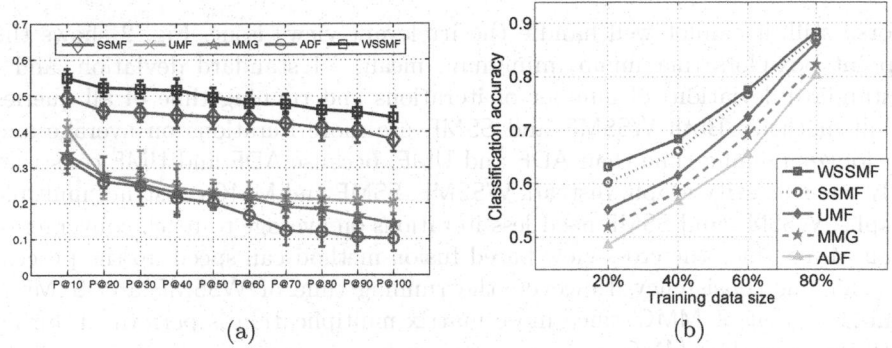

(a)   (b)

Fig. 1. (a) Top-s precision on COREL5K data set. (b) Classification accuracy with respect to sample rate on Caltech-101 image data set.

Unsurprisingly, WSSMF outperforms the other methods in top-s average precision, since it can better achieve the consistency from multi-views than the other methods. In addition, it can effectively address the problem of irrelevant views at each iteration. SSMF is the next after WSSMF. SSMF does not handle the problem of irrelevant views. Like SSMF, UMF **(1)** does not consider the irrelevant view detection, either. Moreover, **(2)** UMF does not fuse label propagation results during the fusion process, **(3)** as such UMF fails to further exploring the graph structure to improve the metric similarity like SSMF and WSSMF as discussed in section 2. Consequently, UMF is inferior to SSMF.

Both MMG and ADF perform worse than the others. MMG outperforms ADF in most cases, since MMG fully explores the graph structure for different views, and it linearly combines the independent label propagation results with different weights. ADF, however, is different from MMG. It assigns the same weight to all views in combining the label propagation results, the single graph associated with averaged metric obstructs the graph structure of original inherent individual views. However, MMG is inferior to SSMF and WSSMF, since such one-combo-fits-all late fusion method is undesirable to achieve the consistency among all views by independently fusing all the label propagation result from all views.

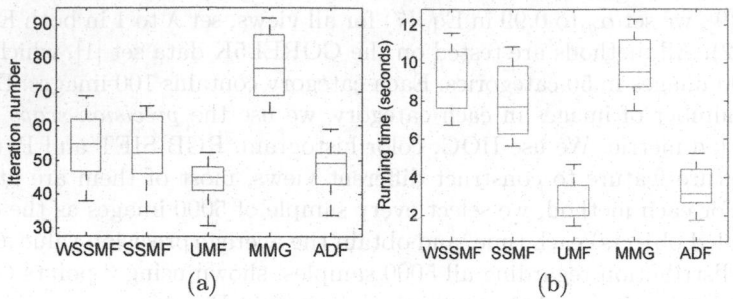

Fig. 2. Comparison of (a) number of iterations. (b) running time.

Worse still, it cannot well handle the irrelevant views issue. Fig. 2 shows the 5-point box-plots (maximum, minimum, mean, +1 standard deviation, and -1 standard deviation) of number of iterations and running time of all queries in all methods. Both WSSMF and SSMF use more iterations on average and sot longer running time than ADF and UMF, because ADF and UMF construct only one similarity graph. Instead, WSSMF, SSMF and MMG construct multiple graphs. WSSMF and SSMF need less iterations on average to reach convergence than MMG, since the cross-view based fusion method can speed up the process of achieving consistency. However, the running time of WSSMF and SSMf is similar to that of MMG, since more matrix multiplication is performed during each iteration than MMG.

5.2   Multi-view Based Multi-label Image Classification

Multi-view based multi-label image classification can be regarded as multi-view based semi-supervised learning with graphs. The Caltech-101 data set (http://www.vision.caltech.edu/Image_Datasets/Caltech101/) is used to test multi-label image classification. It contains 9146 images organized into 101 categories. The number of images in different categories ranges from 40 to 800. We set $c = 101$ and $n = 9146$ in the label relevance matrix $R_g \in \mathbb{R}^{n \times c}$ and $\mathbf{L} \in \mathbb{R}^{n \times c}$ in Eq. (2), along with $k = 20$ in Eq. (1) and $\lambda = 1$ in Eq. (12).

We use the same sample rate to draw a random sample of images from each category as labeled images. The rest of images are treated as unlabeled. Each experiment is repeated 5 times, and the average value is reported. The classification accuracy on all unlabeled images is used to evaluate different methods. HOG, color histogram, pyramid wavelet texture feature and SIFT are used to construct different views. The results are shown in Fig. 1(b).

WSSMF outperforms the other methods. SSMF is the second best method. The results verify the advantages of our iterative fusion methods. We also observe that the difference among different methods decreases as the sample rate increases, since a higher sample rate makes the problem less challenging.

6 Conclusion

In this paper, we propose a novel iterative fusion technique for graph based semi-supervised learning from multi-views. The central idea is to fuse metrics and label propagation results from multi-views iteratively and weight views dynamically. The experimental results clearly show that our new methods outperform the state-of-the-art methods on real data sets. As future work, we will investigate how to fuse selective labeling results from multi-view based graphs rather than tackling all the data points including both informative and noise data points. We will also investigate active learning based methods for better effectiveness and efficiency.

Acknowledgment. Jian Pei's Research is supported in part by an NSERC Discovery Grant and a BCFRST NRAS Endowment Research Team Program Project. Xuemin Lin is supported by ARC DP0987557, ARC DP110102937, ARC DP120104168 and NSFC61021004. Wenjie Zhang is supported by ARC DE120102144 and DP120104168. All opinions, findings, conclusions and recommendations in this paper are those of the authors and do not necessarily reflect the views of the funding agencies.

References

1. Duygulu, P., Barnard, K., de Freitas, J.F.G., Forsyth, D.: Object recognition as machine translation: Learning a lexicon for a fixed image vocabulary. In: Heyden, A., Sparr, G., Nielsen, M., Johansen, P. (eds.) ECCV 2002, Part IV. LNCS, vol. 2353, pp. 97–112. Springer, Heidelberg (2002)
2. Huang, Y., Liu, Q., Zhang, S., Metaxas, D.N.: Image retrieval via probabilistic hypergraph ranking. In: CVPR (2010)
3. Tong, H., He, J., Li, M., Zhang, C., Ma, W.: Graph based multi-modality learning. In: ACM MM (2005)
4. Wang, B., Jiang, J., Wang, W., Zhou, Z., Tu, Z.: Unsupervised metric fusion by cross diffusion. In: CVPR (2012)
5. Wang, M., Hua, X., Hong, R., Tang, J., Qi, G., Song, Y.: Unified video annotation via multigraph learning. IEEE Trans. Circuits Syst. Video Techn. 19(5), 733–746 (2009)
6. Wang, Y., Cheema, M.A., Lin, X., Zhang, Q.: Multi-manifold ranking: Using multiple features for better image retrieval. In: Pei, J., Tseng, V.S., Cao, L., Motoda, H., Xu, G. (eds.) PAKDD 2013, Part II. LNCS, vol. 7819, pp. 449–460. Springer, Heidelberg (2013)
7. Wang, Y., Lin, X., Zhang, Q.: Towards metric fusion on multi-view data: A cross-view based graph random walk approach. In: ACM CIKM (2013)
8. Zhou, D., Bousquet, O., Lal, T.N., Weston, J., Schlkopf, B.: Learning with Local and Global Consistency. In: NIPS (2003)
9. Zhou, D., Weston, J., Gretton, A., Bousquet, O., Schlkopf, B.: Ranking on data manifolds. In: NIPS (2003)
10. Zhu, X.: Semi-supervised learning with graphs. PhD thesis, Carnegie Mellon University (2005)

Crime Forecasting Using Spatio-temporal Pattern with Ensemble Learning

Chung-Hsien Yu[1], Wei Ding[1], Ping Chen[1], and Melissa Morabito[2]

[1] University of Massachusetts Boston,
100 Morrissey Blvd., Boston, MA 02125, USA
{csyu,ding}@cs.umb.edu, Ping.Chen@umb.edu
[2] University of Massachusetts Lowell,
One University Avenue, Lowell, MA 01854, USA
Melissa_Morabito@uml.edu

Abstract. Crime forecasting is notoriously difficult. A crime incident is a multi-dimensional complex phenomenon that is closely associated with temporal, spatial, societal, and ecological factors. In an attempt to utilize all these factors in crime pattern formulation, we propose a new feature construction and feature selection framework for crime forecasting. A new concept of multi-dimensional feature denoted as spatio-temporal pattern, is constructed from local crime cluster distributions in different time periods at different granularity levels. We design and develop the Cluster-Confidence-Rate-Boosting (CCRBoost) algorithm to efficiently select relevant local spatio-temporal patterns to construct a global crime pattern from a training set. This global crime pattern is then used for future crime prediction. Using data from January 2006 to December 2009 from a police department in a northeastern city in the US, we evaluate the proposed framework on residential burglary prediction. The results show that the proposed CCRBoost algorithm has achieved about 80% on accuracy in predicting residential burglary using the grid cell of 800-meter by 800-meter in size as one single location.

Keywords: Spatio-temporal Pattern, Crime Forecasting, Ensemble Learning, Boosting.

1 Introduction

Crime forecasting is notoriously difficult. A crime incident is a multi-dimensional complex phenomenon that is closely associated with temporal, spatial, societal, and ecological factors. In an attempt to utilize all these factors in crime pattern formulation, we propose a new feature construction and feature selection framework for crime forecasting. A new concept of multi-dimensional feature denoted as spatio-temporal pattern, is constructed from local crime cluster distributions in different time periods at different granularity levels.

Crime distributions are of different sizes and shapes with respect to spatial space over time. We use clustering to find local crime distributions in different time periods. The spatial-temporal patterns then are induced from each

V.S. Tseng et al. (Eds.): PAKDD 2014, Part II, LNAI 8444, pp. 174–185, 2014.
© Springer International Publishing Switzerland 2014

crime distribution through classification. Each spatio-temporal pattern uses related crime incidences as indicators to represent a local crime pattern at certain clustered locations during a certain time period. However, these locally learned patterns could be redundant or overlapping at global level.

We design and develop the Cluster-Confidence-Rate-Boosting (CCRBoost) algorithm to efficiently select relevant local spatio-temporal patterns to construct a global crime pattern from a training set. The main idea of this approach is to iteratively pick a set of local patterns which give the least classification error at each boosting round. Each set of local patterns is referred as an ensemble spatio-temporal pattern and is assigned a score (Called confidence-rate in our approach). At the end of boosting, a global pattern is constructed from these ensemble patterns. This global pattern is capable of predicting crime by scaling the total score of an input, a collection of crime indicators, evaluated on each crafted ensemble patterns. The proposed algorithm is depicted in Figure 1.

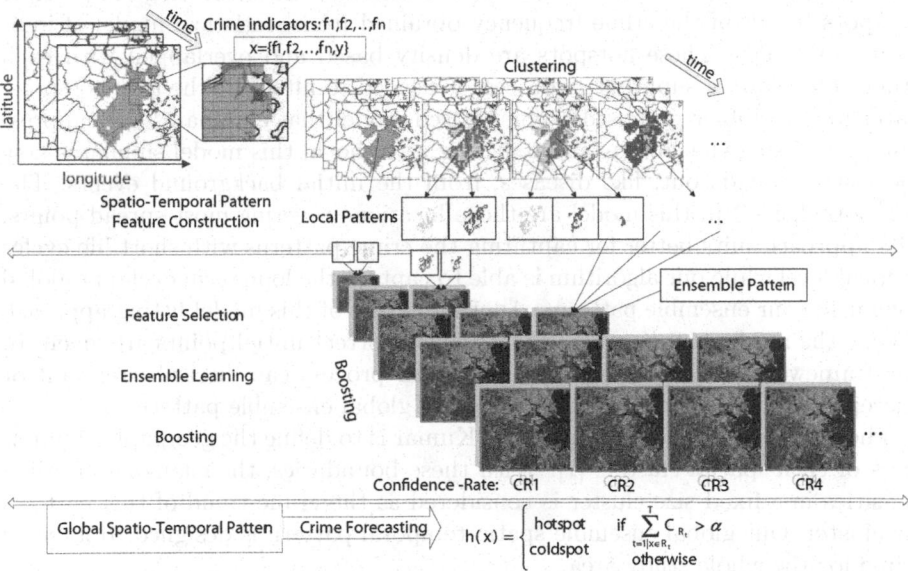

Fig. 1. The flowchart of the proposed CCRBoost algorithm (Better viewed in color)

In our real-world case study, we collaborated with the police department of a northeastern city in the US to collect 4-year historical crime data, from January 2006 to December 2009. These data are used to evaluate the proposed framework on residential burglary prediction. This city is 90 square miles in size and more than 600 thousands in population. The results show that the proposed CCRBoost algorithm has achieved about 80% on accuracy in predicting residential burglary using the grid cell of 800-meter by 800-meter in size as one single location.

This rest of the paper is organized as the followings. The related work is described and compared with our algorithm in Section 2. Our proposed CCRBoost algorithm is thoroughly discussed in Section 3, including its theoretical analysis. The setting and results of our experiments are presented in Section 4. We then conclude our study in Section 5.

2 Related Work

Crime forecasting techniques from feature construction and feature selection point of view, can be categorized into statistic mapping, mathematical modeling, and clustering.

Statistic mapping uses historical statistics of the crime occurred at the same location for forecast[1]. It focuses on seasonality of the crime with the assumption that same type of crime recurs regularly with respect to time, while our approach, in addition to time dimension, also factors in spatial neighborhood and other relevant societal and ecological factors.

In [11], mathematical modeling is used to simulate the formatting of the crime hotspots based on the crime frequency obtained from statistical model of individual criminals. These hotspots are density based and overlapped with each other. Therefore, a suppression process is needed to filter out the true hotspots. Later in [8], Mohler proposed a point-based model that eliminates the suppression step. Using the concept in predicting aftershock, this model simulates how the crime spreads out, like diseases, from the initial background events. The hotspots defined in this model are those locations covering most spread points. The approach suits better for capturing the crime patterns with short life cycles at local level while our algorithm is able to capture the long term cycles at global level using our ensemble patterns. Another caveat of this model fitting approach is that the results can be way off when the incorrect initial points are given. In our framework, the built-in feature selection process can discard irrelevant or misrepresenting patterns when learning the global ensemble pattern.

The clustering approach adapted by Kumar is to define the geographic boundaries of each spatial clusters [7]. With these boundaries, the changing of crime densities in a fixed size cluster is considered as the crime trend of this particular cluster. Our global ensemble spatio-temporal pattern is designed to forecast crime for the whole study area.

3 Crime Forecasting Using Spatio-temporal Patterns

3.1 The Concept of Spatio-temporal Patterns

Our approach is designed to enhance the utility of the near repeat hypothesis formulated in Social Science [12]. This hypothesis suggests that the same type of crime possibly recurs not only at the same neighboring locations but also at a regular interval of time. In addition, crime incidents are closely related with social behaviors and environmental conditions[6]. This implies that crime tends to have similar trends at locations with similar societal and ecological structures. We hypothesize that crime can be foreseen by investigating the trends of its correlated crime incidences. All of these three dimensions (location, time, and correlated incidences) are taking into account when we define a spatio-temporal pattern. A spatio-temporal pattern is a local pattern that represents the crime pattern at certain locations during certain time period using correlated crime incidences as the indicators. These indicators are used to represent societal and ecological factors of different locations.

The challenges are that how many local patterns there are during certain time period and at which locations are unknown. Additionally, crime is not evenly distributed throughout a city and there are areas that are more attractive than others to criminals[3]. In order to find the possible local patterns, we use the unsupervised clustering approach without involving geographical features to group those locations with similar indicators in the same time period. This group of locations is considered as the spatial distribution of a possible crime pattern. By varying the configuration of the clustering method, the clusters with different sizes during different periods of time can be generated. A classifier is then trained from each cluster and will be used to represent a local crime pattern.

Our next task is to use the spatio-temporal patterns as features to construct a global level spatio-temporal pattern. This global crime pattern should be capable of detecting crime incidences at every location. Which of these patterns should be selected to form the global crime pattern? Those locally learned spatio-temporal patterns could be redundant or overlapping. And, how can this global pattern be constructed? To resolve these two issues at the same time, we propose a confidence-rate boosting approach. We will first formulate the problem and then discuss our boosting algorithm in detail.

3.2   Problem Formulation

We denote one crime indicator, a type of relevant crime event, as f_p. Different indicators of the same location in the same period of time are used to form a vector, denoted as $x = [f_1, f_2, \ldots, f_P]$, where P is the number of correlated incidence types. Each vector x has one class label y which tells whether this location is a hotspot. Through the clustering process, the vectors with similar indicators are grouped into one cluster, denoted as c. A local spatio-temporal pattern, denoted as r, is defined as: $r = q(c)$. $q()$, in our case, is a classifier induced from the cluster c and used to extract the crime pattern. This pattern r is used to identify whether a vector is a hotspot. We denote $x \in r$ if x is recognized as hotspot by pattern r. Otherwise, $x \notin r$.

In reality, a crime pattern might not be represented as one single local pattern because this pattern might shift location-wise or change size over time[9]. In order to truly capture the dynamics of crime patterns, we introduce the ensemble spatio-temporal pattern, denoted as $R = [r_1 \wedge r_2 \wedge \ldots \wedge r_i]$, which is the conjunction of selected spatio-temporal patterns. Thus, if $x \in R$, then it must be true that $x \in r_1$ and $x \in r_2$ and $\ldots$ and $x \in r_i$. This ensemble pattern is served as the base learner in our confidence-rate boosting approach.

3.3   Confidence-Rate Boosting

Before boosting, we balance the training data by setting the weight of hotspots as $\frac{1}{2H}$ and $\frac{1}{2C}$ for coldspots, H and C are the numbers of hotspots and coldspots, respectively. Each vector x_i in the training data is denoted as $D(i)$. The main idea of our confidence-rate boosting approach is to repeatedly pick the best hypothesis h_t which yields the least error rate at each boosting round t. In our case, h_t is an ensemble spatio-temporal pattern R_t built at round t. The error function is defined as:

$$E_{i \sim D_t}[y_i h_t(x_i)] = \sum_i D_t(i) y_i h_t(x_i), \tag{1}$$

where D_t is the weight distribution at boosting round t. The theoretical background of the confidence-rate boosting approach is analyzed as follows.

Based on the study in [10], it has been proved that $\sum_i D_t(i) y_i h_t(x_i) \leq \prod_t Z_t$ so the upper bound of the error rate is $\prod_t Z_t$. Z_t is defined as:

$$Z_t = \sum_i D_t(i) exp(-\alpha_t y_i h_t(x_i)) \tag{2}$$

Thus, a smaller Z_t that has a lower error upper bound will lead to a smaller training error at each boosting round. Now, we let $C_R = \alpha_t h_t(x_i)$ and ignore the boosting round t. Then, we define our loss function as:

$$Z = \sum_i D(i) exp(-C_R y_i) \tag{3}$$

and we want to find the minimum value of Z to lower the training error as much as possible. C_R is the confidence-rate for pattern R and $C_R = 0$ if $x_i \notin R$. Here, $x_i \in R$ means that x_i is recognized by pattern R as a hotspot and then set $y_i = 1$. Otherwise, set $y_i = -1$. Since $C_R = 0$ where $x_i \notin R$, we obtain

$$Z = \sum_{i | x_i \notin R} D(i) + \sum_{i | x_i \in R} D(i) exp(-C_R y_i) \tag{4}$$

Equation (4) can be rewritten as:

$$Z = W_0 + W_+ exp(-C_R) + W_- exp(C_R), \tag{5}$$

where $W_0 = \sum_{i | x_i \notin R} D(i)$ so W_0 is the total weights of predicted coldspots. And,

$$W_+ = \sum_{i | x_i \in R \, and \, y=1} D(i), W_- = \sum_{i | x_i \in R \, and \, y=-1} D(i), \tag{6}$$

W_+ is the total weights of true hotspots (true positives), and W_- is the total weights of false hotspots (false positives). By taking the first derivative of Z with respect to C_R and let $\frac{dZ}{dC_R} = 0$, we can find the value of C_R when Z has the maximum or minimum value:

$\frac{dZ}{dC_R} = -W_+ exp(-C_R) + W_- exp(C_R) = 0$

$\implies W_- exp(C_R) = W_+ exp(-C_R)$

$\implies \ln(W_- exp(C_R)) = \ln(W_+ exp(-C_R))$

$\implies \ln(W_-) + C_R = \ln(W_+) - C_R$

$\implies 2C_R = \ln(W_+) - \ln(W_-)$

$\implies C_R = \frac{1}{2}\ln(\frac{W_+}{W_-})$ And then, we take the second derivative of Z, $\frac{dZ}{dC_R^2} = W_+ exp(-C_R) + W_- exp(C_R) > 0$. Since the second derivative of Z is greater than zero, Z has the minimum value of $W_0 + 2\sqrt{W_+ W_-}$ when $C_R = \frac{1}{2}\ln(\frac{W_+}{W_-})$.

To prevent the situation of $W_- = 0$, we adjust the above equation as:

$$\hat{C}_R = \frac{1}{2}\ln(\frac{W_+ + \frac{1}{2n}}{W_- + \frac{1}{2n}}), \tag{7}$$

where n is the total number of vectors. Equation (7) is then used to calculate the confidence-rate $\hat{C}_{R_t}$ for pattern R_t at each round t.

3.4   CCRBoost Algorithm

As described in Algorithm 1, the first task of the CCRBoost algorithm is to identify spatio-temporal patterns of different sizes and shapes with respect to spatial space during each period of time. To add spatio-temporal dimension to our feature, a clustering step is adopted to find the crime distributions at local level in different time periods. K-Means, but not limited to, is chosen to find these patterns. We perform K-Means K times to obtain $1+2+\ldots+K$ clusters and then train classifiers from each cluster to extract local spatio-temporal patterns at different granularity levels. The data is divided into M subsets before clustering by certain length of time interval. For example, if the raw crime data is processed by month, then M equals to 12 when one year worth of data is used. As a result, there are total $M \times (1 + 2 + \ldots + K)$ possible patterns acquired from these M subsets.

Next, the weights of the entire data set is set to be in a probability distribution which makes the total weight equals to 1. The data set is then randomly divided into two subsets, **GrowSet** and **PruneSet**. This split is based on the total weight instead of the number of records. By calling $BuildChain()$, an ensemble spatio-temporal pattern R is built from those local patterns. This R gives the minimum Z value while evaluating R on **GrowSet**. Furthermore, $PruneChain()$ is called to trim the list of R and prevents R from over fitting by using **PruneSet** to reevaluate R and then obtain the final ensemble pattern R_t. The confidence-rate $\hat{C}_{R_t}$ is then calculated by evaluating R_t on the entire data set using Equation (7). Based on $\hat{C}_{R_t}$, the boosting algorithm updated the weights of those vectors that are classified as hotspots. This weight update function is defined as:

$$D_{t+1}(i) = \frac{D_t(i)}{exp(y_i \hat{C}_{R_t})}, \; if \; x_i \in R_t \tag{8}$$

The goal is to exponentially lower the weights on those vectors that are recognized by the current global pattern. This way, the data instances which have not been fitted to the pattern are getting more attentions for the next round.

The boosting process is repeated for T times, where T is a user-defined variable. At the end, T ensemble spatio-temporal patterns, $R_1, R_2, \ldots, R_T$, and T confidence-rates, $\hat{C}_{R_1}, \hat{C}_{R_2}, \ldots, \hat{C}_{R_T}$ are produced. The formula of the final global spatio-temporal pattern is defined as:

$$h(x) = \begin{cases} hotspot & \sum_{R_t | x \in R_t} \hat{C}_{R_t} > \alpha \\ coldspot & otherwise \end{cases}, \tag{9}$$

while α is a user-defined threshold.

By taking an input vector x, this formula evaluates x over each ensemble pattern R_t. If x is recognized by R_t as a hotspot, then $\hat{C}_{R_t}$ is added to the total confidence score $h(x)$. x is predicted as a hotspot if $h(x)$ is greater than the threshold α. Normally, this threshold α is set to zero. This ensemble learning algorithm is inspired by Cohen and Singer's research in [2]. The steps of the CCRBoost algorithm are given in Algorithm 1.

Algorithm 1. CCRBoost Algorithm

1. Given crime data $(x_1, y_1), \ldots, (x_n, y_n)$.
2. K is a user-defined variable and M is the total number of time periods.
3. **for** $k = 1 \ldots K$ **do**
4.   **for** $m = 1 \ldots M$ **do**
5.   Run K-Means using the vectors in period m to generate k clusters. Then, k spatio-temporal patterns are extracted from these clusters.
6.   **end for**
7. **end for**
8. Balance the data set by weights.
9. **for** $t = 1 \ldots T$ **do**
10.   Normalize the weights, let D_t be a probability distribution.
11.   Divide weighted data into two sets, *GrowSet* and *PruneSet*.
12.   Call *BuildChain()* and then *PruneChain()* to obtain R_t.
13.   Calculate $\hat{C}_{R_t}$ using entire data set and Equation (7).
14.   Update the weights based on Equation (8).
15. **end for**
16. The final global spatio-temporal pattern is defined as:
$$h(x) = \begin{cases} hotspot & \sum_{R_t : x \in R_t} \hat{C}_{R_t} > \alpha \\ coldspot & otherwise \end{cases}, \alpha \text{ is a user-defined threshold.}$$

---

4   Case Study: Forecasting Residential Burglary in a Northeastern City of the U.S.A.

Data Configuration: 4-years' (January 2006 to December 2009) crime records have been used for the evaluation. In addition, three different grid resolutions have been applied to generate three data sets from the original crime records.

These three resolutions have the squared cell/block with edge lengths of 800, 600, and 450 meters, respectively.

The targeting crime to be predicted is residential burglary in our experiments. Residential burglary is a particularly interesting crime to study from a prediction perspective since the near repeat hypothesis suggests that proximity to a burgled residence increases the likelihood of victimization of other domiciles in the neighborhood[12].

Based on the criminology theory[6] and after consulting with the domain experts, six categories of incidences are identified having the higher correlation with residential burglary crime than others. These six categories are arrest, commercial burglary, foreclosure, motor vehicle larceny, 911 call, and street robbery. Thus, the aggregations of these six categories from the crime records are used as the crime indicators in our experiments.

The Choice of Pattern Learning Classifier: LADTree[5] has been chosen as the base classifier to identify these patterns in our experiments because LADTree adapts same confidence-rated system to grow a decision tree. However, our algorithm is not limited to LADTree because a spatio-temporal pattern can be represented in any format or model as long as it can tell whether a vector is a hotspot.

Clustering Approach in Finding Spatio-temporal Patterns: In this experiment, the effectiveness of K is evaluated. The other user-defined variables T (The number of boosting iteration) and α (The threshold for $h(x)$) are set to 500 and 0, respectively. The results of this experiment are obtained from three data sets with different grid cell sizes. When $K = 1$, the data is not clustered. Therefore, the results obtained from the setting of $K = 1$ is used as the baseline to compare with others. According to the results shown in Figure 2, the clustering approach yields not only the better overall accuracy but also the better F1-score on hotspots. This is because using clustering enhances the feature with spatial dimension by taking into account the crime distribution at local levels. Moreover, we found that the performance reaches certain level when $K = 4$ and then maintains at this level when $K \geq 5$. This shows that the patterns lose the true representative of local crime distributions when the resolution is set too high and suggests that there are less than or equal to 5 different levels of local crime patterns in our target city. Thus, K is set to 5 in the rest of our experiments.

Comparing Spatio-Temporal Pattern Features with Random Sampling Features: In this experiment, the variable K is used to decide the number of random sampling data sets. This sampling method randomly selects 50% of the data records from a monthly data set for $1 + 2 + \ldots + K$ times without replacement, which means that there is no duplicated records in each sample. This method constructs $1 + 2 + \ldots + K$ samples with unified size and then trains the base classifiers from them using the LADTree algorithm. The purpose is to have same number of features while comparing random sampling with the proposed spatio-temporal pattern.

Next, our confidence-rate boosting algorithm is used to pick features from those patterns generated from random sampled clusters and then build a global

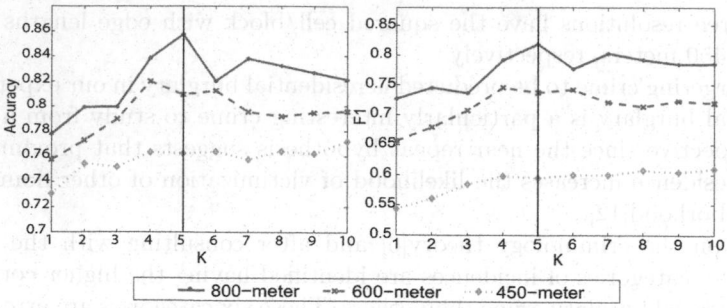

Fig. 2. The results of using different K for clustering on 3 data sets

pattern. By this way, we can tell that which kind of feature delivers the better prediction results. According to Figure 3, using spatio-temporal patterns has better performance regardless the resolution of the data set. Thus, spatio-temperal patterns do have the advantage over random sampling due to their spatio-temporal multi-dimensional characteristic.

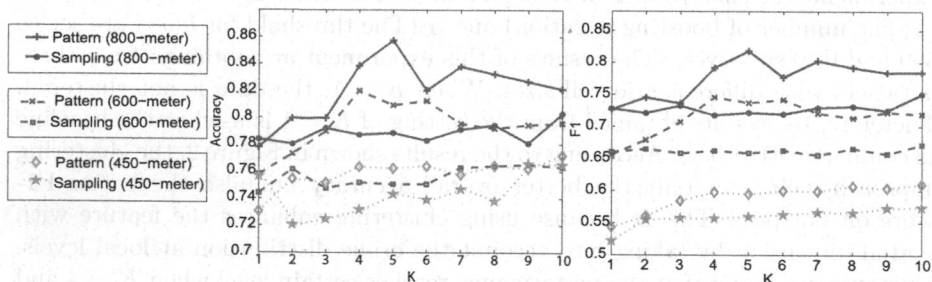

Fig. 3. Comparing spatio-temperal patterns with random sampling on different datasets

Comparing with Classification Approaches: Using the same crime data sets, other commonly used classification methods are adapted to generate the prediction results to compare with our proposed algorithm. Support Vector Machine (SVM), C4.5, Naive Bayes classifier, and LADTree[5] are chosen in this experiment. As shown in Table 1, our proposed CCRBoost algorithm has the best accuracy and F1-score over other classifiers on all three data sets.

Comparing CCRBoost with AdaBoost: During this experiment, the various numbers of iterations, T, are used in comparing our algorithm with the AdaBoost[4] algorithm. LADTree is chosen as the base learning classifier in both algorithms. Shown in Figure 4, the accuracy obtained from the AdaBoost algorithm reaches its ceiling when $T > 50$. However, our CCRBoost algorithm not only can obtain better accuracy but also has better convergence rate throughout

Table 1. The results of comapring CCRBoost with existing classifiers

Data Set	800-meter		600-meter		450-meter	
Method	Accuracy	F1	Accuracy	F1	Accuracy	F1
SVM	0.817	0.801	0.776	0.742	0.651	0.489
C4.5	0.500	0.667	0.500	0.667	0.500	0
NaiveBayes	0.730	0.675	0.703	0.647	0.667	0.592
LADTree	0.772	0.757	0.728	0.702	0.644	0.487
CCRBoost	**0.857**	**0.818**	**0.820**	**0.746**	**0.772**	**0.610**

three data sets. In conclusion, the boosting effect of our algorithm is more efficient than AdaBoost because our algorithm enhanced with new spatio-temporal features has a strong impact in predicting crime.

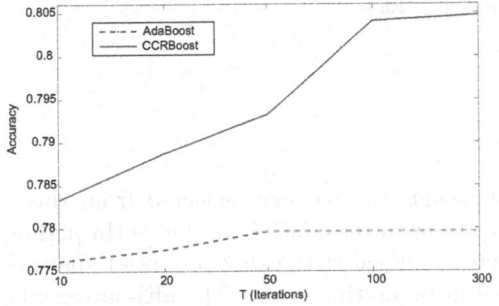

Fig. 4. Comparing AdaBoost with CCRBoost on different iterations T

The Resulting Global Spatio-temporal Pattern: The selection of the local patterns used in the final spatio-temporal pattern has been visualized on the map, which is shown in Figure 5. The red grid cells represent hotspots and blue cells are coldspots. The first pattern chosen by the proposed algorithm is a cluster from September 2007. The locations of this cluster are consistent with known crime pattern of our target city. The second cluster representing August 2009 data identifies crime hotspots that were excluded from the first cluster. More importantly, this second cluster is useful for pinpointing coldspots areas that have some protective factor against residential burglary and other crimes.

As a result, the first two clusters are complementary in identifying locations where we would expect residential burglary across the entire city as well as areas that are coldspots. Interestingly and consistent with criminological literature, both clusters are in the summer months when children are out of school and individuals may take vacations and be less vigilant about protecting their property. It may be that there is an increased likelihood of residential burglary in this city during the summer time. Based on the consistency with actual crime patterns, our algorithm does find the patterns which recognize not only the spatial but also the temporal factors that are useful for criminal justice professionals in predicting the incidence of future crime.

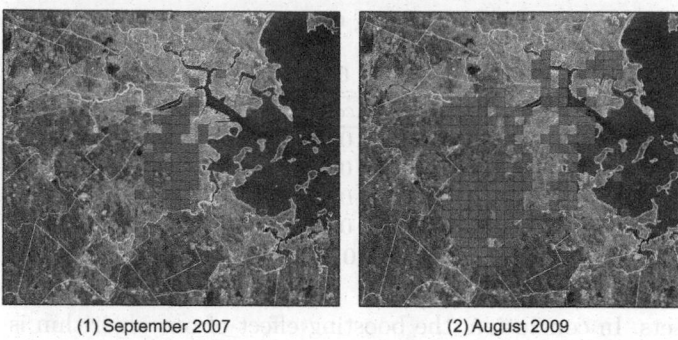

(1) September 2007   (2) August 2009

Fig. 5. The first two local patterns used in the final global spatio-temporal pattern resulting from 800-meter grid data set. The red blocks are hotspots and blues are coldspots. (Better viewed in color).

5   Conclusions

From a practical standpoint, the patterns selected from this algorithm are indicative of the true locations of residential burglaries throughout the target city. This gives the concrete evidence that using proposed spatio-temporal pattern has the great potential in predicting crime. The ultimate goal of our research is to build a crime prediction system with strong predictive power, which is able to provide forecast in a timely manner and requires less amount of data inputs. Ultimately, the law enforcement is able to fight criminals pro-actively instead of passively.

Acknowledgments. The work was partially funded by the National Institute of Justice (No.2009- DE-BX-K219).

References

1. Cohen, J., Gorr, W.L.: Development of crime forecasting and mapping systems for use by police. H. John Heinz III School of Public Policy and Management. Carnegie Mellon University (2005)
2. Cohen, W., Singer, Y.: A simple, fast, and effective rule learner. In: Proceedings of The Sixteenth National Conference on Artificial Intelligence (1999)
3. Eck, J., Chainey, S., Cameron, J., Wilson, R.: Mapping crime: Understanding hotspots (2005)
4. Freund, Y., Schapire, R.: A decision-theoretic generalization of on-line learning and an application to boosting. In: Computational Learning Theory: Eurocolt, pp. 23–37. Springer (1995)
5. Holmes, G., Pfahringer, B., Kirkby, R., Frank, E., Hall, M.: Multiclass alternating decision trees. In: Elomaa, T., Mannila, H., Toivonen, H. (eds.) ECML 2002. LNCS (LNAI), vol. 2430, pp. 161–172. Springer, Heidelberg (2002)

6. Kelling, G., Coles, C.: Fixing Broken Windows: Restoring Order and Reducing Crime in Our Communities. Free Press (1998)
7. Kumar, M.V., Chandrasekar, C.: Spatial clustering simulation on analysis of spatialtemporal crime hotspot for predicting crime activities. International Journal of Computer Science and Information Technologies 2(6), 2864–2867 (2011)
8. Mohler, G.O., Short, M.B., Brantingham, P.J., Schoenberg, F.P., Tita, G.E.: Self-exciting point process modeling of crime. Journal of the American Statistical Association 106(493) (2011)
9. Ratcliffe, J.H.: Aoristic signatures and the spatio-temporal analysis of high volume crime patterns. Journal of Quantitative Criminology 18(1), 23–43 (2002)
10. Schapire, R.E., Singer, Y.: Improved boosting algorithms using confidence-rated predictions. Machine learning 37(3), 297–336 (1999)
11. Short, M.B., Bertozzi, A.L., Brantingham, P.J.: Nonlinear patterns in urban crime: Hotspots, bifurcations, and suppression. SIAM Journal on Applied Dynamical Systems 9(2), 462–483 (2010)
12. Townsley, M., Homel, R., Chaseling, J.: Infectious burglaries. A test of the near repeat hypothesis. British Journal of Criminology 43(3), 615–633 (2003)

NLPMM: A Next Location Predictor with Markov Modeling

Meng Chen[1], Yang Liu[1], and Xiaohui Yu[1,2,*]

[1] School of Computer Science and Technology, Shandong University, Jinan, China, 250101
[2] School of Information Technology, York University, Toronto, ON, Canada, M3J 1P3
`chenmeng114@hotmail.com`, `{yliu,xyu}@sdu.edu.cn`

Abstract. In this paper, we solve the problem of predicting the next locations of the moving objects with a historical dataset of trajectories. We present a Next Location Predictor with Markov Modeling (*NLPMM*) which has the following advantages: (1) it considers both individual and collective movement patterns in making prediction, (2) it is effective even when the trajectory data is sparse, (3) it considers the time factor and builds models that are suited to different time periods. We have conducted extensive experiments in a real dataset, and the results demonstrate the superiority of *NLPMM* over existing methods.

Keywords: moving pattern, next location prediction, time factor.

1 Introduction

The prevalence of positioning technology has made it possible to track the movements of people and other objects, giving rise to a variety of location-based applications. For example, GPS tracking using positioning devices installed on the vehicles is becoming a preferred method of taxi cab fleet management. In many social network applications (e.g., Foursquare), users are encouraged to share their locations with other users. Moreover, in an increasing number of cities, vehicles are photographed when they pass the surveillance cameras installed over highways and streets, and the vehicle passage records including the license plate numbers, the time, and the locations are transmitted to the data center for storage and further processing.

In many of these location-based applications, it is highly desirable to be able to accurately predict a moving object's next location. Consider the following example in location-based advertising. Lily has just shared her location with her friends on the social network website. If the area she will pass by is known in advance, it is possible to push plenty of information to her, such as the most popular restaurant and the products on sale in that area. As another example, if we could predict the next locations of vehicles on the road, then we will be able to forecast the traffic conditions and recommend more reasonable routes to drivers to avoid or alleviate traffic jams.

Several methods have been proposed to predict next locations, most of which fall into one of two categories: (1) methods that use only the historical trajectories of individual objects to discover individual movement patterns [7, 12], and (2) methods that use the

---

* Corresponding Author.

V.S. Tseng et al. (Eds.): PAKDD 2014, Part II, LNAI 8444, pp. 186–197, 2014.

historical trajectories of all objects to identify collective movement patterns [10, 11]. The majority of the existing methods train models based on frequent patterns and/or association rules to discover movement patterns for prediction.

However, there are a few major problems with the existing methods. First, those methods focus on either the individual patterns or the collective patterns, but very often the movements of objects reflect both individual and collective properties. Second, in some circumstances (e.g., social check-in, and vehicle surveillance), the data points are very sparse; the trajectories of some objects may consist of only one record. One cannot construct meaningful frequent patterns with these trajectories. Finally, the existing methods do not give proper consideration to the time factor. Different movement patterns exist in different time, for example, Bob is going to leave his house. If it is 8 a.m. on a weekday, he is most likely to go to work. But if it is 11:30 a.m., he is more likely to go to a restaurant, and he may go shopping if it is 3 p.m on weekends. Failing to take time factor into account would result in higher error rates in predicting the next locations.

To address those problems, we propose a Next Location Predictor with Markov Modeling (*NLPMM*) to predict the next locations of moving objects given past trajectory sequences. *NLPMM* builds upon two models: the Global Markov Model (*GMM*) and the Personal Markov Model (*PMM*). *GMM* utilizes all available trajectories to discover global behaviours of the moving objects based on the assumption that they often share similar movement patterns (e.g., people driving from A to B often take the same route). *PMM*, on the other hand, focuses on modeling the individual patterns of each moving object using its own past trajectories. The two models are combined using linear regression to produce a more complete and accurate predictor.

Another distinct feature of NLPMM lies in its treatment of the time factor. The movement patterns of objects vary from one time period to another (e.g., weekdays vs. weekends). Meanwhile, similarities also exist for different time periods (e.g., this Monday and next), and the movement patterns of moving objects tend to be cyclical. We thus propose to cluster the time periods based on the similarity in movement patterns and build a separate model for each cluster.

The performance of *NLPMM* is evaluated in a real dataset consisting of the vehicle passage records over a period of 31 days (1/1/2013 - 1/31/2013) in a metropolitan area. The experimental results confirm the superiority of the proposed methods over existing methods.

The contributions of this paper can be summarized as follows.

- We propose a Next Location Predictor with Markov Modeling to predict the next location a moving object will arrive at. To the best of our knowledge, *NLPMM* is the first model that takes a holistic approach and considers both individual and collective movement patterns in making prediction. It is effective even when the trajectory data is sparse.

- Based on the important observation that the movement patterns of moving objects often change over time, we propose methods that can capture the relationships between the movement patterns in different time periods, and use this knowledge to build more refined models that are better suited to different time periods.

- We conduct extensive experiments using a real dataset and the results demonstrate the effectiveness of *NLPMM*.

The remainder of this paper is organized as follows. Section 2 reviews related work. Section 3 gives the preliminaries of our work. Section 4 describes our approach of Markov modeling. Section 5 presents methods that take the time factor into consideration. The experimental results and performance analysis are presented in Section 6. Section 7 concludes this paper.

2   Related Work

There have appeared a considerable body of work on knowledge discovery from trajectories, where a trajectory is defined as a sequence of locations ordered by time-stamps. In what follows, we discuss three categories of studies that are most closely related to us.

Route planning: Several studies use GPS trajectories for route planning through constructing a complete route [2, 3, 16]. Chen et al. search the k Best-Connected Trajectories from a database [3] and discover the most popular route between two locations [2]. Yuan et al. find the practically fastest route to a destination at a given departure time using historical taxi trajectories [16].

Long-range prediction: Long-range prediction is studied in [5, 8], where they try to predict the whole future trajectory of a moving object. Krumm proposes a Simple Markov Model that uses previously traversed road segments to predict routes in the near future [8]. Froehlich and Krumm use previous GPS traces to make a long-range prediction of a vehicle's trajectory [5].

Short-range prediction: Short-range prediction has been widely investigated [7, 10–12], which is concerned with the prediction of only the next location. Some of these methods make prediction with only the individual movements [7, 12], while others use the historical movements of all the moving objects [10, 11]. Xue et al. construct a Probabilistic Suffix Tree (PST) for each road using the taxi traces and propose a method based on Variable-order Markov Models (VMMs) for short-term route prediction [12]. Jeung et al. present a hybrid prediction model to predict the future locations of moving objects, which combine predefined motion functions using the object's recent movements with the movement patterns of the object [7]. Monreale et al. use the previous movements of all moving objects to build a T-pattern tree to make future location prediction [10]. Morzy uses a modified version of the PrefixSpan algorithm to discover frequent trajectories and movement rules with all the moving objects' locations [11].

In addition to the three aforementioned categories of work, there has also appeared work on using social-media data for trajectory mining [9, 13]. Kurashima et al. recommend travel routes based on a large set of geo-tagged and time-stamped photographs [9]. Ye et al. utilize a mixed Hidden Markov Model to predict the category of a user's next activity and then predict a location given the category [13].

3   Preliminaries

In this section, we will explain a few terms that are required for the subsequent discussion, and define the problem addressed in this paper.

Definition 1 (Sampling Location). *For a given moving object o, it passes through a set of* sampling locations, *where each sampling location refers to a point or a region (in a two-dimensional area of interest) where the position of o is recorded.*

For example, the positions of the cameras in the traffic surveillance system can be considered as the sampling locations.

Definition 2 (Trajectory Unit). *For a given moving object o, a* trajectory unit, *denoted by u, is the basic component of its trajectory. Each trajectory unit u can be represented by $(u.l, u.t)$, where $u.l$ is the id of the sampling location of the moving object at time-stamp $u.t$.*

Definition 3 (Trajectory). *For a moving object, its* trajectory T *is defined as a time-ordered sequence of trajectory units: $< u_1, u_2, \ldots, u_n >$.*

From Definition 2, T can also be represented as $< (u_1.l, u_1.t), (u_2.l, u_2.t), \ldots, (u_n.l, u_n.t) >$ where $u_i.t < u_{i+1}.t$ $(1 \leqslant i \leqslant n - 1)$.

Definition 4 (Candidate Next Locations). *For the sampling location $u_i.l$, we define a sampling location $u_j.l$ as a* candidate next location *of $u_i.l$ if a moving object can reach $u_j.l$ from $u_i.l$ directly.*

The set of candidate next locations can be obtained either by prior knowledge (e.g., locations of the surveillance cameras combined with the road network graph), or by induction from historical trajectories of moving objects.

Definition 5 (Sampling Location Sequence). *For a given trajectory $< (u_1.l, u_1.t), (u_2.l, u_2.t), \ldots, (u_n.l, u_n.t) >$, its* sampling location sequence *refers to a sequence of sampling locations appearing in the trajectory, denoted as $< u_1.l, u_2.l, \ldots, u_n.l >$.*

Definition 6 (Prefix Set). *For a sampling location $u_i.l$ and a given set of trajectories $\mathcal{T}$, its* prefix set *of size N, denoted by $\mathcal{S}_i^N$, refers to the set of sequences such that each sequence is a length N subsequence that immediately precedes $u_{i+1}.l$ in the sampling location sequence of some trajectory $T \in \mathcal{T}$.*

4 Markov Modeling

We choose to use Markov models to solve the next location prediction problem. Specifically, a state in the Markov model corresponds to a sampling location, and state transition corresponds to moving from one sampling location to the next.

In order to take into consideration both the collective and the individual movement patterns in making the prediction, we propose two models, a Global Markov Model (*GMM*) to model the collective patterns, and a Personal Markov Model (*PMM*) to model the individual patterns and solve the problem of data sparsity. They are combined using linear regression to generate a predictor.

4.1 Global Markov Model

Using historical trajectories, we can train an order-N *GMM* to give a probabilistic prediction over the next sampling locations for a moving object, where N is a user-chosen
parameter. Let $P\left(l_i\right)$ represents a discrete probability of a moving object arriving at
sampling location l_i. The order-N *GMM* implies that the probability distribution $P(l')$
for the next sampling location l' of a given moving object o is independent of all but
the immediately preceding N locations that o has arrived at:

$$P(l' \mid < l_j, \ldots, l_i >) = P(l' | \mathcal{S}_i^N) \tag{1}$$

For a given trajectory dataset, an order-N *GMM* for the sampling location l_i can be
trained in the following way. We first construct the prefix set $\mathcal{S}_i^N$. Next, for every prefix
in $\mathcal{S}_i^N$, we compute the frequency of each distinct sampling location appearing after this
prefix in the dataset. These frequencies are then normalized to get a discrete probability
distribution over the next sampling location.

We start with a first order *GMM*, followed by a second-order *GMM*, etc., until the
order-N *GMM* has been obtained, to train a variable-order GMM. In contrast to the
order-N *GMM*, the variable-order *GMM* learns such conditional distributions with a
varying N and provides the means of capturing different orders of Markov dependencies based on the observed data. There exist many ways to utilize the variable-order
GMM for prediction. Here we adopt the principle of longest match. That is, for a given
sampling location sequence ending with l_i, we find its longest suffix match from the set
of sequences in the prefix set of l_i.

4.2 Personal Markov Model

The majority of people's movements are routine (e.g., commuting), and they often have
their own individual movement patterns. In addition, about 73% of trajectories in our
dataset contain only one point, but they also can reflect the characteristics of the moving
objects' activities. For example, someone who lives in the east part of the city is unlikely
to travel to a supermarket 50 kilo-meters away from his home. Therefore, we propose a
Personal Markov Model (*PMM*) for each moving object to predict next locations.

The training of PMM consists of two parts: training a variable-order Markov model
for every moving object using its own trajectories of length than 1, and a zero-order
Markov model for every moving object using the trajectory units.

For training the variable-order Markov model, we construct the prefix set for every
moving object using its own trajectories, and then we compute the probability distribution of the next sampling locations. Specially, we iteratively train a variable-order
Markov model with order i ranging from 1 to N using the trajectories of one moving
object.

We train a zero-order Markov model using the trajectory units. For a moving object,
let $N(l')$ denotes the number of times a sampling location l' appears in the training
trajectories. Let $L_{l'}$ be the set of distinct sampling locations appearing in the training
trajectories. Then we have

$$P(l') = \frac{N(l')}{\sum_{l \in L_{l'}} N(l)}. \tag{2}$$

The zero-order Markov model can be seamlessly integrated with the variable-order Markov model to obtain the final *PMM*.

4.3 Integration of GMM and PMM

There are many methods to combine the results from a set of predictors. For our problem, we choose to use linear regression to integrate the two models we have proposed.

For the given i-th trajectory sequence, both GMM and PMM can get a vector of probabilities, $\mathbf{p}_i^w = \left(p_1^i, p_2^i, \cdots, p_m^i\right)'$ ($w = 1$ for *GMM* and $w = 2$ for *PMM*), where m is the number of the sampling locations, and p_j^i is the probability of location j being the next sampling location. We also have a vector of indicators $\mathbf{y}_i = (y_1^i, y_2^i, \cdots, y_m^i)'$ for the i-th trajectory sequence, where $y_j^i = 1$ if the actual next location is j and 0 otherwise. We can predict $\mathbf{y}_i$ through a linear combination of the vectors generated by *GMM* and *PMM*:

$$\hat{\mathbf{y}}_i = \beta_0 \mathbf{1} + \sum_{w=1}^{2} \beta_w \mathbf{p}_i^w \tag{3}$$

where $\mathbf{1}$ is a unit vector, and β_0, β_1, and β_2 are the coefficients to be estimated.

Given a set of n training trajectories, we can compute the optimal values of β_i through standard linear regression that minimizes $\sum_{i=1}^{n} ||\mathbf{y}_i - \hat{\mathbf{y}}_i||$, where $|| \cdot ||$ is the Euclidean norm. The β_i values thus obtained can then be used for prediction. For a particular trajectory, we can predict the top k next sampling locations by identifying the k largest elements in the estimator $\hat{\mathbf{y}}$.

5 Time Factor

The movement of human beings demonstrates a great degree of temporal regularity [1, 6]. In this section, we will first discuss how the movement patterns are affected by time, and then show how to improve the predictor proposed in the preceding section by taking the time factor into consideration.

5.1 Observations and Discussions

We illustrate how time could affect people's movement patterns through Figure 1. In this case, for a sampling location l, there are seven candidate next locations, and the distributions over those locations do differ from one period to another. For instance, vehicles are most likely to arrive at the fifth location during the period from 9:00 to 10:00, whereas the most probable next location is the second for the period from 14:00 to 15:00.

Therefore, the prediction model should be made time-aware, and one way to do this is to train different models for different time periods. In what follows, we will explore a few methods to determine the suitable time periods. Here, we choose *day* as the whole time span, i.e., we study how to find movement patterns within a day. However, any other units of time, such as *hour*, *week* or *month*, could also be used depending on the scenario.

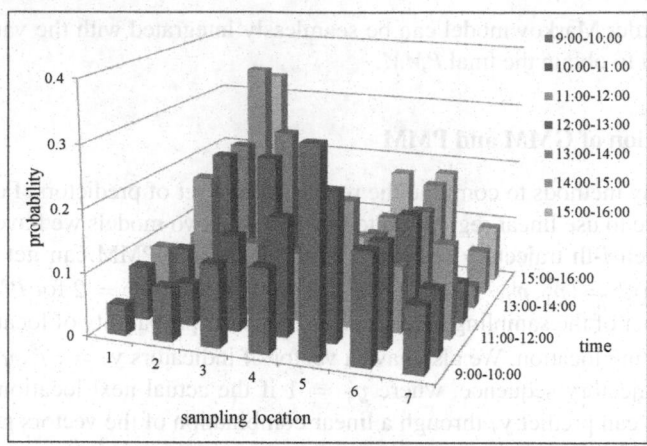

Fig. 1. An example of time affecting people's movement patterns

5.2 Time Binning

A straight-forward approach is to partition the time span into a given number (M) of equi-sized time bins, and all trajectories are mapped to those bins according to their time stamps. A trajectory spanning over more than one bin is split into smaller sub-trajectories such that the trajectory units in each sub-trajectory all fall in the same bin. We then train M independent models, each for a different time bin, using the trajectories falling in each bin. Prediction is done by choosing the right model based on the time-stamp. We call this approach *Time Binning (TB)*.

However, this approach has some limitations: the sizes of all time bins are equal, rendering it difficult to find the correct bin sizes that fit all movement patterns in the time span, as some patterns manifest themselves over longer periods whereas others shorter. One possible improvement to *TB* is to start with a small bin size, and gradually merge the time bins whose distributions are considered similar by some metric.

5.3 Distributions Clustering

We propose a method called Distributions Clustering (*DC*) to perform clustering of the time bins based on the similarities of the probability distributions in each bin. Here, the probability distribution refers to the transition probability from one location to another. Compared with *TB*, the trajectories having similar probability distributions are expected to be put in one cluster, leading to clearer revelation of the moving patterns. Here, we use cosine similarity to measure the similarities between the distributions, but the same methodology still applies when other distance metrics such as the Kullback-Leibler divergence [4] are used.

For an object o appearing at a given sampling location l with a time point falling into the ith time bin, let p_i^m be an m-dimensional vector that represents the probabilities of o moving from l to another location, where m is the total number of sampling locations. We measure the similarity of two time bins i and j (with respect to o) using the cosine similarity, $\cos_{ij} = \left(p_i^m . p_j^m\right) / \left(|p_i^m| . |p_j^m|\right)$. With the similarity metric defined, we can perform clustering for each sampling location l on the time bins. The algorithm is detailed in Algorithm 1. The results will be a set of clusters, each containing a set of time bins, for the sampling location l.

Algorithm 1. DC: Detecting Q clusters for the M time bins

Input: cluster number Q, time bins number M and the probability distributions of trajectories in each time bin;
Output: the clusters;
1. random select Q time bins as the initial cluster centres;
2. **repeat**
3.   calculate the similarity of the probability distributions of trajectories in each time bin and the cluster centres;
4.   assign each time bin to the cluster centre with the maximum similarity;
5.   recalculate the probability distributions of trajectories in the cluster centres;
6. **until** clusters do not change or the maximum number of iterations has been reached
7. return the clusters;

For a given location l_i, we can get Q clusters, defined as $C_i^k, k = 1, 2, \cdots, Q$. Combined with the order-N Markov model, the probability distribution $P(l')$ for the next sampling location l' of a given moving object o can be computed with the formula:

$$P(l' | < (u_j.l, u_j.t), \ldots, (u_i.l, u_i.t) >) = P(l'|C_i^k, \mathcal{S}_i^N) \tag{4}$$

We then train Q models with the trajectories in each cluster to form a new model *NLPMM-DC* (which stands for *NLPMM with Distributions Clustering*). In the new model, the sequence of just-passed locations and the time factor are both utilized by combing distributions clustering and Markov model.

6 Performance Evaluation

We have conducted extensive experiments to evaluate the performance of the proposed *NLPMM* using a real vehicle passage dataset. In this section, we will first describe the dataset and experimental settings, followed by the evaluation metrics to measure the performance. We then show the experimental results.

6.1 Datasets and Settings

The dataset used in the experiments consists of real vehicle passage records from the traffic surveillance system of a major metropolitan area with a 6-million population. The dataset contains 10,344,058 records during a period of 31 days (from January 1, 2013 to January 31, 2013). Each record contains three attributes, the license plate number of the vehicle, the ID of the location of the surveillance camera, and the time of vehicle passing the location. There are about 300 camera locations on the main roads. The average distance between a neighboring pair of camera locations is approximately 3 kilometers.

6.2   Pre-processing

We pre-process the dataset to form trajectories, resulting in a total of 6,521,841 trajectories. According to statistics, the trajectories containing only one point account for about 73% of all trajectories, which testifies to the sparsity of data sampling. We choose a total of 1,760,897 trajectories with the length greater than one to calculate the number of candidate next locations for every sampling location. Due to the sparsity of camera locations, about 86.3% of the sampling locations have more than 10 candidate next sampling locations, and the average number of candidate next locations is about 43. We predict top-k next sampling locations in the experiments.

6.3   Evaluation Metrics

Our evaluation uses the following metrics that are widely employed in multi-label classification studies [14].

Prediction Coverage: It is defined as the percentage of trajectories for which the next location can be predicted based on the model. Let $c(l)$ be 1 if it can be predicted and 0 otherwise. Then $PreCov_\mathcal{T} = \sum_{l \in \mathcal{T}} c(l) / |\mathcal{T}|$, where $|\mathcal{T}|$ denotes the total number of trajectories in the testing dataset.

Accuracy: It is defined as the frequency of the true next location occurring in the list of predicted next locations. Let $p(l)$ be 1 it does and 0 otherwise. Then $accuracy_\mathcal{T} = \sum_{l \in \mathcal{T}} p(l) / |\mathcal{T}|$.

One-error: It is defined as the frequency of the top-1 predicted next location not being the same as the true next location. Let $e(l)$ be 0 if the top-1 predicted sampling location is the same as the true next location and 1 otherwise. Then $one - error_\mathcal{T} = |\sum_{l \in \mathcal{T}} e(l) / \mathcal{T}|$.

Average Precision: Given a list of top-k predicted next locations, the average precision is defined as $AvePrec_\mathcal{T} = \sum_{l \in \mathcal{T}} (p(i)/i) / |\mathcal{T}|$, where i denotes the position in the predicted list, and $p(i)$ takes the value of 1 if the predicted location at the i-th position in the list is the actual next location.

6.4   Evaluation of NLPMM

We evaluate the performance of *NLPMM* and its components, *PMM*, and *GMM*. For each experiment, we perform 50 runs and report the average of the results. First, we study the effect of the order of the Markov model by varying N from 1 to 6. Figure 2(a) shows that the accuracy has an apparent improvement when the order N increases from 1 to 2 for all models. The accuracy reaches the maximum when N is set to 3 and remains stable as N increases further. Therefore, we set N to 3 in the following experiments. Next, we evaluate the effect of top k on *PMM*, *GMM*, and *NLPMM*. From Figure 2(b), we can observe that the accuracy of all three models improves as k increases. Furthermore, the accuracy of GMM and NLPMM is significantly better than that of *PMM*, and the best results are given by *NLPMM*. Since the average number of candidate next

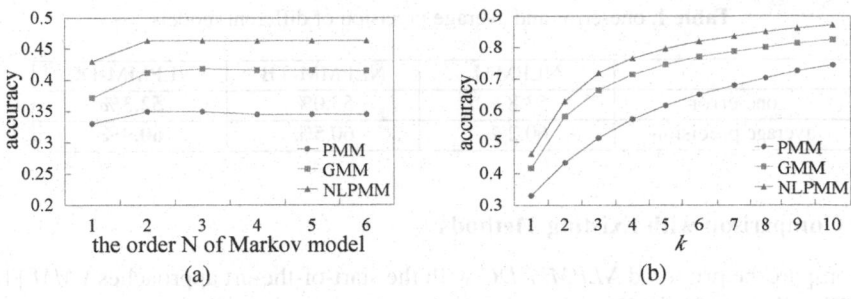

Fig. 2. Performance of PMM, GMM, and NLPMM

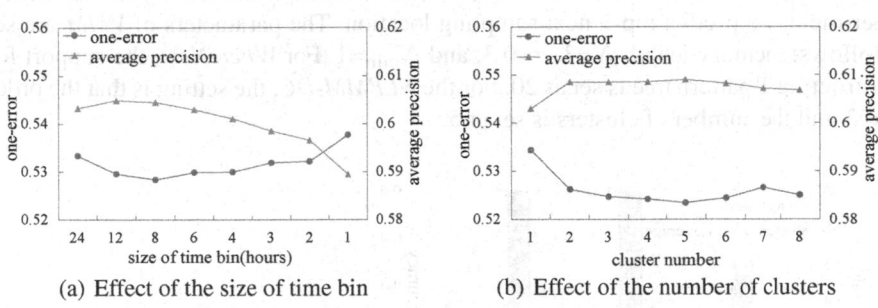

(a) Effect of the size of time bin   (b) Effect of the number of clusters

Fig. 3. Effect of parameters

locations is 43 (meaning there are 43 possibilities), the accuracy of 0.88 is surprisingly good when k is set to 10.

6.5   Effect of the Time Factor

We evaluate the proposed methods that take into consideration of the time factor. Figure 3(a) shows the effect of bin size on *NLPMM-TB* (which stands for *NLPMM with Time Binning*). The performance of *NLPMM-TB* starts to deteriorate when the bin size becomes less than 8, because when the bins get smaller, the trajectories in them become too sparse to generate a meaningful collective pattern. Figure 3(b) shows the effect of the number of clusters on *NLPMM-DC* (which stands for *NLPMM with Distributions Clustering*). When it is set to 1, the model is the same as *NLPMM*. The one-error rate declines and the average precision improves as the number increases from 1 to 5. When it continues to increase, the result starts to get worse. This is because having too many or too few clusters with either hurt the cohesiveness or the separation of the clusters.

We evaluate the performance of *NLPMM*, *NLPMM-TB* and *NLPMM-DC* using one-error and average precision. The results are shown in Table 1. *NLPMM-TB* and *NLPMM-DC* perform better than *NLPMM*, which is because we can get a more refined model by adding the time factor and generate more accurate predictions. *NLPMM-DC* performs best, validating the effectiveness of the method of distributions clustering. It will be used in the following comparison with alternative methods.

Table 1. one-error and average precision of different models

	NLPMM	NLPMM-TB	NLPMM-DC
one-error	53.8%	53.0%	**52.3%**
average precision	60.2%	60.5%	**60.9%**

6.6 Comparison with Existing Methods

We compare the proposed *NLPMM-DC* with the start-of-the-art approaches *VMM* [12] and *WhereNext* [10]. *VMM* uses individual trajectories to predict the next locations, whereas *WhereNext* uses all available trajectories to discover collective patterns. In this experiment, we predict top-1 next sampling location. The parameters of *VMM* are set as follows: memory length N=3, σ=0.3, and N_{min}=1. For *WhereNext*, the support for constructing T-pattern tree is set as 20. For the *NLPMM-DC*, the setting is that the order $N = 3$ and the number of clusters is set at 5.

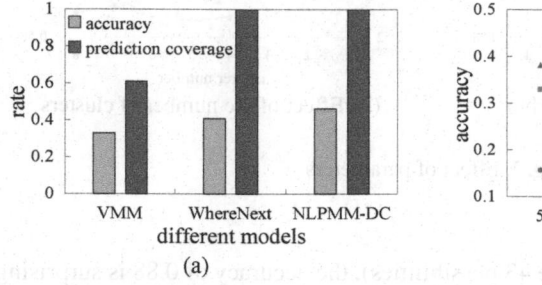 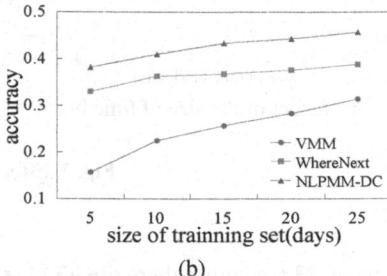

(a)   (b)

Fig. 4. Performance comparison of *NLPMM-DC*, *VMM*, and *WhereNext*

Figure 4 shows the performance comparison of *NLPMM-DC*, *VMM* and *WhereNext* in terms of prediction coverage and accuracy. As shown in Figure 4(a), *NLPMM-DC* performs the best, which can be attributed to the combination of individual and collective patterns as well as the consideration of time factor. Figure 4(b) shows that the accuracy of each model improves as the size of training set increases. It is worth mentioning that *NLPMM-DC* performs better than *VMM* and *WhereNext* in terms of accuracy for any training set size.

7   Conclusions

In this paper, we have proposed a Next Location Predictor with Markov Modeling to predict the next sampling location that a moving object will arrive at with a given trajectory sequence. The proposed *NLPMM* consists of two models: Global Markov Model and Personal Markov Model. Time factor is also added to the models and we propose two methods to partition the whole time span into periods of finer granularities, including Time Binning and Distributions Clustering. New time-aware models are trained accordingly. We have evaluated the proposed models using a real vehicle passage record dataset. The experiments show that our predictor significantly outperforms the state-of-the-art methods (*VMM* and *WhereNext*).

Acknowledgement. This work was supported in part by the National Natural Science Foundation of China Grant (No. 61272092), the Program for New Century Excellent Talents in University (NCET-10-0532), the Natural Science Foundation of Shandong Province of China Grant (No. ZR2012FZ004), the Independent Innovation Foundation of Shandong University (2012ZD012), the Taishan Scholars Program, and NSERC Discovery Grants. The authors would like to thank the anonymous reviewers, whose valuable comments helped improve this paper.

References

1. Ben-Elia, E., Shiftan, Y.: Which road do i take? A learning-based model of route-choice behavior with real-time information. Transportation Research Part A: Policy and Practice 44(4), 249–264 (2010)
2. Chen, Z., Shen, H.T., Zhou, X.: Discovering popular routes from trajectories. In: ICDE, pp. 900–911 (2011)
3. Chen, Z., Shen, H.T., Zhou, X., Zheng, Y., Xie, X.: Searching trajectories by locations: an efficiency study. In: SIGMOD, pp. 255–266 (2010)
4. Ertoz, L., Steinbach, M., Kumar, V.: A new shared nearest neighbor clustering algorithm and its applications. In: SDM, pp. 105–115 (2002)
5. Froehlich, J., Krumm, J.: Route prediction from trip observations. SAE SP 2193, 53 (2008)
6. Gonzalez, M.C., Hidalgo, C.A., Barabasi, A.L.: Understanding individual human mobility patterns. Nature 453(7196), 779–782 (2008)
7. Jeung, H., Liu, Q., Shen, H.T., Zhou, X.: A hybrid prediction model for moving objects. In: ICDE, pp. 70–79 (2008)
8. Krumm, J.: A markov model for driver turn prediction. SAE SP 2193(1) (2008)
9. Kurashima, T., Iwata, T., Irie, G., Fujimura, K.: Travel route recommendation using geotags in photo sharing sites. In: CIKM, pp. 579–588 (2010)
10. Monreale, A., Pinelli, F., Trasarti, R., Giannotti, F.: Wherenext: A location predictor on trajectory pattern mining. In: SIGKDD, pp. 637–646 (2009)
11. Morzy, M.: Mining frequent trajectories of moving objects for location prediction. In: Perner, P. (ed.) MLDM 2007. LNCS (LNAI), vol. 4571, pp. 667–680. Springer, Heidelberg (2007)
12. Xue, G., Li, Z., Zhu, H., Liu, Y.: Traffic-known urban vehicular route prediction based on partial mobility patterns. In: ICPADS, pp. 369–375 (2009)
13. Ye, J., Zhu, Z., Cheng, H.: Whats your next move: User activity prediction in location-based social networks. In: SDM, pp. 171–179 (2013)
14. Ye, M., Shou, D., Lee, W.C., Yin, P., Janowicz, K.: On the semantic annotation of places in location-based social networks. In: SIGKDD, pp. 520–528 (2011)
15. Yin, Z., Cao, L., Han, J., Luo, J., Huang, T.: Diversified trajectory pattern ranking in geotagged social media. In: SDM, pp. 980–991 (2011)
16. Yuan, J., Zheng, Y., Zhang, C., Xie, W., Xie, X., Sun, G., Huang, Y.: T-drive: Driving directions based on taxi trajectories. In: GIS, pp. 99–108 (2010)

Beyond Poisson: Modeling Inter-Arrival Time of Requests in a Datacenter

Da-Cheng Juan[1], Lei Li[2], Huan-Kai Peng[1],
Diana Marculescu[1], and Christos Faloutsos[3]

[1] Electrical and Computer Engineering, Carnegie Mellon University
{dacheng,pumbaapeng,dianam}@cmu.edu
[2] Computer Science Division, University of California, Berkeley
leili@cs.berkeley.edu
[3] School of Computer Science, Carnegie Mellon University
christos@cs.cmu.edu

Abstract. How frequently are computer jobs submitted to an industrial-scale datacenter? We investigate the trace that contains job requests and execution collected in one of large-scale industrial datacenters, which spans near half of a Terabyte. In this paper, we discover and explain two surprising patterns with respect to the inter-arrival time (IAT) of job requests: (a) multiple periodicities and (b) multi-level bundling effects. Specifically, we propose a novel generative process, Hierarchical Bundling Model (HIBM), for modeling the data. HIBM is able to mimic multiple components in the distribution of IAT, and to simulate job requests with the same statistical properties as in the real data. We also provide a systematic approach to estimate the parameters of HIBM.

1 Introduction

What are the major characteristics of job inter-arrival process in a datacenter? Could we develop a tool to create synthetic inter-arrivals that match the properties of the empirical data? Understanding the characteristics of job inter-arrivals is the key to design effective scheduling policies to manage massively-integrated and virtually-shared computing resources in a datacenter. Conventionally, during the development of a cloud-based scheduler, job requests are assumed (1) to be submitted independently and (2) to follow a constant rate λ, which results in a simple and elegant model, Poisson process (PP). PP generates independent and identically distributed (i.i.d.) inter-arrival time (IAT) that follows an (negative) exponential distribution [5]. However, in reality, how much does this inter-arrival process deviate from PP?

To demonstrate how the real inter-arrival process deviates from PP, we use Fig. 1 to present the histogram of the IAT for 668,000 jobs submitted and collected in an industrial, large-scale datacenter. The resolution of IAT is 1 microsecond (μs, 10^{-6} sec). As Fig. 1(a) shows, the IATs "seem" to follow an (negative) exponential distribution. However, in logarithmic scale as Fig. 1(b) shows, surprisingly, *four* distinct clusters (denoted as $\mathcal{A}$, $\mathcal{B}$, $\mathcal{C}$ and $\mathcal{D}$) with either center-or left-skewed shapes can be seen. This distribution (or a mixture of distributions) clearly does not follow an (negative) exponential distribution, which is always right-skewed in logarithmic scale and therefore cannot create such shapes. This phenomenon has confirmed that the i.i.d. assumption

V.S. Tseng et al. (Eds.): PAKDD 2014, Part II, LNAI 8444, pp. 198–209, 2014.

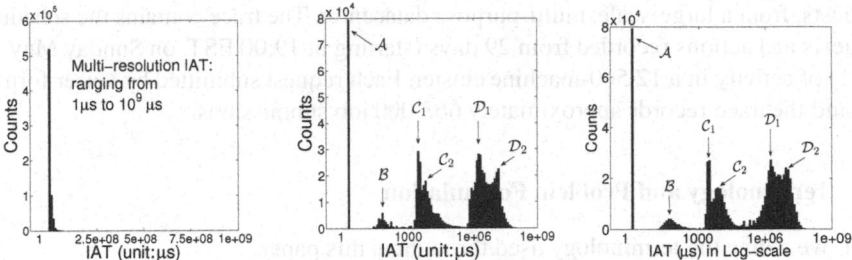

(a) Empirical IAT in lin. scale (b) Empirical IAT in log scale (c) Synthetic IAT in log scale

Fig. 1. Deviation from Poisson Process: (a) Histogram of job IAT ($\approx$ 668,000 jobs) in linear-scale. (b) Same histogram in log-scale. (c) Synthetic IATs from HIBM. In (a), the histogram has limited number of bins to demonstrate IATs of such a fine-resolution, and the marginal distribution may be misidentified as an (negative) exponential distribution. In (b), *four* distinct clusters can be seen: $\mathcal{A}$: $1\mu s$, $\mathcal{B}$: 10-$10^3 \mu s$, $\mathcal{C}$: 10^3-$10^5 \mu s$, and $\mathcal{D}$: 10^6-$10^9 \mu s$. All four clusters are captured by HIBM as shown in (c).

of PP barely holds since certain job requests may depend on one another. For example, a request of disk-backup may immediately be submitted after a request of Gmail service; this dependency violates the i.i.d assumption and thus invalidates conventional statistical analysis. In this paper we aim at solving the following two problems:

- **P1: Find Patterns.** How to characterize this marginal distribution?
- **P2: Pattern-Generating Mechanism.** What is a possible mechanism that can generate such job inter-arrivals?

This work brings the following two contributions:

- **Pattern Discovery.** Two key patterns of job inter-arrivals are provided: (1) multiple periodicities and (2) bundling effects. We show the majority (approximately 78%) of job requests show a regular periodicity with a *log-logistic* noise, a skewed, power-law-like distribution. Furthermore, the submission of a job may depend on the occurrence of its previous job, and we refer to this dependency as the *bundling effect*, since these two associated jobs are considered to belong to the same bundle.
- **Generative Model.** We propose HIBM, a "HIerarchical Bundling Model," that is succinct and interpretative. HIBM's mathematical expression is succinct that requires only a handful of parameters to create synthetic job inter-arrivals matching the characteristics of empirical data, as shown in Fig. 1(c). Furthermore, HIBM has the capability to explain the attribution of the four clusters ($\mathcal{A}$, $\mathcal{B}$, $\mathcal{C}$ and $\mathcal{D}$) and the "spikes" ($\mathcal{A}$, $\mathcal{C}_1$, $\mathcal{C}_2$, $\mathcal{D}_1$, and $\mathcal{D}_2$) in Fig. 1(b).

The remainder of this paper is organized as follows. Section 2 provides the problem definition, Section 3 details the proposed HIBM, Section 4 provides the discussions and Section 5 surveys the previous work. Finally, Section 6 concludes this paper.

2 Problem Definition

In this work, we use the trace from Google's cluster [12], which is the first publicly available dataset that presents the diversity and dynamic behaviors of real-world service requests, from a large-scale, multi-purpose datacenter. The trace contains the scheduler requests and actions recorded from 29 days (starting at 19:00 EST, on Sunday May 1st, 2011) of activity in a 12,500-machine cluster. Each request submitted by a user forms a *job* and the trace records approximately 668,000 job submissions.

2.1 Terminology and Problem Formulation

First, we define the terminology used throughout this paper.

Definition 1 (Job type and job instance). *"Job type" represents a certain type of job that can occur once or multiple times, and "job instance" is the actual occurrence of a job request.*

For example, "disk-backup" is a job type that can instantiate several requests; each request (such as "disk-backup at 1:00P.M. on May 2nd") is a job instance.

Definition 2 (Job bundle). *"Job bundle" represents the association of two job types – if two job types are in the same job bundle, the IATs of their job instances will be correlated.*

Like the example used in Section 1, two job types "disk-backup" and "Gmail" are functionally-associated, and thus they are considered belonging to the same job bundle. In this case, the inter-arrival of each disk-backup instance will depend on the occurrence of each Gmail instance.

Definition 3 (Job class). *"Job class" represents the priority (or latency sensitiveness) of a job type. In the trace, job class is enumerated as $\{0, 1, 2, 3\}$ with a job type of class 3 being the highest priority.*

As mentioned in the Introduction, we have two goals:

- P1: **Find patterns.** Given (1) the job type j, (2) the time stamp of its ith instance (denoted as $t_{j,i}$), and (3) the job class, find the most distinct patterns that are sufficient to characterize the IATs of all job instances in a datacenter.
- P2: **Pattern-generating mechanism.** Given the patterns found in P1, design a model that can generate IATs that match these characteristics of the empirical data and report the model parameters.

2.2 Dataset Exploration

We begin this section by illustrating the number of job instances over time in Fig. 2(a). We collect the time stamp of each job instance when it is first submitted to the datacenter, and then aggregate the total number of job instances within each hour to construct a dataset of one-dimensional time-series. On average, 959.8 job instances are submitted per hour, and in general, less instances are submitted on the weekends whereas more are submitted during weekdays. Interestingly, around 2:00 A.M. on May 19th (Thursday), a burst of 3,152 job instances can be observed, and its amount is approximately three times higher than the amount on typical Thursday midnights.

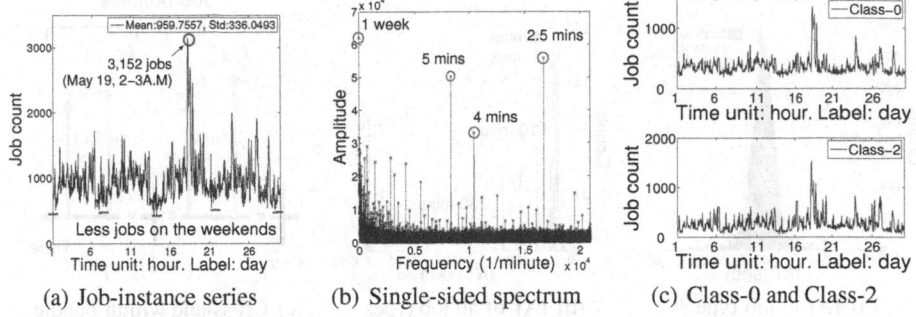

(a) Job-instance series   (b) Single-sided spectrum   (c) Class-0 and Class-2

Fig. 2. A burst and periodicities: (a) Job instances per hour. A burst (indicated by the red circle) at May 19th can be observed. (b) Discrete Fourier Transform (DFT) on the job-instance series. The high-amplitude signals correspond to the periods of 1 week, etc. (c) Class-0 (the lowest priority) and class-2 instance series. Notice their similarity (correlation coefficient is 0.94).

Discrete Fourier Transform (DFT) is also performed on the job-instance series. Fig. 2(b) provides the amplitude of each discrete frequency, on which we denote four frequencies of high power-spectrum amplitudes: 1-week, 5-min, 4-min and 2.5-min. The reason that the 1-week signal has a high amplitude can be explained by the periodic behavior between weekends and weekdays. Later in Section 3.1, we characterize the periodicity and show that both 5-min and 4-min periods can be found during the job inter-arrivals.

2.3 Class Interdependency

Not all jobs are submitted equal: certain job types have higher priority to be scheduled and executed (class-3, *e.g.*, website services), whereas other jobs do not (class-0, *e.g.* MapReduce workloads) [12].

Observation 1. *The spike $\mathcal{A}$ (1 μs) in Fig. 1(b) is attributed to the 1 μs IAT between a class-0 and a class-2 instance.*

As shown in Fig. 2(c), the pattern of class-0 job instances (low priority) is highly similar with the pattern of class-2 instances (high priority), in terms of both trend and quantity. As it can be seen that these instances of class-0 and class-2 contribute to the burst on May 19th observed in Fig. 2(a). Furthermore, the correlation coefficient between class-0 and class-2 instances is 0.94, which makes us think: what is the IAT between a class-0 and a class-2 instance? Surprisingly, this IAT is *exactly 1 μs* , which forms the first cluster in Fig. 1(b). This phenomenon immediately piques our interest: how to characterize and attribute the rest of three clusters ($\mathcal{B}$, $\mathcal{C}$, and $\mathcal{D}$) and the corresponding spikes? The answer lies in the "bundling effect" as we will elaborate in Section 3.

3   HIBM: HIerarchical Bundling Model

In this section, we introduce two major components of HIBM: cross-bundle effects (Section 3.1) and within-bundle effects (Section 3.2). The complete HIBM framework is presented in Section 3.3.

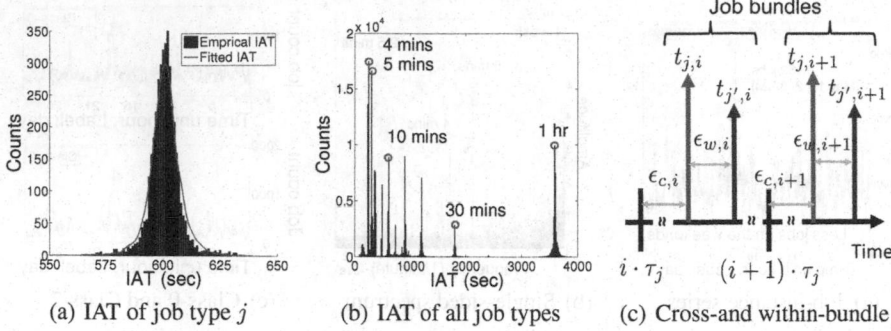

(a) IAT of job type j (b) IAT of all job types   (c) Cross-and within-bundle

Fig. 3. Multiple periodicities: (a) IAT of job type j and fitted PDF by HIBM. (b) IAT of all job types. (c) Illustration of the cross-bundle noise ($\epsilon_{c,i}$) and the within-bundle noise ($\epsilon_{w,i}$) under the period τ_j.

3.1 First Component: Cross-Bundle Effect

Multiple periodicities To characterize the periodicity of each job type, we first calculate the IAT between every two consecutive job instances of that job type as follows:

$$\delta_{j,i} = t_{j,i} - t_{j,i-1}, \text{ for } i = 1 \ldots n_j \tag{1}$$

where $\delta_{j,i}$ is the ith IAT, $t_{j,i}$ represents the occurrence time of the ith instance of job type j, and n_j is the total number of instances of job type j. Fig. 3(a) shows the histogram of such IATs, $\delta_{j,i}$. The histogram is symmetric and has a spike at 600 seconds (10 minutes), which means each instance of job type j arrives approximately every 10 minutes with some noise. Therefore, $t_{j,i}$ can be expressed as:

$$t_{j,i} = i \cdot \tau_j + \epsilon_{c,i} \tag{2}$$

where τ_j stands for the period (*e.g.*, 10 minutes in this case) and $\epsilon_{c,i}$ is a random variable representing the "cross-bundle noise." As illustrated in Fig. 3(c), the cross-bundle noise ($\epsilon_{c,i}$) represents the delay of a job bundle from its scheduled time ($i \cdot \tau_j$) and in this example two job types j and j' are in the same bundle. Here, we focus on only the job type j (the red arrows); the within-bundle noise will be elaborated in Section 3.2. In this work, τ_j is estimated by using the median of IATs of job type j; however, what distribution $\epsilon_{c,i}$ follows remains unclear for now.

Observation 2. *Multiple periodicities are observed: 4-min, 5-min, 10-min, 15-min, 20-min, 30-min, and 1-hr.*

One question may arise: is this periodic job type a special case, or do IATs of many job types behave like this? To find the answer, we further collect the IATs from all job types and illustrate them by using Fig. 3(b). For better visualization, only periods smaller than one hour are demonstrated. In Fig. 3(b), multiple periodicities are observed, and the two highest peaks are 4-min and 5-min, which matches the DFT results in Fig. 2(b): the frequencies with high amplitudes are 4-min and 5-min. 4-min is also the smallest period that exists in the trace. We would like to point out that the "10-min peak" in

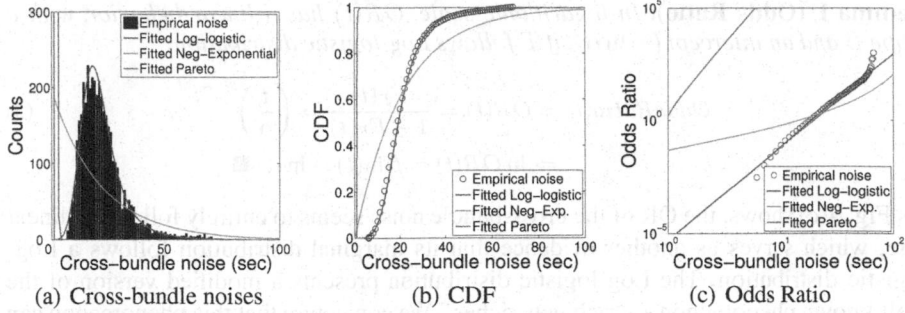

(a) Cross-bundle noises   (b) CDF   (c) Odds Ratio

Fig. 4. Modeling cross-bundle noise: (a) PDF, (b) CDF (c) Odds Ratio are demonstrated by using Log-logistic, negative-exponential and Pareto distribution, respectively

Fig. 3(b) seems sharper than the peak in Fig. 3(a); this is because Fig. 3(b) contains several job types that have the same period (10-min), whereas Fig. 3(a) contains only one such job type.

Now the question is: what random noise $\epsilon_{c,i}$ will create such IAT distribution shown in Fig. 3(a)? Could we use famous "named" distributions, say (negative) exponential or Pareto (power-law), to model this noise?

Modeling cross-bundle noise. Among many statistical distributions, we propose to model the cross-bundle noise $\epsilon_{c,i}$ by using Log-logistic distribution (LL), since it is able to model **both the cross-bundle noise and the within-bundle noise** (Section 3.2), leading to the unified expression in HIBM. Also, it provides intuitive explanations for sporadic, large delays. The Log-logistic distribution has a power-law tail and its definition is as follows.

Definition 4 (Log-logistic distribution). *Let T be a non-negative continuous random variable and $T \sim LL(\alpha, \beta)$; the cumulative density function (CDF) of a Log-logistic distributed variable T is , $CDF(T = t) = F_T(t) = \frac{1}{1+(t/\alpha)^{-\beta}}$, where $\alpha > 0$ is the scale parameter, and $\beta > 0$ is the shape parameter. The support $t \in [0, \infty)$.*

Fig. 4(a) presents the cross-bundle noise $\epsilon_{c,i}$ and three fitted distributions by using Maximum Likelihood Estimate (MLE) [3]. The distribution shows a left-skewed behavior and sporadically, a few job instances suffer from large delays. This phenomenon is difficult to be captured by distributions with tails decaying exponentially fast (*e.g.*, negative-exponential). On the other hand, the Pareto distribution (a power-law probability distribution), which is also a heavy-tail distribution, lacks the flexibility to model a "hill-shaped" distribution. The goodness-of-fit is tested by using Kolmogorov-Smirnov test [11] with the null hypothesis that the cross-bundle noise is from the fitted Log-logistic distribution. The resulting P-value is 0.2441, and therefore we retain the null hypothesis under the 95% confidence level and conclude that the cross-bundle noise follows Log-logistic distribution.

To better examine the distribution behavior both in the head and tail, we propose to use the Odds Ratio (OR) function.

Lemma 1 (Odds Ratio). *In logarithmic scale, $OR(t)$ has a linear behavior, with a slope β and an intercept $(-\ln\alpha)$, if T follows Log-logistic distribution:*

$$OddsRatio(t) = OR(t) = \frac{F_T(t)}{1 - F_T(t)} = \left(\frac{t}{\alpha}\right)^{\beta} \tag{3}$$

$$\Rightarrow \ln OR(t) = \beta\ln(t) - \ln\alpha \quad \blacksquare$$

As Fig. 4(c) shows, the OR of the cross-bundle noise seems to entirely follow the linear line, which serves as another evidence that its marginal distribution follows a Log-logistic distribution. The Log-logistic distribution presents a modified version of the well known phenomenon − "rich gets richer." We conjecture that this phenomenon can be adapted to explain the cross-bundle noise of periodic job instances − "*those delayed long get delayed longer*." If the submission schedule of a job instance is delayed (or preempted) by other jobs with a higher priority, it is likely that this job instance is going to suffer from being further delayed.

3.2 Second Component: Within-Bundle Effect

Bundling effect and within-bundle noise The bundling effect represents the temporal dependency between two job types j and j'. If the instances of two job types (*e.g.*, Gmail and disk-backup, denoted as job type j and j', respectively) are independent from each other, the correlation coefficient of their IATs should be close to zero. However, as Fig. 5(a) shows, IATs of two job types can be highly correlated; the correlation coefficient (CC) is 0.9894. In this context, each $t_{j,i}$ and $t_{j',i}$ must share the same $\epsilon_{c,i}$ due the high correlation. More interestingly, the instances of job type j' always occur after the corresponding instance of j, *i.e.*, $t_{j,i} < t_{j',i}$ as illustrated in Fig. 3(c).

We further examine the IAT between job type j and j', namely, $t_{j',i} - t_{j,i}$, referred as "within-bundle noise" ($\epsilon_{w,i}$). The concept of the within-bundle noise also is illustrated by Fig. 3(c); furthermore, Fig. 5(b) presents a bi-modal distribution of $\epsilon_{w,i}$: one peak at 1.5-sec observed from 2:00P.M. to 6:00A.M. and the other at 16-sec observed from 6:00A.M. to 2:00P.M.

Observation 3. *The spikes $\mathcal{D}_1$ (1.5sec) and $\mathcal{D}_2$ (16sec) in Fig. 1(b) are attributed to HiBM's within-bundle noise in the scale of seconds.*

A possible explanation is that the submissions of job type j' (class 1, latency-insensitive) are delayed or preempted by other high priority job types during the working hours from 6:00A.M. to 2:00P.M., which creates the second mode (the 16-sec peak). Therefore, we model this bi-modal distribution by using a mixture of two Log-logistic distributions. Fig. 5(c) shows the Q-Q plot between the empirical $\epsilon_{w,i}$ and samples drawn from the fitted Log-logistic mixture. As it can be seen, each quantile of simulated samples matches the empirical $\epsilon_{w,i}$ very well.

A highly similar situation can be observed from another job bundle, shown in Fig. 5(d)(e)(f). Instead of seconds, as Fig. 5(e) shows, $\epsilon_{w,i}$ is bi-modal and in the scale of millisecond.

Observation 4. *The spikes $\mathcal{C}_1$ (3ms) and $\mathcal{C}_2$ (5.5ms) in Fig. 1(b) are attributed to HiBM's within-bundle noise in the scale of milliseconds.*

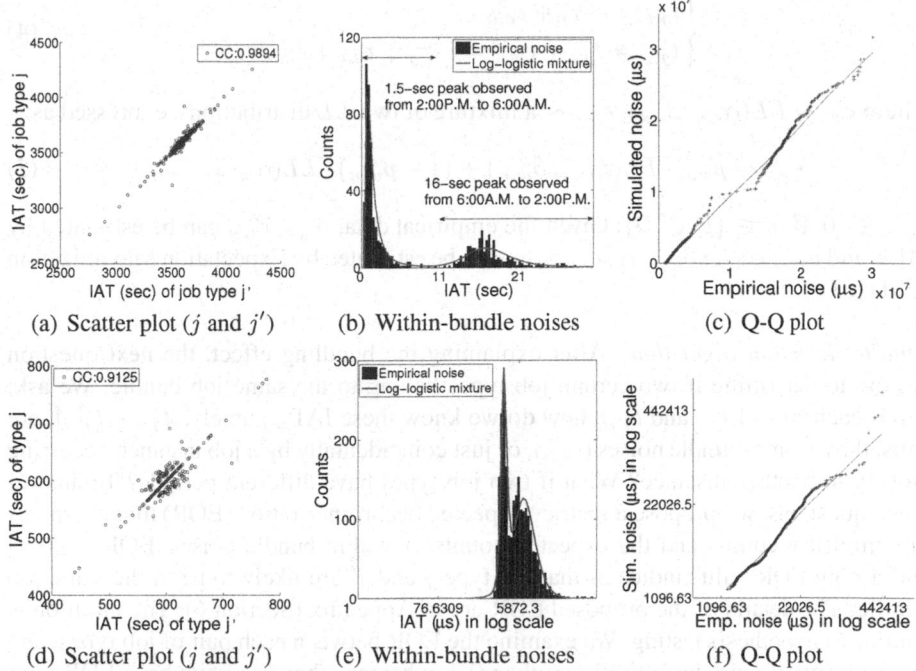

(a) Scatter plot (j and j')   (b) Within-bundle noises   (c) Q-Q plot

(d) Scatter plot (j and j')   (e) Within-bundle noises   (f) Q-Q plot

Fig. 5. HIBM fits real within-bundle noises: (a) IATs of job type j and j' are highly correlated; the correlation coefficient (CC) is 0.9894. Here, both job type j and j' have the period of 1 hour. (b) Within-bundle noise ($\epsilon_{w,i}$) that creates the spikes $\mathcal{D}_1$ and $\mathcal{D}_2$ can be modeled as a mixture of two Log-logistic distributions. (c) Q-Q plot between the empirical $\epsilon_{w,i}$ and the samples drawn from the fitted Log-logistic mixture. (d)(e)(f) demonstrate another $\epsilon_{w,i}$ in millisecond-scale, and have similar explanations. We would like to point out the spikes $\mathcal{C}_1$ and $\mathcal{C}_2$ can be attributed to the within-bundle noise shown in (e).

In this case, $\epsilon_{w,i}$ can also be modeled by a mixture of two Log-logistic distributions as Fig. 5(e)(f) show. For both cases (within-bundle noises in both second-and millisecond-scale), Kolmogorov-Smirnov test is performed; the null hypothesis that $\epsilon_{w,i}$ and the fitted Log-logistic mixture follow the same distribution, is retained under the 95% confidence level. In addition, within-bundle noises are also observed in μs scale, which forms the cluster (and the spike) $\mathcal{B}$ in Fig. 1(b) and can also be modeled by the Log-logistic distribution. This is not shown here due to the space limit. Now we are able to explain and model all the clusters and spikes ($\mathcal{B}$, $\mathcal{C}_1$, $\mathcal{C}_2$, $\mathcal{D}_1$ and $\mathcal{D}_2$) with the Log-logistic distribution, leading to the succinctness of HIBM.

Interestingly, even if $\epsilon_{w,i}$ exists, the IATs of job type j and of j' are still highly correlated. The key to create such a phenomenon lies in the hierarchy that cross-bundle noise is always larger than within-bundle noise, $\epsilon_{c,i} > \epsilon_{w,i}$. In the trace, the scale of $\epsilon_{c,i}$ is approximately in the magnitude of minutes, whereas $\epsilon_{w,i}$ is in the magnitude of seconds, milliseconds or even microseconds. Based on this observation, we propose a unified model to describe the IATs of two job types in the same bundle, which serves as the backbone of the proposed HIBM:

$$\begin{cases} t_{j,i} = i \cdot \tau_j + \epsilon_{c,i} \\ t_{j',i} = t_{j,i} + \epsilon_{w,i} = i \cdot \tau_j + \epsilon_{c,i} + \epsilon_{w,i} \end{cases} \qquad (4)$$

where $\epsilon_{c,i} \sim LL(\alpha_{c,\kappa}, \beta_{c,\kappa})$, $\epsilon_{w,i} \sim$ a mixture of two LL distributions, expressed as:

$$\epsilon_{w,i} \sim p_{w,\kappa} \cdot LL(\alpha_{w,\kappa}, \beta_{w,\kappa}) + (1 - p_{w,\kappa}) \cdot LL(\alpha_{w',\kappa}, \beta_{w',\kappa}) \qquad (5)$$

$p_{w,\kappa} \in [0,1]$, $\kappa \in \{\mathcal{B}, \mathcal{C}, \mathcal{D}\}$. Given the empirical data, $\alpha_{c,\kappa}$, $\beta_{c,\kappa}$ can be estimated by MLE and $p_{w,\kappa}$, $\alpha_{w,\kappa}$, $\beta_{w,\kappa}$, $\alpha_{w',\kappa}$, $\beta_{w',\kappa}$ can be estimated by Expectation Maximization (EM) [3].

Bundle detection algorithm. After explaining the bundling effect, the next question is how to determine if two certain job types belong to the same job bundle. We ask: given each pair of $t_{j,i}$ and $t_{j',i}$, how do we know these IATs, namely, $|t_{j,i} - t_{j',i}|$, are caused by within-bundle noises ($\epsilon_{w,i}$), or just coincidentally by a job instance occurring closely to another instance? What if two job types have different periods? To answer these questions, we propose a metric "expected occurrence ratio" (EOR) that compares the empirical counts and the expected counts of within-bundle noises. EOR $\in [0,1]$ and a high EOR value indicates that job type j and j' are likely to be in the same job bundle. The details of the proposed EOR are in Appendix (Section 6). The intuition is similar to hypothesis testing. We examine the EOR between each pair of job types, and the majority of pairs have EOR less than 0.3, whereas other few pairs have EOR very close to 0.8. In this work, we select an EOR of 0.3 as threshold and therefore two job types are considered unbundled if their EOR is less than 0.3.

3.3 Complete HIBM Framework

By assembling the cross-bundle effect (Section 3.1) and the within-bundle effect (Section 3.2) together, we describe here the complete HIBM framework by using Algorithm 1. The inputs to HIBM are user-defined periods, the total duration $\mathcal{T}$, and the parameters of Log-logistic distributions as described in Eq (4). In our case, the periods are set according to the empirical data as shown in Fig. 3(b), the $\mathcal{T}$ is set to one month as mentioned in Section 2.2, and the parameters described in Eq (4) are estimated by MLE and EM. For each job type j, HIBM calculates its total number of instances by $\left\lfloor \frac{\mathcal{T}}{\tau_j} \right\rfloor$. Next, for the ith instance of job type j, there will be two possible cases: (1) $t_{j,i}$ is bundled with $t_{j',i}$ or (2) $t_{j,i}$ is in its own job bundle (not bundled with any other job type). In the first case, $t_{j,i}$ is estimated according to Eq (2), whereas in the second case, $t_{j,i}$ is estimated according to Eq (4). The estimated $t_{j,i}$ is recorded in JS for all j and i. Finally, JS is sorted in ascending order and then HIBM outputs JS as job inter-arrivals.

4   Experimental Results and Discussion

We validate HIBM by using the empirical data. The comparisons between the synthetic IATs generated by HIBM and empirical IATs are illustrated by Fig. 6. Fig. 6(a)(b) present the histogram of the empirical IATs and the synthetic IATs side by side. As it

---

Algorithm 1. HIBM Generation

Result: Inter-arrival process of job instances, $t_{j,i}$ for all j and i, given periods τ_j for each
job type j, total duration $\mathcal{T}$, $\alpha_{c,\kappa}$, $\beta_{c,\kappa}$, $p_{w,\kappa}$, $\alpha_{w,\kappa}$, $\beta_{w,\kappa}$, $\alpha_{w',\kappa}$, and $\beta_{w',\kappa}$.

initialization: $JS = []$;

for *each* j **do**

    for $i = 1$ *to* $\left\lfloor \frac{\mathcal{T}}{\tau_j} \right\rfloor$ **do**

        if *job type j is bundled with job type j'* **then**

            $t_{j,i} = t_{j',i} + \epsilon_{w,i}$,

            $\epsilon_{w,i} \sim p_{w,\kappa} \cdot LL(\alpha_{w,\kappa}, \beta_{w,\kappa}) + (1 - p_{w,\kappa}) \cdot LL(\alpha_{w',\kappa}, \beta_{w',\kappa})$;

        else

            $t_{j,i} = i \cdot \tau_j + \epsilon_{c,i}, \epsilon_{c,i} \sim LL(\alpha_{c,\kappa}, \beta_{c,\kappa})$;

        $JS = JS$ appending $t_{j,i}$;

Sort JS in ascending order;

return JS;

---

can be seen, the synthetic IATs match the distinct characteristics of the empirical IATs: the job-instance counts (only 0.3% difference), the four clusters, and all the spikes ($\mathcal{A}$, $\mathcal{B}, \mathcal{C}_1, \mathcal{C}_2 \mathcal{D}_1$, and $\mathcal{D}_2$). Fig. 6(c) presents the Q-Q plot, from which we can also observe that each quantile of the synthetic IATs matches the corresponding quantile from the empirical data very well.

We begin the discussion with HIBM's succinctness. HIBM requires only a handful of parameters as described in Algorithm 1 to generate job inter-arrivals that match the characteristics from the empirical data, even when the i.i.d. assumption is violated — the submissions of certain instances depend on one another. Therefore, HIBM can be used as a tool to create more realistic job inter-arrivals to design, evaluate, and optimize the cloud-based scheduler of a datacenter.

Also thanks to HIBM's interpretability, we now understand the four distinct clusters observed from the empirical data can be attributed to both class interdependency ($\mathcal{A}$: $1\mu s$) and within-bundle noises ($\mathcal{B}$: 10-$10^3 \mu s$, $\mathcal{C}$:10^3-$10^5 \mu s$, and $\mathcal{D}$:10^6-$10^9 \mu s$). In addition, the 3ms and 5ms spikes ($\mathcal{C}_1$ and $\mathcal{C}_2$) can be attributed to the within-bundle noise shown in Fig. 5(e), and similarly 1.5sec and 16sec spikes ($\mathcal{D}_1$ and $\mathcal{D}_2$) can be attributed to the within-bundle noise shown in Fig. 5(b). Furthermore, the cross-bundle noises in HIBM provides intuitive explanation — "those delayed long get delayed longer" — for the delays occurred on periodic job instances.

5 Related Work

Many papers have attempted to model the sequential and streaming data. Leland et al. [10], Wang et al. [14], and Kleinberg et al. [8] have addressed the issues of self-similar and bursty internet traffic. Saveski et al. [13] has adapted active learning to model the web services. Benson et al. [2] has proposed a network-level, empirical traffic generator for datacenters. Ihler et al. [7] has proposed a time-varying poisson process for adaptive event detection. However, none of these work has addressed the issue of inter-arrivals with both periodicity and bundling effects.

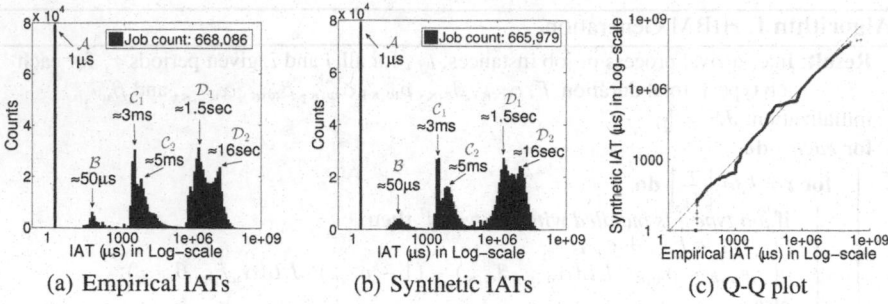

(a) Empirical IATs   (b) Synthetic IATs   (c) Q-Q plot

Fig. 6. Comparisons between Synthetic IATs and the empirical IATs: (a) Histogram of empirical IATs in log scale. (b) Histogram of synthetic IATs in log scale. (c) Q-Q plot. The synthetic IATs generated by HIBM match the characteristics of the empirical IATs: the job-instance counts (only 0.3% difference), the four clusters, and all the spikes ($\mathcal{A}, \mathcal{B}, \mathcal{C}_1, \mathcal{C}_2$ $\mathcal{D}_1$, and $\mathcal{D}_2$). In addition, each quantile of the synthetic IATs matches the corresponding quantile from the empirical data very well.

Regarding to the Log-logistic distribution, it has been developed and used for survival analysis [9,1]. Recently, prior work has demonstrated its use in modeling the duration of telecommunication [4] and software reliability [6]. To the best of our knowledge, this is the first work to use Log-logistic distributions to model the delays of job inter-arrivals in a datacenter.

6 Conclusion

In this work, we investigate and analyze the inter-arrivals of job requests in an industrial, large-scale datacenter. Our paper has two contributions:

- **Pattern Discovery.** We discover two key patterns of job inter-arrivals: (a) multiple periodicities and (b) bundling effects. In addition, we propose to use Log-logistic distributions to model both cross-bundle and within-bundle noises.
- **Generative Model.** We propose HIBM, a succinct and interpretative model. HIBM requires only a handful of parameters to generate job inter-arrivals mimicking the empirical data. In addition, HIBM also attributes the four distinct clusters and the corresponding spikes to both within-bundle noises and class interdependency, and provides intuitive explanation "those delayed long get delayed longer" to the cross-bundle noises of periodic job types.

Acknowledgements. The authors would like to thank Zhen Tang, Chulei Liu, Ilari shafer, Alexey Tumanov and anonymous reviewers for their valuable suggestions. This material is based upon work supported by the National Science Foundation under Grant No. IIS-1247489, IIS-1217559, and CNS-1314632. The research was sponsored by the U.S. Army Research Office (ARO), Defense Advanced Research Projects Agency (DARPA) under Contract Number W911NF-11-C-0088, and Google Focused Research Award. Any opinions, findings, and conclusions or recommendations expressed in this material are those of the author(s) and do not necessarily reflect the views of the National Science Foundation, DARPA, or other funding parties. The U.S. Government is authorized to reproduce and distribute reprints for Government purposes notwithstanding any copyright notation here on.

References

1. Bennett, S.: Log-logistic regression models for survival data. Applied Statistics, 165–171 (1983)
2. Benson, T., Anand, A., Akella, A., Zhang, M.: Understanding data center traffic characteristics. ACM SIGCOMM Computer Communication Review 40(1), 92–99 (2010)
3. Casella, G., Berger, R.L.: Statistical inference, vol. 70. Duxbury Press, Belmont (1990)
4. Vaz de Melo, P.O.S., Akoglu, L., Faloutsos, C., Loureiro, A.A.F.: Surprising patterns for the call duration distribution of mobile phone users. In: Balcázar, J.L., Bonchi, F., Gionis, A., Sebag, M. (eds.) ECML PKDD 2010, Part III. LNCS, vol. 6323, pp. 354–369. Springer, Heidelberg (2010)
5. Fischer, W., Meier-Hellstern, K.: The markov-modulated poisson process (mmpp) cookbook. Performance Evaluation 18(2), 149–171 (1993)
6. Gokhale, S.S., Trivedi, K.S.: Log-logistic software reliability growth model. In: HASE, pp. 34–41. IEEE (1998)
7. Ihler, A., Hutchins, J., Smyth, P.: Adaptive event detection with time-varying poisson processes. In: KDD, pp. 207–216. ACM (2006)
8. Kleinberg, J.: Bursty and hierarchical structure in streams. Data Mining and Knowledge Discovery 7(4), 373–397 (2003)
9. Lawless, J.F.: Statistical models and methods for lifetime data, vol. 362. John Wiley & Sons (2011)
10. Leland, W.E., Taqqu, M.S., Willinger, W., Wilson, D.V.: On the self-similar nature of ethernet traffic. ACM SIGCOMM Computer Communication Review 23, 183–193 (1993)
11. Massey Jr., F.J.: The kolmogorov-smirnov test for goodness of fit. JASA 46(253), 68–78 (1951)
12. Reiss, C., Tumanov, A., Ganger, G.R., Katz, R.H., Kozuch, M.A.: Heterogeneity and dynamicity of clouds at scale: Google trace analysis. In: SOCC, p. 7. ACM (2012)
13. Saveski, M., Grčar, M.: Web services for stream mining: A stream-based active learning use case. ECML PKDD 2011, 36 (2011)
14. Wang, M., Madhyastha, T., Chan, N.H., Papadimitriou, S., Faloutsos, C.: Data mining meets performance evaluation: Fast algorithms for modeling bursty traffic. In: ICDE, pp. 507–516. IEEE (2002)

A Appendix

The expected occurrence ratio (EOR) of job type j and j' can be calculated as:

$$EOR(j, j') = \mathcal{N}_\kappa \cdot \left(\frac{\mathcal{T}}{LCM(\tau_j, \tau_{j'})} \cdot \rho_j \cdot \rho_{j'} \right)^{-1} \tag{6}$$

where $\mathcal{N}_\kappa$ represents the number of the IATs occurred in the range of the cluster $\kappa \in \{\mathcal{B}, \mathcal{C}, \mathcal{D}\}$ in Fig. 1(b), $\mathcal{T}$ is the total duration, $LCM(\tau_j, \tau_{j'})$ is the Least Common Multiple (LCM) between two periods τ_j and $\tau_{j'}$, finally ρ_j and $\rho_{j'}$ are the missing rates of job type j and j', respectively.

Brownian Bridge Model for High Resolution Location Predictions

Miao Lin and Wen-Jing Hsu

School of Computer Engineering
Nanyang Technological University, Singapore
linm0018@e.ntu.edu.sg, hsu@pmail.ntu.edu.sg

Abstract. Given a person's current and historical traces, a useful yet challenging task is to predict the future locations at high spatial-temporal resolution. In this study, we propose a Brownian Bridge model to predict a person's future location by using the individual's historical traces that exhibit similarity with the current trace. The similarity of the traces with the current trace is evaluated based on the notion of edit distance. The predicted location at the future point in time is a weighted result obtained from a modified Brownian Bridge model that incorporates linear extrapolation. Both Brownian Bridge and linear extrapolation aim to capture aspects of the individual's mobility behaviors. Compared to using either historical records or linear extrapolation method alone, the proposed location prediction method shows lower mean prediction error in predicting locations at different time horizons.

Keywords: location prediction, Brownian Bridge, GPS.

1 Introduction

With the wide availability of GPS devices, current location-based services and applications already have high spatial and temporal requirements for predicting individuals' future locations [1]. These requirements pose a nontrivial challenge to location predictions. For instance, in pervasive or mobile computing, the spatial resolution expected of a prediction is in the order of 10 meters and the temporal resolution can be as high as a few seconds. In [3,2] both the theoretical studies and experimental studies show that individual's next locations are highly predictable, which is around 90%, in either cell tower data or GPS data. However, in two respects these two studies are incomplete. Firstly, the predicted results in these two studies are too coarse in terms of both spatial and temporal resolutions. Specifically, the temporal resolution in two datasets are one hour, and the spatial resolution in cell tower data is a few kilometers and in GPS data is around 500 meters [2]. Secondly, the high prediction accuracy indicates the overall predictability during the entire 14 weeks. This high predictability may be dominated by the cases when the individuals stay at home or in the office for most of the time. However, in certain cases, such as, when making moves, it may be difficult to predict individual's next location at high spatial-temporal resolutions.

V.S. Tseng et al. (Eds.): PAKDD 2014, Part II, LNAI 8444, pp. 210–221, 2014.

In this paper, the location prediction problem is defined as follows. Given a set of historical positioning records of an individual and a sequence of samples in the current trace, where all positioning records are GPS readings with a fixed sampling rate, e.g., 60 seconds, the problem is to predict this individual's location in the near future, e.g., 60 seconds, 120 seconds, or 180 seconds later. To achieve such purpose, a location prediction method making use of historical records while observing current mobility behaviors is proposed. Since previous studies suggest that there is a high probability that the individuals will follow previous traces [3,2], the current trace may be similar to some traces in the past. Therefore, we propose to use dynamic time warping [4] to measure the similarity between traces. However, when using similar traces to estimate the future location, there may not be a record that corresponds exactly to the required point of time, and hence there is a need to model the individual's movements in between two sampled positions. A Brownian Bridge model [12], therefore, is proposed to model the variance of an inferred location in between two existing sample points. In the Brownian Bridge model, the location at any time is modeled as a Gaussian distribution to cater for the measurement errors associated with GPS readings. A linear extrapolation method is also used to model the individual's current mobility behaviors. The predicted location is a combination of the results from the Brownian Bridge models and the result from the linear extrapolation method.

In order to evaluate our location prediction method, we compare our method with two baseline methods, namely, the linear extrapolation method and a method that only uses similar traces. In terms of average prediction errors, the experimental results show that our method is much better than either method that uses only linear extrapolation or similar traces alone, in predicting locations at different time horizons.

Our contributions consist of a few parts. Firstly, we propose an algorithm of applying the edit distance for measuring the similarity between mobility traces. Secondly, we present and evaluate the Brownian Bridge model for modeling a person's movements in between any two sample points. Thirdly, we present a prediction method that is capable of high spatial-temporal resolution prediction by exploiting the individual's mobility behaviors in the current trace and historical records.

2   Related Work

The existing approaches for location prediction can be classified into two groups. In the first group, individual's mobility history is constructed based on either significant locations [5,6] or region-of-interest [7], both of which are generated by applying a clustering method, such as DBSCAN [8] or K-Means [9], on historical records from either the given person or a large population. Then, according to the transition records between the significant locations or region-of-interest, a probabilistic model based on either Bayesian theory or Markovian theory is constructed to infer the next location. A major issue here is that, the constructed probabilistic models can be applied to infer the next locations only when individuals visit the significant location; otherwise the model fails. Also, in these studies, the transition time between two locations is generally disregarded.

In the second group of approaches, the next move is predicted entirely based on trajectory patterns [10,11]. In such cases, the trajectories are represented as spatio-temporal items, e.g, a list of locations and the corresponding transition times. The current trace is matched to the historical records by searching through the pattern tree. Two methods [10,11] that make use of the trajectory patterns alone typically suffer from the major limitation in the timing allowed for making predictions, because of the tight coupling between the prediction time and the sampling rate. Similarly, the trajectory pattern mining method is only applicable to inferring the location at a point in time for which the corresponding sample is available in the matched traces; predicting the location at any other time in between the sampling method is impossible.

3   Trace Distance

Table 1. The meanings of the notations used

Symbol	Explanation
$T_i = \{p_1, ...p_{t_n}\}$	Trace T_i contains t_n sample points
p_i, t_i	The coordinate and time of ith point
T_{tr}	The collection of the traces
$T_i(j : k)$	A sub-trace from index j to index k
$M = \{m_1, ...m_k\}$	The collection of portions of historical traces that match the current trace
$m_i.tid, m_i.\delta(tid, p)$	Trace index of a matched result and the distance of the trace with the target trace
$m_i.sid, m_i.eid$	The starting and ending index in the matched trace
$\delta_{init}, \delta_{thd}$	Initial and maximum distance threshold

In this section, we describe our method for measuring the distance between traces based on two ideas. Firstly, there is a high probability for the individuals to follow the same routes [3,2]; however, the recorded traces may be largely different due to different initial recording time or the errors in GPS readings. Therefore, it should be useful to find the most similar subtraces while tolerating certain degree of inexactness. Secondly, individuals often traverse a route in both directions. Thus, we compare the traces in both directions and record the results that satisfy the distance threshold.

According to these ideas, we present a trace matching algorithm *Trace-matching* based on dynamic time warping. Note that a matching need not be starting from the first index, but it should always end with the last index $|t_p|$. This ensures using the last location in finding a match, which is further elaborated in Section 4.3. The matching procedure starts by finding a sequence of

---

Algorithm 1. Trace-matching(T_p, T_{tr}, δ_{init}, δ_{thd})

---

1. **for** each $p_i \in T_p$, $i \geq 2$ **do**
2.   initial trace starts at index $j = 1$;
3.   **while** $j \leq i$ **do**
4.   **for** each trace $T_i \in T_{tr}$ **do**
5.   **for** each point $p_k \in T_i$ **do**
6.   calculate $dist = distance(p_j, p_k)$;
7.   **if** $dist \leq \delta_{init}$ **then**
8.   save the trace index and starting index to M;
9.   **end if**
10.   **end for**
11.   **end for**
12.   **if** M is not empty **then**
13. $k = j + 1$;
14.   **for** $k \leq i$ **do**
15. M =follow-up-matching($T_p(j : k)$,M, T_{tr}, δ_{thd});
16.   **if** M is empty **then**
17. $j = j + 1$;break;
18.   **end if**
19.   **end for**
20.   **end if**
21.   **end while**
22. **end for**
23. **return** M

Algorithm 2. follow-up-matching(T_p,M, T_{tr}, δ_{thd})

---

1. **for** each matching information m_i **do**
2.   Get candidate trace id $j = m_i.tid$, $r = m_i.sid$, $s = m_i.eid$;
3.   **if** $r < s$ **then**
4. $T'_j = T_j(r : s + 1)$;
5.   **if** $|T_p| == 2$ **then**
6.   Add a new match $m_k = \{j, s, r\}$;
7.   **end if**
8.   **else**
9.   Get the inverse partial trace $T'_j = T_j(r : s - 1)$;
10.   **end if**
11.   Calculate $dist = \mathcal{D}(T_p, T'_j)$;
12.   **if** $dist < \delta_{thd}$ **then**
13.   **if** $r < s$ **then**
14.   update $m_i = \{tid, sid, eid + 1\}$;
15.   **else**
16.   update $m_i = \{tid, sid, eid - 1\}$;
17.   **end if**
18.   **end if**
19. **end for**
20. **return** M

---

locations in each trace in T_{tr}, such that each location is within δ_{init} distance to the initial index in T_p (Line 3 to Line 11). Given the traces with the points within δ_{init} distance to the initial index of the target trace, we start matching from the second location of the target trace to the following locations given in Alg. 2. If there is a matched candidate within a given distance threshold δ_{thd}, we output the matched results. Otherwise, we shorten the current trace by removing the most temporally-distant one (Line 19), and the matching process repeats.

Alg. 2 describes the follow-up matching steps given the matched traces from the initial phase. We determine the direction of the matching when calling the *follow-up-matching* for the first time. Specifically, if the starting index *sid* is less than the ending index *eid*, the current candidate trace is from index *sid* to $eid + 1$ (Line 3). Otherwise, the current candidate trace is the reverse of the given trace from index *sid* to $eid - 1$. Also, when calling *follow-up-matching* for the first time, (Line 5), we add an additional piece of information by reversing the previous matching. The reason is that when comparing two line segments, we do not consider the direction of each line, therefore the subsequent matching may be conducted in either direction. The distance of the two traces, given as $\mathcal{D}(T_p, T_j')$, is measured by dynamic time warping [4]. If the distance is less than the threshold δ_{thd}, we update the matching information according to the current direction, otherwise we remove the current matching.

Here the complexity of the two algorithms is analyzed. In Alg. 1, the most computationally expensive part is the initial matching. The complexity of finding the initial matching in each trace is $O(N_p)$, where N_p is the total number of sample points in existing traces. The space requirement is $O(N_p)$. In Alg. 2, the complexity of the matching procedure is $O(|T_p|^2)$, where $|T_p|$ is the length of the target trace. However, once all the current distance values in the matrix are greater than the threshold, the calculation can be terminated and current matching can be removed. The space requirement is upper-bounded by the distance matrix of size $|T_p|^2$. Therefore, the overall time complexity is $O(|T_p|^2)$. In practice, the time complexity is determined by the longest trace in each individual's data. The overall space requirement is $O(|T_p|^2)$.

4   Location Prediction Methods

In this section, we consider the issue of predicting the next location at a specific point in time, given a few previous locations and historical trajectories that match the current trace.

4.1   Method A: Linear Extrapolation

The first location prediction method is based on linear extrapolation without using historical records. We predict the location at time t_x based on the last two locations on the current trace. Let the two locations arise at t_{i-1} and t_i, and there is no record in current trace in between time t_i to t_x. In the linear extrapolation, the location at time t_x is given as

$$\widetilde{p_l} = p_i + (p_i - p_{i-1})\frac{t_x - t_i}{t_i - t_{i-1}} \tag{1}$$

4.2 Method B: Estimation According to the Distance

The predicted location is the weighted mean location from the results of extending the matched traces one step further following the matched sequence, where the weight of each location is proportional to the inverse of the corresponding edit distance to the current trace[1].

4.3 Method C: Estimation by Brownian Bridge

This location prediction method combines the inferred location according to the Brownian Bridge model of each matched trace.

Preliminary. Brownian Bridge is a random process following a specific rule to generate a path between two given locations.

Definition 1. *A one-dimensional Brownian motion [12] $W(t) \in R$ is a continuous-time stochastic process satisfying the following properties: 1)$W_0 = 0$, 2)W_t is almost surely continuous, and 3) for $0 \leq s \leq t$, W_t has independent increment, and $W_t - W_s \sim \mathcal{N}(0, t - s)$, where $\mathcal{N}(\mu, \sigma^2)$ denotes the normal distribution with mean μ and standard deviation σ.*

Definition 2. *Let $W(t)$ be a one-dimensional Brownian motion. Given $T > 0$, the Brownian Bridge [12] from $W_0 = 0$ to $W_T = 0$ on $[0, T]$ is the process $X(t) = W(t) - \frac{t}{T}W(T)$, where $t \in [0, T]$.*

The Brownian Bridge from $W_0 = a$ to $W_T = b$ on $[0, T]$ is the process

$$X^{a \to b}(t) = a + \frac{(b - a)t}{T} + X(t) \tag{2}$$

where $X(t) = X^{0 \to 0}$ is the Brownian Bridge given in Def. 2.

According to Eq. (2), $EX^{a \to b}(t) = a + \frac{(b-a)t}{T}$, and $var(X^{a \to b}(t)) = \frac{t(T-t)}{T}$.

The Brownian Bridge from a to b is described in the following way. At time t, the estimated location is chosen from a Gaussian distribution, where the mean of the Gaussian distribution is given by the linear estimation and the variance varies with respect to time. Specifically, the variance increases at time $T/2$ to the maximum, and decreases till time T. The property of the variance is very suitable for modeling the uncertainties about a person's locations between two

---

[1] Note that this prediction method is only suitable for the case in which the current trace has the same sampling rate as that of the historical records. Otherwise, this method may lead to a large error.

known locations. When the predicted time is very close to the time obtained from the existing position samples, the mobility may not change greatly and the prediction can be of high accuracy. When the time of interest lies in between the two actual samples, the accuracy of the prediction result decreases.

In our case, the one-dimensional Brownian Bridge is extended to the two-dimensional version, where the x and y coordinates vary independently. An additional parameter, the variance of each trace σ_m^2, is introduced to model the mobility behaviors in each trace. Therefore, the variance of the estimated location is adjusted to [2]

$$var(X^{a \to b}(t)) = \frac{t(T-t)}{T} \sigma_m^2 \qquad (3)$$

According to the Brownian Bridge model, the method of estimating the location at time t_x based on the current trace is given as follows. In the current trace, two existing points p_k and p_{k+1} satisfy that $t_k < t_x < t_{k+1}$. When considering the error of each point, the locations at time t_k and t_{k+1} are generated from $p_k \sim \mathcal{N}(\mu_k, \sigma_g^2)$, $p_{k+1} \sim \mathcal{N}(\mu_{k+1}, \sigma_g^2)$, where $\sigma_g = 0.01 km$ is the standard error of the GPS reading. The location at t_x is given as $\widetilde{p}_x \sim \mathcal{N}(\widetilde{\mu}_x, \widetilde{\sigma}_x^2)$, where

$$\widetilde{\mu}_x = \alpha_k \mu_k + (1 - \alpha_k) \mu_{k+1} \qquad (4)$$

$$\widetilde{\sigma}_x^2 = t'_x \alpha_k (1 - \alpha_k) \sigma_m^2 + \alpha_k^2 \sigma_g^2 + (1 - \alpha_k)^2 \sigma_g^2 \qquad (5)$$

where σ_m is the standard deviation of the trace. $\alpha_k = \frac{t_x - t_k}{t_{k+1} - t_k}$ and $t'_x = t_{k+1} - t_k$.

Location Prediction. Given individual's current records in the given trace $T_p = \{p_1, p_2, ..., p_i\}$, similar historical traces and a time value t_x, where $t_x > t_i$, we focus on predicting the location at the time t_x.

Algorithm 3. location-prediction(M, t_x, T_p,i_p)

---

1. Normalize the time in trace T_p and the predicted time t_x to t'_x according to the initial matched index i_p;
2. **for** each matched trace information m_i **do**
3.   Get trace index $j = m_i.tid$, initial index $r = m_i.sid$, ending index $s = m_i.eid$, and the edit distance to the current trace $\delta(p, j)$;
4.   Normalize the time in the matched trace based on index r;
5.   Get the location with index i_s and i_e in T_j, such that $t_{i_s} < t'_x < t_{i_e}$;
6.   Estimate the location $\widetilde{p}_i$ at time t'_x according to the current trace and the locations at index i_s and i_e;
7. **end for**
8. Get the mean location from the list of estimated location $\widetilde{p}_i \sim \mathcal{N}(\widetilde{\mu}_i, \widetilde{\sigma}_i^2)$, and the one $\widetilde{p}_l$ estimated by the linear model given in Eq. (1), where the weight for each $\widetilde{p}_i$ is the inverse of the corresponding distance value, and the weight for $\widetilde{p}_l$ is the inverse of the initial distance δ_{init};

---

[2] Due to the space limit, the estimation of the parameter σ_m is given in [13].

Alg. 3 presents the details of the location prediction method. Firstly, we normalize the time in both predicted trace and the matched trace according to the initial matched index (Line 1 and Line 4). This is because the estimated results for the Brownian Bridge model is highly time dependant. Secondly, the estimated location $\tilde{p}_i$ based on current matched trace is given as a Gaussian distribution $\mathcal{N}(\tilde{\mu}_i, \tilde{\sigma}_i^2)$, where the parameters are inferred according to Eq. (4) and Eq. (5), respectively. Lastly, we combine the estimated results from each Brownian Bridge and also the linear extrapolation result from the current trace. The estimated location $\tilde{p}_b$ is given by $\tilde{p}_b \sim \mathcal{N}(\tilde{\mu}_b, \tilde{\sigma}_b^2)$, where $\tilde{\mu}_b = \frac{\sum_{k=1}^{|M|+1} w_k' \tilde{\mu}_k}{\sum w_k'}$ and $\tilde{\sigma}_b = \sum_{k=1}^{|M|} (\frac{w_k'}{\sum w_k'})^2 \tilde{\sigma}_k^2$. When $\tilde{\mu}_k = \tilde{\mu}_i$, $\tilde{\sigma}_k = \tilde{\sigma}_i$ and $w_i' = 1/\delta(p, j)$, for $1 \leq k \leq |M|$, and j is the index of the matched trace. For $k = |M| + 1$, $\tilde{\mu}_{|M|+1} = \tilde{p}_l$ and $w_{|M|+1}' = 1/\delta_{init}$, which is the result from linear extrapolation.

5 Experimental Results

In this section, we compare three location prediction methods.

5.1 Mobility Data

The mobility data used in this study is a subset of the GPS dataset released by Zheng Yu [14,15]. From this dataset, we extract 40 individual's trajectories over 14 weeks. The following preprocessing is conducted. Firstly, each individual's trajectory is divided into a series of traces based on the time gap between two GPS points. If the gap is longer than 300 seconds, the sequence will be divided into different traces. Each trace is resampled at the rate of 60 seconds, and the dataset is denoted as $T_{tr,60}$. Each individual's traces are randomly separately into a training set and predicting set, where the training set contains 70% of all the traces and the remaining traces are used to evaluate the location prediction methods. To cater for the characteristics of individual's mobility behavior, the training and prediction steps are separately evaluated on the individual's own data. The location of each point indicated by a pair of latitude and longitude is converted to x and y coordinates according to $0°$ in both latitudinal and longitudinal direction [16].

When estimating the standard deviation of each trace fitted by Brownian Bridge model according to our method [13], among the 3225 training traces in our dataset for 82.5% of the traces show a standard deviation of less than 0.05 km, with only one trace presents a standard deviation of 0.069 km, and the mean standard deviation is 0.021 km, indicating that the Brownian Bridge model is a suitable choice for predicting the location given the locations before and after the predicting time.

5.2 Location Prediction

In this subsection, we evaluate the location prediction methods based on $T_{tr,60}$ in two respects. Here, firstly the three location prediction methods are compared.

For each prediction, all the previous locations in the current trace and the time for prediction are given. The performance is evaluated by the distance of the predicted location to the actual location in the current trace. Note that when predicting using the Brownian Bridge model, the estimated mean location is the result. Also the parameter δ_{thd} is tested with the values of 0.3 km, 0.2 km, 0.1 km and 0.05 km, with the initial threshold set to be $\delta_{init} = 0.05$ km in all these cases.

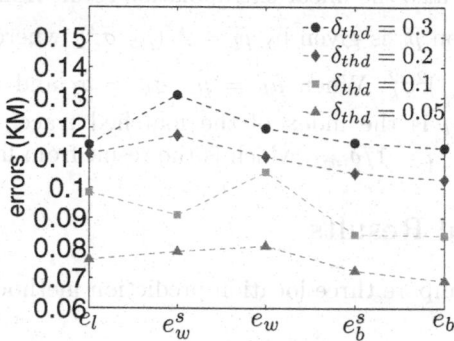

Fig. 1. The mean prediction errors in one step location prediction based on three different methods, where e_b, e_w and e_l corresponds, respectively, with respect to the mean prediction error given by the Brnownian Bridge method, prediction results using similar traces, and the linear extrapolation method. e_w^s and e_b^s are the results only using the most similar trace according to the matched results.

In all the cases tested, the prediction errors given by Method C, the combination of linear extrapolation and Brownian Bridge model, is always lower than either Method A or Method B that uses linear extrapolation or similar traces alone, respectively. Secondly, in Method B the choice of using either the most similar trace or the list of similar traces depends on the distance threshold. Specifically, in Method B, when the similarity threshold is large, e.g., $\delta_{thd} = 0.3$ or 0.2, the prediction result is adversely influenced by the most similar trace, since using only the most similar trace results in a larger mean prediction error than that by using all the matched traces. However, when the distance threshold is small, e.g., $\delta_{thd} = 0.1$ or 0.05, using the most similar trace is more effective than using all the matched traces in Method B, which may be due to the stricter distance criterion in measuring similarity. However, for the prediction results by Method C, in all the cases mentioned before, making use of the most similar trace is slightly worse than that obtained by using all the similar traces. This is because the Brownian Bridge model is able to adjust the prediction with respect to the given point of time. Thirdly, given $\delta_{thd} = 0.1$, Method C is much more discriminative than both the linear extrapolation method (Method A) and the prediction using only similar traces (Method B). Specifically, the mean prediction error for Method C is 0.084 km, while that of the Method A and Method B is 0.099 km and 0.105 km, respectively.

Note that the prediction results may be affected by δ_{thd} when using either method B or method C according to the most similar trace. This is because varying δ_{thd} the most similar trace may be different. For instance, when δ_{thd} is small, the most similar trace may have only two or three samples. When increasing δ_{thd}, a longer matched trace or subtrace may be found. Therefore, the distance threshold δ_{thd} also affects the prediction results obtained by the most similar trace.

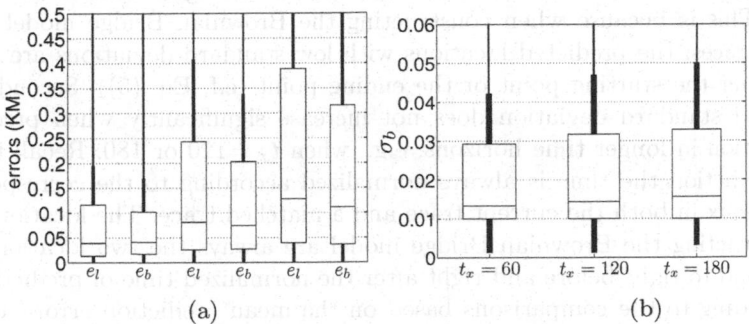

(a)   (b)

Fig. 2. (a) The distribution of the prediction errors according to linear extrapolation method e_l and Method C e_b in the following 60 seconds, 120 seconds and 180 seconds. The two boxplots on the left show the distribution of the one step location prediction results, i.e., predicting the location 60 seconds into the future. The middle two plots are the corresponding results given by predicting the location 120 seconds into the future. The last two are the prediction results for 180 seconds into the future. (b) The distributions of the standard deviation of the results given by Method C in predicting the location at 60 seconds, 120 seconds and 180 seconds later.

Secondly, we evaluate our method for predicting location with time horizons of 120 seconds and 180 seconds. Note that the location prediction by using similar traces is only suitable for the case to predict the location at a point in the similar traces. This is because the longer the prediction time is, the less likely to find a location at roughly the same time in the similar traces. Therefore, the comparison is only made between the linear extrapolation method and Method C. The prediction results, which are given in terms of the distribution of errors shown in Figure 2 (a). The mean errors for the consecutive steps are 0.099 km, 0.214 km, and 0.345 km for the linear extrapolation method, while the corresponding result for the Brownian Bridge model are 0.084 km, 0.173 km, 0.281 km, respectively. In all these cases, the Brownian Bridge model shows lower mean errors than the linear extrapolation, especially for predicting the location in longer time horizons.

Moreover, Figure 2 (b) shows the distributions of the estimated standard deviation of the predictions in Method C in the case of predicting the future location at $t_x = 60$, $t_x = 120$, and $t_x = 180$. A few observations can be made from this figure. Firstly, in all the cases, the estimated standard deviation is very small, which is less than 0.03 in the most cases. This is due to that the estimated standard deviation is calculated according to each Brownian Bridge model given by Eq. (5).

The mean standard deviation σ_m for the traces is 0.021 km and the error of GPS measurement is $\sigma_g = 0.01$ km. Therefore, the maximum standard deviation given by Eq. (5) is when $\alpha_x = \frac{1}{2}$, indicating inferring a location in between two samples, which is roughly 0.082 when $t'_x = 60$ in our case. However, a much smaller standard deviation given in Figure 2 (b) indicates that the predictions are mostly near either the starting location or the ending location in the similar traces. This is because when constructing the Brownian Bridge model in the similar traces, the predicted locations with low standard deviations are mostly near either the starting point or the ending point (c.f. Eq. (5)). Secondly, the estimated standard deviation does not increase significantly when predicting the location in longer time horizons, e.g., when $t_x = 120$ or 180. Recall that in each prediction the time is always normalized according to the corresponding initial index in both the current trace and a matched trace. The locations used in constructing the Brownian Bridge model are always the two locations that correspond to right before and right after the normalized time of prediction.

According to the comparisons based on the mean prediction errors and the estimated standard deviations, combining the linear extrapolation method and individual's historical records is a suitable choice for location prediction.

5.3   Discussions

For the few location prediction methods discussed before, the Brownian Bridge model fails to predict if there is no similar trace within a given distance threshold to the current trace. Because the linear extrapolation method shows decent results while using only two location samples. A possible remedy to the Brownian Bridge method is to apply the linear extrapolation in the case of failing to find a match in the records.

One useful method for location prediction is the Kalman filter. We do not choose Kalman filter for the following reasons. Firstly, a Kalman filter needs to be specifically designed for each trace since the mobility behaviors may change greatly even for the same individual. Secondly, learning the parameters in each trace needs sufficient number of sample points, implying that the method is unable to make any prediction for traces with few records. In contrast, with our Method C, since the matching can be based on a subsequence of the current trace with a few existing samples, it will be able to predict the future location even with a few records.

6   Conclusion

In this paper, we have presented a new algorithm for location prediction by making use of similar traces and individual's current mobility information. The similar traces with current target trace are found on the basis of the edit distance. In order to predict individual's location at any given point of time, we use a Brownian Bridge to model the uncertainties about person's movements in between any two locations. The final prediction combine both the current mobility behaviors described by the linear extrapolation method and also the estimated results from the individual's Brownian Bridge models from similar traces. Experimental results show that our location prediction method by using Brownian Bridge model is better than that using only the historical records or the linear extrapolation method.

There are a few issues unresolved in this study. For instance, we have shown in some cases, the linear extrapolation method is better than using the historical records or vice versa. Therefore, how to choose the right method at each predicting time is a relevant question for future research.

References

1. Gruteser, M., Grunwald, D.: Anonymous Usage of Location-Based Services Through Spatial and Temporal Cloaking. In: MobiSys, pp. 31–42. ACM, New York (2003)
2. Lin, M., Hsu, W.-J., Lee, Z.Q.: Predictability of Individuals' Mobility with High-Resolution Positioning Data. In: UbiComp, pp. 381–390. ACM, New York (2012)
3. Song, C., Qu, Z., Blumm, N., Barabái, A.L.: Limits of Predictability in Human Mobility. Science 327, 1018–1021 (2010)
4. Kim, S.-W., Park, S., Chu, W.W.: An index-based approach for similarity search supporting time warping in large sequence databases. In: ICDE, pp. 607–614. IEEE Computer Society, Washington (2001)
5. Ashbrook, D., Starner, T.: Using GPS to Learn Significant Locations and Predict Movement across Multiple Users. Personal Ubiquitous Comput. 7, 275–286 (2003)
6. Krumm, J., Horvitz, E.: Predestination: Inferring destinations from partial trajectories. In: Dourish, P., Friday, A. (eds.) UbiComp 2006. LNCS, vol. 4206, pp. 243–260. Springer, Heidelberg (2006)
7. Giannotti, F., Nanni, M., Pinelli, F., Pedreschi, D.: Trajectory pattern mining. In: KDD, pp. 330–339. ACM, New York (2007)
8. Ester, M., Kriegel, H., Jörg, S., Xu, X.: A density-based algorithm for discovering clusters in large spatial databases with noise. In: KDD, pp. 226–231. AAAI (1996)
9. Han, J., Kamber, M.: Data Mining: Concepts and Techniques. Morgan Kaufmann, San Francisco (2006)
10. Chen, L., Lv, M., Chen, G.: A system for destination and future route prediction based on trajectory mining. Pervasive Mobile Computing 6, 657–676 (2010)
11. Monreale, A., Pinelli, F., Trasarti, R., Giannotti, F.: WhereNext: A location predictor on trajectory pattern mining. In: KDD, pp. 637–646. ACM, New York (2009)
12. Shreve, S. E.: Stochastic Calculus for Finance II - Contunuous-Time Models. Springer (2004)
13. The Inference of the Parameter of the Brownian Bridge Model, https://dl.dropboxusercontent.com/u/6694774/BB_varianceInference.pdf
14. Zheng, Y., Zhang, L., Xie, X., Ma, W.-Y.: Mining interesting locations and travel sequences from GPS trajectories. In: WWW, pp. 791–800. ACM, New York (2009)
15. Zheng, Y., Li, Q., Chen, Y., Xie, X., Ma, W.-Y.: Understanding mobility based on GPS data. In: Ubicomp, pp. 312–321. ACM, New York (2008)
16. Zaidi, Z.R., Mark, B.L.: Mobility Tracking Based on Autoregressive Models. IEEE Transactions on Mobile Computing 10, 32–43 (2011)

Mining Correlation Patterns among Appliances in Smart Home Environment

Yi-Cheng Chen[1], Chien-Chih Chen[2], Wen-Chih Peng[2], and Wang-Chien Lee[3]

[1] Department of Computer Science and information engineering, Tamkang University, Taiwan
[2] Department of Computer Science, National Chiao Tung University, Taiwan
[3] Department of Computer Science and Engineering, The Pennsylvania State University, USA
ycchen@mail.tku.edu.tw, {flykite,wcpeng}@cs.nctu.edu.tw,
wlee@cse.psu.edu

Abstract. Since the great advent of sensor technology, the usage data of appliances in a house can be logged and collected easily today. However, it is a challenge for the residents to visualize how these appliances are used. Thus, mining algorithms are much needed to discover appliance usage patterns. Most previous studies on usage pattern discovery are mainly focused on analyzing the patterns of single appliance rather than mining the usage correlation among appliances. In this paper, a novel algorithm, namely, *Correlation Pattern Miner* (*CoPMiner*), is developed to capture the usage patterns and correlations among appliances probabilistically. With several new optimization techniques, CoPMiner can reduce the search space effectively and efficiently. Furthermore, the proposed algorithm is applied on a real-world dataset to show the practicability of correlation pattern mining.

Keywords: correlation pattern, smart home, sequential pattern, time interval-based data, usage representation.

1 Introduction

Recently, due to the advance of sensor technology, the electricity usage data of in-house appliances can be collected easily. In particular, an increasing number of smart power meters, which facilitates data collection of appliance usage, have been deployed. With the usage data, residents could supposedly visualize how the appliances are used. Nonetheless, with an anticipated huge amount of appliance usage data, subtle information may exist but hidden. Therefore it is necessary to devise data mining algorithms to discover appliance usage patterns in order to make representative usage behavior of appliances explicit. Appliance usage patterns not only help users to better understand how they use the appliances at home but also detect abnormal usages of appliances. Moreover, it facilitates appliance manufacturers to design intelligent control of smart appliances.

Most prior studies focus on knowledge extraction for a single appliance instead of the correlation among appliances in a house. In our daily life, we usually use different appliances simultaneously. For example, while the night, air conditioner and

V.S. Tseng et al. (Eds.): PAKDD 2014, Part II, LNAI 8444, pp. 222–233, 2014.

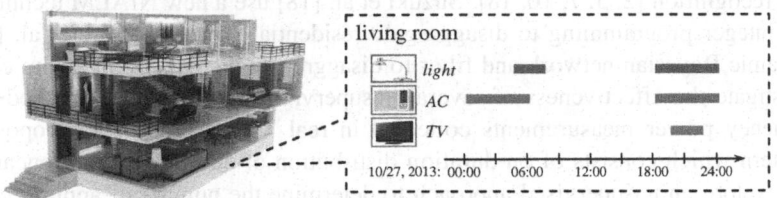

Fig. 1. An example of daily usage sequence

television in the living room may be turned on in the evening (as shown in Fig. 1). The correlation among the usage of some appliances can provide valuable information to assist residents better understand how they use appliances.

So far, little attention has been paid to the issue of mining correlation among appliances, which undoubtedly is more complex and arduous than mining the usage patterns of an appliance alone, and thus requires new mining techniques. In this paper, a new framework fundamentally different from previous work is proposed to discover the usage correlation patterns.

The contributions of our work are as follows: (1) We define the notion of *correlation pattern* based on time interval-based sequence including probability concept. Since the usage of a device can be regarded as a usage interval (duration between *turn-on* and *turn-off*), interval-based sequences can depict users' daily behaviors unambiguously. (2) The relation between any two usage intervals is intrinsically complex which may lead to more candidate sequences and heavier workload for computation. We propose a method, called *usage representation*, to simplify the processing of complex relations among intervals by considering the global information of intervals in the sequence. (3) We develop an efficient algorithm, called *Correlation Pattern Miner* (abbreviated as *CoPMiner*), to capture the usage patterns implying the correlations among appliances with several optimized techniques to reduce the search space effectively. (4) The readability of patterns is also an essential issue. A large number of patterns may become an obstacle for users to understand their actual behaviors. A spatial constraint is introduced to prune off non-promising correlation and reduce the number of generated correlation patterns. (5) To demonstrate the practicability of correlation pattern mining, we apply CoPMiner on a real dataset and analyze the results to show the discovered patterns are not just an anecdote.

The rest of the paper is organized as follows. Sections 2 and 3 provide the related works and preliminaries, respectively. Section 4 introduces the proposed CoPMiner algorithm. Section 5 reports the experimental results in a performance study, and finally Section 6 concludes the paper.

2   Related Work

In this section, we discuss some previous works extracted useful knowledge and patterns of a single device applying on energy disaggregation [3, 6, 11, 13, 18] or appliance recognition [2, 5, 7, 10, 18]. Suzuki et al. [18] use a new NIALM technique based on integer programming to disaggregate residential power use. Lin et al. [13] use a dynamic Bayesian network and filter to disaggregate the data online. Kim et al. [11] investigate the effectiveness of several unsupervised disaggregation methods on low frequency power measurements collected in real homes. They also propose a usage pattern which consists of on-duration distribution of all appliances. Goncalves et al. [6] explore an unsupervised approach to determine the number of appliances in the household, including their power consumption and state, at any given moment. Chen et al. [3] disaggregate utility consumption from smart meters into specific usage associated with certain human activities. They propose a novel statistical framework for disaggregation on coarse granular smart meter readings by modeling fixture characteristic, household behavior, and activity correlations. Ito et al. [7] extract features from the current (e.g., amplitude, form, timing) to develop appliance signatures. For appliance recognition, Kato et al. [10] use Principal Component Analysis to extract features from electric signals and classify them using Support Vector Machine. Aritoni et al. [2] develop a software prototype to understand the behaviors of household appliances. Chen et al. [5] introduce two types of usage patterns to describe users' representative behaviors. Based on these two types of patterns, an intelligent system, Jakkula et al. [8, 9] propose an Apriori-based algorithm for activity prediction and anomaly detection from sensor data in a smart home. All aforementioned studies focus on knowledge extraction for a single appliance instead of the correlation among appliances in a house. In this paper, we propose a mining algorithm to extract patterns including correlation among appliances and probability concept.

date	appliance symbol	turn-on time	turn-off time	interior location	pictorial example	usage representation (usage sequence, time sequence)
1	A	02:10	07:30	(1, 1, 1)		
1	B	05:20	10:00	(1, 2, 1)		$\begin{pmatrix} A^+ & (B^+ & C^+) & A^- & B^- & C^- & D^+ & E^+ & E^- & D^- \\ 2 & 5 & 5 & 7 & 10 & 12 & 16 & 18 & 20 & 22 \end{pmatrix}$
1	C	05:20	12:30	(3, 4, 2)		
1	D	16:10	22:40	(1, 3, 1)		
1	E	18:00	20:00	(3, 4, 1)		
2	B	00:40	05:30	(1, 2, 1)		
2	D	08:00	14:00	(1, 3, 1)		$\begin{pmatrix} B^+ & B^- & D^+ & (E^+ & F^+) & (E^- & F^-) & D^- \\ 0 & 5 & 8 & 10 & 10 & 13 & 13 & 14 \end{pmatrix}$
2	E	10:20	13:10	(3, 4, 2)		
2	F	10:20	13:10	(2, 2, 1)		
3	A	06:00	12:20	(1, 1, 1)		
3	B	07:20	14:00	(1, 2, 1)		$\begin{pmatrix} A^+ & B^+ & A^- & (B^- & D^+) & E^+ & E^- & D^- \\ 6 & 7 & 12 & 14 & 14 & 17 & 19 & 20 \end{pmatrix}$
3	D	14:00	20:30	(1, 3, 1)		
3	E	17:30	19:00	(3, 4, 1)		
4	B	08:30	10:00	(1, 2, 1)		
4	A	13:20	16:00	(1, 1, 1)		$\begin{pmatrix} B^+ & B^- & A^+ & A^- & D^+ & E^+ & E^- & D^- \\ 8 & 10 & 13 & 16 & 20 & 21 & 22 & 23 \end{pmatrix}$
4	D	20:00	23:30	(1, 3, 1)		
4	E	21:30	22:40	(3, 4, 1)		

Fig. 2. An example of usage database

3   Preliminaries

Definition 1 (Usage-interval and usage-interval sequence). Let $A = \{a_1, a_2,..., a_k\}$ be a set of k appliances. Without loss of generality, we define a set of uniformly spaced location and time points based on natural numbering N. A function, $Loc: A \rightarrow N^3$, specifies the location of each appliance in A. Let the triplet $(a_i, o_i, f_i) \in A \times N \times N$ denote a usage-interval of a_i, where $a_i \in A$, $o_i, f_i \in N$ and $o_i < f_i$. The two time points

o_i, f_i denote the *using times*, where o_i and f_i are the turn-on time and the turn-off time of appliance a_i, respectively. A usage-interval sequence is a series of usage-intervals $\langle (a_1, o_1, f_1), (a_2, o_2, f_2), ..., (a_n, o_n, f_n) \rangle$, where $o_i \leq o_{i+1}$, and $o_i < f_i$. $Loc(a_i)$ is the interior location of appliance a_i in a smart home environment.

Definition 2 (Usage-interval database). Considering a database $DB = \{r_1, r_2, ..., r_m\}$, each record r_i, where $1 \leq i \leq m$, consists of a date, a usage-interval and an interior location of appliance. DB is called a *usage-interval database*. If all records in DB with the same date are grouped together and ordered by nondecreasing turn-on time, turn-off time and appliance symbol, actually, DB can be transformed into a collection of daily usage-interval sequences. Note that the location information can be viewed as attachment to appliances. Fig. 2 shows a usage database which consists of 17 usage intervals and 4 daily usage-interval sequences.

Definition 3 (Usage-point and usage sequence). Given a usage-interval sequence $Q = \langle (a_1, o_1, f_1), (a_2, o_2, f_2), ..., (a_n, o_n, f_n) \rangle$, the set $TS_Q = \{o_1, f_1, o_2, f_2, ..., o_i, f_i, ..., o_n, f_n\}$ is called a *time set corresponding to Q*. By ordering all the elements of TS_Q in nondecreasing order, we can derive a sequence $T_Q = \langle t_1, t_2, ..., t_{2n} \rangle$ where $t_i \in TS_Q$, $t_i \leq t_{i+1}$. T_Q is called a *time sequence corresponding to Q*. A function Φ that maps a usage interval (a_i, o_i, f_i) into two usage-points a_i^+ and a_i^- is defined as follows,

$$\Phi(t_j, Q) = \begin{cases} a_i^+ & \text{if } t_j = o_i \\ a_i^- & \text{if } t_j = f_i \end{cases}, \tag{1}$$

where a_i+ and a_i- are called *on*-point and *off*-point of interval (a_i, o_i, f_i), respectively. The usage-points $a_k^*, ..., a_\ell^*$ (* can be $+$ or $-$) are collected in brackets as a pointset if they occur at the same time in T_Q, denoted as $(a_k^*, ..., a_\ell^*)$. A usage sequence S_Q of Q is denoted by $\langle s_1, ..., s_i, ..., s_{2n} \rangle$ where s_i is a usage-point. For example, in Fig. 2, the database collects 4 daily usage-interval sequences. The usage sequence of date 2 is $\langle B + B - D + (E + F +)(E - F -)D - \rangle$, and $(E + F +)$ and $(E - F -)$ are two pointsets because they occur at the same time, respectively.

Definition 4 (Usage representation). Given a usage-interval sequence $Q = \langle (a_1, o_1, f_1), ..., (a_n, o_n, f_n) \rangle$ and corresponding time sequence $T_Q = \langle t_1, ..., t_i, ..., t_{2n} \rangle$, by Definition 3, we can derive a usage sequence $S_Q = \langle s_1, ..., s_i, ..., s_{2n} \rangle$. The usage representation of Q is defined as a pair,

$$(S_Q, T_Q) = \begin{pmatrix} s_1 & ... & s_i & ... & s_{2n} \\ t_1 & ... & t_i & ... & t_{2n} \end{pmatrix}. \tag{2}$$

Note that the using time of usage point s_i in S_Q is t_i in T_Q. Take the database in Fig. 2 as an example. Without leading into ambiguity, we consider the turn-on and turn-off times by hour. The usage representation of DB is shown in the last column in Fig. 2. For the rest of this paper, we assume the usage database has already been transformed into usage representation.

Let $S_1 = \langle x_1, \ldots, x_i, \ldots, x_n \rangle$ and $S_2 = \langle x_1', \ldots, x_j', \ldots, x_m' \rangle$ be two usage sequences, where x_i, x_j' are pointsets and $n \leq m$. S_1 is called a subsequence of S_2, denoted as $S_1 \sqsubseteq S_2$, if there exist integers $1 \leq k_1 \leq k_2 \leq \ldots \leq k_n \leq m$ such that $x_1 \subseteq x_{k_1}', x_2 \subseteq x_{k_2}', \ldots, x_n \subseteq x_{k_n}'$. Given a usage-interval database DB in usage representation, the tuple $(date, S, T)$ $\in DB$ is said to contain a usage sequence S' if $S' \sqsubseteq S$. The support of a usage sequence S' in DB, denoted as $support(S')$, is the number of tuples in the database containing S'. More formally, $support(S') = |\{ (date, S, T) \in DB \mid S' \sqsubseteq S \}|.$ (3)

As mentioned above, each appliance in a house has its own location. For an appliance a in A, the function, $Loc: A \rightarrow N \times N \times N$, gives the locations (a_x, a_y, a_z) of a. The similarity between two appliances a_1 and a_2 is defined as follows:

$$similarity(a_1, a_2) = \begin{cases} 1 & \text{if } Loc(a_1) = Loc(a_2) \\ \dfrac{1}{Loc(a_1) - Loc(a_2)} & \text{if } Loc(a_1) \neq Loc(a_2) \end{cases}, \text{ where } Loc(a_1) - Loc$$

$(a_2) = |a_{1x} - a_{2x}| + |a_{1y} - a_{2y}| + |a_{1z} - a_{2z}|.$ (4)

For example, in Fig. 2, the similarity of appliances B and C is $\dfrac{1}{2+2+1} = \dfrac{1}{5} = 0.2.$ We use a support threshold, min_sup and min_sim, to filter out insignificant usage sequences. A usage sequence $S = \langle s_1, \ldots, s_n \rangle$ in DB is called a *frequent sequence*, if $support(S) \geq min\_sup$ and $\forall$ s_i, s_j in S where $i, j \leq n$, $similarity(s_i, s_j) \geq min\_sim$.

Definition 5 (Correlation pattern). Given DB in usage representation and two thresholds, min_sup and min_sim, the set of frequent sequences, FS, includes all frequent usage sequences in DB. A correlation pattern P is defined as,

$$P = (S, f(S)) = \begin{pmatrix} s_1 & \cdots & s_i & \cdots & s_n \\ f_1 & \cdots & f_i & \cdots & f_n \end{pmatrix}, \text{ where } S = \langle s_1, \ldots, s_n \rangle \in FS \text{ and } f_i \text{ is the}$$ (5)

probability function of s_i in DB.

We modify the idea of multivariate kernel density estimation [14, 17] to estimate the probability function of each s_i in S. Suppose the time information of s_i in DB is $\{t_{i1}, t_{i2}, \ldots t_{im}\}$, the probability function is defined as,

$$f_i(x) = (K(x), h, \{t_{i1}, t_{i2}, \ldots, t_{im}\}) = \frac{1}{mh} \sum_{j=1}^{m} K(\frac{x - t_{ij}}{h}), \text{ where } K \text{ is Gaussian Normal,}$$

i.e., $K(x) = \dfrac{1}{\sqrt{2\pi}} e^{-\frac{1}{2}x^2}$, and $h = \dfrac{range(\{t_{i1}, t_{i2}, \ldots, t_{im}\})}{\sqrt{m}}$. (6)

For example, in Fig. 2, with $min\_sup = 2$ and $min\_sim = 0.3$, $\langle A+A-D-D- \rangle$ is a frequent sequence since its support is $3 \geq 2$ and similarity $(A, D) = 0.5 \geq 0.3$. The correlation pattern with respective to $\langle A+A-D-D- \rangle$ is $\begin{pmatrix} A^+ & A^- & D^+ & D^- \\ f_{A^+} & f_{A^-} & f_{D^+} & f_{D^-} \end{pmatrix}$.

We only discuss f_{A+} as an example. The time information of $A+$ is $\{2, 6, 13\}$; hence $f_{A+}(x)$

$$= \frac{1}{3h\sqrt{2\pi}}(e^{-\frac{1}{2}(\frac{x-2}{h})^2} + e^{-\frac{1}{2}(\frac{x-6}{h})^2} + e^{-\frac{1}{2}(\frac{x-13}{h})^2}) \text{ with } h = \frac{range(\{2,6,13\})}{\sqrt{3}} = \frac{13-2}{\sqrt{3}} =$$
6.35.

4   Mining Appliance Usage Patterns

We focus our study on correlation pattern mining in smart home due to its wide applicability and the lack of research on this topic. In this paper, we develop a new algorithm, called *Correlation Pattern Miner* (abbreviated as **CoPMiner**), to discover correlation patterns effectively and efficiently. CoPMiner utilizes the arrangement of endpoints to accomplish the mining of correlation among appliances' usage. We also propose four pruning strategies to effectively reduce the search space and speedup the mining process.

4.1   Merits of Correlation Pattern and Usage Representation

Extracting correlation patterns from data collected in smart homes can provide resident useful information to better understand the relation among usage of appliances. Given a correlation pattern, as defined in Definition 5, a user can know the distribution of usage time of appliances. With a turn-on/off time of an appliance, we can derive the usage probability of other appliances. Consider the correlation pattern in aforementioned example. Suppose appliances A and D are the light and the coffee machine, respectively. Given the turn-on/off times of light and coffee machine, we can derive the usage probability for them, i.e., the probability for the light and coffee machine to be on/off at that time. This probability information is very useful for several applications, such as abnormal detection and activity prediction.

Obviously, the correlation pattern mining is an arduous task. Since the time period of the two usage-intervals may overlap, the relation between them is intrinsically complex. Allen's 13 temporal logics [1], in general, can be adopted to describe the relations among intervals. However, Allen's logics are binary relations. When describing relationships among more than three intervals, Allen's temporal logics may suffer several problems.

A suitable representation is very important for describing a correlation pattern. In this paper, a new expression, called *usage representation*, is proposed to effectively address the ambiguous and scalable issue [19] for describing relationships among intervals. Given two different usage-intervals A and B, the usage representation of Allen's 13 relations between A and B is categorized as in Fig. 3. Several merits of usage representation are discussed as follows: (1) **Lossless:** Usage representation not only implies the temporal relation among intervals, but also includes the accurate usage time of each interval. This concept can achieve a lossless representation to express the nature of the interval sequence. (2) **Nonambiguity:** According to [19], we can find that the usage representation has no ambiguous problem. First, by Definition 3, we can transform every usage-interval sequence to a unique usage sequence.

In other words, the temporal relations among intervals can be mapped to a usage sequence. Second, in a usage sequence, the order relation of the starting and finishing endpoints of *A* and *B* can be depicted easily. Hence, we can infer the original temporal relationships between intervals *A* and *B* nonambiguously. (3) **Simplicity:** Obviously, the complex relations between intervals are the major bottleneck of correlation pattern mining. However, the relation between two usage points is simple, just "*before*," "*after*" and "*equal*." The simpler the relations, the less number of intermediate candidate sequences are generated and processed.

4.2   CoPMiner Algorithm

Before introducing the algorithm, we modify the idea in [16] and define the projected database first. Let α be a usage sequence in a database *DB* with usage representation. The α - projected database, denoted as $DB_{|\alpha}$, is the collection of postfixes of sequences (including usage sequences and corresponding time sequence) in *DB* with regards to prefix α.

Algorithm 1 illustrates the main framework of CoPMiner. It first transforms the usage database to usage representation and calculates the count of each usage-point concurrently (line 2, algorithm 1). CoPMiner removes infrequent usage-points under given support threshold, *min_sup* (line 3, algorithm 1). For each frequent starting usage-point *s*, we find all its time information $\{t_{s1}, t_{s2}, \ldots t_{sm}\}$ in *DB* and estimate the probability function f_s by Definition 5 (lines 6-7, algorithm 1).

Algorithm 1: CoPMiner (*DB*, *min_sup*, *min_sim*)
Input: a usage-interval database *DB*, the support threshold *min_sup*, the similarity threshold *min_sim*
Output: all correlation patterns *P*
01:   $P \leftarrow \varnothing$;
02:   transform *DB* into usage presentation by Definition 4;
03:   find all frequent usage-points and remove infrequent usage-points in *DB*;
04:   *FS* $\leftarrow$ all frequent "on-points";
05:   **for** each $s \in FS$ **do**
06:        find all corresponding usage time information of *s* in *DB*;
07:        $f_s \leftarrow$ calculate the probability function of *s* by Definition 5;
08:        construct $DB_{|s}$ only with each usage-point *v*, where   *similarity*(*s*, *v*) $\geq$ *min_sim*;   // **spatial-pruning strategy**
09:        **UPrefixSpan**($DB_{|s}$, *s*, f_s, *min_sup*, *P*);
10:   output all correlation patterns *P*;

As mentioned above, the spatial distance may conflict with the correlation dependency between two appliances. When building the projected database $DB_{|s}$, CoPMiner collects the postfixes by using **spatial pruning strategy**. We eliminate the usage-points which have the similarity with regard to *s* smaller than *min_sim* in collected postfix sequences (line 8, algorithm 1). Finally, CoPMiner calls UPrefixSpan recursively and output all correlation patterns (lines 9-10, algorithm 1).

By borrowing the idea of the PrefixSpan [16], UPrefixSpan is developed with two search space pruning methods. The pseudo code is shown in Algorithm 2. For a prefix α, UPrefixSpan scans its projected database $DB_{|\alpha}$ once to discover all local frequent usage-points and remove infrequent ones (line 1, algorithm 2). For frequent usage-point s, we can append it to original prefix to generate a new frequent sequence α' with the length increased by 1. We also use the time information of s in $DB_{|\alpha}$ to estimate the probability function f_s by Definition 5, and then include f_s into $f(\alpha')$. As such, the prefixes are extended (lines 3-7, algorithm 2). If all usage-points in a frequent sequence appear in pairs, i.e., every on(off)-point has corresponding off(on)-point, we can output this frequent sequence and its probability function as a correlation pattern (lines 8-9, algorithm 2). Finally, we can discover all correlation patterns by constructing the projected database with the frequently extended prefixes and recursively running until the prefixes cannot be extended (lines 10-11, algorithm 2).

Taking into account the property of usage-point, we propose two pruning strategies, point-pruning and postfix-pruning to reduce the searching space efficiently and effectively. Firstly, the on-points and the off-points definitely occur in pairs in a usage sequence. We only require projecting the frequent on-points or the frequent off-points which have the corresponding on-points in their prefixes. For example, if we scan the projected database $DB_{|\langle A+\rangle}$ with respective to prefix $\langle A+\rangle$ and find three frequent local usage-points, $A-$, $B+$ and $B-$. We only require extending prefix $\langle A+\rangle$ with $A-$ and $B+$ (i.e., $\langle A+A-\rangle$ and $\langle A+B+\rangle$), since $B-$ has no corresponding on-points in its prefix. It is because that sequence $\langle A+B-\rangle$ has no chance to grow to a frequent sequence. This strategy is called **point-pruning strategy** (line 2 and lines 12-19, algorithm 2) which can prune off non-qualified patterns before constructing projected database.

Second, when we construct a projected database, some usage-points in postfix sequences need not be considered. With respect to a prefix sequence $\langle \alpha \rangle$, an off-point in a projected postfix sequence is insignificant, if it has no corresponding on- points in $\langle \alpha \rangle$. Hence, when collecting postfix sequences to construct $DB_{|\langle \alpha \rangle}$, we can eliminate all insignificant off-points since they can be ignored in the discovery of correlation patterns. This pruning method is called **postfix-pruning strategy** which can shrink the length of postfix sequence and further reduce the size of projected database effectively (line 14 and lines 20-25, algorithm 2).

Algorithm 2: UPrefixSpan $(DB_{|\alpha}, \alpha, f(\alpha), min\_sup, P)$

Input: a projected database $DB_{|\alpha}$, an usage sequence α, the support threshold min_sup, a similarity threshold min_sim, and a set of correlation patterns P
Output: a set of correlation patterns P

01: scan $DB_{|\alpha}$ once, remove infrequent usage-points and find every frequent usage-point v such that:
    (i) v can be assembled to the last pointset of α to form a frequent sequence; or
    (ii) $\langle v \rangle$ can be appended to α to form a frequent sequence;
02: $FS \leftarrow$ all frequent usage-points;
03: $FS \leftarrow point\_pruning(FS, \alpha)$; // **point-pruning strategy**
04: **for** each $s \in FS$ **do**
05:    find all corresponding usage time information of s in $DB_{|\alpha}$;
06:    $f_s \leftarrow$ calculate the probability function of s by Definition 5;
07:    append s to α to form α';
08:    $f(\alpha') \leftarrow f(\alpha) + f_s$;
09:    **if** α' is a correlation pattern **then**
10:        $P \leftarrow P \cup (\alpha', f(\alpha'))$;
11:    $DB_{|\alpha} \leftarrow DB\_construct(DB_{|\alpha}, \alpha')$; // **prefix-pruning strategy**
12:    $UPrefixSpan(DB_{|\alpha}, \alpha', f(\alpha'), min\_sup, P)$;

Procedure point_pruning (FS, α)
13: $temp\_point \leftarrow \emptyset$;
14: **for** each $s \in FS$ **do**
15:    **if** s is a "off-point"   then   // **point-pruning strategy**
16:        **if** exist corresponding "on-point" in α **then**
17:            $temp\_point \leftarrow temp\_point \cup s$;
18:    **if** s is a "on-point" **then**
19:        $temp\_point \leftarrow temp\_point \cup s$;
20: **return** $temp\_point$;

Procedure DB_construct $(DB_{|\alpha}, \alpha')$
21: $temp\_seq \leftarrow \emptyset$;
22: find all postfix sequences of α' in $DB_{|\alpha}$ to form $DB_{|\alpha'}$;
23: **for** each postfix sequence $q \in DB_{|\alpha'}$ **do**
24:    eliminate the "off-points" in q which has no corresponding "on-point" in α'; // **postfix-pruning strategy**
25:    $temp\_seq \leftarrow temp\_seq \cup q$;
26: **return** $temp\_seq$;

5   Experimental Results

To best of our knowledge, CoPMiner is the first algorithm discussing the correlation among appliances included probability concept. Three interval-pattern mining algorithms, *CTMiner* [4], *IEMiner* [15] and *TPrefixSpan* [19] have been implemented for performance discussion. For fair comparison, when comparing the execution time of CoPMiner with other interval-pattern mining algorithms, we only discuss the part of usage sequence mining (i.e., exclusive of computation of probability function). All algorithms were implemented in Java language and tested on a workstation with Intel i7-3370 3.4 GHz with 8 GB main memory. First, we compare the execution time using synthetic datasets at different minimum support. Second, we conduct an experiment to observe the memory usage and the scalability on execution time of CoPMiner. Finally, CoPMiner is applied in real-world dataset [12] to show the performance and the practicability of mining correlation patterns. The synthetic datasets in the experiments are generated using synthetic generator in [4] and the parameter setting is shown in Fig. 3.

Parameters	Description
$\lvert D \rvert$	Number of event sequences
$\lvert C \rvert$	Average size of event sequences
$\lvert S \rvert$	Average size of potentially frequent sequences
N_S	Number of potentially frequent sequences
N	Number of event symbols

Fig. 3. Parameters of synthetic data generator

5.1   Performance and Scalability on Synthetic Dataset

In all the following experiments, two parameters are fixed, i.e., $\lvert S \rvert = 4$ and $N_S = 5,000$. The other parameters are configured for comparison. Note that, for fair comparison, when comparing the performance of CoPMiner with other interval-pattern mining algorithms, we only discuss the part of usage sequence mining (i.e., exclusive of computation of probability function). Fig. 4(a) shows the running time of the four algorithms with minimum supports varied from   1 % to 5 % on the dataset $D100k–C20–N10k$. Obviously, when the minimum support value decreases, the processing time required for all algorithms increases. We can see that when we continue to lower the threshold, the runtime for IEMiner and TPrefixSpan increase drastically compared to CTMiner and CoPMiner. This is partly because these two algorithms still process interval-based data with complex relationship which may lead to generate more number of intermediate candidate sequences.

Then, we study the scalability of CoPMiner. Here, we use the data set $C = 20$, $N = 10k$ with varying different database size. Fig. 4(b) shows the results of scalability tests of four algorithms with the database size growing from 100K to 500K sequences. We fix the min_sup as 1%. Fig. 4(c) depicts the results of scalability tests of CoPMiner under different database size growing with different minimum support threshold varying from 1% to 5%. As the size of database increases and minimum support decreases, the processing time of all algorithms increase, since the number of patterns also increases. As can be seen, CoPMiner is linearly scalable with different minimum support threshold. When the number of generated patterns is large, the runtime of CoPMiner still increases linearly with different database size.

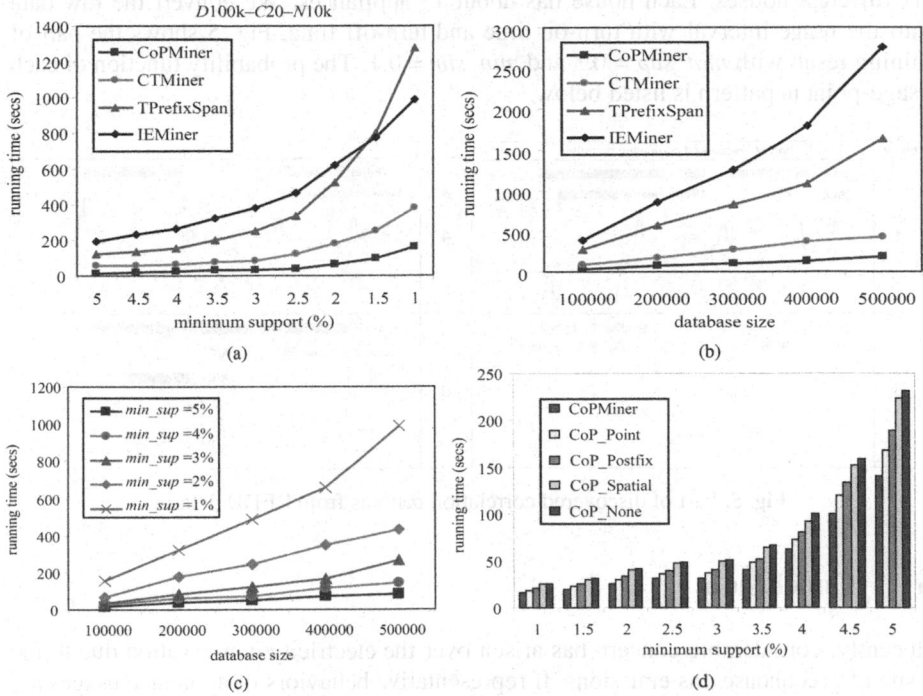

Fig. 4. Experimental results on synthetic datasets

5.2   Influence of Proposed Pruning Strategies

To reflect the speedup of proposed pruning methods, we measure CoPMiner with pruning strategies and without pruning strategy on time performance. We compare five algorithms, *CoPMiner* (includes all pruning strategies), *CoP_Point* (only point-pruning strategy), *CoP_Postfix* (only postfix-pruning strategy), *CoP_Spatial* (only spatial-pruning strategy) and *CoP_None* (without any pruning strategy). The experiment is performed on the data set D100k–C20–N10k. Fig. 4(d) is the results of varying minimum support thresholds from 0.5% to 1%. As shown in figure, point-pruning can improve about 25% performance. Because of removing non-qualified usage-points before database projection, point-pruning can efficiently speedup the execution time. As can be seen from the graph, postfix-pruning can improve about 11% performance. Postfix-pruning can improve the performance by effectively eliminating all useless usage-points for correlation pattern construction. We also can observe that spatial-pruning constantly ameliorate the performance about 2.5.

5.3   Real-World Dataset Analysis

In addition to using synthetic datasets, we also have performed an experiment on real-world dataset to indicate the applicability of correlation pattern mining. The dataset REDD [12] used in the experiment is the power reading of appliances collected from six different houses. Each house has about 15 appliances. We convert the raw data into the usage interval with turn-on time and turn-off time. Fig. 5 shows the part of mining result with $min\_sup = 0.3$ and $min\_sim = 0.1$. The probability function of each usage-point in pattern is listed below.

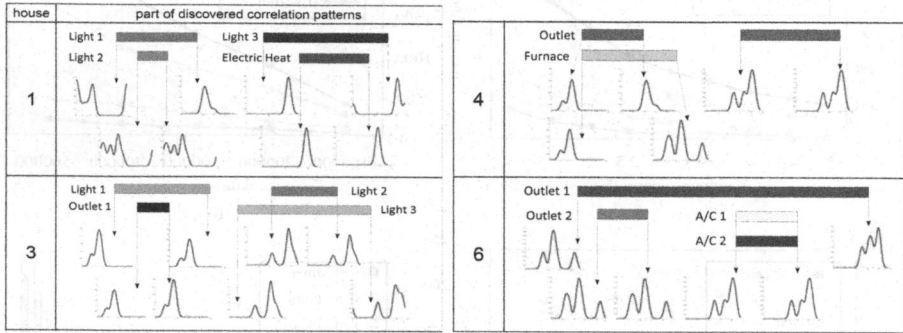

Fig. 5. Part of discovered correlation patterns from REDD dataset

6   Conclusion

Recently, considerable concern has arisen over the electricity conservation due to the issue of greenhouse gas emissions. If representative behaviors of appliance usages are available, residents may adapt their usage patterns to conserve energy effectively. However, previous studies on usage pattern discovery are mainly focused on analyzing single appliance and ignore the usage correlation. In this paper, we introduce a new concept, correlation pattern, to capture the usage patterns and correlations among appliances probabilistically. An efficient algorithm, CoPMiner is developed to discover patterns based on proposed usage representation. The experimental studies indicate that CoPMiner is efficient and scalable. Furthermore, CoPMiner is applied on a real-world dataset to show the practicability of correlation pattern mining.

References

1. Allen, J.: Maintaining Knowledge about Temporal Intervals. Communications of ACM 26(11), 832–843 (1983)
2. Aritoni, O., Negru, V.: A Methodology for Household Appliances Behavior Recognition in AmI Systems Integration. In: 7th International Conference on Automatic and Autonomous Systems (ICAS 2011), pp. 175–178 (2011)
3. Chen, F., Dai, J., Wang, B., Sahu, S., Naphade, M., Lu, C.T.: Activity Analysis Based on Low Sample Rate Smart Meters. In: 7th ACM SIGKDD International Conference on Knowledge Discovery and Data Mining (KDD 2011), pp. 240–248 (2011)

4. Chen, Y., Jiang, J., Peng, W., Lee, S.: An Efficient Algorithm for Mining Time Interval-based Patterns in Large Databases. In: Proceedings of 19th ACM International Conference on Information and Knowledge Management (CIKM 2010), pp. 49–58 (2010)
5. Chen, Y.-C., Ko, Y.-L., Peng, W.-C., Lee, W.-C.: Mining Appliance Usage Patterns in a Smart Home Environment. In: Pei, J., Tseng, V.S., Cao, L., Motoda, H., Xu, G. (eds.) PAKDD 2013, Part I. LNCS (LNAI), vol. 7818, pp. 99–110. Springer, Heidelberg (2013)
6. Goncalves, H., Ocneanu, A., Bergés, M.: Unsupervised disaggregation of appliances using aggregated consumption data. In: KDD Workshop on Data Mining Applications in Sustainability, SustKDD 2011 (2011)
7. Ito, M., Uda, R., Ichimura, S., Tago, K., Hoshi, T., Matsushita, Y.: A method of appliance detection based on features of power waveform. In: 4th IEEE Symposium on Applications and the Internet (SAINT 2004), pp. 291–294 (2004)
8. Jakkula, V., Cook, D.: Using Temporal Relations in Smart Environment Data for Activity Prediction. In: Proceedings of the 24th International Conference on Machine Learning (ICML 2007), pp. 1–4 (2007)
9. Jakkula, V., Cook, D., Crandall, A.: Temporal pattern discovery for anomaly detection in a smart home. In: Proceedings of the 3rd IET Conference on Intelligent Environments (IE 2007), pp. 339–345 (2007)
10. Kato, T., Cho, H.S., Lee, D., Toyomura, T., Yamazaki, T.: Appliance recognition from electric current signals for information-energy integrated network in home environments. In: Mokhtari, M., Khalil, I., Bauchet, J., Zhang, D., Nugent, C. (eds.) ICOST 2009. LNCS, vol. 5597, pp. 150–157. Springer, Heidelberg (2009)
11. Kim, H., Marwah, M., Arlitt, M., Lyon, G., Han, J.: Unsupervised disaggregation of low frequency power measurements. In: 11th SIAM International Conference on Data Mining (SDM 2011), pp. 747–758 (2011)
12. Kolter, J.Z., Johnson, M.J.: REDD: A public data set for energy disaggregation research. In: KDD Workshop on Data Mining Applications in Sustainability, SustKDD 2011 (2011)
13. Lin, G., Lee, S., Hsu, J., Jih, W.: Applying power meters for appliance recognition on the electric panel. In: 5th IEEE Conference on Industrial Electronics and Applications (ISIEA 2010), pp. 2254–2259 (2010)
14. Liu, B., Yang, Y., Webb, G.I., Boughton, J.: A Comparative Study of Bandwidth Choice in Kernel Density Estimation for Naive Bayesian Classification. In: Theeramunkong, T., Kijsirikul, B., Cercone, N., Ho, T.-B. (eds.) PAKDD 2009. LNCS, vol. 5476, pp. 302–313. Springer, Heidelberg (2009)
15. Patel, D., Hsu, W., Lee, M.: Mining Relationships Among Interval-based Events for Classification. In: Proceedings of the 2008 ACM SIGMOD International Conference on Management of Data (SIGMOD 2008), pp. 393–404 (2008)
16. Pei, J., Han, J., Mortazavi-Asl, B., Pito, H., Chen, Q., Dayal, U., Hsu, M.: PrefixSpan: Mining Sequential Patterns Efficiently by Prefix-Projected Pattern Growth. In: Proceedings of 17th International Conference on Data Engineering (ICDE 2001), pp. 215–224 (2001)
17. Silverman, B.: Density Estimation for Statistics and Data Analysis. Chapman and Hall (1986)
18. Suzuki, K., Inagaki, S., Suzuki, T., Nakamura, H., Ito, K.: Nonintrusive appliance load monitoring based on integer programming. In: International Conference on Instrumentation, Control and Information Technology, pp. 2742–2747 (2008)
19. Wu, S., Chen, Y.: Mining Nonambiguous Temporal Patterns for Interval-Based Events. IEEE Transactions on Knowledge and Data Engineering 19(6), 742–758 (2007)

Shifting Hypergraphs by Probabilistic Voting

Yang Wang[1,2], Xuemin Lin[1], Qing Zhang[2,1], and Lin Wu[1]

[1] The University of New South Wales, Sydney, Australia
{wangy,lxue,linw}@cse.unsw.edu.au
[2] Australian E-Health Research Center
qing.zhang@csiro.au

Abstract. In this paper, we develop a novel paradigm, namely hypergraph shift, to find robust graph modes by probabilistic voting strategy, which are semantically sound besides the self-cohesiveness requirement in forming graph modes. Unlike the existing techniques to seek graph modes by shifting vertices based on pair-wise edges (i.e, an edge with 2 ends), our paradigm is based on shifting high-order edges (hyperedges) to deliver graph modes. Specifically, we convert the problem of seeking graph modes as the problem of seeking maximizers of a novel objective function with the aim to generate good graph modes based on sifting edges in hypergraphs. As a result, the generated graph modes based on dense subhypergraphs may more accurately capture the object semantics besides the self-cohesiveness requirement. We also formally prove that our technique is always convergent. Extensive empirical studies on synthetic and real world data sets are conducted on clustering and graph matching. They demonstrate that our techniques significantly outperform the existing techniques.

Keywords: Hypergraphs, Mode Seeking, Probabilistic Voting.

1 Introduction

Seeking graph based modes is of great importance to many applications in machine learning literature, e.g., image segmentation [9], feature matching [3]. In order to find the good modes of graphs, Pavan et al. [16] converted the problem of mode seeking into the problem of discovering dense subgraphs, and proposed a constrained optimization function for this purpose. Liu et al. [14] proposed another method, namely graph shift. It generalized the idea of non-parametric data points shift paradigms (i.e., Mean Shift [4] and Medoid Shift [17,18,19] to graph shift for graph mode seeking). An iterative method is developed to get the local maximizers, of a constrained objective function, as the good modes of graphs. While the graph (vertices) shift paradigm may deliver good results in many cases for graph mode seeking, we observe the following limits. Firstly, the graph modes generated based on shifting vertices only involve the information of pair-wise edges between vertices. As a result, the generated graphs modes may not always be able to precisely capture the overall semantics of objects. Secondly, the graph shift algorithm is still not strongly robust to the existence

V.S. Tseng et al. (Eds.): PAKDD 2014, Part II, LNAI 8444, pp. 234–246, 2014.

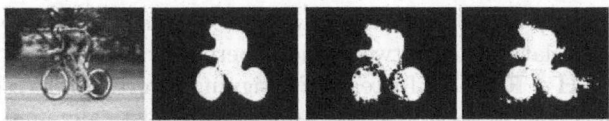

Fig. 1. Comparison between graph shift and hypergraph shift on saliency detection, from left to right: input image, ground truth, graph shift, hypergraph shift

of a large number of outliers. Besides, no theoretical studies are conducted to show the convergence of iteration of shifting.

Our Approach: Observing the above limits, we propose a novel paradigm, namely hypergraph shift, aimed at generating graph modes with high order information. Different from graph shift paradigms that only shift vertices of graphs based on pair-wise edges, our technique shifts high order edges (hyperedges in hypergraphs). Our technique consists of three key phases, 1) mode seeking (section 4.1) on subhypergraphs, 2) probabilistic voting (section 4.2) to determine a set of hyperedges to be expanded in mode seeking, and 3) iteratively perform the above two stages until convergence.

By these three phases, our approach may accurately capture the overall semantics of objects. Fig. 1 illustrates an example where the result of our approach for hypergraph shift can precisely capture the the the scene of a person riding on a bicycle. Nevertheless, the result performed by graph shift method in [14] fails to capture the whole scene; instead, by only focusing on the requirement of self-cohesiveness, three graph modes are generated.

Contributions: To the best of our knowledge, this is the first work based on shifting hyperedges to conduct graph mode seeking. Our contributions may be summarized as follows. (1) We specify the similarities on hyperedges, followed by an objective function for mode seeking on hypergraphs. (2) An effective hypergraph shift paradigm is proposed. Theoretical analysis for hypergraph shift is also provided to guarantee its convergence. The proposed algorithm is naturally robust to outliers by expanding modes via the probabilistic voting strategy. (3) Extensive experiments are conducted to verify the effectiveness of our techniques over both synthetic and real-world datasets.

Roadmap: We structure our paper as follows: The preliminaries regarding hypergraph are introduced in section 2, followed by our technique for hypergraph shift in sections 3 and 4. Experimental studies are performed in section 5, and we conclude this paper in section 6.

2   Probabilistic Hypergraph Notations

Different from simple graph, each edge of hypergraph (known as hyperedge) can connect more than two vertices. Formally, we denote a weighted hypergraph as $\mathbf{G} = (\mathcal{V}, \mathcal{E}, \mathcal{W})$, with vertex set as $\mathcal{V} = \{v_1, v_2, \ldots, v_{|\mathcal{V}|}\}$, hyperedge set as

$\mathcal{E} = \{e_1, e_2, \ldots, e_{|\mathcal{E}|}\}$, and $\mathcal{W} = \{w(e_1), w(e_2), \cdots, w(e_{|\mathcal{E}|})\}$, where $w(e_i)$ is the weight of e_i. The relationship between the hyperedges and vertices is defined by incidence matrix $\mathbf{H} \in \mathbb{R}^{|\mathcal{V}| \times |\mathcal{E}|}$. Instead of assigning a vertex v_i to a hyperedge e_j with a binary decision, we establish the values probabilistically [5,8]. Specifically, we define the entry h_{v_i,e_j} of $\mathbf{H}$ as Eq. (1).

$$h_{v_i,e_j} = \begin{cases} p(v_i|e_j), & \text{if } v_i \in e_j; \\ 0, & \text{otherwise.} \end{cases} \tag{1}$$

where $p(v_i|e_j)$ describes the likelihood that a vertex v_i is connected to hyperedge e_j. Then we define a diagonal matrix $\mathbf{D}_e$ regarding the degree of all hyperedges, with $\mathbf{D}_e(i,i) = \delta(e_i) = \sum_{v \in \mathcal{V}} h_{v,e_i}$, and a diagonal matrix $\mathbf{D}_v$ regarding the degree of all vertices, with $\mathbf{D}_v(i,i) = \sum_{e \in \mathcal{E}} h_{v_i,e} w(e)$. Based on that, to describe the similarity between hyperedges, we define a novel **hyperedge-adjacency matrix** $\mathbf{M} \in \mathbb{R}^{|\mathcal{E}| \times |\mathcal{E}|}$ in the context of hypergraph. Specifically, we have

$$\mathbf{M}(i,j) = \begin{cases} w(e_i)\frac{|e_i \cap e_j|}{\delta(e_i)} + w(e_j)\frac{|e_i \cap e_j|}{\delta(e_j)} & i \neq j \\ 0, & \text{otherwise} \end{cases} \tag{2}$$

Example 1. Consider the case in Fig.2, for e_2 and e_3, the only common vertex is v_2, then, we have $|e_2 \cap e_3|=1$, and the affinity value between e_2 and e_3 is $\mathbf{M}(2,3) = w(e_2) \cdot \frac{1}{2} + w(e_3) \cdot \frac{1}{2} = \frac{w(e_2)+w(e_3)}{2}$.

Now, we describe the modes of hypergraph.

3 Modes of Hypergraph

We consider the mode of a hypergraph as a dense subhypergraph consisting of hyperedges with high self-compactness. We first define the **hypergraph density**, then formulate the modes of a hypergraph, which leads to our hypergraph shift algorithm in section 4.

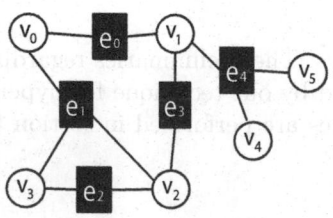

	e_0	e_1	e_2	e_3	e_4
V_0	h(v₀,e₀)	h(v₀,e₁)			
V_1				h(v₁,e₃)	h(v₁,e₄)
V_2		h(v₂,e₁)	h(v₂,e₂)	h(v₂,e₃)	
V_3		h(v₃,e₁)	h(v₃,e₂)		
V_4					h(v₄,e₄)
V_5					h(v₅,e₄)

Fig. 2. A toy example on hypergraph. Left: a hypergraph. Right: The incidence matrix of hypergraph.

Hypergraph Density. We describe hypergraph $\mathbf{G}$ with n hyperedges by probabilistic coordinates fashion as $\mathbf{p} \in \Delta^n$, where $\Delta^n = \{\mathbf{p}|\mathbf{p} \geq 0, |\mathbf{p}|_1 = 1\}$, $|\mathbf{p}|_1$ is the L_1 norm of vector $\mathbf{p}$, and $\mathbf{p} = \{\mathbf{p}_1, \mathbf{p}_2, \dots, \mathbf{p}_n\}$. Specifically, $\mathbf{p}_i$ indicates the probability of e_i contained by the probabilistic cluster of $\mathbf{G}$. Then the affinity value between any pair-wise $\mathbf{x} \in \Delta^n$ and $\mathbf{y} \in \Delta^n$ is defined as $\mathbf{m}(\mathbf{x}, \mathbf{y}) = \sum_{i,j} \mathbf{M}(i,j)\mathbf{x}_i\mathbf{y}_j = \mathbf{x}^T\mathbf{M}\mathbf{y}$. The **hypergraph density** or self-cohesiveness of $\mathbf{G}$, is defined as Eq. (3).

$$\mathcal{F}(\mathbf{p}) = \mathbf{p}^T\mathbf{M}\mathbf{p}. \tag{3}$$

Intuitively, **hypergraph density** can be interpreted by the following principle. Suppose hyperedge set $\mathcal{E}$ is mapped to $\mathbf{I} = \{i_m|m = 1, \dots, |\mathcal{E}|\}$, which is the representation in a specific feature space regarding all hyperedges in $\mathcal{E}$, where we define a kernel function $\mathcal{K} : \mathbf{I} \times \mathbf{I} \to \mathbb{R}$. Specifically, $\mathcal{K}(i_m, i_n) = \mathbf{M}(m, n)$. Thus, the probabilistic coordinate $\mathbf{p}$ can be interpreted to be a probability distribution, that is, the probability of i_m occurring in a specific subhypergraph is $\mathbf{p}_m$. Assume that the distribution is sampled $\mathcal{N}$ times, then the number of data i_m is $\mathcal{N}\mathbf{p}_m$. For i_m, the density is $d(i_m) = \frac{\sum_n \mathcal{N}\mathbf{p}_m\mathcal{K}(m,n)}{\mathcal{N}}$, then we have the average density of the data set:

$$\bar{d} = \frac{\sum_m \mathcal{N}\mathbf{p}_m d(i_m)}{\mathcal{N}} = \sum_{m \neq n} \mathbf{p}_m\mathcal{K}(i_m, i_n)\mathbf{p}_n = \mathbf{p}^T\mathbf{M}\mathbf{p} \tag{4}$$

Definition 1. *(Hypergraph Mode)* *The mode of a hypergraph $\mathbf{G}$ is represented as a dense subhypergraph that locally maximizes the Eq. (3).*

Given a vector $\mathbf{p} \in \Delta^n$, the **support** of $\mathbf{p}$ is defined as the set of indices corresponding to its nonzero components: $\theta(\mathbf{p}) = \{i \in |\mathcal{E}| : \mathbf{p}_i \neq 0\}$. Thus, its corresponding subhypergraph is $\mathbf{G}_{\theta(\mathbf{p})}$, composed of all vertices whose indices are in $\theta(\mathbf{p})$. If $\mathbf{p}^*$ is a local maximizer i.e., the mode of $\mathcal{F}(\mathbf{p})$, then $\mathbf{G}_{\theta(\mathbf{p}^*)}$ is a dense subhypergraph. Hence, the problem of mode seeking on a hypergraph is equivalent to maximizing the density measure function $\mathcal{F}(\mathbf{p})$, which is taken as the criterion to evaluate the goodness of any subhypergraph.

To find the modes, i.e., the local maximizers of Eq. (3), we classify it into the standard quadratic program (StQP) [16,1]:

$$\max \mathcal{F}(\mathbf{p}), s.t. \mathbf{p} \in \Delta^n, \tag{5}$$

According to [16,1], a local maximizer $\mathbf{p}^*$ meets the Karush-Kuhn-Tucker(KKT) condition. In particular, there exist $n + 1$ real Lagrange multipliers $\mu_i \geqslant 0 (1 \leq i \leq n)$ and λ, such that:

$$(\mathbf{M}\mathbf{p})_i - \lambda + \mu_i = 0 \tag{6}$$

for all $i = 1, \dots, n$, and $\sum_{i=1}^n \mathbf{p}_i^*\mu_i = 0$. Since $\mathbf{p}^*$ and μ_i are nonnegative, it indicates that $i \in \theta(\mathbf{p}^*)$ implies $\mu_i = 0$. Thus, the KKT condition can be rewritten as:

$$(\mathbf{M}\mathbf{p}^*)_i \begin{cases} = \lambda, i \in \theta(\mathbf{p}^*); \\ \leqslant \lambda, \text{otherwise}. \end{cases} \tag{7}$$

where $(\mathbf{M}\mathbf{p}^*)_i$ is the affinity value between $\mathbf{p}^*$ and e_i.

4 Hypergraph Shift Algorithm

Commonly the hypergraph can be very large, a natural question is how to perform modes seeking on a large hypergraph? To answer this question, we perform mode seeking on subhypergraph, and determine whether it is the mode of the hypergraph. If not, we shift to a new subhypergraph by expanding the neighbor hyperedges of the current mode to perform mode seeking. Prior to that, we study the circumstances that determine whether the mode of a subhypergraph is the mode of that hypergraph.

Assume $\mathbf{p}_\mathcal{S}^*$ is the mode of subhypergraph $\mathcal{S}$ containing $m = |\theta(\mathbf{p}_\mathcal{S}^*)|$ hyperedges, then we expand the m dimensional $\mathbf{p}_\mathcal{S}^*$ to $|\mathcal{E}|$ dimensional $\mathbf{p}^*$ by filling zeros into the components, whose indices are in the set of $\mathbf{G} - \mathcal{S}$. Based on that, Theorem 1 is presented to determine whether $\mathbf{p}_\mathcal{S}^*$ is the mode of hypergraph $\mathbf{G}$.

Theorem 1. *A mode* $\mathbf{p}_\mathcal{S}^*$ *of the subgraph* $\mathcal{S}$ *is also the mode of hypergraph* $\mathbf{G}$ *if and only if for all hyperedge* e_j, $\mathbf{m}(\mathbf{p}^*, \mathbf{I}_j) \leqslant \mathcal{F}(\mathbf{p}^*) = \mathcal{F}_\mathcal{S}(\mathbf{p}_\mathcal{S}^*)$, $j \in \mathbf{G} - \mathcal{S}$, *where* $\mathbf{p}^*$ *is computed from* $\mathbf{p}_\mathcal{S}^*$ *by filling zeros to the elements whose indices are in* $\mathbf{G} - \mathcal{S}$ *and* $\mathbf{I}_j$ *is the vector containing only hyperedge* e_i *where its i-th element is 1 with others 0.*
Proof. Straightforwardly, $\theta(\mathbf{p}^*) = \theta(\mathbf{p}_\mathcal{S}^*)$, $\mathcal{F}(\mathbf{p}^*) = \mathcal{F}_\mathcal{S}(\mathbf{p}_\mathcal{S}^*)$. Due to $\mathbf{m}(\mathbf{p}^*, \mathbf{I}_j) \leqslant \mathcal{F}(\mathbf{p}^*) = \mathcal{F}_\mathcal{S}(\mathbf{p}_\mathcal{S}^*) = \lambda$, $\forall j \in \mathbf{G} - \mathcal{S}$, $\mathbf{p}^*$ is the mode of hypergraph $\mathbf{G}$. Otherwise if $\mathbf{m}(\mathbf{p}^*, \mathbf{I}_j) > \mathcal{F}(\mathbf{p}^*) = \lambda$, which indicates that $\mathbf{p}^*$ violates the KKT condition, thus it is not the mode of $\mathbf{G}$. ∎

Next, we introduce our hypergraph shift algorithm, which consists of two steps: The first step performs mode seeking on an initial subhypergraph. If the mode obtained in the first step is not the mode of that hypergraph, it shifts to a larger subhypergraph by expanding the support of the current mode to its neighbor hyperedges using the technique, namely probabilistic voting. The above steps alternatively proceed until the mode of hypergraph is obtained.

4.1 Higher-Order Mode Seeking

Given an initialization of $\mathbf{p}(0)$, we find solutions of Eq. (5) by using the replicator dynamics, which is a class of continuous and discrete-time dynamical systems arising in evolutionary game theory [20]. In our setting, we use the following form:

$$\mathbf{p}_i(t+1) = \mathbf{p}_i(t) \frac{(\mathbf{M} \cdot \mathbf{p}(t))_i}{\mathbf{p}(t)^T \mathbf{M} \mathbf{p}(t)}, i = 1, \dots, |\mathcal{E}| \qquad (8)$$

It can be seen that the simplex Δ^n is invariant under these dynamics, which means that every trajectory starting in Δ^n will remain in Δ^n for all future times. Furthermore, according to [20], the objective function of Eq. (3) strictly increases along any nonconstant trajectory of Eq. (8), and its asymptotically stable points are in one-to-one with local solutions of Eq. (5).

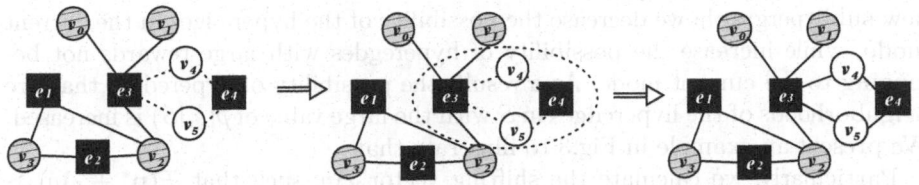

Fig. 3. Probabilistic voting strategy. The hyperedge e_3 is selected to be the dominant seed because of its higher closeness, as measured by Eq. (12). We start the expansion from e_3 and then include its nearest neighbor e_4 into the mode.

4.2   Probabilistic Voting

We propose to find the **dominant seeds** of the subhypergraph, from which we perform hypergraph shift algorithm. Before presenting the formal definition of dominant seeds, we start with the intuitive idea that the assignment of hyperedge-weights induces an assignment of weights on the hyperedges. Therefore, the average weighted degree of a hyperedge e_k from subhypergraph $\mathcal{S}$ is defined as:

$$g_{\mathcal{S}}(e_k) = \frac{1}{|\mathcal{S}|} \sum_{e_j \in \mathcal{S}} \mathbf{M}(k, j) \tag{9}$$

Note that $g_{e_k}(e_k) = 0$ for any $e_k \in \mathcal{S}$. Moreover, if $e_j \not\subseteq \mathcal{S}$, we have:

$$\psi_{\mathcal{S}}(e_i, e_j) = \mathbf{M}(i, j) - g_{\mathcal{S}}(e_i) \tag{10}$$

Intuitively, $\psi_{\mathcal{S}}(e_i, e_j)$ measures the relative closeness between e_j and e_i with respect to the average closeness between e_i and its neighbors in $\mathcal{S}$.

Let $\mathcal{S} \subseteq \mathcal{E}$ be a nonempty subset of hyperedges, and $e_i \in \mathcal{S}$. The weight of e_i is given as

$$w_{\mathcal{S}}(e_i) = \begin{cases} 1, & \text{if } |\mathcal{S}| = 1; \\ \sum_{e_j \in \mathcal{S} - \{e_i\}} \psi_{\mathcal{S} - \{e_i\}}(e_j, e_i) w_{\mathcal{S} - \{e_i\}}(e_j), & \text{otherwise.} \end{cases} \tag{11}$$

$w_{\mathcal{S}}(e_i)$ measures the overall closeness between hyperedge e_i and other hyperedges of $\mathcal{S} - \{e_i\}$. Moreover, the total weight of $\mathcal{S}$ is defined as $W(\mathcal{S}) = \sum_{e_i \in \mathcal{S}} w_{\mathcal{S}}(e_i)$. Finally, we formally define the dominant seed of subhypergraph $\mathcal{S}$ as follows.

Definition 2. *(Dominant Seed) The dominant seed of a subhypergraph $\mathcal{S}$ is the subset of hyperedges with higher closeness than others.*

Besides, the closeness of the dominant seed is evaluated as follows:

$$p(e_i|\mathcal{S}) = \begin{cases} \frac{w_{\mathcal{S}}(e_i)}{W(\mathcal{S})}, & \text{if } e_i \in \mathcal{S} \\ 0, & \text{otherwise.} \end{cases} \tag{12}$$

We utilize dominat seeds to expand the current subhypergraph, which is named **probabilistic voting** that works by the following priciple. To expand $\mathcal{S}$ to a new subhypergraph, we decrease the possibility of the hyperedges in the current mode, while increase the possibility of hyperegdes with large rewards not belonging to the current mode. As a result, the possibility of hyperedges that are neighborhoods of the hyperedges in $\mathcal{S}$ with the large value of $p(e_i|\mathcal{S})$ is increased. We present an example in Fig.3 to illustrate that.

Particularly, we calculate the shifting vector Δp, such that $\mathcal{F}(\mathbf{p}^* + \Delta p) > \mathcal{F}(\mathbf{p}^*)$. According to Theorem 1, there exist some hyperedges e_i, such that $\mathbf{m}(\mathbf{p}^*, \mathbf{I}_i) > \mathcal{F}(\mathbf{p}^*)$, $i \in \mathbf{G} - \mathcal{S}$. We define a direction vector h as $h_i = \mathbf{p}_i^* - 1$ if $i \in \theta(\mathbf{p}^*)$, otherwise, $h_i = \max\{\sum_{e_j \cap e_i \neq \emptyset} p(e_j|\mathcal{S})(\mathbf{m}(\mathbf{p}^*, \mathbf{I}_i) - \mathcal{F}(\mathbf{p}^*)), 0\}$. The above definition of h_i for $i \in \theta(\mathbf{p}^*)$, decreases the possibility of e_i in the current mode. However, we try to preserve the dominant seeds with a larger value of $\mathbf{p}_i^* - 1$, and increase the possibility of the hyperedges $e_j \in \mathbf{G} - \mathcal{S}$ that are the neighborhoods of dominant seeds of the current mode.

Assume $\mathcal{F}(h) = \eta$, then we have:

$$Q(c) = \mathcal{F}(\mathbf{p}^* + ch) - \mathcal{F}(\mathbf{p}^*) \tag{13}$$
$$= \eta c^2 + 2c(\mathbf{p}^*)^T \mathbf{M} h$$

We want to maximize Eq. (13), which is the quadratic function of c. Since $\Delta = 4(p^*\mathbf{M}h)^2 > 0$, if $\eta < 0$, then we have $c = \frac{p^*\mathbf{M}h}{\lambda}$. Otherwise, for $i \in \theta(\mathbf{p}^*)$, we have $\mathbf{p}_i^* + c(\mathbf{p}_i^* - 1) \geq 0$, then $c \leq \min_i\{\frac{p_i^*}{1-p_i^*}\}$. Thus, $c^* = \min\{\frac{p^*\mathbf{M}h}{\lambda}, \min_i\{\frac{p_i^*}{1-p_i^*}\}$, and $\Delta p = c^* h$, which is the expansion vector.

We summarize the procedure of hypergraph shift in Algorithm 1.

Algorithm 1. Hypergraph Shift Algorithm.

Input: The hyperedge-adjacency matrix $\mathbf{M}$ of hypergraph $\mathbf{G}$, the start vector $\mathbf{p}$ (a cluster of hyperedges).
Output: The mode of hypergraph $\mathbf{G}$.
while $\mathbf{p}$ *is not the mode of* $\mathbf{G}$ **do**

    Evolve $\mathbf{p}$ towards the mode of subhypergraph $\mathbf{G}_\theta(\mathbf{p})$ by Eq. (8);
    if $\mathbf{p}$ *is not the mode of hypergrah* G **then**
        Expand $\mathbf{p}$ by using expansion vector Δp;
        Update $\mathbf{p}$ by mode seeking;
    else
        return;

One may wonder whether Algorithm 1 converges, we answer this question in theorem 2.

Theorem 2. *Algorithm 1 is convergent.*
Proof. The mode sequence set $\{(\mathbf{p}^*)(t)\}_{t=1}^\infty \subset U$ generated by Algorithm 1 is compact. We construct $-\mathcal{F}(\mathbf{p})$, which is a continuous and strict decreasing function over the trajectory of sequence set. Assume the solution set is Γ, then the mode sequence generated by Algorithm 1 is closed on $U - \Gamma$. The above three

conclusions are identical to the convergence conditions of Zangwill convergence theorem [21].   ∎

5   Experimental Evaluations

In this section, we conduct extensive experiments to evaluate the performance of hypergraph shift. Specific experimental setting are elaborated in each experiment.

Competitors. We compare our algorithm against a few closely related methods, which are introduced as follows.

For clustering evaluations, we consider the following competitors:

- The method proposed by Liu *et al.* in [13], denote by Liu *et al.* in follows.
- The approach presented by Bulo *et al.* in [2], denote by Bulo *et al.* in follows.
- Efficient hypergraph clustering [12] (EHC) aims to handle the higher-order relationships among data points and seek clusters by iteratively updating the cluster membership for all nodes in parallel, and converges relatively fast.

For graph matching, we compare our method to the state-of-the-arts below:

- Graph shift (GS).
- Two hypergraph matching methods (TM) [6] and (PM) [22].
- SC+IPFP. The algorithm of spectral clustering [10] (SC), enhanced by the technique of integer projected fixed point [11], namely SC+IPFP is an effective method in graph matching. Thus, it is suitable to compare our method against SC+IPFP in terms of graph matching.

5.1   Clustering Analysis

Consider that hypergraph shift is a natural clustering tool, and all the hyperedges shifting towards the same mode should belong to a cluster. To evaluate the clustering performance, we compare HS against Liu *et al.* , Bulo *et al.* and EHC over the data set of five crescents, as shown in Fig.4.

We performed extensive tests including clustering accuracy and noise robustness on five crescents gradually decreasing sampling density from 1200 pts to 100 pts. We used the standard clustering metric, normalized mutual information (NMI). The NMI accuracy are computed for each method in Fig.5 (a), with respect to decreasing sample points and increasing outliers. It shows that hypergraph shift has the best performance even in sparse data, whereas EHC quickly degenerates from 600 pts. The accuracies of methods in Liu *et al.* and Bulo *et al.* are inferior to EHC, which is consistent to the results in [12]. To test the robustness against noises, we add Gaussian noise ϵ, such that $\epsilon \sim \mathcal{N}(0, 4)$, in accordance with [14], to the five crescents samples, and re-compute the NMI values. As illustrated in Fig.5 (b), the three baselines of Liu *et al.* , Bulo *et al.* and EHC drop faster than hypergraph shift. This is because the eigenvectors required by Liu *et al.* are affected by all weights, no matter they are deteriorated or not;

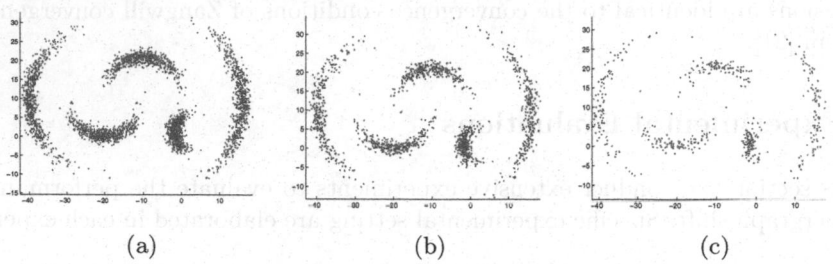

(a)   (b)   (c)

Fig. 4. "Five crescents" examples with decreasing sample points from 600 pts in (a) to 300 pts in (b) and 200 pts in (c)

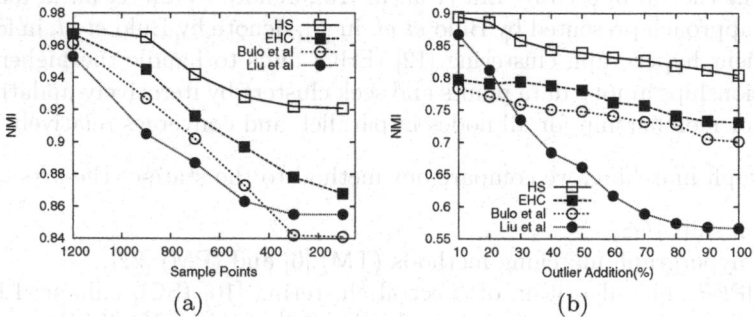

(a)   (b)

Fig. 5. Clustering comparisons on different sample sizes. We illustrate the averaged NMI with respect to the number of sample points and the ratio of outlier addition in (a) and (b), respectively.

EHC is better than Liu *et al.* and Bulo *et al.* , however, it performs clustering by only considering the strength of affinity relationship within a hyperedge, which is not as robust against noises as the mode with high-order constraints; Hypergraph shift, in contrast, can find a dense high-order subhypergraph, which is more robust to noises.

We are interested in another important aspect: speed of convergence, under varying number of data points. In Fig.6, we present the evaluation of the computational cost of the four methods with varying number of data points. Fig.6 (a) shows the average computational time per iteration of each method against the number of samples. We can see that the computation time per step for each method varies almost linearly with the number of data points. As expected, the least expensive method per step is Liu *et al.* , which performs update in sequence. And our method proceeds with expansion and dropping strategy, in the expense of more time. However, the drawback of Liu *et al.* is its large iterations to convergence. In contrast, both ours and EHC are relatively stable w.r.t. the number of samples. Our method converges very fast, requiring on average 10

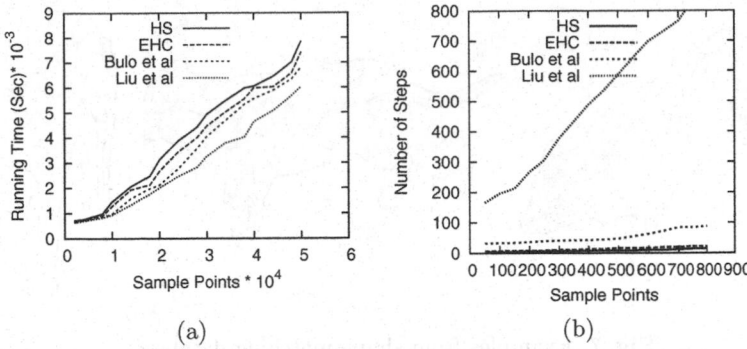

Fig. 6. The computational cost as a function of the sample size. (a) The running time against increasing number of sample points. (b) The number of steps to convergence against increasing number of sample points.

iterations. This figure experimentally show that our method, by taking larger steps towards a maximum, has significantly better speed of convergence with slightly better accuracy.

5.2   Graph Matching

In this part, we present some experiments on graph matching problems. We will show that this graph matching problem is identical to mode seeking on a graph with certain amount of noises and outliers. Following the experiment setup of [14], the equivalence of graph matching problem to mode seeking can be described as follows. Suppose there are two sets of feature points, P and Q, from two images. For each point $p \in P$, we can find some similar points $q \in Q$, based on local features. Each pair of (p, q) is a possible correspondence and all such pairs form the correspondence set $C = \{(p, q)|p \in P, q \in Q\}$. Then a graph G is constructed based on C with each vertex of G representing a pair in C. Edge $e(v_i, v_j)$ connecting v_i and v_j reflects the relation between correspondence c_i and c_j. Due to space limitations, we refer the interested readers to [14] for details. Afterwards, the hyperedge construction and weight calculation are conducted according to our technique section. We use the PASCAL 2012 [7] database as benchmark in this evaluation. The experiments are difficult due to the large number of outliers, that are, large amount of vertices and most of them represent incorrect correspondence, and also due to the large intra-category variations in shape present in PASCAL 2012 itself. Under each category, we randomly select two images as a pair and calculate the matching rate by each method. We run 50 times on each category and the averaged results are report in Table 1. The final matching rate is averaged over rate values of all categories.

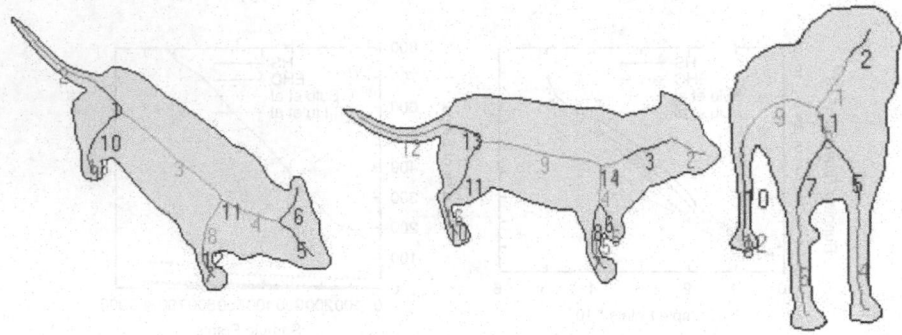

Fig. 7. Examples from shape matching database

Table 1. Average matching rates for the experiments on PASCAL 2012 database

	SC+IPFP	GS	TM	PM	HS
Car	62.5%	60.1%	60.7%	59.2%	**66.4%**
Motorbike	62.3%	60.1%	63.5 %	62.7%	**67.3%**
Person	**57.6%**	55.7%	54.2%	48.7%	53.1%
Animal	46.7%	49.2%	44.9%	40.3%	**54.3%**
Indoor	30.6%	28.5%	26.6%	24.3%	**36.9%**
All-averaged	51.8%	50.7%	50.1%	47.0%	**55.8%**

We also conducted shape matching [15] on the affinity data on the database from ShapeMatcher[1], which contains 21 objects with 128 views for each object. A few examples of dog's shape are shown in Fig.7. For each shape, we compute the matching score as the affinity value using the shape matching method [15], thus obtain a 2688×2688 affinity matrix. We compare our method with EHC and GS. The results are shown in Table 2. Both GS and HS can specify the number of objects, however, HS outperforms GS in terms of precision due to the fact that HS considers high-order relationship among vertices rather than pair-wise relation.

Table 2. Precision results for EHC, GS and HS on the shape matching affinity data

	EHC	GS	HS
Objects recognized	18	21	21
Precision	72.7%	83.5%	89.82 %

---

[1] http://www.cs.toronto.edu/~dmac/ShapeMatcher/index.html

6   Conclusion

In this paper, we propose a novel hypergraph shift algorithm aimed at finding the robust graph modes by probabilistic voting strategy, which are semantically sound besides the self-cohesiveness. Experimental studies show that our paradigm outperforms the state-of-the-art clustering and matching approaches observed from both synthetic and real-world data sets.

Acknowledgment. Xuemin Lin's research is supported by ARC DP0987557, ARC DP110102937, ARC DP120104168 and NSFC61021004.

References

1. Bomze, I.M.: Branch-and-bound approahces to standard quadratic optimization problems. Journal of Global Optimization 22(1-4), 17–37 (2002)
2. Bulo, S., Pellilo, M.: A game-theoretic approach to hypergraph clustering. In: NIPS (2009)
3. Cho, M., Lee, K.M.: Progressive graph matching: Making a move of graphs via probabilistic voting. In: CVPR (2012)
4. Comaniciu, D., Meer, P.: Mean shift: A robust approach toward feature space analysis. TPAMI 24(5), 603–619 (2002)
5. Ding, L., Yilmaz, A.: Interactive image segmentation using probabilistic hypergraphs. Pattern Recognition 43(5), 1863–1873 (2010)
6. Duchenme, O., Bach, F., Kweon, I., Ponce, J.: A tensor-based algorithm for high-order graph matching. In: CVPR (2009)
7. Everingham, M., Van Gool, L., Williams, C.K.I., Winn, J., Zisserman, A.: The PASCAL Visual Object Classes Challenge 2012 (VOC 2012) Results, http://www.pascal-network.org/challenges/VOC/voc2012/workshop/index.html
8. Huang, Y., Liu, Q., Zhang, S., Metaxas, D.N.: Image retrieval via probabilistic hypergraph ranking. In: CVPR (2010)
9. Kim, S., Nowozin, S., Kohli, P., Yoo, C.D.: Higher-order correlation clustering for image segmentation. In: NIPS (2011)
10. Leoradeanu, M., Hebert, M.: A spectral technique for correspondence problems using pairwise constraints. In: ICCV (2005)
11. Leordeanu, M., Hebert, M., Sukthankar, R.: Integer projected fixed point for graph matching and map inference. In: NIPS (2009)
12. Leordeanu, M., Sminchisescu, C.: Efficient hypergraph clustering. In: AISTATS (2012)
13. Liu, H., Latecki, L., Yan, S.: Robust clustering as ensembles of affinity relations. In: NIPS (2010)
14. Liu, H., Yan, S.: Robust graph mode seeking by graph shift. In: ICML (2010)
15. Macrini, D., Siddiqi, K., Dickinson, S.: From skeletons to bone graphs: Medial abstrations for object recognition. In: CVPR (2008)
16. Pavan, M., Pelillo, M.: Dominant sets and pairwise clustering. TPAMI 29(1), 167–171 (2007)
17. Sheikh, Y., Khan, E.A., Kanade, T.: Mode seeking by medoidshifts. In: ICCV (2007)

18. Wang, Y., Huang, X.: Geometric median-shift over riemannian manifolds. In: Zhang, B.-T., Orgun, M.A. (eds.) PRICAI 2010. LNCS, vol. 6230, pp. 268–279. Springer, Heidelberg (2010)
19. Wang, Y., Huang, X., Wu, L.: Clustering via geometric median shift over riemannian manifolds. Information Science 20, 292–305 (2013)
20. Weibull, J.W.: Evolutionary game theory. MIT Press (1995)
21. Zangwill, W.: Nonlinear programming: a unified approach. Prentice-Hall (1969)
22. Zass, R., Shashua, A.: Probabilistic graph and hypergraph matching. In: CVPR (2008)

Extensions to Quantile Regression Forests for Very High-Dimensional Data

Nguyen Thanh Tung[1], Joshua Zhexue Huang[2], Imran Khan[1], Mark Junjie Li[2], and Graham Williams[1]

[1] Shenzhen Key Laboratory of High Performance Data Mining. Shenzhen Institutes of Advanced Technology, Chinese Academy of Sciences, Shenzhen 518055, China
[2] College of Computer Science and Software Engineering, Shenzhen University
tungnt@wru.vn, {zx.huang,jj.li}@szu.edu.cn, imran.khan@siat.ac.cn, Graham.Williams@togaware.com

Abstract. This paper describes new extensions to the state-of-the-art regression random forests *Quantile Regression Forests* (QRF) for applications to high-dimensional data with thousands of features. We propose a new subspace sampling method that randomly samples a subset of features from two separate feature sets, one containing important features and the other one containing less important features. The two feature sets partition the input data based on the importance measures of features. The partition is generated by using feature permutation to produce raw importance feature scores first and then applying p-value assessment to separate important features from the less important ones. The new subspace sampling method enables to generate trees from bagged sample data with smaller regression errors. For point regression, we choose the prediction value of Y from the range between two quantiles $Q_{0.05}$ and $Q_{0.95}$ instead of the conditional mean used in regression random forests. Our experiment results have shown that random forests with these extensions outperformed regression random forests and quantile regression forests in reduction of root mean square residuals.

Keywords: Regression Random Forests, Quantile Regression Forests, Data Mining, High-dimensional Data.

1 Introduction

Regression is a task of learning a function $f(\mathbf{X}) = E(Y|\mathbf{X})$ from a training data $\mathbb{L} = \{(\mathbf{X}, Y) = (\mathbf{X_1}, Y_1), ..., (\mathbf{X_N}, Y_N)\}$, where N is the number of objects in $\mathbb{L}$, $X \in \mathbb{R}^M$ are predictor variables or features and $Y \in \mathbb{R}^1$ is a response variable or feature. The regression model has the form

$$Y = E(Y|\mathbf{X}) + \epsilon \tag{1}$$

where error $\epsilon \sim N(0, \sigma^2)$.

A parametric method assumes that a formula for conditional mean $E(Y|\mathbf{X})$ is known, for instance, linear equation $Y = \beta_0 + \beta_1 X_1, ..., \beta_M X_M$. The linear

V.S. Tseng et al. (Eds.): PAKDD 2014, Part II, LNAI 8444, pp. 247–258, 2014.

regression model is solved by estimating parameters $\beta_0, \beta_1, \ldots, \beta_M$ from $\mathbb{L}$ with least squares method to minimize the sum of square residuals. A nonparametric method does not require that a model form be known. Instead, a model structure is specified, such as a neural network and $\mathbb{L}$ is used to learn the model. Linear regression models do not perform on nonlinear domains and suffer the problem of curse of dimensionality. Neural networks are not scalable to big data.

Decision tree is a nonparametric regression model that works on nonlinear situations. A decision tree model partitions the training data $\mathbb{L}$ into subsets of leaf nodes and the prediction value in each leaf node is taken as the mean of Y values of the objects in that leaf node. Decision tree model is unstable in high-dimensional data because of the large prediction variance. This problem can be remedied by using an ensemble of decision trees or random forests [3] built from the bagged samples of $\mathbb{L}$ [2]. Regression random forests takes the average of multiple decision tree predictions to reduce the prediction variance and increase the accuracy of prediction.

Quantile regression forests (QRF) represents the state-of-the-art technique for nonparametric regression [7]. Instead of modeling $Y = E(Y|\mathbf{X})$, QRF models $F(y|X = x) = P(Y < y|X = x)$, i.e., the conditional distribution function. Given a continuous distribution function and a probability α, the α-quantile $Q_\alpha(x)$ can be computed as

$$P(Y < Q_\alpha(x)|X = x) = \alpha \tag{2}$$

where $0 < \alpha < 1$. Given two quantile probabilities α_l and α_h, QRF enables to predict the range $[Q_{\alpha_l}(x), Q_{\alpha_h}(x)]$ of Y with a given probability τ that $P(Q_{\alpha_l}(x) < Y < Q_{\alpha_h}(x)|X = x) = \tau$. Besides the range prediction, quantile regression forests can perform well in situations where the conditional distribution function is not in normal distribution.

Both regression random forests and quantile regression forests suffer performance problems in high-dimensional data with thousands of features. The main cause is that in the process of growing a tree from the bagged sample data, the subspace of features randomly sampled from the thousands of features in $\mathbb{L}$ to split a node of the tree is often dominated by less important features, and the tree grown from such randomly sampled subspace features will have a low accuracy in prediction which affects the final prediction of the random forests.

In this paper, we propose a new subspace feature sampling method to grow trees for regression random forests. Given a training data set $\mathbb{L}$, we first use feature permutation method to measure the importance of features and produce raw feature importance scores. Then, we apply p-value assessment to separate important features from the less important ones and partition the set of features in $\mathbb{L}$ into two subsets, one containing important features and one containing less important features. We independently sample features from the two subsets and put them together as the subspace features for splitting the data at a node. Since the subspace always contains important features which can guarantee a better split at the node, this subspace feature sampling method enables to generate trees from bagged sample data with smaller regression errors.

For point regression, we choose the prediction value of Y from the range between two quantiles $Q_{0.05}$ and $Q_{0.95}$ instead of the conditional mean used in regression random forests. Our experiment results have shown that random forests with these extensions outperformed regression random forests and quantile regression forests in reduction of root mean square residuals (RMSR).

2   Random Forests for Regression

2.1   Regression Random Forests

Given a training data $\mathbb{L} = \{(\mathbf{X_1}, Y_1), ..., (\mathbf{X_N}, Y_N)\}$, where N is the number of objects in $\mathbb{L}$, a regression random forests model is built as follows.

- Step 1: Draw a bagged sample $\mathbb{L}_k$ from $\mathbb{L}$.
- Step 2: Grow a regression tree T_k from $\mathbb{L}_k$. At each node t, the split is determined by the decrease in impurity that is defined as $\sum_{x_i \in t}(Y_i - \bar{Y}_t)/N(t)$, where $N(t)$ is the number of objects and $\bar{Y}_t$ is the mean value of all Y_i at node t. At each leaf node, $\bar{Y}_t$ is assigned as the prediction value of the node.
- Step 3: Let $\hat{Y}^k$ be the prediction of tree T_k given input X. The prediction of regression random forests with K trees is

$$\hat{Y} = \frac{1}{K} \sum_{k=1}^{K} \hat{Y}^k$$

Since each tree is grown from a bagged sample, it is grown with only two-third of objects in $\mathbb{L}$. About one-third of objects are left out and these objects are called *out-of-bag (OOB)* samples which are used to estimate the prediction errors.

2.2   Quantile Regression Forests

Quantile Regression Forests (QRF) uses the same method as described above to grow trees [7]. However, at each leaf node, it retains all Y values instead of only the mean of Y values. Therefore, QRF keeps the raw distribution of Y values at leaf node.

To describe QRF with notation by Breiman [3], we compute a positive weight $w_i(x, \theta_k)$ by each tree for each case $X_i \in \mathbb{L}$, where θ_k indicates the kth tree for a new given x. Let $l(x, \theta_k)$ be a leaf node t. All $X_i \in l(x, \theta_k)$ are assigned to an equal weight $w_i(x, \theta_k) = 1/N(t)$ and $X_i \notin l(x, \theta_k)$ are assigned to 0 otherwise, where $N(t)$ is the number of objects in $l(x, \theta_k)$. For single tree prediction, given $X = x$, the prediction value is

$$\hat{Y}^k = \sum_{i=1}^{N} w_i(x, \theta_k)Y_i = \sum_{x, X_i \in l(x, \theta_k)} w_i(x, \theta_k)Y_i = \frac{1}{N(t)} \sum_{x, X_i \in l(x, \theta_k)} Y_i \qquad (3)$$

The weight $w_i(x)$ assigned by random forests is the average of weights by all trees, that is

$$w_i(x) = \frac{1}{K} \sum_{k=1}^{K} w_i(x, \theta_k) \tag{4}$$

The prediction of regression random forests is

$$\hat{Y} = \sum_{i=1}^{N} w_i(x) Y_i \tag{5}$$

We note that $\hat{Y}$ is the average of conditional mean values of all trees in the regression random forests.

Given an input X, we can find the leaf node $l_k(x, \theta_k)$ from all trees and the set of Y_i in these leaf nodes. Given all Y_i and the corresponding weights $w(i)$, we can estimate the conditional distribution function of Y given X as

$$\hat{F}(y|\mathbf{X} = x) = \sum_{i=1}^{N} w_i(x) \mathcal{I}(Y_i \leq y) \tag{6}$$

where $\mathcal{I}(\cdot)$ is the indicator function that is equal to 1 if $Y_i \leq y$ and 0 if $Y_i > y$. Given a probability α, we can estimate the quantile $Q_\alpha(X)$ as

$$\hat{Q}_\alpha(\mathbf{X} = x) = inf\{y : \hat{F}(y|\mathbf{X} = x) \geq \alpha\}. \tag{7}$$

For range prediction, we have

$$[Q_{\alpha_l}(X), Q_{\alpha_h}(X)] = [inf\{y : \hat{F}(y|\mathbf{X} = x) \geq \alpha_l\}, inf\{y : \hat{F}(y|\mathbf{X} = x) \geq \alpha_h\}] \tag{8}$$

where $\alpha_l < \alpha_h$ and $(\alpha_h - \alpha_l) = \tau$. Here, τ is the probability that prediction Y will fall in the range of $[Q_{\alpha_l}(X), Q_{\alpha_h}(X)]$.

For point regression, the prediction can choose a value in a range such as the mean or median of Y_i values. The median surpasses the mean in robustness towards extreme values/outliers. We use the median of Y values in the range of two quantiles as the prediction of Y given input $X = x$.

3   Feature Weighting Subspace Selection

3.1   Importance Measure of Features by Permutation

Given a training data set $\mathbb{L}$ and a regression random forests model RF, Breiman [3] described a permutation method to measure the importance of features in the prediction. The procedure for computing the importance scores of features consists of the following steps.

1. Let $\mathbb{L}_k^{oob}$ be the *out-of-bag* samples of the kth tree. Given $X_i \in \mathbb{L}_k^{oob}$, use the tree T_k to predict $\hat{Y}_i^k$, denoted as $\hat{f}_i^k(\mathbf{X_i})$.

2. Choose a predictor feature j and randomly permute the value of feature j in X_i with another case in $\mathbb{L}_k^{oob}$. Use tree T_k to obtain the new prediction on the permuted X_i as $\hat{f}_i^{k,p,j}(\mathbf{X_i})$. Repeat the permutation process P times.

3. For M_i trees grown without X_i, compute the out-of-bag prediction by RF in the pth permutation of the jth predictor feature as

$$\hat{f}_i^{p,j}(\mathbf{X_i}) = \frac{1}{M_i} \sum_{X_i \in \mathbb{L}_k^{oob}} \hat{f}_i^{k,p,j}(\mathbf{X_i})$$

4. Compute the two *mean square residuals* (MSR) with and without permutations of predictor feature j on X_i as $MSR_i = \frac{1}{M_i} \sum_{k \in M_i} (\hat{f}_i^k(\mathbf{X_i}) - Y_i)^2$ and $MSR_i^j = \frac{1}{P} \sum_{p=1}^{P} (\hat{f}_i^{p,j}(\mathbf{X_i}) - Y_i)^2$, respectively.

5. Let $\Delta MSR_i^j = max(0, MSR_i^j - MSR_i)$. The importance of feature j is $IMP_j = \frac{1}{N} \sum_{i \in \mathbb{L}} \Delta MSR_i^j$. To normalize the importance measures, we have the raw importance score as

$$VI_j = \frac{IMP_j}{\sum_l IMP_l} \tag{9}$$

With the raw importance scores by (9) we can rank the features on the importance.

3.2 p-Value Feature Assessment

Permutation method only gives the importance ranking of features. We need to identify important features from less important ones. To do so, we use Welch's two-sample t-test that compares the importance score of a feature with the maximum importance scores of generated noisy features called shadows. The shadow features do not have prediction power to the response feature. Therefore, any feature whose importance score is smaller than the maximum importance score of noisy features, it is less important. Otherwise, it is considered as important. This idea was introduced by Stoppiglia et al. [10], and were further developed in [5], [11].

Table 1. The importance scores matrix of all real features and shadows with R replicates

Iteration	VI_{X_1}	VI_{X_2}	$\cdots$	VI_{X_M}	$VI_{A_{M+1}}$	$VI_{A_{M+2}}$	$\cdots$	$VI_{A_{2M}}$
1	$VI_{x_{1,1}}$	$VI_{x_{1,2}}$	$\cdots$	$VI_{x_{1,M}}$	$VI_{a_{1,(M+1)}}$	$VI_{a_{1,(M+2)}}$	$\cdots$	$VI_{a_{1,2M}}$
2	$VI_{x_{2,1}}$	$VI_{x_{2,2}}$	$\cdots$	$VI_{x_{2,M}}$	$VI_{a_{2,(M+1)}}$	$VI_{a_{2,(M+2)}}$	$\cdots$	$VI_{a_{2,2M}}$
$\vdots$	$\vdots$	$\vdots$						$\vdots$
R	$VI_{x_{R,1}}$	$VI_{x_{R,2}}$	$\cdots$	$VI_{x_{R,M}}$	$VI_{a_{R,(M+1)}}$	$VI_{a_{R,(M+2)}}$	$\cdots$	$VI_{a_{R,2M}}$

We build a random forests model RF from this extended data set. Following the importance measure by permutation procedure, we use RF to compute $2M$ importance scores for $2M$ features. We repeat the same process R times to compute R replicates. Table 1 shows the importance measure of M features in input data and M shadow features generated by permutating the values of the corresponding feature in data.

From the replicates of shadow features, we extract the maximum value from each row and put it into the comparison sample $V^* = max\{A_{ri}\}, (r = 1, ..R; i = M + 1, ..2M)$. For each data feature X_i, we compute t-statistic as:

$$t_i = \frac{\overline{X}_i - \overline{V}^*}{\sqrt{(s_1^2 + s_2^2)/R}} \tag{10}$$

where s_1^2 and s_2^2 are the unbiased estimators of the variances of the two samples. For significance test, the distribution of t_i in (10) is approximated as an ordinary Student's distribution with the degrees of freedom df calculated as

$$df = \frac{(s_1^2/n_1 + s_2^2/n_2)^2}{(s_1^2/n_1)^2/(n_1 - 1) + (s^2{}_2/n_2)^2/(n_2 - 1)} \tag{11}$$

where $n_1 = n_2 = R$.

Having computed the t statistic and df, we can compute the p-value for the feature and perform hypothesis test on $\overline{X}_i > \overline{V}^*$. Given a statistical significance level, we can identify important features. This test confirms that if a feature is important, it consistently scores higher than the shadow over multiple permutations.

3.3 Feature Partition and Subspace Selection

The p-value of a feature indicates the importance of the feature in prediction. The smaller the p-value of a feature, the more correlated the predictor feature to the response feature, and the more powerful the feature in prediction.

Given all p values for all features, we set a significance level as the threshold λ for instance $\lambda = 0.05$. Any feature whose p-value is smaller than λ is added to the important feature subset X_{high}, and it is added to the less important feature subset X_{low} otherwise. The two subsets partitions the set of features in data. Given X_{high} and X_{low}, at each node, we randomly select some features from X_{high} and some from X_{low} to form the feature subspace for splitting the node. Given a subspace size, we can form the subspace with 80% of features sampled from X_{high} and 20% sampled from X_{low}.

4 A New Quantile Regression Forests Algorithm

Now we can extend the quantile regression forests with the new feature subspace sampling method to generate splits at the nodes of decision trees and select prediction value of Y from the range of low and high quantiles with a high probability. The new quantile regression forests algorithm eQRF is summarized as follows.

1. Given $\mathbb{L}$, generate the extended data set $\mathbb{L}^e$ in $2M$ dimensions by permutating the corresponding predictor feature values for shadow features.
2. Build a regression random forests model RF^e from $\mathbb{L}^e$ and compute R replicates of raw importance scores of all predictor features and shadows with RF^e. Extract the maximum importance score of each replicate to form the comparison sample V^* of R elements.
3. For each predictor feature, take R importance scores and compute t statistic as (10).
4. Compute the degree of freedom df as (11).
5. Given t statistic and df, compute all p-values for all predictor features.
6. Given a significance level threshold λ, separate important features from less important features in two feature subsets X_{low} and X_{high}.
7. Sample the training set $\mathbb{L}$ with replacement to generate bagged samples $\mathbb{L}_1, \mathbb{L}_2, .., \mathbb{L}_K$.
8. For each sample $\mathbb{L}_k$, grow a regression tree T_k as follows:
   (a) At each node, select a subspace of $m = \lfloor \sqrt{M} \rfloor$ $(m > 1)$ features randomly and separately from X_{low} and X_{high} and use the subspace features as candidates for splitting the node.
   (b) Each tree is grown nondeterministically, without pruning until the minimum node size n_{min} is reached. At each leaf node, all Y values of the objects in the leaf node are kept.
   (c) Compute the weights of each X_i by individual trees and the forests with out-of-bag samples.
9. Given a probability τ, α_l and α_h for $\alpha_h - \alpha_l = \tau$, compute the corresponding quantile Q_{α_l} and Q_{α_h} with (8) (We set default values $[\alpha_l = 0.05, \alpha_h = 0.95]$ and $\tau = 0.9$).
10. Given a $\mathbf{X}$, estimate the prediction value from a value in the quantile range of Q_{α_l} and Q_{α_h} such as mean or median.

5   Simulation Analysis

5.1   Simulation Data

We used three models as listed in Table 2 to generate synthetic data for simulation analysis. Each model has 5 predictor variables or features. With each model, we first created 200 objects in 5 dimensions plus a response feature. After this, we expanded the data set with different numbers of noisy features and obtained 5 data sets named as {LM5, LM50, LM500, LM2000, LM5000} where the number in the data name indicates dimensions of the data set. Similarly, we generated extra 5 data sets with 1000 objects from each model as test data sets named {HM5, HM50, HM500, HM2000, HM5000}.

Table 2. Three simulation models for synthetic data generation. Each model uses 5 iid predictor features from $U(0,1)$ and ϵ from *Exp(1)* (exponential mean 1) distribution.

Model	Error Distribution	Simulation models
1	*Exp(1)*	$Y = 10(X_1 + X_2 + X_3 + X_4 + X_5 - 2.5)^2 + \epsilon$
2	*Exp(1)*	$Y = 10sin(\pi X_1 X_2) + 20(X_3 - 0.5)^2 + 10X_4 + 5X_5 + \epsilon$
3	*Exp(1)*	$Y = 0.1e^{4X_1} + 4/[1 + e^{-20(X_2-0.5)}] + 3X_3 + 2X_4 + X_5 + \epsilon$

5.2   Evaluation Measure

The performance of a model was evaluated on test data with the *root mean of square residuals (RMSR)* computed as

$$RMSR = \sqrt{\frac{1}{\|\mathbb{H}\|} \sum_{\mathbf{X}_i \in \mathbb{H}} [\hat{f}_{\mathbb{H}}(\mathbf{X}_i) - Y_i]^2}. \tag{12}$$

where $\hat{f}_{\mathbb{H}}(\mathbf{X}_i)$ is the prediction given X_i, $\mathbb{H}$ is a test data set and $\|\mathbb{H}\|$ is the number of objects in test data $\mathbb{H}$.

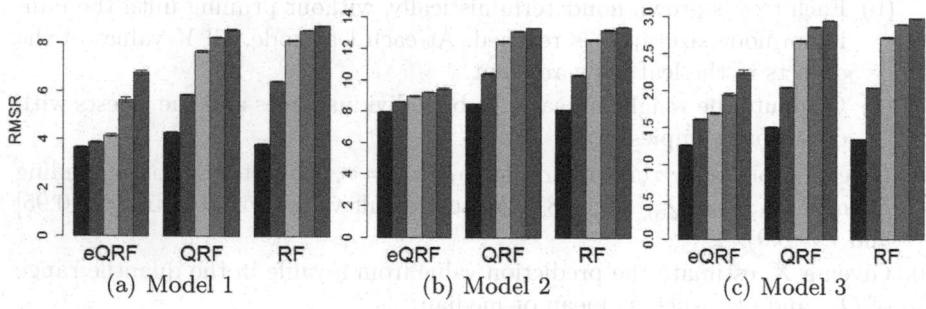

Fig. 1. Comparisons of three regression forests algorithms on 5 test data sets generated with the simulation models in Table 2

5.3   Evaluation Results

We used regression random forests RF, quantile regression forests QRF and our algorithm eQRF to build regression models from the training data sets and used evaluation measure (12) to evaluate the models with the test data sets. We used the latest RF and QRF packages *randomForest, quantregForest* in R in these experiments [6], [8]. For each training data set, we built 100 regression models, each with 500 trees and tested the 100 models with the corresponding test data. Then, the result was evaluated with (12) and the average of 100 models was computed.

Figure 1 shows the evaluation results of three random forests regression methods in RMSR measures. Each random forests method produced 5 test results on 5 simulation data sets from the left to right as {HM5, HM50, HM500, HM2000, HM5000}. We can see that the more noisy features in the data, the lower accuracy in the prediction model. Clearly, in all simulated data generated by the three models in Table 2, eQRF performed the best and its RMSR was significantly lower than those of QRF and RF.

6 Experiments on Real Datasets

6.1 Real-World Data

Five real-world data sets were used to evaluate the performance of our new regression random forests algorithm. The general characteristics of these data sets are presented in Table 3.

The *computed tomography (CT)* data was taken from the UCI[1] which was used to build a regression model to calculate the relative locations of CT slices on the axial axis. The data set was generated from 53,500 images taken from 74 patients (43 males and 31 females). Each CT slice was described by two histograms in a polar space. The first histogram describes the location of bone structures in the image and the second represents the location of air inclusions inside of the body. Both histograms are concatenated to form the feature vector.

TFIDF-2006 [2] is a text data set containing financial reports. Each document is associated with an empirical measure of financial risk. These measures are log transformed volatilities of stock returns.

The *Microarray data "Diffuse Large B-cell Lymphoma"* (DLBCL) was collected from Rosenwald et al. [9]. The DLBCL data consisted of measurements of 7399 genes from 240 patients with diffuse large B-cell lymphoma. The outcome was survival time, which was either observed or censored. We used observed survival time as the response feature because censored data only indicates two states, dead or alive. A detailed description can be found in [9].

"Leukemia" and *"Lung cancer"* are two gene data sets taken from NCBI [3]. Each of those data sets contains two classes. We changed one class label to 1 and another label to 0. We treat 0 and 1 as continuous values and consider this problem as a regression problem. We built a regression random forests model to estimate the outcome and used a defined threshold to divide the outcomes into two classes.

6.2 Experiments and Results

For each real-world data set, we used two-third of data for training and one-third for testing. We generated 10 models from each training data and each model

---

[1] The data are available at http://archive.ics.uci.edu/

[2] http://www.csie.ntu.edu.tw/~cjlin/libsvmtools/datasets

[3] http://www.ncbi.nlm.nih.gov

Table 3. Description of the real data sets sorted by the number of features and RMSR performance of three regression algorithms

Dataset	Name	#training	#testing	#features	eQRF	RF	QRF
1	CT Data	35,700	17,800	385	**0.29**	1.33	2.09
2	Leukemia	48	24	7,129	**0.17**	0.22	0.24
3	DLBCL Data	160	80	7,399	**3.77**	4.28	4.55
4	Lung cancer	114	58	54,675	**0.21**	0.32	0.36
5	TFIDF-2006	16,087	3,308	150,361	**0.41**	0.68	0.69

contained 200 trees. We computed the average of RMSRs of the 10 models with (12). The average RMSRs of three regression random forests models on five real-world data sets are shown in Table 3 on the right. We can see that eQRF had the lowest average RMSR. RF performed better than QRF.

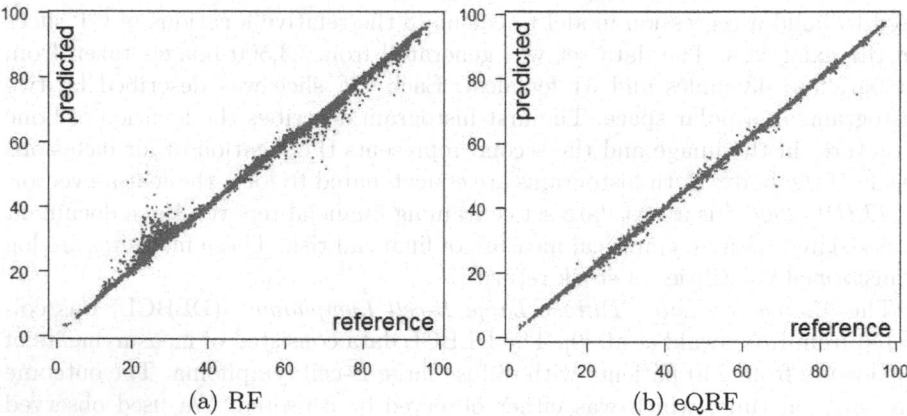

(a) RF   (b) eQRF

Fig. 2. Plots of predicted response values against the true values of CT test data. (a) Result of RF. (b) Result of eQRF.

Figure 2 plots the predicted values by RF and eQRF against the true values of the response feature in CT test data. We can see from Figure 2 (a) that there are some regions that RF predicted higher than the true value, for instance [25 cm, 35 cm], and some regions that RF predicted lower than the true value, for instance > 70 cm. These are the prediction error regions including shoulder [20-30cm] and abdomen [60-75cm]. On the contrast, the predicted values of eQRF were more close to the true values as shown in Figure 2 (b) and the prediction results are consistent and more stable in all regions of human body.

Figure 3 shows the average RMSR box plots of three regression models from the real-world data sets. Figure 3 (a) is the result of CT data and Figure 3 (b) is the result of DLBCL data. We can see that eQRF produced less RMSR than QRF and RF and the variance is also small.

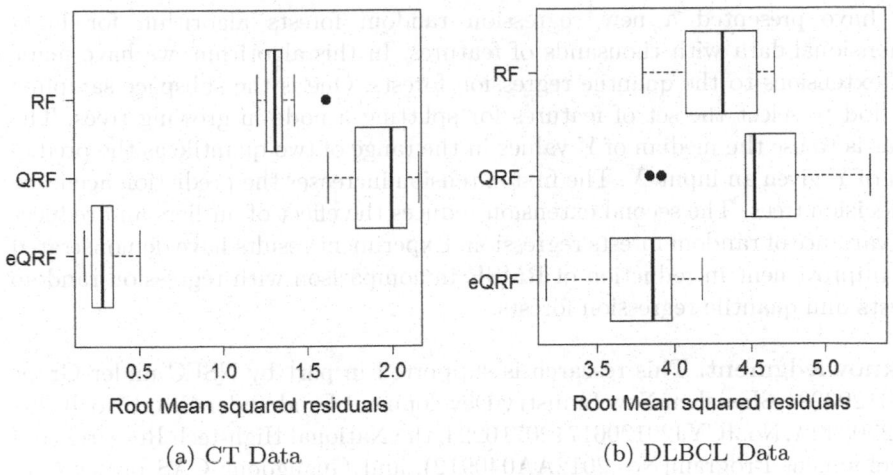

(a) CT Data (b) DLBCL Data

Fig. 3. Boxplots of RMSR of three models RF, QRF and eQRF. (a) Result of CT test data. (b) Result of DLBCL test data.

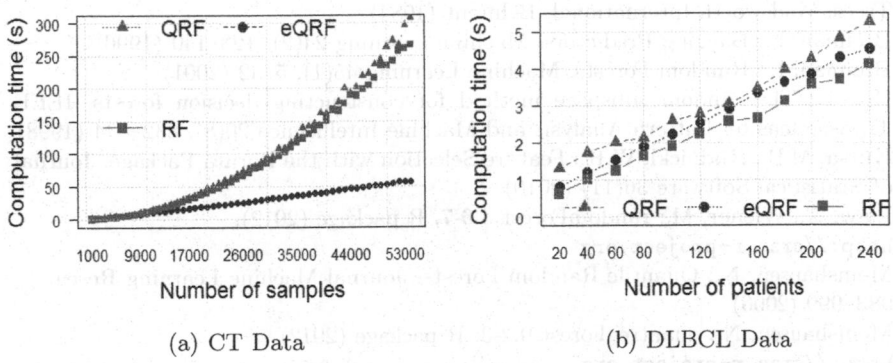

(a) CT Data (b) DLBCL Data

Fig. 4. .Plots of computational time of three algorithms against the number of objects in data. The experiments were conducted on a computer with 2.13 Ghz Intel Core 2 Quad processor and 24GB RAM. (a) Result of CT data. (b) Result of DLBCL data.

Figure 4 shows the computational time of three regression models on the two data sets. We can see that the computational time of the three models linearly increases as the number of objects increases if the size is small, such as DLBCL data. However, for data set with a large number of objects as CT data, the computational times of RF and QRF increase exponentially as shown in Figure 4 (a) but eQRF still maintains a linear increase as shown in Figure 4 (b).

7 Conclusions

We have presented a new regression random forests algorithm for high-dimensional data with thousands of features. In this algorithm, we have made two extensions to the quantile regression forests. One is the subspace sampling method to select the set of features for splitting a node in growing trees. The other is to use the median of Y values in the range of two quantile as the prediction of Y given an input X. The first extension increases the prediction accuracy of decision trees. The second extension reduces the effect of outliers and reduces the variance of random forests regression. Experiment results have demonstrated the improvement in reduction of RMSR in comparison with regression random forests and quantile regression forests.

Acknowledgment. This research is supported in part by NSFC under Grant No.61203294, Shenzhen New Industry Development Fund under Grant No.JC201 005270342A, No.JCYJ20120617120716224, the National High-tech Research and Development Program(No. 2012AA040912), and Guangdong-CAS project(No. 2011B090300025).

References

1. Breiman, L., Friedman, J.H., Olshen, R.A., Stone, C.: Classification and Regression Trees. Wadsworth International, Belmont (1984)
2. Breiman, L.: Bagging Predictors. Machine Learning 24(2), 123–140 (1996)
3. Breiman, L.: Random Forests. Machine Learning 45(1), 5–32 (2001)
4. Ho, T.: The random subspace method for constructing decision forests. IEEE Transactions on Pattern Analysis and Machine Intelligence 20(8), 832–844 (1998)
5. Kursa, M.B., Rudnicki, W.R.: Feature Selection with the Boruta Package. Journal of Statistical Software 36(11) (2010)
6. Liaw, A., Wiener, M.: randomForest 4.6-7. R package (2012), http://cran.r-project.org
7. Meinshausen, N.: Quantile Random Forests. Journal Machine Learning Research, 983–999 (2006)
8. Meinshausen, N.: quantregForest 0.2-3. R package (2012), http://cran.r-project.org
9. Rosenwald, A., et al.: The use of molecular profiling to predict survival after chemotherapy for diffuse large-b-cell lymphoma. N. Engl. J. Med. 346, 1937–1947 (2002)
10. Stoppiglia, H., Dreyfus, G.: Ranking a random feature for variable and feature selection. The Journal of Machine Learning Research 3, 1399–1414 (2003)
11. Tuv, E., Borisov, A., Runger, G., Torkkola, K.: Feature selection with ensembles, artificial variables, and redundancy elimination. The Journal of Machine Learning Research 10, 1341–1366 (2009)

Inducing Controlled Error over Variable Length Ranked Lists

Laurence A.F. Park and Glenn Stone

School of Computing, Engineering and Mathematics,
University of Western Sydney, Australia
{l.park,g.stone}@uws.edu.au
http://www.scem.uws.edu.au/~lapark

Abstract. When examining the robustness of systems that take ranked lists as input, we can induce noise, measured in terms of Kendall's tau rank correlation, by applying a set number of random adjacent transpositions. The set number of random transpositions ensures that any ranked lists, induced with this noise, has a specific expected Kendall's tau. However, if we have ranked lists of varying length, it is not clear how many random transpositions we must apply to each list to ensure that we obtain a consistent *expected* Kendall's tau across the collection. In this article we investigate how to compute the number of random adjacent transpositions required to obtain an expected Kendall's tau for a given list length, and find that it is infeasible to compute for lists of length more than 9. We also investigate an alternate and more efficient method of inducing noise in ranked lists called Gaussian Perturbation. We show that using this method, we can compute the parameters required to induce a consistent level of noise for lists of length 10^7 in just over six minutes. We also provide an approximate solution to provide results in less than 10^{-5} seconds.

1 Introduction

The robustness of a modeling or prediction system is defined as the system's ability to handle noise or error applied to its input, and operate within certain limits. For example, if we have a classification system that predicts a state, given a set of observations, we can measure its accuracy by examining how many of its predictions are correct. We can then measure the robustness of the system by applying a specified level of random noise to the observations, and then examine its change in accuracy. Of course, if we increased the level of the added noise, we would expect the accuracy of the system to decrease, but a more robust system would provide a slower decrease.

To determine the robustness level of a system, we must often perform a simulation experiment, where we can control the noise. Noise is usually randomly sampled from a predefined distribution with carefully chosen parameters to ensure the noise is consistent for the experiment. For example, if our observations are elements of the real number set, we may generate noise using a Normal distribution with zero mean and variance of one. If our observations are frequency values, we may generate noise using a Poisson distribution with mean 1. Further experiments can then be performed by adjusting the variance of the noise distribution and measuring the change in accuracy.

V.S. Tseng et al. (Eds.): PAKDD 2014, Part II, LNAI 8444, pp. 259–270, 2014.

Many systems, such as collaborative filtering systems, meta search engines, query expansion systems, and rank aggregation systems require a set of ranked items as input.

Therefore, to measure the robustness of such systems, we can randomly permute the lists to obtain a given *expected* Kendall's τ between the permuted and unpermuted lists. We can achieve this by performing a set number of randomly chosen adjacent transpositions, as long as *all* of the list lengths are the same. If the lists lengths are not the same, it is not clear how many adjacent transpositions we should apply to obtain a given expected Kendall's τ across all lists.

In this article, we investigate how to compute the number of random adjacent transposition required to obtain a given expected τ for a given list length. We find that this is a very computationally expensive task. We also propose an alternate method of inducing error in ranked lists called Gaussian Perturbation. We show that its parameter is a function of the rank correlation reduction caused by the error, and therefore can be used to induce controlled noise in ranked lists. We provide the following contributions:

- An analysis of the relationship between the expected Kendall's τ, the number of random adjacent transpositions and the list length (Section 3.1),
- A novel ranked list noise induction method called Gaussian Perturbation that can be computed for larger lists in reasonable time, (Section 3.2),
- An approximate version of Gaussian Perturbation to compute the required parameters in minimum time (Section 3.3).

The article will proceed as follows: Section 2 reviews how we measure the robustness of a system, and examines the form of Kendall's τ. Section 3 examines the noise induced using random adjacent transpositions, introduces and analyses Gaussian Perturbation, and provides a faster approximation to Gaussian Perturbation.

2   Robustness Using Ranked Lists

To assess the robustness of a prediction system using simulation, we must define a method of inducing controlled noise into the observation space. In this section, we will examine how to measure robustness, given a noise distribution. We will then examine how to use Kendall's τ rank correlation to measure the induced noise.

2.1   Measuring System Robustness

A robust system is one that can function within a set of predefined limits in a noisy (error prone) environment. Therefore a robust classification system would be able to provide a certain level of accuracy, when making predictions based on observations, with a given level of noise. For example, if our observation space is the one dimensional real line $\mathbb{R}$, we might define an unknown true value as $a \in \mathbb{R}$, and an observation with added noise as $\hat{a} = a + \epsilon$ where $\epsilon \sim N(0, \sigma)$, a sample from a Normal distribution, with mean 0 and standard deviation σ.

In this situation the level of noise is controlled by the parameter σ. As the difference between the noisy and noise free observations increase, we would expect the system prediction accuracy to change, but we would expect a more robust system's behaviour to change less.

Analysis of robust system behaviour has provided us with strategies that allow prediction models to provide good accuracy using a wide range of observation input data. Some well known general strategies are to use cross validation when training a model [12] and including a regularisation term in the optimisation function (used in SVMs [13] and Lasso/Ridge regression [7]). Such systems introduce bias into the optimisation, in order to reduce the variance of the classification, and hence increase the robustness.

Similar methods can be used in clustering and dimension reduction. Compressive sampling has been used in image analysis [2], clustering [11] and outlier detection [1]. Each of these methods use l_1 regularisation to obtain sparse solutions to the dimension reduction, clustering, and outlier detection problems. These methods have also been applied to Principal Component Analysis [3], to obtain biased, but more robust principal components.

Now consider the observation space as the set of all ranked lists of multiple lengths. There are many systems that have an observation space of ranked lists, therefore it is important that we provide a method of measuring the robustness of this space. Systems using Collaborative filtering [9] use an observation space of ranked lists. These systems obtain ranked responses from the user community and aggregate them to assist in the decision making process. Query expansion [4] also requires ranked lists, where ranked documents are used to extract potential query terms. Meta search over multiple databases [8,6] obtain ranked results from various databases (e.g. airline travel prices, book prices, Web search results) and combines them to form a single result. Also, Rank aggregation systems [5,10] are used to combine multiple ranked lists into a single ranked list (e.g. results from tennis competitions to form an overall player ranking).

To test the robustness of methods applied in such systems, we need a method to induce noise into ranked lists. Furthermore, we must be able to control the *amount* of noise induced. Ranked lists contain ordinal numbers, therefore the error term ϵ would be the application of a random permutation of the list elements (clearly, simply adding an error variate to the *true* ranking will not be appropriate for ordinal numbers).

2.2 Measuring Error in Ranked Lists

The ranking of n items can be identified with an ordering of the integers $1, \ldots, n$. Thus the sample space for ranked lists can be considered S_n, the set of permutations of the integers 1 to n. We will talk about elements $\boldsymbol{x} = (x_1, \ldots, x_n)$ of S_n, where x_i is the rank of item i. Note that we cannot induce error on lists of length $n = 1$, therefore, we will only consider $n > 1$ in this article.

Given two rankings $\boldsymbol{x}$ and $\boldsymbol{y} \in S_n$, one way to measure the similarity of the two rankings is using Kendall's τ rank correlation. Kendall's τ is defined as:

$$\tau(\boldsymbol{x}, \boldsymbol{y}) = \frac{\sum_{1 \leq i < j \leq n} c_{ij} - d_{ij}}{n(n-1)/2}$$

where the sum is over the set of all possible pairs of items being ranked, and

$$c_{ij} = \begin{cases} 1, & \text{if } ((x_i < x_j) \text{ and } (y_i < y_j)) \text{ or } ((x_i > x_j) \text{ and } (y_i > y_j)) \\ 0, & \text{otherwise} \end{cases}$$

and $d_{ij} = 1 - c_{ij}$. A pair of items are concordant ($c_{ij} = 1$) if their ordering in each of the two lists matches, otherwise they are discordant ($d_{ij} = 1$). We can see that $\tau = 1$ if and only if $x = y$ (all items are concordant), implying perfect correlation. Also $\tau = -1$ if and only if $x = \text{reverse}(y)$ (all items are discordant), implying perfect anti-correlation. The result of $\tau = 0$ implies no correlation between x and y. Therefore, Kendall's τ can be simplified to:

$$\tau(x, y) = \frac{2\sum_{1 \leq i < j \leq n} c_{ij}}{n(n-1)/2} - 1$$

since $\sum_{1 \leq i < j \leq n} c_{ij} + d_{ij} = n(n-1)/2$.

We can use Kendall's τ to measure the noise induced in a ranked list. Given ranked lists $x, y \in S_n$, the level of noise between x and y is measured using $\tau(x, y)$. The greater the value of τ, the lower the amount of noise.

Since Kendall's τ is a measure of correlation, it is comparable across lists of different lengths. This means that if we induce noise on a list of length n_1 and on another of length n_2, where the measure of noise using Kendall's τ is the same for both, then we have induced the same level of noise for both lists.

3   Inducing Controlled Noise in Ranked Lists

To examine the robustness of a system, we must examine how it performs when noise is introduced. If the system takes a set of ranked lists as input, we must induce controlled noise in the ranked lists. We showed that error in ranked lists can be measured using Kendall's τ, which is a function of the permutation required to remove error from the erroneous ranked list. Therefore, to induce controlled noise, we must perform controlled permutations.

Ideally, supposing the *truth* to be $x \in S_n$, we would weight all elements $y \in S_n$ such that the expected τ is as desired. We can then sample from the set S_n with probability proportional to the assigned weights. However, there are $n!$ elements in S_n and enumeration rapidly becomes impractical.

In this section, we examine two methods of sampling from S_n to obtain an expected τ; one controlled by the number of transpositions t, the other controlled by the standard deviation σ.

3.1   Using Adjacent Transpositions to Induce Controlled Noise

When measuring error using Kendall's τ, an obvious choice of inducing error is to perform adjacent transpositions. The set of adjacent transpositions is a generating set for the symmetric group, therefore we can obtain every possible ranking of n items using a finite sequence of adjacent transpositions.

To induce error in a ranked list x of length n, we randomly select x_i, where $i \in \{1, 2, \ldots, n-1\}$ and transpose it with the adjacent item x_{i+1}. By performing t random adjacent transpositions, we obtain a permuted list y, which when compared to the original list x, gives a value of τ. If we repeat this random process many times, we find that we obtain a distribution over τ, that is dependent on n and t.

To compute the expected Kendall's τ after t random adjacent transpositions, we first construct the probability transition matrix containing the probability of moving from one state to another in one adjacent transposition. The state of the list is the order of the items in the list. Therefore, if there are n items in the list, there are $n!$ possible states. The probability transition matrix will be of size $n! \times n!$, but each column will contain only $n-1$ nonzero elements (since for any list of n items, we can only move to $n-1$ other states using a single adjacent transposition). This gives us a probability transition matrix containing $n! \times (n-1)$ nonzero elements.

For example, if $n = 3$, we have the $n! = 6$ possible states $x_1 = (1, 2, 3)$, $x_2 = (1, 3, 2)$, $x_3 = (3, 1, 2)$, $x_4 = (3, 2, 1)$, $x_5 = (2, 3, 1)$ and $x_6 = (2, 1, 3)$ giving a probability transition matrix T with $n!(n-1) = 12$ nonzero values:

$$T = \begin{bmatrix} 0 & 0.5 & 0 & 0 & 0 & 0.5 \\ 0.5 & 0 & 0.5 & 0 & 0 & 0 \\ 0 & 0.5 & 0 & 0.5 & 0 & 0 \\ 0 & 0 & 0.5 & 0 & 0.5 & 0 \\ 0 & 0 & 0 & 0.5 & 0 & 0.5 \\ 0.5 & 0 & 0 & 0 & 0.5 & 0 \end{bmatrix} \tag{1}$$

where $t_{i,j}$, the elements of T, contain the probability of moving from state j to state i. If we begin in state 1: $x_1 = (1, 2, 3)$, our initial state probability vector is $p_0 = [\,1\ 0\ 0\ 0\ 0\ 0\,]'$. By taking a random walk of length 1, we compute our new state probability as $p_1 = Tp_0 = [\,0\ 0.5\ 0\ 0\ 0\ 0.5\,]'$. A random walk of length 2 is computed as $p_1 = Tp_1 = T^2 p_0 = [\,0.5\ 0\ 0.25\ 0\ 0.25\ 0\,]'$. A random walk of length n gives us the state distribution $T^n p_0$. Once we have the probability of each state after t random adjacent transpositions, we can compute the expected Kendall's τ using:

$$\mathbb{E}[\tau] = \sum_{i=1}^{n!} p_{t,i} \tau(x_i, x_1) \tag{2}$$

where $p_{t,i}$ is the ith element of p_t. If we perform two random transpositions (a random walk of length 2), on a list of length $n = 3$, we can obtain a Kendall's τ of 0 or $-1/3$, but the expected Kendall's τ is:

$$\mathbb{E}[\tau] = 0.5 \times 1 + 0 \times 1/3 + 0.25 \times (-1/3)$$
$$+ 0 \times (-1) + 0.25 \times (-1/3) + 0 \times 1/3 = 1/3$$

By representing the problem as a random walk on an undirected graph, we obtain additional information about its stationary distribution, being proportional to the degree of each vertex over an undirected graph. The degree of each vertex

Table 1. The number of nonzero elements (Nonzero) in the adjacent transposition adjacency matrix and the computation time required (Time) to compute the 50 expected values

List length	2	3	4	5	6	7	8	9
Nonzero	2	12	72	480	3600	30240	282240	2903040
Time (sec)	0.055	0.099	0.311	1.322	7.628	56.034	9.78 min	10.38 hr

over the set of permutations is equal $(n - 1)$, therefore the stationary distribution is the Uniform. This implies that as the number of adjacent transpositions approaches infinity, each state is equally likely. Kendall's τ is symmetric about 0 over all permutation states, therefore the expected Kendall's τ approaches 0 as t approaches infinity for all lists of length $n > 1$.

Given the task of randomly sampling ranked lists of length n with a given expected τ error, it is not obvious how we should choose t. In fact there is no way to directly compute t, other than trial and error (set t and n and examine the associated expected τ). To achieve this task, we have provided Table 2 containing the computed expected τ values of lists of length 2 to 9, using 1 to 50 random adjacent transpositions. So given $\mathbb{E}[\tau] = 0.5952$, the table shows that we are required to perform 13 random adjacent transpositions for $n = 8$. If we induce error in lists of length 7 and 9, we find that 9 and 18 random adjacent transpositions will provide the wanted $\mathbb{E}[\tau]$ respectively.

We have also provided the computation time for each of the lists in Table 1. It is interesting to note how fast the number of nonzero elements and computation time grows as n increases. Based on the trend, we found that the time is increasing double exponentially, meaning that the expected τ for $n = 10$ would take approximately 100 days to compute. Clearly, it is not feasible to compute the expected τ using this method for lists of length 10 or more.

3.2   Using Gaussian Perturbation to Induce Controlled Noise

Rather than performing adjacent transpositions, controlled noise can be induced in ranked lists by perturbing the rank of the elements. Perturbation requires the introduction of a latent value for each item i as a sample from a Normal distribution $X_i \sim N(\mu = x_i, \sigma)$, with mean equal to its rank x_i and constant standard deviation σ.

Once we have sampled a value for each item, we generate the new rank, by ordering the latent values. An example of this process is shown in Figure 1. We can see in this example four items $(1, 2, 3, 4)$, each having a Normal distribution with equal standard deviation, centred on its rank. A sample from these four distributions gives $(1.2, 3.1, 2.7, 4.4)$, providing us with the noise induced ranked list $(1, 3, 2, 4)$. Using this method of noise induction, it is more likely that each item moves a smaller number of ranks than one item move a large number, which is the typical form of error seen in ranked lists.

Table 2. The expected Kendall's τ after t transpositions over lists is length 2 to 9

t	List length							
	2	3	4	5	6	7	8	9
1	-1.0000	0.3333	0.6667	0.8000	0.8667	0.9048	0.9286	0.9444
2	1.0000	0.3333	0.5556	0.7000	0.7867	0.8413	0.8776	0.9028
3	-1.0000	0.0000	0.4198	0.6125	0.7216	0.7901	0.8361	0.8685
4	1.0000	0.1667	0.3580	0.5469	0.6681	0.7469	0.8007	0.8389
5	-1.0000	-0.0833	0.2716	0.4883	0.6216	0.7091	0.7695	0.8127
6	1.0000	0.1250	0.2318	0.4395	0.5807	0.6754	0.7414	0.7890
7	-1.0000	-0.1042	0.1759	0.3955	0.5440	0.6449	0.7158	0.7673
8	1.0000	0.1146	0.1501	0.3571	0.5107	0.6170	0.6922	0.7472
9	-1.0000	-0.1094	0.1139	0.3223	0.4803	0.5911	0.6703	0.7285
10	1.0000	0.1120	0.0972	0.2913	0.4523	0.5672	0.6498	0.7109
11	-1.0000	-0.1107	0.0738	0.2633	0.4264	0.5447	0.6306	0.6943
12	1.0000	0.1113	0.0630	0.2381	0.4023	0.5237	0.6124	0.6786
13	-1.0000	-0.1110	0.0478	0.2153	0.3798	0.5038	0.5952	0.6637
14	1.0000	0.1112	0.0408	0.1947	0.3587	0.4850	0.5789	0.6494
15	-1.0000	-0.1111	0.0309	0.1761	0.3390	0.4672	0.5633	0.6358
16	1.0000	0.1111	0.0264	0.1593	0.3205	0.4503	0.5483	0.6227
17	-1.0000	-0.1111	0.0200	0.1441	0.3030	0.4342	0.5341	0.6102
18	1.0000	0.1111	0.0171	0.1303	0.2865	0.4188	0.5204	0.5981
19	-1.0000	-0.1111	0.0130	0.1179	0.2710	0.4041	0.5072	0.5864
20	1.0000	0.1111	0.0111	0.1066	0.2564	0.3900	0.4945	0.5752
21	-1.0000	-0.1111	0.0084	0.0964	0.2426	0.3765	0.4823	0.5643
22	1.0000	0.1111	0.0072	0.0872	0.2295	0.3636	0.4705	0.5538
23	-1.0000	-0.1111	0.0054	0.0789	0.2171	0.3511	0.4591	0.5436
24	1.0000	0.1111	0.0046	0.0714	0.2055	0.3391	0.4481	0.5337
25	-1.0000	-0.1111	0.0035	0.0645	0.1944	0.3276	0.4374	0.5241
26	1.0000	0.1111	0.0030	0.0584	0.1840	0.3166	0.4271	0.5148
27	-1.0000	-0.1111	0.0023	0.0528	0.1741	0.3059	0.4171	0.5057
28	1.0000	0.1111	0.0019	0.0478	0.1648	0.2956	0.4074	0.4968
29	-1.0000	-0.1111	0.0015	0.0432	0.1559	0.2857	0.3979	0.4882
30	1.0000	0.1111	0.0013	0.0391	0.1476	0.2761	0.3887	0.4798
31	-1.0000	-0.1111	0.0010	0.0353	0.1397	0.2669	0.3798	0.4716
32	1.0000	0.1111	0.0008	0.0320	0.1322	0.2580	0.3711	0.4636
33	-1.0000	-0.1111	0.0006	0.0289	0.1251	0.2494	0.3627	0.4558
34	1.0000	0.1111	0.0005	0.0262	0.1184	0.2411	0.3545	0.4482
35	-1.0000	-0.1111	0.0004	0.0237	0.1120	0.2331	0.3465	0.4407
36	1.0000	0.1111	0.0003	0.0214	0.1060	0.2254	0.3387	0.4335
37	-1.0000	-0.1111	0.0003	0.0194	0.1003	0.2179	0.3311	0.4263
38	1.0000	0.1111	0.0002	0.0175	0.0950	0.2106	0.3236	0.4193
39	-1.0000	-0.1111	0.0002	0.0158	0.0899	0.2037	0.3164	0.4125
40	1.0000	0.1111	0.0001	0.0143	0.0851	0.1969	0.3094	0.4058
41	-1.0000	-0.1111	0.0001	0.0130	0.0805	0.1904	0.3025	0.3993
42	1.0000	0.1111	0.0001	0.0117	0.0762	0.1841	0.2958	0.3928
43	-1.0000	-0.1111	0.0001	0.0106	0.0721	0.1780	0.2892	0.3865
44	1.0000	0.1111	0.0001	0.0096	0.0682	0.1721	0.2828	0.3804
45	-1.0000	-0.1111	0.0000	0.0087	0.0646	0.1664	0.2766	0.3743
46	1.0000	0.1111	0.0000	0.0078	0.0611	0.1609	0.2705	0.3684
47	-1.0000	-0.1111	0.0000	0.0071	0.0578	0.1556	0.2645	0.3626
48	1.0000	0.1111	0.0000	0.0064	0.0547	0.1505	0.2587	0.3568
49	-1.0000	-0.1111	0.0000	0.0058	0.0518	0.1455	0.2530	0.3512
50	1.0000	0.1111	0.0000	0.0053	0.0490	0.1407	0.2474	0.3457

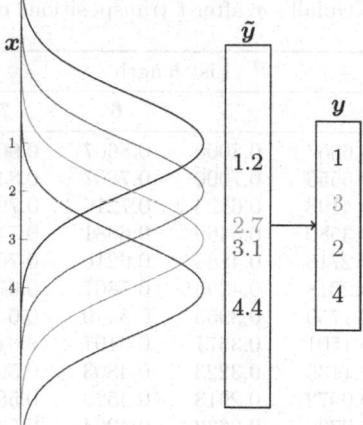

Fig. 1. Inducing error in ranked list x containing four items, using Gaussian Perturbation. Each item is treated as a Normal distribution with mean equal to its rank and constant standard deviation (in this case, $\sigma = 1$). A ranked list with error y is obtained by sampling from the Normal distributions (to obtain the latent values $\tilde{y}$), then ordering by the sample value. We can see that a value of 3.1 was sampled from the Normal distribution with mean 2, and 2.7 was sampled from the distribution with mean 3, pushing item 3 to the 2nd rank and item 2 to the 3rd rank.

Using Gaussian Perturbation, we can compute the expected τ as a function of n and σ. The expected value of τ is given as:

$$\mathbb{E}[\tau] = \mathbb{E}\left[\frac{2\sum_{1 \le i < j \le n} c_{ij}}{n(n-1)/2} - 1\right]$$

$$= \frac{2\sum_{1 \le i < j \le n} \mathbb{E}[c_{ij}]}{n(n-1)/2} - 1 \tag{3}$$

showing that it is dependent on the expected concordance of each pair of items. Let us consider x the ranked list of length n, and the permuted list y, based on latent values $\tilde{y}$, which are realisations of Normal random variables $\tilde{Y}$. Given two items i and j, where $x_i < x_j$ in ranked list x, then the pair is concordant if $y_i < y_j$ which happens if and only if $\tilde{y}_i < \tilde{y}_j$. Now;

$$\mathbb{E}[c_{ij}] = 1 \times P(c_{ij} = 1) + 0 \times P(c_{ij} = 0)$$
$$= P(\tilde{y}_i < \tilde{y}_j)$$
$$= P(\tilde{y}_i - \tilde{y}_j < 0)$$

and $\tilde{y}_i$ is a sample from a distribution $N(\mu = x_i, \sigma)$. Therefore, $\tilde{y}_i - \tilde{y}_j$ is a sample from a Normal distribution with mean $\mu = x_i - x_j$, and standard deviation $\sqrt{2}\sigma$. After standardising, we obtain:

$$\mathbb{E}[c_{ij}] = P\left(Z < \frac{x_j - x_i}{\sqrt{2}\sigma}\right) \tag{4}$$

Table 3. The computation time to compute σ when given $\mathbb{E}[\tau]$ and n for Gaussian Perturbation for lists of length 10^1 to 10^6

List Length	10^1	10^2	10^3	10^4	10^5	10^6
Computation Time (sec)	0.002	0.020	0.254	3.083	34.259	379.931

where $Z \sim N(0,1)$ is the standard Normal distribution. Substituting equation 4 back into 3 gives:

$$\mathbb{E}[\tau] = \sum_{1 \leq i < j \leq n} P\left(Z < \frac{x_j - x_i}{\sqrt{2}\sigma}\right) \frac{4}{n(n-1)} - 1$$

By noticing that $x_j - x_i$ is an integer from 1 to $n-1$, and the sum involves several copies of each such integer, we can simplify the equation further:

$$\mathbb{E}[\tau|n,\sigma] = \sum_{k=1}^{n-1} P\left(Z < \frac{k}{\sqrt{2}\sigma}\right) \frac{4(n-k)}{n(n-1)} - 1 \tag{5}$$

Since $\sigma > 0$ and $n > 1$, each term of the sum in equation 5 is non-negative, showing that $\mathbb{E}[\tau|n,\sigma] \geq 0$. $\mathbb{E}[\tau|n,\sigma]$ can only be zero when all terms of the sum are zero, implying that $\mathbb{E}[\tau|n,\sigma] \to 0$ if and only if $\sigma \to \infty$.

Using equation 5, we can compute the value of σ for a given $\mathbb{E}[\tau]$ and n using a one-dimensional optimisation function. For example, if we want to induce noise so that $\mathbb{E}[\tau] = 0.6$, we find that we must assign $\sigma = 22.46$, 44.48 and 110.54 for $n = 100$, 200 and 500 respectively.

Table 3 provides us with the computation time to perform the one-dimensional optimisation to compute σ. The table shows us that the time to compute σ increases linearly with n. When comparing this to Table 1, we see that Gaussian Perturbation has benefit over the Adjacent Transposition sampling method in terms of the parameter computation time.

3.3 Estimating σ for Gaussian Perturbation

In the previous section, we derived the equation for $\mathbb{E}[\tau]$, dependant on n and σ, and stated that we had to run a one dimensional optimisation over the function to compute σ when given $\mathbb{E}[\tau]$ and n. The computation of σ would be faster if we had a formula that gives the appropriate σ in terms of n and $\mathbb{E}[\tau]$. Unfortunately, σ is embedded in the Normal cumulative density function (CDF) and summed $n-1$ times.

To allow inversion of equation 5 we have made use of the Shah piece-wise approximation [14] to the Standard Normal CDF:

$$P(Z < x) = \begin{cases} x(4.4 - x)/10 + 1/2 & \text{if } x \leq 2.2 \\ 0.99 & \text{if } 2.2 < x < 2.6 \\ 1 & \text{if } x \geq 2.6 \end{cases} \tag{6}$$

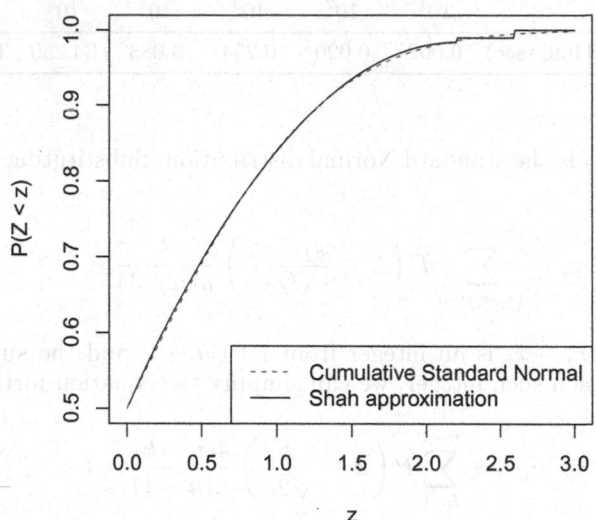

Fig. 2. The Cumulative Standard Normal function and the Shah piece-wise approximation

for $x \geq 0$. The similarity of the Shah approximation to the Standard Normal CDF is shown in Figure 2.

If we assume that $(n - 1)/(\sqrt{2}\sigma) \leq 2.2$, we can substitute $P(Z < x)$ with the first piece of the Shah approximation $x(4.4 - x)/10 + 1/2$. By making this substitution into equation 5 and simplifying, we obtain the expected τ approximation:

$$\mathbb{E}[\tau|n, \sigma] \approx \frac{4.4\sqrt{2}(n + 1)}{30\sigma} - \frac{(n + 1)n}{60\sigma^2}$$

which is a quadratic equation in terms of $1/\sigma$. By solving for σ, we obtain:

$$\sigma \approx \frac{n}{4.4\sqrt{2} \pm \sqrt{2 \times 4.4^2 - \frac{60\mathbb{E}[\tau]n}{n+1}}} \tag{7}$$

providing a solution under the condition $\mathbb{E}[\tau] < 4.4^2 \frac{n+1}{n}/30$, meaning that we can compute an approximate σ for all values of n when $0 < \mathbb{E}[\tau] < 0.6453$, and greater values of $\mathbb{E}[\tau]$ as n decreases. Also note that $\sigma \to \infty$ as $\mathbb{E}[\tau] \to 0$, as shown in equation 5.

Of the two solutions, the negative form provides accurate estimates for most of the τ, n combinations, but the positive form provides poor estimates. To examine the estimate's accuracy, we chose a desired value of $\mathbb{E}[\tau]$, used equation 7 to compute the approximate σ, then used equation 5 to compute the obtained $\mathbb{E}[\tau]$.

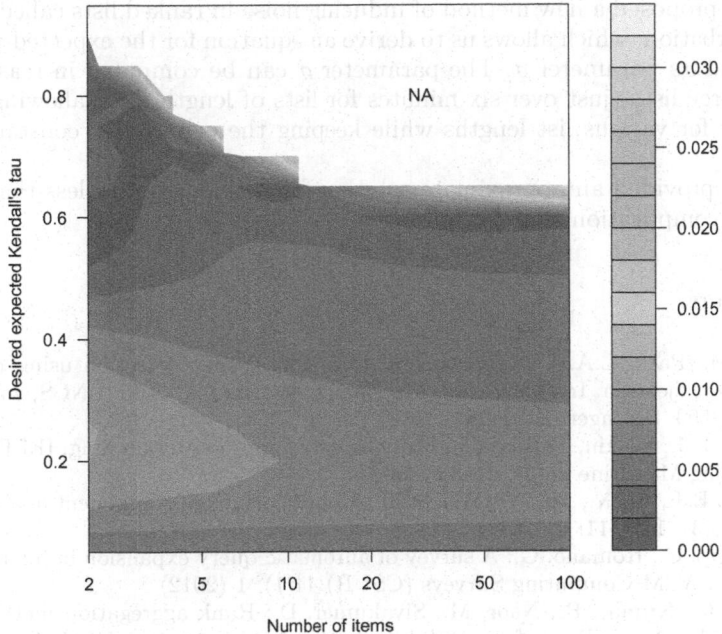

Fig. 3. The absolute difference when comparing the desired expected Kendall's τ, to the obtained expected Kendall's τ, when using the approximate σ computed from equation 7. The difference is computed for varying number if items (n) and desired expected Kendall's τ ($\mathbb{E}[\tau]$).

Figure 3 shows the absolute difference in desired $\mathbb{E}[\tau]$ compared to the obtained $\mathbb{E}[\tau]$, when using an approximate σ. We can see that that the σ estimate is a good estimate since most of the error is smaller than 0.01.

When using equation 7 to compute σ the computation time is less that 10^{-5} seconds for all n.

4   Conclusion

To examine the robustness of systems that take ranked lists as input, we can induce noise by applying a set number of random adjacent transpositions, where the error is measured using Kendall's τ rank correlation. For a fixed list length n, we can ensure a fixed expected τ by keeping the number a random adjacent transpositions constant, allowing a consistent level of noise to be applied to all lists. If the lists are of varying length, it is not clear how many adjacent transposition should be applied to each list to obtain a consistent expected τ over all lists.

In this article, we examined the relationship between the expected τ, the list length n and the number of random adjacent transpositions t. We found that it is possible to compute the expected τ, given n and t, but the computation time rapidly increases with n, making the computation infeasible for $n > 9$.

We also proposed a new method of inducing noise in ranked lists called Gaussian Perturbation, which allows us to derive an equation for the expected τ when given n and its parameter σ. The parameter σ can be computed in reasonable time for large lists (just over six minutes for lists of length 10^6), allowing us to compute σ for various list lengths while keeping the expected τ constant over all lists.

We also provided an approximate solution for σ that requires less that 10^{-5} seconds of computation time for all n.

References

1. Aouf, M., Park, L.A.F.: Approximate document outlier detection using random spectral projection. In: Thielscher, M., Zhang, D. (eds.) AI 2012. LNCS, vol. 7691, pp. 579–590. Springer, Heidelberg (2012)
2. Candès, E.J., Wakin, M.B.: An introduction to compressive sampling. IEEE Signal Processing Magazine 25(2), 21–30 (2008)
3. Candès, E.J., Li, X., Ma, Y., Wright, J.: Robust principal component analysis? J. ACM 58(3), 11:1–11:37 (2011)
4. Carpineto, C., Romano, G.: A survey of automatic query expansion in information retrieval. ACM Computing Surveys (CSUR) 44(1), 1 (2012)
5. Dwork, C., Kumar, R., Naor, M., Sivakumar, D.: Rank aggregation methods for the web. In: Proceedings of the 10th International Conference on World Wide Web, pp. 613–622. ACM (2001)
6. Farah, M., Vanderpooten, D.: An outranking approach for rank aggregation in information retrieval. In: Proceedings of the 30th Annual International ACM SIGIR Conference on Research and Development in Information Retrieval, SIGIR 2007, pp. 591–598. ACM, New York (2007)
7. Friedman, J., Hastie, T., Tibshirani, R.: The elements of statistical learning. Springer Series in Statistics, vol. 1 (2001)
8. Krüpl, B., Holzinger, W., Darmaputra, Y., Baumgartner, R.: A flight meta-search engine with metamorph. In: Proceedings of the 18th International Conference on World Wide Web, pp. 1069–1070. ACM (2009)
9. Linden, G., Smith, B., York, J.: Amazon. com recommendations: Item-to-item collaborative filtering. IEEE Internet Computing 7(1), 76–80 (2003)
10. Liu, Y.T., Liu, T.Y., Qin, T., Ma, Z.M., Li, H.: Supervised rank aggregation. In: Proceedings of the 16th international Conference on World Wide Web, pp. 481–490. ACM (2007)
11. Park, L.A.F.: Fast approximate text document clustering using compressive sampling. In: Gunopulos, D., Hofmann, T., Malerba, D., Vazirgiannis, M. (eds.) ECML PKDD 2011, Part II. LNCS, vol. 6912, pp. 565–580. Springer, Heidelberg (2011)
12. Ronchetti, E., Field, C., Blanchard, W.: Robust linear model selection by cross-validation. Journal of the American Statistical Association 92(439), 1017–1023 (1997)
13. Schölkopf, B., Smola, A.J.: Learning with kernels: Support vector machines, regularization, optimization, and beyond. MIT press (2001)
14. Shah, A.K.: A simpler approximation for areas under the standard normal curve. The American Statistician 39(1), 80–80 (1985)

Com2: Fast Automatic Discovery
of Temporal ('Comet') Communities

Miguel Araujo[1,5], Spiros Papadimitriou[2], Stephan Günnemann[1],
Christos Faloutsos[1], Prithwish Basu[3], Ananthram Swami[4],
Evangelos E. Papalexakis[1], and Danai Koutra[1]

[1] iLab & School of Computer Science, Carnegie Mellon University, Pittsburgh, USA
{maraujo,sguennem,christos,epapalex,danai}@cs.cmu.edu
[2] Rutgers University, New Brunswick, USA
spapadim@business.rutgers.edu
[3] BBN Technologies, Cambridge, USA
pbasu@bbn.com
[4] Army Research Laboratory, Adelphi, USA
ananthram.swami.civ@mail.mil
[5] University of Porto, Porto, Portugal

Abstract. Given a large network, changing over time, how can we find
patterns and anomalies? We propose COM2, a novel and fast, incremental tensor analysis approach, which can discover both transient and pe-
riodic/repeating communities. The method is (a) scalable, being linear
on the input size (b) general, (c) needs no user-defined parameters and
(d) effective, returning results that agree with intuition.

We apply our method on real datasets, including a phone-call network
and a computer-traffic network. The phone call network consists of *4
million* mobile users, with *51 million* edges (phonecalls), over 14 days.
COM2 spots intuitive patterns, that is, temporal communities (*comet
communities*).

We report our findings, which include large 'star'-like patterns, near-
bipartite-cores, as well as tiny groups (5 users), calling each other hundreds of times within a few days.

Keywords: community detection, temporal data, tensor decomposition.

1 Introduction

Given a large time-evolving network, how can we find patterns and communities?
How do the communities change over time? One would expect to see strongly
connected communities (say, groups of people, calling each other) with near-
stable behavior—possibly a weekly periodicity. Is this true? Are there other
types of patterns we should expect to see, like stars? How do they evolve over
time? Is the central node fixed with different leaves every day or are they fixed
over time? Perhaps the star appears on some days but not others?

Here we focus on exactly this problem: how to find time-varying communities, in a scalable way without user-defined parameters. We analyze a large,

V.S. Tseng et al. (Eds.): PAKDD 2014, Part II, LNAI 8444, pp. 271–283, 2014.
© Springer International Publishing Switzerland 2014

million-node graph, from an anonymous (and anonymized) dataset of mobile customers of a large population and a bipartite computer network with hundreds of thousands of connections, available to the public. We shall refer to time-varying communities as *comet communities*, because they (may) come and go, like comets.

Spotting communities and understanding how they evolve are crucial for forecasting, provisioning and anomaly detection. The contributions of our method, Com2, are the following:

- **Scalability**: Com2 is linear on the input size, thanks to a careful, incremental tensor-analysis method, based on fast, iterated rank-1 decompositions.
- **No User-Defined Parameters**: Com2 utilizes a novel Minimum Description Length (MDL) based formulation of the problem, to automatically guide the community discovery process.
- **Effectiveness**: We applied Com2 on real and synthetic data, discovering time-varying communities that agree with intuition.
- **Generality**: Com2 can be easily extended to handle higher-mode tensors.

2   Background and Related Work

In this section, we summarize related work on graph patterns, tensor decomposition methods, and general anomaly detection algorithms for graphs.

Tensor Decomposition. An n-mode tensor is a generalization of the concept of matrices: a 2-mode tensor is just a matrix, a 3-mode tensor looks like a data-cube, and a 1-mode tensor is a vector. Among the several flavors of tensor decompositions (see [1]), the most intuitive one is the so called Canonical Polyadic (CP) or PARAFAC decomposition [2]. PARAFAC is the generalization of SVD (Singular Value Decomposition) in higher modes.

Tensors have been used for anomaly detection in computer networks [3] and Facebook interactions [4] and for clustering of web pages [5].

Static Community Detection. Static community detection methods are closely related to graph partitioning and clustering problems. Using a more algebraic approach, community detection can also be seen as a feature identification problem in the adjacency matrix of a graph and several algorithms based on spectral clustering have been developed. Santo Fortunato wrote a detailed report on community detection [6].

Time Evolving Graphs. Graph evolution has been a topic of interest for some time, particularly in the context of web data [7,8]. MDL-based approaches for detecting overlapping communities in static graphs [9] as well as non-overlapping communities in time-evolving graphs [10] have been previously proposed. However, the former cannot be easily generalized to time-evolving graphs, whereas the latter focuses on incremental, streaming community discovery, imposing segmentation constraints over time, rather than on discovering *comet* communities. Other work, e.g. [11], studies the problem of detecting changing communities, but requires selection of a small number of parameters. Furthermore, broadly related work uses tensor-based methods for analysis and prediction of time-evolving "multi-aspect" structures, e.g., [12].

Table 1 compares some of the most common static and temporal community detection methods.

Table 1. Comparison of common (temporal) community detection methods

	Scalable	Temporal	Non-consecutive[*]	Parameter free[†]	Interpretability[‡]
COM2	✓	✓	✓	✓	✓
Graphscope[10]	✓	✓	✗	✓	✓
CP	✗	✓	✓	✗	✗
SDP + Rounding[13]	✗	✓	✓	✗	✓
Eigenspokes[14]	✓	✗	N/A	✓	✓
METIS[15]	✓	✗	N/A	✗	✓

[*] Temporal communities do not need to be contiguous.
[†] No user-defined parameter.
[‡] Results are easy to interpret; elements of the community can be identified easily.

3 Proposed Method

In this section, we formalize our problem, present the proposed method and analyze it. We first describe our MDL-based formalization which guides the community discovery process. Next, we describe a novel, fast, and efficient search strategy, based on iterated rank-1 tensor decompositions which can discover time varying communities in a fast and effective manner.

3.1 Formal Objective

We are given a temporal directed network consisting of sources $\mathcal{S}$, destinations $\mathcal{D}$, and time stamps $\mathcal{T}$. We represent this network via a 3-mode tensor $\mathbf{X} \in \{0,1\}^{|\mathcal{S}| \times |\mathcal{D}| \times |\mathcal{T}|}$ where $\mathbf{X}_{i,j,t} = 1$ if source i is connected to destination j at time t. As abbreviations we use $N = |\mathcal{S}|$, $M = |\mathcal{D}|$, and $K = |\mathcal{T}|$. The goal is to automatically detect communities:

Definition 1. *Community*
A community is a triplet $C = (S, D, T)$ with $S \subseteq \mathcal{S}$, $D \subseteq \mathcal{D}$, and $T \subseteq \mathcal{T}$ such that each triplet describes an 'important' time-varying aspect.

We propose to measure the 'importance' of a community via the principle of compression, i.e. by the community's ability to help us compress the 3-mode tensor: if most of the sources are connected to most of the destinations during most of the indicated times, then we can compress this 'comet-community' easily.

By finding the set of communities leading to the best compression of the tensor, we get the overall most important communities.

More specifically, we use MDL (Minimum Description Length) [16]. That is, we aim to minimize the number of bits required to encode the detected patterns (i.e. the model) and to describe the data given these patterns (corresponding to the effects of the data which are not captured by the model). Thus, the overall description cost automatically trades off the model's complexity and its goodness of fit. In the following, we provide more details about the description cost:

Description cost. The first part of the description cost accounts for encoding the detected patterns $\mathcal{C} = \{C_1, \ldots, C_l\}$ (where l is part of the optimization and not a priori given). Each pattern $C_i = (S_i, D_i, T_i)$ can completely be described by the cardinalities of the three included sets and by the information which vertices and time stamps belong to these sets. Thus, the coding cost for a pattern C_i is

$$L_1(C_i) = \log^* |S_i| + \log^* |D_i| + \log^* |T_i| + |S_i| \cdot \log N + |D_i| \cdot \log M + |T_i| \cdot \log K$$

The first three terms encode the cardinalities of the sets via the function $\log^*$ using the universal code length for integers [17][1]. The last three terms encode the actual membership information of the sets: e.g., since the original graph contains N sources, each source included in the pattern can be encoded by $\log N$ bits, which overall leads to $|S_i| \cdot \log N$ bits to encode all sources included in the pattern.

Correspondingly, a set of patterns $\mathcal{C} = \{C_1, \ldots, C_l\}$ can be encoded by the following number of bits:

$$L_2(\mathcal{C}) = \log^* |\mathcal{C}| + \sum_{C \in \mathcal{C}} L_1(C)$$

That is, we encode the number of patterns and sum up the bits required to encode each individual pattern.

The second part of the description cost encodes the data given the model. That is, we have to provide a lossless reconstruction of the data based on the detected patterns. Since in real world data we expect to find overlapping communities, our model should not be restricted to disjoint patterns. But how to reconstruct the data based on overlapping patterns? As an approach, we refer to the principle of Boolean algebra: multiple patterns are combined by a logical disjunction. That is, if an edge occurs in at least one of the patterns, it is also present in the reconstructed data. This idea related to the paradigm of Boolean tensor factorization. More formally, the reconstructed tensor is given by:

Definition 2. *Tensor reconstruction*
Given a pattern $C = (S, D, T)$. We define the indicator tensor $\mathbf{I}^C \in \{0,1\}^{N \times M \times K}$ to be the 3-mode tensor with $\mathbf{I}^C_{i,j,k} = 1 \Leftrightarrow i \in S \wedge j \in D \wedge k \in T$.
Given a set of patterns $\mathcal{C}$, the reconstructed tensor $\mathbf{X}^{\mathcal{C}}$ is defined as $\mathbf{X}^{\mathcal{C}} = \bigvee_{C \in \mathcal{C}} \mathbf{I}^C$ where $\vee$ denotes element-wise disjunction.

---

[1] Not to be confused with the *iterated logarithm* (log*). $\log^*$ is defined as $\log^* x = \log x + \log \log x + \ldots$, where only the positive terms are included in the sum.

The tensor $\mathbf{X}^{\mathcal{C}}$ might not perfectly reconstruct the data. Since MDL, however, requires a lossless compression, a complete description of the data has to encode the 'errors' made by the model. Here, an error might either be an edge appearing in $\mathbf{X}$ but not in $\mathbf{X}^{\mathcal{C}}$, or vice versa. Since we consider a binary tensor, the number of errors can be computed based on the squared Frobenius norm of the residual tensor, i.e. $\left\| \mathbf{X} - \mathbf{X}^{\mathcal{C}} \right\|_F^2$.

Since each 'error' corresponds to one triplet (source, destination, time stamp), the description cost of the data can now be computed as

$$L_3(\mathbf{X}|\mathcal{C}) = \log^* \left\| \mathbf{X} - \mathbf{X}^{\mathcal{C}} \right\|_F^2 + \left\| \mathbf{X} - \mathbf{X}^{\mathcal{C}} \right\|_F^2 \cdot (\log N + \log M + \log K)$$

Technically, we also have to encode the cardinalities of the set $\mathcal{S}$, $\mathcal{D}$, and $\mathcal{T}$ (i.e. the size of the original tensor). Given a specific dataset, however, these values are constant and thus do not influence the detection of the optimal solution.

Overall model. Given the functions L_2 and L_3, we are now able to define the communities that minimize the overall number of bits required to describe the model and the data:

Definition 3. *Finding comet communities*
Given a tensor $\mathbf{X} \in \{0,1\}^{|\mathcal{S}| \times |\mathcal{D}| \times |\mathcal{T}|}$. The problem of finding comet communities is defined as finding a set of patterns $\mathcal{C}^* \subseteq (\mathcal{P}(\mathcal{S}) \times \mathcal{P}(\mathcal{D}) \times \mathcal{P}(\mathcal{T}))$ such that

$$\mathcal{C}^* = \arg \min_{\mathcal{C}} [L_2(\mathcal{C}) + L_3(\mathbf{X}|\mathcal{C})]$$

Again, it is worth mentioning that the patterns detected based on this definition are not necessarily disjoint, thus better representing the properties of real data.

Obviously, computing the optimal solution to the above problem is infeasible as it is NP-hard. In the following, we present an approximate but scalable solution based on an iterative processing scheme.

3.2   Algorithmic Solution

We approximate the optimal solution via an iterative algorithm, i.e., we *sequentially* detect important communities. However, given the extremely large search space of the patterns (with most of the patterns leading to only low compression), the question is how to spot the 'good' communities?

Our idea is to exploit the paradigm of tensor decomposition [2]. Tensor decomposition provides us with a principled solution to detect patterns in a tensor while simultaneously considering the global characteristics of the data. It is worth mentioning that tensor decomposition cannot directly be used to solve our problem: (1) Tensor decomposition methods usually require the specification of the number of components in advance, while we are interested in a parameter-free solution. (2) Traditional tensor decomposition does not support the idea of Boolean disjunctions as proposed in our method, and Boolean tensor factorization methods [18] are still limited and a new field to explore. (3) Tensor decomposition does not scale to large datasets if the number of components is large as many local maxima exist. In our case, we expect to find many communities in the data.

Thus, in this work, we propose a novel, incremental tensor analysis for the detection of temporal communities. The outline of our method is as follows:

- **Step 1: Candidate 'comet' community**: We spot candidates by using an efficient rank-1 tensor decomposition. This step provides 3 vectors that represent the score of each source, destination and time stamp.
- **Step 2: Ordering and community construction**: The scores from step 1 are used to guide the search for important communities. We order the candidates and use MDL to determine the correct community size.
- **Step 3: Tensor deflation**: Based on the communities already detected, we deflate the tensor so that the rank-1 approximation is steered to find novel communities in later iterations.

In the following, we discuss each step of the method.

Candidate Generation. As explained, exhaustive search of all candidate communities is not possible. We propose to find a good initial candidate community using a fast implementation of rank-1 tensor decomposition. We aim at finding vectors $\mathbf{a} \in \mathbb{R}^N$, $\mathbf{b} \in \mathbb{R}^M$, and $\mathbf{c} \in \mathbb{R}^K$ providing a low rank approximation of the community. Intuitively, sources connected to highly-connected destinations at highly active times get a higher score in the vector $\mathbf{a}$ and similarly for the other two vectors. Specifically, to find these vectors, a scalable extension of the matrix-power-method only needs to iterate over the equations:

$$
\mathbf{a}_i \leftarrow \sum_{j=1,k=1}^{M,K} \mathbf{X}_{i,j,k}\mathbf{b}_j\mathbf{c}_k \quad , \quad \mathbf{b}_j \leftarrow \sum_{i=1,k=1}^{N,K} \mathbf{X}_{i,j,k}\mathbf{a}_i\mathbf{c}_k \quad , \quad \mathbf{c}_k \leftarrow \sum_{i=1,j=1}^{N,M} \mathbf{X}_{i,j,k}\mathbf{a}_i\mathbf{b}_j
$$

$$(1)$$

where $\mathbf{a}_i$, $\mathbf{b}_j$ and $\mathbf{c}_k$ are the scores of source i, destination j and time k. These vectors are then normalized and the process is repeated until convergence.

Lemma 1. *ALS [19] reduces to Equation 1, when we ask for rank-1 results.*

Proof. Substituting vectors $\mathbf{a}$, $\mathbf{b}$, $\mathbf{c}$, instead of matrices $(\mathbf{A}, \mathbf{B}, \mathbf{C})$, and carefully handling the Khatri-Rao products, we obtain the result.

Notice that the complexity is linear in the size of the input tensor: Let E be the number of non zeros in the tensor, we can easily show that each iteration has complexity $O(E)$ as we only need to consider the non zero $\mathbf{X}_{i,j,k}$ values. In practice, we select an ϵ and compare two consecutive iterations in order to stop the method when convergence is achieved. In our experimental analysis in Section 4 (using networks with millions of nodes) we saw that a relatively small number of iterations (about 10) is sufficient to provide reasonable convergence.

We can now use the score vectors $\mathbf{a}$, $\mathbf{b}$ and $\mathbf{c}$ as a heuristic to guide our community construction.

Community Construction Using MDL. Since the tensor decomposition provides numerical values for each node/time stamp, its result cannot be directly used to specify the communities. Additionally, there might be no clear threshold to distinguish between the nodes/time stamps belonging to the community and the rest. Our goal is to find a single community $C' \in (\mathcal{P}(\mathcal{S}) \times \mathcal{P}(\mathcal{D}) \times \mathcal{P}(\mathcal{T}))$ leading to the best compression, based on a local (i.e. community-wise) evaluation based on MDL (see Definition 3).

The definition of $L_3(\mathbf{X}|\mathcal{C})$ can be adapted to represent the MDL of this single community. By using the Hadamard product $(\mathbf{X} \circ \mathbf{I}^{C'})$, we restrict the tensor to the edges of the pattern:

$$\hat{L}_3(\mathbf{X}|C') = \log^* \left\| \mathbf{X} \circ \mathbf{I}^{C'} - \mathbf{I}^{C'} \right\|_F^2 + \left\| \mathbf{X} \circ \mathbf{I}^{C'} - \mathbf{I}^{C'} \right\|_F^2 \cdot (\log|S| + \log|D| + \log|T|)$$
$$+ \log^* \left\| \mathbf{X} - \mathbf{X} \circ \mathbf{I}^{C'} \right\|_F^2 + \left\| \mathbf{X} - \mathbf{X} \circ \mathbf{I}^{C'} \right\|_F^2 \cdot (\log N + \log M + \log K)$$

Even though we now only have to find a single community, minimizing this equation is still hard. Therefore, we exploit the result of the tensor decomposition to design a good search strategy.

We first sort the sources, destination, and time stamps according to the scores provided by the tensor decomposition. Let $\mathcal{S}'=(s_1,\ldots,s_N)$, $\mathcal{D}'=(d_1,\ldots,d_M)$ and $\mathcal{T}'=(t_1,\ldots,t_K)$ denote the lists storing the sorted elements. We start constructing the community by selecting the most promising triplet first, i.e., we form the community using the most promising edge and we evaluate its description cost.

Given the current community, we incrementally let the community grow. For each mode, we randomly select an element that is not currently part of the community using the score vectors as sampling bias. For each of these elements, we calculate the description length considering that we would add it to the community. The lowest description length is then selected, and the corresponding element is added to the community. If none of these elements decreases the overall description length, we reject them, proceed with the old community and repeat this process. If we observe l consecutive rejections, the method stops. It can be shown that the probability that an element that should have been included in the community was not included decreases exponentially as a function of l and of its initial score, thus a relatively small value of l is sufficient to identify a vast majority of the elements in the community. In our experimental analysis, a default value of $l = 20$ was seen to be enough, i.e. larger values have not led to the addition of further elements even when considering communities with thousands of elements. Therefore, we consider this parameter to be general and it does not need to be defined by the user of the algorithm.

Tensor Deflation. The output of the previous two steps is a *single* community. To detect multiple communities, multiple iterations are performed. The challenge of such an iterative processing is to avoid generating the same community repeatedly: we have to explore different regions of the search space.

As a solution, we propose the principle of tensor deflation. Informally, we remove the previously detected communities from the tensor, to steer the tensor decomposition to different regions. More formally: Let $\mathbf{X}^{(1)} = \mathbf{X}$ be the original tensor. In iteration i of our method we analyze the tensor $\mathbf{X}^{(i)}$ leading to the community C_i. The tensor used in iteration $i + 1$ is recursively computed as

$$\mathbf{X}^{(i+1)} = \mathbf{X}^{(i)} - \mathbf{I}^{C_i} \circ \mathbf{X}^{(i)}$$

where $\circ$ is once again the Hadamard product. This deflated tensor might either be used in both the candidate generation and community construction stages, in case we want to penalize overlapping communities, or in the candidate generation stage alone if overlapping communities are not to be penalized.

The method might terminate when the tensor is fully deflated (if possible), or when a specific number of communities has been found, or when some other measure of community quality was not achieved in the most recent communities (e.g. community size).

Complexity Analysis

Lemma 2. *Our algorithm has a runtime complexity of $O(M \cdot (k \cdot E + N \cdot \log N))$, where M is the number of communities we obtain, E is the number of non-zeros of the tensor, N is the length of the biggest mode, and k the number of iterations to obtain convergence. Thus, our method scales linearly w.r.t. the input E.*

Proof. Omitted for brevity.

4   Experiments

We tested our method on a variety of synthetic tensors to assess it's quality and scalability. We also applied COM2 on two realworld datasets: a large phone call network and a public computer communications network, demonstrating that it can find interesting patterns in challenging, real-world scenarios. This section details the experiments on the datasets summarized in Table 2.

Table 2. Networks used: Two small, synthetic networks; two large real networks

Abbr	Nodes	#Non zeros	Time	Description
OLB	10-20	1000-2000	100	Overlapping blocks.
DJB	1 000	50000	500	Disjoint blocks.
LBNL	1 647 + 13 782	113 030	30	Bipartite Internet traces from LBNL.
PHONE	3 952 632	51 119 177	14	Phone call network.

4.1   Quality of the Solutions

The characterization of the temporal communities identified by the method is important. In particular we want to answer the following questions: How are "overlapping blocks" identified? How "dense" are the communities found?

Impact of overlap. A tensor with two disjoint communities was constructed and, iteratively, elements from each of the modes of one of the communities were replaced with elements of the other. Our tests show that the communities are reported as independent until there is an overlap of about 70% of the elements in each mode, in which case they start being reported as a single community. This corresponds to an overlapping of slightly over 20% of the non-zero values of the two communities and the global community formed has 63% of non-zeros.

This clearly demonstrates that COM2 has high discriminative power: it can detect the existence of communities that share some of their members and it is able to report them independently, regardless of their size (the method is scale-free).

Impact of block density. We also performed experiments to determine how density impacts the number of communities found. Fifty disjoint communities were created in a tensor and non-zeros were sampled without repetition from each community with different probabilities and random noise was then added. We analyzed the number of non-zeros in the first fifty communities reported by our method in order to calculate its accuracy. As we show in Figure 1a, COM2 has high discriminative power even with respect to varying density.

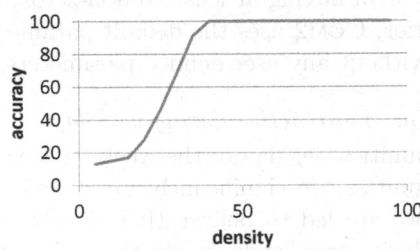

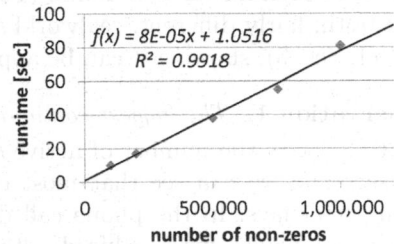

(a) Tensor with disjoint blocks - **Com2 identifies communities even at low densities.**

(b) **Com2 scales linearly** with input size: Running time versus number of non-zeros for random tensors.

Fig. 1. Experiments on synthetic data

4.2 Scalability

As detailed before, COM2's running time is linear on the number of communities and in the number of non-zero values in the tensor. We constructed a tensor of size $10\,000 \times 10\,000 \times 10\,000$ and randomly created connections between sources and destinations at different timesteps. Figure 1b shows the runtime versus the number of non-zeros in the tensor when calculating the first 200 communities of the tensor. We consider random insertion to be a good worst-case scenario for many real-life applications, as the lack of pre-défined structure will force many small communities to be found, effectively penalizing the running time of COM2.

In addition to its almost linear runtime, COM2 is also easily parallelizable. By selecting different random seeds in the tensor decomposition step, different communities can be found in parallel.

4.3 Discoveries on Real Data

We applied COM2 to a dataset from a european mobile carrier, to characterize the communities found in real phone call data. We considered the network formed by calls between clients of this company over a period of 14 days.

During this period, 3 952 632 unique clients made 210 237 095 phone calls, 51 119 177 of which formed unique (caller, callee, day) triplets. The tensor is very sparse, with density in the order of 10^{-7}. We extracted 900 communities using COM2. These communities contain a total of 229 287 unique non-zeros. 293 unique callers and 97 677 unique callees are represented, so the first observation is that the temporal communities are usually heavy on one side with large outgoing stars.

We also applied COM2 to a public computer network dataset captured in 1993, made available by the Lawrence Berkeley National Laboratory. 30 days of TCP connections between 1 647 IP addresses inside the laboratory and 13 782 external IP addresses were recorded. This tensor was totally deflated and a total of 19 046 communities were found (1 930 of them having at least 10 non-zeros).

In both, fairly different, realworld scenarios, COM2 uses the default parameters (cf. Sec. 3), showing it can be applied without any user-defined parameters.

Observation 1. *The biggest communities are more active during weekdays.*
Figure 2 shows the number of active communities per day of the week on both datasets and we can see that most communities are significantly more active during weekdays. In the phone call data, we are led to believe that these are mostly companies with reduced activity during weekends, while the reduced activity during the weekends in the research laboratory is to be expected.

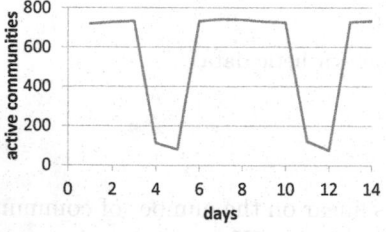

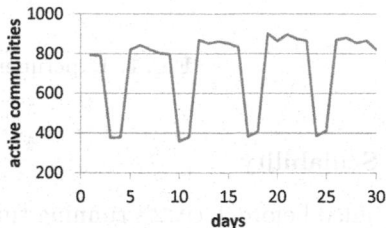

(a) **Weekly periodicity** phone call data.

(b) **Weekend activity** computer network data.

Fig. 2. Weekly periodicity: number of active communities vs time. Notice the weekend dives on a) days 4, 5 and 11, 12 and b) days 3, 4, 10, 11, 17, 18, 24, 25

Observation 2. *A typical pattern is the "Flickering stars".*
When analyzing a phone call network, a pattern to be expected is the marketeer pattern in which a number calls many others a very small number of times (1 or 2). Surprisingly, the stars reported by COM2 were not of this type. Two callers stand out in an analysis of the communities reported: one participated in 78 279 (source, destination, time) triplets as a caller but only in 10 triplets as a receiver, while the other participated in 8 909 triplets as a caller and in none as a receiver. These two nodes are centers of two distinct outgoing stars and were detected by the algorithm. However, the time component of these stars was not a single day but rather spanned almost all the weekdays. This behavior does not seem

typical of a marketeer, so we hypothesize that it is a big company communicating with employees. Many of the reported communities are stars of this type: a caller calling a few hundred people in a subset of the weekdays - we call them flickering because there is still some activity during the rest of the weekdays, only reduced so that those days are not considered part of the community.

In the LBNL dataset, one star was particularly surprising. It received connections from over 750 different IP addresses inside the laboratory but only on a single day. One of the other big stars corresponded to 40 connections on a single day to an IP address attributed to the Stanford Research Institute, which is not surprising given the geographical proximity.

We define *Flickering stars* as a common temporal-community that has a varying number of receivers. These communities are active on different days, not necessarily consecutive. Stars active on many days (e.g. every weekday) are more common than single day stars.

Observation 3. *A typical pattern is the "Temporal Bipartite Cores".*
Several near-bipartite cores were detected as communities in the phone call dataset. These are communities with about 5 callers and receivers that are active on nearly each day under analysis. These communities represent between 75 and 150 of the non-zeros of the original tensor, with a block density of around 40%.

An example of such communities can also be shown for the LBNL data. 7 machines of the laboratory communicated with 6 external IP addresses on every weekday of the month. After analyzing the IP addresses, the outside machines were found to be part of the Stanford National Accelerator Laboratory, the University of California in San Francisco, the UC Davis, the John Hopkins University, and the U.S. Dept. of Energy. COM2 was able to detect this research group (possibly in particle physics) using communications data alone.

5   Conclusions

We focused on deriving patterns from time-evolving graphs, and specifically on spotting *comet* communities, that come and go (possibly periodically). The main contributions are the following:

- **Scalability**: Our method, COM2, is linear on the input size; instead of relying on a complete tensor factorization, we carefully leverage rank-1 decompositions to incrementally guide the search process for community detection.
- **No user-defined parameters**: In addition to the above, efficient, incremental search process, we also proposed a novel MDL-based stopping criterion, which finds such *comet* communities in a parameter-free fashion.
- **Effectiveness**: We applied COM2 on real and synthetic data, where it discovered communities that agree with intuition.
- **Generality**: COM2 can be easily extended to handle higher-mode tensors.

COM2 can also be applied on edge-labeled graphs, by considering the labels as the third mode of the tensor. Future work could focus on exploiting side information, like node-attributes (for example, demographic data for each node). COM2 is available at http://cs.cmu.edu/~maraujo/publications.html.

Acknowledgments. This material is based upon work supported by the National Science Foundation under Grant No. IIS-1247489. Research was sponsored by the Defense Threat Reduction Agency and was accomplished under contract No. HDTRA1-10-1-0120. Also, sponsored by the Army Research Laboratory and was accomplished under Cooperative Agreement Number W911NF-09-2-0053. Additional funding was provided by the U.S. Army Research Office (ARO) and Defense Advanced Research Projects Agency (DARPA) under Contract Number W911NF-11-C-0088. This work is also partially supported by a Google Focused Research Award, by the Fundação para a Ciência e a Tecnologia (Portuguese Foundation for Science and Technology) through the Carnegie Mellon Portugal Program, and by a fellowship within the postdoc-program of the German Academic Exchange Service (DAAD). Any opinions, findings, and conclusions or recommendations expressed in this material are those of the author(s) and do not necessarily reflect the views of the National Science Foundation, DARPA, or other funding parties. The U.S. Government is authorized to reproduce and distribute reprints for Government purposes notwithstanding any copyright notation here on.

References

1. Kolda, T., Bader, B.: Tensor decompositions and applications. SIAM Review 51(3) (2009)
2. Harshman, R.: Foundations of the PARAFAC procedure: Models and conditions for an "explanatory" multimodal factor analysis (1970)
3. Maruhashi, K., Guo, F., Faloutsos, C.: Multiaspectforensics: Pattern mining on large-scale heterogeneous networks with tensor analysis. In: Proceedings of the Third International Conference on Advances in Social Network Analysis and Mining (2011)
4. Papalexakis, E.E., Faloutsos, C., Sidiropoulos, N.D.: Parcube: Sparse parallelizable tensor decompositions. In: Flach, P.A., De Bie, T., Cristianini, N. (eds.) ECML PKDD 2012, Part I. LNCS, vol. 7523, pp. 521–536. Springer, Heidelberg (2012)
5. Kolda, T.G., Bader, B.W., Kenny, J.P.: Higher-order web link analysis using multilinear algebra. In: ICDM, pp. 242–249. IEEE Computer Society (2005)
6. Fortunato, S.: Community detection in graphs. Physics Reports 486(35), 75–174 (2010)
7. Kumar, R., Novak, J., Raghavan, P., Tomkins, A.: On the bursty evolution of blogspace. In: WWW (2003)
8. Leskovec, J., Kleinberg, J., Faloutsos, C.: Graph evolution: Densification and shrinking diameters. IEEE TKDD (2007)
9. Gionis, A., Mannila, H., Seppänen, J.K.: Geometric and combinatorial tiles in 0–1 data. In: Boulicaut, J.-F., Esposito, F., Giannotti, F., Pedreschi, D. (eds.) PKDD 2004. LNCS (LNAI), vol. 3202, pp. 173–184. Springer, Heidelberg (2004)
10. Sun, J., Papadimitriou, S., Faloutsos, C., Yu, P.S.: Graphscope: Parameter-free mining of large time-evolving graphs. In: KDD (2007)
11. Liu, Z., Yu, J., Ke, Y., Lin, X., Chen, L.: Spotting significant changing subgraphs in evolving graphs. In: ICDM (2008)
12. Sun, J., Tao, D., Faloutsos, C.: Beyond streams and graphs: Dynamic tensor analysis. In: KDD (2006)

13. Tantipathananandh, C., Berger-Wolf, T.Y.: Finding communities in dynamic social networks. In: ICDM (2011)
14. Prakash, B.A., Sridharan, A., Seshadri, M., Machiraju, S., Faloutsos, C.: Eigenspokes: Surprising patterns and scalable community chipping in large graphs. In: Zaki, M.J., Yu, J.X., Ravindran, B., Pudi, V. (eds.) PAKDD 2010. LNCS, vol. 6119, pp. 435–448. Springer, Heidelberg (2010)
15. Karypis, G., Kumar, V.: Metis: unstructured graph partitioning and sparse matrix ordering system. Technical report (1995)
16. Grünwald, P.D.: The minimum description length principle. The MIT Press (2007)
17. Rissanen, J.: A universal prior for integers and estimation by minimum description length. The Annals of Statistics, 416–431 (1983)
18. Miettinen, P.: Boolean tensor factorizations. In: ICDM (2011)
19. Takane, Y., Young, F.W., De Leeuw, J.: Nonmetric individual differences multidimensional scaling: an alternating least squares method with optimal scaling features. Psychometrika 42(1), 7–67 (1977)

A Graphical Model for Collective Behavior Learning Using Minority Games

Farhan Khawar and Zengchang Qin

Intelligent Computing and Machine Learning Lab
School of ASEE, Beihang University, Beijing, China
farhan.khawar@asee.buaa.edu.cn, zcqin@buaa.edu.cn

Abstract. The Minority Game (MG) is a simple game theory model
for the collective behavior of agents in an idealized situation where they
compete for some finite resource. In this paper, we assume that collective
behavior is determined by the aggregation of individual actions of agents.
This causal relation between collective behavior and individual actions
is investigated. A graphical model is proposed to model the generative
process of collective behavior using a group of agents whose actions are
modeled by minority games. In this model, we can infer the individual behavior of the agents by training on the global information, and
then make predictions about the future collective behavior. Experimental results on a set of stock indexes from the Chinese market and foreign
exchange (FX) rates show that the new proposed model can effectively
capture the rises and falls of market and be significantly better than
a random predictor. This framework also provides a new data mining
paradigm for analyzing collective data by modeling micro-level actions
of agents using game theory models.

Keywords: Collective Intelligence, Minority Game, Probabilistic Graphical Models.

1 Introduction

Collective intelligence is a shared or group intelligence that emerges from the interactions (both collaborative and competitive) of many individuals and appears
in consensus decision making of agents. Collective behaviors can be modeled by
agent-based games where each individual agent follows its own local rules. Agent-
based experimental games have attracted much attention in different research
areas, such as psychology [1], economics [2,3], financial market modeling [4,5]
and market mechanism designs [6,7]. Agent-based models (ABM) of complex
adaptive systems (CAS) provide invaluable insight into the highly non-trivial
collective behavior of a population of competing agents. These systems are universal and researchers aim to model the systems where involving agents with
similar capability are competing for a limited resource. Agents may share global
information and learn from past experience. In such a complex system, if we
assume that every agent in the market knows the history data, the key problem
is how to decide to act based on this global information.

V.S. Tseng et al. (Eds.): PAKDD 2014, Part II, LNAI 8444, pp. 284–295, 2014.

The minority game [8] is a simple ABM originating from the El-Farol Bar [9] problem. An odd number of agents compete with each other to be in the minority by making one of two choices; the players who end up in the minority side win the game. The collective recognizable patterns generated by the minority game are due to the interaction of many individual agents. There are two main features for the minority game: first, the minority rule, which makes complete steady state in the population impossible and secondly, every agent has its own way of perceiving the available global information about the game and using it into its own strategy. Each agent is aware of the global information and can use this information to make decisions based on its own unique strategy, as it is unrealistic that all agents follow the same deterministic strategy [10]. Due to its simplicity and attractive properties, the minority game model has attracted much attention in different research communities [11].

In [12] the authors purpose that collective behaviors of MG can be decomposed into several micro-level behaviors from different group of agents and they use Genetic Algorithm (GA) to optimize the agent behavior parameters in order to get the best guess of the original system. Also, different game theory model can be used to model the individual behavior. More accurate a model is for individual behavior, more accurate collective behavior we can obtain [13]. Uncertainty is present in all real-world scenarios, especially when it comes to the task of predicting the outcomes of aggregated actions. Probabilistic graphical models (PGM) [14] have the capability to deal with uncertainty by incorporating prior beliefs about the domain and updating these beliefs as new evidence is obtained. Using PGMs we can construct richly structured models to understand hidden relations.

In this paper, we propose a novel generative probabilistic graphical model for modeling the process of generation of the collective behavior from the individual behavior. We use the proposed PGM to infer the behavior of individual agents from the available global information and use the learned agent behavior to predict the future collective behavior. The main contribution of our work is four fold: (1) We use a game theory model, the minority game, to model individual behavior. It also has the flexibility of using any appropriate game theories to model individual behavior; (2) A PGM is used to model generative process of collective data from a group of individual actions. We can infer the individual strategy from the observable collective data, and this can be used for predicting future collective data; (3) We use this novel framework to show how it can be applied to time-series data mining of financial market data; (4) Comprehensive experiments on real world stock market data and foreign exchange rates demonstrate the effectiveness of the new model.

2   Behavior Modeling Using Minority Game

In a MG, there are an odd number of players and each player must independently choose one of two options at each round of the game, the players who end up at the minority side are winners and the choice of the minority players is referred to as the winning choice r. There is no prior communication among players; the only global information available is numbers of players corresponding to the two choices from the previous rounds.

2.1   Terminology

We begin by first introducing the notation and the terminology used in this paper: *Agent*: A player of the game is referred to as an "agent" and it is the entity that makes decisions based on its "strategy". The number of agents that participate in the MG is N, which is an odd number. Agent is indexed by an integer A where $A \in \{1, 2, \ldots, N\}$.

Choice: An action made by an agent. Choice C has two possible values: $C \in \{0, 1\}$. The total number of choices are N.

Game: Every run of the MG will be referred to as a "game". In every game the choices are represented as a vector of N elements of binary value $\{0, 1\}$. It can also be viewed as a sequence of choices $C_1, C_2, \ldots, C_N$ where C_n is the nth agent's choice in the game. The total number of games is denoted by G.

Minority Choice: For every game the choice of the agents on the minority side is the winning outcome and is called the "minority choice". Formally, let t denote the current game, then the minority choice in game t is defined by:

$$r(t) = \begin{cases} 0 & \text{If } \sum_{n=1}^{N} C_n(t) > \frac{N}{2} \\ 1 & \text{Otherwise} \end{cases} \tag{1}$$

Memory and **History**: In the minority game we assume that the agent's actions are governed by its strategy and previous minority choices of the game. If the agents have an m-bit memory which means that the agent will take into the information (minority choices) in the previous m rounds. The minority choice for the last m round of games is defined as the "history" $H \in \{1, 2, 3, \ldots, K\}$ and $K = 2^m$, where K is the maximum value of history. History is normally a binary string of the past m minority choices, but without loss of generality, for the representational convenience of our model we have defined history as a decimal number. For example if the agents have a 3-bit memory then the history belongs to the set $H = \{1, 2, 3, \ldots, 8\}$.

Strategy: We assume that each agent's action is governed by a strategy which can be regarded as a set of rules or functions taking the previous minority choices as inputs. Given a history, the agent makes its decision based on its own predefined rules named "strategy" S [8,15,16]. The strategy is a mapping from each possible m-bit memory(each possible history) to a corresponding choice of making choice-0 or choice-1. Therefore, there are 2^{2^m} possible strategies in the strategy space and we assume that at one time each agent has exactly one strategy. A strategy can be regarded as a particular set of decisions on all the permutations of previous history of minority choices. Fo r example, the first 3 rows of Table 1 show the 3-bit memory, history and a sample strategy.

Probabilistic Strategy: For each agent, "probabilistic strategy (PS)" is a strategy that maps the history to a probability distribution over the two choices

Table 1. Sample strategy and PS of one agent for 3-bit memory and history

Memory	$(000)_2$	$(001)_2$	$(010)_2$	$(011)_2$	$(100)_2$	$(101)_2$	$(110)_2$	$(111)_2$
History	1	2	3	4	5	6	7	8
Strategy	1	0	1	1	0	0	0	1
PS	$P_B(0.1)$	$P_B(0.8)$	$P_B(0.3)$	$P_B(0.1)$	$P_B(0.6)$	$P_B(0.8)$	$P_B(0.6)$	$P_B(0.3)$

instead of mapping directly to one choice only. We use a Bernoulli distribution to represent probabilistic strategy. The bottom row of Table 1 shows a sample PS where $P_B(p)$ represents a Bernoulli distribution, p is the probability of making choice-0 and $q = 1 - p$ is the probability of making choice-1. The advantage of using such a strategy is that it is able to incorporate uncertainty in the learning process of the PS via PGMs, because for the next game this PS provide a prior to the PGM.

2.2   Probabilistic Graphical Model for Collective Behavior Learning

Previous works show that collective behavior can be decomposed into the aggregation of individual agents' actions [12,13]. In this paper we assume that the collective behavior is generated by agents with probabilistic strategies. Our task is two fold: first, to decompose the collective behavior by inferring individual agent behaviors, and the second, to use these learned individual behaviors to predict the future collective behavior. PGMs provide us the ability to deal with uncertainty and incorporate prior knowledge. Moreover, the proposed PGM will provide a unified framework for both the inference of individual behaviors and the prediction of the global behavior.

With this premise, the motivation behind our purposed PGM is to model the procedure of an agent making a choice. We first start by drawing a distribution over agents from a Dirichlet prior, then we randomly select an agent from this agent distribution to make a choice. After selecting this agent we observe the history for the present game. Then the agent's choice, corresponding to the observed history, is sampled from its PS. This is repeated N times to generate all the choices of a game. Formally this generative process can be outlined as:

- For each agent and each history:
  - Draw a vector of distribution over the two choices $\phi_{n,k} \sim Dir_2(\beta)$ from a Dirichlet prior.
- Observe the history H for the current game.
- For each choice
  - Draw a vector of agent distribution from a Dirichlet prior i.e. $\psi_n \sim Dir_N(\alpha)$.
  - Draw an agent index $A_n \sim Mult(\psi_n)$, $A_n \in \{1, \ldots, N\}$.
  - Draw a choice $C_n \sim Bernoulli(\phi_{A_n,H})$, $C_n \in \{0, 1\}$.

where α and β are scalars that parameterize the symmetric prior Dirichlet distributions, $Dir_2(\beta)$ denotes a 2-dimensional Dirichlet with the scalar parameter

β and $Dir_N(\alpha)$ denotes an N dimensional Dirichlet parameterized by α. A symmetric Dirichlet is a Dirichlet distribution where every component of the parameter is equal to the same scalar value. The Dirichlet distribution is a distribution over discrete distributions and it is conjugate to the Multinomial distribution; each component in the sampled random vector is the probability of drawing the item associated with that component. $Mult(.)$ denotes a discrete Multinomial distribution.

The "choice distribution" $\phi_{n,k}$ is a 2-dimensional random vector that corresponds to the probability of making choice-0 or choice-1 for agent n and history k. For a single agent n, the set of these distributions for all values of history corresponds to that agent's PS i.e. $\{\phi_{n,m=1,...,K}\}$ is the PS of agent n. The distribution ψ_n is the "agent distribution" and is a N-dimensional random vector where each component gives the probability of selecting the agent index associated with that component. The corresponding directed graphical model is shown

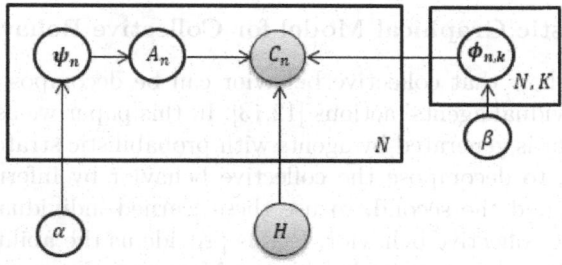

Fig. 1. Graphical model representation

in Figure 1. It is worth mentioning here that the choices that are observed are unordered choices: the N choices for every game are provided as a string of 0's followed by 1's. If the actual choices made by the $N=5$ agents were $[1,0,1,0,0]$ in one game and $[0,1,1,0,0]$ in the other game then in both cases the observed choices would be $[0,0,0,1,1]$. These unordered choices are referred to as choices throughout the paper.

The joint distribution corresponding to the PGM in Figure 1 is:

$$p(\psi_{1:N}, A_{1:N}, \phi_{1:N,1:K}, C_{1:N}, H | \alpha, \beta, H) = \prod_{i=1}^{N} \prod_{k=1}^{K} p(\phi_{i,k}|\beta)$$

$$p(H) \left[\prod_{n=1}^{N} p(\psi_n|\alpha)p(A_n|\psi_n)p(C_n|A_n, \phi_{1:N,1:K}, H) \right] \qquad (2)$$

Notice that since H is observed and does not depend on another variable. It is a deterministic variable and $p(H)$ can be omitted from the joint of Eq. (2). The model specifies a number of dependencies between random variables: the agent index A_n depends on agent distribution ψ_n, the choices C_n depends on the agent index A_n , history H and all of the choice distributions $\phi_{1:N,1:K}$. Here the notation $\phi_{1:N,1:K}$ denotes the set of distributions $\{\phi_{n,k}|0 < n \le N ,0 < k \le K\}$. This is equivalent to saying that $p(C_n|A_n, \phi_{1:N,1:K}, H) = \phi_{A_n,H}$ i.e. from the PS of the agent denoted by A_n we find the choice distribution corresponding to the observed history H. From Figure 1 we can also see that once choices are observed the agent distributions ψ_n and choice distributions $\phi_{n,k}$ are conditionally dependent according to d-separation.

3   Inference and Prediction

In generative probabilistic modeling, we treat our data as arising from a generative process that includes hidden and observed variables. This generative process defines a joint probability distribution over both the observed and latent random variables. We perform data analysis by using that joint distribution to compute the conditional distribution of the latent variables given the observed variables. The decomposition of collective behavior to individual behavior corresponds to observing the choices and history and inferring the posterior distribution of the hidden variables. The choices and the history are the global behavior and the posterior of the agent distributions $p(\psi_{1:N}|C_{1:N}, H)$ and choice distributions $p(\phi_{1:N,1:K}|C_{1:N}, H)$ are the local behaviors. Here the notation $\psi_{1:N}$ denotes the set of distributions $\{\psi_n|0 < n \le N\} = \{\psi_1, \psi_2, \dots, \psi_N\}$. The posterior distribution can be written as:

$$p(\psi_{1:N}, A_{1:N}, \phi_{1:N,1:K}|C_{1:N}, H, \alpha, \beta) = \frac{p(\psi_{1:N}, A_{1:N}, \phi_{1:N,1:K}, C_{1:N}, H|\alpha, \beta)}{p(C_{1:N}|\alpha, \beta)} \quad (3)$$

where the numerator of Eq. (3) is defined in Eq. (2). In order to normalize this posterior distribution we need to marginalize over the latent variables to give the denominator as shown in Eq. (4):

$$p(C_{1:N}|\alpha, \beta) = \prod_{i=1}^{N} \prod_{k=1}^{K} \int \prod_{n=1}^{N} \int p(\psi_n|\alpha) \sum_{A_n} p(A_n|\psi_n)$$
$$p(C_n|A_n, \phi_{1:N,1:K}, H) d\psi_n p(\phi_{i,k}|\beta) d\phi_{i,k} \quad (4)$$

Posterior inference of our model is done using approximate message passing algorithm [17], specifically we are using Variational Message Passing algorithm (VMP) [18] in the Infer.Net [19] package. VMP is deterministic approximate inference algorithm that is guaranteed to converge to some solutions and it works by using only local message passing operations. Our goal is to infer the individual behaviors, then use these individual behaviors to predict the choices for the next game and calculate our accuracy of prediction. This procedure is explained in the Algorithm 1. Here $C_n(t + 1)$ denotes the actual choices of the next game. $PC_n(t+1)$ denote predicted choices of the next game and the predicted minority choice of the next game $\hat{r}(t + 1)$ is defined as:

$$\hat{r}(t+1) = \begin{cases} 0 & \text{If } \sum_{n=1}^{N} PC_n(t+1) > \frac{N}{2} \\ 1 & \text{Otherwise} \end{cases} \quad (5)$$

Algorithm 1. Model Inference and Prediction

    Parameters: α, β, Number of Agents N, History size K, Number of games G
1: Construct Bayesian inference engine E using Variational Message Passing:
2: **for** $t = 1 \rightarrow G$ **do**
3:     **if** $t = 1$ **then**
4:         Assign symmetric *Dirichlet* priors i.e. $\psi_{1:N} \sim Dir_N(\alpha)$ and $\phi_{1:N,1:K} \sim Dir_2(\beta)$.
5:     **else**
6:         Assign posterior distributions from the last game as the current prior i.e.
$p(\psi_{1:N}(t)) = p(\psi_{1:N}(t-1)|C_{1:N}(t-1), H(t-1))$ and $p(\phi_{1:N,1:K}(t)) = p(\phi_{1:N,1:K}(t-1)|C_{1:N}(t-1), H(t-1))$
7:     **end if**
8:     Observe the global data of the current game (choices $C_{1:N}(t)$ and history $H(t)$) and input them to E.
9:     Execute E and calculate the posterior distributions i.e. $p(\psi_{1:N}(t)|C_{1:N}(t), H(t))$ and $p(\phi_{1:N,1:K}(t)|C_{1:N}(t), H(t))$
10:    Given the posteriors inferred in Line 9 and history of next game $H(t+1)$, execute the engine E to infer *Bernoulli* distributions of the predicted choices of the next game $PC_{1:N}(t+1)$ i.e.

$$p(PC_{1:N}(t+1)|\psi_{1:N}(t), A_{1:N}(t+1), \phi_{1:N,1:K}(t), H(t+1)) \sim P_B(\phi_{A_{1:N}(t+1),H(t+1)})$$

11:    To get the predicted choices of the next game, sample from the distributions inferred in Line 10 i.e. $PC_{1:N}(t+1) \sim P_B(\phi_{A_{1:N}(t+1),H(t+1)})$, $PC_{1:N} \in \{0, 1\}$
12:    Predict the minority choice of the next game $\hat{r}(t+1)$ from Eq. (5)
13:    Calculate the prediction accuracy from Eq. (6)
14: **end for**

Then the prediction accuracy $Acc(t)$ after observing the game t is defined as:

$$Acc(t) = \frac{\#(\hat{r}(t+1)=r(t+1))}{\#(\hat{r}(t+1)=r(t+1))+\#(\hat{r}(t+1)\neq r(t+1))} \tag{6}$$

where the $\#(.)$ is an incremental counter, initialized with 0 at $t = 1$, that increments by 1 each time its argument (.) is true.

4   Experimental Studies

4.1   Test on Artificial Data

To test the validity of our model we performed experiments on an artificial dataset generated according to the assumptions of the MG and used our model to learn the individual behavior and then make predictions. To make our experiment more realistic we assumed that some agents follow a random strategy because in the real-world there are always certain trends in the data that cannot be either modeled or captured. The random agent makes random choices between 0 and 1 following a uniform distribution. We further assume that some agents are adaptive agents and they might divert from their original strategy during the experiment. Every adaptive agent maintains its loosing probability for every history; if this loosing probability is greater than a threshold (0.6 in our experiments) then the agent changes its strategy (by random picking another strategy from the strategy space, e.g. [10]). Moreover, the change in strategy can only occur after every 200 games. We set $\alpha = 0.25$, $\beta = 0.25$, $m = 3$, $N = 31$ and use our framework of Algorithm 1 to make predictions on this data. Figure 2 shows the prediction accuracy and error bars for 1000 games with 31 agents, 10% random and 20% adaptive agents, along with the prediction accuracy for the case of 10% random and no adaptive agents. To obtain averaged results, the experiment is repeated 10 times. Strategies of adaptive agents change every 200 games resulting in a dip in prediction accuracy and after 1000 games the accuracy is around 67% with adaptive agents and 84% without adaptive agents.

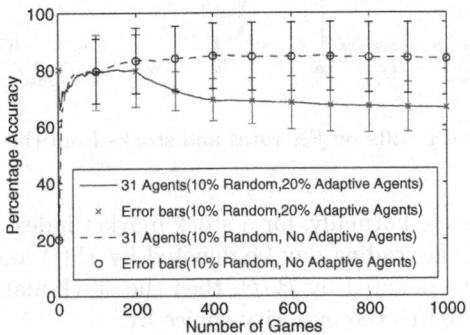

Fig. 2. Experimental results on artificial data

4.2 Experiments on Real-World Market Data

The minority game is related to many real-world complex scenarios [5,20,8] including financial markets. In the following experiments, stock market index data and the foreign exchange (FX) rate of U.S. Dollar (USD) against Renminbi (RMB) and Japanese Yen (JPY) against RMB are tested (Table 2). On the macro-level, the global behavior of these real-world market data appear random and unpredictable, but in our experiments we assume the global behavior of these markets as being generated according to the minority game and then use the PGM to infer local behaviors to predict possible future trends of these markets. This framework provides a new way of understanding the relationship between macro-trends and micro-trends; the combination of these individual behavior have the potential of generating very complex and apparently random global behaviors.

In our experiments we predict whether a stock market index or FX rate will rise or fall the next day. Thus, rise and fall are the two values corresponding to the two possible minority choices (choice-0 and choice-1) and one trading day

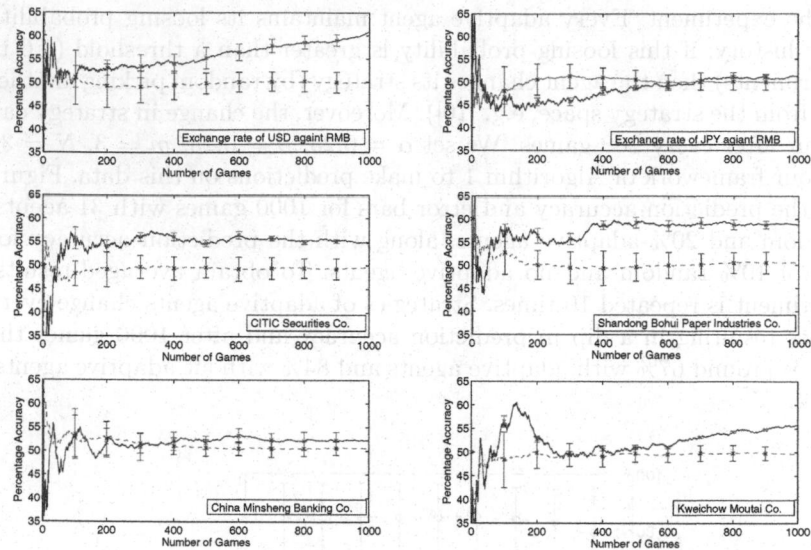

Fig. 3. Experimental results on FX rates and stocks from the Chinese market

corresponds to one game. Formally, for a stock market index (or a FX rate), let the opening price of that trading day be denoted by $P_o(t)$ and the closing price for that trading day be denoted by $P_c(t)$, then the stock market fluctuation for that day can be encoded to the minority choice by:

$$r(t) = \begin{cases} 1 & \text{If } P_c(t) \geq P_o(t) \\ 0 & \text{Otherwise} \end{cases} \tag{7}$$

The proposed PGM has two observed variables for the history and choices, respectively. To obtain the choices from the market data, we first need to give an appropriate number of agents. Given the market data, the number of agents making choice-1, denoted by $\#C_1$, is obtained according to:

$$\#C_1 \propto P_o(t) - P_c(t) \tag{8}$$

Based on the training data, the value of $\#C_1$ is scaled appropriately. The quantity $P_o(t) - P_c(t)$ is not always positive and is roughly centered around zero, so we shift it appropriately to make $\#C_1$ non-negative. This shifting factor is greater than the magnitude of its minimum or maximum value and it becomes the new mean. We then set N equal to twice the shifting factor. Then the choices are obtained by forming a string of $[N - \#C_1]$ zeros followed by $\#C_1$ ones.

4.3 Experimental Results

The parameters for the PGM used in our experiments are the following: the hyper-parameters α and β governing the prior Dircihlet distribution of agent

Table 2. Descriptions of the real market data

Name	Stock index	Test date	Acc after 1000 games	(# Agents) N
USD–RMB Exchange Rate	-	12/04/00-12/02/04	60.78%	21
JPY–RMB Exchange Rate	-	12/04/00-12/02/04	51.28%	29
CITIC Securities Co.	600030	08/11/08-01/09/13	54.82%	31
Shandong Bohui Paper Industrial Co.	600966	06/08/04-04/10/08	58.79%	31
China Minsheng Banking Co.	600016	12/19/00-10/21/04	52.21%	21
Kweichou Moutai Co.	600519	08/27/01-06/29/05	55.76%	31

distributions $\psi_{1:N}$ and of choice distributions $\psi_{1:N,1:K}$, respectively, were both set to 0.25 as from our multiple experiments we found that these values work well for most real markets, moreover, the history was assumed to be generated by agents having a memory length of 3 i.e. $m = 3$.

In order to get stable and unbiased results we ran the algorithm (described in Algorithm 1) 10 times for each data set and then calculated the corresponding error bars. We tested the proposed PGM on 6 datasets, 4 are from real sock markets and 2 datasets are for RMB exchange rate (shown in Table 2). All the data was downloadable by searching for each company by its stock index[1] and the exchange rate data is also available online[2]. Figure 3 shows the prediction accuracy defined in Eq. (6) for each of the data sets for 1000 games. It also shows another dashed curve for each dataset that corresponds to random prediction, based on discrete uniform distribution over $[0, 1]$, for the future trends of the real datasets. This dashed curve represents the base line as it corresponds to no learning and just randomly predicting $\hat{r}(t + 1) \sim U(0, 1)$. Therefore as long as our prediction accuracy can be above this curve we can consider that our proposed inference and prediction technique has learned some local behaviors that can predict the future trends with more than random accuracy.

The accuracy of USD-RMB exchange rate after 1000 games is around 60.78% which is high compared to the accuracy obtained for other real-world data sets. This may suggest that this exchange rate data has some prominent patterns and these findings are consistent with the findings of [12]. Conversely, for JPY the exchange rate against RMB for the same time period was analyzed and the result show that our algorithm does not perform much better than random prediction. In fact for the first 700 games the random prediction works better than our proposed algorithm, however after that our algorithm performs slightly better with an accuracy of 51.28% after 1000 games. For the other markets the accuracy is between 52% and 58%. The prediction accuracy on Shandong Bohui Paper Industrial Co. and CITIC Securities Co. datasets is above random prediction but for China Minsheng Banking Co. random prediction performs better for the first 350 games, after that the proposed algorithm performs better on average although the the error bars overlap suggesting that on some occasions the performance of the proposed technique is comparable to random prediction. The prediction accuracy for Kweichow Moutai increases to around 60% in the

---

[1] http://finance.yahoo.com/

[2] http://bbs.jjxj.org/thread-69632-1-7.html

first 130 games after which it drops indicating a change in the trend for this
market, then after game 280 the accuracy begins to increase again indicating
another major change in the trend, but the increase in accuracy suggests that
the new trend is similar to the trend previously observed by the model.

Table 2 provides the details of the data sets we used in experiments and the
number of agents that were set for each data set based on Eq. (8). For our
simulations we used C# along with Matlab on a 32-bit computer with 3GB of
RAM and two 2.93GHz processors. And for one iteration of Algorithm 1 it takes
around 1.1 sec if the number of agents N is 31. Therefore the total duration for
10 iterations of Algorith 1 for 1000 games is around 3 hours.

5　Conclusions and Future Work

In this paper we modeled the process of generation of the collective behavior of
the minority game with a PGM and showed that we can use the proposed PGM
to decompose the collective behavior by inferring individual agent behavior and
then use them to predict the future trends of real world market data. We first
performed experiments on artificial data to validate our model and then tested
it on the real-world market data. Although finding patterns of real-world market
data has always been a controversial topic as it violates the efficient-market hypothesis (EMH) [21], however, based on our empirical studies, we indeed found
statistical significant patterns by training the new proposed model on history
data. Especially for the USD-RMB exchange rate, we present quantitative evidence that there are some stronger patterns comparing to other FX rate and
stock index. Our future work will focus on applying the framework of Bayesian
learning to learn the hyper-parameters α and β instead of setting them by experimental evaluation and also to test our framework on more real-world data
but not limited to stock market indexes and FX rates.

Acknowledgements. This research is funded by the NSFC under the Grant
No. 61305047.

References

1. Rapoport, A., Chammah, A., Orwant, C.: Prisoner's Dilemma: A Study in Conict
   and Cooperation. University of Michigan Press, Ann Arbor, Michigan (1965)
2. Smith, V.L.: An experimental study of competitive market behavior. The Journal
   of Political Economy 70(2), 111–137 (1962)
3. Farmer, J.D., Foley, D.: The economy needs agent-based modelling. Nature 460,
   685–686 (2009)
4. Gode, D., Sunder, S.: Allocative efficiency of markets with zero-intelligence traders:
   Market as a partial substitute for individual rationality. Journal of Political Economy 101(1), 119–137 (1993)
5. Johnson, N.F., Jeffries, P., Hui, P.M.: Financial market complexity. Oxford
   University Press, Oxford (2003)

6. Qin, Z.: Market mechanism designs with heterogeneous trading agents. In: Proceedings of Fifth International Conference on Machine Learning and Applications (ICMLA 2006), pp. 69–74 (2006)
7. Qin, Z., Dong, Y., Wan, T.: Evolutionary models for agent-based complex behavior modeling. In: Yang, X.-S. (ed.) Artificial Intelligence, Evolutionary Computing and Metaheuristics. SCI, vol. 427, pp. 601–631. Springer, Heidelberg (2013)
8. Challet, D., Zhang, Y.C.: Emergence of cooperation and organization in an evolutionary game. Physica A 246, 407–418 (1997)
9. Arthur, W.B.: Bounded rationality and inductive behavior (the el farol problem). American Economic Review 84, 406–411 (1994)
10. Li, G., Ma, Y., Dong, Y., Qin, Z.: Behavior learning in minority games. In: Guttmann, C., Dignum, F., Georgeff, M. (eds.) CARE 2009 / 2010. LNCS, vol. 6066, pp. 125–136. Springer, Heidelberg (2011)
11. Challet, D., Marsili, M., Zhang, Y.C.: Minority Games: Interacting Agents in Financial Markets. Oxford University Press, USA (2004)
12. Ma, Y., Li, G., Dong, Y., Qin, Z.: Minority game data mining for stock market predictions. In: Cao, L., Bazzan, A.L.C., Gorodetsky, V., Mitkas, P.A., Weiss, G., Yu, P.S. (eds.) ADMI 2010. LNCS, vol. 5980, pp. 178–189. Springer, Heidelberg (2010)
13. Du, Y., Dong, Y., Qin, Z., Wan, T.: Exporing market behaviors with evolutionary mixed-game learning model. In: Jędrzejowicz, P., Nguyen, N.T., Hoang, K. (eds.) ICCCI 2011, Part I. LNCS, vol. 6922, pp. 244–253. Springer, Heidelberg (2011)
14. Bishop, C.M.: Pattern Recognition and Machine Learning. Springer, New York (2006)
15. Challet, D., Marsili, M., Zhang, Y.C.: Stylized facts of financial markets and market crashes in minority games. Physica A 294, 514–524 (2001)
16. Challet, D., Marsili, M., Zecchina, R.: Statistical mechanics of systems with heterogeneous agents: Minority games. Phys. Rev. Lett. 84, 1824–1827 (2000)
17. Koller, D., Friedman, N.: Probabilistic Graphical Models: Principles and Techniques. MIT Press (2009)
18. Winn, J., Bishop, C.M.: Variational message passing. J. Mach. Learn. Res. 6, 661–694 (2005)
19. Minka, T., Winn, J., Guiver, J., Knowles, D.: Infer.NET 2.4, Microsoft Research Cambridge (2010), http://research.microsoft.com/infernet
20. Lo, T.S., Hui, P.M., Johnson, N.F.: Theory of the evolutionary minority game. Phys. Rev. E 62, 4393–4396 (2000)
21. Fama, E.: Efficient capital markets: A review of theory and empirical work. Journal of Finance 25, 383–417 (1970)

Finding Better Topics: Features, Priors and Constraints

Xiaona Wu, Jia Zeng, Jianfeng Yan, and Xiaosheng Liu

School of Computer Science and Technology, Soochow University,
Suzhou 215006, China
zengja@gmail.com

Abstract. Latent Dirichlet allocation (LDA) is a popular probabilistic topic modeling paradigm. In practice, LDA users usually face two problems. First, the common and stop words tend to occupy all topics leading to bad topic interpretability. Second, there is little guidance on how to improve the low-dimensional topic features for a better clustering or classification performance. To find better topics, we re-examine LDA from three perspectives: continuous features, asymmetric Dirichlet priors and sparseness constraints, using variants of belief propagation (BP) inference algorithms. We show that continuous features can remove the common and stop words from topics effectively. Asymmetric Dirichlet priors have substantial advantages over symmetric priors. Sparseness constraints do not improve the overall performance very much.

Keywords: Latent Dirichlet allocation, belief propagation, continuous features, asymmetric Dirichlet priors, sparseness constraints.

1 Introduction

Latent Dirichlet allocation (LDA) [1] is a widely-used probabilistic topic modeling paradigm, which has found many important applications in natural language processing and computer vision areas. LDA represents documents as mixtures over latent topics, where each topic is a distribution over a fixed vocabulary. Using approximate inference techniques like variational Bayes (VB) [1], Gibbs sampling (GS) [2] or belief propagation (BP) [3], LDA automatically learns the topic-word and document-topic distributions from a large collection of documents. In practice, LDA users usually encounter two problems. First, the common and stop words tend to occupy all topics. For example, if we use LDA to extract topics from a machine learning corpus like NIPS, we find that the common words "learning" and "model" dominate (having very high likelihood) almost all topic-word distributions. This phenomenon makes the interpretability of topics undesirable [4]. Second, there is relatively little guidance on how to improve the lower-dimensional topic features for a better retrieval, clustering and classification performance. Therefore, we explore LDA from three perspectives: continuous features, asymmetric Dirichlet priors and sparseness constraints to find better topics.

V.S. Tseng et al. (Eds.): PAKDD 2014, Part II, LNAI 8444, pp. 296–310, 2014.

LDA has long been used for discrete features such as word tokens and counts. Continuous features or term weighting schemes have been rarely discussed such as term frequency-inverse document frequency (TF-IDF) [5] and LTC [6]. One major concern is that LDA cannot generate continuous observations in its probabilistic modeling process. So, in practice users have to manually remove stop words having little contribution to the meaning of the text [7]. But, removing common words requires contextual knowledge of the entire corpus, which is often a big challenge to users without prior knowledge. Recently, continuous features for LDA have gained intensive research interests. A simple term-frequency feature scheme [8] has been used for tagged document within the framework of LDA. Point-wise mutual information (PMI) features [9] have been incorporated into the GS inference algorithm referred to as pmiGS. The PMI feature gives common and stop words some lower weights. Then, pmiGS infers topic-word distributions from weighted word counts. The results show that the PMI feature not only lowers the likelihood of common and stop words in the topic-word distribution, but also gains a no-trivial improvement in cross-language retrieval tasks. This line of research inspires us to consider continuous features for LDA to improve the topic interpretability.

Most LDA algorithms [2, 3, 7] consider fixed symmetric Dirichlet priors over document-topic and topic-word distributions for simplicity. Although it is possible to automatically learn Dirichlet hyperparameters from training data according to the maxumum-likelihood criterion [10], the extensive empirical studies [11] confirm that the inferred symmetric priors do not significantly improve the topic modeling performance than the fixed ones. However, asymmetric Dirichlet priors over document-topic and symmetric Dirichlet priors over topic-word distributions have substantial advantages on removing the common words and choosing the number of topics [12]. The asymmetric prior over document-topic distribution can guide common or stop words to be grouped into a few topics with higher likelihoods because these words often occupy the larger proportion of each document. So, asymmetric priors are also effective in finding better topics.

If we can control the sparseness of document-topic and topic-word distributions, we can possibly control the quality and interpretability of lower-dimensional topic features. Sparse topic coding (STC) [13] can directly control the sparsity of the inferred representations by relaxing the normalization constraint, which can be integrated with any convex loss function. STC identifies sparse topic meanings of words and improves time efficiency and classification accuracy. Also, sparse coding can be directly combined with LDA's extensions [14] for computer vision applications. In sparse coding, each document or word only has a few salient topical meanings or senses. Sparse distributions carry salient information for a better interpretability, so that the low-dimensional sparse topic features may be more distinguishable. Therefore, we will consider adding sparse constrains [15] on LDA's document-topic and topic-word distributions.

Although continuous features, asymmetric priors and sparseness constraints for LDA have been studied either by GS [2] or by VB [1] inference algorithms, we re-examine these three perspectives within the novel BP inference framework [3], which is very competitive in both speed and accuracy. As a result, we incoporate continuous features, asymmetric Dirichlet priors and sparseness constraints into BP algorithms to find better topics than traditional GS and VB algorithms. Besides, most of previous studies focus only on one of three aspects, and lack a comprehensive comparison in terms of generalization performance, document clustering/classification and topic interpretability. Here, we compare these three aspects on different data sets, and provide evidence on which one can produce high-quality topics.

2   Background

We begin by reviewing batch BP algorithms for learning collapsed LDA [3,16,17]. The probabilistic topic modeling task can be interpreted as a labeling problem, in which the objective is to assign a set of thematic topic labels, $\mathbf{z}_{W \times D} = \{z_{w,d}^k\}$, to explain the observed elements in document-word matrix, $\mathbf{x}_{W \times D} = \{x_{w,d}\}$. The notations $1 \leq w \leq W$ and $1 \leq d \leq D$ are the word index in vocabulary and the document index in corpus. The notation $1 \leq k \leq K$ is the topic index. The nonzero element $x_{w,d} \neq 0$ denotes the number of word counts at the index $\{w, d\}$. For each word token $x_{w,d,i} = \{0,1\}, 1 \leq i \leq x_{w,d}$, there is a topic label $z_{w,d,i}^k = \{0,1\}, \sum_{k=1}^K z_{w,d,i}^k = 1, 1 \leq i \leq x_{w,d}$, so that the soft topic label for the word index $\{w, d\}$ is $z_{w,d}^k = \sum_{i=1}^{x_{w,d}} z_{w,d,i}^k / x_{w,d}$.

The collapsed LDA [18] has joint probability $p(\mathbf{x}, \mathbf{z} | \alpha v_k, \beta u_w)$, where the Dirichlet hyperparameters $\{\alpha v_k, \beta u_w\}, \sum_k v_k = 1, \sum_w u_w = 1, \alpha, \beta > 0$. In practice, we may use the fixed symmetric hyperparameters $\{v_k = 1/K, u_w = 1/W\}$ and the concentration parameters $\{\alpha, \beta\}$ are provided by users for simplicity [2]. To maximize the joint probability in terms of $\mathbf{z}$, the BP algorithm [3] computes the posterior probability, $\mu_{w,d}(k) = p(z_{w,d,i}^k = 1 | \mathbf{z}_{-(w,d,i)}^k, \mathbf{x})$, called *message*, which can be normalized by local computation, i.e., $\sum_{k=1}^K \mu_{w,d}(k) = 1$. The approximate message update equation is

$$\mu_{w,d}(k) \propto \frac{[\hat{\theta}_{-w,d}(k) + \alpha v_k] \times [\hat{\phi}_{w,-d}(k) + \beta u_w]}{[\sum_w x_{w,d} + \alpha] \times [\hat{\phi}_{-(w,d)}(k) + \beta]}, \tag{1}$$

where the *sufficient statistics* for LDA model are

$$\hat{\theta}_{-w,d}(k) = \sum_{-w} x_{w,d} \mu_{w,d}(k), \tag{2}$$

$$\hat{\phi}_{w,-d}(k) = \sum_{-d} x_{w,d} \mu_{w,d}(k), \tag{3}$$

where $-w$ and $-d$ denote all word indices except w and all document indices except d. Obviously, the message update equation (1) depends on all other neighboring messages $\boldsymbol{\mu}_{-(w,d)}$ excluding the current message $\mu_{w,d}$. Two multinomial parameters, the document-topic distribution θ and the topic-word distribution

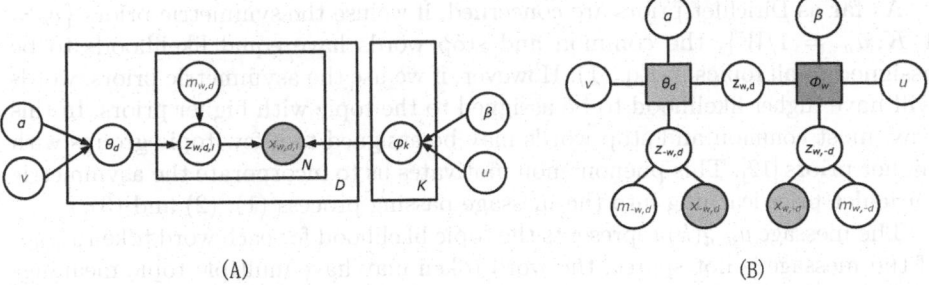

Fig. 1. (A)Generative graphical representation of LDA based on continuous features, asymmetric Dirichlet priors and sparseness constraints, (B)Factor graph and message passing

ϕ, can be calculated from *sufficient statistics* $\hat{\theta}_d(k)$ and $\hat{\phi}_w(k)$ by normalization. Message passing process will iterate Eqs. (1), (2) and (3) until all messages converge to a local stationary point [3].

As mentioned in Section 1, LDA users often use the document-topic distribution in (2) as the lower-dimensional features for document retrieval, clustering and classification. The word-topic distribution in (3) is used to find the hot words in each topic. Usually, users will inspect the hot words with higher likelihood in each topic to understand the topic's semantic meaning. Observing (2) and (3), we find that these two distributions are determined by three factors:

1. The features or observations: the word counts $x_{w,d}$.
2. The Dirichlet priors or hyperparameters: the base vectors $\{v_k, u_w\}$ and the concentration parameters $\{\alpha, \beta\}$ in Eq. (1).
3. The message: the K-tuple vector $\mu_{w,d}(k)$ for the topic likelihood at index $\{w, d\}$.

In this paper, we will regulate these three factors to find better topics including document-topic (2) and topic-word distributions (3).

3   Finding Better Topics

The major reason that the common and stop words occupy almost all topics is that LDA uses word counts as features. The bigger the word counts, the higher the influence to the topic distributions. In Eqs. (2) and (3), the normalized message $\mu_{w,d}(k)$ is multiplied by the nonzero word count $x_{w,d}$. Thus, $x_{w,d}$ can be regarded as the weight of $\mu_{w,d}(k)$ in estimating document-topic and topic-word distributions. In this way, the topics may be dominated by those high-frequent common and stop words. We see that the bigger word count $x_{w,d}$ corresponds to the greater influence of the estimated distributions in (2) and (3). This phenomenon motivates us to use the continuous features such as TF-IDF or LTC to lower the weights of common and stop words during message passing.

As far as Dirichlet priors are concerned, if we use the symmetric priors $\{v_k = 1/K, u_w = 1/W\}$, the common and stop words have equal likelihoods to be assigned to all topics in Eq. (1). However, if we use the asymmetric priors, words will have higher likelihood to be assigned to the topic with higher priors. In this way, most common and stop words may be assigned to a few topic groups with higher priors [12]. This phenomenon motivates us to incorporate the asymmetric Dirichlet prior learning into the message passing process (1), (2) and (3).

The message $\mu_{w,d}(k)$ represents the topic likelihood for each word token $x_{w,d,i}$. If the message is not sparse, the word token may have multiple topic meanings leading to unclear explanations. So, we encourage passing those sparse messages by adding a weight proportional to the sparseness of the message. This weighted message passing strategy can strengthen the sparseness of document-topic and topic-word distributions in (2) and (3). According to [13] and [14], the sparseness will make the lower-dimensional topic features more distinguishable for clustering or classification purposes. This motivates us to add sparseness constraints on messages during their passing process.

Fig. 1(A) shows the continuous features, asymmetric Dirichlet priors and sparseness constraints denoted by red colors in the generative graphical representation of LDA. The asymmetric Dirichlet priors are divided into the connection parameters $\{\alpha, \beta\}$ and the base measure vectors $\{\mathbf{v}, \mathbf{u}\}$, and $m_{w,d}$ is the sparseness constraints for the message $\mu_{w,d}(k) \sim z_{w,d}^k$. Note that if $x_{w,d} = \sum_i x_{w,d,i}$ becomes continuous observations like TF-IDF, the generative model in Fig. 1(A) cannot generate such observations. However, the factor graph representation of the collapsed LDA [3] shows that it is possible to describe the continuous features using the undirected factor graph, which does not need to encode the generative relations between variables. In this way, we may think that the factor graph is a close approximation to LDA [3]. Fig. 1(B) shows the factor graph representation and the message passing process based on continuous features, asymmetric Dirichlet priors and sparseness constraints. We see that the message $\mu_{w,d}(k) \sim z_{w,d}^k$ can be inferred by its neighboring messages including $\{(\mathbf{x}_{-w,d}, \mathbf{z}_{-w,d}^k, \mathbf{m}_{-w,d}), \alpha v_k\}$ and $\{(\mathbf{x}_{w,-d}, \mathbf{z}_{w,-d}^k, \mathbf{m}_{w,-d}), \beta u_w\}$ via factor nodes θ_d and ϕ_w, respectively. We group the variables $(\mathbf{x}_{w,d}, \mathbf{z}_{w,d}^k, \mathbf{m}_{w,d})$ together because they work together to influence the neighboring messages according to (1). From the message passing over factor graphs, we can derive the similar message update equation to (1) that considers continuous features, asymmetric priors and sparseness constraints within the unified BP framework.

3.1   Continuous Features

In linguistics, the high frequent stop words like "the, and, of" which occur in most of the documents do not contribute to the topic formation. To avoid stop words dominating every topic, we have to remove stop words before running LDA according to a corpus-specific stop word list. However, even if the stop words have been removed, there still are many common words such as "model, learning, data" in the machine learning corpus. In such cases, we may use the

continuous features such as TF-IDF [5] and LTC [6] that give the lower weights
to the "common word" messages in (1). Let $x_{w,d}/\sum_w x_{w,d}$ be the frequency of
word w in document d, and $\sum_d x_{w,d}$ be the total number of times that the word
w occurs in all documents. We get the continuous TF-IDF feature as

$$x_{w,d}^{tfidf} = \frac{x_{w,d}}{\sum_w x_{w,d}} \times \log\left(\frac{D}{\sum_d x_{w,d}}\right),$$

(4)

and the LTC feature as

$$x_{w,d}^{ltc} = \frac{\log(\frac{x_{w,d}}{\sum_w x_{w,d}} + 1) \times \log\left(\frac{D}{\sum_d x_{w,d}}\right)}{\sqrt{\sum_{d=1}^{D}\left[\log(\frac{x_{w,d}}{\sum_w x_{w,d}} + 1) \times \log\left(\frac{D}{\sum_d x_{w,d}}\right)\right]^2}}.$$

(5)

The difference between (5) and (4) is that (5) uses the logarithm of word frequency and is normalized by the geometric mean of the numerator. This normalization makes LTC features more distinguishable than TF-IDF features.

We simply replace the discrete word count feature $x_{w,d}$ by the continuous
features $x_{w,d}^{tfidf}$ and $x_{w,d}^{ltc}$ in Eqs. (2), (3) and (1). Without loss of generality,
we focus on LTC features for topic modeling. We refer to the message passing
algorithms for LTC feature as ltcBP. Obviously in (2) and (3), the higher TF-IDF
and LTC values will have the bigger influence to the topic formation. Generally,
the stop and common words have lower TF-IDF and LTC weights, so that they
will be automatically removed from hot word list in each topic during the message
passing process.

3.2   Asymmetric Priors

There are several approaches to learn Dirichlet priors from training data. Here,
we choose to place Gamma priors on the hyperparameters $\alpha \sim G[C, S]$, where C
and S are shape and scale parameters of Gamma distribution. Generally, these
parameters are fixed by users during learning Dirichlet priors. We adopt the
improved method of Minka's fixed point iteration [10,12]. However, this method
is based on discrete counts on topic labels rather than messages in BP (1). To
solve this problem, we sample the topic label $z_{w,d,i}^k$ for each word token $x_{w,d,i}$
from the conditional probability $\mu_{w,d}(k)$. From the sampled $[z_{w,d,i}^k = 1]$, we get
two topic count matrices

$$\gamma_d(k) = \sum_{w=1}^{W}\sum_{i=1}^{x_{w,d}} [z_{w,d,i}^k = 1],$$

(6)

$$\eta_w(k) = \sum_{d=1}^{D}\sum_{i=1}^{x_{w,d}} [z_{w,d,i}^k = 1].$$

(7)

$$
\begin{array}{ll}
\textbf{input} : \mathbf{x}_{W \times D}, \mathbf{K}, \mathbf{T}, \alpha\mathbf{v}, \beta\mathbf{u}, \mathbf{C}, \mathbf{S}. \\
\textbf{output} : \theta_d, \phi_w. \\
1 \quad \mu^1_{w,d}(k) \longleftarrow \text{initialization and normalization;} \\
2 \quad \hat{\theta}^1_{-w,d}(k) \leftarrow \sum_{-w} x_{w,d}\mu^1_{w,d}(k); \\
3 \quad \hat{\phi}^1_{w,-d}(k) \leftarrow \sum_{-d} x_{w,d}\mu^1_{w,d}(k); \\
4 \quad \alpha \leftarrow 50, v_k \leftarrow 50/K, \beta u_w \leftarrow 0.01, C \leftarrow 1.001, S \leftarrow 1; \\
5 \quad \textbf{for } t \leftarrow 1 \textbf{ to } T \textbf{ do} \\
6 \qquad \mu^{t+1}_{w,d}(k) \propto \frac{[\hat{\theta}^t_{-w,d}(k) + \alpha v^t_k] \times [\hat{\phi}^t_{w,-d}(k) + \beta u^t_w]}{[\sum_w x_{w,d} + \alpha^t] \times [\hat{\phi}^t_{-(w,d)}(k) + \beta^t]}; \\
7 \qquad \hat{\theta}^{t+1}_{-w,d}(k) \leftarrow \sum_{-w} x_{w,d}\mu^{t+1}_{w,d}(k); \\
8 \qquad \hat{\phi}^{t+1}_{w,-d}(k) \leftarrow \sum_{-d} x_{w,d}\mu^{t+1}_{w,d}(k); \\
9 \qquad \text{sampling } z \text{ from } \mu^{t+1}_{w,d}(k); \\
10 \qquad \gamma_d(k) \leftarrow \sum_{w=1}^{W} \sum_{i=1}^{x_{w,d}} z^k_{w,d,i}; \\
11 \qquad \eta_w(k) \leftarrow \sum_{d=1}^{D} \sum_{i=1}^{x_{w,d}} z^k_{w,d,i}; \\
12 \qquad \alpha v^{t+1}_k \leftarrow \alpha v^t_k \frac{\sum_{n=1}^{b1} I_{n,k} \sum_{f=1}^{n} \frac{1}{f-1+\alpha v^t_k} + C}{\sum_{n=1}^{b2} I_n \sum_{f=1}^{n} \frac{1}{f-1+\alpha^t} - \frac{1}{S}}; \\
13 \qquad \text{using } \eta_w(k) \text{ to learn symmetric } \beta^{t+1}; \\
14 \qquad \beta u^{t+1}_w \leftarrow \beta^{t+1}/W; \\
15 \quad \textbf{end} \\
16 \quad \theta_d(k) \leftarrow \frac{\theta_d(k) + \alpha v_k}{\sum_k \theta_d(k) + \alpha}; \phi_w(k) \leftarrow \frac{\phi_w(k) + \beta u_w}{\sum_w \phi_w(k) + \beta}.
\end{array}
$$

Fig. 2. The asBP algorithm for LDA

Based on these two count matrices, we can directly use the Minka's fixed point iteration

$$
\alpha v_k \leftarrow \alpha v_k \frac{\sum_{n=1}^{b1} I_{n,k} \sum_{f=1}^{n} \frac{1}{f-1+\alpha v_k} + C}{\sum_{n=1}^{b2} I_n \sum_{f=1}^{n} \frac{1}{f-1+\alpha} - \frac{1}{S}}, \tag{8}
$$

where

$$
I_{n,k} = \sum_{d=1}^{D} \delta(\gamma_d(k) - n), \tag{9}
$$

$$
I_n = \sum_{d=1}^{D} \delta(len(d) - n), \tag{10}
$$

where $b1 = \max_d \gamma_d(k), b2 = \max_d len(d)$, and $len(d)$ is the total number of observations in document d, and n and f are positive integers. The value αv_k acts as an initial set for the topic k in all documents. $I_{n,k}$ is the number of documents in which the topic k has been seen exactly n times. I_n is the number of documents that contain a total of n observations. $I_n(\cdot) = \sum_{k=1}^{K} I_{n,k}$ is the total number of documents whose topics $(1, \ldots, K)$ has been seen exactly n times. For the symmetric Dirichlet priors, the base measure is fixed as $v_k = 1/K$ and the concentration parameter α is updated as

$$
\alpha v_k \leftarrow \frac{\alpha}{K} \times \frac{\sum_{n=1}^{b3} I_n(\cdot) \sum_{f=1}^{n} \frac{1}{f-1+\alpha/K}}{\sum_{n=1}^{b2} I_n \sum_{f=1}^{n} \frac{1}{f-1+\alpha}}, \tag{11}
$$

where $b3 = max_{d,k}\gamma_d(k)$. It is the same way to learn asymmetric or symmetric βu_w according to the count matrix $\eta_w(k)$.

Symmetric and asymmetric Dirichlet priors over $\{\theta, \phi\}$ play different roles in topic modeling. Similar to [12], we implement an asymmetric prior over θ and a symmetric prior over ϕ, which is referred to as the asBP algorithm. In practice, this implementation performs the best than other combinations of priors [12]. Fig. 2 summaries the asBP algorithm for learning LDA, where T is the total number of learning iterations. The asymmetric prior αv_k can be learned by Eqs. (9), (10), (8). At the first $t \leq 100$ iterations, asBP is the same with the batch BP which updates and normalizes all messages for all topics. For $t > 100$, we learn the asymmetric prior αv_k and the symmetric prior βu_w every 20 iterations.

3.3 Sparseness Constraints

In addition to the continuous features and asymmetric Dirichlet priors, sparseness constraints over messages also has an effect on the topic interpretability. In this paper, we adopt a sparseness measure based on the L_1 norm and the L_2 norm [15],

$$m_{w,d} = \frac{\sqrt{K} - (\sum_k |\mu_{w,d}(k)|)/\sqrt{\sum_k [\mu_{w,d}(k)]^2}}{\sqrt{K} - 1}, \tag{12}$$

where K is the number of topics and the dimensionality of $\mu_{w,d}(k)$. The quantity $m_{w,d}$ is the sparseness of $\mu_{w,d}$. Usually, the messages of stop and common words have relatively lower sparseness because they often occupy many topics for a lower interpretability. For example, when the number of topics is 10 in CORA data set, the meaningful words such as "reinforcement", "Bayesian" have relatively higher sparseness values 0.9999 and 0.9615 than 0.8663 and 0.8417 of the common words such as "learning" and "model". Our intuition is that we need to encourage passing those messages with higher sparseness values, so we use the sparseness value (12) as the weight of message during message update (1). More specifically, we simply use the weighted sum $m_{w,d}x_{w,d}\mu_{w,d}(k)$ in Eqs. (2), (3) and (1). Such a weighted message passing strategy will encourage sparse messages with higher weights in topic formation. We refer this message passing algorithm as conBP. If all sparseness constraints $m_{w,d} = 1$, conBP will become the standard BP algorithm for learning LDA [3].

4 Experiments

In this section, we evaluate the effectiveness of the proposed ltcBP, asBP, and conBP algorithms on six publicly available data sets. Table 1 summarizes the statistics of six data sets, where D is the total number of documents, $\overline{N}_d$ the average document length, N the total number of tokens, W the vocabulary size, and "stop" indicates whether there are stop words. All algorithms are evaluated by five performance metrics. Lower perplexity [3,11] indicates better generalization performance. The lower-dimensional document-topic distributions can be fed into standard SVM classifiers for document classification. The higher classification accuracy implies the more distinguishable ability of the lower-dimensional

Table 1. Data set statistics

Data sets	D	$\overline{N}_d$	N	W	STOP
CORA	2410	57	136394	2961	no
WEK	2785	127	352647	7061	no
NIPS	1740	1323	2301375	13649	no
20NEWS	2000	200	399669	36863	no
NIPS (STOP)	1740	2939	5114634	70629	yes
20NEWS (STOP)	2000	372	743180	37370	yes

Algorithm	NIPS(STOP)	20NEWS(STOP)
pmiGS	training set the and test performance error class classification on network neural networks the recurrent control output to systems of learning the in a on reinforcement task learn to control the of in cells cell and cortex direction neurons cortical	you jpeg if file gif image it from on this comp windows edu ibm os sys misc ms mac hardware space gov nasa sci at au access digex jpl on edu rutgers christian not are religion may all who mit
asGS	data and error prediction set training model validation regression selection the network input output networks neural a i is to state a and learning q policy reinforcement the value for and in model of cells cell j neurons system c the of in a to is by are this with	jpeg image you file images color files gif format comp graphics x video sys mac monitor hardware card screen space sci doc launch shuttle nasa mission toronto henry orbit rutgers christian edu god he of religion geneva jesus church the is to a of in and that it this
BP	the of a and in for to is learning r with generalization the network of neural a input networks to output is the a of and to learning state in is for q s reinforcement the of and in to a model cells by is	the image is it jpeg to graphics of a from windows comp os ms edu i to the misc a the space nasa gov to and of sci s on rutgers edu of the christian in god to that is
ltcBP	classifier classifiers classification nearest classes neighbor classify class classified classifying associative memory capacity hopfield memories neuron stored neurons recall retrieval robot controller arm control trajectory plant motor trajectories controllers robotics cortex receptive orientation cortical cells visual selectivity tuning dominance spatial	graphics x comp file windows code image program files motif windows os ms comp de dos nl ui apps win nasa space jpl gov elroy sci alaska launch orbit moon god jesus christians faith bible his christ he paul religion
asBP	classification training class classifier the set data performance classes classifiers network units hidden input layer output the networks unit training learning state q action s value reinforcement policy optimal time visual motion cells direction field spatial model receptive orientation response the of a and is in i for to we	image jpeg file graphics images color files gif format bit windows comp os ms dos x microsoft unix window program space nasa sci launch shuttle venus gov station mission orbit rutgers christian god geneva religion athos church jesus soc may the of in to and a on for was by
conBP	the of and classification training class classifier to for in the network units of to input hidden output layer unit the of and control to in model is motor trajectory the to learning and is robot s goal environment task	image jpeg file you it from graphics images the files windows comp os ms edu i misc cs for dos edu gov com nasa apr stratus usenet indiana ucs jpl rutgers edu christian of in that god we religion i

Fig. 3. Top ten words of four topics when $K = 50$. Blue and black colors denote stop and common words, respectively. Red color denotes meaningful key words in each topic.

topic features. We can also use the document-topic distribution as the soft document clustering results. Normalized mutual information (NMI) [19] evaluates the performance of clustering by comparing predicted clusters with true class labels of a corpus. When displaying topics to users, each topic is generally represented as a list of the most probable words (for example, top ten hot words in each topic). Topic "coherence" [20] evaluates the topic quality. Point-wise mutual information (PMI) [21] is very similar to coherence. The higher coherence and PMI values correspond to the better topic interpretability.

For a fair comparison, we implement all algorithms using the MATLAB C/C++ MEX platform publicly available at [22] and run experiments on the Sun fire X4270 M2 server. The initial hyperparameters is set as $\alpha = 50/K, \beta = 0.01$, where K is the number of topics. We use the same $T = 1000$ training iterations for all algorithms. We compare our algorithms with the four benchmark topic modeling algorithms such as BP [3], asGS [12], pmiGS [9] and STC [13]. Since STC outputs the word-topic distribution containing negative values, we only compare our algorithms with STC in terms of document clustering and classification tasks.

Fig. 3 shows the top ten words of four topics when $K = 50$. The meaningful key words of each topic are highlighted with the red color, and the stop and

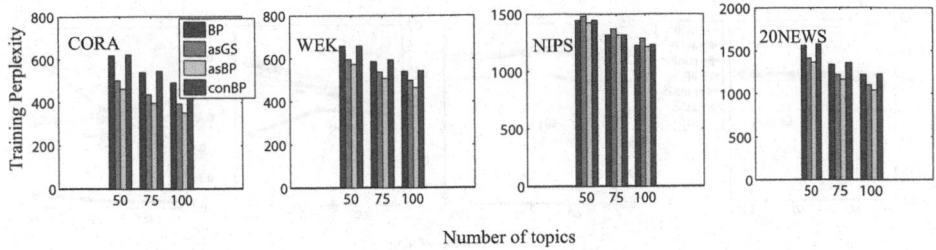

Fig. 4. Training perplexity as a function of the number of topics

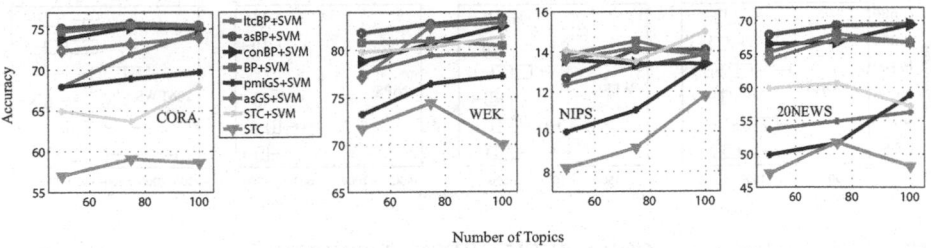

Fig. 5. Document classification accuracy as a function of the number of topics

common words are highlighted with blue and black colors, respectively. We use the subjective "word intrusion" [4] to evaluate the topic interpretability, i.e., the number of conflict stop and common words in each topic. It is easy to see that ltcBP performs the best to remove almost all stop and common words in each topic, which demonstrates the effectiveness of the continuous LTC features in topic modeling. Note that asBP can also remove the most stop words by clustering them such as "the of a and is in i for we" in a separate topic on both NIPS (STOP) and 20NEWS (STOP). This result shows that the asymmetric prior has an effect on allocating the most frequent stop words to a specific topic with a higher prior value v_k. But asBP still has difficulty in handling some common words like "learning" and "model". Note that asGS can also cluster stop words in one topic, but some topics contain more common words than those of asBP. BP performs the worst since its extracted topics are influenced by those high-frequent stop and common words. Although pmiGS uses the continuous PMI feature in topic modeling, it performs significantly worse than ltcBP because it cannot remove most stop and common words in each topic. The underlying reason is that LTC features are more effective in lowering the weights of stop and common words in topic modeling. We see that using sparseness constraints cannot effectively remove stop and common words from each topic. The conBP is only slightly better than BP, but significantly worse than both asBP and ltcBP. So, to find more interpretable topic-word distributions, the continuous features and asymmetric priors provide the best performance.

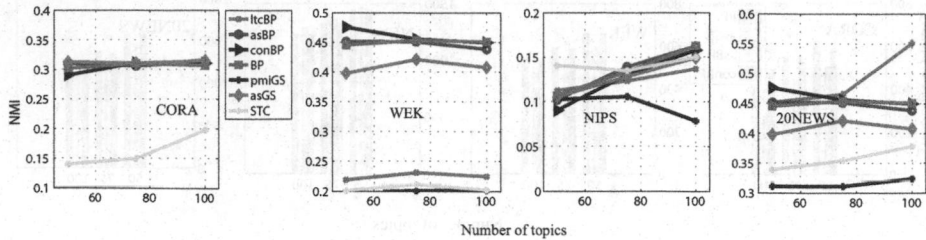

Fig. 6. The NMI as a function of the number of topics

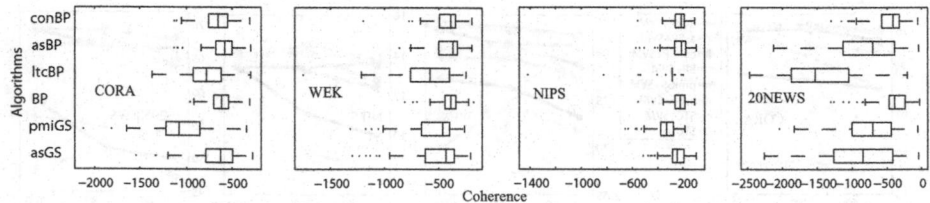

Fig. 7. The coherence of CORA, WEK, NIPS and 20NEWS datasets when $K = 100$

Fig. 4 shows the training perplexity as a function of the number of topics on CORA, WEK, NIPS and 20NEWS for $K = \{50, 75, 100\}$. Note that ltcBP, pmiGS and STC do not describe how to generate word tokens, so that they cannot be measured by the perplexity metric. Except on NIPS, asGS yields a lower perplexity value than BP. We see that conBP has almost the same perplexity of BP, which implies that sparseness constraints do not improve the likelihood of word generation. On all data sets, we see that the training perplexity of asBP is the lowest, showing the highest topic modeling accuracy. The result shows that learning asymmetric Dirichlet prior of αv_k and the symmetric prior βu_w can improve the topic modeling accuracy. The training perplexity has a smaller difference on the NIPS data set. One possible reason is that each document in NIPS contains more word tokens, so that the prior has a smaller impact on the message update (1). To summarize, learning an asymmetric Dirichlet prior over the document-topic distributions and an symmetric Dirichlet prior over the topic-word distributions still has substantial advantages on improving the document-topi and topic-word distributions to generate word tokens.

Fig. 5 shows the document classification accuracy as a function of the number of topics on CORA, WEK, NIPS and 20NEWS for $K = \{50, 75, 100\}$. In our experiments, we randomly divide each data set into half as training and test sets. Then, we use the standard linear SVM classifier to classify the lower-dimensional document-topic features produced by the topic modeling algorithms. As far as STC is concerned, it can directly output the class predictions. Also, we can use STC to generate lower-dimensional topic features and use SVM to do the classification.

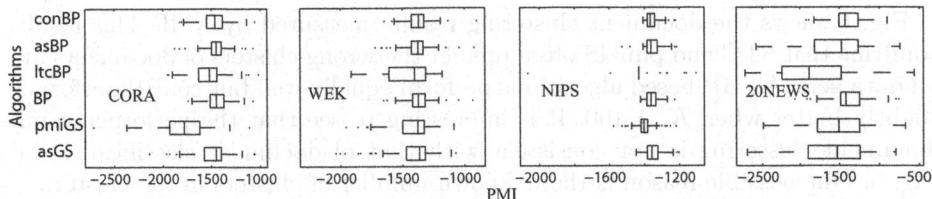

Fig. 8. The PMI of CORA, WEK, NIPS and 20NEWS datasets when $K = 100$

Table 2. Performance on CORA, WEK, NIPS and 20NEWS datasets when $K = 100$

Datasets	CORA					WEK				
	Perplexity	Accuracy	NMI	PMI	Coherence	Perplexity	Accuracy	NMI	PMI	Coherence
ltcBP	–	74.58	**0.3168**	−1536.1	−781.42	–	79.51	0.2251	−1458.8	−620.45
asBP	**352.29**	**75.42**	0.3150	−1447.6	**−609.72**	**462.66**	83.32	0.2469	−1380.6	−426.72
conBP	496.16	75.00	0.3107	−1466.7	−673.98	541.94	**84.46**	0.2347	−1381.4	−428.86
BP	491.31	75.33	0.3069	**−1444.5**	−631.86	539.41	80.45	0.2297	**−1370.9**	**−403.68**
pmiGS	–	69.68	0.2161	−1788.9	−1075.10	–	77.21	0.2015	−1418.9	−553.56
asGS	393.79	73.92	0.2852	−1485.5	−677.81	497.56	82.82	**0.2532**	−1425.0	−526.96
STC	–	67.94	0.1981	–	–	–	81.38	0.2027	–	–

Datasets	NIPS					20NEWS				
	Perplexity	Accuracy	NMI	PMI	Coherence	Perplexity	Accuracy	NMI	PMI	Coherence
ltcBP	–	13.84	0.1365	**−1254.1**	−415.10	–	56.30	**0.5511**	−1837.3	−1461.1
asBP	**1215.72**	14.07	**0.1632**	−1357.2	−227.49	**1039.56**	**69.50**	0.4386	−1549.2	−793.0
conBP	1230.30	13.38	0.1577	−1357.5	**−225.55**	1222.99	69.40	0.4498	−1386.9	−474.63
BP	1226.43	13.73	0.1626	−1358.3	−226.54	1219.04	66.70	0.4511	**−1374.8**	**−411.8**
pmiGS	–	13.38	0.0785	−1389.7	−279.01	–	58.90	0.3242	−1518.1	−772.23
asGS	1288.02	13.84	0.1489	−1349.9	−249.15	1096.89	66.80	0.4079	−1604.7	−906.59
STC	–	**14.99**	0.1449	–	–	–	57.20	0.3785	–	–

We see that BP and asBP performs comparably, and outperform other methods. Their classification performance is relatively stable as the number of topics changes. Although ltcBP can effectively remove stop and common words, it does not perform the best in document classification. On possible reason is that the distributions of stop and common words also provide useful information for classification. Surprisingly, STC cannot predict the class label very well when compared with other methods. But STC works well on the lower-dimensional topic features. As we see, conBP works slightly better than BP on classification when $K = 100$, which implies that sparseness constraints do not provide useful information in this task. Overall, asBP performs the best in document classification. For example, asBP outperforms BP and asGS by around 0.6% and 3.7% on CORA for $K = 50$, and by around 4.0% and 3.9% on 20NEWS data set for $K = 100$ in terms of classification accuracy. This result shows that the asymmetric priors play an important role in regulating document-topic features for classification. When the dimensionality of latent space is small, learning an asymmetric Dirichlet prior over the document-topic distributions and symmetric Dirichlet prior over the topic-word distributions is worse than heuristically set symmetric Dirichlet priors on NIPS. One reason is that the Dirichlet prior have more effects on shorter documents than longer documents.

Fig. 6 shows the document clustering results measured by NMI. This result confirms that STC and pmiGS often predict the wrong clusters of documents on all data sets. All BP-based algorithms perform equally well but conBP performs slightly better when $K = 100$. It is interesting to see that the performance of document clustering is not consistent with that of document classification in Fig. 5. One possible reason is the unknown number of clusters in the clustering task.

Fig. 7 shows the coherence on all data sets when $K = 100$. Because STC has no topic-word distributions, it cannot be measured by the coherence metric. The plot produces a separate box for $K = 100$ coherence values of each algorithm. On each box, the central mark is the median, the edges of the box are the 25th and 75th percentiles, the whiskers extend to the most extreme data points not considered outliers, and outliers are plotted individually by the black dot sign. We see that asBP and conBP have higher coherence median values with smaller variances. BP also yields a stable coherence value. However, ltcBP and pmiGS have lower coherence values. The major reason is that they remove most common words, which contribute much to the coherence metric.

Fig. 8 shows the PMI values of all algorithms when $K = 100$. Because STC has no topic-word distributions, it cannot be measured by the PMI metric. The plot produces a separate box for $K = 100$ PMI values of each algorithm. On each box, the central mark is the median, the edges of the box are the 25th and 75th percentiles, the whiskers extend to the most extreme data points not considered outliers, and outliers are plotted individually by the black dot sign. We see that most results are consistent with those of Fig. 7. For example, asBP, conBP and BP have relatively smaller variances and median values, while ltcBP and pmiGS have relatively bigger variances and median values. Both Fig. 7 and 8 confirm that asBP provide more coherent and related word groups. Note that asBP clusters stop and common words in a separate topic, which enhances coherence and PMI when compared with ltcBP.

Table 2 summarizes the overall performance of all algorithms on four data sets when $K = 100$. We mark the best performance by the bold face. We see that asBP wins 8/20 columns and all variants of BP win around 18/20 columns. This result confirms that BP and its variants find better document-topic and topic-word distributions. As far as perplexity is concerned, asBP is always the best method, which means that it is very likely to recover the observed words from the document-topic and topic-word distributions. We see that ltcBP and asBP learns better document-topic distributions for soft document clustering with relatively higher NMI values. Moreover, both ltcBP and asBP can effectively remove stop and common words as shown in Fig. 3. Although STC uses sparse coding for document classification, it performs relatively worse than conBP partly because conBP incorporates the sparseness constraints naturally. Note that conBP often provides a stable clustering and classification performances though it is not the best. On CORA and 20NEWS, conBP outperforms BP with a large margin, which reflects that sparseness constraints can improve clustering and classification performance. When compared with pmiGS, ltcBP wins all columns, confirming the effectiveness of LTC features for topic modeling as well as BP framework for learning LDA. Form Table 2, we suggest continuous features and asymmetric priors for topic modeling because sparseness constraints do not provide significant improvement. The underlying reason is that the estimated document-topic and topic-word distributions are already very sparse so that any sparseness constraints can give only marginal improvement.

5   Conclusions

In this paper, we extensively explore three factors to find better topics: continuous features, asymmetric priors, and sparseness constraints within the unified BP framework. We develop several novel BP-based algorithms to study the three perspectives. Through extensive experiments, we advocate asymmetric priors for topic modeling because they can enhance the overall performance in terms of several metrics. Also, the continuous features can improve the interpretability of topic-word distributions by effectively remove almost all stop and common words. Finally, we find that sparseness constraints do not improve the topic modeling performance very much, partly because the sparse nature of document-topic and topic-word distributions of LDA.

Acknowledgements. This work is supported by NSFC (Grant No. 61003154, 61373092, 61033013, 61272449 and 61202029), Natural Science Foundation of the Jiangsu Higher Education Institutions of China (Grant No. 12KJA520004), Innovative Research Team in Soochow University (Grant No. SDT2012B02), and Guangdong Province Key Laboratory Project(Grant No. SZU-GDPHPCL-2012-09).

References

1. Blei, D.M., Ng, A.Y., Jordan, M.I.: Latent Dirichlet allocation. J. Mach. Learn. Res. 3, 993–1022 (2003)
2. Griffiths, T.L., Steyvers, M.: Finding scientific topics. Proc. Natl. Acad. Sci. 101, 5228–5235 (2004)
3. Zeng, J., Cheung, W.K., Liu, J.: Learning topic models by belief propagation. IEEE Trans. Pattern Anal. Mach. Intell. 33(5), 1121–1134 (2013)
4. Chang, J., Boyd-Graber, J., Gerris, S., Wang, C., Blei, D.: Reading tea leaves: How humans interpret topic models. In: NIPS, pp. 288–296 (2009)
5. Salton, G., McGill, M.J.: Introduction to modern information retrieval. McGraw-Hill, New York (1983)
6. Buckley, C.: Automatic query expansion using SMART: Trec 3. In: Proceedings of The Third Text REtrieval Conference (TREC-3), pp. 69–80 (1994)
7. Hoffman, M., Blei, D., Bach, F.: Online learning for latent Dirichlet allocation. In: NIPS, pp. 856–864 (2010)
8. Ramage, D., Heymann, P., Manning, C.D., Garcia-Molina, H.: Clustering the tagged web. In: Web Search and Data Mining, pp. 54–63 (2009)

9. Wilson, A.T., Chew, P.A.: Term weighting schemes for latent Dirichlet allocation. In: North American Chapter of the Association for Computational Linguistics: Human Language Technologies, pp. 465–473 (2010)
10. Minka, T.P.: Estimating a Dirichlet distribution. Technical report, Microsoft Research (2000)
11. Asuncion, A., Welling, M., Smyth, P., Teh, Y.W.: On smoothing and inference for topic models. In: UAI, pp. 27–34 (2009)
12. Wallach, H., Mimno, D., McCallum, A.: Rethinking LDA: Why priors matter. In: NIPS, pp. 1973–1981 (2009)
13. Zhu, J., Xing, E.P.: Sparse topical coding. In: UAI (2011)
14. Zhu, W., Zhang, L., Bian, Q.: A hierarchical latent topic model based on sparse coding. Neurocomputing 76(1), 28–35 (2012)
15. Hoyer, P.O.: Non-negative matrix factorization with sparseness constraints. Journal of Machine Learning Research 5, 1457–1469 (2004)
16. Zeng, J., Cao, X.-Q., Liu, Z.-Q.: Residual belief propagation for topic modeling. In: Zhou, S., Zhang, S., Karypis, G. (eds.) ADMA 2012. LNCS, vol. 7713, pp. 739–752. Springer, Heidelberg (2012)
17. Zeng, J., Liu, Z.Q., Cao, X.Q.: A new approach to speeding up topic modeling, arXiv:1204.0170 [cs.LG] (2012)
18. Heinrich, G.: Parameter estimation for text analysis. Technical report, University of Leipzig (2008)
19. Zhong, S., Ghosh, J.: Generative model-based document clustering: A comparative study. Knowl. Inf. Syst. 8(3), 374–384 (2005)
20. Mimno, D.M., Wallach, H.M., Talley, E.M., Leenders, M., McCallum, A.: Optimizing semantic coherence in topic models. In: EMNLP, pp. 262–272 (2011)
21. Newman, D., Karimi, S., Cavedon, L.: External evaluation of topic models. In: Australasian Document Computing Symposium, pp. 11–18 (2009)
22. Zeng, J.: TMBP: A topic modeling toolbox using belief propagation. J. Mach. Learn.Res. 13, 2233–2236 (2012)

Finding Well-Clusterable Subspaces for High Dimensional Data
A Numerical One-Dimension Approach

Chuanren Liu[1], Tianming Hu[2,*], Yong Ge[3], and Hui Xiong[1]

[1] Rutgers University, New Jersey, USA
{chuanren.liu,hxiong}@rutgers.edu
[2] Dongguan University of Technology, Guangdong, China
tmhu@ieee.org
[3] UNC Charlotte, North Carolina, USA
yong.ge@uncc.edu

Abstract. High dimensionality poses two challenges for clustering algorithms: features may be noisy and data may be sparse. To address these challenges, subspace clustering seeks to project the data onto simple yet informative subspaces. The projection process should be fast and the projected subspaces should be well-clusterable. In this paper, we describe a numerical one-dimensional subspace approach for high dimensional data. First, we show that the numerical one-dimensional subspaces can be constructed efficiently by controlling the correlation structure. Next, we propose two strategies to aggregate the representatives from each numerical one-dimensional subspace into the final projected space, where the clustering problem becomes tractable. Finally, the experiments on real-world document data sets demonstrate that, compared to competing methods, our approach can find more clusterable subspaces which align better with the true class labels.

Keywords: numerical one-dimension, clusterable subspace, subspace learning.

1 Introduction

People often face a dilemma when analyzing high dimensional data. On one hand, more features imply more information available for the learning task. On the other hand, irrelevant/contradicting features introduce noise and may mislead the learning algorithms. This difficulty has been studied extensively in the literature from different perspectives including dimension reduction, feature selection, model ensembling, etc.

Among them, multiple subspace learning is a promising paradigm to address the high dimensional difficulty. In this approach, we construct multiple simple

* Corresponding Author. This research was partially supported by National Science Foundation via grant CCF-1018151 and IIS-1256016. Also, it was supported in part by Natural Science Foundation of China (No. 61100136 and 71329201).

V.S. Tseng et al. (Eds.): PAKDD 2014, Part II, LNAI 8444, pp. 311–323, 2014.

yet informative subspaces of the original high dimensional data. For example, principle component analysis (PCA) chooses the subspaces that best preserve the variance of the data. Then we can either build learning models in the aggregated space, or build models collaboratively in each of the subspaces. This paradigm brings several desirable advantages. First, we can construct the subspaces by grouping related features together and separating contradicting features simultaneously. This is superior to simple feature reduction which may lose information carried by contradicting features. Second, such collaborative learning mode in the aggregated space is superior to separately learning one submodel at a time and finally combining them. In fact, this mode share some spirit with multi-source learning [3] in the literature. In the language of multi-source learning, directly learning the original high dimensional data is actually the early-source-combination based approach, which might be too difficult for a single model. At the other extreme, directly assembling the separately learned submodels is actually the late-source-combination based approach, which might make very limited or even no information to be shared among different submodels. The aggregated/collaborative learning mode is actually the intermediate-source-combination approach, which can balance between the learning difficulty of too many features for individual models and the ensemble difficulty of many too isolated and non-cooperative models.

Along this line, in this paper, we focus on the task of subspace learning for clustering high dimensional data. Specifically, we first construct numerical one-dimensional subspaces consisting of highly related features. In theory, such subspaces can substantially alleviate the unstable difficulties often encountered by clustering algorithms such as K-means. In practice, we show such subspaces can be efficiently constructed by leveraging correlation coefficients. Next, by further exploiting the one-dimension nature, we propose strategies to aggregate the representatives from the numerical one-dimensional subspaces into the final projected space. Finally, we use real-world document data sets to compare our approach with several competing methods in terms of performance lift and clustering separability. The experimental results demonstrate that our approach can find more clusterable subspaces which align better with the true class labels.

The rest of the paper is organized as follows. Section 2 summarizes recent works related to subspace learning. In Section 3, we show the numerical one-dimensional subspaces can be constructed by controlling the correlation structure. In Section 4, we propose strategies to build the final projected subspace by aggregating the representatives from the numerical one-dimensional subspaces. Section 5 validates the effectiveness of our idea on real-world document data sets. Section 6 concludes this paper with some remarks on future work.

2   Related Work

Our work can be categorized as dimension reduction for clustering. Although there have been extensive studies of dimension reduction techniques in the literature, few of them are designed specially for the general clustering problems.

In [10], the idea of grouping correlated features was exploited for the regression of the DNA microarray data. Specifically, the authors defined the "supergenes" by averaging the genes within the correlated feature subspaces and then used them to fit the regression models. In our case of unsupervised clustering, however, we do not have response for learning, which was used in [10] to analyze the accuracy improvement of the regression with the averaged features. Instead, we show that the subspaces of correlated features are actually of numerical one-dimension, which speaks to the improved clustering stability. Furthermore, empirical studies on real-world data sets suggest that they enjoy higher clustering separability which aligns better with the true class labels. In [1], another approach of dimension reduction, random projection, was exploited for the clustering problems. It is shown that any set of N points in D dimensions can be projected into $O(K/\epsilon^2)$ dimensions, for $\epsilon \in (0, 1/3)$, where optimal K-means can be preserved. In the later experiments, we will compare our methods with this baseline approach.

Another category of related work includes the validation measures of the clustering results. [17] gave an organized study of the external validation measures. Normalization solutions and major properties of several measures were provided. Later, [9] investigated more widely used internal clustering validation measures. Recently, [5] studied the effectiveness of the validation measures with respect to different distance metrics. It is shown that the validation measures might biasedly prefer some distance metrics. Thus, we should be careful with the choice of validation measures involving distance computation.

3   Numerical 1-Dimensional Subspace Construction

In this section, we first use the simpleness of 1-dimensional clustering to introduce the motivation of our work. Then we show how to construct numerical 1-dimensional subspaces by controlling the correlation structure of the features.

3.1   1-Dimensional Clustering

The clustering problem can be formulated as:

Problem 1. Given a set of observations $\mathbf{X}$, and the number of clusters K, the optimal clustering solution $C = \{C_1, \cdots, C_K\}$ minimizes the so-called within-cluster sum of squares (WCSS):

$$\text{WCSS}(\mathbf{X}|C) = \sum_{k=1}^{K} \sum_{x \in \mathbf{X} \cap C_k} \|x - \mu_k\|^2$$

where μ_k is the centroid of cluster C_k.

The most common solver for this problem, K-means [18], can only achieve local optima, which are not stable. Indeed, we might have more than one solutions, which are often inconsistent with one another. However, there is a special place where K-means yields more stable clustering results: 1-dimensional space.

Proposition 1. *For any two K-means clustering solutions on a 1-dimensional data set, $C^1 = \{C_1^1, C_2^1, \cdots\}$ and $C^2 = \{C_1^2, C_2^2, \cdots\}$, with cluster centers $c_i^j \in C_i^j$ where $c_1^j < c_2^j < \cdots$ for $j = 1, 2$, there are no data points x_1 and x_2 such that $x_1 \in C_1^1, x_2 \in C_2^1$ but $x_1 \in C_2^2, x_2 \in C_1^2$.*

The proof is straightforward and is omitted due to space limit. In other words, K-means clustering is very simple in 1-dimensional space, which is equivalent to finding the cut points. This can also be intuitively visualized in the *clustergram* [12], as we will see later in Figure 1. In short, the *clustergram* examines how data points in each cluster are assigned to new clusters in the next round as the number of clusters increase. When Proposition 1 holds, it is expected that there are few cross lines connecting the consecutive solutions. However, few data are so perfectly "1-dimensional" in reality. Hence, in the following, we seek 1-dimension-like subspaces, where Proposition 1 can be preserved approximately.

3.2   Numerical 1-Dimensional Subspace

In 1-dimension-like subspaces (subset of features), it is observed that, if most of the variation of the data can be captured by the first principle component, then K-means is roughly equivalent to clustering in 1-dimensional space (along the first principle direction). In this case, Proposition 1 will still hold under the mild assumption that all cluster centers can be roughly connected by a line parallel to the first principle direction. Specifically, note that, if data point x is closer to cluster center c, its projection $\langle x, v \rangle$ is also closer to c on the axis of the first principle direction v. Formally, this notion is captured by the numerical 1-dimensional space define below [11, 7]:

Definition 1. *A data set $\mathbf{X}$ is numerical 1-dimensional with error ϵ, if and only if $\sigma^2 \leq \epsilon\sigma^1$, where $\sigma_1 \geq \sigma_2 \geq \cdots$ are singular values of $\mathbf{X}$ (standardized to be of zero-mean and unit-variance along each feature).*

At first glance, we need to perform singular value decomposition many times to find such subspaces, which is expensive in high dimensional space. Nevertheless, as we will show below, the error ϵ is bounded with a term of correlation among features, which can be leveraged to construct the desired subspaces efficiently.

Theorem 1. *If the average correlation of different features in the d-dimensional data set $\mathbf{X}$ is $\rho > 0$, then $\mathbf{X}$ is numerical 1-dimensional with error*

$$\epsilon \leq \sqrt{\frac{(1-\rho)d - 1 + \rho}{\rho d + 1 - \rho}} < \sqrt{\frac{1 - \rho}{\rho}}.$$

Proof. Suppose matrix $\mathbf{X} \in \mathbb{R}^{N \times d}$ is already standardized to be of zero-mean and unit-variance along each feature (column). Then the feature correlations of $\mathbf{X}$ can be expressed by $\mathbf{C} = \frac{1}{N}\mathbf{X}'\mathbf{X}$ where the diagonal coefficients are all 1. With the singular value decomposition (SVD) $\mathbf{X} = U\Sigma V'$ where U, V are unitary matrices and the diagonal coefficients of Σ are $\sigma_1, \sigma_2, \cdots$, we have $\mathbf{C} = \frac{1}{N} V \Sigma' \Sigma V'$ where $\Sigma' \Sigma = \text{diag}(\sigma_1^2, \sigma_2^2, \cdots)$. It follows that

$$\frac{1}{N}(\sigma_1^2 + \sigma_2^2) \leq \text{tr}(\mathbf{C}) = d.$$

Let J be the column vector with 1 as all coefficients, then on one hand we have

$$N J' \mathbf{C} J = (V'J)' \Sigma' \Sigma (V'J) = \sum_i (\sum_j v_{ji})^2 \sigma_i^2$$

$$\leq \sigma_1^2 \sum_i (\sum_j v_{ji})^2 = \sigma_1^2 (V'J)'(V'J) = \sigma_1^2 J' J = d\sigma_1^2.$$

On the other hand, with the average of non-diagonal coefficients in $\mathbf{C}$, ρ, we have $J' \mathbf{C} J = \sum_{i,j} c_{ij} \geq (d^2 - d)\rho + d$. Hence, it follows that

$$\frac{1}{N}\sigma_1^2 \geq \rho d + 1 - \rho$$

$$\frac{1}{N}\sigma_2^2 \leq (1 - \rho)d - 1 + \rho$$

and this concludes our proof.

Theorem 1 suggests that, with a proper threshold of average correlation, the agglomerative hierarchical clustering over the feature set with average linkage can unambiguously group the original space into numerical 1-dimensional subspaces with error lower than the desired level. The standard Euclidean distance between features can be used as the linkage when the data matrix is of zero-mean and unit-variance along each feature. In the general case, the computational complexity of the agglomerative average linkage algorithm for D-dimensional data is $O(D^3)$, which is not efficient for big data applications. However, we note that Theorem 1 still holds if we denote ρ as the minimal correlation between features. This leads to the complete linkage clustering for which the computational complexity can be reduced to roughly $O(D^2)$. We will use this procedure in our experiments and denote it by $\mathcal{F} = N1dSpaces(\mathbf{X}, \epsilon)$ in the following discussions, where $\mathbf{X}$ is the data matrix, ϵ is the maximal error of numerical 1-dimensional subspaces, and $\mathcal{F}$ is the constructed subspaces.

The effectiveness of the subspace construction algorithm can be visualized in Figure 1, as mentioned earlier. Specifically, for a given high dimensional data set $\mathbf{X}$, we can produce a *clusgtergram* by directly applying a clustering algorithm, such as K-means with increasing number of clusters. Then we can construct the numerical 1-dimensional subspaces $\mathcal{F}$, and produce the same *clusgergram* in each subspace $\mathbf{S}$ in $\mathcal{F}$. The results show that, in the subspaces, there are few cross lines connecting the consecutive solutions.

4   Collaborative Ensemble of Subspaces

Now we have constructed subspaces where the clustering problem can be approached stably. However, clustering algorithms directly applied to the isolated

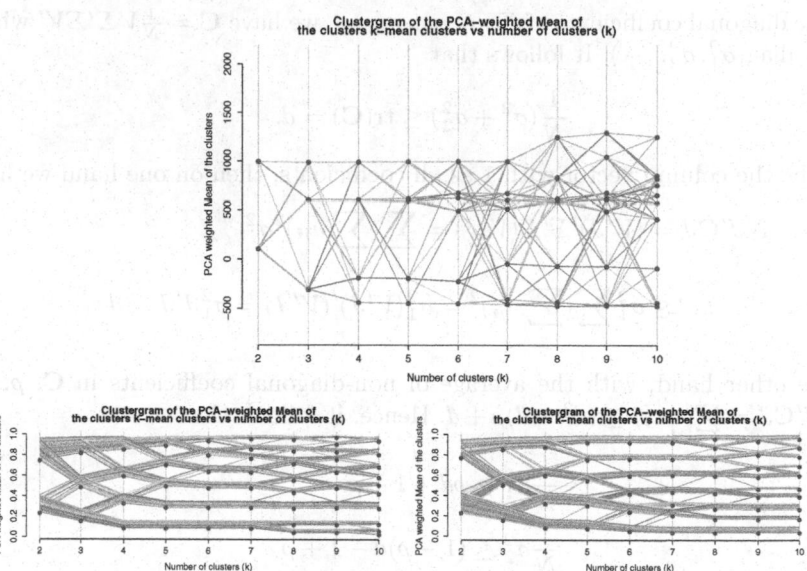

Fig. 1. Comparison of *clustergram*, where cluster means of consecutive cluster solutions are connected with parallelograms whose widths are proportional to the size of data assigned from the previous clusters. The top figure shows the *clustergram* of the high dimensional space. The bottom figures show the *clustergram* of two numerical 1-dimensional subspaces.

subspaces might produce degenerated solutions, since no information is shared between the subspaces. On the other hand, since each subspace $\mathbf{S}$ is numerically only of 1 dimension, it can be approximated by a few observation features. A natural way to this end is to investigate the SVD $\mathbf{S} = \mathbf{U}\Sigma\mathbf{V}'$, where $\Sigma = \text{diag}(\sigma_1, \cdots, \sigma_s)$ is a diagonal matrix consisting of s positive singular values of $\mathbf{S}$: $\sigma_1 \geq \cdots \geq \sigma_s$. In general, we can transform $\mathbf{S}$ to $\mathbf{S}\mathbf{V}$ by the principal directions in $\mathbf{V}$. Then, guaranteed by Theorem 1, we can use only the first principal component $\mathbf{S}\mathbf{v}$ where $\mathbf{v}$ is the first principal direction in $\mathbf{V}$ corresponding to σ_1. Note that, this is often computationally more efficient, since we only need the first singular vector and it is not necessary to fully decompose $\mathbf{S}$. Also, when the number of features are small in $\mathbf{S}$, the computation can be further boosted by decomposing $\mathbf{S}'\mathbf{S}$ as in Theorem 1. This collaborative strategy of subspace ensemble is detailed in Algorithm 1, where $mSpace(\mathcal{F})$ denotes the combination of the projected components of the multiple subspaces in $\mathcal{F}$, and $mCluster$ denotes the clustering problem solver applied to $mSpace(\mathcal{F})$.

In addition to the above strategy of aggregating projected components, we can also progressively approximate the subspaces in the light of [8]. Specifically, suppose we have the approximation $\widehat{\mathbf{S}}^d$ for the first d subspaces $\mathbf{S}^1, \mathbf{S}^2, \cdots, \mathbf{S}^d$. To approximate the next new subspace $\mathbf{S}^{d+1}$, we compute the SVD $\mathbf{S} = \mathbf{U}\Sigma\mathbf{V}'$, where $\mathbf{S} = (\widehat{\mathbf{S}}^d, \mathbf{S}^{d+1})$ is concatenation of $\widehat{\mathbf{S}}^d$ and $\mathbf{S}^{d+1}$. Then the new approximation $\widehat{\mathbf{S}}^{d+1} = \mathbf{S}\mathbf{P}$ where $\mathbf{P}$ are the top $d+1$ principal directions in $\mathbf{V}$.

Algorithm 1. The multiple subspaces clustering algorithm

Signature: $C = mCluster(\mathbf{X}, K, \epsilon)$
Input: The data matrix $\mathbf{X}$; The number of clusters K; The maximal error of numerical
1-dimensional subspaces ϵ.
Output: The clustering C.

1. Construct subspaces $\mathcal{F} \leftarrow N1dSpaces(\mathbf{X}, \epsilon)$.
2. **for** Each subspace $\mathbf{S} \in \mathcal{F}$ **do**
3.   Compute the first singular vector $\mathbf{v}$ of $\mathbf{S}$.
4.   Replace $\mathbf{S}$ in $\mathcal{F}$ by $\mathbf{Sv}$.
5. **end for**
6. Construct $\widehat{\mathbf{X}} = mSpace(\mathcal{F})$ by combining the approximated subspaces in $\mathcal{F}$.
7. Solve Problem 1 in the space $\widehat{\mathbf{X}}$ with the parameter K, e.g., compute the Kmeans
   clustering $C \leftarrow kmeans(\widehat{\mathbf{X}}, K)$.

Table 1. The characteristics of data sets

data	fbis	kla	la1	re0	re1	wap
#doc	2463	2340	3204	1504	1657	1560
#term	2000	4707	6188	2886	3758	8460
#class	17	20	6	13	25	20
MinClass	38	9	273	11	10	5
MaxClass	506	494	943	608	371	341
Min/Max	0.075	0.018	0.290	0.018	0.027	0.015

The details are given in Algorithm 2, where $pSpace(\mathcal{F})$ denotes the approximation described above for the subspaces in $\mathcal{F}$, and $pCluster$ denotes the clustering problem solver applied to $pSpace(\mathcal{F})$.

5   Experimental Evaluation

5.1   Experimental Data Sets

For evaluation, we used six real data sets from different domains, all of which are available at the website of CLUTO [4]. Some characteristics of these data sets are shown in Table 1. One can see diverse characteristics in terms of size (#doc), dimension (#term), number of clusters (#class) and cluster balance are covered by the investigated data sets. The cluster balance is measured by the ratio MinClass/MaxClass, where MinClass and MaxClass are the sizes of the smallest class and the largest class, respectively.

5.2   Comparison of Performance Lift

To see how much improvements can be achieved by the subspaces, regardless which solver of Problem 1 is used, we compute the performance lift [14, 5] in

---

Algorithm 2. The progressive subspaces clustering algorithm

Signature: $C = pCluster(\mathbf{X}, K, \epsilon)$
Input: The data matrix $\mathbf{X}$; The number of clusters K; The maximal error of numerical
    1-dimensional subspaces ϵ.
Output: The clustering C.

1. Construct subspaces $\mathcal{F} = \{\mathbf{S}^1, \mathbf{S}^2, \cdots\} \leftarrow N1dSpaces(\mathbf{X}, \epsilon)$.
2. Order subspaces in $\mathcal{F}$ by the descending order of numerical 1-dimensional error.
3. Initialize the $pSpace(\mathcal{F})$ as $\widehat{\mathbf{X}} \leftarrow ()$, i.e., empty space.
4. $d \leftarrow 0$.
5. **repeat**
6. $d \leftarrow d + 1$.
7. $\mathbf{S} \leftarrow (\widehat{\mathbf{X}}, \mathbf{S}^d)$.
8.   Compute the first d singular vectors $\mathbf{P}$ of $\mathbf{S}$.
9. $\widehat{\mathbf{X}} \leftarrow \mathbf{SP}$.
10. **until** d reaches the number of subspaces in $\mathcal{F}$
11. Solve Problem 1 in the space $\widehat{\mathbf{X}}$ with the parameter K, e.g., compute the Kmeans
    clustering $C \leftarrow kmeans(\widehat{\mathbf{X}}, K)$.

---

the approximated subspaces. Specifically, the performance lift can be defined by the expectation: $\text{lift}(\mathbf{X}|Y) = E[\frac{\text{WCSS}(\mathbf{X}|C)}{\text{WCSS}(\mathbf{X}|Y)}]$ where C is a random clustering assignments for the data set $\mathbf{X}$ and Y is the true class labels. The performance lift actually represents the difference between the ground truth of the clustering structure and the random clustering solution. The higher the lift is, the easier it will be for the solver of Problem 1 to find the optimal solutions. Thus, we can use this lift to see which subspaces help most. To estimate the $\text{lift}(\mathbf{X}|Y)$, we can generate T (e.g., 10) random clustering assignments $\{C_1, \cdots, C_T\}$, and compute the average: $\frac{1}{T} \sum_{t=1}^{T} \frac{\text{WCSS}(\mathbf{X}|C_t)}{\text{WCSS}(\mathbf{X}|Y)}$. In Figure 2, we show the performance lifts in different approximated subspaces for all the data sets.

Specifically, we generate $T = 10$ random clustering assignments to estimate the performance lift. By controlling the error ϵ used in $\mathcal{F} = N1dSpaces(\mathbf{X}, \epsilon)$, we can construct approximation $mSpace(\mathcal{F})$ and $pSpace(\mathcal{F})$ with different dimensions, e.g., $d = 100, 200, \cdots, 1000$. For comparison, we also compute the performance lifts with top d principal components constructed by simple PCA, as denoted by "PC" in Figure 2. The line denoted by "RP" stands for Random Projection [1], which constructs the low dimensional approximation of $\mathbf{X} \in \mathbb{R}^{N \times D}$ by $\mathbf{X}\Omega$ where $\Omega \in \mathbb{R}^{D \times d}$ is random matrix with entries $+1/\sqrt{d}$ or $-1/\sqrt{d}$ with equal probability. We can see that $mSpace$, $pSpace$, and PC are all effective to boost the performance lift. Also, while $mSpace$ and $pSpace$ outperform others consistently, $mSpace$ achieves significantly higher lift of performance.

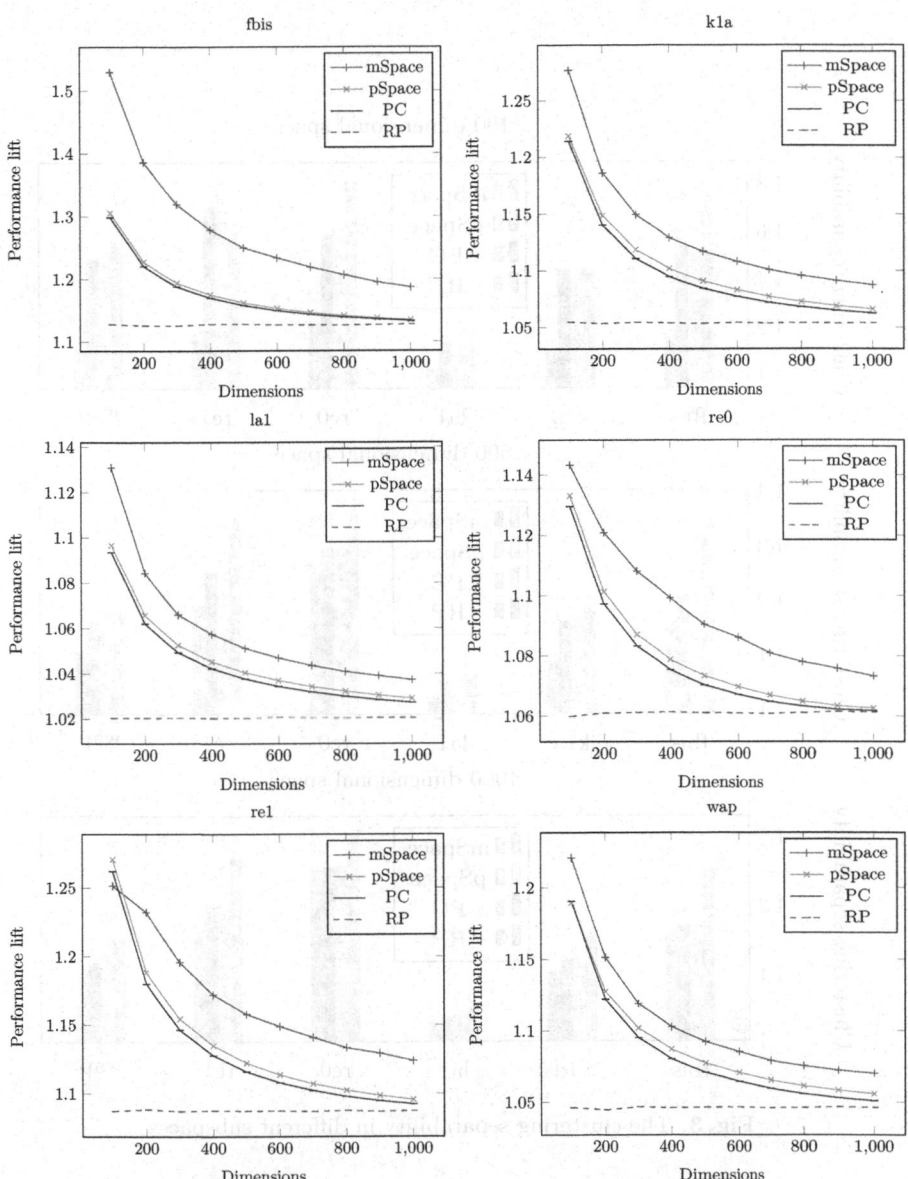

Fig. 2. The performance lift in different subspaces

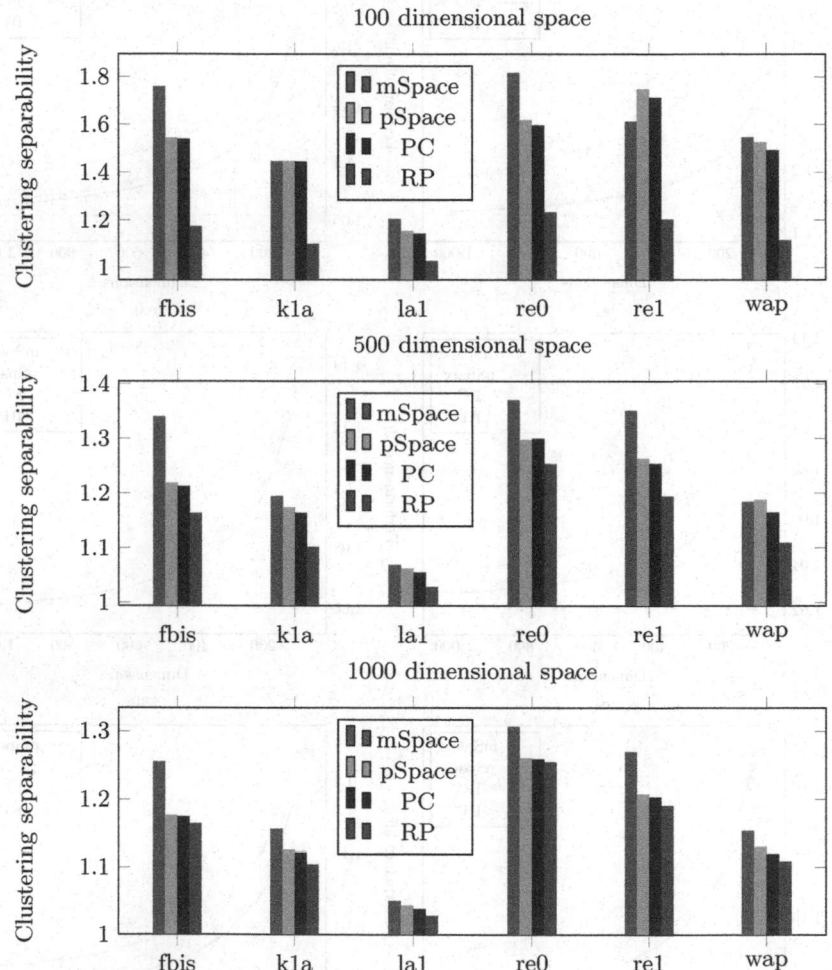

Fig. 3. The clustering separability in different subspaces

5.3 Comparison of Clustering Separability

Table 2. The clustering separability in 500 dimensional subspaces constructed with different methods on 'la1'

Cluster ID	Cluster Label	mSpace	pSpace	PC	RP
1	Entertainment	1.1596	**1.2010**	1.1482	1.0328
2	Financial	**1.0628**	1.0287	1.0372	1.0299
3	Foreign	1.0160	1.0176	1.0223	**1.0355**
4	Metro	1.0287	**1.0442**	1.0395	1.0136
5	National	1.0225	**1.0418**	1.0347	1.0191
6	Sports	**1.1200**	1.0410	1.0440	1.0390
Average		**1.0684**	1.0624	1.0543	1.0283

Adopted in [4, 5], one can investigate the data separability for the unsupervised clustering problem. Specifically, for each cluster C_i in the clustering solution $\{C_1, \cdots, C_K\}$, we can compute the ratio $\frac{EDis(C_i)}{IDis(C_i)}$ of the average external distance, $EDis(C_i)$, over the average internal distance, $IDis(C_i)$. The average internal distance $IDis(C_i)$ is the average distance between the instances in C_i, and the average external distance $EDis(C_i)$ is the average distance between the instances in C_i and the instances in the rest of the clusters C_j where $j \neq i$. The higher the ratio is for a cluster, the more compact and isolated the cluster will be, which, in turn, makes it easier for a clustering solver to identify the cluster. The ratio results of data set 'la1' are listed in Table 2, which clearly indicates that *mSpace* and *pSpace* provide better clustering separability. The last row also reports the average separability for the 6 clusters, where *mSpace* performs best. Besides, Figure 3 shows the average cluster separability for all of the six data sets. One can see that *mSpace* performs best on all data sets with few exceptions where *pSpace* performs better.

5.4 Analysis of Computational Cost

To reduce the dimensionality of the data matrix $\mathbf{X} \in \mathbb{R}^{N \times D}$ to d, our methods first construct the numerical 1-dimensional spaces. In the general case, the complexity of this step is $O(D^2)$, as we discussed in Section 3. To construct *mSpace*, we need to perform SVD further d times in the subspaces, each of $O(N)$ time, since most of the subspaces are of very low dimensions and we only need the first principal component. Thus the total computational cost for *mSpace* is $O(D^2 + dN)$. For *pSpace*, the computational complexity of the progressive SVD is costly $O(d^2 N)$ and the total cost is $O(D^2 + d^3 N)$. For PCA, we have the computational cost of $O(N^2 D)$ when $N \leq D$ or $O(ND^2)$ when $D \leq N$. In our experiments, since most of the data sets are very high dimensional, we have the order of computational cost for the evaluated methods: $RP < PC < mSpace < pSpace$, which aligns with the order of performance.

6   Concluding Remarks

In this paper, we proposed a numerical one-dimension approach to high dimensional data clustering. An efficient correlation-based method was provided to construct the numerical one-dimensional subspace, which is well-clusterable and thus makes the clustering stable. Also, we discussed two strategies to collaboratively aggregate them into the final projected space. The experiments on real-world data sets demonstrated that such transformed data aligns better with the true class labels with respect to clustering.

This paper focused on the collaborative ensembling of the one-dimensional subspaces. For the future work, we plan to investigate collaboratively building clustering submodels directly in each of these one-dimensional subspaces. In the literature, this is related to the areas of multiple clustering [19, 6], clustering kernel [16] and clustering ensembles [13, 2, 15], where there are still many open problems to be answered.

References

1. Boutsidis, C., Zouzias, A., Drineas, P.: Random projections for k-means clustering. In: NIPS, pp. 298–306 (2010)
2. Fred, A.L.N., Jain, A.K.: Combining multiple clusterings using evidence accumulation. IEEE Transactions on Pattern Analysis and Machine Intelligence 27(6), 835–850 (2005)
3. Gönen, M., Alpaydın, E.: Multiple kernel learning algorithms. The Journal of Machine Learning Research 12, 2211–2268 (2011)
4. Karypis, G.: CLUTO: Data clustering software,
   http://glaros.dtc.umn.edu/gkhome/views/cluto
5. Liu, C., Hu, T., Ge, Y., Xiong, H.: Which distance metric is right: An evolutionary k-means view. In: SDM, pp. 907–918 (2012)
6. Liu, C., Xie, J., Ge, Y., Xiong, H.: Stochastic unsupervised learning on unlabeled data. Journal of Machine Learning Research - Proceedings Track 27, 111–122 (2012)
7. Liu, G., Lin, Z., Yu, Y.: Robust subspace segmentation by low-rank representation. In: ICML, vol. 3 (2010)
8. Liu, J., Chen, S., Zhou, Z.-H.: Progressive principal component analysis. In: Yin, F.-L., Wang, J., Guo, C. (eds.) ISNN 2004. LNCS, vol. 3173, pp. 768–773. Springer, Heidelberg (2004)
9. Liu, Y., Li, Z., Xiong, H., Gao, X., Wu, J.: Understanding of internal clustering validation measures. In: ICDM, pp. 911–916. IEEE (2010)
10. Park, M.Y., Hastie, T., Tibshirani, R.: Averaged gene expressions for regression. Biostatistics 8(2), 212–227 (2007)
11. Rangan, A.V.: Detecting low-rank clusters via random sampling. Journal of Computational Physics 231(1), 215–222 (2012)
12. Schonlau, M.: The clustergram: A graph for visualizing hierarchical and non-hierarchical cluster analyses. The Stata Journal 3, 316–327 (2002)
13. Strehl, A., Ghosh, J.: Cluster ensembles—a knowledge reuse framework for combining multiple partitions. The Journal of Machine Learning Research 3, 583–617 (2003)

14. Strehl, A., Ghosh, J., Mooney, R.: Impact of similarity measures on web-page clustering. In: Workshop on Artificial Intelligence for Web Search (AAAI 2000), pp. 58–64 (2000)
15. Topchy, A., Jain, A.K., Punch, W.: Clustering ensembles: Models of consensus and weak partitions. IEEE Transactions on Pattern Analysis and Machine Intelligence 27(12), 1866–1881 (2005)
16. Weston, J., Leslie, C., Zhou, D., Elisseeff, A., Noble, W.S.: Semi-supervised protein classification using cluster kernels. In: Thrun, S., Saul, L., Schölkopf, B. (eds.) NIPS (2003)
17. Wu, J., Xiong, H., Chen, J.: Adapting the right measures for k-means clustering. In: SIGKDD, pp. 877–886. ACM (2009)
18. Wu, X., et al.: Top 10 algorithms in data mining. Knowledge and Information Systems 14(1), 1–37 (2008)
19. Zhang, J., et al.: Pattern classification of large-scale functional brain networks: Identification of informative neuroimaging markers for epilepsy. PloS One 7(5), e36733 (2012)

Determine Optimal Number of Clusters
with an Elitist Evolutionary Approach

Lydia Boudjeloud-Assala and Ta Minh Thuy

Laboratory of Theoretical and Applied Computer Science
LITA EA 3097, University of Lorraine
Ile du Saulcy, Metz, F-57045, France
{lydia.boudjeloud-assala,minh-thuy.ta}@univ-lorraine.fr

Abstract. This article proposes an elitist evolutionary approach to determine the optimal number of clusters for clustering data sets. The
proposed method is based on the cluster number optimization and in
the same time, finds the potential clusters seeds. This method can be
used as an initialization of k-means algorithm or directly as a clustering
algorithm without prior knowledge of the clusters number. In this approach, elitist population is composed of the individuals with potential
clusters seeds. We introduce a new mutation strategy according to the
neighborhood search and new evaluation criteria. This strategy allows us
to find the global optimal solution or near-optimal solution for clustering
tasks, precisely finding the optimal clusters seeds. The experimental results show that our algorithm performs well on multi-class and large-size
data sets.

Keywords: Elitist approach, Evolutionary algorithm, Clustering, Optimal number of clusters, Cluster seed initialization.

1 Introduction

Clustering is a challenging research area in data mining. A common form of clustering is partitioning the data set into homogeneous clusters such that members
of the same cluster are similar and members of distinct clusters are dissimilar.
Determining the optimal clusters number is one of the most difficult issues in
clustering data. In this work, we deal with the clustering problem without prior
knowledge on the appropriate clusters number and we try to propose the global
optimal or near-optimal cluster seeds. Clustering algorithms can be broadly classified into two groups: hierarchical and partitional [1]. Hierarchical algorithms
recursively find nested clusters either in a divisive or agglomerative method. In
contrast, partitional algorithms find all the clusters simultaneously as a partition
of the data and do not impose a hierarchical structure. Common formulation of
the clustering problem is assuming S is the given data set include n data points:
$S = \{x_1, x_2, ..., x_n\}$ where x_i $(i = 1, ..., n)$ is a real vector $d - dimensions$ and an
integer k. The goal of clustering is to determine a set of k clusters $C_1, C_2, ..., C_k$
such that the points belonging to the same cluster are similar, while the points

V.S. Tseng et al. (Eds.): PAKDD 2014, Part II, LNAI 8444, pp. 324–335, 2014.

belonging to different clusters are dissimilar in the sense of the given metric. The problem of finding an optimal solution to the partition of n data into k clusters is $NP-complete$, and heuristic methods are widely effective on $NP-complete$ global optimization problems and they can provide good sub-optimal solutions in reasonable time. We propose a clustering algorithm that can detect compact and hyperspherical clusters that are well separated using an Euclidean distance. To detect hyperellipsoidal clusters, we can use a more general distance function such as the Mahalanobis distance for example [2]. Some recent research has shown that the problem of searching efficient initialization methods for k-means clustering algorithm is a great challenge. Numerous initialization methods have been proposed to address this problem. Celebi and al. [3] present an overview of these methods with an emphasis on their computational efficiency. In their study, they investigate some of the most popular initialization methods developed for the k-means algorithm. They describe and compare initialization methods that can be used to initialize other partitional clustering algorithms such as fuzzy c-means and its variants and expectation maximization, and they conclude that most of these methods can be used independently of k-means as standalone clustering algorithms. Some others methods are proposed, based on metaheuristics such as simulated annealing [4] and genetic algorithms [5]. These algorithms start from a random initial configuration (population) and use k-means to evaluate their solutions in each iteration (generation). There are two main drawbacks associated with these methods. First, they involve numerous parameters that are difficult to tune [6]. Second, due to the large search space, they often require a large number of iterations, which renders them computationally prohibitive for most of the data sets. This paper presents a method that proposes, in the same time, the optimal cluster number with the initial clusters seeds. We don't use k-means to evaluate our solution because it depends on several parameters itself (k, initial seeds, ...), or any other clustering algorithm, so, our method can be used as standalone clustering algorithms without a prior knowledge of cluster number. Generally metaheuristics approaches involve numerous parameters that are difficult to tune, to deal with this problem, we propose an Elitist Evolutionary Approach that involves numerous evolutionary algorithms (EAs). The difference between them is implemented by the parameters and we select only the best concurrent solution. The remainder of this paper is organized as follows: Section 2 describes the related work on initialization and elitist methods. Section 3 gives the details of the new proposed algorithm, including the algorithm description, motivation of population initialization, as well as the mutation process based on neighborhood search. Section 4 shows the experimental results on data sets clustering. Finally, we draw the conclusions.

2   Related Work

In this section, we briefly review some of the commonly used initialization methods and elitist methods.

2.1   Initialization Methods

Celebi and al. [3] investigate some of the most popular initialization methods developed for the k-means algorithm. Their motivation is threefold. First, a large number of initialization methods have been proposed in the literature. Second, these initialization methods can be used to initialize other partitional clustering algorithms such as fuzzy c-means and its variants and expectation maximization. Third, most of these initialization methods can be used independently of k-means as standalone clustering algorithms. They review some of the commonly used initialization methods with an emphasis on their time complexity. They conclude that the super linear methods often have more elaborate designs when compared to linear ones. An interesting feature of the super linear methods is that they are often deterministic, which can be considered as an advantage especially when dealing with large data sets. In contrast, linear methods are often non-deterministic and/or order-sensitive. A frequently cited advantage of the more elaborate methods is that they often lead to faster k-means convergence, i.e. require fewer iterations, and as a result the time gained during the clustering phase can offset the time lost during the initialization phase. This may be true when a standard implementation of k-means is used. However, convergence speed may not be as important when a fast k-means variant is used as such methods often require significantly less time compared to a standard k-means implementation. Some other initialization methods such as the binary-splitting method [7] takes the mean of data as the first center. In iteration t, each of the existing 2^{t-1} centers is split into two new centers by subtracting and adding a fixed perturbation vector. These 2^t new centers are then refined using k-means. There are two main disadvantages associated with this method. First, there is no guidance on the selection of a proper value for the vector, which determines the direction of the split [8]. Second, the method is computationally demanding since after each iteration k-means has to be run for the entire data set. Some other methods based on metaheuristics such as simulated annealing [4] and genetic algorithms [5]. These algorithms start from a random initial configuration and use classical algorithm such as k-means to evaluate their solutions in each iteration. There are two main disadvantages associated to these methods. First, they involve numerous parameters that are difficult to tune (initial temperature, cooling schedule, population size, crossover/mutation probability, etc.) [6]. Second, due to the large search space, they often require a large number of iterations, which renders them computationally prohibitive for all but the smallest datasets. Interestingly, with the recent developments in combinatorial optimization algorithms, it is now feasible to obtain globally minimum clusterings solution for small data sets without resorting to metaheuristics [9].

2.2   Elitist Methods

Prevent promising individuals from being eliminated from the population during the application of genetic operators is a very important task. To ensure that the best chromosome is preserved, elitist methods copy the best individual found so far into the new population. Different EAs variants achieve this goal of preserving the best solution in different ways. However, elitist strategies tend to make the search more exploitative rather than explorative and may not work for problems in which one is required to find multiple optimal solutions [10]. Elitist methods are widely applied on different domains, Qasem and Shamsuddin [11] developed a Mimetic Elitist Pareto Differential Evolution algorithm, in order to deal with the hybrid learning problem (unsupervised and supervised learning), they use the multi-elitist approach to help the learning algorithm to get out of local minimum, therefore improving the accuracy of the proposed learning model. Das and al. [12] proposed a method based on a modified version of classical Particle Swarm Optimization algorithm, known as the Multi-Elitist Particle Swarm Optimization model. The proposed algorithm has been shown to meet or beat the other state of the art clustering algorithms in a statistically meaningful way over several benchmark datasets. The unique disadvantage of this algorithm is choosing the best suited parameters to find optimal solution. Gou and al. [13] apply Multi Elitist approach on quantum clustering problems. They use these methods to avoid getting stuck in local extremes. They used the mechanism of cluster center updating with a property of k-means clustering, that can influence the clustering results. This is one of disadvantages of this method, adding parameters that the method is based on. According to elitist strategy that work for problems in which one is required to find multiple optimal solutions, we introduce a new approach where multi evolutionary algorithms run together at the same time to compare their proposed solutions, and we select only the best one which is the optimal or nearest optimal solution.

3   Proposed Approach

We propose an Elitist Evolutionary Clustering Algorithm (EECA) (figure 1) that combine different techniques such as evolutionary algorithm with elitist approach and local search approach. This approach allows us to determine the optimal cluster number as well as finding cluster seeds.

3.1   Evolutionary Algorithm

In this section, we try to explain succinctly the evolutionary algorithm which is shown in right of the figure 1.

Gene Representation. A potential solution of our problem is a combination of potential optimal seeds, we try to find optimal number of these seeds. According to this, we consider a genetic individual (chromosome) as a combination of k_{max} potential optimal seeds, k_{max} being parameter of the algorithm. Each gene in genetic individual is an integer number, which takes values from $\{1, 2, .., n\}$. This value indicates the data point identification. The gene with value 0 indicates that there is no point selected as center. The number of genes in the optimal solution different from 0 represents the optimal cluster number. Each data point is a

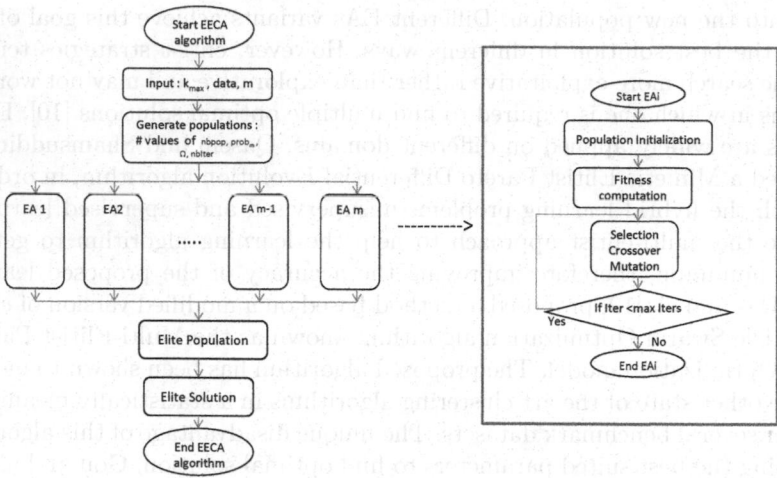

Fig. 1. EECA schema

$d - dimensions$ vector containing the d real values. We want to diversify the population, for this, we associate a frequency rate at each data point, each new genetic individual will be composed by the data point had not been used before, that have frequencies equal or close to zero.

Population Initialization. The population initialization is created by nb_{pop} genetic individuals with nb_{pop} given before. Each gene in the genetic individual is selected randomly on : $\{0, 1, 2, ..., n\}$ data set identification. Each genetic individual corresponds to a specific clustering solution in terms of cluster seeds. We impose in the genetic individual to have a different gene to obtain specific and different clusters seeds. We also impose to have an initial population without redundant genetic individuals. Studying individual is applied in each genetic individual which is created. And then, we verify that is no identical genetic individual, in the population which has the same gene. At this step, the first population is ready. Once the population is evaluated and sorted according to the fitness function we operate the genetic operator described bellow.

Genetic Operators. The crossover operation produces new offspring genetic individuals from parent individual. Two new genetic individuals are created by exchanging genes from two parent chromosomes. This exchange may start from one or several positions in the chromosomes called cut point. We can use two types of crossover: a randomly determined cut point or an optimized cut point. In the latter case, we determine the best point before the cut, this implies an evaluation of each possible cut for the individual. For our case, we first use a randomly chosen cut point which is modified to obtain optimized cut point, which implies evaluation of each new created individual. The first child resulting from by repeated randomly crossover which is better than initial child will determine the optimal cut point. The mutation makes gene inversion in the genetic individual. This genetic operator is used to avoid the degeneration of the population in order to prevent a too fast convergence of the genetic algorithm. If implemented appropriately, the operator can make the algorithm able to leave from local optimum. The mutation is usually applied with a small probability ($prob_m$, algorithm parameter). The mutated gene is chosen according to neighborhood search (section 3.2), we should verify if a new gene value is not in neighborhood of other genes composing the genetic individual. It is evident that we study the composition of each creation of new individual to have a different gene in a specific genetic individual.

Evaluation Criteria. The first objective is to find the optimal number of clusters, where clusters are compact and separated between them. Several measures have been proposed, Milligan and Cooper [14] presented a survey and comparison of 30 internal validity indexes for clustering algorithms and out-perform that $CH(k)$ [15] is one of the best solutions.

This index represents a ratio of the sum of between-cluster and the sum of within-cluster, which is expressed as follows:

$$CH(k) := \frac{[traceB/(k-1)]}{[traceW/(n-k)]}$$

Where n is the number of data points, k is clusters number.

$$traceB := \sum_{i=1}^{k} |C_i| \, \|\overline{C_i} - \overline{x}\|^2$$

$$traceW := \sum_{i=1}^{k} \sum_{j \in C_i} \|x_j - \overline{C_i}\|^2$$

With $|C_i|$ is the number of assigned objects to the cluster C_i ($i = 1, \ldots, k$); $\overline{C_i}$ is a center of class C_i and $\overline{x} = \frac{1}{n} \sum_{i=1}^{n} x_i$ is the global center of all data points. The optimal clusters number is found by maximizing Calinski and Harabasz index ($CH - index$).

The second objective is to find clusters that are compact and separate, so we introduce a measure that minimizes the overlapping between clusters, this criteria is defined by :

$$OP = \sum_{i \neq j}^{k} Card(C_i \cap C_j)$$

We compare our method with two fitness functions, first test series focus on maximizing $CH(k)$ in each EAi and in the elite population, where second test series focus on minimizing OP in each EAi and sort the elite population with $CH(k)$.

3.2 Neighborhood Search

When we operate the mutation we should firstly verify if all genes are different; secondly, according to the obtained potential clusters seeds, we introduce a new verification process where we impose that a new gene must not be in the neighborhood of others genes composing the genetic individual. For this, we introduce an automatic method to detect the limit of the cluster.

Cluster limit detection. In order to obtain the neighborhood seed, we search the data points contained in the cluster, we select the ones close enough to the seed. We choose the threshold Ω (algorithm parameter) by computing the distance between all the data points and the cluster seed and ordering them from the closest object to the farthest. We then try to find an abrupt increasing of distance that will indicate the cluster limit. We choose to use the peak detection method presented by Palshikar [16] applied on differential distances. Other cluster limit detection might be used, but this one is fast and gives the algorithm a complexity of $O(n \times d \times (p + g))$, where n is the number of data points, d the number of dimensions, p the population size and g the number of generations.

Then, we operate mutation by changing gene value by new value which is not in the neighborhood of other genes composing the genetic individual, and if we don't find this new gene, we change the value by 0 (no center selected).

3.3 Elite Population

Generally, the goal of the adaptive elitist-population search method is to adaptively adjust the population size according to the features of our technique to achieve, firstly, a single elitist individual searching for each solution; and secondly, all the individuals in the population searching for different solutions in parallel. For satisfying multi modal optimization search, we define the elitist individuals in the population as the individual with the best fitness on different solutions of the multiple domain. These elitist individuals define the elite population. Then we propose the elitist genetic operators that can maintain and even improve the diversity of the population and performing different evolutionary algorithms. A major advantage of using EAs over traditional learning algorithms is the ability to escape from local minimum using genetic operators [17]. The evolutionary algorithm depend on 4 parameters : nb_{pop}, $prob_m$, nb_{iter} and Ω. We perform as much as possible many evolutionary algorithms according to different values of these parameters and then we select the best solution, which compose elite population, without focusing on setting parameters. From this elite population we select the best solution (elite individual), which is the optimal solution for our problem.

4 Experimental Results

To evaluate the performance of the proposed method, we proceed several experiments on data sets from University of California at Irvine (UCI) machine

Table 1. Datasets description

Dataset	No. Points	No. Attributes	No. Clusters
Iris	150	4	3
Vehicule	846	18	4
Haberman	306	3	2
Synthetic	500	2	5
Wine	178	13	3
Blood Tranfusion	748	4	2
Seed	210	7	3
Ecoli	336	7	8

learning benchmark repository [18]. Data sets information are summarized in Table 1.

In our experiments, we perform an Elitist Algorithm with varying different parameters values in each Evolutionary Algorithm. We vary Evolutionary Algorithm parameters nb_{pop}, $prob_m$, nb_{iter} and Ω as follow :

- $nb_{pop} \in \{50, 100, 150\}$
- $prob_m \in \{0.1, 0.2, 0.3\}$
- $nb_{iter} \in \{200, 300, 500\}$
- $\Omega \in \{1, 2, 3\}$

For each data set we perform the elitist algorithm with 18 EAs and $k_{max} = 10$. The results of finding optimal k are illustrated in table 2. We compare our method with the two fitness functions presented before. First test series focus on maximizing $CH(k)$ in each EAi and in the elite population ($EACH$ corresponding column in table 2). The second test series focus on minimizing OP in each EAi and sort the elite population with $CH(k)$ ($EECA$ corresponding column in table 2).

Table 2. Results description

Dataset	k-real	$EACH$	$EECA$
Iris	3	2	3
Vehicule	4	2	4
Haberman	2	2	2
Synthetic	5	5	5
Wine	3	2	3
Blood Tranfusion	2	2	2
Seed	3	2	3
Ecoli	8	7	8

As we can see in the table 2, we find exactly the same number of clusters as in real data sets using overlapping fitness function combined with $CH(k)$ index. When only $CH(k)$ index is used, we always find $k = 2$, (except for 2 data sets, Ecoli and Synthetic data sets). This result can be explained by the formula of $CH(k)$ index, when we try to maximize only this index, we converge to small number of clusters. To find the data set partitions, we use a cluster limit detection method based on hyperspherical cluster forms. For Synthetic data set which presents prefect hyperspherical clusters, the two fitness functions performs perfectly. But in other data sets it seems important to consider the overlapping fitness in the first, and then to select the better solution according to the $CH(k)$ index. To confirm our results, we visualize some data sets using scatter plot methods [19] which represent all 2D projection of the data set. The figure 2 represents the projection of iris data set, and the figure 3 represents the projection of Haberman data set, as we can see in the figures the data points colors (different forms) represent the real clusters, and in red (square form) we can see the clusters seeds that are detected by our method. We can note that each of them corresponds to the real clusters, and can be considered as the center of the different clusters or as initial seed for any clustering algorithm.

These results and the visualizations showed the effectiveness of our methods on data sets that have compactness clusters structures. As we can see in the figure 4, that represent a synthetic data set, composed by five clusters, with Gaussian distribution. We can easily adapt our method with changing and adapting distance measures to extract other cluster structures. In fact, for this we need to

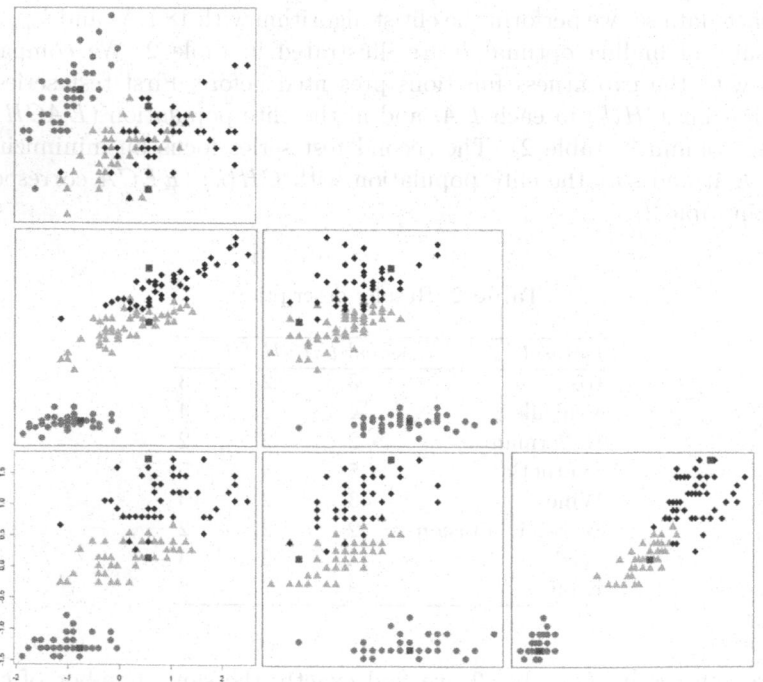

Fig. 2. Scatter plot visualization of Iris data set with detected seeds

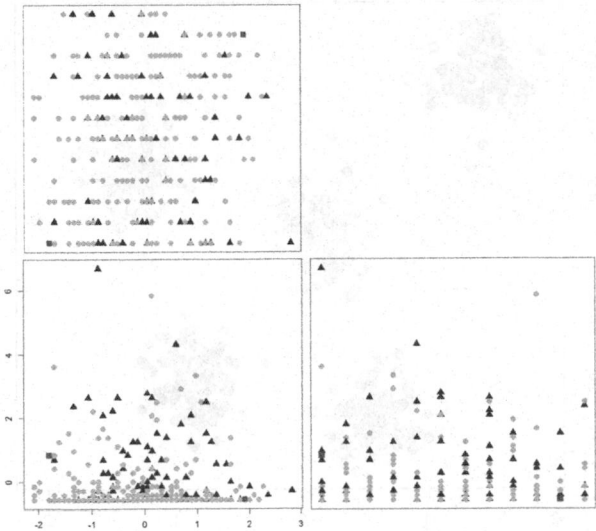

Fig. 3. Haberman data set with detected seeds

build a good benchmark data bases. We can also improve our method with allowing an overlapping degree between different extracted clusters. This approach allows us to apply our method on more different data sets structures that can be added on the benchmark.

5 Conclusion

This article proposes a new method that finds at the same time the optimal cluster number and proposes the initial clusters seeds without a prior knowledge of cluster number. We don't use any other clustering algorithm to evaluate our solution, so, our method can be used as standalone clustering algorithms dealing with hyperspherical clusters. Unlike other algorithms such as k-means, our approach do not need to fix a priori the clusters number. We propose a new mutation process using neighborhood search and we use, for this, an automatic cluster limit detection method. We also introduce a new combined fitness function to evaluate our solution. To deal with the problem of involving numerous parameters, we propose an Elitist Evolutionary Approach that involve numerous evolutionary algorithms (EAs). The difference between them is implemented through parameters and we select only the best concurrent solution. This proposition can also address the problem of exploration of large search space. The initial population of numerous evolutionary algorithms (EAs) is substantially different, so, we can deal with the large data sets in few number of iterations. First results are encouraging, as meaningful clusters seeds have been found from different data sets. Results showed the effectiveness of our methods on data sets that have compactness clusters structures. We can easily adapt our method with changing

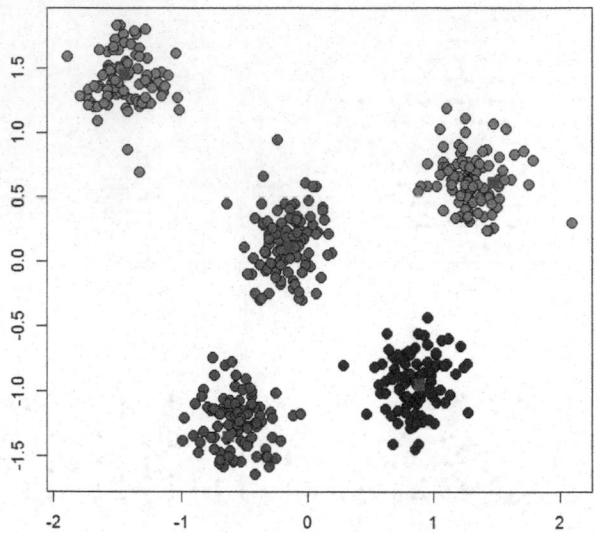

Fig. 4. Synthetic data set with detected seeds

and adapting distance measures to extract other cluster structure. Concerning further work, we plan to test our approach to different benchmark data bases to detect hyperellipsoidal clusters or other data sets structures and we think improving our methods with allowing an overlapping degree between different extracted clusters. This approach allows us to apply our method on more different data sets structures that can be added on the benchmark. We want also to apply our approach on different subspaces, we think that the optimal clustering can be different according to the subspace data projection. And then, our method can be applied on multiview clustering or on subspace clustering.

References

1. Jain, A.K.: Data clustering: 50 years beyond k-means. Pattern Recognition Letters 31(8), 651–666 (2010)
2. Mao, J., Jain, A.K.: A self-organizing network for hyperellipsoidal clustering (HEC). IEEE Transacations on Neural Networks 7(1), 16–29 (1996)
3. Celebi, M.E., Kingravi, H.A., Vela, P.A.: A comparative study of efficient initialization methods for the k-means clustering algorithm. Journal Expert Systems with Applications: An International Journal Archive 1(40), 200–210 (2013)
4. Babu, G.P., Murty, M.N.: Simulated annealing for selecting optimal initial seeds in the k-means algorithm. Indian Journal of Pure and Applied Mathematics 25(1-2), 85–94 (1994)
5. Babu, G.P., Murty, M.N.: A near-optimal initial seed value selection in k-means algorithm using a genetic algorithm. Pattern Recognition Letters 14(10), 763–769 (1993)

6. Jain, A.K., Murty, M.N., Flynn, P.J.: Data clustering: A review. ACM Computing Surveys 31(3), 264–323 (1999)
7. Linde, Y., Buzo, A., Gray, R.: An algorithm for vector quantizer design. IEEE Transactions on Communications 28(1), 84–95 (1980)
8. Huang, C.M., Harris, R.W.: A comparison of several vector quantization code-book generation approaches. IEEE Transactions on Image Processing 2(1), 108–112 (1993)
9. Aloise, D., Hansen, P., Liberti, L.: An improved column generation algorithm for minimum sum-of-squares clustering. Mathematical Programming, 1–26 (2010)
10. Sarma, J., De, J.: Generation gap methods. In: Handbook of Evolutionary Computation, vol. 2(7), pp. 1–5 (1997)
11. Qasem, S.N., Shamsuddin, S.M.: Memetic Elitist Pareto Differential Evolution algorithm based Radial Basis Function Networks for classification problems. Original Research Article Applied Soft Computing 8(11), 5565–5581 (2011)
12. Das, S., Abraham, A., Konar, A.: Automatic kernel clustering with a Multi-Elitist Particle Swarm Optimization Algorithm. Pattern Recognition Letters 5(29), 688–699 (2008)
13. Gou, S., Zhuang, X., Li, Y., Xu, C., Jiao, L.C.: Multi-elitist immune clonal quantum clustering algorithm. Neurocomputing 101(4), 275–289 (2013)
14. Milligan, G., Cooper, M.: An examination of procedures for determining the number of clusters in a data set. Psychometrika 50, 159–179 (1985)
15. Calinski, T., Harabasz, J.: A dendrite method for cluster analysis. Communications in Statistics Simulation and Computation 3(1), 1–27 (1974)
16. Palshikar, G.: Simple algorithms for peak detection in time-series. In: Proceedings of 1st International Conference on Advanced Data Analysis Business Analytics and Intelligence (2009)
17. Radcliffe, N.J.: Equivalence class analysis and presentation of strong rules. In: Knowledge Discovery in Database, vol. 11, pp. 229–248 (1991)
18. Blake, C.L., Merz, C.J.: UCI repository of machine learning databses. University of California, Irvine, Dept. of Information and Computer Sciences (1998), http://archive.ics.uci.edu/ml/datasets.html (accessed on May 2013)
19. Carr, D.B., Littlefield, R.J., Nicholson, W.L.: Scatter-plot matrix techniques for large N. Journal of the American Statistical Association 82(398), 424–436 (1987)

Crowdordering

Toshiko Matsui[1], Yukino Baba[1], Toshihiro Kamishima[2], and Hisashi Kashima[1,3]

[1] The University of Tokyo
matsui@sr3.t.u-tokyo.ac.jp,
{yukino_baba,kashima}@mist.i.u-tokyo.ac.jp
[2] National Institute of Advanced Industrial Science and Technology (AIST)
mail@kamishima.net
[3] JST PRESTO

Abstract. Crowdsourcing is a promising solution to problems that are difficult for computers, but relatively easy for humans. One of the biggest challenges in crowdsourcing is quality control, since high quality results cannot be expected from crowdworkers who are not necessarily very capable or motivated. Several statistical crowdsourcing quality control methods for binary and multinomial questions have been proposed. In this paper, we consider tasks where crowdworkers are asked to arrange multiple items in the correct order. We propose a probabilistic generative model of crowd answers by extending a distance-based order model to incorporate worker ability, and propose an efficient estimation algorithm. Experiments using real crowdsourced datasets show the advantage of the proposed method over a baseline method.

1 Introduction

Crowdsourcing offers online marketplaces where specific tasks can be outsourced to a large group of people. With the recent expansion of the use of crowdsourcing platforms, such as Amazon Mechanical Turk, various professional and non-professional tasks, including audio transcription, article writing, language translation, program coding, and graphic designing, can now easily be outsourced. The popularity of crowdsourcing is increasing exponentially in computer science as well, and researchers exploit it as an efficient and inexpensive way to process a large number of tasks that humans can perform much more easily than computers, such as image annotation and web content categorization. Crowdsourcing has been successfully applied to such fields as natural language processing, computer vision, and human computer interaction [1–4].

One of the most challenging problems in crowdsourcing research is achieving *quality control* to ensure the quality of crowdsourcing results, because there is no guarantee that the ability of all workers is sufficient to complete the offered tasks at a satisfactory level of quality. Moreover, it is known that some untrustworthy workers try to receive remuneration while expending as little effort as possible, which results in outputs of no value. Most crowdsourcing platforms allow requesters to check the submitted results and to reject low-quality results; however, if their volume is large, realistically, they cannot all be checked manually.

One popular approach to the quality control problem is to use tasks with known correct answers to evaluate the ability of each worker. This approach has been implemented

V.S. Tseng et al. (Eds.): PAKDD 2014, Part II, LNAI 8444, pp. 336–347, 2014.
© Springer International Publishing Switzerland 2014

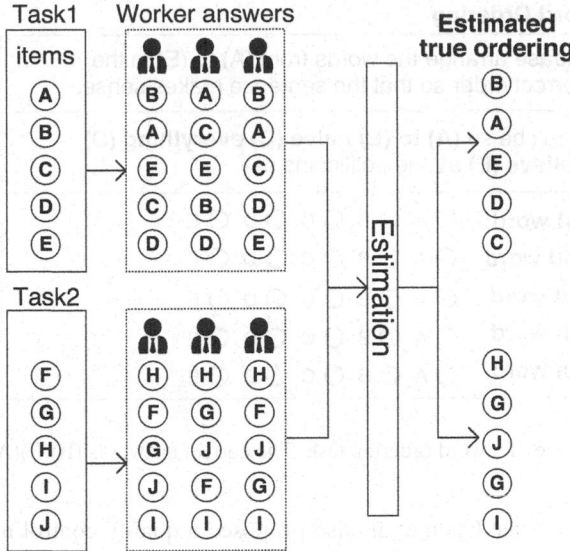

Fig. 1. Overview of quality control problem for item ordering tasks in crowdsourcing. The objective is to estimate true ordering of given items from crowd-generated answers for each item ordering task.

on several commercial crowdsourcing platforms such as CrowdFlower; however, its usage is limited because of the high cost of preparing the correct answers or the difficulty of determining one unique answer. Another promising approach is to introduce *redundancy*. A single task is assigned to multiple workers, and their responses are aggregated by majority voting [5] or more sophisticated statistical aggregation techniques that consider the characteristics of each worker or task, such as the ability of each worker and the difficulty of each task [6–8].

In most existing approaches, it is assumed that the tasks are binary questions to which binary answers (e.g., "yes" or "no") are expected, or multiple-choice questions. Only a few methods have been proposed that extend the applicability of the aggregation-based quality control approach to more general crowdsourcing tasks [9]. Following the the same line, we consider *item ordering tasks*, where workers are asked to arrange multiple items in the correct order. Item ordering tasks, typical examples of which are the ranking of web search results and ordering of items in a to-do list in according to their dependencies [10], are frequently posted on crowdsourcing sites.

In this paper, we propose an aggregation-based statistical quality control method for item ordering tasks. We model the generative process of a worker response (i.e., an ordering of items) using a distance-based probabilistic ordering model [11]. The ability of each worker is naturally incorporated into the concentration parameter of the distance-based model. We also present an effective algorithm for estimating the true ordering, which is particularly efficient because the Spearman distance [12] is employed as the distance measure between two different orderings of items.

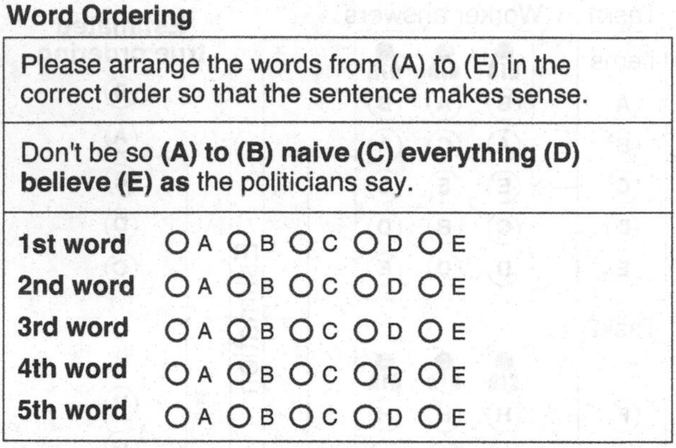

Word Ordering

Please arrange the words from (A) to (E) in the correct order so that the sentence makes sense.				
Don't be so **(A) to (B) naive (C) everything (D) believe (E) as** the politicians say.				
1st word	○ A ○ B ○ C ○ D ○ E			
2nd word	○ A ○ B ○ C ○ D ○ E			
3rd word	○ A ○ B ○ C ○ D ○ E			
4th word	○ A ○ B ○ C ○ D ○ E			
5th word	○ A ○ B ○ C ○ D ○ E			

Fig. 2. Example of a word ordering task. The correct answer is (B)(E)(A)(D)(C).

It should be noted that Chen et al. also proposed a quality control method for item ordering tasks [13] based on a pairwise ranking model; however, their method focuses on finding the correct ordering of a single (large) set of items, whereas our method focuses on solving multiple different (relatively small) ordering tasks simultaneously. Additionally, since their method is based on pairwise comparisons, it is not always suitable for tasks where more than two items are needed to determine their correct order. Fig. 2 shows an example of such a task.

We describe our experiments in which word and sentence ordering tasks were posted on a commercial crowdsourcing marketplace. We compare our quality control method to an aggregation method that does not consider the abilities of workers. The experimental results show that our method achieves answers that are more accurate than those of baseline method.

In summary, this paper makes three main contributions:

1. We address the quality control problem for a set of item ordering tasks (Section 2).
2. We propose a generative model of worker responses to item ordering tasks that extend a distance-based probabilistic ordering model to incorporate the ability of each worker (Section 3).
3. We introduce an efficient algorithm to estimate the true ordering from multiple worker responses (Section 4).

2 Crowdsourcing Quality Control for Item Ordering Tasks

We first define the crowdsourcing quality control problem related to item ordering tasks, where each ordering task requires crowdworkers to place given items in the correct order. We then present a model for aggregating the answers collected from multiple workers to obtain answers that are more accurate.

Let us assume I ordering tasks, whose i-th task has M_i items to be ordered. The true order is represented as a *rank vector* $\boldsymbol{\pi}_i = (\pi_{i,1}, \pi_{i,2}, \ldots, \pi_{i,M_i})$, where $\pi_{i,j}$

indicates the position of item j of task i in the true order of the items of M_i [11]. For example, for a task with five items indexed as $1, 2, 3, 4$, and 5, whose true order is given as $(2, 4, 1, 3, 5)$, the true rank vector is $(3, 1, 4, 2, 5)$. Note that π_i is a permutation of $(1, 2, \ldots, M_i)$.

We resort to crowdsourcing to obtain estimates for the true rank vectors. It is assumed that a total of K crowdworkers is employed. In the following, $\mathcal{I}^{(k)}$ denotes the indices of tasks on which the k-th worker works, and $\mathcal{K}_i$ denotes the indices of the workers who work on the i-th task. $\pi_i^{(k)} = (\pi_{i,1}^{(k)}, \pi_{i,2}^{(k)}, \ldots, \pi_{i,M_i}^{(k)})$ denotes the rank vector that the k-th worker gives to the i-th item ordering task.

Our goal is to estimate the true rank vectors $\{\pi_i\}_{i \in \{1,2,\ldots,I\}}$ given the (unreliable) rank vectors $\{\pi_i^{(k)}\}_{k \in \{1,2,\ldots,K\}, i \in \mathcal{I}^{(k)}}$ collected using crowdsourcing.

3 Model

To resolve the issue of the aggregation problem of the crowd-generated answers to item ordering tasks, we present a statistical model of the generative process of worker responses, so that we apply statistical inference to estimate the true order from the observed responses.

3.1 Distance-Based Model for Orders

We first review the probabilistic ordering model on which our generative model of crowdworker responses is based. We chose a distance-based model [11] from several variations of the ordering models. A distance-based model gives the probability of a rank vector $\tilde{\pi}$, given a modal order π and a concentration parameter λ, namely,

$$\Pr[\tilde{\pi} \mid \pi, \lambda] = \frac{1}{Z(\lambda)} \exp\left(-\lambda d(\tilde{\pi}, \pi)\right),$$

where $d(\cdot, \cdot)$ denotes a distance between two rank vectors, and $Z(\lambda)$ is a normalizing constant given as

$$Z(\lambda) = \sum_{\tilde{\pi}} \exp\left(-\lambda d(\tilde{\pi}, \pi)\right).$$

Specifically, we employ the Euclidean distance (also referred to as the *Spearman distance* in the ranking model literature) due to its convenience for deriving an effective parameter estimation method, which will be described later. The distance-based model in which the Spearman distance is applied is called the Mallows θ model [12].

3.2 Extension of the Distance-Based Model for the Crowdsourcing Setting

In crowdsourcing, some workers may have sufficient abilities to provide accurate orders, while some are unskilled and often submit wrong orders. To capture such worker characteristics, we incorporate the worker dependent concentration parameters into the distance-based ordering model. Namely, it is assumed that the k-th worker has his/her own personal concentration parameter $\lambda^{(k)}$, and the generative model for the worker is then given as

$$\Pr[\tilde{\pi} \mid \pi, \lambda^{(k)}] = \frac{1}{Z(\lambda^{(k)})} \exp\left(-\lambda^{(k)} d(\tilde{\pi}, \pi)\right).$$

In this model, the answer of a worker who has a high concentration parameter $\lambda^{(k)}$ is likely to be an accurate order whose distance from the true order (i.e., the modal order π) is small. Therefore, we can interpret the personal concentration parameter $\lambda^{(k)}$ as the ability parameter of the k-th worker.

4   Estimation

Based on the distance-based crowd-ordering model introduced in the previous section, we introduce a maximum likelihood estimation method to obtain estimates for the true rank vectors as well as the worker ability parameters. Our strategy for optimization is to repeat two optimization steps: optimizing the true rank vector and optimizing the worker ability.

4.1   Objective Function

We apply the maximum likelihood estimation to estimate the true rank vector $\{\pi_i\}_i$ and the worker ability parameters $\{\lambda^{(k)}\}_k$, given the crowd-generated rank vectors $\{\pi_i^{(k)}\}_{i,k}$. The objective function for the maximization problem is the log-likelihood function L, given as

$$L(\{\lambda^{(k)}\}_k, \{\pi_i\}_i) = \sum_k \sum_{i \in \mathcal{I}^{(k)}} \log \frac{1}{Z(\lambda^{(k)})} \exp\left(-\lambda^{(k)} d(\pi_i^{(k)}, \pi_i)\right)$$

$$= -\sum_k \sum_{i \in \mathcal{I}^{(k)}} \left\{\lambda^{(k)} d(\pi_i^{(k)}, \pi_i) + \log \sum_{\tilde{\pi}} \exp\left(-\lambda^{(k)} d(\tilde{\pi}, \pi_i)\right)\right\}. \quad (1)$$

4.2   Optimization

Our strategy for optimizing the objective function (1) w.r.t. $\{\lambda^{(k)}\}_k$ and $\{\pi_i\}_i$ is to repeat the two optimization steps, that w.r.t. $\{\lambda^{(k)}\}_k$ and that w.r.t. $\{\pi_i\}_i$. Since L is *not* a convex function, and therefore, its solution depends on the initial parameters, we start with the solution assuming all workers have equal abilities, specifically, $\lambda^{(k)} = \lambda^{(\ell)}$ for an arbitrary pair of k and ℓ.

One major virtue of our model is that the optimization problem is decomposable with respect to each worker and task, that is, each small optimization problem solved at each iteration step depends always on one single variable (a worker ability or a mode order), so that the computational cost linearly is dependent on the numbers of workers and tasks.

Optimization w.r.t. True Rank Vectors. Given that all the worker ability parameters $\{\lambda^{(k)}\}_k$ are fixed, the true rank vectors $\{\pi_i\}_i$ are obtained by maximizing the first term of the objective function (1). Optimization with respect to $\{\pi_i\}_i$ is a combinatorial optimization problem that is often computationally hard to solve; however, we are able to solve it efficiently by employing the Spearman distance as the distance measure $d(\cdot, \cdot)$.

The optimal true rank vector π_i for task i is given as follows[1]. First, for each item $m(= 1, \ldots, M_i)$, we calculate a *weighted rank* $w_{i,m}$, which is a weighted mean of the ranks given by workers weighted by the worker abilities,

$$w_{i,m} = \frac{1}{|\mathcal{K}_i|} \sum_{k \in \mathcal{K}_i} \lambda^{(k)} \pi_{i,m}^{(k)}.$$

The maximum likelihood estimator of the true item ordering is given by sorting the items by $w_{i,1}, w_{i,2}, \ldots, w_{i,M_i}$ in ascending order. It should be noted that each $\{\pi_i\}_i$ is obtained independently of the others.

Optimization w.r.t. Worker Ability Parameters. Optimization with respect to the worker ability parameters $\lambda^{(k)}$ with fixed true rank vectors $\{\pi_i\}_i$ is performed by numerical optimization. The objective function (1) is represented as the sum of the different objective functions $\{J^{(k)}\}_k$, where $J^{(k)}$ for each k is defined as

$$J^{(k)}(\lambda^{(k)}) = \lambda^{(k)} d(\pi_i^{(k)}, \pi_i) + \log \sum_{\tilde{\pi}} \exp\left(-\lambda^{(k)} d(\tilde{\pi}, \pi_i)\right).$$

Noting that $J^{(k)}(\lambda^{(k)})$ depends only on $\lambda^{(k)}$, we can consider K independent optimization problem with only one variable.

Since only a single variable function $J^{(k)}(\lambda^{(k)})$ needs to be considered to optimize $\lambda^{(k)}$, the optimization is easily performed by applying a standard optimization method. In the experiments, we employed a simple gradient descent method.

5 Experiments

We collected two crowdsourced datasets, one for word ordering tasks, and the other for sentence ordering tasks. We experimentally evaluated the advantages of our model as compared to a baseline method.

5.1 Datasets

We collected two datasets using Lancers[2], which is a general purpose crowdsourcing marketplace. Table 1 gives the general statistics of the datasets.

Word Ordering. Word ordering is a task whose objective is to order given English words into a grammatically correct sentence. The word ordering problem can be a

---

[1] Due to the space limitation, we omit the proof of the optimality.
[2] http://lancers.jp

Table 1. Statistics about the datasets

	#tasks	#workers	Avg. #items per task	Reward for each task	#all obtained orderings
Word ordering	20	15	5.2	$0.05	300
Sentence ordering	13	15	5.1	$0.07	195

subproblem of machine translation between languages with different grammatical word ordering, such as English to Japanese translation. Although several methods have been proposed to solve this ordering problem [14], computer programs still cannot easily perform this task. However, humans, especially the native speakers of the target language, can skillfully perform the word ordering tasks. The workers were given an English sentence with five or six randomly shuffled words, and asked to correct the order of the words. An example of the task is given in Fig. 2. Since we had the correct order of each sentence as the ground truth, we could evaluate the accuracies of our estimation results.

Sentence Ordering. Sentence ordering is a task in which given sentences are ordered such that the aligned texts logically make sense. It emulates several tasks that we presume are posted in crowdsourcing marketplaces, for example, to revise a piece of writing such that its focal point is emphasized more clearly ,or ordering items in a to-do list by their dependencies [10]. In each sentence ordering task, a paragraph consisting of five or six sentences whose order was permuted was presented to the workers, and they were requested to arrange the sentences correctly. Fig. 3 shows an example of the sentence ordering task.

5.2 Results

We applied our method to the two crowd-generated datasets, and calculated the Spearman distance (i.e., the squared error) between each estimated rank vector and the ground truth rank vector. We also tested a baseline method that does not consider the workers' ability. Concretely, we fixed the worker ability parameter $\lambda^{(k)} = 1$ for all workers k, and then optimized the objective function (1) with respect only to $\{\pi_i\}_i$. It should be noted that our proposed method uses the solution of this baseline method as the initial parameters.

The number of workers involved in each task directly affects the monetary cost of posting tasks to an actual crowdsourcing marketplace. In order to investigate the impact on the estimation accuracy engendered by the number of workers assigned to each task, we randomly selected n (ranging from 3 to 15) workers from the all workers for each task, and only used the responses of the selected workers for the estimation. We examined the averaged estimation errors of 50 trials. The results are shown in Fig. 4.

In the word ordering task, our proposed method drastically reduced the estimation error of the baseline method when the number of workers assigned to each task was more than four. It is worth mentioning that the averaged squared error of our method was

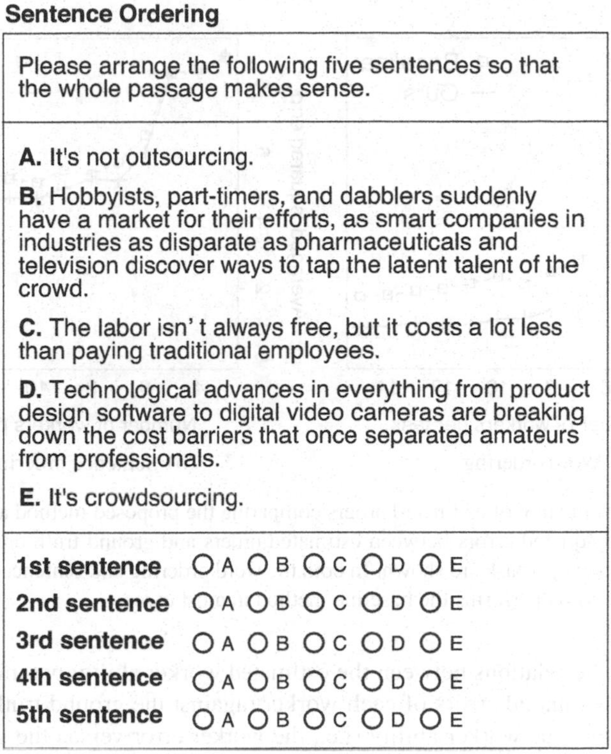

Sentence Ordering

Please arrange the following five sentences so that
the whole passage makes sense.

A. It's not outsourcing.

B. Hobbyists, part-timers, and dabblers suddenly
have a market for their efforts, as smart companies in
industries as disparate as pharmaceuticals and
television discover ways to tap the latent talent of the
crowd.

C. The labor isn't always free, but it costs a lot less
than paying traditional employees.

D. Technological advances in everything from product
design software to digital video cameras are breaking
down the cost barriers that once separated amateurs
from professionals.

E. It's crowdsourcing.

1st sentence   ○ A ○ B ○ C ○ D ○ E

2nd sentence   ○ A ○ B ○ C ○ D ○ E

3rd sentence   ○ A ○ B ○ C ○ D ○ E

4th sentence   ○ A ○ B ○ C ○ D ○ E

5th sentence   ○ A ○ B ○ C ○ D ○ E

Fig. 3. Example of a sentence ordering task. The passage in this example is from *The Rise of Crowdsourcing* by Jeff Howe. The correct answer is (D)(B)(C)(A)(E).

only 0.902 when all the collected responses were used, while the squared error easily reached 2. When the order of a pair of items that were adjacent in the correct order were incorrectly estimated, the squared error was 2. For example, a rank vector was estimated as $(2, 1, 3, 4, 5)$, when the correct one was $(1, 2, 3, 4, 5)$. This result implies that our method reduces the number of such errors by approximately half.

Our method outperformed the baseline method in the sentence ordering task as well, when the number of workers assigned to each task was more than five. Since the sentence ordering task is generally more difficult than the word ordering task, the averaged estimation errors of both the proposed and baseline methods in the sentence ordering task increased as compared with those in the word ordering task. The best result achieved by our method was a squared error of 4.25, which is relatively large; however, considering the expected squared errors when using random guessing is 21.2, it can be said the result is acceptable. In addition to the Spearman distance, we compared our method and the baseline method in three different measures shown in Table 2. The results in all the measures demonstrated the performance improvement of our method in both the word ordering and sentence ordering tasks.

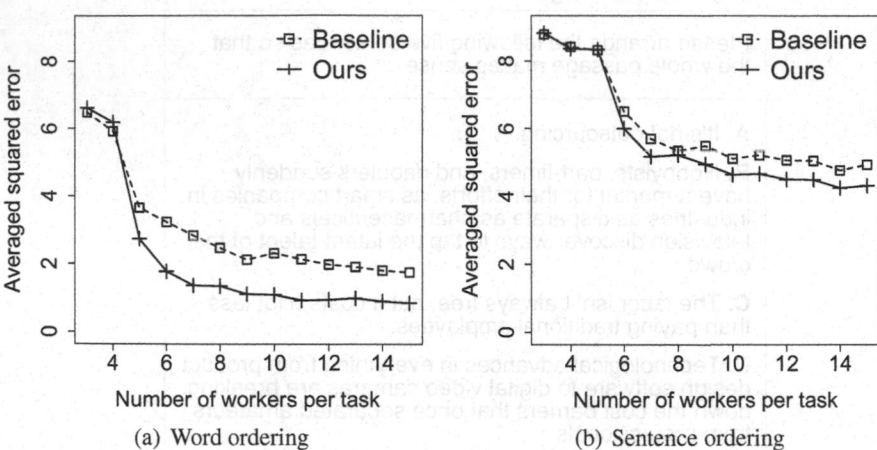

Fig. 4. Accuracy evaluation of estimated orders comparing the proposed method and the baseline method. Averaged squared errors between estimated orders and ground truth orders along with the number of workers per task are shown. In both the word ordering and sentence ordering tasks, the proposed method outperforms the baseline method in most cases.

Fig. 5 shows the relations between the estimated worker ability parameters $\{\lambda^{(k)}\}_k$ and the averaged squared errors of each worker (against the ground truths). These results show that the true worker ability (i.e., the worker error versus the ground truths) certainly varies from person to person, and that the proposed method gives higher weights to superior workers, which explains its improved performance. In fact, the estimated worker abilities and the worker errors showed strong negative correlations of -0.853 and -0.695 for the word ordering tasks and the sentence ordering tasks, respectively.

Finally, we mention the scalability of our proposed method. Generally, estimated orders show convergence after five to ten iterations. The ability parameters for good workers require more iterations than those for inferior workers. As discussed before, the complexity of each iteration depends linearly on the numbers of workers and tasks.

In summary, we verified that the proposed method shows clear advantages as compared to the baseline method for estimating the correct orders in both the word ordering and sentence ordering tasks. We also confirmed that the proposed method precisely estimates worker ability.

6   Related Work

One of the fundamental challenges in crowdsourcing is controlling the quality of the obtained data. Crowdworkers are rarely trained and they do not necessarily have adequate ability to complete the tasks [3]. There also exist large differences between the

Table 2. Evaluation of estimated orders in several measures. *Error rate* is the fraction of tasks where the estimated order did not exactly match with the ground truth order, *Hamming distance* counts the number of items whose position is different from the ground truth, and *Kendall distance* counts the number of item pairs who are in the opposite order of the ground truth. Average Hamming distance and Kendall distance of task are shown. Number of workers per task is fixed to 15. The results in all of these measures clearly indicate that the proposed method is superior to the baseline in both the word ordering and sentence ordering tasks.

Task	Method	Error rate	Avg. Hamming distance	Avg. Kendall distance
Word ordering	Baseline	0.350	0.800	0.600
	Ours	0.200	0.500	0.350
Sentence ordering	Baseline	0.769	2.231	1.692
	Ours	0.615	1.923	1.385

skills of individual workers. Moreover, a number of malicious workers participate in crowdsourcing [15]. They are motivated by financial rewards and try to complete the tasks as quickly as possible with the minimum effort by providing illogical submissions.

A widely used approach is to obtain multiple submissions from different workers and aggregate them by applying a majority vote [5] or other rules. Dawid and Skene addressed the problem of aggregating the medical diagnoses of multiple doctors to achieve more accurate decisions [6]. Smyth et al. applied the method to the problem of inferring the true labels of images from multiple noisy labels [16]. Whitehill et al. explicitly modeled the difficulty of each task [7], and Welinder et al. introduced the difficulty of each task for each worker [8]. The usage of these methods is limited to the tasks that constitute binary or multiple-choice questions; however, the tasks in crowdsourcing comprise varied types of questions. A few methods have been proposed to extend the applicability of the aggregation-based quality control approach to more general crowdsourcing tasks [9].

Although the probabilistic models for ranking have been widely studied [11], only a few studies in the literature focused on item ordering tasks in the context of crowdsourcing. Chen et al. proposed a quality control method for item ordering tasks [13] based on a pairwise ranking model; however, their method aims to find the correct ordering of a single, large set of items, while our method focuses on solving multiple different (relatively small) ordering tasks simultaneously. Additionally, since their method is based on pairwise comparisons, it is not always suitable for tasks where more than two items are needed to decide their correct positions. Wu et al. also employed the general distance-based model in the context of *learning to rank* from multiple annotators [17], while our approach employs a more specific distance measure, i.e., Spearman distance, so that the inference is more simple and efficient.

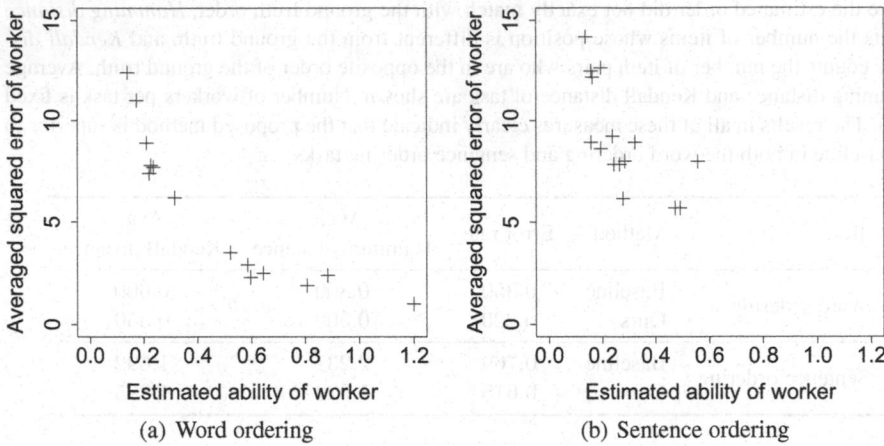

(a) Word ordering   (b) Sentence ordering

Fig. 5. Accuracy evaluation of estimated worker abilities. Relations between averaged squared error of each worker's responses to ground truths and estimated worker ability are shown. Strong negative correlations (-0.853 and -0.695 for the word ordering tasks and the sentence ordering tasks, respectively) are confirmed.

7 Conclusion

We addressed the problem of quality control for item ordering tasks in crowdsourcing, where multiple workers are asked to perform each task, which consists of positioning given items in the correct order. By extending a distance-based probabilistic ordering model to incorporate the ability of each worker, we built our proposed method for aggregating the collected orders to obtain more accurate orders in a setting where variability in the workers' abilities exists. We also introduced an efficient algorithm to estimate the true orders that employs the Spearman distance as the distance measure in a distance-based ordering model. Experimental results on two kinds of crowdsourcing tasks, word ordering tasks and sentence ordering tasks, showed that our method successfully achieved more accurate orders than the baseline method, which does not consider the worker's ability.

Acknowledgments. Y. Baba was supported by the Funding Program for World-Leading Innovative R&D on Science and Technology (FIRST Program).

References

1. Bernstein, M., Little, G., Miller, R., Hartmann, B., Ackerman, M., Karger, D., Crowell, D., Panovich, K.: Soylent: A word processor with a crowd inside. In: Proceedings of the 23rd Annual ACM Symposium on User Interface Software and Technology, UIST (2010)
2. Bigham, J., Jayant, C., Ji, H., Little, G., Miller, A., Miller, R., Miller, R., Tatarowicz, A., White, B., White, S., et al.: VizWiz: Nearly real-time answers to visual questions. In: Proceedings of the 23nd Annual ACM Symposium on User Interface Software and Technology, UIST (2010)

3. Snow, R., O'Connor, B., Jurafsky, D., Ng, A.Y.: Cheap and fast – but is it good? evaluating non-expert annotations for natural language tasks. In: Proceedings of the Conference on Empirical Methods in Natural Language Processing, EMNLP (2008)
4. Sorokin, A., Forsyth, D.: Utility data annotation with Amazon Mechanical Turk. In: Proceedings of the 1st IEEE Workshop on Internet Vision (2008)
5. Sheng, V.S., Provost, F., Ipeirotis, P.G.: Get another label? improving data quality and data mining using multiple, noisy labelers. In: Proceeding of the 14th ACM SIGKDD International Conference on Knowledge Discovery and Data Mining, KDD (2008)
6. Dawid, A.P., Skene, A.M.: Maximum likelihood estimation of observer error-rates using the em algorithm. Journal of the Royal Statistical Society. Series C (Applied Statics) 28(1), 20–28 (1979)
7. Whitehill, J., Ruvolo, P., Wu, T., Bergsma, J., Movellan, J.: Whose vote should count more: Optimal integration of labels from labelers of unknown expertise. In: Advances in Neural Information Processing Systems, vol. 22 (2009)
8. Welinder, P., Branson, S., Belongie, S., Perona, P.: The multidimensional wisdom of crowds. In: Advances in Neural Information Processing Systems, vol. 23 (2010)
9. Lin, C., Mausam, M., Weld, D.: Crowdsourcing control: Moving beyond multiple choice. In: Proceedings of the 28th Conference on Uncertainty in Artificial Intelligence, UAI (2012)
10. Zhang, H., Law, E., Miller, R., Gajos, K., Parkes, D., Horvitz, E.: Human computation tasks with global constraints. In: Proceedings of the 2012 ACM Annual Conference on Human Factors in Computing Systems, CHI (2012)
11. Marden, J.I.: Analyzing and Modeling Rank Data, vol. 64. CRC Press (1995)
12. Mallows, C.L.: Non-null ranking models. I. Biometrika 44, 114–130 (1957)
13. Chen, X., Bennett, P.N., Collins-Thompson, K., Horvitz, E.: Pairwise ranking aggregation in a crowdsourced setting. In: Proceedings of the 6th ACM International Conference on Web Search and Data Mining, WSDM (2013)
14. Chang, P.C., Toutanova, K.: A discriminative syntactic word order model for machine translation. In: Proceedings of the 47th Annual Meeting of the Association for Computational Linguistics: Human Language Technologies, ACL (2007)
15. Eickhoff, C., de Vries, A.: How crowdsourcable is your task? In: Proceedings of the Workshop on Crowdsourcing for Search and Data Mining, CSDM (2011)
16. Smyth, P., Fayyad, U., Burl, M., Perona, P., Baldi, P.: Inferring ground truth from subjective labelling of venus images. In: Advances in Neural Information Processing Systems, vol. 7 (1995)
17. Wu, O., Hu, W., Gao, J.: Learning to rank under multiple annotators. In: Proceedings of the 22nd International Joint Conference on Artificial Intelligence (IJCAI), pp. 1571–1576 (2011)

Document Clustering with an Augmented Nonnegative Matrix Factorization Model

Zunyan Xiong, Yizhou Zang, Xingpeng Jiang, and Xiaohua Hu

College of Computing and Informatics, Drexel University, Philadelphia, USA
{zunyan.xiong,yizhou.zang,xingpeng.jiang,
xiaohua.hu}@drexel.edu

Abstract. In this paper, we propose an augmented NMF model to investigate the latent features of documents. The augmented NMF model incorporates the original nonnegative matrix factorization and the local invariance assumption on the document clustering. In our experiment, first we compare our model to baseline algorithms with several benchmark datasets. Then the effectiveness of the proposed model is evaluated using datasets from CiteULike. The clustering results are compared against the subject categories from Web of Science for the CiteULike dataset. Experiments of clustering on both benchmark data sets and CiteULike datasets outperforms many state of the art clustering methods.

Keywords: nonnegative matrix factorization, graph Laplacian, regularization, clustering, social tagging.

1 Introduction

Document clustering is an unsupervised machine learning technique aims to discover the classification of documents according to their similarities. So far many document clustering methods have been proposed, such as k-means [1], spectral clustering [2], non-negative matrix factorization (NMF) [3][4], and Probabilistic Latent Semantic Analysis (PLSA) [5] etc.

Besides, traditional clustering methods mostly apply linear dimensional reduction to extract features of the data set. However, recent research has shown that most data structures are nonlinear. To deal with this problem, researchers referred to the idea of manifold learning. Manifold learning [6] is a nonlinear matrix dimensionality reduction approach that tries to discover the low dimensional structure for the data in the high dimension. There are several popular manifold learning methods, such as Isomap [7][8][9], Locally Linear Embedding [10], and Laplacian Eigenmaps [11] etc. In [12], Cai et al. proposed a graph regularized non-negative matrix factorization (GNMF). GNMF embedded manifold learning into nonnegative matrix factorization by means of local invariance assumption. Experiments proved that the manifold learning can significantly improve the clustering results.

One defect of the aforementioned clustering methods is that they only focus on one dimension of the data. However, to better study the cluster dataset, it's important to explore the data structure from two dimensions, because the geometrical structures of

V.S. Tseng et al. (Eds.): PAKDD 2014, Part II, LNAI 8444, pp. 348–359, 2014.

vectors in two dimensions of the matrix are independent to each other. For example, in a document-term matrix, the similarities of the document vectors are independent of the term vectors.

Motivated by addressing this problem, we propose a novel model named augmented nonnegative matrix factorization (ANMF), which incorporates both matrix factorization and manifold learning on both dimensions of the data matrix. Then, we applied the method in a social tagging system, CiteULike. One of the biggest challenges here is how to establish a reliable clustering evaluation method. Since most contents of the social tagging systems are created by users, rarely any criterion exists to classify the contents, not to mention a gold standard for evaluation. In this paper, the dataset applied for the experiment is from CiteULike. In this paper, we solve the problem by using the subject classification from Web of Science for the CiteULike dataset. The classification provides an objective and reliable standard to test the effectiveness of algorithms.

2   Related Work

2.1   NMF

Nonnegative Matrix Factorization (NMF)[3][4][13] is widely applied in recent years and proved to be efficient and robust in various situations. NMF decomposes the original matrix to the product of two nonnegative matrices consisting of latent vectors. The factorized matrices consist of nonnegative components, which is convenient for data analysis. The relationship between the data points can be regarded as their distance of similarity on the graph.

Given a data matrix $X = [X_1, ..., X_N] \in \mathbb{R}^{M \times N}$, the goal of nonnegative matrix factorization is to find two low-rank nonnegative matrices U and V whose product can best approximate the original matrix X:

$$X \approx UV^T$$

Xu et al. apply NMF method in document clustering, and the experiment results indicate that NMF method outperforms the latent semantic indexing and the spectral clustering methods in document clustering accuracies [14]. However, one defect of NMF is that it does not preserve the relational structure of the data during the factorization process. Two documents that are originally similar in their learned latent represents maybe dissimilar after factorization.

2.2   GNMF

In[12], Cai et al. introduced Graph Regularized Nonnegative Matrix Factorization (GNMF) model, which is an extension of Nonnegative Matrix Factorization (NMF). The two nonnegative matrices represent the latent factors of the original matrix from the two dimensions respectively. GNMF model applies the local invariance assumption on the one of the two nonnegative matrices, which results in a regularization term to the NMF objective function. The main idea of the local invariance assumption is that if two data points are close to each other in the original geometry, they should still be close in the new representation after factorized [12]. This is formulated as follows:

$$O = \|X - UV^T\|^2 + \lambda Tr(V^T LV)$$

where $Tr(\cdot)$ denotes the trace of a matrix. L is a Laplacian matrix, which is defined by $L = D - W$ [15]. D is a diagonal matrix and its diagonal entries are the sum of columns or rows of the weight matrix W (for W is a symmetric matrix), i.e., $d_{ii} = \sum_{j=1}^{M} w_{ij}$.

In the experiment, two image data sets and one document corpus were selected for clustering and showed GNMF outperforms NMF model, k-means and SVD methods. A flaw of this model is that it only considers the local invariance assumption from one dimension of the data.

2.3   Document Clustering on the Web

Many document clustering methods including the aforementioned ones are developed on the relation between documents and terms [14][5][12]. However, for web applications other than textual resources, it's difficult to get direct content information. Sometimes even the textual documents are no longer available due to the instability of webpages. Therefore, it is necessary to exploit other information sources to improve the clustering effectiveness. Moreover, studies in [1] and [2] proved that compared to textual contents or keywords of Web pages, social tag information is more reliable for clustering. Besides, most traditional content-based clustering algorithms such as k-means and NMF ignore the semantic relations among terms. As a result, two documents with no common terms will be regarded as dissimilar even though they have many synonymic or semantically related terms. Therefore, it's natural to incorporate other useful information to benefit the document clustering research.

Ramage and Heymann proposed two methods in Web clustering that include both term and tag information [16]. One applied k-means in an extended vector that includes both term and tag information; the other used the term and tag information in a generative clustering algorithm based on latent Dirichlet Allocation. Their study shows that including tagging data can significantly improve the clustering quality.

Lu et al. exploited the tripartite information, i.e. resources, users and tags, for webpage clustering [17]. The authors integrated the tripartite information together for better clustering performance. Three different methods are proposed in the paper. The first method applies the structure of the tripartite network to cluster Web pages; the second method uses the tripartite information in k-means by combining two or three different vectors together; the third method utilizes the Link k-means algorithm in the tripartite information. Results indicated that all clustering methods incorporating tagging information significantly outperform the content-based clustering. Furthermore, compared to the other two methods, the tripartite network has better performance.

3   Augmented Nonnegative Matrix Factorization Model

In this section, we first propose the augmented nonnegative matrix factorization model, which simultaneously incorporates the geometric structures of both the data manifold and the feature manifold. Then, we introduce the model and its iterative algorithm in detail.

The objective function of the ANMF model is:

$$O = \|X - UV^T\|^2 + \lambda Tr(V^T LV) + \mu Tr(U^T \tilde{L} U)$$

3.1   Notations and Problem Formalization

Before describing the model, some useful definitions are introduced. Given a data set, $\mathcal{D} = \{d_1, d_2, \ldots, d_M\}$, and $\mathcal{F} = \{f_1, f_2, \ldots, f_N\}$ be the set of data features. Their relational matrix is denoted by $X = (x_{ij})_{M \times N}$, where x_{ij} denotes the weighting of the data feature f_j for the data point d_i.

For convenience, the meaning of notations used in the paper is summarized in Table 1.

Table 1. Important notation used in this paper

Notations	Description	Notations	Description
$\mathcal{D}$	data set	$\mathcal{F}$	data feature set
M	number of data points	N	number of features
X	data matrix of size $M \times N$	x_{ij}	data point in the matrix
U	data partition of size $M \times K$	V	feature partition of size $N \times K$
u_k	kth column of U	v_k	kth column of V
L	data graph Laplacian	D	data degree matrix
W	feature adjacency matrix	$\tilde{W}$	data adjacency matrix
$w_{jj'}$	a cell in the feature adjacency matrix	$\tilde{w}_{ii'}$	a cell in the data adjacency matrix
K	the number of latent component		

To analyze the similarities between the data points and the features respectively, the nonnegative matrix factorization can be applied to decompose the matrix X

$$X \approx UV^T,$$

such that the matrix U and V consists of the latent vectors associated to data points and features, i.e. the row vectors of U and V represent the latent vectors for the data d_i and feature f_j respectively. In this approach, the latent vectors are supposed to represent the factorized meaningful parts (or topic) of the data set and their features. In order to quantize the similarities of data and features, we next use the idea of local invariance assumption to obtain two regularization terms, Regularizer I and Regularizer II.

3.2   Local Invariance Assumption

The local invariance assumption is a general principle that can be interpreted in this context as following. If the data are close in some sense, after the NMF decomposition, they should still be close in the latent space. To measure the similarities between points of the original data, we construct two adjacency matrix W and $\widetilde{W}$ from X for both feature and data vectors. The metric of the adjacency (closeness) between the vectors $w_{jj'}$ (or $\widetilde{w}_{ii'}$) can be defined in different ways, such as 0-1 weighting, heat kernel weighting and dot-product weighting, the definitions of the three weighting modes can be referenced in [18]. For each data point, only the p nearest neighbors are considered.

In the NMF decomposition, let K be the number of latent component, and $K \ll M$, $K \ll N$. The data and features are mapped to points in a lower K dimensional Euclidean space. From a geometric point of view, their similarity can be easily compared by Euclidean distance.

Consider the matrix $X = (x_{ij})_{M \times N}$, let x_j be the jth column vector, i.e. $X = [x_1, x_2, \dots x_N]$. Then x_j can be regarded as the coordinates of feature f_j in the standard basis. Under the matrix decomposition,

$$X \approx UV^T,$$

Let the kth column vector of U be $u_k^{(K)} = \langle u_{ik}^{(K)} | i = 1, 2, \dots, |M| \rangle$. Then the original vector x_j is approximated by the linear combination of vectors u's:

$$x_j \approx \sum_{k=1}^K u_k v_{jk} \tag{1}$$

In this expression, now u_k's can be regarded as new basis vectors for the latent space, and the new coordinates for feature f_j hence are $v_j^{(K)} = \langle v_{jk}^{(K)} | k = 1, 2, \dots, |K| \rangle$. According to the local invariance assumption, if feature vector f_j and $f_{j'}$ are close in the original coordinates, they should still be close in the new coordinates. To quantize this information, we use the Euclidean distance in the latent space $\|f_j - f_{j'}\|$, weighted by their original closeness $w_{jj'}$. As stated previously, $w_{jj'}$ can be calculated by 0-1 weighting, heat kernel weighting or dot-product weighting. Also as in [12], we define the Regularizer I as

$$\mathcal{R}_1 = \frac{1}{2} \sum_{j,j'=1}^N \|v_j - v_{j'}\|^2 w_{jj'}$$

It can be seen heuristically that for $\mathcal{R}_1$ bounded, if $w_{jj'}$ is large, meaning features j, j' are close in the adjacency, the Euclidean distance is forced to be small, which implies the factored feature v_j, $v_{j'}$ are close. For computational convenience, we simplify the regularizer as following.

$$\mathcal{R}_1 = \frac{1}{2} \sum_{j,j'=1}^N \|v_j - v_{j'}\|^2 w_{jj'}$$

$$= \sum_{j=1}^{N} \boldsymbol{v}_j^T \boldsymbol{v}_j w_{jj} - \sum_{j,j'=1}^{N} \boldsymbol{v}_j^T \boldsymbol{v}_{j'} w_{jj'}$$

$$= Tr(V^T DV) - Tr(V^T WV) = Tr(V^T LV),$$

where $Tr(\cdot)$ denotes the trace of a matrix. D is a diagonal matrix and its diagonal entries are the sum of columns or rows of W (for W is a symmetric matrix), i.e., $d_{ii} = \sum_{j=1}^{N} w_{ij}$. The Laplacian matrix L is defined by $L = D - W$ [15].

Incorporating this information to the NMF model, the objective function now becomes

$$O = \|X - UV^T\|^2 + \lambda Tr(V^T LV) \tag{2}$$

Here λ is a regularization parameter that balances the effects of local invariance. This is the model considered in [1], called the graph regularized NMF method (GNMF).

At this point, it is important to notice that for our problem, the local invariance assumption applies to the other piece of data, the features.

To reflect the local invariance of the data, a second regularization term is added:

$$\mathcal{R}_2 = \frac{1}{2}\sum_{i,i'=1}^{N}\|\boldsymbol{u}_i - \boldsymbol{u}_{i'}\|^2 \widetilde{w}_{ii'}$$

$$= \sum_{i=1}^{N} \boldsymbol{u}_i^T \boldsymbol{u}_i d_{ii} - \sum_{i,i'=1}^{N} \boldsymbol{u}_i^T \boldsymbol{v} \boldsymbol{u}_{i'} \widetilde{w}_{ii'}$$

$$= Tr(U^T \widetilde{D} U) - Tr(U^T \widetilde{W} U)$$
$$= Tr(U^T \widetilde{L} U)$$

where $\widetilde{W}$ is the adjacency matrix for the data and $\widetilde{L} = \widetilde{D} - \widetilde{W}$. Now the final cost function can be defined as

$$O = \|X - UV^T\|^2 + \lambda Tr(V^T LV) + \mu Tr(U^T \widetilde{L} U) \tag{3}$$

We call this new model the augmented NMF (ANMF). Here the two regularization parameters λ and μ are positive numbers to be chosen later. They balance the effects of local invariance and the original NMF. Heuristically, the larger the parameters, the stronger will the local invariance be reflected in the results. The optimal solution is obtained by minimizing O over all non-negative matrices U and V. We will discuss the algorithms in next section.

3.3   Iterative Algorithm

As in the original matrix factorization model [19] or the GNMF model [12], the cost function O is not convex in W and H jointly. Thus it is not possible to find global minima. However, it is convex in W for fixed H and vice versa. In fact, the Lagrange multiplier method used in [2] is also applicable here to give an iterative algorithm. However the updating rules could only be expected converge to a local (not global) minima.

The cost function can be rewritten as

$$O = Tr(X - UV^T)(X - UV^T)^T + \lambda Tr(V^T LV) + \mu Tr(U^T \tilde{L} U)$$

$$= Tr(XX^T) - 2Tr(XVU^T) + Tr(UV^T VU^T) + \lambda Tr(V^T LV) + \mu Tr(U^T \tilde{L} U)$$

Here the basic properties $Tr(A) = Tr(A^T)$ and $Tr(AB) = Tr(BA)$ are used for any matrices A and B. Next let ψ_{ik} be the Lagrange multiplier for the condition $u_{ik} \geq 0$, and ϕ_{jk} be the multiplier for the condition $v_{jk} \geq 0$. The augmented Lagrangian is

$$\mathcal{L} = Tr(XX^T) - 2Tr(XVU^T) + Tr(UV^T VU^T) + \lambda Tr(V^T LV) + \mu Tr(U^T \tilde{L} U) + Tr(\Psi U) + Tr(\Phi V^T) \tag{4}$$

where $\Psi = (\psi_{ik})_{M \times K}$ and $\Phi = (\phi_{jk})_{N \times K}$. The partial derivatives are

$$\frac{\partial \mathcal{L}}{\partial U} = -2XV + 2UV^T V + 2\mu \tilde{L} U + \Psi \tag{5}$$

$$\frac{\partial \mathcal{L}}{\partial V} = -2XU + 2VU^T U + 2\lambda LV + \Phi \tag{6}$$

The derivatives vanish at local minima. Using the Karush-Kuhn-Tucker (KKT) condition, $\psi_{ik} u_{ik} = 0, \phi_{ik} v_{ik} = 0$. The equations (5) and (6) become

$$-(XV)_{ik u_{ik}} + (UV^T V)_{ik u_{ik}} + \mu(\tilde{L} U)_{ik u_{ik}} = 0 \tag{7}$$

$$-(X^T U)_{jk v_{jk}} + (VU^T U)_{jk v_{jk}} + \lambda(LU)_{jk u_{jk}} = 0 \tag{8}$$

These equations give the following updating rules

$$u_{ik} \leftarrow u_{ik} \frac{(XV + \mu \hat{W} U)_{ik}}{(UV^T V + \mu \tilde{D} U)_{ik}} \tag{9}$$

$$v_{jk} \leftarrow v_{jk} \frac{(X^T U + \lambda WV)_{jk}}{(VU^T U + \lambda DV)_{jk}} \tag{10}$$

The updating rules of our model actually lead to convergence sequences, which are justified by Theorem 1 and its proof below.

Theorem 1. Under the updating rules (9) and (10), the objective function (3) is non-increasing.

As in [12] and [18], the proof of Theorem 1 is essentially based on the existence of a proper auxiliary function for the ANMF. We give a simple proof on the ground of the following results from [12].

Lemma 2. Under the updating rule

$$v_{jk} \leftarrow v_{jk} \frac{(X^T U + \lambda WV)_{jk}}{(VU^T U + \lambda DV)_{jk}} \tag{11}$$

The cost function O_1 in GNMF, i.e.

$$O_1 = \|X - UV^T\|^2 + \lambda Tr(V^T LV) \tag{12}$$

is non-increasing.

Proof of Theorem 1. Consider the objective function O under the updating of V by (11). Then the last term $\mu Tr(U^T \tilde{L} U)$ in O will not change. It suffices to prove $O_V = O - \mu Tr(U^T \tilde{L} U) = O_1$ is non-increasing, which is exactly given by Lemma 2. Next consider O under the updating of U. Since H is not changed, it suffices to consider

$$O_U = O - \lambda Tr(V^T LV) = \|X - UV^T\|^2 + \mu Tr(U^T \tilde{L} U)$$
$$= \|X^T - VU^T\|^2 + \mu Tr(U^T \tilde{L} U)$$

Now interchange U, V and replace μ by λ, X by X^T in Lemma 2, O_U is not increasing under the updating of W by (9). ∎

One problem of the objective function of ANMF is that the solutions U and V are not unique. If U and V are the solutions, then UD and VD^{-1} can also be the solutions of the objective function. To obtain unique solutions, we refer to the approach from [12] that enforces the Euclidean distance of the column vectors in matrix U as one. This approach can be achieved by

$$u_{ik} \leftarrow \frac{u_{ik}}{\sqrt{\sum_i u_{ik}^2}} \tag{13}$$

$$v_{jk} \leftarrow v_{jk}\sqrt{\sum_i u_{ik}^2} \tag{34}$$

Table 2 shows the simple algorithm of ANMF model.

Table 2. Algorithm of ANMF

Input: the data matrix X, regularization parameter λ and μ.
Output: the data-topic matrix U, and the topic-feature matrix V.
Method:
Construct weighting matrix W and $\tilde{W}$, compute the diagonal matrix D and $\tilde{D}$;
Random initialize U and V;
Repeat (9) and (10) **until** convergence;
Normalize U and V using (13) and (14).

3.4   Complexity Analysis

In this section, the computational cost of NMF, GNMF and ANMF algorithms are discussed. Supposing the algorithm stops after t iterations, the overall cost for NMF is $O(tMNK)$. For GNMF, the adjacency matrix needs $O(N^2M)$ to construct, so the overall cost for GNMF is $O(tMNK + N^2M)$. As ANMF adds one more adjacency matrix on the other dimension, so the overall cost for ANMF is $O(tMNK + N^2M + M^2N)$.

4   Experiments

4.1   Data Sets and Evaluation Metrics

Before applying CiteULike data set, four data sets were chosen as the benchmark, which were Coil20, ORL, TDT2, and Reuters-21578. Two of them are image data and the other two are text data.

The results of our experiments were evaluated by Clustering Accuracy [20] and normalized mutual information (NMI) [21]. Both of the evaluation metrics range from zero to one, and a high value indicates better clustering result.

4.2   Parameter Settings

In this section, we compared our proposed method with the following methods, K-means [22], NMF [4], and GNMF [12]. For both GNMF and ANMF, we normalized the vectors on columns of W and H.

To fairly compare algorithms, each algorithm was run under different parameter settings, and the best results were selected to compare with each other. The number of clusters was set equal to the true number of standard categories for all the data sets and clustering algorithms.

The 0-1 weighting was applied in GNMF and ANMF algorithms for convenience. Here we set the nearest neighborhood p as 7 for both the algorithms. The value of p determines the construction of the adjacency matrix for both GNMF and ANMF, which lies on the assumption that the neighboring data points share the same topic. So the performance of GNMF and ANMF are supposed to decrease as p increases, which was verified by [12] for GNMF. There is only one regularization parameter in GNMF, the parameter was set by the grid $\{10^{-3}, 10^{-2}, 10^{-1}, 1, 10, 10^2, 10^3, 10^4\}$. For ANMF algorithm, there are two regularization parameters λ and μ. Both of them were set by the grid $\{10^{-3}, 10^{-2}, 10^{-1}, 1, 10, 10^2, 10^3, 10^4\}$.

The aforementioned algorithms were repeated 20 times for each parameter setting, the average results were computed. The best average results are shown in Table 3 and Table 4.

4.3   Clustering Results

Table 3 and 4 display the Accuracy and NMI of all algorithms on the four data sets respectively. We can see that overall both GNMF and ANMF performed much better than K-means and NMF algorithms. Note that both GNMF and ANMF consider the geometrical structure of the data through the local invariance assumption, the results imply the importance of the geometrical structure in mining the latent features of the data. Besides, ANMF shows the best performance in all the four data sets, which indicates that by adding the geometrical structure for the two dimensions of the data, the algorithm can achieve better performance.

Table 3. Clustering Accuracy (%)

Data Sets	k-means	NMF	GNMF	ANMF
Coil20	95.56%	95.90%	97.80%	**97.82%**
ORL	96.40%	96.01%	96.61%	**97.19%**
TDT2	90.92%	90.11%	95.09%	**95.35%**
Reuters-21578	74.04%	73.68%	74.68%	**75.13%**

Table 4. Normalized Mutual Information (%)

Data Sets	k-means	NMF	GNMF	ANMF
Coil20	73.86%	74.36%	89.17%	**90.14%**
ORL	71.82%	66.80%	72.01%	**75.24%**
TDT2	64.54%	58.75%	83.49%	**84.65%**
Reuters-21578	33.90%	29.98%	34.41%	**36.31%**

5   Study on the CiteULike Data Set

5.1   Data Processing

CiteULike is a social bookmarking platform that allows researchers to share scientific references, so nearly all the bookmarks in CiteULike are academic papers. The CiteULike data was crawled during January-December 2008. We extracted the article id, journal name of the articles, user id and tag information from the original data. The journal name of the articles was used for setting evaluation standard. Before processing the dataset, we unified the format of the tags. Tags such as "data_mining", "data-mining", "data.mining", "datamining", etc. were all considered as the same one. Here we excluded the articles, users and tags with less than four bookmarks. To evaluate the CiteULike dataset, we utilized the subject categories in Web of Science [23]. There are a total of 176 top-level subject categories for science journals. Under each subject category, they display a list of the afflicted journals. By overlapping the journals of all articles from CiteULike with the journals under the categories in Web of Science, we could discover the subject categories of the articles in CiteULike dataset. Under the 176 subject categories, we only kept the 44 biggest subject categories with the largest articles numbers. Finally, we had 3,296 bookmarks with 2406 articles, 1220 users and 4593 tags.

5.2   Clustering Results

We construct two matrices for CiteULike data set, article-user matrix and article-tag matrix. Besides, in order to test if combining the article vectors from article-user and article-tag vectors can get a better performance, we also construct a new matrix that consists of the linear combination of the article-user vectors and article-tag vectors. Just as the experiments in section 4, we compare the clustering results of ANMF with GNMF, NMF and k-means based on the Clustering Accuracy and NMI. The settings for the parameters and the value of the nearest neighborhood p are all the same as in section 4.

Table 5. The Evaluation Results for CiteULike Data Set

Data Sets	k-means	NMF	GNMF	ANMF
	Clustering Accuracy (%)			
article-user matrix	30.04%	44.22%	86.57%	**87.17%**
article -tag matrix	73.42%	76.24%	**88.46%**	88.43%
the combination matrix	68.65%	68.94%	85.48%	**87.60%**
	Normalized Mutual Information (%)			
article -user matrix	10.83%	19.03%	27.24%	**28.55%**
article -tag matrix	25.00%	27.72%	32.07%	**36.85%**
the combination matrix	26.24%	23.85%	36.91%	**42.35%**

Table 5 displays the evaluation scores of the four algorithms with CiteULike dataset. The experiments reveal several interesting points:

- ANMF still performs the best among the four algorithms. Specifically, the improvement is significant in NMI results. This shows that ANMF is efficient not only in image and text data, but also in the data from social tagging systems, which suggests the potential of ANMF in collaborative filtering area.
- The evaluation results of the combination matrix are rather poor for k-means and NMF algorithms for the article-user matrix and article-tag matrix. For GNMF and ANMF, their NMI scores are better than the other two matrices, while the Clustering Accuracy scores are a little lower.

6   Conclusion

In this paper, we have explored a graph regularized nonnegative matrix factorization model for document clustering. First, we applied our algorithm in four benchmark data sets, and compared it with three canonical algorithms to evaluate its performance in clustering. Then the algorithm was used in CiteULike dataset by applying user and tag information for analysis. The experiment results demonstrate that our algorithm outperforms GNMF, NMF and k-means models in both benchmark data sets and CiteULike data set.

References

1. Jain, A.K., Murty, M.N., Flynn, P.J.: Data clustering: a review. ACM Comput. Surv. CSUR 31(3), 264–323 (1999)
2. Guy, I., Carmel, D.: Social recommender systems. In: Proceedings of the 20th International Conference Companion on World Wide Web, pp. 283–284 (2011)
3. Lee, D.D., Seung, H.S.: Learning the parts of objects by non-negative matrix factorization. Nature 401(6755), 788–791 (1999)
4. Seung, D., Lee, L.: Algorithms for non-negative matrix factorization. Adv. Neural Inf. Process. Syst. 13, 556–562 (2001)

5. Hofmann, T.: Probabilistic latent semantic indexing. In: Proceedings of the 22nd Annual International ACM SIGIR Conference on Research and Development in Information Retrieval, pp. 50–57 (1999)
6. Law, M.H., Jain, A.K.: Incremental nonlinear dimensionality reduction by manifold learning. Pattern Anal. Mach. Intell. IEEE Trans. 28(3), 377–391 (2006)
7. Balasubramanian, M., Schwartz, E.L.: The isomap algorithm and topological stability. Science 295(5552), 7–7 (2002)
8. Bengio, Y., Paiement, J.-F., Vincent, P., Delalleau, O., Le Roux, N., Ouimet, M.: Out-of-sample extensions for lle, isomap, mds, eigenmaps, and spectral clustering. Adv. Neural Inf. Process. Syst. 16, 177–184 (2004)
9. Samko, O., Marshall, A.D., Rosin, P.L.: Selection of the optimal parameter value for the Isomap algorithm. Pattern Recognit. Lett. 27(9), 968–979 (2006)
10. Tenenbaum, J.B., De Silva, V., Langford, J.C.: A global geometric framework for nonlinear dimensionality reduction. Science 290(5500), 2319–2323 (2000)
11. Belkin, M., Niyogi, P.: Laplacian eigenmaps for dimensionality reduction and data representation. Neural Comput. 15(6), 1373–1396 (2003)
12. Cai, D., He, X., Han, J., Huang, T.S.: Graph regularized nonnegative matrix factorization for data representation. Pattern Anal. Mach. Intell. IEEE Trans. 33(8), 1548–1560 (2011)
13. Gaussier, E., Goutte, C.: Relation between PLSA and NMF and implications. In: Proceedings of the 28th Annual International ACM SIGIR Conference on Research and Development in Information Retrieval, pp. 601–602 (2005)
14. Xu, W., Liu, X., Gong, Y.: Document clustering based on non-negative matrix factorization. In: Proceedings of the 26th Annual International ACM SIGIR Conference on Research and Development in Informaion Retrieval, pp. 267–273 (2003)
15. Chung, F.R.: Spectral graph theory, vol. 92. AMS Bookstore (1997)
16. Ramage, D., Heymann, P., Manning, C.D., Garcia-Molina, H.: Clustering the tagged web. In: Proceedings of the Second ACM International Conference on Web Search and Data Mining, pp. 54–63 (2009)
17. Lu, C., Hu, X., Park, J.: Exploiting the social tagging network for Web clustering. Syst. Man Cybern. Part Syst. Humans IEEE Trans. 41(5), 840–852 (2011)
18. Matlab Codes and Datasets for Feature Learning,
    http://www.cad.zju.edu.cn/home/dengcai/Data/data.html (accessed: September 18, 2013)
19. Koren, Y., Bell, R., Volinsky, C.: Matrix factorization techniques for recommender systems. Computer 42(8), 30–37 (2009)
20. Gu, Q., Zhou, J.: Co-clustering on manifolds. In: Proceedings of the 15th ACM SIGKDD International Conference on Knowledge Discovery and Data Mining, pp. 359–368 (2009)
21. Strehl, A., Ghosh, J.: Cluster ensembles—a knowledge reuse framework for combining multiple partitions. J. Mach. Learn. Res. 3, 583–617 (2003)
22. MacQueen, J.: Some methods for classification and analysis of multivariate observations. In: Proceedings of the Fifth Berkeley Symposium on Mathematical Statistics and Probability, vol. 1, p. 14 (1967)
23. Journal Search - IP & Science - Thomson Reuters,
    http://www.thomsonscientific.com/cgi-bin/jrnlst/jlsubcatg.cgi?PC=D (accessed: October 01, 2013)

Visualization of PICO Elements for Information Needs Clarification and Query Refinement

Wan-Tze Vong and Patrick Hang Hui Then

Faculty of Engineering, Computing and Science, Swinburne University of Technology,
Sarawak Campus, Jalan Simpang Tiga, 93350 Kuching, Sarawak, Malaysia
{wvong,pthen}@swinburne.edu.my

Abstract. The UMLS semantic types and natural language processing techniques were collectively utilized to extract PICO elements from the titles and abstracts of 114 MEDLINE articles. 24 sets of PICO elements were generated from the articles based on the derivation of, and the tokenization methods and weighting schemes applied to the elements. The similarity of the I and C elements (called jointly the "Interventions") between pairs of documents was calculated using 42 similarity/distance measures. Similar interventions were grouped together using complete-/average-/ward-link hierarchical clustering. The similarity measure, Yule, performed significantly better than other measures in identifying paired interventions derived from the titles and which had been pre-processed into single term and weighted by binary term-occurrence. The clustering algorithm, complete-link, provides the most appropriate structure for the visualization of interventions. Similarity-based clustering gave a higher mean average precision than random-baseline clustering (MAP = 0.4298 vs. 0.2364) over the 25 queries evaluated.

Keywords: Hierarchical Clustering, PICO Element, Query Refinement, Similarity Measure, Distance Measure.

1   Introduction

A focused, well-defined question warrants a high quality answer. The quality of answers returned by a question-answering (QA) system depends on the quality of questions posed by users. Doctors have difficulty in generating high quality questions that unambiguously and comprehensively defined their information needs [1]. The use of PICO (an acronym for Problem/Population, Intervention, Comparison and Outcome) framework has been widely accepted for the formulation of answerable clinical questions. However, a study by [2] reported that not all clinical questions have all four PICO elements present. Two examples of questions maintained by the National Library of Medicine (NLM) [3] are "What is the best treatment for external otitis?" (Question 1) and "I have a lady with graves' disease (33 years old). She was trying to get pregnant when she was diagnosed with graves. So the question is, what is the best treatment for graves in someone who is trying to get pregnant, and if we use radioactive iodine, how long does she need to wait?" (Question 2). Both of the questions are

V.S. Tseng et al. (Eds.): PAKDD 2014, Part II, LNAI 8444, pp. 360–372, 2014.
© Springer International Publishing Switzerland 2014

categorized under "Treatment and Prevention". Question 1 represents a definitional question that contains only the P element ("external otitis"). Question 2 is described in paragraph format and contains both the P ("graves' disease", "lady") and I ("radioactive iodine") elements. An alternative intervention to the clinical condition and the expected treatment outcome, which denote the C and O elements respectively, are not stated in both Questions 1 and 2. As reported in [4], questions with the I/C and O elements are unlikely to go unanswered. Therefore, the visualization of PICO elements in documents relevant to a user's input query has the potential to assist the user in refining his/her information needs.

The paper presents a case study of utilizing similarity-based clustering to aid the visualization and exploration of interventions (i.e. the I and C elements) for the refinement of questions relating to treatments and drugs. The proposed user interface is illustrated in Fig. 1. As shown in the figure, the natural language (NL) question entered by the user contains only the P element ("breast cancer"). To assist the user in refining or clarifying his/her information needs, the user is allowed to explore a particular subject domain by browsing through the interventions which have been pre-clustered into a hierarchical structure. Simultaneously, the user can identify the interventions encompassed in each cluster and discover the relationships between the interventions. The most potent sets of PICO queries are produced by returning the interventions selected by the user (circled by black line in Fig. 1), accompanied by the P and O elements identified from the titles or abstracts. It is expected that through this process, a user can understand his/her information needs and obtain a more comprehensive knowledge about the domain of interest. An ambiguous query can also be refined by selecting the PICO query that best described the information needs.

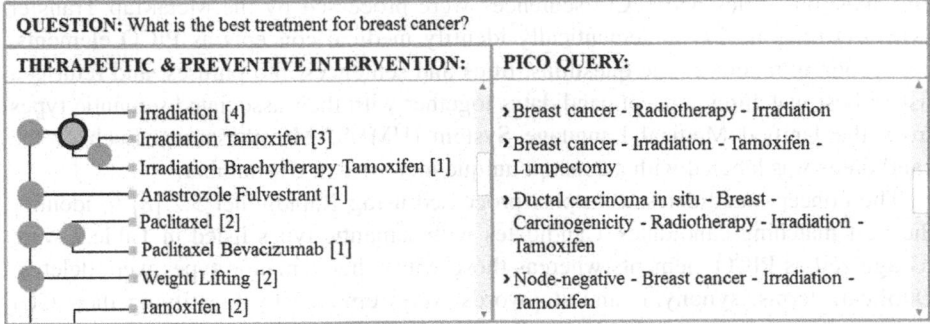

Fig. 1. The proposed user interface

2   Methodology

2.1   Collection of MEDLINE Documents

The processing of the NL question in Fig. 1 as described in Section 2.3 returns the medical concepts: "breast cancer" and "breast neoplasms". The concepts were used as the main search terms and the following filters were activated to retrieve relevant documents from the MEDLINE database: randomized controlled trial, abstract available, publication date from 2002/10/01 to 2012/10/04, humans and English. The documents were limited to those published in 7 core journals: *N Engl J Med, JAMA, Ann Inter Med, Lancet, Br Med J, BMJ* and *BMJ (Clin Res Ed)*. The titles and abstracts of the documents were collected for the extraction of PICO elements.

2.2   Generation of PICO Sentences

Based on previous studies, the position of a sentence within an abstract is useful in determining the PICO elements that the sentence carries [5,6]. Two types of abstracts were identified: structured abstracts with internal section headings such as METHODS and RESULTS, and unstructured abstracts written in paragraph format without the headings. Both structured and unstructured abstracts were cut into three segments respectively based on the headings and the position of the sentences in the abstracts (Table 1). The extracted sentences were called in the remainder of this paper the "PICO sentences".

Table 1. Derivation of PICO sentences

Representation	Internal Section Heading	Position of Sentence
P	Introduction, Background, Objective	First 3 sentences
I/C	Method	Sentences in between the first and the last 3 sentences
O	Result, Conclusion	Last 3 sentences

2.3   Generation of PICO Elements

NL questions, titles and PICO sentences were processed by the MetaMap Transfer (MMTx) program [7] to semantically identify medical concepts as PICO elements. The program tokenizes the questions, titles and sentences into phrases, and returns a list of best matching concept candidates together with their associated semantic types from the Unified Medical Language System (UMLS) Metathesaurus. Each of the candidates was labeled with a concept unique identifier (CUI) number.

The concept candidates were post-processed using Rapidminer 5.2 [8] to identify the best matching candidates. Candidates with semantic types listed in Table 2 were recognized as PICO elements whereas those with other semantic types were deleted. Duplicate terms, synonyms and stopwords were removed by identifying their CUI numbers. For instance, "blood sugar" and "blood glucose" are synonyms with the same CUI number (i.e. C0005802). Examples of stopwords are "find", "release", "peer support", "still", "little" and "inform". If candidate terms of different lengths were identified at the same location in a document, candidates with the highest number of words were selected. For example, the processing of the phrase "management of orbital cellulitis" returns the concept candidates "orbital", "cellulitis" and "orbital cellulitis". "Orbital cellulitis" is selected and the rest are removed. For each document, a list of best matching medical concepts was collected respectively from the titles and the abstracts as PICO elements.

Table 2. Identification of PICO elements by semantic types (adapted from [6])

Representation	Semantic Type
P/O	Age group, Family group, Group, Human, Patient or disabled group, Population group, Acquired abnormality, Anatomical abnormality, Cell or molecular dysfunction, Congenital abnormality, Disease or syndrome, Experimental model of disease, Finding, Injury or poisoning, Mental or behavioral dysfunction, Neoplastic process, Pathologic function, Sign or symptom.
I/C	Daily or recreational activity, Amino acid, peptide, or protein, Antibiotic, Clinical drug, Eicosanoid, Enzyme, Hormone, Inorganic chemical, Lipid, Neuroreactive substance or biogenic amine, Nucleic acid, nucleoside, or nucleotide, Organic chemical, Organophosphorus compound, Pharmacologic substance, Receptor, Steroid, Vitamin, Diagnostic procedure, Therapeutic or preventive procedure.

2.4   Preprocessing of PICO Elements

The preprocessing involves three steps: (1) the I and C elements, i.e. the "interventions", were collected from titles, abstracts or a combination from both sections ("Tile + Abstract"), (2) the interventions were tokenized using "Loose" (LO) or "Strict" (ST) method, and (3) the interventions were weighted using normalized term frequency (TF), binary term occurrence (BI), term occurrence (TO) or term frequency-inverse document frequency (TF-IDF). The tokenization methods and weighting schemes are detailed as follow:

— LO: The interventions were tokenized into single term. For instance, the phrase "ascorbic acid" is tokenized into "ascorbic" and "acid"; ST: The interventions were not tokenized. For example, the phrase "breast radiotherapy" remains unchanged.
— TF: The ratio of the frequency of a term to the maximum term frequency of any term in a document, producing a numerical value between 0 and 1; BI: The occurrence of a term in a document with a binary value of 0 or 1; TO: A nominal value obtained by calculating the number of times a term occurs in a document; TF-IDF: A numerical value calculated by multiplying the frequency of a term in a document to the inverse of the number of documents in a collection that contains the term.

The three steps described above were achieved using Rapidminer 5.2 [8]. 24 sets of baseline data were generated based on the derivation of, and the tokenization methods and weighting schemes applied to the interventions.

2.5   Inter-document Similarity Tests

The baseline data were assembled into pairs of interventions. The similarity between each pair of interventions was computed using the "dist" and "simil" functions available in the R package "proxy" [9]. A total of 42 similarity/distance measures (Table 3) were utilized to compute the similarity or distance between the pairs of interventions. A distance measure was converted to a similarity measure using (1). The similarity values were normalized to a scale of 0 to 1. The normalized similarity value of each pair of interventions S_i was calculated using (2). S_{min} is the minimum similarity value and S_{max} is the maximum similarity value among all pairs of interventions.

Table 3. Similarity and distance measures

Data Type	Similarity Measure	Distance Measure
Numerical	Correlation, Cosine, eJaccard, fJaccard	Bhattacharyya, Bray, Canberra, Chord, Divergence, Euclidean, Geodesic, Hellinger, Manhattan, Soergel, Supremum, Wave, Whittaker
Binary	Braun-blanquet, Dice, Fager, Faith, Hamman, Jaccard, Kulczynski1, Kulczynski2, Michael, Mountford, Mozley, Ochiai, Phi, Russel, Simple Matching, Simpson, Stiles, Tanimoto, Yule, Yule2	-
Nominal	Chi-squared, Cramer, Pearson, Phi-square, Tschuprow	-

$$Similarity = \frac{1}{Distance + 1} \quad (1) \qquad S_{norm} = \frac{S_i - S_{min}}{S_{max} - S_{min}} \quad (2)$$

A retrospective analysis was conducted manually to judge the similarity of the pairs of interventions. Interventions which are highly similar were identified as paired interventions whereas those with low similarity were identified as unpaired interventions. Histograms and boxplots were created to assess the effectiveness of the similarity/distance measures in separating paired interventions from unpaired interventions. A one-way ANOVA was performed to compare the means of similarity values between paired and unpaired interventions. The mean difference (MD) was calculated to compare the performance of the 42 measures on the 24 sets of baseline data using (3). $\overline{S_{paired}}$ is the mean of similarity values of paired interventions and $\overline{S_{unpaired}}$ is the mean of similarity values of unpaired interventions.

$$Mean\ Difference\ (MD) = \overline{S_{paired}} - \overline{S_{unpaired}} \quad (3)$$

2.6 Cluster Structure Analysis

Similar interventions were clustered together using agglomerative hierarchical clustering methods: average-link (AL), complete-link (CL) and ward-link (WL). The three types of clusterings were generated using the "hclust" function available in the R package "stats" [9].

A sample of clustering was shown in Fig. 2a. Based on the figure, each interventions (e.g. "Tamoxifen Bevacizumab") represents a single document. The number of levels and the number of documents that a user will need to explore to obtain all the relevant documents for a topic of interest were identified. For instance, a user will need to explore two levels to discover two documents representing the topic "Irradiation Tamoxifen".

The precision (P), recall (R) and F-measure (F) of each cluster were calculated. P is the ratio of relevant documents retrieved for a given topic (N_{Rel}) over the total number of relevant and irrelevant documents retrieved ($N_{Rel} + N_{Irrel}$) (4). R is the ratio of relevant documents retrieved for a given topic (N_{Rel}) over the total number of relevant documents retrieved and not retrieved ($N_{Rel} + M_{Rel}$) (5). F is the harmonic mean of P and R (6).

$$Precision\ (P) = \frac{N_{Rel}}{N_{Rel} + N_{Irrel}} \qquad (4) \qquad Recall\ (R) = \frac{N_{Rel}}{N_{Rel} + M_{Rel}} \qquad (5)$$

$$F - measure\ (F) = 2 \times \frac{P \times R}{P + R} \qquad (6)$$

Random-baseline clusterings were constructed to evaluate the information retrieval performance of similarity-based clusterings. A random-baseline clustering was created by randomly assigning the interventions into a clustering that has the same number of clusters and the same number of documents in each cluster of a similarity-based clustering. An example is given in Fig. 2. The P, R and mean average precision (MAP) over 25 topics were computed using the TREC_EVAL program [10].

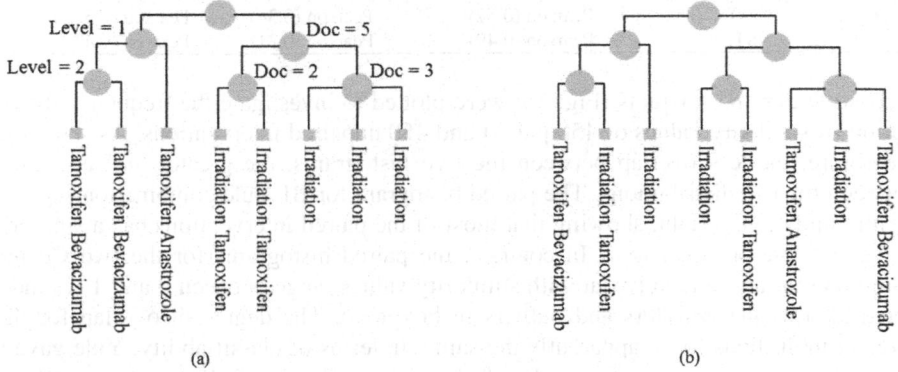

Fig. 2. (a) Similarity-based clustering and (b) random-baseline clustering. Doc = number of documents

3   Results

3.1   The Inter-document Similarity Tests

42 types of similarity/distance measures were employed to calculate the similarities between pairs of interventions. A value close to 1 indicates strong similarity whereas a value close to 0 means low similarity. The MD was calculated to indicate the difference between the mean of paired and the mean of unpaired similarities. The larger the MD, the greater the differentiation and the less overlap between the distributions of paired and unpaired similarities. Table 4 summarizes the measures that produced the highest MD over the 24 sets of baseline data. The table revealed that: (1) BI is better than other weighting schemes, (2) the tokenization method, LO, is superior to ST, (3) interventions derived from title are better than those derived from abstract or "title + abstract", and (4) Yule gives the highest MD compared to other measures.

One of the most popular distance measures between two document vectors is the Cosine similarity. Fig. 3a compares the MD of Yule to the MD of Cosine with an increase in number of pairs of interventions. The figure shows that the number of pairs has little influence on the performance of Yule and Cosine. The MD of Yule (average MD = 0.86 ± 0.04) is evidently higher than the MD of Cosine (average MD = 0.50 ± 0.02 and 0.40 ± 0.02 respectively when tokenized by TF and TF-IDF). For both measures, a significant difference in similarity values between paired and unpaired interventions was found ($p < 0.005$).

Table 4. Similarity/distance measures that produced the highest mean difference (MD)

Weighting Scheme	Tokenization Method	Derivation of Interventions		
		Title (MD)	Abstract (MD)	Title + Abstract (MD)
TF	LO	eJaccard (0.53)	Cosine (0.34)	Cosine (0.45)
	ST	eJaccard (0.44)	Cosine (0.28)	Cosine (0.36)
TF-IDF	LO	eJaccard (0.43)	Cosine (0.24)	Cosine (0.31)
	ST	eJaccard (0.36)	Cosine (0.20)	Cosine (0.27)
BI	LO	Yule (0.86)	Yule (0.67)	Yule (0.74)
	ST	Yule (0.66)	Yule (0.53)	Yule (0.63)
TO	LO	Pearson (0.57)	Pearson (0.34)	Pearson (0.53)
	ST	Pearson (0.49)	Pearson (0.31)	Pearson (0.43)

Histograms and boxplots (Fig. 3b) were plotted to investigate the frequency distribution of similarity values of 450 paired and 450 unpaired interventions. As shown in the figure, the less overlap between the two histograms, the greater the separation between the two distributions. The paired histogram for BI-Yule combination skewed significantly to the right, showing that most of the paired interventions has a similarity value close or equal to 1. In contrast, the paired histograms for the two Cosine combinations are relatively flat with similarity values range between 0 and 1 (as indicated also by the whiskers and outliers in boxplots). The degree of overlap for the three combinations looks apparently the same. In terms of classifiability, Yule gave a more clear-cut separation of paired and unpaired similarities in histograms and boxplots than Cosine.

In summary, the similarity measure, Yule, performed better than other measures at identifying paired interventions or at differentiating between paired and unpaired interventions derived from titles and which had been weighted and tokenized respectively using BI and the LO method.

3.2 The Clustering Tests

The similarities between the interventions that occurred in a collection of 114 documents were calculated using the BI-LO-Title-Yule combination. Hierarchical clusterings were computed using AL, CL and WL algorithms. As shown in Fig. 4a (1st column), the structure of AL and CL clusterings are wide with many branches at the top of the hierarchies, whereas for WL clustering, branches are located mainly at the bottom of the hierarchy. The CL algorithm produced a structure with lower number of levels (the highest number of levels = 5 compared to 10 for WL and 15 for AL). A comparison of the structures obtained by calculating the similarities using the TF-LO-Title-Cosine combination (Fig. 4a, 2nd column) revealed a higher number of levels in the clusterings (the highest number of levels = 7 for CL, compared to 13 for WL and

18 for AL). The higher the number of levels, the longer it takes for a user to browse and search for a topic in a hierarchy. The Yule-based CL clustering, compared to other clusterings, provides a better structure in terms of the number of levels that a user needs to explore to reach a topic of interest.

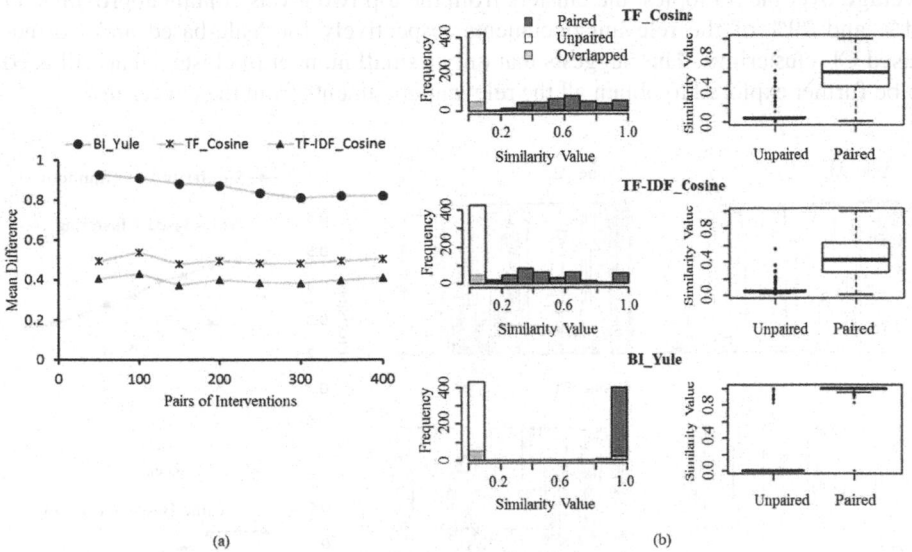

(a)   (b)

Fig. 3. (a) Mean difference against number of pairs of interventions; (b) Histograms and box-plots of the distribution of paired and unpaired similarities. Derivation of interventions: title, tokenization method: LO

The clusters that best represent 33 topics that covered in the 114 documents were identified by computing the P, R and F of each cluster in a hierarchy. Table 5 shows the level of the clustering hierarchy (Lev), the number of documents (Doc), the number of relevant documents (Rel) and the P, R and F values of a cluster in a Yule-based CL clustering. A good cluster is supposed to contain as many relevant documents as possible with high P and high R. The F-measure quantifies the balance between P and R. The higher the F value, the higher the quality of a cluster. It can be seen from Table 5 that: (1) relevant documents are grouped in one (e.g. Level 1 of Topic 1, R = 1.0) or two clusters (e.g. Level 1 of Topic 2, R = 0.5 and 0.5 respectively), (2) the best clusters appear at the top of the structure with high P, R and F for Topics 1, 2 and 4, (3) the best clusters occur at the bottom of the structure with high P, R and F for Topic 3, (4) Topic 4 can be identified without exploring the structure (Level = 0), and (5) some of the relevant documents are grouped in different clusters with irrelevant documents (e.g. Level 1 of Topic 3, P = 0.6 and 0.3 respectively). The results indicate that the best clusters located at different levels of the structure.

Table 6 shows the average number of levels that a user needs to explore to discover the best clusters for the 33 topics. The Yule-based CL clustering provides the best hierarchical structure for the exploration of different topics, followed by the

Cosine-based CL clustering (Average No. of Level = 1.70 ± 1.10 and 2.33 ± 1.95 respectively). The findings suggest that the best clusters appear on average at the top two levels of a CL clustering. This was evaluated by identifying the clusters with the highest F-measure (F_{Max}) from the top two levels. The percentages of relevant documents covered by the clusters were then calculated and are shown in Table 7. On average over the 33 topics, the clusters from the top two levels contain approximately 81% and 79% of the relevant documents respectively for Yule-based and Cosine-based CL clusterings. This suggests that only a small number of clusters that will need to be further explored to obtain all the relevant documents from the clusterings.

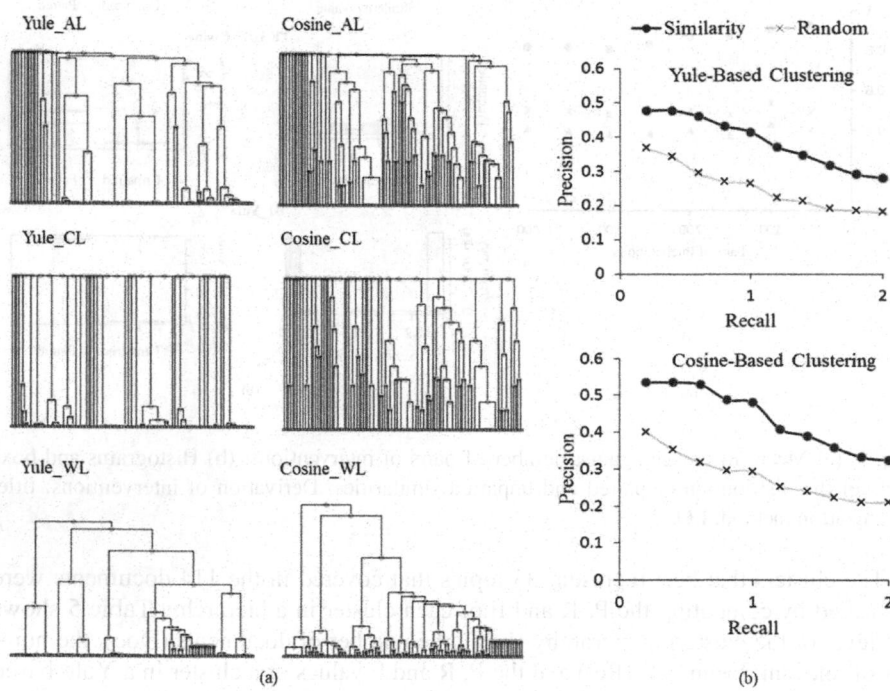

Fig. 4. (a) A comparison of clusterings by similarity measures and clustering methods; (b) The precision-recall performance of similarity-based and random-baseline CL clusterings

The effectiveness of similarity-based clusterings in grouping similar interventions to the same or small number of clusters were evaluated by comparing with random-baseline clusterings. Interventions were grouped into different clusters without similarity constraint to produce a random-baseline clustering. A total of 25 topics were created for the evaluation. Each topic was treated as a query. Similarity-based clusterings outperform random-baseline clusterings in terms of mean average precision (MAP = 0.43 vs. 0.24 and 0.48 vs. 0.25 respectively for Yule-based and Cosine-based CL clusterings). This is further indicated in the P-R curves shown in Fig. 4b.

The overall clustering results indicate that the top two levels of a Yule-based CL clustering provide the most appropriate hierarchical clustering for the exploration and visualization of interventions.

Table 5. Examples of the distribution of the best cluster in a Yule-based CL clustering

Topic	Lev	Doc	Rel	P	R	F
1	1	10	10	1.0	1.0	1.0
	2	2	2	1.0	0.2	0.3
	2	2	8	1.0	0.8	0.9
	3	3	7	1.0	0.7	0.8
2	1	15	4	0.3	0.5	0.4
	1	4	4	1.0	0.5	0.7
	⋮	⋮	⋮	⋮	⋮	⋮
	4	3	1	0.3	0.1	0.2
3	1	6	1	0.6	0.3	0.2
	1	11	3	0.3	0.8	0.4
	2	2	1	0.5	0.3	0.3
	2	10	3	0.3	0.8	0.4
	3	3	3	1.0	0.8	0.9
4	0	1	1	1.0	0.5	0.5
	0	1	1	1.0	0.5	0.5

Table 6. Location of the best cluster by the average number of level over 33 topics

Clustering Method	Similarity Measure	Average ± SD No. of Level
CL	Cosine	2.33 ± 1.95
CL	Yule	1.70 ± 1.10
AL	Cosine	10.12 ± 4.84
AL	Yule	8.21 ± 3.56
WL	Cosine	7.67 ± 3.91
WL	Yule	5.97 ± 2.49

Table 7. Percentages of relevant documents in top clusters in CL clusterings

	Cosine_CL				Yule_CL			
Topic	F_{Max}	Doc	Rel	%	F_{Max}	Doc	Rel	%
Q1	0.6	8	11	73	0.6	8	11	73
Q2	0.1	1	2	50	0.2	1	2	50
⋮	⋮	⋮	⋮	⋮	⋮	⋮	⋮	⋮
Q33	0.6	2	2	100	0.6	2	2	100
Mean ± SD				79 ± 23				81 ± 23

4   Discussion and Conclusions

What types of clinical information do doctors need? Where do they search for information? An early study by [11] reported that approximately 33% of information needs related to treatment of specific conditions, 25% to diagnosis and 14% to drugs. Similar findings were reported in [12] that the top categories of information needs were treatment/therapy (38%), diagnosis (24%) and drug therapy/ information (11%). Studies by [13,14] further supported that one of the doctors' greatest information needs is for information about treatments and drugs. The primary electronic resource used by doctors for evidence-based clinical decision making is MEDLINE [15,16]. Junior doctors accessed MEDLINE (44%), UpToDate (42%), internet search engines (5%), MDCONSULT (3%) and the Cochrane Library (2%) for clinical information [17]. The findings support the use of MEDLINE in this study as the preferred information source for PICO elements.

The Use of PICO Framework for Query Refinement. One of the obstacles that prevents physicians from answering patient-care question is the tendency to formulate unanswerable question [1]. To formulate an answerable question, physicians are recommended to change their search strategies by rephrasing their questions [17]. Other studies recommended the use of question frameworks such as PICO, PICOT, PICOS and PESICO for the formulation of clinical question [18,19,20,21]. A study evaluating the use of PICO as a knowledge representation reported that the framework is primarily centered on therapy question [2]. This supports the focus of the present study on refining questions relating to treatments and drugs. An earlier study by [4] found that questions that contain a proposed intervention and a relevant outcome were unlikely to go unanswered. It is recommended by [22] that at least 3 of the PICO elements are needed to formulate an answerable question. In summary, the completeness of PICO elements determines whether a clinical question is likely to be answered.

Visualization of Interventions Using Similarity-Based Clustering. Current medical QA (MedQA) systems focus on providing direct and precise answers to doctors' questions. A recent review by [23] concluded that current MedQA systems have limitations in terms of the types and formats of questions that they can process. The Info-Bot [24] and the EpoCare [25] systems can only handle structured queries in PICO format but not in NL. An example of PICO query is *"Atrial Fibrillation AND Warfarin AND Aspirin AND Secondary Stroke"*. The use of the system may be limited by the ability of users to apply Boolean operators (e.g. AND and OR) and by the lack of vocabulary due to limited knowledge of a particular domain. The AskHermes system [26,27], on the other hand, accepts both well-structured and ill-formed NL questions. For example, *"What is the best treatment for a needle stick injury after a human immunodeficiency virus exposure?"* A poorly formulated question cannot be refined. This can in turn lead to the discovery of irrelevant documents. Current MedQA systems assume that users are aware of their knowledge deficit. Little research has focused on assisting users in formulating high quality questions, supporting them in exploring a problem domain and clarifying their information needs.

The present study adopted the concept of system-mediated information access, introduced by [28], to assist users in refining an ill-defined question. It is expected that users can clarify or refine their information needs through browsing and searching interventions which have been pre-clustered into a hierarchical structure. The inter-document similarity and cluster structure analysis revealed that the combination of BI-LO-Title-Yule-CL produced the most appropriate hierarchical clustering for the visualization of interventions. The Yule measure appeared to be slightly better than the Cosine measure at contributing to the identification of similar interventions. The Cosine similarity, which measures the cosine of the angle between two vectors, has been applied to both document clustering [29] and short text clustering [30]. The Yule similarity calculates the strength of association between binary variables. Though not as well studied as the Cosine similarity, [31] reported an improvement in clustering performance using the Yule measure. The cluster structure analysis revealed that documents with similar interventions are likely to be grouped into the same cluster. The top two levels of a CL clustering provide the most appropriate structure for the exploration of different topics. Previous work by [32] reported that AL produces a more effective clustering than CL for information retrieval. However, in the present study, the AL clustering requires users to explore a higher number of levels to discover a problem domain. Doctors often have very tight schedules. When seeking information for patient care, they are more likely to look for information that can be accessed quickly with minimal effort [11]. Therefore, it is argued that CL provides a quicker and more appropriate clustering than AL for the visualization and exploration of interventions.

Limitations. The study was conducted only on MEDLINE articles relevant to a single question. The single source of documents may restrict the applicability of the findings from this study to documents from other resources such as the Cochrane Library and UpToDate. The question tested was posed with only the P element. Further analysis should be undertaken with higher number of questions addressed with different combinations of PICO elements. Compared to the titles, a higher number of interventions were collected from the abstracts. The case study however shows that the title-based approach superior to the abstract-based approach. The study can be improved by evaluating the effectiveness of the methodologies used for PICO extraction and the effects of different numbers of interventions between two documents on the measurement of similarity. Despite of the limitations, the experimental results show that the similarity-based clustering approach has the potential to aid the visualization and exploration of interventions for the applications of clinical information needs clarification and query refinement.

References

1. Ely, J.W., Osheroff, J.A., Chambliss, M.L., Ebell, M.H., Rosenbaum, M.E.: Answering physicians' clinical questions: obstacles and potential solutions. J. Am. Med. Inform. Assoc. 12(2), 217–224 (2005)
2. Huang, X., Lin, J., Demner-Fushman, D.: Evaluation of PICO as a knowledge representation for clinical questions. In: AMIA Annual Symposium Proceedings 2006, pp. 359–363 (2006)
3. Cao, Y., Liu, F., Simpson, P., Antineau, L., Bennet, A., Cimino, J.J., Ely, J., Yu, H.: AskHERMES: an online question answering system for complex clinical questions. J. Biomed. Inform. 44(2), 227–288 (2011)
4. Bergus, G.R., Randall, C.S., Sinift, S.D., Rosenthal, D.M.: Does the structure of clinical questions affect the outcome of curbside consultations with specialty colleagues? Arch. Fam. Med. 9(6), 541–547 (2000)
5. Demner-Fushman, D., Lin, J.: Answering clinical questions with knowledge-based and statistical techniques. Association for Computational Linguistics 33(1), 63–103 (2007)
6. Boudin, F., Nie, J.Y., Bartlett, J.C., Grad, R., Pluye, P., Dawes, M.: Combining classifiers for robust PICO element detection. BMC Med. Inform. Decis. Mak. 10(1), 29–36 (2010)
7. Aronson, A.R.: Effective mapping of biomedical text to the UMLS Metathesaurus: the MetaMap program. In: Proceedings of AMIA Annual Symposium 2001, pp. 17–21 (2001)
8. RapidMiner: Report the future, http://rapid-i.com/content/blogcategory/38/69/ (Assessed: August 2013)
9. R: The R project for statistical computing, http://www.r-project.org (Assessed: August 2013)
10. Trec_eval, http://trec.nist.gov/trec_eval/ (Assessed: August 2013)
11. Smith, R.: What clinical information do doctors need? BMJ 313(7064), 1062–1068 (1996)
12. Davies, K., Harrison, J.: The information-seeking behavior of doctors: a review of the evidence. Health Info. Libr. J. 24(2), 78–94 (2007)
13. Schwartz, K., Northrup, J., Crowell, K., Lauder, N., Neale, A.V.: Use of on-line evidence-based resources at the point of care. Family Medicine 35(4), 251–256 (2003)

14. Yu, H., Cao, Y.G.: Automatically extracting information needs from ad hoc clinical questions. In: AMIA Annual Symposium Proceedings 2008, pp. 96–100 (2008)
15. Davies, K.: UK doctors awareness and use of specified electronic evidence-based medicine resources. Inform. Health Soc. Care 36(1), 1–19 (2011)
16. Schilling, L.M., Steiner, J.F., Lundahl, K., Anderson, R.J.: Residents' patient-specific clinical questions: opportunities for evidence-based learning. Academic Medicine 80(1), 51–56 (2005)
17. Ely, J.W., Osheroff, J.A., Maviglia, S.M., Rosenbaum, M.E.: Patient-care questions that physicians are unable to answer. J. Am. Med. Inform. Assoc. 14(4), 407–414 (2007)
18. Schardt, C., Adams, M.B., Owens, T., Keitz, S., Fontelo, P.: Utilization of the PICO framework to improve searching PubMed for clinical questions. BMC Med. Inform. Decis. Mak. 7(1), 16 (2007)
19. Rios, L.P., Ye, C., Thabane, L.: Association between framing of the research question using the PICOT format and reporting quality of randomized controlled trials. BMC Med. Res. Methodol. 10(1), 11–18 (2010)
20. Robinson, K.A., Saldanha, I.J., Mckoy, N.A.: Frameworks for determining research gaps during systematic reviews. In: Methods Future Research Needs Reports, vol. 2. Agency for Healthcare Research and Quality, Rockville (MD) (2011)
21. Schlosser, R.W., Koul, R., Costello, J.: Asking well-built questions for evidence-based practice in augmentative and alternative communication. Journal of Communication Disorders 40(3), 225–238 (2007)
22. Staunton, M.: Evidence-based radiology: steps 1 and 2 – asking answerable questions and searching for evidence. Radiology 242(1), 23–31 (2007)
23. Athenikos, S.J., Han, H.: Biomedical question answering: a survey. Comput. Methods Programs Biomed. 99(1), 1–24 (2010)
24. Demner-Fushman, D., Seckman, C., Fisher, C., Hauser, S.E., Clayton, J., Thoma, G.R.: A prototype system to support evidence-based practice. In: AMIA Annual Symposium Proceedings 2008, pp. 151–155 (2008)
25. Niu, Y., Hirst, G., McArthur, G., Rodriguez-Gianolli, P.: Answering clinical questions with role identification. In: Proceedings, Workshop on Natural Language Processing in Biomedicine, 41st Annual Meeting of the Association for Computational Linguistics, pp. 73–80 (2003)
26. Yu, H., Kaufman, D.: A cognitive evaluation of four online search engines for answering definitional questions posed by physicians. In: Pacific Symposium on Biocomputing 2007, pp. 328–339 (2007)
27. Cao, Y.G., Cimino, J.J., Ely, J., Yu, H.: Automatically extracting information needs from complex clinical questions. J. Biomed. Inform. 43(6), 962–971 (2010)
28. Muresan, G., Harper, D.J.: Topic modelling for mediated access to very large document collections. J. Am. Soc. Inf. Sci. Technol. 55(10), 892–910 (2004)
29. Subhashini, R., Kumar, V.: Evaluating the performance of similarity measures used in document clustering and information retrieval. In: Proceedings of the First International Conference on Integrated Intelligent Computing 2010, pp. 27–31 (2010)
30. Rangrej, A., Kulkarni, S., Tendulkar, A.: Comparative study of clustering techniques for short text documents. In: Proceedings of the 20th International Conference Comparison on World Wide Web 2011, pp. 111–112 (2011)
31. Malik, H.H., Kender, J.R.: High quality, efficient hierarchical document clustering using closed interesting itemsets. In: Proceedings of the Sixth International Conference on Data Mining 2006, pp. 991–996 (2006)
32. Aljaber, B., Stokes, N., Bailey, J., Pei, J.: Document clustering of scientific texts using citation contexts. Information Retrieval 13(2), 101–131 (2010)

Mining Biomedical Literature
and Ontologies for Drug Repositioning Discovery

Chih-Ping Wei, Kuei-An Chen, and Lien-Chin Chen

Department of Information Management, National Taiwan University, Taipei, Taiwan, R.O.C.
{cpwei,r00725020,lcchen101}@ntu.edu.tw

Abstract. Drug development is time-consuming, costly, and risky. Approximate 80% to 90% of drug development projects fail before they ever get into clinical trials. To reduce the high risk of failure for drug development, pharmaceutical companies are exploring the drug repositioning approach for drug development. Previous studies have shown the feasibility of using computational methods to help extract plausible drug repositioning candidates, but they all encountered some limitations. In this study, we propose a novel drug-repositioning discovery method that takes into account multiple information sources, including more than 18,000,000 biomedical research articles and some existing ontologies that cover detailed relations between drugs, proteins and diseases. We design two experiments to evaluate our proposed drug repositioning discovery method. Overall, our evaluation results demonstrate the capability and superiority of our proposed drug repositioning method for discovering potential, novel drug-disease relationships.

Keywords: Drug repositioning, Drug repurposing, Literature-based discovery, Medical literature mining.

1 Introduction

Drug development is time-consuming, costly, and risky. The development process to bring a new drug to market requires about 10-15 years and costs between 500 million and 2 billion U.S. dollars [1]. However, as the U.S. National Institutes of Health reported, 80 to 90 percent of drug development projects fail before they ever get tested in human [2]. To reduce the high risk of failure for *de novo* drug development, pharmaceutical companies have been evaluating alternative paradigms for drug development, e.g., drug repositioning. The goal of drug repositioning is to find new indications (i.e., treatment for diseases) for existing drugs. Because existing drugs already have their preclinical properties and established safety profiles, many experiments, analyses and tests can therefore be bypassed [3]. Thus, drug repositioning can reduce significant time and cost in the discovering and preclinical stage. Moreover, drug repositioning helps a company exploit its intellectual property portfolio by extending its old or expiring patents, or getting new method-of-use patents [3].

One notable example of repositioned drug is Thalidomide. It was originally marketed as a sedative and antiemetic for pregnant women to treat morning sickness,

V.S. Tseng et al. (Eds.): PAKDD 2014, Part II, LNAI 8444, pp. 373–384, 2014.

but was completely withdrawn from the market after the drug was found responsible for severe birth defects [4]. After Celgene Corporation's repositioning works, FDA approved Thalidomide for use in the treatment of Erythema Nodosum Leprosum (ENL) in 1998 [5]. The company further discovered that Thalidomide is also effective against several other diseases, including multiple myeloma. Accordingly, Celgene gets several utility patents for the repositioned Thalidomide, and it brings in over 300 million U.S. dollars in revenue annually since 2004 [6-8].

Several computational drug repositioning approaches have been developed to help medical researchers sift the most plausible drug-disease pairs from a wide range of combinations. Dudley et al. [9] summarized several computational drug repositioning methods and categorized them as either drug-based, where discovery from the chemical or pharmaceutical perspective, or disease-based, where discovery from the perspective of disease management, symptomatology, or pathology. Another excellent review in [10] highlights computational techniques for systematic analysis of transcriptomics, side effects and genetics data to generate new hypotheses for additional indications. Several network based approaches use various heterogeneous data resources to discovery drug repositioning opportunity [11, 12]. Moreover, Wu et al. [13] summarized 26 different sources of databases related to disease, genes, proteins, and drugs for drug repositioning. In summary, existing drug-repositioning methods can broadly be classified into two approaches: *literature-based* and *ontology-based*. The literature-based approach assumes that if a drug frequently co-occurs with some biomedical concepts (such as enzymes, genes, pathological effects, and proteins) and many of these concepts also frequently co-occur with a disease in biomedical literature (e.g., MEDLINE), it is likely that the disease is a new indication for the focal drug [14]. In contrast, the ontology-based approach relies on existing ontologies (or knowledge bases) to discover hidden relationships between drugs and diseases. These existing methods have shown their feasibility for drug repositioning. However, most existing methods rely only on single information source, i.e., literature or ontologies.

In this study, we propose a drug repositioning method that exploits multiple information sources to discover hidden relationships between drugs and diseases. Specifically, a comprehensive network of biomedical concepts is first constructed by combining and integrating relations of biomedical concepts extracted from literature and existing ontologies. Subsequently, we follow Swanson's ABC model [14] to obtain links between a focal drug (A) and intermediate terms (Bs) and then between Bs and diseases (Cs) from the comprehensive concept network. A novel link weighting method and two target term ranking measures are proposed to effectively rank candidate diseases that are likely to be new indications of the focal drug A. To evaluate the proposed method, we collect the literature from MEDLINE and three ontologies (i.e., DrugBank [15], Online Mendelian Inheritance in Man (OMIM) [16], and Comparative Toxicogenomics Database (CTD) [17]), and follow the evaluation procedure proposed in [18] to conduct a series of experiments. According to our empirical evaluation results, our proposed method outperforms the existing method for drug repositioning.

The remainder of this paper is organized as follows: Section 2 reviews existing methods related to this study, and discuss their limitations to justify our research motivation. In Section 3, we describe the design of our proposed drug repositioning discovery method. Section 4 reports on our evaluation of our proposed method. Finally, we conclude our study in Section 5.

2 Related Work

In this section, we review existing drug repositioning methods, which can be classified into two major approaches: literature-based and ontology-based.

2.1 Literature-Based Approach

Swanson [19] first introduced the idea of discovering hidden relationships from biomedical literatures in the mid-1980s. He examined across disjoint literatures, manually identified plausible new connections, and found fish oil might be beneficial to the treatment of Raynaud's syndrome [19]. Furthermore, Swanson and Smalheiser developed a computational model, namely "*ABC model*" or "*undiscovered public knowledge (UPK) model*" [14]. The basic assumption of ABC model is that if a biomedical concept *A* relates to intermediate concept *B* and intermediate concept *B* relates to another concept *C*, there is a logically plausible relation between *A* and *C*. The ABC model generally consists of three major phases: *term selection* to extract textual terms (concepts) from the literature, *link weighting* to assess the link strength between two concepts, and *target term ranking* to rank target terms by assigning a score to each target term on the basis of the connections and link weights between the starting term and the target term. This approach is often referred to as the literature-based discovery.

Weeber et al. [20] followed Swanson's idea of co-occurrence analysis and mapped words from titles and abstracts extracted from MEDLINE articles to Unified Medical Language System (UMLS) concepts to filter link candidates with the help of semantic information. Similarly, Wren et al. [21] mapped full text from articles into OMIM concepts. They measured link weights between concepts by mutual information. Lee et al. [22] further combined multiple thesauruses to better translate text into biomedical concepts. These studies employed the full text for concept extraction with the help of thesauri. On the other hand, some other studies used Medical Subject Headings (MeSH) as keywords to annotate each article in MEDLINE [18] [23-24]. They applied tf-idf, association rule, and z-score as the measurement of link weights. All of them reported the metadata-only approach is feasible, though Hristovski et al. [24] noted some shortcoming of using MeSH such as insufficient information of involving genes. Based on the ABC model, several drug repositioning methods [20-21] [25-26] were proposed to find undiscovered relations between drugs and diseases through selecting different semantic groups of intermediate terms such as adverse effects, genes, and proteins.

For evaluating the performance of the literature-based approach, Yetisgen-Yildiz and Pratt [18] developed an evaluation methodology. They used two literature sets collected from separated time spans, and trained systems by using the older set to predict novel relations in the newer set. They compared the effectiveness of various link weighting methods. According to their study, association rule appears to achieve the best performance over tf-idf, mutual information measure, and z-score. They also compared different target term ranking algorithms and suggested that the use of link term count with average minimum weight can achieve the best effectiveness.

2.2   Ontology-Based Approach

Campillos et al. [27] constructed a network of side-effect driven drug-drug relations from UMLS ontology by measuring side-effect similarity between drugs. Assuming that similar side effects of unrelated drugs may be caused by common targets, they can be used to predict new drug-target interactions. They also experimentally validated their results, and thus reported the feasibility of using phenotypic information to infer unexpected biomedical relations. Yang and Agarwal [28] also based on side effect likelihood between drugs, but they constructed Naïve Bayes models to make predictions. They took PharmGKB and SIDER knowledge bases, rather than phenotype database, as their information sources. Cheng et al. [29] built a bipartite network by extracting known drug-target interaction data from DrugBank, and used the network similarity to predict new targets of drugs. Li and Lu [30] built a network similar to Cheng et al.'s work, but added the similarity of drug chemical structure into consideration.

Qu et al. [31] and Lee et al. [26] both attempted to increase the size and scope of semantic data by constructing integrated network or database of ontologies. However, to our best knowledge, few prior studies, if any, take both ontologies and literature into account, which may be a good way to acquire deeper and broader biomedical knowledge for making predictions of drug-disease relations. Li et al. [32] tried to incorporate more knowledge by using protein-protein interactions extracted from Online Predicted Human Interaction Database (OPHID) to expand disease-related proteins, and built disease-specific drug-protein connectivity maps based on literature mining. His work inspires us to build a network over multiple information sources.

3   The Proposed Method

We propose a drug repositioning discovery method that is based on Swanson's ABC model [14] but takes both biomedical literature and existing ontologies into account. Fig. 1 illustrates our proposed method, which consists of four main phases: *comprehensive concept network construction, related concept retrieval, link weighting,* and *target term ranking.*

3.1   Phase 1: Comprehensive Concept Network Construction

The goal of this phase is to construct a literature-based concept network from the biomedical literature and an ontology-based concept network from existing ontologies (i.e., DrugBank, OMIM, and CTD in this study) and, subsequently, integrate them into a comprehensive concept network.

For the literature-based concept network construction, we collect the biomedical literature from MEDLINE 2011 baseline. U.S. National Library of Medicine (NLM) indexes the publication type for each article. We follow Yetisgen-Yildiz's preprocessing procedure [18] to remove 18 irrelevant types (e.g., address, bibliography, comment, etc.) from the 61 publication types in MEDLINE 2011 baseline. NLM also indexes several MeSH terms for each biomedical article. As a result, our literature database consists of 18,712,338 biomedical articles. The number of MeSH terms per article ranges from 1 to 97, and its average is 9.44. We further select several MeSH subcategories related to drug repositioning, as shown in Table 1. Next, association rule mining is applied on the collected literature where biomedical articles and MeSH terms are considered as transactions and items, respectively. We follow Yetisgen-Yildiz and Pratt's experiment [18] by setting the minimum support threshold to 2.6 and the minimum confidence threshold to 0.0055 in this study. After filtering, we extract 12,278 MeSH terms and 2,623,222 relations to construct a literature-based concept network.

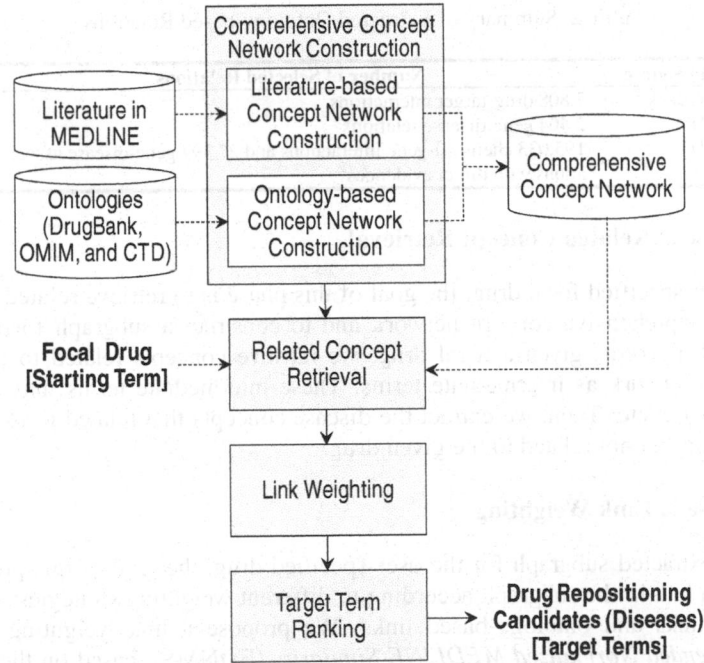

Fig. 1. Overall Process of the Proposed Drug Repositioning Discovery Method

For the ontology-based concept network construction, the ontologies we use in this study include DrugBank [15], OMIM [16], and CTD [17]. DrugBank (http://www.drugbank.ca/) is a richly annotated database which provides extensive information about targets, pathways, indications, adverse effects, and related proteins of various drugs, whereas OMIM (http://www.omim.org/) is a comprehensive and

Table 1. Selected MeSH Subcategories

Semantic Group	MeSH Subcategories
Drugs	D01-D05, D09, D10, D20, D26, D27
Genes, Proteins, and Enzymes	D06, D08, D12, D13, D23
Pathological Effects	G03-G16
Diseases	C01-C23

authoritative knowledge base of human genes and genetic phenotypes. CTD (http://ctdbase.org/) is a database that integrates data from scientific literature to describe chemical interactions with genes and proteins, and diseases and genes/proteins, and others. The collected information includes various types of relations, such as drug-target, gene-disease, gene-gene, protein-protein, chemical-gene, gene-disease relations, etc. And, we also try to translate those terms to MeSH term and select only MeSH subcategories shown in Table 1. The summary of these collected ontology-based relations is shown in Table 2.

Table 2. Summary of Extracted Ontology-based Relations

Data Source	Number of Selected Relations
DrugBank	7,808 drug-target interactions
OMIM	2,404 gene-disease relations
CTD	195,033 chemical-gene interactions and 27,397 gene-disease associations with direct evidences

3.2   Phase 2: Related Concept Retrieval

Given a user-specified focal drug, the goal of this phase is to retrieve related relations from the comprehensive concept network and to construct a subgraph for the focal drug. In other words, given a focal drug, we retrieve concepts related to the given drug in the network as intermediate terms. These intermediate terms may be gene, protein, disease, etc. Then, we extract the disease concepts that related to these intermediate terms but not related to the given drug.

3.3   Phase 3: Link Weighting

Given the extracted subgraph for the user-specified drug, the goal of this phase is to weight each link in the subgraph according to different weighting strategies for literature-based links and ontology-based links. We propose a link weighting method, namely *Extended Normalized MEDLINE Similarity* (ExtNMS), based on the Normalized Google Distance (NGD) [33] to calculate similarity between MeSH term A and B as follows:

$$NMD(A,B) = \frac{\max\{\log|D_A|,\log|D_B|\}-\log|D_A \cap D_B|}{\log M - \min\{\log|D_A|,\log|D_B|\}}, NMD = 1 \text{ if } NMD > 1, \tag{1}$$

$$NMS(A,B) = 1 - NMD(A,B), \tag{2}$$

$$ExtNMS(A,B) = \begin{cases} NMS(A,B), & if\,(A,B) \in Literature\ and \notin Ontology \\ 1, & if\,(A,B) \in Ontology \end{cases}, \quad (3)$$

where D_A is the set of articles including MeSH term A, and M denotes the total number of articles in MEDLINE. This weighting measure ranges from 0 to 1, in which 0 being completely unrelated and 1 being credibly related. If link (A,B) is from the literature-based concept network and is not found in the ontology-based concept network, we weight it by calculating its *Normalized MEDLINE Distance* (NMD) and subtracting from 1, called *Normalized MEDLINE Similarity* (NMS); otherwise, if the link appears in the ontology-based concept network, we assign its weight as 1 since this relation should have been validated.

3.4   Phase 4: Target Term Ranking

Yetisgen-Yildiz & Pratt [18] suggested that using *Link Term Count with Average Minimum Weight* (LTC-AMW) can achieve the best effectiveness for target term ranking. LTC-AMW takes the number of intermediate terms between starting term and target term as the primary ranking criteria, i.e. the number of paths. The average minimum weight of paths is used only when two target terms are identical in their number of paths. In this study, we propose two target term ranking measures, *Summation of Minimum Weight* (SumMW) and *Summation of Average Weight* (SumAW), as follows:

$$SumMW(A,C) = \sum_{B \in N(A) \cap N(C)} \min\{Wt(A,B), Wt(B,C)\}, \quad (4)$$

$$SumAW(A,C) = \sum_{B \in N(A) \cap N(C)} \frac{Wt(A,B) + Wt(B,C)}{2}, \quad (5)$$

where $N(A)$ denotes the neighbor concepts of term A, and $Wt(A,B)$ is the weight of link between term A and B. The above measures differentiate the importance of each path according to their minimum or average weight of constituent links, and assign an importance score to each target term according to the cumulative information of all paths between the starting term and the target term. We then order target terms according to their importance scores.

4   Evaluation and Results

In this study, we conduct two experiments to evaluate our proposed method for drug repositioning. The first experiment is to evaluate our proposed comprehensive concept network and link weighting measure (i.e., ExtNMS). The second experiment is to evaluate our proposed target term ranking measure.

4.1   Evaluation Design

We follow the evaluation procedure proposed by Yetisgen-Yildiz and Pratt [18]. Specifically, we describe our experiment procedure step by step in the following.

Given a starting term A (i.e., drug):

1. We set cut-off date as January 1, 2000 and divide MEDLINE 2011 baseline into two datasets:
   (a) *Pre-cut-off* set (S_{t1}) includes the documents prior to 1/1/2000.
   (b) *Post-cut-off* set (S_{t2}) includes the documents on and after 1/1/2000.
2. The documents in the pre-cut-off set are used along with ontologies as the input to construct the comprehensive concept network.
3. We define a gold-standard set G_A, which contains terms that satisfy the following rules:
   (a) Terms are within our specified target semantic group, i.e., disease.
   (b) Terms co-occur with term A in the post-cut-off set, but do not co-occur with term A in the pre-cut-off set. In other words, these terms co-occur with term A in literature only after the cut-off date (i.e., 1/1/2000).
   (c) Terms are not related to term A in the ontology-based concept network.
4. The discovery effectiveness is estimated by using the information retrieval metrics as follows:
   (a) Precision:

$$P_A = \frac{|T_A \cap G_A|}{|T_A|},$$ (6)

   (b) Recall:

$$R_A = \frac{|T_A \cap G_A|}{|G_A|},$$ (7)

where T_A is the set of target terms generated by our discovery method.

Table 3 shows the list of semantic groups used in our experiments. In this study, we randomly select 100 terms from the semantic group of drugs as the starting terms, i.e., the focal drugs.

Table 3. Selected Semantic Groups for Our Experiments

Selected Intermediate Terms	Selected Target Terms
Drugs	Diseases
Genes, Proteins, and Enzymes	
Pathological Effects	
Diseases	

4.2   Exp. 1: Evaluation of the Comprehensive Concept Network and Link Weighting Method

The performance benchmark is the original ABC model over only the literature which uses association rules as link weighting algorithm [14]. We evaluate three versions of our discovery method, i.e., one is over only the literature-based concept network, another one is over only the ontology-based concept network, and the third one is over the comprehensive concept network. All methods (including the benchmark) under investigation apply LTC-AMW as the target term ranking measure. Table 4 shows the evaluation results, and the Area Under Curve of Precision and Recall (AUC-PR) represents the overall performance. A higher AUC-PR value represents a greater effectiveness. As Table 4 illustrates, our proposed ExtNMS measure outperforms the benchmark link weighting measure (i.e., association rules) when the information source is the literature only or the integrated information sources (i.e., the comprehensive concept network that contains literature and ontologies). Moreover, using both literature and ontologies as information sources improves the overall performance, especially precisions on higher ranks. This would better help researchers sift plausible drug-disease relations for the purpose of drug repositioning.

Table 4. Comparative Evaluation Results of the Link Weighting Measures Under Different Concept Networks

Recall	Precision			
	Association Rules (Literature)	ExtNMS (Literature)	ExtNMS (Ontology)	ExtNMS (Comprehensive Concept Network)
0%	62.61%	57.72%	39.01%	59.33%
10%	29.72%	29.93%	20.16%	30.54%
20%	22.07%	23.75%	15.96%	23.89%
30%	17.80%	18.95%	14.22%	19.01%
40%	15.13%	16.27%	12.58%	16.26%
50%	11.73%	13.80%	12.25%	13.62%
60%	9.52%	11.69%	13.77%	11.53%
70%	7.61%	9.66%	0%	9.61%
80%	7.17%	7.76%	0%	7.69%
90%	2.31%	6.01%	0%	5.97%
100%	0.60%	3.80%	0%	3.84%
AUC-PR	**15.47%**	**16.86%**	**10.84%**	**16.97%**

4.3   Exp. 2: Evaluation of Target Term Ranking Measure

In this experiment, we evaluate the proposed target term ranking measures, *Summation of Minimum Weight* (SumMW) and *Summation of Average Weight* (SumAW). Based on the experiment 1, we apply ExtNMS as the link weighting measure in this experiment. In this experiment, LTC-AMW is used as the benchmark ranking measure. Table 8 shows the comparative evaluation results of the three target term ranking measures using the comprehensive concept network as the information sources. Both SumMW and SumAW outperform the benchmark measure, LTC-AMW. These results show that our link weighting measure, ExtNMS, is a more effective measure to weight links, and considering both the number and weights of paths between the starting term and target terms can improve the discovery effectiveness.

Table 5. Comparative Evaluation Results of Target Term Ranking Measures (Using the Comprehensive Concept Network)

Recall	Precision		
	LTC-AMW	SumMW	SumAW
0%	59.33%	59.14%	61.70%
10%	30.54%	33.46%	33.16%
20%	23.89%	24.86%	24.52%
30%	19.01%	20.91%	20.69%
40%	16.26%	17.32%	17.01%
50%	13.62%	14.74%	14.56%
60%	11.53%	12.42%	12.17%
70%	9.61%	10.50%	10.29%
80%	7.69%	8.41%	8.22%
90%	5.97%	6.35%	6.22%
100%	3.84%	3.84%	3.84%
AUC-PR	16.97%	18.05%	17.96%

5   Conclusions

In this study, we develop a drug repositioning discovery method that uses both biomedical literature and ontologies as information sources for constructing a comprehensive network of biomedical concepts. We also develop a link weighting method (i.e., ExtNMS) and two target term ranking measures. We experimentally evaluate our proposed method and show that taking both literature and ontologies into account and using our ExtNMS measure can improve the effectiveness of predicting novel drug-disease relationships. Besides, our proposed target term ranking measures can better infer plausible drug-disease relations. Overall, our proposed drug repositioning discovery method can help researchers sift most plausible unknown drug-disease relationships, i.e., potential drug repositioning candidates.

References

1. Adams, C.P., Brantner, V.V.: Estimating the cost of new drug development: Is it really $802 million? Health Aff. 25(2), 420–428 (2006)
2. National Institutes of Health. NIH Announces New Program to Develop Therapeutics for Rare and Neglected Diseases (May 20, 2009),
   http://rarediseases.info.nih.gov/files/TRND%20Press%20Release.pdf
3. Ashburn, T.T., Thor, K.B.: Drug repositioning: identifying and developing new uses for existing drugs. Nat. Rev. Drug Discov. 3(8), 673–683 (2004)
4. McBride, W.G.: Thalidomide and congenital abnormalities. Lancet 278(7216), 1358 (1961)
5. Stephens, T.D., Brynner, R.: Dark Remedy: The Impact of Thalidomide and Its Revival as a Vital Medicine. Perseus Publishing, Cambridge (2001)

6. Celgene Corporation: 2005 Annual Report. Celgene Corporation, Summit, NJ (2006)
7. Celgene Corporation: 2008 Annual Report on Form 10-K. Celgene Corporation, Summit, NJ (2009)
8. Celgene Corporation: 2012 Annual Report on Form 10-K. Celgene Corporation, Summit, NJ (2013)
9. Dudley, J.T., Deshpande, T., Butte, A.J.: Exploiting drug-disease relationships for computational drug repositioning. Brief. Bioinformatics 12, 303–311 (2011)
10. Hurle, M.R., Yang, L., Xie, Q., Rajpal, D.K., Sanseau, P., Agarwal, P.: Computational drug repositioning: From data to therapeutics. Clin. Pharmacol. Ther. 93(4), 335–341 (2013)
11. Emig, D., Ivliev, A., Pustovalova, O., Lancashire, L., Bureeva, S., Nikolsky, Y., Bessarabova, M.: Drug target prediction and repositioning using an integrated network-based approach. PLoS One 8(4), e60618 (2013)
12. Kim, S., Jin, D., Lee, H.: Predicting Drug-Target Interactions Using Drug-Drug Interactions. PLoS One 8(11), e80129 (2013)
13. Wu, Z., Wang, Y., Chen, L.: Network-based drug repositioning. Mol. Biosyst. 9, 1268–1281 (2013)
14. Swanson, D.R., Smalheiser, N.R.: An interactive system for finding complementary literatures: A stimulus to scientific discovery. Artif. Intell. 91(2), 183–203 (1997)
15. Knox, C., Law, V., Jewison, T., Liu, P., Ly, S., Frolkis, A., Wishart, D.S.: DrugBank 3.0: A comprehensive resource for 'omics' research on drugs. Nucleic Acids Res. 39(suppl. 1), D1035–D1041 (2011)
16. Hamosh, A., Scott, A.F., Amberger, J.S., Bocchini, C.A., McKusick, V.A.: Online Mendelian Inheritance in Man (OMIM), a knowledgebase of human genes and genetic disorders. Nucleic Acids Res. 33(suppl. 1), D514–D517 (2005)
17. Davis, A., King, B.L., Mockus, S., Murphy, C.G., Saraceni-Richards, C., Rosenstein, M., Mattingly, C.J.: The Comparative Toxicogenomics Database: Update 2013. Nucleic Acids Res. 39(suppl. 1), D1067–D1072 (2013)
18. Yetisgen-Yildiz, M., Pratt, W.: A new evaluation methodology for literature-based discovery systems. J. Biomed. Inform. 42(4), 633–643 (2009)
19. Swanson, D.R.: Fish oil, Raynaud's syndrome, and undiscovered public knowledge. Perspect. Biol. Med. 30(1), 7–18 (1986)
20. Weeber, M., Klein, H., de Jong-van den Berg, L.T., Vos, R.: Using concepts in literature-based discovery: Simulating Swanson's Raynaud–fish oil and migraine–magnesium discoveries. J. Am. Soc. Inf. Sci. Technol. 52(7), 548–557 (2001)
21. Wren, J.D., Bekeredjian, R., Stewart, J.A., Shohet, R.V., Garner, H.R.: Knowledge discovery by automated identification and ranking for implicit relationships. Bioinformatics 20(3), 389–398 (2004)
22. Lee, S., Choi, J., Park, K., Song, M., Lee, D.: Discovering context-specific relationships from biological literature by using multi-level context terms. BMC Medical Informatics and Decision Making (BMC Med. Inform. Decis. Mak.) 12(suppl. 1), S1 (2012)
23. Srinivasan, P.: Text mining: Generating hypotheses from MEDLINE. J. Am. Soc. Inf. Sci. Technol. 55(5), 396–413 (2004)
24. Hristovski, D., Peterlin, B., Mitchell, J.A., Humphrey, S.M.: Using literature-based discovery to identify disease candidate genes. Int. J. Med. Inform. 74, 289–298 (2005)
25. Frijters, R., van Vugt, M., Smeets, R., van Schaik, R., de Vlieg, J., Alkema, W.: Literature mining for the discovery of hidden connections between drugs, genes and diseases. PLoS Comput. Biol. 6(9), e1000943 (2010)

26. Lee, H.S., Bae, T., Lee, J.-H., Kim, D., Oh, Y., Jang, Y., Kim, S.: Rational drug reposi-
    tioning guided by an integrated pharmacological network of protein, disease and drug.
    BMC Syst. Biol. 6(1), 80 (2012)
27. Campillos, M., Kuhn, M., Gavin, A.-C., Jensen, L.J., Bork, P.: Drug target identification
    using side-effect similarity. Science 321(5886), 263–266 (2008)
28. Yang, L., Agarwal, P.: Systematic drug repositioning based on clinical side-effects. PLOS
    ONE 6(12), e28025 (2011)
29. Cheng, F., Liu, C., Jiang, J., Lu, W., Li, W., Liu, G., Tang, Y.: Prediction of drug-target in-
    teractions and drug repositioning via network-based inference. PLoS Comput. Biol. 8(5),
    e1002503 (2012)
30. Li, J., Lu, Z.: A new method for computational drug repositioning. In: IEEE International
    Conference on Bioinformatics and Biomedicine, pp. 1–4. IEEE Press, Philadelphia (2012)
31. Qu, X.A., Gudivada, R.C., Jegga, A.G., Neumann, E.K., Aronow, B.J.: Inferring novel
    disease indications for known drugs by semantically linking drug action and disease me-
    chanism relationships. BMC Bioinformatics 10(suppl. 5), S4 (2009)
32. Li, J., Zhu, X., Chen, J.Y.: Building disease-specific drug-protein connectivity maps
    from molecular interaction networks and PubMed abstracts. PLoS Comput. Biol. 5(7),
    e1000450 (2009)
33. Cilibrasi, R.L., Vitányi, P.M.: The Google similarity distance. IEEE Trans. Knowl. Data
    Eng. 19(3), 370–383 (2007)

Inferring Metapopulation Based Disease Transmission Networks

Xiaofei Yang[1], Jiming Liu[1],
William Kwok Wai Cheung[1], and Xiao-Nong Zhou[2]

[1] Department of Computer Science, Hong Kong Baptist University
Hong Kong SAR, China
{xfyang09,jiming,william}@comp.hkbu.edu.hk
[2] National Institute of Parasitic Diseases, China CDC, Shanghai, China
zhouxn1@chinacdc.cn

Abstract. To investigate how an infectious disease spreads, it is most desirable to discover the underlying disease transmission networks based on surveillance data. Existing studies have provided some methods for inferring information diffusion networks, where nodes correspond to individual persons. However, in the case of disease transmission, to effectively develop intervention strategies, it would be more realistic and reasonable for policy makers to study the diffusion patterns at the metapopulation level, that is, to consider disease transmission networks where nodes represent subpopulations, and links indicate their interrelationships. Such networks are useful to: (i) investigate hidden factors that influence epidemic dynamics, (ii) reveal possible sources of epidemic outbreaks, and (iii) practically develop and improve strategies for disease control. Therefore, based on such a real-world motivation, we aim to address the problem of inferring disease transmission networks at the metapopulation level. Specifically, we propose an inference method called NetEpi (Network Epidemic), and evaluate the method by utilizing synthetic and real-world datasets. The experiments show that NetEpi can recover most of the ground-truth disease transmission networks based only on the surveillance data. Moreover, it can help detect and interpret patterns and transmission pathways from the real-world data.

Keywords: Network inference, disease transmission networks, metapopulation, Bayesian learning, partial correlation networks.

1 Introduction

Infectious disease transmission has been studied with a network based approach and at an individual level [1]. However, existing studies often assume network structures are given in advance (e.g., air travels for the spread of H1N1 [2]), suggesting that it is possible to know which individual could be infected next. In reality, what is possible to observe is only the spatiotemporal surveillance data, containing infection times and locations of reported infection cases. This data

V.S. Tseng et al. (Eds.): PAKDD 2014, Part II, LNAI 8444, pp. 385–399, 2014.

provides no knowledge of the hidden transmission pathways that denote the routes of disease propagation among geographical locations. This real-world situation directly poses a significant challenge to the policy makers in applying intervention strategies at appropriate times and locations. In this regard, inferring disease transmission networks (DTNs) becomes an important and urgent research problem in epidemiological studies, which is our key objective.

The network inference problem has been widely studied in the information diffusion domain and is usually conducted at an individual level. Based on empirical time-series data of when people get informed, the static network inference is transformed into a combinatorial optimization problem [3]. Formulating it as a MAX-k-COVER problem, Rodriguez et al. have proven that selecting the top k edges that maximize the likelihood of the static information diffusion network (IDN) structure is NP-hard. They introduced a greedy algorithm based on the submodularity properties to approximate the optimal solution. Myers and Leskovec formulated a similar problem with heterogeneous edge weights into a convex optimization problem, and proposed a maximum likelihood method to solve it [4]. In addition, having noticed that the structure of a social network is sparse, they introduced penalty functions into the objective function to improve the accuracy. In a recently published study on inferring DTNs at the individual level [5], Teunis and Heijne used a pairwise kernel likelihood function to incorporate disease related information, and trained and applied the model using a real-world dataset collected from a university hospital.

The above work has provided insights into solving network inference problems at an individual level. However, inferring DTNs is more meaningful and practical at a metapopulation level, where nodes and edges represent patches with subpopulations (e.g., cities) and transmission pathways among them (e.g., transportation) rather than individual persons and their pairwise connections (e.g., social contacts). This is due to the considerations of: (i) the appropriateness of simulating disease transmission in both spatial and temporal scales [6], (ii) difficulties in simulating complicated human behaviors and collecting a huge amount of personal information [1], and (iii) the practice of controlling disease transmission from the view point of policy makers [7]. However, this treatment leads to two additional challenges: (i) nodes within metapopulation based DTNs can in addition connect to themselves, indicating susceptible people get infected by infected people within the same subpopulation, and (ii) metapopulation based disease transmission follows Directed Cyclic Graphs (DCGs) rather than Directed Acyclic Graphs (DAGs) as in information diffusion or individual based DTNs. Even if a large proportion of a certain subpopulation is infected, the remaining susceptible persons that have not been temporally infected will still have chances of being infected later.

Inferring metapopulation based DTNs is not only desirable but also challenging. As far as we know, there has not been such work done before. In this paper, we will address this problem, and more specifically, make three contributions: (i) to build a generalized linear disease transmission model that considers all possible transmission pathways at the metapopulation level, (ii) to develop an inference method, called NetEpi, that infers hidden DTNs based only on the spatiotemporal surveillance data, and (iii) to solve the network inference problem over DCGs rather than DAGs. We believe such work is also practically meaningful since it helps computationally predict large-scale infectious disease spread and provide policy makers with insights into optimizing intervention strategies.

The paper is organized as follows. The metapopulation based DTN inference problem is formulated in Section 2. A two-step inference method (NetEpi) is introduced in Section 3. NetEpi is evaluated by using both synthetic and real-world datasets in Section 4. Finally, we make conclusions in Section 5.

2   Problem Statement

A ground-truth DTN is defined as $G = <V, E>$, where the set of nodes is denoted as $V = \{v_i \mid i = 0, 1, 2, ..., N\}$. i is the index of a specific node. v_0 represents the external source node for the imported cases that would potentially cause local epidemics [8] (Imported cases are the laboratory-confirmed infection cases where people have traveled to disease endemic regions within days before the onset of the disease [8]). v_i $(i = 1, 2, ..., N)$ correspond to the rest of nodes within the target region. $E = \{e_i \mid i = 1, 2, ..., N\}$ denotes the set of edges with weights $W = \{w_i \mid i = 1, 2, ..., N\}$. e_i is the set of incoming links for node i, and w_i is the corresponding weight vector. Source node v_0 has no incoming links. The physical meanings of these edges that have non-zero weights describe the generalized transmission pathways that *temporally correlate* subpopulations in terms of their infection observations. In reality, G cannot be directly obtained. What is often collected is surveillance data, which can be represented as $D = \{<v_i, ic_i, t_i> \mid i = 0, 1, 2, ..., N, t \in T\}$ after aggregating infection cases based on locations and infection times. v_i corresponds to a geographical location (e.g., a city, or a township), ic_i is the aggregated number of infection cases, and t_i indicates a time step. T is the considered time period of disease transmission.

We refer to the estimated DTN as G^*, and consider three types of transmission pathways: (i) internal transmission component (ITC), which indicates that infected people, directly (e.g., in the air-borne disease of influenza) or indirectly (e.g., in the vector-borne disease of malaria), infect susceptible people within the same subpopulation, (ii) neighborhood transmission component (NTC), where disease transmits, through physically connected highways, adjacent borders, etc., among several subpopulations (it signifies the interactions happening between infected people in different subpopulations), and (iii) external influence component (EIC), which represents the source of imported cases from distant endemic regions or countries. In G, it is an external node connected to all the other nodes.

To characterize a disease transmission process over G, we integrate both of the internal transmission component and the external influence component with the neighborhood transmission component. The total number of infection cases can be written as a linear combination of the above three components plus an error term ε that captures the unpredicted biases, as follows:

$$ic_i^t = itc_i^t + ntc_i^t + eic_i^t + \varepsilon$$

$$= w_{ii} \times ic_i^{t-1} + \sum_j^{N_i} w_{ji} \times ic_j^{t-1} + w_{0i} \times ic_0^{t-1} + \varepsilon \tag{1}$$

where itc_i^t, ntc_i^t, and eic_i^t refer to the numbers of infection cases from ITC, NTC, and EIC to node i ($i \neq 0$) at time step t, respectively. N_i is the number of the neighbors of node i. The error term ε follows a zero-mean normal distribution, $\varepsilon \sim N(0, \beta)$. Eq. 1 characterizes the temporal dynamics of infection cases at each location. To be noticed, in the real world, once a patient is diagnosed to be infected, treatments and interventions (e.g., medication and isolation) would be taken by the physicians or hospitals. Thus, the infection cases at the current time step would be set to be isolated in the following time steps.

Given an observed surveillance dataset $D = \{<v_i, ic_i, t_i> \mid i = 0, 1, 2, ..., N, t \in T\}$, we intend to infer E of G and their corresponding weights W. The likelihood function for a specific node i based on Eq. 1 is:

$$\mathcal{L}(\boldsymbol{w_i}, \beta \mid ic_i) = \prod_{t=1}^{T} \frac{1}{(2\pi\beta)^{(1/2)}} e^{-\frac{1}{2\beta}(ic_i^t - w_{ii} \times ic_i^{t-1} - \sum_j^{N_i^*} w_{ji} \times ic_j^{t-1} - w_{0i} \times ic_0^{t-1})} \tag{2}$$

where N_i^* indicates the number of the estimated neighbors of node i within G^*. This set of neighbors can be written as $V_i^* = \{v_j \mid j = 0, 1, 2, ..., N \text{ and } w_{ji} \neq 0\}$. β is the variance of the normal distribution for ε. Therefore, we transform the network inference problem into an optimization problem, which is to find an optimal combination of neighbors with accurate weights for a specific node i. Specifically, to infer network G^*, we aim to maximize the likelihood function, given as:

$$\mathcal{L}(W, \beta \mid D) = \prod_{i=1}^{N} \mathcal{L}(\boldsymbol{w_i}, \beta \mid ic_i) \tag{3}$$

3 The Proposed Network Inference Method

3.1 Partial Correlation Network Construction

Given D, we first hope to construct an approximate network structure. It will reduce the trivial computations for our second step as well as filter out a proportion of false positive edges. Using the pearson correlation to build such networks is intuitive but not workable in the case of disease transmission. As shown in Fig. 1(a), disease transmission may follow a path from i to k, then to j. Even though i and j are not directly connected, they may still be correlated. Therefore, in the approximate network structure, denoted as G^p, they may be connected

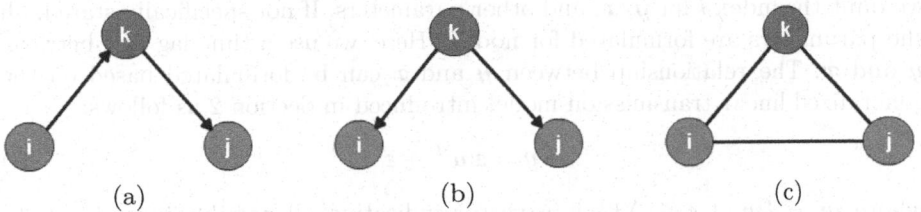

Fig. 1. Possible transmission relationships among three nodes [9]. Blue nodes are the targets of correlation analysis. The red one is the intermediate node. (a) shows that there is no directed edge between nodes i and j. Disease transmission follows a path from node i to k, then to j. (b) shows node k transmits to nodes i and j simultaneously and independently. (c) shows the pearson correlation results for (a) and (b).

as illustrated in Fig. 1(c). The same problem also exists in the case of Fig. 1(b), where i and j are the children of k considering the disease transmission dynamics.

To avoid such situations, we carry out first-order partial correlation analysis, which measures the dependence between two variables, while removing or fixing a third variable. In this regard, to compute it between nodes i and j, we remove or fix the impact of another node k, where $k = 0, 1, 2, ..., N$ and $k \neq i, j$. From the results, we choose coefficients that indicate strong correlations with significant p-values. It should be mentioned that partial correlation usually does not provide edge directions [10]. Therefore, to infer directed edges, we analyze the time-series data with a time lag (e.g., one day or one week). Then, the direction is from the node using the previous time-series data to the node using the current one. The partial correlation coefficient between nodes i and j after fixing the variable of node k is $\rho_{ij.k} = (\rho_{ij} - \rho_{ik}\rho_{jk})(\sqrt{1 - \rho_{ik}^2}\sqrt{1 - \rho_{jk}^2})$, where ρ_{ij}, ρ_{ik}, and ρ_{jk} are the covariances. This method removes many false positive edges as well as generates an approximate partial correlation network (PCN), G^p.

3.2 Back-Tracking Bayesian Learning

It should be noted that some edges in G^p still do not exist in G. A possible solution is to set the weights of these false positive edges with values of zero during the inference process. This is similar to the removal of irrelevant basis components as in basis pursuit for dimensionality reduction [11]. In our proposed inference method, we base our second step on the Sparse Bayesian Learning (SBL) framework [12]. To be noticed, if two components are similar, SBL only chooses one of them in order to compress the relevant information. However, in our case, even two nodes are similar, we aim to find both of them.

For node i, we divide preprocessed surveillance dataset D into two subsets: an $M \times 1$ vector of $y = \{<v_i, ic_i, t_i> \mid t_i = 2, 3, ..., M+1, M \in T\}$ and an $M \times |N^p|$ matrix of $x = \{<v_j, ic_j, t_j> \mid j \in N^p, t_j = 1, 2, ..., M, M \in T-1\}$. M is the size of output variable y and input variable x. N^p represents the indices of the possible neighbors that node i has based on G^p. $T-1$ is the previously considered time period of disease transmission. For the sake of presentation, in the following, we omit the index i for $\boldsymbol{y}$, $\boldsymbol{x}$, and other parameters. If not specifically stated, all the parameters are formulated for node i. Here, we use a time lag of 1 between $\boldsymbol{y}$ and $\boldsymbol{x}$. The relationship between $\boldsymbol{y}$ and $\boldsymbol{x}$ can be formulated based on the generalized linear transmission model introduced in Section 2 as follows:

$$y = xw^T + \varepsilon \tag{4}$$

where $\boldsymbol{w} = \{w_j \mid j \in N^p\}$ is a vector indicating all possible incoming links estimated based on G^p. ε is an error term. Under the framework of SBL, both $\boldsymbol{w}$ and ε follow a zero-mean Gaussian distribution with variances of $\boldsymbol{\alpha}$ and β, respectively. They are defined respectively as: $p(\boldsymbol{w}|\boldsymbol{\alpha}) = \prod_{j=1}^{N^p} N(w_i|0, \alpha_j^{-1})$ and $p(\varepsilon) = N(0, \beta)$. Because we have no prior knowledge about $\boldsymbol{w}$ and ε, it is reasonable to set them with non-informative prior distributions (e.g., Gamma distribution). $\boldsymbol{\alpha}$ and β are assumed to have the same hyperparameters for all nodes.

Given the observation data of $\boldsymbol{y}$ and the prior distribution of $\boldsymbol{\alpha}$ and β, the posterior distribution of $\boldsymbol{w}$ is:

$$p(\boldsymbol{w}|\boldsymbol{y}, \boldsymbol{\alpha}, \beta) = \frac{likelihood \times prior}{normalize\ factor} = \frac{p(\boldsymbol{y}|\boldsymbol{w}, \beta)p(\boldsymbol{w}|\boldsymbol{\alpha})}{p(\boldsymbol{y}|\boldsymbol{\alpha}, \beta)} \tag{5}$$

which is a Gaussian distribution $N(\boldsymbol{\mu}, \boldsymbol{\Sigma})$ with $\boldsymbol{\mu} = \beta^{-1}\boldsymbol{\Sigma}\boldsymbol{x}^T\boldsymbol{y}$, $\boldsymbol{\Sigma} = (\boldsymbol{\Lambda} + \beta^{-1}\boldsymbol{x}^T\boldsymbol{x})^{-1}$ where $\boldsymbol{\Lambda} = diag(\alpha_1, \alpha_2, ..., \alpha_{N^p})$. "type-II maximization likelihood" maximization combined with a maximum a posteriori probability (MAP) estimate transforms the whole problem into that of maximizing the marginal likelihood function of:

$$p(\boldsymbol{y}|\boldsymbol{\alpha}, \beta) = \int p(\boldsymbol{y}|\boldsymbol{w}, \beta)p(\boldsymbol{w}|\boldsymbol{\alpha})d\boldsymbol{w} \tag{6}$$

Writing Eq. 6 into a logarithm form $\mathcal{L}(\boldsymbol{\alpha})$, we have:

$$\begin{aligned} \mathcal{L}(\boldsymbol{\alpha}) = \log p(\boldsymbol{y}|\boldsymbol{\alpha}, \beta) &= \log \int p(\boldsymbol{y}|\boldsymbol{w}, \beta)p(\boldsymbol{w}|\boldsymbol{\alpha})d\boldsymbol{w} \\ &= -\frac{1}{2}[M \log 2\pi + \log|\boldsymbol{C}| + \boldsymbol{y}^T\boldsymbol{C}^{-1}\boldsymbol{y}] \end{aligned} \tag{7}$$

with $\boldsymbol{C} = \beta I + \boldsymbol{x}\boldsymbol{\Lambda}^{-1}\boldsymbol{x}^T$. The derivatives with respect to α_j and β are [13]:

$$\frac{\partial \mathcal{L}(\boldsymbol{\alpha})}{\partial \log \alpha_j} = \frac{1}{2}(1 - \alpha_j \Sigma_{jj} - \alpha_j \mu_j^2) \tag{8}$$

$$\frac{\partial \mathcal{L}(\boldsymbol{\alpha})}{\partial \log \beta} = \frac{1}{2}[\frac{M}{\beta} - \|\boldsymbol{y} - \boldsymbol{x}\boldsymbol{\mu}\|^2 - trace(\boldsymbol{\Sigma}\boldsymbol{x}^T\boldsymbol{x})] \tag{9}$$

Setting Eqs. 8 and 9 to zero, the estimations of α_j and β become:

$$\alpha_j^{new} = \frac{1 - \alpha_j \Sigma_{jj}}{\mu_j^2} \tag{10}$$

$$\beta^{new} = \frac{M - \sum_{j=1}(1 - \alpha_j \Sigma_{jj})}{\|\boldsymbol{y} - \boldsymbol{x}\boldsymbol{\mu}\|^2} \tag{11}$$

The above iterative estimation procedure is solved by using the Expectation-Maximization algorithm. In each iteration, we estimate the contributions to the marginal likelihood function for all nodes in G^p. The one with maximum contribution is selected as the candidate neighbor. Then, its corresponding weight is computed. As to be noted, in G, we only have positive links. However, the prior distribution may cause w to become negative. To avoid this, a constraint of limiting w to be positive is introduced. To incorporate this constraint, we use a back-tracking technique. During the EM learning procedure, the update of marginal likelihood function and other parameters proceeds sequentially. Consequently, each time μ, Σ, α_j, and β are updated, we select those α_j that fail the constraint, and put their corresponding indices into a blacklist. The program is rolled back to the previous step and proceeds by selecting nodes that do not exist in the blacklist. The algorithm is shown in Alg. 1.

Algorithm 1. Back-Tracking Bayesian Learning

Require: D: Preprocessed surveillance dataset; G^p: Partial correlation network;
Ensure: G^*: Inferred disease transmission network;

---

1. Divide D into two subsets with time lag of one time unit;
2. **for all** node $i = 1, 2, ..., N$ **do**
3.   Initialize parameters for prior distributions;
4.   Construct marginal likelihood function $p_i(y|\alpha, \beta)$ (shown in Eq. 6);
5.   **while** not reaching stopping criteria **do**
6.   **for all** node $j \in N^p$, and $i \neq j$ **do**
7.   Compute contributions to $p_i(y|\alpha, \beta)$;
8.   **end for**
9.   Select node with maximum contribution;
10.   Re-estimate all weights of current neighbors of node i;
11.   **if** all weights are not less than zero **then**
12.   Update neighborhood list;
13.   **else**
14.   Remove neighbors with weights less than zero, and put them into blacklist;
15.   Roll back $p_i(y|\alpha, \beta)$;
16.   **end if**
17.   **end while**
18. **end for**
19. Combine all neighborhood lists to construct G^*;
20. return G^*;

3.3   Discussions

As stated in [3], it is not trivial nor practical to find all the edges within G, or the exact time required to stop the inference program. Thus, once the program iterates to the maximum permitted iteration steps, or the update of the marginal likelihood function converges to a small value, we will stop the learning procedure. To compute the PCN, the time complexity is $O(N^3)$. To speed up this process, we use dynamic programming to recursively compute the first-order partial correlation based on the result of zeroth-order partial correlation. As for the back-tracking Bayesian learning, the complexity of Bayesian learning is mainly distributed over the computation of parameters of Σ, which requires $O(N^3)$. An efficient incremental algorithm proposed in [12] can optimize this computation. Besides, the computation based on G^p can also reduce this computational time. After integrating the back-tracking algorithm, the time complexity becomes exponential. However, based on our experiments, the algorithm usually converges fast. That is to say, the algorithm seldom tracks back to the nodes that are selected at the very beginning. This is caused by the previous Bayesian learning; it selects those significantly contributing nodes at the very beginning, making the marginal likelihood function converge to a near optimum solution without large space to increase, and stable until reaching the stopping criteria.

4   Experiments

4.1   Experiments Based on Synthetic Data

The synthetic data generation proceeds as follows: we first use Kronecker Graphs model [14] to generate a basic network structure. Then, we link all the nodes with an external node v_0, and generate self-connected edges with predefined probabilities. We iteratively run the transmission model, as given in Eq. 1, for a sufficient number of time steps to generate the disease surveillance data.

Experimental Setting. We construct 3 types of network structures: (i) core-periphery networks (CPNs), which have a cluster of nodes in the core of the network, (ii) hierarchical community networks (HCNs), where a proportion of the nodes form several small communities, and (iii) random graphs (RGs), which have no obvious pattern. Then, for each structure, we generate networks with different sizes: 64n with 100e, 150e ("n"and "e" are the abbreviations of "nodes" and "edges", respectively); 128n with 180e, 200e; 256n with 350e, 400e; 512n with 720e, 800e. For each of them 10 datasets are produced. Specifically within each generation process, we make sure that the transmission process cover all the edges in G. In total, there are 3 types of network topologies $\times$ 8 different sizes $\times$ 10 independent transmission processes = 240 datasets.

The Baseline Method. To our best knowledge, there have not been much prior work on inferring network structures over DCGs. Therefore, we utilize a probability based baseline method. At two adjacent time steps $t = n$ and $t = n + 1$, all the nodes that have infection cases at $t = n$ will have connections to those nodes have infection cases at $n + 1$. The edge weight is affected by the number of infection cases and the number of infected nodes at the previous time step. We select the top k edges with the highest weights, and form the estimated disease transmission network G^* accordingly. The mathematical formula to compute the baseline edge weight is $w_{ij} = ic_i{}^t ic_j{}^{t+1} / \sum_{i=1}^{N} ic_i{}^t$.

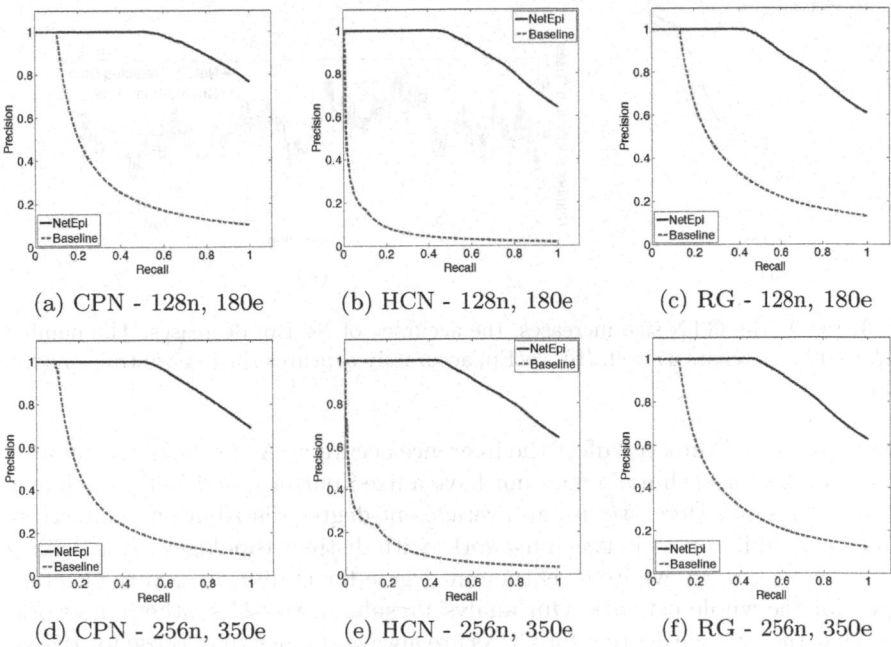

(a) CPN - 128n, 180e   (b) HCN - 128n, 180e   (c) RG - 128n, 180e

(d) CPN - 256n, 350e   (e) HCN - 256n, 350e   (f) RG - 256n, 350e

Fig. 2. The precision-recall curves for synthetic DTNs. It is obvious that NetEpi outperforms the baseline method in all cases. The average degree for the above six networks are 1.3876, 1.4496, 1.4651, 1.3619, 1.5175, and 1.5097, respectively, from (a) to (f).

Result Evaluation. To evaluate the inference results, we compute the precision-recall curves as shown in Fig. 2. For the sake of space, we display only part of our experimental results here. The precision and recall are defined as "what fraction of edges in G^* is also present in G", and "what fraction of edges of G appears in G^*", respectively [3]. For nodes i and j, if both ground-truth edge e_{ij} and inferred edge e_{ij}^* exist, and the difference between the corresponding weights $|w_{ij} - w_{ij}^*|$ is less than a threshold, then we say the inferred edge is accurate.

In our experiments, NetEpi outperforms the baseline method in all 240 datasets. Specifically, for networks that have the same sizes but different topologies, NetEpi performs the best on the CPNs. Nodes located in the core region have more connections as compared with those in the periphery region. Therefore, to achieve an optimal solution, core-located nodes will have higher probabilities to possess a more number of combinations of neighbors. In other words, the probabilities to find a globally optimal solution for a single node will decrease as the number of its incoming edges increases. The accuracies of NetEpi over CPNs are consequently biased by the tradeoff between core-located and periphery-located nodes. In comparison, for HCNs, there is no longer a single core. In contrast, there are several sub-cores that individually form a sub-community. This structure makes the average number of combinations for each

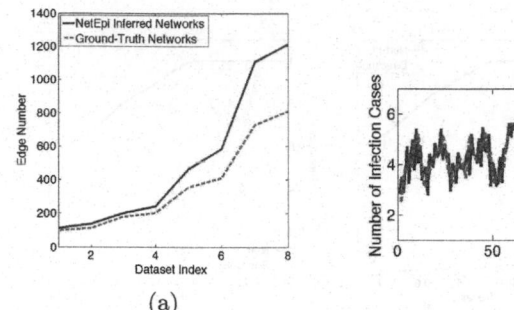
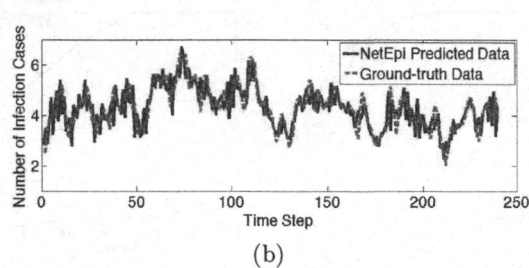

(a)   (b)

Fig. 3. (a) As the GTN size increases, the accuracy of NetEpi decreases. The number of false edges increases as well. (b) NetEpi accurately captures the disease transmission trend.

node increase and directly affect the inference accuracy. As for RGs, the number of connections for each node does not have a fixed pattern, and NetEpi achieves oscillating results. Here, we use an average out-degree distribution to illustrate the accuracy differences between networks with distinct topologies. It is defined as $d_{avg} = \sum_{i=1}^{N} d_i/N$ where d_i is the out degree for node i, d_{avg} is the average degree for the whole network. Our analysis results of the 24 synthetic networks show that the average degrees for CPNs are always smaller than those for HCNs. And, the average degrees for RGs present oscillating patterns (Fig. 2).

For networks with the same topologies but different sizes, NetEpi achieves better results on inferring smaller ones as shown in Fig. 2. During the inference process, the whole Ground-Truth Network (GTN) is treated as a complete network. Even given the approximate structure G^p, the complexity quadratically increases as the number of nodes increases. Meanwhile, as the edge number increases, the number of combinations of neighbors for each node to achieve optimal solutions increases as well, which directly interferes the inference results as in Fig. 3(a). However, the network sizes of metapopulation based DTNs are usually small at the administrative level. For example, for a global epidemic disease, WHO publishes statistical reports at the country level. Therefore, a possible solution to infer large-size networks is to perform hierarchical clustering based on geographical information. NetEpi is conducted from the highest level where each node represents a cluster of lower-level nodes. Then, within each high-level node, NetEpi is performed again to infer lower-level transmission networks. This process is repeatedly and sequentially conduced in order to get a whole picture of large-size networks. Fig. 3(b) shows an example of the prediction results of NetEpi. It is obvious that the predicted epidemic trend happening in the GTN is well captured by the inferred network. This validates that NetEpi converges to an optimal solution, although this may not be the global one.

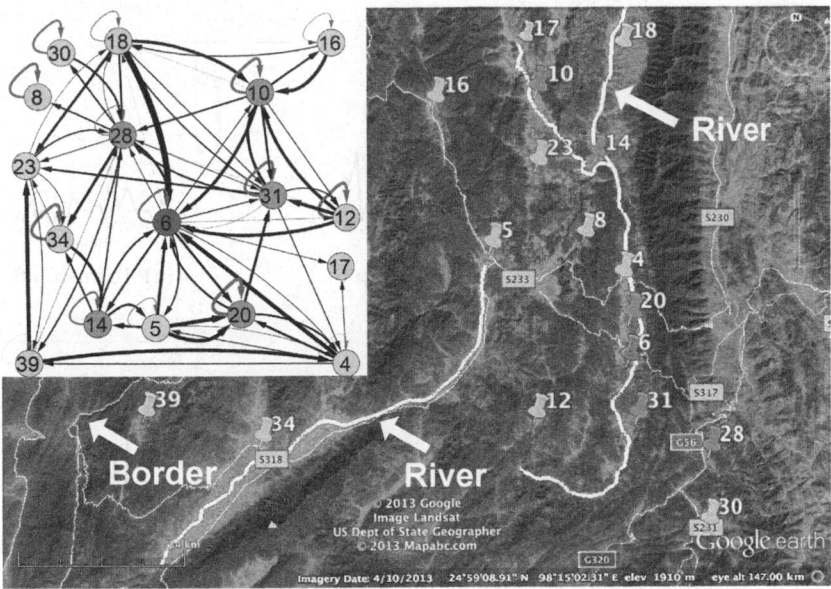

Fig. 4. Townships that form a community in the inferred malaria transmission network

4.2 Experiments Based on a Real-World Dataset

Experimental Setting. The real-world dataset was provided by Chinese Center for Disease Control and Prevention. It contains reported malaria cases in Yunnan province, China. In total, there are 2928 cases reported in 51 townships in 2005. These townships are distributed along the border between China and Myanmar (a high malaria-endemic country) and classified into 5 categories based on the numbers of infection cases: $(200, +\infty)$ (red), $(150, 200]$ (purple), $(100, 150]$ (green), $(50, 100]$ (yellow), and $(0, 50]$ (blue). The dataset is very sparse, with missing data. Moreover, there is no complete information about the sources and identifications of imported cases. Thus, a fixed external node cannot be set up before the inference procedure. Like the periodical pattern of the Internal Transmission Component, the External Influence Component also presents regular pattern because of the frequent human mobility motivated by cross-border trade and business. We consequently merge EIC with ITC, and represent either of them, or their combination, by self-connected edges. This is reasonable because it has been recorded that most of these imported cases were due to working, trading, and/or visiting in/with Myanmar regularly. Therefore, self-connected edges are able to capture these regular patterns. We are informed that there exist imported cases, and expect that the inferred malaria transmission network contain many self-connected edges. It has been widely reported that the incubation time for *Plasmodium vivax* is $12 \sim 17$ days [15]. However, studies have also reported that the incubation time could be longer from several months to several years [15]. Therefore, we choose 21 days as the time window when inferring

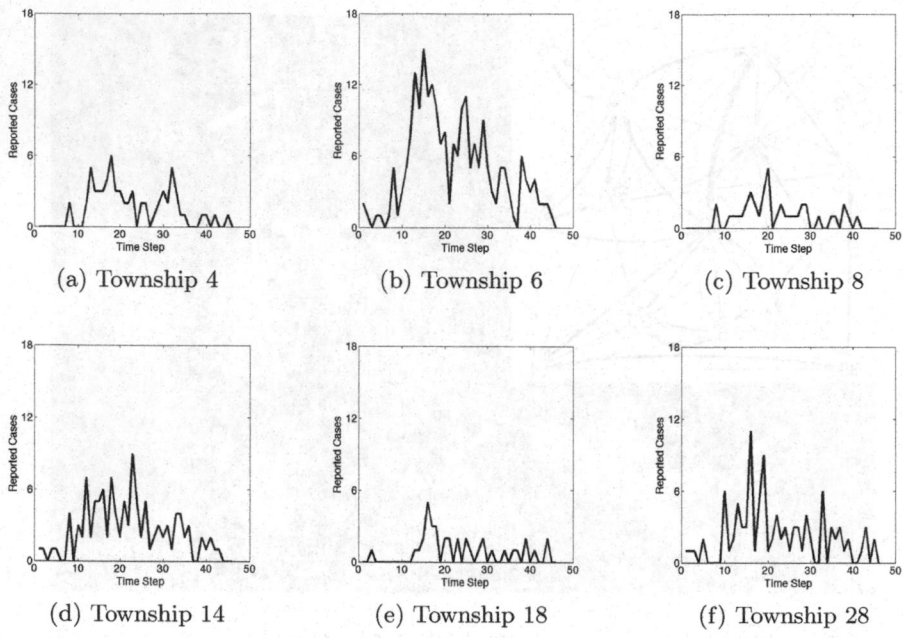

Fig. 5. The reported cases for the selected nodes in 2005. In order to present them clearly, we aggregate the reported cases on an eight-day basis.

the underlying malaria transmission network, so that it compromises both the reported incubation time and the sensitivity analysis that we have conducted previously.

Result Interpretation. The inferred network contains two classes of nodes. Some of them connect to themselves as we expected, while the others form communities. Self-connected nodes occupy 50.98% of the whole network. This caters to our previous expectation. These nodes are located adjacent to the border between China and Myanmar, or connected with the border by highways, or situated close to rivers, which provide suitable environments for the vectors of malaria to reproduce. Therefore, the malaria endemic within these self-connected nodes are possibly caused by EIC, ITC, or their combinations.

As shown in Fig. 4, there is a community found in the inferred network. It contains most nodes that have severe endemic situations. Many townships are distributed along two rivers. Besides providing suitable habitat there, rivers also bring the larva of vectors from the upstream to the downstream. Therefore, the inferred edges of these nodes possibly represent partial influences from rivers, and impact of vectors' movements. In addition, the severest township 6 has connections to all the other second level severity townships (green nodes), indicating that their disease transmission interactions may be the dominant reason for the local malaria endemic in the region. Other townships 16, 28, and 30 are connected with the others by highways (e.g., S231, S233, S317, and S318), indicating that their transmission pathways are possibly caused by transportation.

It can readily be noted from Fig. 4 that some inferred edges are thicker than others, denoting higher transmission influences. For example, e_{18-6} (the dash in the index is used for separation) is thicker than e_{14-6}, e_{4-6}, and e_{28-6}. We interpret this based on Fig. 5 (a) - (f) where reported cases are aggregated on an eight-day basis for clear presentation. As shown, although township 18 (Fig. 5(e)) has fewer reported cases than other example townships and contains many zero-case intervals, its temporal trend does not significantly violate the trend of township 6 (Fig. 5(b)). In comparison, the "mountain-valley-mountain" pattern of township 6 can only be partially matched with other townships (e.g., townships 4 (Fig. 5(a)), 14 (Fig. 5(d)) and 28 (Fig. 5(f))). The influence from township 6 to 4 is much less than that from the reverse direction. This is because the second highest peak appearing between time steps 20 to 30 in the trend of township 6 cannot contribute to the valley appearing at the same time interval in the trend of township 4. However, the reverse contribution is reasonable. Intuitively, the pair of townships 4 and 8 (Fig. 5(c)) and the pair of townships 14 and 28 have similar trends respectively, but NetEpi only finds edges between townships 14 and 28. This is due to that, for townships 4 and 8, their trends before time step 20 seem to be similar, but those after step 20 present a time lag of around 8*8 days.

There are totally 47 rather than 51 townships contained in the inferred network. The 4 missing nodes have neither self-connected edges nor neighborhood connected edges. The sum of their infection cases is 81, which is a very small proportion of all infection cases. Therefore, we think their disease transmission dynamics are caused by accidentally imported cases. In addition, although some townships are located very close to each other, and in the positions of the upstream or the downstream of the same river, they are not connected in the inferred network (e.g., townships 10 and 17). We believe this is because their transmission pathways are not significant or their malaria endemic is mainly affected by imported cases that overtook the impact of other factors. To interpret them, currently available information about highways, rivers, and geographical locations may not be fully adequate, because they represent the transmission pathways that are the *comprehensive results of all impact factors*. In addition, the roads that are locally formed and managed are not displayed in the map, which may also play significant roles in malaria transmission. Missing reports and data sparsity may affect the results as well. However, our method can still detect some hidden connections that may draw the attention of policy makers.

5   Conclusion

In this study, based on the need for real-world disease transmission pattern discovery, we have defined and addressed an inverse network inference problem. Given only the surveillance data, we have proposed a two-step network inference algorithm, called NetEpi. Having highlighted the major differences between the individual based network inference and the metapopulation based network inference problems, we defined a linear disease transmission model over a Directed Cyclic Graph (DCG) containing three types of transmission pathways as often found in the real-world situations, namely, internal transmission component, neighborhood transmission component, and external influence component. We performed partial correlation analysis to construct an approximate network structure for the underlying disease transmission network, and then conducted back-tracking Bayesian learning to iteratively infer edges and estimate their corresponding weights. We have evaluated the proposed method by using synthetic data. The experimental results have shown that NetEpi outperforms a probability based baseline inference method, and performs well over a relatively small-scale network, which is sufficient for metapopulation based disease transmission network modeling in practice. Meanwhile, NetEpi achieves a reasonable accuracy over different network topologies. In addition, we have applied NetEpi to a real-world disease transmission dataset and have discovered certain meaningful community patterns as well as transmission pathways. Our future work will focus on inferring disease transmission networks in which there exist various underlying, sometimes dynamically-changing network structures. We will also consider other impact factors that may be disease-dependent. This work will further be applied to the real-world situations for policy makers to develop and implement intervention strategies for controlling disease transmission.

Acknowledgement. The authors would like to acknowledge the funding support from Hong Kong Research Grants Council (HKBU211212) and from National Natural Science Foundation of China (NSFC81273192) for the research work being presented in this paper.

References

1. Keeling, M.J., Eames, K.T.: Networks and Epidemic Models. J. R. Soc. Interface. 2, 295–307 (2005)
2. Bajardi, P., Poletto, C., Ramasco, J.J., Tizzoni, M., Colizza, V., Vespignani, A.: Human Mobility Networks, Travel Restrictions, and Global Spread of 2009 H1N1 Pandemic. PLoS ONE 6, e16591 (2011)
3. Gomez-Rodriguez, M., Leskovec, J., Krause, A.: Inferring Networks of Diffusion and Influence. In: 16th ACM SIGKDD International Conference on Knowledge Discovery and Data Mining, pp. 1019–1028. ACM Press, New York (2010)
4. Myers, S., Leskovec, J.: On the Convexity of Latent Social Network Inference. In: Advances in Neural Information Processing Systems, pp. 1741–1749 (2010)
5. Teunis, P., Heijne, J.C.M., Sukhrie, F., van Eijkeren, J., Koopmans, M., Kretzschmar, M.: Infectious Disease Transmission as a Forensic Problem: Who Infected Whom? J. R. Soc. Interface 10(81) (2013)
6. Arino, J.: Diseases in Metapopulations. In: Ma, Z., Zhou, Y., Wu, J. (eds.) Modeling and Dynamics of Infectious Diseases. Series in Contemporary Applied Mathematics, vol. 11, pp. 65–123. World Scientific (2009)
7. Ndeffo, M.M.L., Gilligan, C.A.: Resource Allocation for Epidemic Control in Metapopulations. PLoS ONE 6, e24577 (2011)

8. Shang, C.S., Fang, C.T., Liu, C.M., Wen, T.H., Tsai, K.H., King, C.C.: The Role of Imported Cases and Favorable Meterorological Conditions in the Onset of Dengue Epidemics. PLoS Negl. Trop. Dis. 4, e775 (2010)
9. Yuan, Y., Li, C.T., Windraw, O.: Directed Partial Correlation: Inferring Large-Scale Gene Regulatory Network through Induced Topology Disruptions. PLoS ONE 6, e16835 (2011)
10. Lasserre, J., Chung, H.R., Vingron, M.: Finding Associations among Histone Modifications Using Sparse Partial Correlation Networks. PLoS Comput. Biol. 9, e1003168 (2013)
11. David, P.W., Bhaskar, D.R.: Sparse Bayesian learning for basis selection. IEEE Trans. Signal Processing. 52(8), 2153–2164 (2004)
12. Tipping, M.E., Faul, A.: Fast Marginal Likelihood Maximization for Sparse Bayesian Models. In: 9th International Workshop on Artificial Intelligence and Statistics, pp. 3–6 (2003)
13. Tzikas, D., Likas, C., Galatsanos, N.: Sparse Bayesian Modeling with Adaptive Kernel Learning. IEEE Trans. Neural Networks. 20(6), 926–937 (2009)
14. Leskovec, J., Faloutsos, C.: Scalable Modeling of Real Graphs using Kronecker Multiplication. In: 24th International Conference on Machine Learning, pp. 497–504. ACM Press, New York (2007)
15. Brasil, P., de Pina Costa, A., Pedro, R., da Silveira Bressan, C., da Silva, S., Tauil, P., Daniel-Ribeiro, C.: Unexpectedly Long Incubation Period of Plasmodium vivax Malaria, in the Absence of Chemoprophylaxis, in Patients Diagnosed outside the Transmission Area in Brazil. Malar. J. 10(1), 122 (2011)

A Novel Framework to Improve siRNA Efficacy Prediction

Bui Thang Ngoc

Japan Advanced Institute of Science and Technology
1-1 Asahidai, Nomi City, Ishikawa, 923-1211 Japan
thangbn@jaist.ac.jp

Abstract. Short interfering RNA sequences (siRNAs) can knockdown target genes and thus have an immense impact on biology and pharmacy research. The key question of which siRNAs have high knockdown ability in siRNA research remains challenging as current known results are still far from expectation. This work aims to develop a generic framework to enhance siRNA knockdown efficacy prediction. The key idea is first to enrich siRNA sequences by incorporating them with rules found for designing effective siRNAs and representing them as transformed matrices, then to employ the bilinear tensor regression to do prediction on those matrices. Experiments show that the proposed method achieves results better than existing models in most cases.

1 Introduction

In 2006, Fire and Mello received their Nobel Prize for their contributions to research on RNA interference (RNAi) that is the biological process in which RNA molecules inhibit gene expression, typically by causing the destruction of specific mRNA molecules. Their work and that of others on discovery of RNAi have had an immense impact on biomedical research and will most likely lead to novel medical applications. On RNAi research, designing of siRNAs (short interfering RNAs) with high efficacy is one of the most crucial RNAi issues. Highly effective siRNAs can be used to design drugs for viral-mediated diseases such as Influenza A virus, HIV, Hepatitis B virus, RSV viruses, cancer disease and so on. As a result, siRNA silencing is considered one of the most promising techniques in future therapy. Finding highly effective siRNAs among thousands of potential siRNAs for an mRNA remains a great challenge.

Various siRNA design rules have been found by empirical processes since 1998. The first rational siRNA design rule was detected by Elibalshir *et al.* [2]. They suggested that siRNAs having 19–21 nt (nucleotide) in length with 2 nt overhangs at 3' end can efficiently silence mRNAs. Scherer *et al.* reported that the thermodynamic properties (G/C content of siRNA) to target specific mRNAs are important characteristics [11]. Soon after these works, many rational design rules for effective siRNAs have been found, typically those in [10], [15], [1], [4], [7], [14]. For example, Reynolds *et al.* [10] analyzed 180 siRNAs and found eight criteria for improving siRNA selection: (1) G/C content 30−52%, (2) at least 3

V.S. Tseng et al. (Eds.): PAKDD 2014, Part II, LNAI 8444, pp. 400–412, 2014.

As or Us at positions from 15 to 19, (3) absence of internal repeats, (4) an A at position 19, (5) an A at position 3, (6) a U at position 10, (7) a base other than G or C at position 19, (8) a base other than G at position 13.

However, most of siRNA design tools using the above-mentioned design rules have low accuracy, because about 65% of the siRNAs predicted as high effective was failed when tested experimentally as they were 90% in inhibition and near 20% of them were found to be inactive [9]. One reason is the previous empirical analyses only based on small datasets and focused on specific genes. Therefore, each of these rules certainly is poor to individually design effective siRNAs.

Since nearly a decade, machine learning techniques have alternatively been applied to predict knockdown efficacy of siRNAs. The first predictive model was proposed by Huesken et al. in which motifs for effective and ineffective siRNA sequences were detected basing on the significance of nucleotides by using a neural network to train 2,182 scoring siRNAs (scores are real numbers in $[0, 1]$, the high score the higher knockdown efficacy) and test on 249 siRNAs [5]. This data set was consequently used to build other predictive models [6], [13], [16]. Recently, Qui et al. used multiple support vector regression with RNA string kernel for siRNA efficacy prediction [8], and Sciabola et al. applied three dimension structural information of siRNA to increase predictability of the regression model [12]. However, most of those methods suffer from some drawbacks. Their correlations between predicted values and experimental values of dependent variable ranging from 0.60 to 0.68 were considerably decreased when testing on independent data sets. It may be caused by the fact that the Huesken dataset may not be representative of the siRNA population having about 4^{19} siRNAs and the sample size is small. Besides the scoring siRNA dataset, the labelled siRNA datasets, e.g. siRecord database [9] with labels such as 'very high", 'high', 'medium', 'low' for the knockdown ability were also exploited by classification methods.

Our work aims to develop a novel framework for better prediction of the siRNA knockdown ability. The key idea is not only focusing on learning algorithms but also exploiting results of the empirical process to enrich the data. To this end, we first learn transformation matrices by incorporating existing siRNA design rules with labelled siRNAs in siRecord database. We then use the transformation matrices to enrich scoring siRNAs as transformed matrices and do prediction with them by bilinear tensor regression where the Frobenius norm is appropriately replaced by L_2 regularization norm for an effective computation. Experiments show that the proposed method achieves results better than most existing models. The contributions of this work are summarized as follows

1. A novel generic framework to predict siRNA efficacy by enriching siRNA sequences with domain knowledge and appropriately using bilinear tensor regression.
2. An optimization method to enrich siRNAs using siRNA design rules found by empirical works.
3. The use of L_2 norm instead of Frobenius norm in bilinear tensor regression that allows effectively learning the set of model parameters.

2   The Framework to Improve siRNA Efficacy Prediction

The problem of siRNA knockdown efficacy prediction using siRNA design rules is formulated as follows:

- **Given:** Two sets of labelled siRNA and scoring siRNA sequences of length n, and a set of K siRNA design rules.
- **Find:** A function that assigns a right score to a given siRNA.

The proposed framework consists of four steps in two phases. The first phase is to encode siRNAs and learn transformation matrices. The second phase is to use transformation matrices to enrich siRNAs as transformed matrices and learn model parameters of the bilinear tensor regression to predict the score of siRNAs using transformed matrices. The steps of the framework are summarized in Table 1.

Table 1. Framework for siRNA knockdown efficacy prediction

---

1. To encode each siRNA sequence as an encoding matrix X representing the nucleotides A, C, G, and U at n positions in the sequence. Thus, siRNA sequences are represented as $n \times 4$ encoding matrices.
2. To learn transformation matrices $T_k, k = 1, ..., K$, each characterizes the knockdown ability of nucleotides A, C, G, and U at n positions in the siRNA sequence regarding the kth design rule. Each T_k is learned from the set of labelled siRNAs and the kth design rule. This incorporation of each design rule with siRNAs leads to solve a newly formulated optimization problem.
3. To transform siRNA (encoding matrices) to transformed matrices by K transformation matrices. The transformed matrices of size $K \times n$ are considered as second order tensor representations of the siRNA sequences.
4. To build a bilinear tensor regression model that uses transformed matrices of scoring siRNAs to predict the knockdown ability of new siRNAs.

2.1   Encoding siRNA and Transformation Matrix Learning

Step 1 of the framework can be easily done where each siRNA sequence with n nucleotides in length is encoded as a binary encoding matrix of size $n \times 4$. In fact, four nucleotides A, C, G, or U are encoded by encoding vectors $(1,0,0,0)$, $(0,1,0,0)$, $(0,0,1,0)$ and $(0,0,0,1)$, respectively. If a nucleotide from A, C, G, and U appears at the jth position in a siRNA sequence, $j = 1, ..., n$, its encoding vector will be used to encode the jth row of the encoding matrix.

Step 2 is to learn transformation matrices T_k regarding the kth design rule, $k = 1, ..., K$. T_k has size of $4 \times n$ where the rows correspond to nucleotides A, C, G, and U and the columns correspond to n positions on sequences. T_k are learned one by one from the set of siRNAs and the kth design rule, thus we use T instead of T_k for simplification. Each cell $T[i, j], i = 1, ..., 4, j = 1, ..., n$, represents the knockdown ability of nucleotide i at position j regarding the kth

Sequence	Encoding matrix X	Transformation matrix T	Transformed data vector $x = T \circ X$
AUGCU	1 0 0 0 0 0 0 1 0 0 1 0 0 1 0 0 0 0 0 1	$\overline{0.5}$ 0.7 0.32 0.2 0.5 0.3 0.1 0.6 $\overline{0.6}$ 0.3 0.1 0.1 $\overline{0.08}$ 0.1 0.1 0.1 $\overline{0.1}$ 0 0.1 $\overline{0.1}$	$(\overline{0.5}, \overline{0.1}, \overline{0.08},$ $\overline{0.6}, \overline{0.1})$

Position	Knockdown ability	Nucleotides	Mapping to T	Constraints on T
19	Effective	A,U	$T[1,19]$, $T[4,19]$	$T[3,19] - T[1,19] < 0$ $T[3,19] - T[4,19] < 0$
	Ineffective	C	$T[2,19]$	$T[2,19] - T[1,19] < 0$ $T[2,19] - T[3,19] < 0$ $T[2,19] - T[4,19] < 0$

Fig. 1. The left table shows an example of encoding matrix, transformation matrix, and transformed vector (the values $\overline{0.5}, \overline{0.1}$ etc. are taken to the transformed vector). The right table is an example of incorporating the condition of a design rule at position 19 to a transformation matrix T by designing constraints.

design rule. Each cell $T[i,j]$ to be learned have to satisfy a number of constraints. First, they are basic and normalization constraints on elements of T

$$T[i,j] \geq 0, \qquad i = 1, ..., 4; \quad j = 1, 2, \ldots, n \tag{1}$$

$$\sum_{i=1}^{4} T[i,j] = 1, \quad j = 1, \ldots, n \tag{2}$$

The second kind of constraints related to design rules. Each design rule propositionally describes the occurrence or absence of nucleotides at different positions of effective siRNA sequences. Therefore, if a design rule shows the occurrence (absence) of some nucleotides on jth position, then their corresponding values in the matrix T would be greater (smaller) than other values at column j. For example, the design rule in the right table in Figure 1 illustrates that at position 19, nucleotides A/U are effective and nucleotide C is ineffective. It means that knockdown ability of nucleotides A/U are bigger than that of nucleotides G/C and knockdown ability of nucleotide C is smaller than that of the other nucleotides. Thus, values $T[1,19], T[2,19], T[3,19]$ and $T[4,19]$ show the knockdown ability of nucleotides A, C, G and U at position 19, respectively. Therefore, five constraints at column 19 of T are formed. Generally, we denote the set of R trick inequality constraints on T by the design rule under consideration by

$$\{g_r(T) < 0\}_{r=1}^{R} \tag{3}$$

The third kind of constraints relating to preservation of the siRNA classes after being transformed by using transformation matrices T_k, it means that siRNAs belonging to the same class should be more similar to each other than siRNAs belonging to the other class.

Let vector x_l of size $1 \times n$ denote the transformed vector of the lth siRNA sequence using the transformation matrix T. The jth element of x_l is the element of T at column j and the row corresponds to the jth nucleotide in the siRNA sequence. To compute x_l, new column-wise inner product is defined as follows

$$x_l = T \circ X_l = (\langle X_l[1,.], T[.,1] \rangle, \langle X_l[2,.], T[.,2] \rangle, \ldots, \langle X_l[n,.], T[.,n] \rangle) \tag{4}$$

where $X_l[j, .]$ and $T[., j]$ are the jth row vector and the jth column of the matrix X_l and T, respectively, and $\langle x, y \rangle$ denotes the inner product of vectors x and y.

The left table in Figure 1 shows an example of encoding matrix X, transformation matrix T and transformed vector x of the given sequence AUGCU. The rows of X represent encoding vectors of nucleotides in the sequence. Given transformation matrix T of size 4×5. The sequence AUGCU is represented by the vector $x = (T[1, 1], T[4, 1], T[3, 3], T[2, 4], T[4, 5]) = (0.5, 0.1, 0.08, 0.6, 0.1)$. Therefore, transformed data can be computed by the column-wise inner product $x = T \circ X$.

The problem of transformation matrix learning is now formulated as finding T under constraints (1), (2) and (3) so that the similarity of transformed vectors x_l in the same class is minimum and the dissimilarity of x_l in different classes is maximum. The learning problem then leads to solve the optimization problem with the following objective function

$$Min \sum_{p,q \in N_1} d^2(x_p, x_q) + \sum_{p,q \in N_2} d^2(x_p, x_q) - \sum_{\substack{p \in N_1 \\ q \in N_2}} d^2(x_p, x_q) \tag{5}$$

Subject to
$$T[i, j] \geq 0, \ \sum_{i=1}^{4} T[i, j] = 1, g_r(T) < 0, \ i = 1, ..., 4; j = 1, ..., n; r = 1, .., R.$$

In the objective function, the two first components are the sum of similarity of sequence pairs belonging to the same class and the last one is similarity of sequence pairs belonging to two different classes; $d(x, y)$ is the similarity measure between x and y (in this work we use Euclidean distance and L_2 norm); N_1 and N_2 are the two index sets of high and low efficacy siRNAs, respectively. Constraints $g_i(T)$ can also help to avoid the trivial solution of the objective function.

This optimization problem is solved by the following Lagrangian form

$$E = \sum_{p,q \in N_1} d^2(x_p, x_q) + \sum_{p,q \in N_2} d^2(x_p, x_q) - \sum_{\substack{p \in N_1 \\ q \in N_2}} d^2(x_p, x_q) + \sum_{j=1}^{n} \lambda_j \left(\sum_{i=1}^{4} T[i, j] - 1 \right) + \sum_{r=1}^{R} \mu_r g_r(T)$$

$$= \sum_{\substack{p \in N_1 \\ q \in N_1}} \| x_p - x_q \|_2^2 + \sum_{\substack{p \in N_2 \\ q \in N_2}} \| x_p - x_q \|_2^2 - \sum_{\substack{p \in N_1 \\ q \in N_2}} \| x_p - x_q \|_2^2 + \sum_{j=1}^{n} \lambda_j \left(\sum_{i=1}^{4} T[i, j] - 1 \right) + \sum_{r=1}^{R} \mu_r g_r(T)$$

$$= \sum_{p,q \in N_1} \sum_{j=1}^{n} (\langle X_p[j, .], T[., j] \rangle - \langle X_q[j, .], T[., j] \rangle)^2 + \sum_{p,q \in N_2} \sum_{j=1}^{n} (\langle X_p[j, .], T[., j] \rangle - \langle X_q[j, .], T[., j] \rangle)^2$$

$$+ \sum_{j=1}^{n} \lambda_j \left(\sum_{i=1}^{4} T[i, j] - 1 \right) + \sum_{r=1}^{R} \mu_r g_r(T) - \sum_{\substack{p \in N_1 \\ q \in N_2}} \sum_{j=1}^{n} (\langle X_p[j, .], T[., j] \rangle - \langle X_q[j, .], T[., j] \rangle)^2$$

where $\mu_r, r = 1, ..., R$ and $\lambda_j, j = 1, ..., n$ are Lagrangian multipliers. To solve the minimization problem, an iterative method is applied. For each pair of (i, j), $T[i, j]$ is solved while keeping the other elements of T. The Karush-Kuhn-Tucker conditions are

- Stationarity: $\frac{\partial E}{\partial T[i,j]} = 0, i = 1, ..., 4$ and $j = 1, ..., n$.
- Primal feasibility: $T[i, j] \geq 0, \sum_{i=1}^{4} T[i, j] = 1, g_r(T) < 0, i = 1, ..., 4;$ $j = 1, ..., n; r = 1, ..., R$.

– Dual feasibility: $\mu_r \geq 0, r = 1, \ldots, R$.
– Complementary slackness: $\mu_r g_r(T) = 0, r = 1, \ldots, R$.

From the last three conditions, we have $\mu_r = 0, r = 1, \ldots, R$. Therefore, the stationarity condition can be derived as follows

$$\frac{\partial E}{\partial T[i,j]} = 2 \sum_{p,q \in N_1} (\langle X_p[j,.], T[.,j] \rangle - \langle X_q[j,.], T[.,j] \rangle)(X_p[j,i] - X_q[j,i])$$

$$+ 2 \sum_{p,q \in N_2} (\langle X_p[j,.], T[.,j] \rangle - \langle X_q[j,.], T[.,j] \rangle)(X_p[j,i] - X_q[j,i])$$

$$- 2 \sum_{p \in N_1, q \in N_2} (\langle X_p[j,.], T[.,j] \rangle - \langle X_q[j,.], T[.,j] \rangle)(X_p[j,i] - X_q[j,i]) + \lambda_j = 0$$

Set $Z_{p,q} = (X_p - X_q)^T$ and A_{ij} is the vector resulting from the column j of matrix A by removing the element $A[i,j]$. Therefore, the above formulation is derived as follows

$$\frac{\partial E}{\partial T[i,j]} = 2(\sum_{p,q \in N_1} \langle (Z_{p,q})_{ij}, T_{ij} \rangle Z_{p,q}[i,j] + \sum_{p,q \in N_2} \langle (Z_{p,q})_{ij}, T_{ij} \rangle Z_{p,q}[i,j]$$

$$- \sum_{p \in N_1, q \in N_2} \langle (Z_{p,q})_{ij}, T_{ij} \rangle Z_{p,q}[i,j])$$

$$+ 2T[i,j] \left(\sum_{p,q \in N_1} Z_{p,q}^2[i,j] + \sum_{p,q \in N_2} Z_{p,q}^2[i,j] - \sum_{p \in N_1, q \in N_2} Z_{p,q}^2[i,j] \right) + \lambda_j = 0$$

We define the following equations

$$S(i,j) = \sum_{p,q \in N_1} Z_{p,q}^2[i,j] + \sum_{p,q \in N_2} Z_{p,q}^2[i,j] - \sum_{p \in N_1, q \in N_2} Z_{p,q}^2[i,j] \tag{6}$$

$$B(i,j) = \sum_{p,q \in N_1} \langle (Z_{p,q})_{ij}, T_{ij} \rangle Z_{p,q}[i,j] + \sum_{p,q \in N_2} \langle (Z_{p,q})_{ij}, T_{ij} \rangle Z_{p,q}[i,j]$$

$$- \sum_{p \in N_1, q \in N_2} \langle (Z_{p,q})_{ij}, T_{ij} \rangle Z_{p,q}[i,j]. \tag{7}$$

Substitute (6) and (7) to $\frac{\partial E}{\partial T[i,j]}$, we have

$$T[i,j] = \frac{\frac{-\lambda_j}{2} - B(i,j)}{S(i,j)} \tag{8}$$

At a column j, T has to satisfy

$$\sum_{i_1=1}^{4} T(i_1, j) = 1 \Leftrightarrow \sum_{i_1=1}^{4} \frac{\frac{-\lambda_j}{2} - B(i_1,j)}{S(i_1,j)} = 1 \Rightarrow \frac{-\lambda_j}{2} = \frac{1 + \sum_{i_1=1}^{4} \frac{B(i_1,j)}{S(i_1,j)}}{\sum_{i_1=1}^{4} \frac{1}{S(i_1,j)}} \tag{9}$$

Substitute (9) to (8), equation (8) can be derived as

$$T[i,j] = \frac{\frac{1+\sum_{i_1=1}^{4} \frac{B(i_1,j)}{S(i_1,j)}}{\sum_{i_1=1}^{4} \frac{1}{S(i_1,j)}} - B(i,j)}{S(i,j)} = \frac{1 + \sum_{i_1 \neq i} \frac{B(i_1,j) - B(i,j)}{S(i_1,j)}}{\sum_{i_1=1}^{4} \frac{S(i,j)}{S(i_1,j)}} \tag{10}$$

In this task, K design rules are used to learn K transformation matrices. The main steps are summarized in Algorithm 1. For each siRNA design rule, the algorithm will update each element of the transformation matrix according to equation (10). In each iterative step, the transformation matrix without trick inequality constraints is updated to reach the global optimal solution. If updated elements in a column satisfy the trick inequality constraints characterizing the condition at the corresponding position of the rule, that column will be updated to the target solution. The transformation matrix is updated until meeting the convergence criteria. $\| \cdot \|_{Fro}$ is the Frobenious norm of a matrix.

Algorithm 1. Transformation matrices learning

Input: A data set $S = \{(s_l, y_l)\}_1^N$ where s_l are siRNA sequences and y_l are their labels, a set DR of K design rules, the length n of siRNA sequences.
Output: K transformation matrices $T_1, T_2, \ldots, T_K$.
Encoding siRNA sequences in S.
for $rule_k$ in DR **do**
    Form the set of constraints C_k based on $rule_k$
    Initialize the transformation matrix T_k satisfying C_k.
    $t = 0$ { Iterative step}
    repeat
        $t \leftarrow t + 1$
        for $j = 1$ to n **do**
            $v = T_k^{(t-1)}[.,j]$ { A temporary vector}
            for $i = 1$ to 4 **do**
                Compute $v[i]$ using equation (10)
            end for
            if (v satisfies the constraints at the position j in C_k) **then**
                $T_k^{(t)}[.,j] \leftarrow v$
            end if
        end for
        until $(\frac{\|T_k^{(t)} - T_k^{(t-1)}\|_{Fro}}{\|T_k^{(t-1)}\|_{Fro}} \leq \epsilon)$ or $(t > t_{Max})$
end for

2.2 Tensor Regression Model Learning

Given a siRNA data set $D = \{(s_l, y_l)\}_1^N$ where s_l is the lth siRNA sequence of size n and $y_l \in \mathbb{R}$ is the knockdown efficacy score of s_l. Let X_l denotes the encoding matrix of s_l. Each encoding matrix X is transformed to K representations by K transformation matrices, $(T_1 \circ X, T_2 \circ X, \ldots, T_K \circ X)$. $R(X) = (T_1 \circ X, T_2 \circ X, \ldots, T_K \circ X)^T$ denotes the second order tensor of size $K \times n$.

The regression model can be defined as the following bilinear form

$$f(x) = \alpha R(X) \beta \tag{11}$$

where $\alpha = (\alpha_1, \alpha_2, \ldots, \alpha_K)$ is a weight vector of the K representations of X and $\beta = (\beta_1, \beta_2, \ldots, \beta_n)^T$ is a parameter vector of the model, and $\alpha R(X)$ component is the linear combination of representations $T_1 \circ X, T_2 \circ X, \ldots, T_K \circ X$. It also shows the relationship among elements on each column of the second order tensor or each dimension of $T_k \circ X, k = 1, 2, \ldots, K$. Equation (11) can be derived as follows

$$f(X) = \alpha R(X)\beta = \left(\beta \otimes \alpha^T\right)^T vec(R(X)) = \left(\beta^T \otimes \alpha\right) vec(R(X)) \qquad (12)$$

where $A \otimes B$ is the Kronecker product of two matrices A and B, and $vec(A)$ is the vectorization of matrix A. The weight vector α and the parameter vector β are learned by minimizing the following regularized risk function

$$L(\alpha, \beta) = \sum_{l=1}^{N} (y_l - \alpha R(X_l)\beta)^2 + \lambda \parallel \beta^T \otimes \alpha \parallel_{Fro}^2 \qquad (13)$$

where λ is the turning parameter to tradeoff between bias and variance, and $\parallel \beta^T \otimes \alpha \parallel_{Fro}$ is the Frobenius norm of the first order tensor $\beta^T \otimes \alpha$. $L(\alpha, \beta)$ can be derived as follows

$$L(\alpha, \beta) = \sum_{l=1}^{N} (y_l - \alpha R(X_l)\beta)^2 + \lambda \sum_{k=1}^{K} \sum_{j=1}^{n} (\alpha_k \beta_j)^2 = \sum_{l=1}^{N} (y_l - \alpha R(X_l)\beta)^2 + \lambda \sum_{k=1}^{K} \alpha_k^2 \sum_{j=1}^{n} \beta_j^2$$

$$= \sum_{l=1}^{N} (y_l - \alpha R(X_l)\beta)^2 + \lambda \sum_{k=1}^{K} \alpha_k^2 \parallel \beta \parallel_2^2 = \sum_{l=1}^{N} (y_l - \alpha R(X_l)\beta)^2 + \lambda \parallel \alpha \parallel_2^2 \parallel \beta \parallel_2^2 \qquad (14)$$

The risk function with Frobenius norm is converted to equation (14) with L_2 norm. In order to solve this optimization problem, an alternative iteration method is used. At each iteration, the parameter vector β is effectively solved by keeping the weight vector α and vice versa.

$$\frac{\partial L(\alpha, \beta)}{\partial \alpha} = -2 \sum_{l=1}^{N} (y_l - \alpha R(X_l)\beta)(R(X_l)\beta)^T + 2\lambda \alpha \parallel \beta \parallel_2^2 = 0$$

$$\Leftrightarrow \qquad \sum_{l=1}^{N} \alpha (R(X_l)\beta)(R(X_l)\beta)^T - \sum_{l=1}^{N} y_l (R(X_l)\beta)^T + \lambda \alpha \parallel \beta \parallel_2^2 = 0$$

$$\Rightarrow \qquad \alpha = \sum_{l=1}^{N} y_l (R(X_l)\beta)^T \left(\sum_{l=1}^{N} (R(X_l)\beta)(R(X_l)\beta)^T + \lambda \parallel \beta \parallel_2^2 I \right)^{-1} \qquad (15)$$

$$\frac{\partial L(\alpha, \beta)}{\partial \beta} = -2 \sum_{l=1}^{N} (y_l - \alpha R(X_l)\beta)(\alpha R(X_l))^T + 2\lambda \beta \parallel \alpha \parallel_2^2 = 0$$

$$\Leftrightarrow \qquad \sum_{l=1}^{N} \alpha R(X_l)\beta (\alpha R(X_l))^T - \sum_{l=1}^{N} y_l (\alpha R(X_l))^T + \lambda \beta \parallel \alpha \parallel_2^2 = 0$$

$$\Leftrightarrow \qquad \sum_{l=1}^{N} \left((\alpha R(X_l))^T \otimes (\alpha R(X_l)) \right) \beta - \sum_{l=1}^{N} y_l (\alpha R(X_l))^T + \lambda \beta \parallel \alpha \parallel_2^2 = 0$$

$$\Rightarrow \qquad \beta = \left(\sum_{l=1}^{N} \left((\alpha R(X_l))^T \otimes (\alpha R(X_l)) \right) + \lambda \parallel \alpha \parallel_2^2 I \right)^{-1} \sum_{l=1}^{N} y_l (\alpha R(X_l))^T \qquad (16)$$

Our proposed tensor regression model learning is summarized in Algorithm 2. In this algorithm, siRNA sequences are firstly represented as encoding matrices. The encoding matrices are then transformed to tensors by using K transformation matrices. After that, the weight vector α and the coefficient vector β are updated until meeting the convergence criteria, where t_{Max} denotes the maximum iterative step to update α and β, and ϵ_1 and ϵ_2 are thresholds for vectors α and β.

Algorithm 2. Tensor Regression Model Learning

Input: A data set $S = \{(s_i, y_i)\}_1^N$ where s_i are scoring siRNA sequences and $y_i \in \mathbb{R}$. K transformation matrices $R_1, R_2, \ldots, R_k$, and the length n of siRNA sequence.

Output: Weight vector $\alpha = (\alpha_1, \alpha_2, \ldots, \alpha_k)$ and parameter vector $\beta = (\beta_1, \beta_2, \ldots, \beta_n)$ that minimize the regularized risk function

– Represent siRNA sequences in S as enconding matrices.
– Transform encoding matrices to tensors using K transformation matrices.
– Initialize α and β randomly.
– $t = 0$ { Iterative step}

repeat

$t \leftarrow t + 1$

Compute $\alpha^{(t)}$ using equation (15)

Compute $\beta^{(t)}$ using equation (16)

until $((\frac{\|\alpha^{(t)} - \alpha^{(t-1)}\|_2}{\|\alpha^{(t-1)}\|_2} \leq \epsilon_1)$ and $(\frac{\|\beta^{(t)} - \beta^{(t-1)}\|_2}{\|\beta^{(t-1)}\|_2} \leq \epsilon_2))$ or $(t > t_{Max})$

---

3   Experimental Evaluation

This section presents experimental evaluation in comparing the proposed method TRM (stands for 'tensor regression model') with the most recent reported methods for siRNA knockdown efficacy prediction on commonly used datasets. Discussion on the framework and methods will follow the experiment report.

Comparative Evaluation. The comparison is carried out using four data sets

– The Huesken dataset of 2431 siRNA sequences targeting 34 human and rodent mRNAs, commonly divided into the training set HU_train of 2182 siRNAs and the testing set HU_test of 249 siRNAs [5].
– The Reynolds dataset of 240 siRNAs [10].
– The Vicker dataset of 76 siRNA sequences targeting two genes [17].
– The Harborth dataset of 44 siRNA sequences targeting one gene [3].

TRM is compared to most state-of-the-art methods for siRNA knockdown efficacy prediction recently reported in the literature. As experiments in those methods cannot be repeated directly, we employed the results reported in the literature and carried out experiments on TRM in the same conditions of the other works. Concretely, the comparative evaluation is done as follows

1. Comparison of TRM with Multiple Kernel Support Vector Machine proposed by Qui *et al.* [8]. The author of [8] reported their Pearson correlation coefficient (R) of 0.62 obtained by 10-fold cross validation on the whole Huesken

Table 2. The R values of 18 models and TRM on three independent data sets

Algorithm	$R^{Reynolds}$ (244si/7g)	R^{Vicker} (76si/2g)	$R^{Harborth}$ (44si/1g)	Algorithm	$R^{Reynolds}$ (244si/7g)	R^{Vicker} (76si/2g)	$R^{Harborth}$ (44si/1g)
GPboot	0.55	0.35	0.43	Stockholm 1	0.05	0.18	0.28
Uitei	0.47	0.58	0.31	Stockholm 2	0.00	0.15	0.41
Amarzguioui	0.45	0.47	0.34	Tree	0.11	0.43	0.06
Hsieh	0.03	0.15	0.17	Luo	0.33	0.27	0.40
Takasaki	0.03	0.25	0.01	i-score	0.54	0.58	0.43
Reynolds 1	0.35	0.47	0.23	Biopredsi	0.53	0.57	0.51
Reynolds 2	0.37	0.44	0.23	DSIR	0.54	0.49	0.51
Schawarz	0.29	0.35	0.01	Katoh	0.40	0.43	0.44
Khvorova	0.15	0.19	0.11	SVM	0.54	0.52	0.54
				TRM	**0.60**	**0.58**	**0.55**

dataset. The Pearson correlation coefficient (R) is carefully evaluated by TRM by 10 times of 10-fold cross validation with the average value of 0.64.

2. Comparison of TRM with four state-of-the-art methods of BIOPREDsi [5], DSIR [16], Thermocomposition21 [13], SVM [12] by HU_train and HU_test. The Pearson correlation coefficients of the four models BIOPREDsi, DSIR, Thermocomposition21 and SVM are 0.66, 0.67, 0.66 and 0.80, respectively. The performance of TRM estimated on HU_test is 0.68 that is slightly higher than that of the first three models but lower than that of the last model.

3. Comparison of TRM with 18 methods including BIOPREDsi, DSIR, Thermocomposition21, SVM when training on HU_train and testing on three independent datasets of Reynolds, Vicker and Harborth as reported in the recent article [12]. As shown in Table 2 (taken from [12] with the added last row of the TRM result), TRM considerably achieved results higher than all of 18 methods on the all three independent testing datasets.

In running Algorithm 2, the thresholds for the weight vector α and the coefficient vector β are set up as 0.001 and the maximum iterative step is 1000. The turning parameter λ is chosen by minimizing the risk function when testing on validation dataset. Particularly, we do 10–fold cross validation on the training set for each λ belonging to $[0, \log 50]$ and compute the risk function

$$R(\lambda) = \frac{1}{F} \sum_{i=1}^{F} \left(\frac{1}{\| fold_i \|} \sum_{x_j \in fold_i} (y_j - f(x_j))^2 \right)$$

where $fold_i$ is validation set, $f(x)$ is a tensor predictor learnt from training set except validation set $fold_i$. F is the number of folds to do cross validation on training set. In our work, we do F-fold cross validation thus F equals to 10.

In the transformation matrices learning task, we use the labelled dataset collected from siRecord database [9]. This data set has 2470 siRNA sequences in 'very high' class and 2514 siRNA sequences in 'low' and 'medium' classes. Each siRNA sequence has 19 nucleotides. Seven design rules used to learn matrices are Reynolds rule, Uitei Rule, Amarzguioui rule, Jalag rule, Hsieh rule, Takasaki

Table 3. The learnt transformation matrix containing characteristics of Reynolds rule

	1	2	3	4		10	11	12		19
A	0.29704	0.217977	0.423469	0.266597	...	0.363636	0.246021	0.224727	...	0.393939
C	0.231159	0.235744	0.255102	0.226922	...	0	0.252513	0.267744	...	0.0757576
G	0.155341	0.211418	0.0459184	0.237968	...	0.229437	0.221336	0.260756	...	0.161616
U	0.31646	0.33486	0.27551	0.268513	...	0.406926	0.28013	0.246773	...	0.368687

rule and Huesken rule. The convergence criteria in Algorithm 1 are set up as following: threshold ϵ for transformation matrices is $2.5E^{-8}$ and the maximum iterative step is 5000.

Discussion. As reported in the experimental comparative evaluation, the proposed TRM achieved higher results than most other methods for prediction of siRNA knockdown efficacy. There are some reasons of that. First, it is expensive and hard to analyze the knockdown efficacy of siRNAs, and thus most available datasets are of relatively small size leading to limited results. Second, TRM has its advantages by incorporating domain knowledge (siRNA design rules) found from different datasets in experiments. Third, TRM is generic and can be easily exploited when new design rules are discovered or more analyzed siRNAs be obtained. Four, one drawback of TRM is its transformation matrices are learned using positional features of available design rules, and thus they lack some characteristics effecting to knockdown efficacy of siRNA sequences such as GC content, thermodynamic properties, GC stretch, etc. It may be one of reasons that at this moment TRM cannot get higher performance when testing on HU_test set than the best current model SVM [12].

Table 3 shows the learned transformation matrix capturing positional characteristics of Reynolds rule. One of characteristics is described as "An nucleotide 'A' at position 19". That characteristic means that at column 19, the cell (1,19) has to be the maximum value. In the matrix, the value at this cell is 0.393939 and is the highest value of this column. In this column, we also know knockdown efficacy of each nucleotide at position 19. Therefore, nucleotides can be arranged by the decreasing order of their efficacy: A, U, G, and C. In the order, nucleotide U has efficacy of 0.368687 that also can be used to design effective siRNAs. In addition, if a position on siRNAs is not described in characteristics of the design rules, values at the column corresponding to this position is learned to satisfy classification assumption and property to get knockdown efficacy of each nucleotide such as values at columns 1, 2, 4 and so on.

4   Conclusion

In this paper, we have proposed a novel framework to predict knockdown efficacies of siRNA sequences by successfully enriching the siRNA sequences into transformed matrices incorporating the effective siRNA design rules and predicting the siRNA knockdown efficacy by bilinear tensor regression. The experimental comparative evaluation on commonly used datasets with standard evaluation procedure in different contexts shows that the proposed framework and corresponding methods achieved better results than most existing methods for doing the same task. One significant feature of the proposed framework is it can be easily extended when new design rules are discovered as well as more siRNAs are analyzed by empirical works.

References

1. Amarzguioui, M., Prydz, H.: An algorithm for selection of functional siRNA sequences. Biochem. Biophys. Res. Commun. 316(4), 1050–1058 (2004)
2. Elbashir, S.M., Lendeckel, W., Tuschl, T.: RNA interference is mediated by 21– and 22–nucleotide RNAs. Genes Dev. 15, 188–200 (2001)
3. Harborth, J., Elbashir, S.M., Vandenburgh, K., Manninga, H., Scaringe, S.A., Weber, K., Tuschl, T.: Sequence, chemical, and structural variation of small interfering RNAs and short hairpin RNAs and the effect on mammalian gene silencing. Antisense Nucleic Acid Drug Dev. 13, 83–105 (2003)
4. Hsieh, A.C., Bo, R., Manola, J., Vazquez, F., Bare, O., Khvorova, A., Scaringe, S., Sellers, W.R.: A library of siRNA duplexes targeting the phosphoinositide 3-kinase pathway: Determinants of gene silencing for use in cell-based screens. Nucleic Acids Res. 32(3), 893–901 (2004)
5. Huesken, D., Lange, J., Mickanin, C., Weiler, J., Asselbergs, F., Warner, J., Mellon, B., Engel, S., Rosenberg, A., Cohen, D., Labow, M., Reinhardt, M., Natt, F., Hall, J.: Design of a Genome-Wide siRNA Library Using an Artificial Neural Network. Nature Biotechnology 23(8), 955–1001 (2005)
6. Ichihara, M., Murakumo, Y., Masuda, A., Matsuura, T., Asai, N., Jijiwa, M., Ishida, M., Shinmi, J., Yatsuya, H., Qiao, S., et al.: Thermodynamic instability of siRNA duplex is a prerequisite for dependable prediction of siRNA activities. Nucleic Acids Res. 35(8), e123 (2007)
7. Jagla, B., Aulner, N., Kelly, P.D., Song, D., Volchuk, A., Zatorski, A., Shum, D., Mayer, T., De Angelis, D.A., Ouerfelli, O., Rutishauser, U., Rothman, J.E.: Sequence characteristics of functional siRNAs. RNA 11(6), 864–872 (2005)
8. Qiu, S., Lane, T.: A Framework for Multiple Kernel Support Vector Regression and Its Applications to siRNA Efficacy Prediction. IEEE/ACM Trans. Comput. Biology Bioinform. 6(2), 190–199 (2009)
9. Ren, Y., Gong, W., Xu, Q., Zheng, X., Lin, D., et al.: siRecords: An extensive database of mammalian siRNAs with efficacy ratings. Bioinformatics 22, 1027–1028 (2006)
10. Reynolds, A., Leake, D., Boese, Q., Scaringe, S., Marshall, W.S., Khvorova, A.: Rational siRNA design for RNA interference. Nat. Biotechnol. 22(3), 326–330 (2004)
11. Scherer, L.J., Rossi, J.J.: Approaches for the sequence-specific knockdown of mRNA. Nat. Biotechnol. 21, 1457–1465 (2003)
12. Sciabola, S., Cao, Q., Orozco, M., Faustino, I., Stanton, R.V.: Improved nucleic acid descriptors for siRNA efficacy prediction. Nucl. Acids Res. 41(3), 1383–1394 (2013)

13. Shabalina, S.A., Spiridonov, A.N., Ogurtsov, A.Y.: Computational models with thermodynamic and composition features improve siRNA design. BMC Bioinformatics 7, 65 (2006)
14. Takasaki, S.: Methods for Selecting Effective siRNA Target Sequences Using a Variety of Statistical and Analytical Techniques. Methods Mol. Biol. 942, 17–55 (2013)
15. Ui-Tei, K., Naito, Y., Takahashi, F., Haraguchi, T., Ohki-Hamazaki, H., Juni, A., Ueda, R., Saigo, K.: Guidelines for the selection of highly effective siRNA sequences for mammalian and chick RNA interference. Nucleic Acids Res. 32, 936–948 (2004)
16. Vert, J.P., Foveau, N., Lajaunie, C., Vandenbrouck, Y.: An accurate and interpretable model for siRNA efficacy prediction. BMC Bioinformatics 7, 520 (2006)
17. Vickers, T.A., Koo, S., Bennett, C.F., Crooke, S.T., Dean, N.M., Baker, B.F.: Efficient reduction of target RNAs by small interfering RNA and RNase H-dependent antisense agents. A Comparative Analysis. J. Biol. Chem. 278, 7108–7118 (2003)

A Selectively Re-train Approach Based on Clustering to Classify Concept-Drifting Data Streams with Skewed Distribution

Dandan Zhang[1], Hong Shen[2,3], Tian Hui[4], Yidong Li[1],
Jun Wu[1], and Yingpeng Sang[1]

[1] School of Computer and Information Tech., Beijing Jiaotong University, China
[2] School of Information Science and Technology, Sun Yat-sen University, China
[3] School of Computer Science, University of Adelaide, Australia
[4] School of Electronics and Info. Engineering, Beijing Jiaotong University, China
hongsh01@gmail.com, {11120501,htian,ydli,wuj,ypsang}@bjtu.edu.cn

Abstract. Classification is an important and practical tool which uses a model built on historical data to predict class labels for new arrival data. In the last few years, there have been many interesting studies on classification in data streams. However, most such studies assume that those data streams are relatively balanced and stable. Actually, skewed data streams (e.g., few positive but lots of negatives) are very important and typical, which appear in many real world applications. Concept drifts and skewed distributions, two common properties of data streams, make the task of learning in streams particularly difficult and the traditional data mining algorithms no longer work. In this paper, we propose a method (Selectively Re-train Approach Based on Clustering) which can deal with concept-drifting and skewed distribution simultaneously. We evaluate our algorithm on both synthetic and real data sets simulating skewed data streams. Empirical results show the proposed method yields better performance than the previous work.

Keywords: data stream, skewed distribution, concept-drifting, selectively Re-train.

1 Introduction

Data stream classification has been widely studied, and there are many successful algorithms for coping with this problem [6,11]. However, most of these studies assume that data streams are relatively balanced and stable, which results in a failure to handle rather skewed distributions. Actually, skewed data streams are very important in many real-life data stream applications, such as credit-card fraud detection, diagnosis of rare diseases, network traffic analysis etc. In such skewed data streams, the probability that we observe positive instances is much less than the probability that we observe negative ones. For imbalanced data streams, instances from the minority class are more costly and thus are of more interest. For example, online credit-card fraud rate of US was just 2 percent in

V.S. Tseng et al. (Eds.): PAKDD 2014, Part II, LNAI 8444, pp. 413–424, 2014.

2006 [4] which means that in credit card transactions stream, there are a lot of genuine transactions and very few fraud transactions. Hence, it's very necessary and important to classify such minority instances from all streams because classifying an instance of credit-card fraud (positive) as a normal transaction (negative) is very costly.

Concept-drift occurs in the stream when the underlying concepts of the stream change over time. It is difficult to predict when and how the concept changes. When any of concept changes occurs, a decrease in classification accuracy usually occurs because the training data the model is built on would be carrying out-of-date concepts. Many of the previous work aimed to effectively update the classification model when stream data flows in. After some period, those approaches throw out the out-of-date examples or fade them out by decreasing their weights as time elapses. Skewed distribution and concept-drifting are two challenges for traditional classification algorithms. Classification algorithms are required to deal with data streams with skewed distributions, at the same time, when concept-drifting occurs, classification algorithms should be convergent to the up-to-date concept with high accuracy and speed. This method will be described in detail in section 3.

The rest of the paper is organized as follows. Section 2 gives the introduction on the related works. Section 3 shows our proposed algorithm in detail. In Section 4, we give experimental results by using a synthetic data set and real data sets which own the properties of skewed data streams. Section 5 summarizes our work and introduces the future work.

2   Related Work

Though there are many stream classification algorithms available, most of them assume that the streams have relatively balanced distribution of classes. In recent years, there have been several algorithms proposed for coping with skewed data stream classification. In these algorithms, data set usually are balanced by sampling, which is the most common method [7].

The basic sampling methods for dealing with the skewed data streams include under-sampling, over-sampling and clustering-sampling [12]. Drummond et al. [1] concluded that under-sampling outperforms over-sampling through detailed experiments. However, under-sampling may lead to loss of useful information. In order to use more majority class data to reduce information loss, clustering-sampling is preferred. This is because clustering can maintain more useful information, which may be thrown away by under-sampling.

Wang et al.[12] proposed an clustering-sampling based ensemble algorithm for classifying data streams with skewed distribution. Empirical results show that clustering-sampling outperforms under-sampling. Particularly in the case of ensemble model, the proposed ensemble based algorithm gives better performance. However, this method can not handle the problem of concept-drifting.

Nguyen et al.[9] proposed a new method for online learning from imbalanced data streams. In this method, a small training set T used to initialize a classification model is first collected. The classification model includes several base models. It is gradually updated by new training instance arriving one by one. If the class of the new training instance is positive, all the base models are updated. If the incoming instance is in negative class, these base models are updated with a certain probability. This approach also can not deal with the problem of concept-drifting.

Gao et al.[3] proposed a framework which employs both sampling and ensemble methods to classify skewed data streams. This paper analyzes the source of concept drifts, classification model selection reason and why ensemble methods reduce classification errors. In the sampling phase, it randomly under-samples the negative instances from the most up-to-date chunk Q_m, at the same time, it collects all the positive instances from data chunk $Q_1, Q_2, \cdots, Q_m$ and keeps them in the training set. Though the training set is balanced through sampling, all collected positive examples may not be consistent with the current concept and affect classification results if regarded as positive instances.

A variety of techniques have also been proposed in the literature for addressing concept-drifting in data stream classification. Wang et al.[11] proposed a general framework for mining concept-drifting data streams using weighted ensemble classifiers. The classifiers in the ensemble are judiciously weighted based on their expected classification accuracy on the test data under the time-evolving environment. Thus, the ensemble approach improves both the efficiency in learning the model and the accuracy in performing classification.

Kolter et al.[13] proposed *AddExp* algorithm, which is similar to expert prediction algorithms, for discrete classes and continuous classes, respectively. This algorithm bounds its performance over changing concepts, not relative to the performance of any abstract expert, but relative to the actual performance of an online learner trained on each concept individually. During the online learning process, new experts can be added.

Although data stream classification algorithms coping with imbalanced distribution problem and concept-drifting problem respectively have been researched for several years, dealing with the two problems in a system framework is still challenging. In this paper, we propose the *SeRt* framework to address the nonstationary stream data with skewed distribution and concept-drifting.

Many available algorithms for classifying data streams measure their performance by considering overall classification accuracy. However, such assessment metric is not suitable for assessing the performance of those algorithms which classify data streams with skewed distributions. This is because that, in such case, the overall accuracy is dominated by the majority class. Even if the classification model performs well only for negative instances (eg. majority examples), the overall classification accuracy may also be high. For example, if 98% of the data is from the majority class and only 2% from the minority class, a classification model classifies every instance as majority, and thus the overall accuracy reaches up to 98% at the cost of no minority class instances correctly classified. Those literatures [2] [10] [5] gives some measurements, like *AUROC* and $G - Mean$, to evaluate the performance of algorithms for classifying skewed data streams.

3   Classifying Data Streams with Skewed Distributions and Concept-Drifting

Imbalanced learning problem and concept drifting phenomenon are quite popular in recent years with a increased number of reports on the difficulties in many practical applications. When concept drifting occurs or class distribution changes, it definitely leads to a sudden drop of classification accuracy if we still use the classifier trained on old data points to classify new instances.

3.1   Basic Idea and Main Framework

The basic idea of our method is that when there is no concept-drifting occurring in data streams with unbalanced distribution, we use the most up-to-date chunk to train a base classifier and use it to update the ensemble classifier. This may improve the accuracy of ensemble classifier, because the most up-to-date chunk contains information about the current target concepts. When concept drifting occurs, we use these data points which consist with the current target concepts to re-train those base classifiers in the previous ensemble classifier E_{j-1}. This way can make the algorithm converge to target concepts with high accuracy efficiently.

When concept-drifting occurs, for each base classifier C_i in the classifier ensemble E, we inject those data points which can represent the up-to-data target concepts into the training set T_i with the *injection probability* p. Here T_i is the training set corresponds to the base classifier C_i . But the value of p has relationship with the overall accuracy and the accuracy of each base classifier. It is obvious that for greater p, more training instances are given in the model which may produce more error reduction. So, when $p = 1$, we may have maximum reduction in prediction error for a single model. However, if the same set of instances are injected in all models, the correlation among them may increase, which reduces the accurate prediction rate of the ensemble. So, we have to choose a value of p to balance the overall accuracy and the accuracy of each classification model [8]. Then we update the classification model C_i.

In stream applications, the training data are read in consecutive data chunks. Suppose the incoming data stream is partitioned into sequential chunks of equal size $S_1, S_2, \cdots, S_j, \cdots$, where S_j is the most up-to-date data chunk. Each data chunk is considered as a conventional imbalanced data set, which makes it easy to apply sampling methods to balance those skewed data chunks. In order to deal with skewed data streams, most of the available algorithms have either used oversampling or under-sampling approach for ensemble of classifiers. In our algorithm, we use clustering-sampling to balance the skewed distribution and selectively re-train approach to solve the problem of concept-drifting. Figure 1 shows the framework of our proposed algorithm.

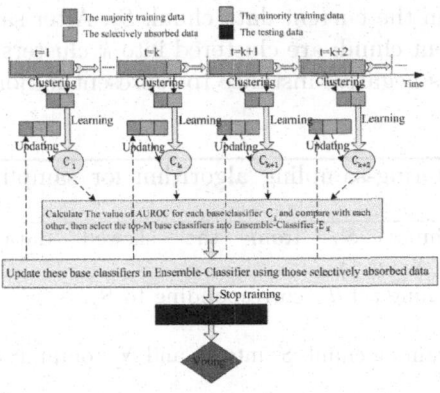

Fig. 1. The system framework of our method

3.2   How to Choose Data Point for Re-training

We illustrate the idea of re-training through a simple hyperplane example in Figure 2. Figure 2(a) shows two true models of the evolving hyperplane at different time. A instance is positive (+) if it is above the hyperplane, otherwise it is negative (−). We assume that *hyperplane₁* is the previous concept and *hyperplane₂* is the current concept, C_1 and C_2 stand for their true models respectively. In Figure 2(b) we draw the optimal model $E_1(ensemble\ classifier)$ for *hyperplane₁*, which are interpolated by straight lines. In Figure 2(c), those data points in the shaded areas can be used to re-train those base classifiers in previous ensemble classifier E_1.

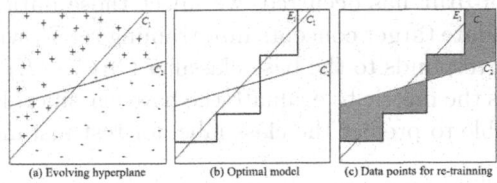

Fig. 2. How to choose data points for re-trainning

3.3   Sampling Skewed Data Streams

The basic sampling methods often used are under-sampling and over-sampling. Although the study result shows that these two sampling methods can somehow address the problem of skewed distribution, they have several drawbacks[1].

In our method, we employ $K-Means$ clustering algorithm, for selecting negative examples to represent majority class, to balance the skewed distribution. Algorithm 1 [12] gives clustering-sampling algorithm for sampling imbalanced data streams. In this algorithm, we set the number of clusters k to the number of positive instances in the current data chunk S_j. After sampling, the negative examples in the current chunk are clustered into k clusters, and the centroid of each cluster is used as negative instance to represent majority class.

Algorithm 1. Clustering-sampling algorithm for sampling imbalanced data streams

---

Input : a   data   chunk S_j from   the   skewed   data   streams, S_j = $\{(X_1, y_1), (X_2, y_2), \cdots, (X_n, y_n)\}$.
Output : balanced training set T_j corresponding to S_j.
begin

    split the current training chunk S_j into P_j and N_j containing positive and negative examples respectively.
1. $k = |P_j|$;
2. $Cluster = Kmeans(N_j, k)$;
3. $NEG_j = \{centroid(c)|c \in Cluster\}$;
4. obtain balanced training set T_j corresponding to S_j, $T_j = P_j \cup NEG_j$;

3.4   Selectively Re-train Algorithm Based on Clustering

The pseudo code of our method for learning from imbalanced data streams is formulated in Algorithm 2. After receiving a data chunk S_j, we firstly balance it using clustering-sampling method, thus we get the corresponding balanced training set T_j. Then we train a base classifier C_j, by an arbitrary successful binary learner from the training set T_j. Before updating the classification ensemble E_{j-1}, we first check if concept-drifting has occurred by checking if $E_{j-1}(X_i) \neq y_i$ holds. If the inequation holds, it means concept-drifting has occurred, or vice verse. If the concept-drift has occurred, we inject those data points which can represent the up-to-date target concepts into training set T_i with a certain probability, where T_i corresponds to the base classifier $C_i(C_i \in E_{j-1})$. The $AUROC$ value is employed as the metric to evaluate the base classifier. Finally, we use the classification ensemble to predict the class label for test instances with weighted majority voting.

In Algorithm2, $Learn(T_i)$ is an arbitrary learning algorithm; $class(X_i)$ returns the class label of instance X_i; $AUC(C_i)$ returns the AUROC value of base classifier C_i; $NormalizeWeight(E_j, C_i)$ returns the normalized weight of base classifier C_i, which satisfies $C_i \in E_j$; $GetRandomNumber()$ returns a random number in $[0, 1]$; $numBaseClassifier$ is the number of the base classifier in Ensemble-classifier; and $balanceRatio$ is a parameter that is used to judge whether the training set T_h needed to be balanced again.

4   Experiments

In our experiment, we evaluate our proposed algorithm which can be abbreviated to *SeRt* and compare it with the approach proposed in [3]which is denoted as

---

Algorithm 2. Selectively Re-train algorithm Based on Clustering towards Concept-drifting Data streams with Skewed Distribution

---

Input : skewed data streams $S_1, S_2, \cdots, S_j, \cdots$;

    the ensemble of classifiers trained on previous data chunks, E_{j-1};

    the maximum size of classifier ensemble, M.

Output : the classifier ensemble, E_j.

begin

    read a data chunk S_j from the skewed data streams, $S_j = \{(X_1, y_1), \cdots, (X_n, y_n)\}$.

1. $T_j = Sampling(S_j)$;
2. built a new base classifier $C_j = Learn(T_j)$;
3. for i from 1 to n do

    if $E_{j-1}(X_i) \neq y_i$ then for h from 1 to $numBaseClassifier$ do

      if $class(X_i) = majority$ && $GetRandomNumber() < injection\ probability$

        put data point (X_i, y_i) into NEG_h; $|NEG_h|$++;

      else   put data point (X_i, y_i) into T_h; $|P_h|$++;

4. for each $T_h (h$ from 1 to $numBaseClassifier)$

    if ($\frac{|P_h|}{|NEG_h|} > \rho$) where $\rho \in [balanceRatio, 1]$ then following step 1 to 4;

5. if $|E_{j-1}| < M$ then $C_j.weight = 1; E_j = E_{j-1} \cup \{C_j\}$;

    else $h = \arg\min(AUC(C_h))$ where $C_h \in E_{j-1}$;

    if $AUC(C_h) < AUC(C_j)$ then $E_j = E_{j-1} \cup \{C_j\} - \{C_h\}$;

    for each $C_h \in E_j$ do $C_h.weight = AUC(C_h)$;

    for each $C_h \in E_j$ do $C_h.weight = NormalizeWeight(E_j, C_h)$;

end

SE in the simulation results on both synthetic and real data sets. To simulate data streams, we partitioned these data sets into several chunks.

4.1 Data Sets

Synthetic Data Set. We create synthetic data sets with drifting concepts based on a rotating hyperplane by reprogramming the software *MOA Task Launcher*. The rotating hyperplane is widely used for experiments [6,11, ??] . A hyperplane in a d-dimensional space is the set of data instance which is denoted by the equation:

$$\sum_{i=1}^{d} a_i x_i = a_0 \tag{1}$$

Here, x_i is the $i - th$ coordinate of instance x. We label instances with $\sum_{i=1}^{d} a_i x_i \geq a_0$ as positive instances, and instances with $\sum_{i=1}^{d} a_i x_i < a_0$ as negative instances. Weights $a_i (1 < i < d)$ in (1) are initialized randomly in the range of $[0, 1]$. Hyperplanes have been used to simulate time-changing concepts because the orientation and the position of the hyperplane can be changed in a smooth manner by changing the magnitude of the weights[6]. We choose the value of a_0 so that the hyperplane cuts the multi-dimensional space into two parts of the different volume, *eg.* $a_0 = r \sum_{i=1}^{d} a_i$ where r is the skewness ratio.

when $r = \frac{1}{2}$, the hyperplane cuts the multidimensional space into two parts of the same volume. In our experiment, $r \neq \frac{1}{2}$.

In our study, we simulate the phenomenon of concept drifting through a series of parameters. Parameter $n(n \in N)$ specifies the number of examples in each batch, and parameter $k(k \in N)$ specifies the number of dimensions whose weights are involved in concept drifting. Parameter $t(t \in R)$ indicates the magnitude of the changing of weights $a_1, a_2, \cdots, a_k$ and $s_i \in \{-1, 1\}$ $(1 \leq i \leq k)$ indicates the direction of change for each weight a_i. Weight $a_i(1 \leq i \leq k)$ is adjusted by $s_i \cdot t$ after each instance is generated.

Real Data Set. We use the data set *Optical Recognition of Handwritten Digits* from the University of California, Irvine's Machine Learning Repository (http://archive.ics.uci.edu/ml/datasets.html). Although this data set doesn't correspond directly to skewed data mining problem, we can convert it into skewed distribution problem by taking one small class as the minority class and the remaining records as the majority class. So, from the original data sets, we can get 10 skewed data sets and average the results over these data sets.

4.2   Evaluation Metrics

Traditionally, overall accuracy is the most commonly used measure for evaluating the performance of classifier. However, for classification with skewed distribution, overall accuracy is no longer proper since the minority class has very little impact on it as compared to the majority class. This measurement is meaningless to some applications where the learning target is to identity the rare instances. For any classifier, there is always a trade-off between true positive rate (TPR) and true negative rate (TNR). For this end, some evaluation metrics associated with confusion matrix are used to validate the effectiveness of those algorithms dealing with class imbalance problem. Table 2 illustrates a confusion matrix of a two-class problem.

TP and TN denote the number of positive and negative instances that are classified correctly , while FP and FN denote the number of examples which are misclassified respectively.

- **Overall Accuracy (OA):**

$$OA = \frac{TP + TN}{TP + FP + TN + FN} \tag{2}$$

- **TPR (ACC^+):**

$$ACC^+ = \frac{TP}{TP + FN} \tag{3}$$

- **TNR (ACC^-):**

$$ACC^- = \frac{TN}{TN + FP} \tag{4}$$

– **G-mean:**

$$G - mean = \sqrt{\frac{TP}{TP + FN} \times \frac{TN}{TN + FP}} \tag{5}$$

– **Mean Squared Error (MSE):**

$$MSE = \frac{1}{|T|} \sum_{x_i \in T} (f(x_i) - p(+ \mid x_i))^2 \tag{6}$$

Where T is the set of testing examples, $f(x_i)$ is the output of the Ensemble-Classifier, which is the estimated posterior probability of testing instance x_i; $p(+ \mid x_i)$ is the true posterior probability of x_i.

4.3 Experiment Configuration and Results

Experiment Configuration. The synthetic data set (Moving Hyperplane Data Set) is used to validate the effectiveness and superiority of our proposed method. In our experiments, the dimensionality of the data stream is set to be 50. Only two dimensionalities are changing with time and the magnitude of the change for every example is set to be 0.1. The percentage of probability that the direction of change is reversed is set to be 10%. At the same time, we set the *balanceRatio* to be 0.85 and the method applied to build base classifier is *Naive Bayes*.

When concept-drifting occurs, for each base classifier C_i in the classifier ensemble E, we inject those data points which can represent the up-to-date target concepts into the training set T_i, where T_i corresponds to the base classifier C_i, with the probability p. But the value of p has relationship with the overall accuracy and the accuracy of each base classifier. If the same set of instances are injected in all the models, then the correlation among them may increase, resulting in reduced prediction accuracy of the ensemble. So, we have to choose a value of p to balance the overall accuracy and the accuracy of each model [8]. In order to find the relationship between p and the performance of our approach, the number of chunks in the data stream is set to be 101, each of which carries 1000 examples and the last chunk contains the testing instances. In the data stream, there are 98064 negative instances and 2936 positive instances. 956 negative examples and 44 positive examples are used for testing.

Figure 3 shows the relationship between p and $TPR, TNR, G - mean$ and MSE, where NS denotes selectively re-train approach based on No-Sampling, US denotes selectively re-train approach based on Under-Sampling and CS denotes selectively re-train approach based on Clustering-Sampling (*eg.*our proposed method *SeRt*). Through experiments, we get that, when p is between 0.5 and 0.75,$TPR, TNR, G - mean$ and MSE are near optimal. And the point where $p = 0.75$ is the turning point, so we set $p = 0.75$.

Experiment Results. After experiment configuration, we can conduct experiments based on it. The following are the experiment results.

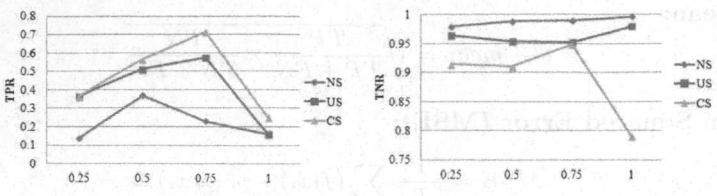

(a) Changes of TPR with differ- (b) Changes of TNR with different
ent p p

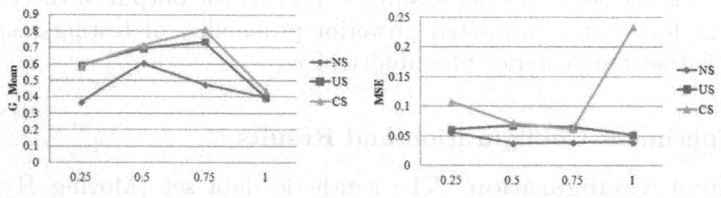

(c) Changes of G-mean with dif- (d) Changes of MSE with different
ferent p p

Fig. 3. The relationship between p and TPR, TNR, G_Mean and MSE

– The performance comparison with different sampling methods

In order to solve the problem of imbalanced distribution of data stream, we should balance it by using sampling method. We compare the performance of different sampling methods, as shows Figure 4, where CS represents Clustering-sampling, US represents Under-sampling, and NS represents that we do not use any sampling method. From this figure, it is obvious that CS outperforms NS and US in TPR and $G - Mean$. Though CS is slightly lower than other two methods in the case of TNR, the results are very competitive, because the cost of misclassifying a positive instance is much larger than the cost of misclassifying a negative instance .

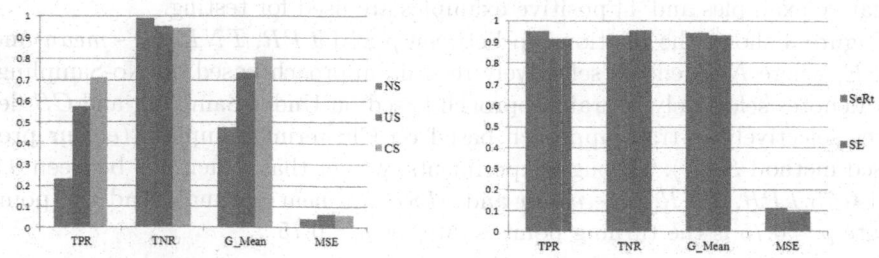

Fig. 4. The performance comparison with **Fig. 5.** The average performance of $SeRt$
different sampling methods   and SE on real data set

- The comparison between our method and the methode which was proposed in [3] based on real dataset just with skewed distribution

Except for synthetic data set, we also employ the real data set *Optical Recognition of Handwritten Digits* to compare our algorithm with the approach proposed in [3] which is denoted as SE in the results. Figure 5 gives the average result of these 10 skewed data streams. From Figure 5, we find that $SeRt$ can get higher TPR and $G - Mean$ with the cost of TNR and MSE than SE. But it is worth, because in the case of skewed distribution, minority class is the most significant one, and the cost of misclassifying of a positive example is more huge than the cost of misclassifying a negative example.

- The comparison between our method and the methode which was proposed in [3] based on synthetic dataset with concept-drifting and skewed distribution

We also use synthetic data set with concept-drifting to validate the performance of $SeRt$ and SE, as follows Figure 6. The data set contains 101000 instances in total, 99817 negative examples and 1183 positive examples respectively. We partition the data set into 34 chunks to simulate data stream. The last chunk for testing contains 2000 examples, and the rest of each chunk for training contains 3000 examples. There are 1929 majority instances and 71 minority instances in the testing set. From Figure 6, we know that when the data stream is changing with time continuously, it is obvious that our proposed algorithm $SeRt$ outperforms SE.

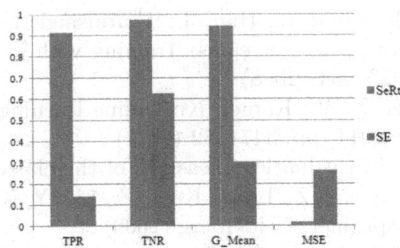

Fig. 6. The comparison between $SeRt$ and SE on synthetic data set with concept-drifting

5   Conclusion and Future Work

In this paper, we propose a new algorithm for mining data streams with skewed distribution, which can also tackle the problem of concept drifting.

The proposed algorithm could only deal with binary classification tasks. However, many real-life applications, such as network intrusion detection, are characterized as multi-class classification tasks. In the future, we will study the problem of classifying multi-class with skewed data streams.

Acknowledgement. This work is supported by National Science Foundation of China under its General Projects funding # 61170232 and 61100218, Fundamental Research Funds for the Central Universities # 2012JBZ017 and # 2011JBM206, Research Initiative Grant of Sun Yat-Sen University (Project 985), State Key Laboratory of Rail Traffic Control and Safety Research Grant # RCS2012ZT011. The corresponding author is Hong Shen.

References

1. Drummond, C., Holte, R.C.: C4.5, class imbalance, and cost sensitivity: Why under-sampling beats over-sampling, pp. 1–8 (2003)
2. Fawcett, T.: Roc graphs: Notes and practical considerations for re-searchers. Technical report, HP Laboratories (2004)
3. Gao, J., Ding, B., Fan, W., Han, J., Yu, P.S.: Classifying data streams with skewed class distributions and concept drifts. IEEE Internet Computing 12(6), 37–49 (2008)
4. Gao, J., Fan, W., Han, J., Yu, P.S.: A general framework for mining concept-drifting data streams with skewed distributions. In: Proc. 2007 SIAM Int. Conf. Data Mining (SDM 2007), Minneapolis (MN2007)
5. He, H., Garcia, E.A.: Learning from imbalanced data. IEEE Transactions on Knowledge and Data Engineering 21(9), 1263–1284 (2009)
6. Hulten, G., Spencer, L., Domingos, P.: Mining time-changing data streams. In: Proceedings of the Seventh ACM SIGKDD International Conference on Knowledge Discovery and Data Mining, KDD 2001, pp. 97–106. ACM, New York (2001)
7. Kotsiantis, S.B., Pintelas, P.E.: Mixture of expert agents for handling imbalanced data sets (2003)
8. Masud, M.M., Gao, J., Khan, L., Han, J., Thuraisingham, B.M.: A practical approach to classify evolving data streams: Training with limited amount of labeled data. In: ICDM, pp. 929–934 (2008)
9. Nguyen, H.M., Cooper, E.W., Kamei, K.: Online learning from imbalanced data streams. In: SOCPAR 2011, pp. 347–352 (2011)
10. Gu, Q., Zhu, L., Cai, Z.: Evaluation measures of the classification performance of imbalanced data sets. In: Cai, Z., Li, Z., Kang, Z., Liu, Y. (eds.) ISICA 2009. CCIS, vol. 51, pp. 461–471. Springer, Heidelberg (2009)
11. Wang, H., Fan, W., Yu, P.S., Han, J.: Mining concept-drifting data streams using ensemble classifiers. In: Proceedings of the Ninth ACM SIGKDD International Conference on Knowledge Discovery and Data Mining, KDD 2003, pp. 226–235. ACM, New York (2003)
12. Wang, Y., Zhang, Y., Wang, Y.: Mining data streams with skewed distributions by static classifier ensemble. In: Chien, B.-C., Hong, T.-P. (eds.) Opportunities and Challenges for Next-Generation Applied Intelligence. SCI, vol. 214, pp. 65–71. Springer, Heidelberg (2009)
13. Kolter, J.Z., Maloof, M.A.: Using additive expert ensembles to cope with concept drift. In: Proceedings of the 22nd International Conference on Machine Learning, pp. 449–456. ACM Press (2005)

Recurring and Novel Class Detection in Concept-Drifting Data Streams Using Class-Based Ensemble

Mohammad Raihanul Islam

Bangladesh University of Engineering and Technology

Abstract. Over the recent years concept-evolution has received a lot of attention because of its importance in the context of mining data streams. Mining data stream has become an important task due to its wide range of applications such as network intrusion detection, credit card fraud protection, identifying trends in the social networks etc. Concept-evolution means introduction of novel class in the data stream. Many recent works address this phenomenon. In addition, a class may appear in the stream, disappears for a while and then reemerges. This scenario is known as recurring classes and remained unaddressed in most of the cases. As a result, generally where a novel class detection system is present, any recurring class is falsely detected as novel class. This results in unnecessary waste of human and computational resources. In this paper, we have proposed a class-based ensemble of classification model addressing the issues of recurring and novel class in the presence of concept drift and noise. Our approach has shown impressive performance compared to the state-of-art methods in the literature.

Keywords: Novel Class, Recurring Class, Concept Evolution, Stream Classification.

1 Introduction

The problem of data stream classification has been studied among the research community over the recent years. One of the major characteristics of data stream mining is that, the classification is a continuous process thus the size of the training data can be considered infinite. So it is almost impossible to store all the examples to train the classifiers. Some methods regarding incremental learning are proposed in [4, 9] to address this problem. Moreover, it is a common scenario that, the underlying concept may changes overtime; a characteristic known as *concept-drift*.

However, another significant phenomenon of the data stream is *concept-evolution*, which is considered as the emergence of novel classes in the stream. For example, a new topic may appear in social network or a new type of intrusion may be identified in the network. If the number of classes in the classifiers is fixed and no novel class detection system is present, then the novel class is falsely identified as existing class. Concept Evolution has become a new research

V.S. Tseng et al. (Eds.): PAKDD 2014, Part II, LNAI 8444, pp. 425–436, 2014.

direction for the researchers recently because of its practical importance. For example, if a new type of attack occurs in the network, it is imperative to identify it and take actions as soon as possible. Several approaches regarding this issue have been studied in the literature [5, 7].

A special case of concept-evolution is *recurring class* where a class reemerges after its long disappearance from the stream. For example, a popular topic may appear in a social network at a particular time of the year (i.e. festivals or elections). This result in a change of topics in the discussion on the social network over the time period and then when the event ends the topic disappears again. A recurring class creates several discrepancies if not properly handled. If it is not properly identified, then it is erroneously considered as a novel class or an existing one. As a result, a significant amount of human resources is wasted to detect its reappearance. Some studies regarding the problem of recurring class are present in [1, 6].

The classification model for data stream can be constructed by ensemble of classifiers. In an ensemble approach, multiple base classifiers learn the decision boundary on the learning patterns and their decisions on test example ares fused to reach the final verdict. The ensemble approach is more popular among the research community because of their higher accuracy, efficiency and flexibility [5].

The contributions of this paper are as follows. In this paper, we propose a new technique to generate ensemble of classifiers to detect novel and recurring class in the data stream which reduces overall classification errors. Moreover, we have observed the phenomenon that, if the class boundary between two classes is very close, then it is possible to get a false prediction if the instances falls closely to boundary region. In our approach, we have employed several strategies to mitigate this problem. Finally, we have also used the falsely predicted instances to update our model. Our proposed method has outperformed the state-of-the-art techniques in the literature.

The rest of the paper is organized as follows. In Section 2, we discuss the previous works regarding data stream classification in the literature. We present our approach in Section 3. We discuss the experimental results in Section 4. We briefly conclude in Section 5.

2 Previous Works

Several studies are present in the literature on data stream classification [1–4, 6, 8–10]. Due to page limitation we have highlighted studies only related to novel and recurring class detection. It has been observed that, existing approaches can be divided into two categories. First one is single model approach where one classification model is used and periodically updated for new data. On the other hand, batch-incremental method constructs each model using batch learning. When older model can no longer give satisfactory results, it is replaced by newer models [9]. The advantage of ensemble model is that, updating the classification model is much simpler in this case. However, these techniques generally do not include novel or recurring class detection.

An approach to identify recurring class is presented in [6]. Here in addition to primary ensemble model, an auxiliary ensemble of classifiers is present. The auxiliary ensemble model is responsible for storing all the classes even after they disappear from the data stream. When an instance is detected as outlier in the primary ensemble, but falls within the decision boundary of auxiliary ensemble, the instances is identified as recurrent class. Any test data outside the decision boundary of both ensembles are analyzed for novel class.

The approaches described in [6] are considered as *chunk-based* method. A *class-based* ensemble approach is presented in [1]. Here an ensemble model is constructed for each class C of the data stream. Each ensemble has K micro-classifiers. Initially, micro-classifiers are trained from the data chunk. When a latest labeled chunk of data arrives, a separate micro-classifier is trained for each class. Then the newly trained micro-classifier replaces the one with highest prediction error of the respective class. An instance falls outside the decision boundary of all the micro-classifiers of all the classes is considered as an outlier and saved in a buffer. The buffer is checked periodically to detect novel class. Authors of [1] have shown theoretically and experimentally that, class-based approach is better than the chunk-based technique.

In this paper, we propose a more sophisticated approach to construct a class-based ensemble of classifiers. We have also present a better way to update and maintain the ensemble model. Moreover, we propose two types of outliers to update the classifiers and novel class detection and also take the wrongly predicted data into account to modify the classifiers. Experiments show the effectiveness of our methods compared to other techniques.

3 Our Approach

First, we discuss the fundamental concept of data stream classification. Then we describe our approach for stream classification subsequently.

3.1 Preliminaries

Each data in the stream arrives in the following format:

$$D_1 = < x_1, ...x_S >,$$
$$D_2 = < x_{S+1}, ...x_{2S} >,$$
$$............$$
$$D_\Gamma = < x_{(\Gamma-1)S+1}, ...x_{\Gamma S} >$$

where x_i is the ith instance in the stream and S is the size of the stream. D_i is the ith data chunk and D_Γ is the latest data chunk. The problem is to predict the class of each data point. Let l_i and $\hat{l}_i$ be the actual and predicted label of instance x_i. If $l_i = \hat{l}_i$ then the prediction is correct otherwise it is incorrect. The goal is to minimize the prediction error.

Stream classification can be used in various applications such as labeling message in social network or identify intrusion in the network traffic. For example, in credit card fraud detection system, each transaction can be considered as an instance or data point and can be predicted either as *authentic* or *fraud* by any classification technique. If the transaction is predicted as *fraud*, then immediate action can be taken to withhold the transaction. Sometimes, the predicted decision can be wrong (authentic transaction predicted as fraud or vice versa). This can be verified from the cardholder later. The feedback can be considered as "labeling" the instance and used to refine the classification model.

The major task in the data stream classification is to keep the classification model up-to-date by modifying it periodically with the most recent concept. The overview of our proposed approach is shown in Figure 1(a). The major parts of the algorithm will be described step-by-step.

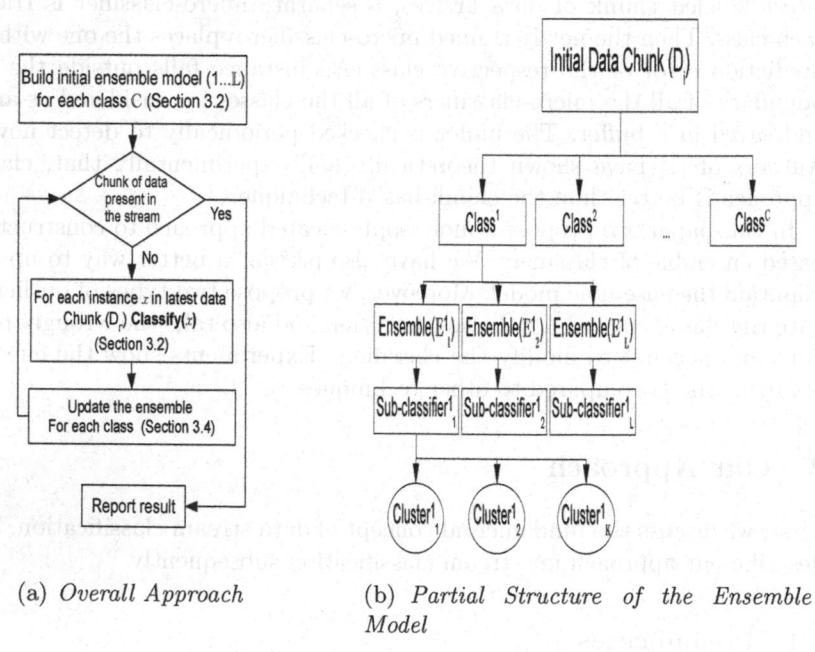

(a) *Overall Approach*

(b) *Partial Structure of the Ensemble Model*

Fig. 1. Overall approach and structure of the classification model

3.2   Ensemble Construction and Training

Now, we present the approach for generating the ensemble model. We will refer our model as **R**ecurring and **N**ovel **C**lass **D**etector **E**nsemble (RNCDE).

Initially, the data chunk is partitioned into $\mathbb{C}$ disjoint groups $(\mathcal{G}^1, \mathcal{G}^2, ..., \mathcal{G}^{\mathbb{C}})$ according to the true class labels, where $\mathbb{C}$ is number of classes in the chunk. Therefore, each group contains the instances of one class only. Then an ensemble of size $\mathbb{L}$ is constructed for each class i using $\mathcal{G}^i$. Each ensemble $\mathbb{E}_l^i$, $i \in \mathbb{C}$, $l \in \mathbb{L}$ is composed of a sub-classifier $\mathbb{S}_l^i$. Each sub-classifier $\mathbb{S}_l^i$ is trained on the instance of class i ($\mathcal{G}^i$). We apply K-means clustering to generate $\mathbb{K}$ clusters on the instances of each class i. For each cluster $\mathbb{H}_{l_j}^i$ of ensemble l of class i, where $j \in \mathbb{K}$ we keep a summary of the cluster i.e. μ, the centroid, r, the cluster radius (distance between centroid and the farthest data point of the cluster) and η, the number of points belonging to the cluster. This way we do not need every data point of the cluster. Therefore, each sub-classifier $\mathbb{S}_l^i$ is composed of all the clusters built from the instances of class i ($\mathbb{S}_l^i = \bigcup_{j=1}^{\mathbb{K}} \mathbb{H}_{l_j}^i$). This process for generating sub-classifiers $\mathbb{S}_l^i$ is repeated $\mathbb{L}$ times to construct the ensemble model $\mathbb{E}^i$ for class i ($\mathbb{E}^i = \bigcup_{l=1}^{\mathbb{L}} \mathbb{S}_l^i$). Finally, the overall model is the union of all the ensemble built for each class i ($\mathbb{E} = \bigcup_{i=1}^{\mathbb{C}} \mathbb{E}^i$). For visual purpose, the partial structure of the ensemble model is shown in hierarchical form in Figure 1(b). It should be noted that, each ensemble for class i has only one sub-classifier.

Note that, each sub-classifier $\mathbb{S}_l^i$ of an ensemble $\mathbb{E}^i$ is trained on the same data $\mathcal{G}^i$. We vary the seed parameters ($\gamma_1, \gamma_2, ...\gamma_{\mathbb{L}}$) of K-means clustering to diversify the sub-classifier. We have shown our method using a hypothetical example in Figure 2. In Figure 2(a), the instances of the same class are shown. The K-means clustering is applied to construct sub-classifier 1 using seed parameter γ_1 (Figure 2(b)), where $\mathbb{K} = 3$. Then again sub-classifier 2 is constructed by K-means clustering initialized by the seed parameter γ_2 shown in Figure 2(c). We can see that, identical instances belong to different clusters at each sub-classifier. This process is repeated $\mathbb{L}$ times to construct $\mathbb{L}$ alternating sub-classifiers $\mathbb{S}_1^i....\mathbb{S}_{\mathbb{L}}^i$ for class i which is shown in Figure 2(d).

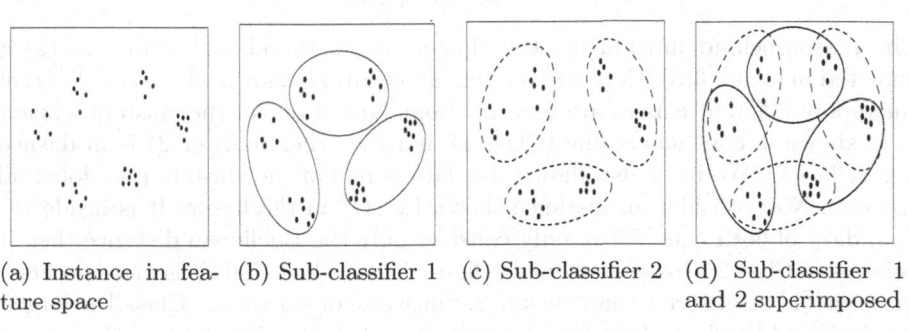

(a) Instance in fea-   (b) Sub-classifier 1   (c) Sub-classifier 2   (d) Sub-classifier 1
ture space   and 2 superimposed

Fig. 2. A hypothetical example of layer for 2-dimensional search space

The advantages of K-means clustering is that, its lower time complexity will allow to built classifiers in reduced time which is a critical requirements for data stream mining. Another benefit is that, after construction of the clusters, it is easy to modify them compared to other types of classifiers.

3.3   Classification

Here we describe our classification procedure and outlier detection. Each data point in the most recently arrived chunk is first checked for whether it is an outlier. We have maintained two types of outlier i.e. class-outlier ($\mathbb{C}$-outlier) and universal-outlier (U-outlier). If any instance is outside the decision-boundary of all the sub-classifiers of all the ensembles($\bigcup_{i=1}^{C} \mathbb{E}^i$), then it considered as a U-outlier. If a data point is a U-outlier, then it is saved in *buffer* to analyze it further. If an instance x_i is not a U-outlier then, it is inside the decision boundary of any class. It is possible that, x_i may be inside of more than one class due to noise and the curse of dimensionality. Let $\mathcal{E}_{x_i}$ be the set of such classes. We decide which class x_i belongs to by computing a coefficient (m-value). We called this coefficient *membership coefficient*. The m-value ($\tau_{l_j}^i$) for cluster $\mathbb{H}_{l_j}^i$, where $i \in \mathcal{E}_{x_i}$, $l \in \mathbb{L}$ and $j \in \mathbb{K}$ can be computed using the equation below,

$$\tau_{l_j}^i = \left(\frac{\eta_{l_j}^i}{\max\limits_{m \in \mathcal{E}_{x_i}, n \in \mathbb{L}, o \in \mathbb{K}} \eta_{n_o}^m} \right) \Big/ \left(\frac{d_{l_j}^i}{\max\limits_{m \in \mathcal{E}_{x_i}, n \in \mathbb{L}, o \in \mathbb{K}} d_{n_o}^m} \right)^{\beta}, \quad (1)$$

where $d_{l_j}^i$ is the Euclidean distance between the instance x_i and the centroid of cluster $\mathbb{H}_{l_j}^i$ where $\eta_{l_j}^i$ is the size of the cluster. Here β is the relative importance of the inverse of distance over the size of the classifier. We refer this constant as ξ-coefficient. The max size and max distance is used for normalization. After computing m-value for each cluster of all the sub-classifiers, the class label for instance x_i is computed using the equation below,

$$c = \arg \max\limits_{i \in \mathcal{E}_{x_i}, l \in \mathbb{L}, j \in \mathbb{K}} \tau_{l_j}^i \quad (2)$$

The reason behind introducing the cluster size in the classification process is depicted in Figure 3(a). Here a hypothetical scenario is shown where two different clusters of different classes are present. Boundary of one of the clusters (cluster 1) is shown in continuous line (Class 1) and the other (cluster 2) is in dashed line (Class 2). We have also shown the data points of the clusters (i.e. dots and crosses). Now consider an instance shown by "O" in the figure. It is inside the boundary of both class. If we only consider only the Euclidean distance then it belongs to Class 2. However, from the figure it is evident that, it is more prone to the centroid of cluster 1 than cluster 2. Since size of cluster for Class 1 is larger, the decision boundary of cluster 1 is more expanded. Considering only the nearest neighbor to label the instance may result in erroneous prediction. However, if we make the assumption that, all the data points of a cluster are uniformly distributed, then the number of points in the overlapped region (common region between two clusters) will be greater for cluster 1 than cluster 2. In this case, the test instance will be labeled as Class 1. Therefore, a more sophisticated measurement can be possible if we take account the size of the cluster in the classification process.

3.4   Ensemble Update

When the labels for data points of a chunk are available (labeled by human expert), the incorrectly predicted data ($\mathcal{W}$) by the ensemble model is identified. Then the wrongly predicted data are separated according to their correct

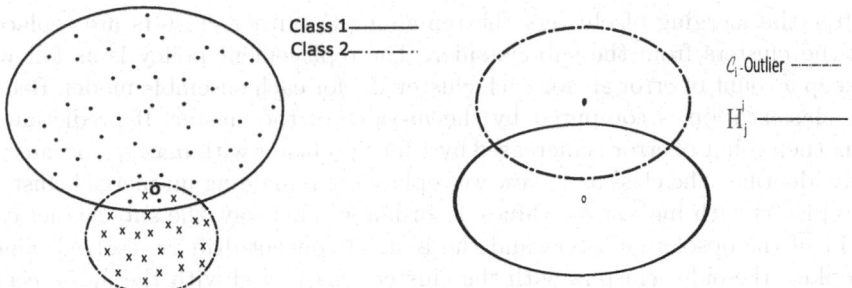

(a) *Hypothetical decision boundary of 2*   (b) *A scenario for cluster merging*
clusters of two different classes

Fig. 3. $\mathbb{K}$ vs ERR

label. As a result, the all the inaccurately predicted data are partitioned into
disjoint sets $(\mathcal{W}^1, \mathcal{W}^2, ... \mathcal{W}^{\mathbb{C}})$. Then the data in $\mathcal{W}^i$ are clustered using K-means
clustering. The number of clusters $\mathcal{K}$ is computed using the following equation:

$$\mathcal{K} = \frac{|D_\Gamma|}{ChunkSize} \cdot \mathbb{K} \tag{3}$$

Here $ChunkSize$ is a constant which can be initialized manually. These newly
formed clusters can be called $\mathcal{C}_i$-outlier clusters where $i \in \mathbb{C}$. The union of $\mathcal{C}_i$-
outlier is the $\mathbb{C}$-outlier. After the formation of $\mathcal{C}_i$-outlier clusters, the Euclidean
distance from each $\mathcal{C}_i$-outlier clusters to each $\mathbb{H}^i_{l_j}$ is computed. Now based on the
distance among the clusters we make two types of modifications. One is cluster
merge and the other is cluster replacement.

If the distance between a $\mathcal{C}_i$-outlier clusters and one of the clusters ($\mathbb{H}^i_{l_j}$) in the
ensemble is less than the radius of $\mathbb{H}^i_{l_j}$ ($r^i_{l_j}$), then the two clusters are merged.
Recall that, the data points of $\mathcal{C}_i$-outlier are actually the wrongly predicted
instances clustered according to the actual class label i. So it is normal that,
any cluster from $\mathcal{C}_i$-outlier will tend to very remain very close to the $\mathbb{H}^i_{l_j}$ in
the ensemble model. A possible scenario depicting the condition for merging the
clusters is shown in Figure 3(b). Here the distance between $\mathbb{C}_i$-oulier cluster and
the centroid of $\mathbb{H}^i_{l_j}$ is less than the radius of $\mathbb{H}^i_{l_j}$ ($r^i_{l_j}$).

Now to merge the cluster, we have to calculate the new centroid, the cluster
size and the radius. To calculate the position of new centroid we have used the
the equation below:

$$\mu^i_{l_j} = \frac{\eta^i_{l_j} \cdot \mu^i_{l_j} + \eta_{\mathcal{C}_i-\text{outlier}} \cdot \mu_{\mathcal{C}_i-\text{outlier}}}{\eta^i_{l_j} + \mu_{\mathcal{C}_i-\text{outlier}}}, \tag{4}$$

where $\eta_{\mathcal{C}-\text{outlier}}$ and $\mu_{\mathcal{C}-\text{outlier}}$ are the size and centroid of the $\mathcal{C}_i$-outlier. Since
two clusters are merged, size is addition of the size of two clusters. The radius is
computed by combining the radii of two clusters with the distance between the
centroids.

After the merging of clusters the remaining C_i-outlier clusters are replaced with the clusters from the sub-classifier. The replacement policy is as follows. We keep a count of error $\varepsilon_{l_j}^i$ for each cluster $\mathbb{H}_{l_j}^i$ for each ensemble model. Recall that, classification is computed by the m-value of the cluster. If prediction is wrong then count of error is increased by 1 for the cluster with max $\tau_{l_j}^i$, because it falsely identified the class as i. Now we replace the remaining un-merged clusters with clusters with highest $\varepsilon_{l_j}^i$ values accordingly. This way, the sub-classier can get rid of the obsolete clusters and the issue of concept-drift is resolved. Since we replace the older clusters with the cluster constructed with the most recent data points, the ensemble model remains up-to-date with the latest concept.

3.5   Novel Class Detection

We have extended and generalized the idea of novel class detection in [1]. The primary assumption behind the novel class detection in [1] was, data points of the same class should be closer to each other (*cohesion*) and farther apart from the other classes (*separation*). However, first assumption (i.e. cohesion) may prove different in some complex cases. It may be possible that, data points of the same class may be clustered together in various groups where these groups may be scattered through the feature space.

If the data points of a novel class emerge in the stream, we can assume that, the instances belonging to novel class will be far from the decision boundary of existing classes. Since data points of U-outlier are outside the decision boundary of all the existing classes, these data are analyzed for novel classes. Recall that, the U-outliers are stored in a buffer, if the size of the buffer reaches a threshold then they are analyzed for novel class. We have modified the metric called q-NSC authors of [1] used and called it **q-mNSC**. In this method, another metric called q,c-*neighborhood* is used. We modify the definition of q,c-neighborhood also, which we called q,h-neighborhood. We define it as follows:

q,h-neighborhood: The q,h-neighborhood (q,h(x) in short) of an U-outlier x is the set of q clusters that are nearest to x. (q-nearest cluster h neighbor of instance x).

Here q is a user defined parameter which can be initialized at the beginning. In summary, we compute the nearest q number of clusters from instance x regardless of the class the clusters belong to.

Now suppose, $\bar{D}_{h_{out},q}(x)$ be the mean distance of a U-outlier instance x to its q nearest U-outlier neighbors. Moreover, let $\bar{D}_{h,q}(x)$ be the mean distance from x to its $q,h(x)$ and $\bar{D}_{h_{min},q}(x)$ be the minimum value among all $\bar{D}_{h,q}(x)$. Here, h is the set of clusters from the existing classes. Then the q-mNSC of x can be computed according our definition:

$$q-\mathrm{mNSC(x)} = \frac{\bar{D}_{h_{min},q}(x) - \bar{D}_{h_{out},q}(x)}{\max(\bar{D}_{h_{min},q}(x), \bar{D}_{h_{out},q}(x))} \tag{5}$$

The value of q-mNSC(x) ranges between -1 to +1. When the value is positive x is closer to $\mathbb{U}$-outlier instances and away from the existing classes resulting more cohesion and vice versa.

Now we explain how we can utilize the metric to detect novel class. First, we apply K-means clustering on $\mathbb{U}$-outliers to partition the data to $\mathcal{K}_0$ number of clusters, where $\mathcal{K}_0 = \mathbb{K} \cdot \frac{|\text{buffer}|}{ChunkSize}$. The reason for applying clustering is to reduce time complexity(reduces from $O(n^2)$ to $O(\mathcal{K}_0^2)$), where n is the total number of data points in $\mathbb{U}$-outlier). For each $\mathbb{U}$-cluster we compute q nearest cluster h_n for all the sub-classifiers of all the class. After that, for each $\mathbb{U}$-cluster we compute q-nearest neighbor cluster of that $\mathbb{U}$-cluster. Then we apply the Equation 5 to compute the q-mNSC for each $\mathbb{U}$-cluster. This way we get a q-mNSC value for each $\mathbb{U}$-cluster in the ensemble. If the positive value of q-mNSC is greater than a fixed number (q_α) than we can conclude a novel class has emerged at the stream.

4   Experimental Findings

First we discuss about the data set and then the parameter settings. Later, we describe the results and our remarks.

We apply the procedure described in [5] to generate synthetic datasets with concept evolution and drift. We generate three types of datasets as described in [5]. Each dataset contain 2.5×10^5 instances with 40 real value attribute. We refer each set as SynNCX having X classes (i.e. SynNC10 where total 10 classes are present).

We have also taken the real-life dataset *Forest* from UCI database and the *10 percent* version of KDD CUP 1999 intrusion detection challenge. First dataset contains 581000 instances with 7 classes and 54 numeric attributes while the second datasets have 490000 instances having 23 classes and 34 numeric attributes. We randomly permutate the instances and construct 10 sequences and report the average results. We have made adjustments to have novel instances in the sequences.

We have compared our approach (RNCDE) with class-based approach (CL) [1], ECSMiner (EM) [5], the clustered-based method presented in [8] (OW) and chunk-based approach (SC) described in [6].

4.1   Parameter Settings

We have set the size of the ensemble $\mathbb{L} = 3$, number of clusters per sub-classifier $\mathbb{K} = 20$. The minimum number of instances to detect novel class $q_\alpha = 20$. Moreover, ξ is varied between 3 to 8 and size of the *buffer* is set to the 20% of the size of the chunk. These parameters are set either according to the parameters of the previous works or by running preliminary experiments.

4.2   Evaluation

We have used the following evaluation criteria for performance measurements. $M_{new} = \%$ of novel class instances misclassified as existing class, $F_{new} = \%$ of existing class instances misclassified as novel class, $OTH = \%$ of existing class instances misclassified as another existing class and ERR = average misclassfication error (average of three types of error).

Initially, we construct the ensemble model from first three data chunks. Then we begin our performance evaluation from the chunk four. Table 1 summarizes the results from all the methods. We have taken the summary results on other methods from [1] and compared with our approach. OTH can be calculated from the other errors, so we do not show it. From the table, we can see that, OW has the highest error rate, because it can not detect majority of the novel class instances. Therefore, the F_{new} rate is also high in case of OW.

EM can identify novel class but it can not detect recurring class. As a result, recurring classes are detected as novel class and it has a high F_{new} rate also. SC maintains an auxiliary ensemble model which contains classifiers for all the class including recurring class. Therefore, it has comparatively lower F_{new} rate than EM. CL uses class- based ensemble to detect novel and recurring class and it has a lower error rate than the approaches above. Our proposed method RNCDE also have shown comparatively lower error rate than other methods. In *Forest* dataset, the ERR is slightly higher than CL, but in other case RNCDE shows better performance than other approaches.

Table 1. Summary results on all the datasets

Performance Criteria	Methods	SynNC10	SynNC20	SynNC40	Forest	KDD
F_{new}	OW	0.9	1.0	1.4	0.0	0.0
	EM	24.0	23.0	20.9	5.8	16.4
	SC	14.6	13.6	11.8	3.1	12.6
	CL	0.01	0.05	0.13	2.3	5.0
	RNCDE	0.01	0.03	0.03	5.8	4.8
M_{new}	OW	3.3	5.0	7.1	89.5	100
	EM	0.0	0.0	0.0	34.5	63.0
	SC	0.0	0.0	0.0	30.1	61.4
	CL	0.0	0.0	0.0	17.5	59.5
	RNCDE	0.0	0.0	0.0	14.4	60.1
ERR	OW	7.5	7.7	8.0	30.3	37.6
	EM	8.2	7.9	7.2	13.7	28
	SC	5.1	4.8	4.3	11.5	26.7
	CL	0.01	0.02	0.05	7.3	26.0
	RNCDE	**0.019**	**0.02**	**0.02**	**10.57**	**24.76**

In Figure 4, ERR rates for both Synthetic and Real Data are shown. In each case X axis represents number of data points and Y axis represents the ERR. For example from the Figure 4(a) and 4(b), we can see that, ERR rates after 300000 data points are 20% for *forest*, 10% in *KDD*. For synthetic data ERR remains almost constant. In case of KDD we can see at the beginning ERR fluctuates, but the ERR decreases afterwards. This occurs because the at first the class boundary among classes are not accurately drawn so misclassification among existing classes (OTH) raises ERR. When the concept is learned comprehensively then ERR decreases. On the other hand, in forest ERR rises gradually. This is because M_{new} increases continuously when more data points arrive.

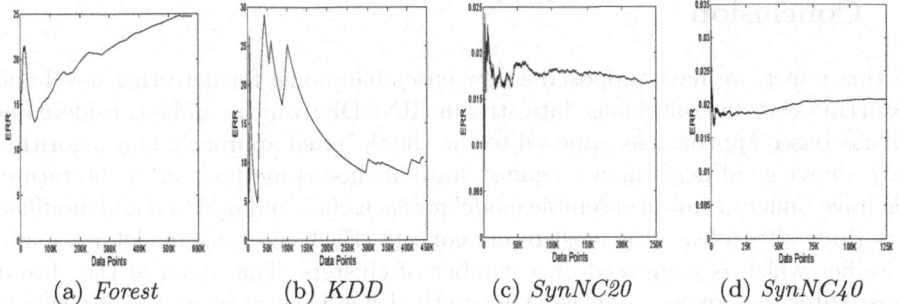

(a) *Forest*   (b) *KDD*   (c) *SynNC20*   (d) *SynNC40*

Fig. 4. ERR for Datasets

4.3   Parameter Sensitiveness

We have observed the effect of a number of parameters on our algorithm. Due
to page limitation we describe only one parameter number of clusters per sub-
classifier $\mathbb{K}$. The $\mathbb{K}$ is varied between 10 to 50. The impact of varying $\mathbb{K}$ for
synthetic dataset is shown in Figure 5. We can see from the figure that, ERR
decreases, if the number of cluster $\mathbb{K}$ increases. The reason behind this is when
the number of clusters increases more accurate decision boundary can be drawn
among the classes. When the value of $\mathbb{K}$ is increased, more clusters will be formed
on the same instances. Therefore, the size of the clusters will be comparatively
lower and each cluster will learn the small portion of the total concept. If the
boundary between two classes is noisy then more and smaller clusters will perform better than fewer and larger clusters. In other words, the boundary of the
class will be more accurate constructed if an increased number clusters is formed.
That is why ERR deceases if $\mathbb{K}$ is increased. However, it should be noted that,
if the value of $\mathbb{K}$ is high, then it would result in high space requirements and
increased time complexity, which has a detrimental effect on the performance
of the model. So the value of $\mathbb{K}$ should be adjusted to balance between the
performance and accuracy.

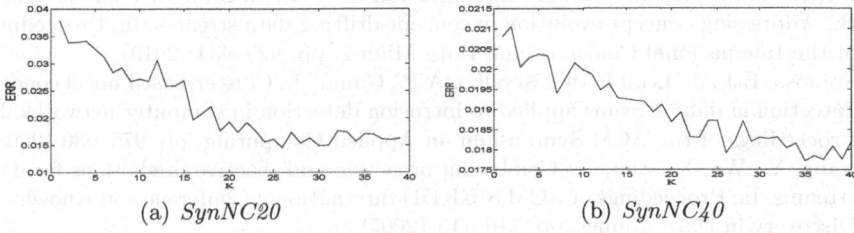

(a) *SynNC20*   (b) *SynNC40*

Fig. 5. $\mathbb{K}$ vs ERR

5 Conclusion

In this paper, we have proposed a new ensemble model for detecting novel and recurring class in continuous data stream (RNCDE) which can be considered as a class-based approach as opposed to the chunk-based approach. Our algorithm have shown good performance against state-of-the-art methods in the literature. We have built our initial ensemble model for each class and updated and modified it periodically to learn the most recent concept. Each ensemble model has a sub-classifier which is composed of a number of clusters. The union of the cluster constitutes the concept of class. Our method has been proven very effective in data stream mining. Inspired by the promising results, we will concentrate on more efficient techniques for data stream classification. We are also planning to experiment our method on other real life data.

References

1. Al-Khateeb, T., Masud, M.M., Khan, L., Aggarwal, C., Han, J., Thuraisingham, B.: Stream classification with recurring and novel class detection using class-based ensemble. In: Proceedings of International Conference on Data Mining, pp. 31–40 (2012)
2. Gao, J., Fan, W., Han, J.: On appropriate assumptions to mine data streams: Analysis and practice. In: Proceedings of International Conference on Data Mining, pp. 143–152 (2007)
3. Hashemi, S., Yang, Y., Mirzamomen, Z., Kangavari, M.: Adapted one-versus-all decision trees for data stream classification. IEEE Transactions on Knowledge and Data Engineering 21(5), 624–637 (2009)
4. Hulten, G., Spencer, L., Domingos, P.: Mining time-changing data streams. In: Proceedings of ACM SIGKDD International Conference on Knowledge Discovery and Data Mining, pp. 97–106 (2001)
5. Masud, M., Gao, J., Khan, L., Han, J., Thuraisingham, B.: Classification and novel class detection in concept-drifting data streams under time constraints. IEEE Transactions on Knowledge and Data Engineering 23(6), 859–874 (2011)
6. Masud, M.M., Al-Khateeb, T.M., Khan, L., Aggarwal, C., Gao, J., Han, J., Thuraisingham, B.: Detecting recurring and novel classes in concept-drifting data streams. In: Proceedings of International Conference on Data Mining (2011)
7. Masud, M.M., Chen, Q., Khan, L., Aggarwal, C., Gao, J., Han, J., Thuraisingham, B.: Addressing concept-evolution in concept-drifting data streams. In: Proceedings of the International Conference on Data Mining, pp. 929–934 (2010)
8. Spinosa, E.J., de Leon F. de Carvalho, A.P., Gama, J.: Cluster-based novel concept detection in data streams applied to intrusion detection in computer networks. In: Proceedings of the ACM Symposium on Applied Computing, pp. 976–980 (2008)
9. Yang, Y., Wu, X., Zhu, X.: Combining proactive and reactive predictions for data streams. In: Proceedings of ACM SIGKDD International Conference on Knowledge Discovery in Data Mining, pp. 710–715 (2005)
10. Zhang, P., Zhu, X., Guo, L.: Mining data streams with labeled and unlabeled training examples. In: Ninth IEEE International Conference on Data Mining, pp. 627–636 (2009)

Detecting Changes in Rare Patterns
from Data Streams

David Tse Jung Huang[1], Yun Sing Koh[1], Gillian Dobbie[1], and Russel Pears[2]

[1] Department of Computer Science, University of Auckland, New Zealand
{dtjh,ykoh,gill}@cs.auckland.ac.nz
[2] School of Computing and Mathematical Sciences, AUT University, New Zealand
rpears@aut.ac.nz

Abstract. Current drift detection techniques in data streams focus on finding changes in streams with labeled data intended for supervised machine learning methods. Up to now there has been no research that considers drift detection on item based data streams with unlabeled data intended for unsupervised association rule mining. In this paper we address and discuss the current issues in performing drift detection of rare patterns in data streams and present a working approach that enables the detection of rare pattern changes. We propose a novel measure, called the M measure, that facilitates pattern change detection and through our experiments we show that this measure can be used to detect changes in rare patterns in data streams efficiently and accurately.

Keywords: Data Stream, Drift Detection, Rare Pattern.

1 Introduction

Mining data streams for knowledge discovery, using techniques such as clustering, classification, and frequent pattern discovery, has become increasingly important. A data stream is an ordered sequence of instances that arrive at a high rate. This characteristic imposes additional constraints on the mining algorithms to be efficient enough to keep up with the fast rate of arrival and also requires an efficient memory usage as not all data instances can be stored in memory. Many techniques that find frequent patterns from data streams have been proposed, such as CPS-Tree [17] and FPStream [6]. Frequent patterns have been widely considered to be informative and useful but in some domains and scenarios rare patterns may be more interesting. Rare patterns are patterns that do not occur frequently and can sometimes be considered as exceptions. Rare patterns often represent irregular behaviors such as frauds. The detection of rare patterns can benefit a wide range of domains such as fraud detection in credit card transactions and auctions. The mining of rare patterns from data streams has been considered in some previous research, such as SRP-Tree [9].

An important characteristic of data streams is that changes in the underlying distribution can signal important changes in the data stream. Many drift detection techniques have been proposed to detect these changes. However,

V.S. Tseng et al. (Eds.): PAKDD 2014, Part II, LNAI 8444, pp. 437–448, 2014.

these techniques are designed with the focus of detecting drift in data streams that contain class labels which are intended for supervised machine learning methods such as classification. These drift detection techniques (*e.g.* ADWIN2 [3]) take in binary inputs that are derived from the error rates of a classifier run on the labeled data stream. Because these techniques are designed for use in labeled data streams, they cannot be applied directly onto unlabeled data streams that frequent and rare pattern mining techniques take in as input.

A naive method for detecting changes in patterns would be to mine the stream for a set of patterns at given intervals and then compare the sets of patterns. This is not the suitable approach, especially in the case of rare patterns, where it is often harder and more costly to discover rare patterns from data streams. A more enlightened scheme would be to apply drift detection techniques at an item level. Therefore, instead of running one instance of the drift detection technique (*e.g.* ADWIN2) on the stream, multiple instances of the technique is run on each separate item (or a subset of the items) found in the unlabeled stream. The binary inputs into ADWIN2 can be derived from the presence or absence of the item in a series of transactions where the binary input 1 would represent that the item occurs in the transaction and the binary 0 would represent the item did not occur in the transaction. For example, consider three transactions: $T_1 : \{a,b,c\}, T_2 : \{a\}, T_3 : \{a,b\}$. The binary inputs used for item a would be $\{111\}$ as item a appears in all three transactions and the input for item b would be $\{101\}$ as it does not occur in transaction T_2. Essentially this monitors the support change of the individual items in the stream and a detected change in this case would represent that the item is either occurring in more transactions or occurring in fewer transactions than it did previously. The issue with this method is that only pattern changes caused by support variations in items will be detected. If there is a change in pattern, but without an accompanying change in support of items, the change will not be detected. For example, consider 4 items $\{a, b, c, d\}$ where $\{a, b\}$ always occur together forming a pattern and $\{c, d\}$ always occur together forming another pattern. If the support of these items do not change but now item a occurs with item c and item b occurs with item d, then this change will not be picked up by simply monitoring the support of items using current drift detection techniques. In this paper, we describe this form of pattern change as a change in item association.

Motivated by the difficulties and the lack of methods for detecting rare pattern changes from unlabeled data streams, the aim of this paper is to address this problem by proposing a novel approach that enables such detection. We propose a novel M measure that enable the detection of both rare pattern changes caused by support change and item association change. The M measure consolidates the state of association of items into one numerical value and in our approach, instead of monitoring the supprot of items as described earlier, we monitor the M measure. Through our evaluations, we demonstrate that this overall approach is capable of detecting rare pattern changes.

There are several scenarios where detecting a change in item associations can be useful. For example, consider a stream of data for recording a series of traceroutes where an item represents a host in the route. Through monitoring a subset of the items (hosts), the user can identify whether there are changes in the relationship of the hosts. A change in the relationship of the hosts could represent a major change in the routing behavior of the network and signal a possible congestion in the network. The prompt identification of these changes can allow the user to quickly respond to these situations.

The major contributions of this paper are as follows:

1. We present a new approach that enables the detection of rare pattern changes from unlabeled data streams. To the best of our knowledge there has been no previous work on this topic.
2. We propose a novel M measure that is a consolidated numerical measure and represents the state of item associations for an item at any point in the stream. Through monitoring this measure using techniques like the Page-Hinkley Test, changes in rare patterns can be detected in an efficient manner. The M measure enables the discovery of pattern changes relating to changes in item associations that was previously not possible.

The rest of the paper is as follows: in Section 2 we detail the current state of research in the areas of drift detection and pattern mining in data streams. Section 3 describes the preliminaries and definitions of the problem we address. In Section 4 we introduce our overall framework for solving the problem of drift detection of rare patterns and introduce our novel M measure. In Section 5 we present the experimental evaluations and analysis of our algorithm and lastly Section 6 concludes the paper.

2   Related Work

There has been intense research in the area of rare pattern mining. Most of the research was designed for a static database environment and can generally be divided into level-wise exploration or tree based approaches. Level-wise approaches are similar to the Apriori algorithm [2] developed by Agrawal. In the Apriori algorithm, k-itemsets (itemsets of cardinality k) are used to generate $k+1$-itemsets. These new $k+1$-itemsets are pruned using the downward closure property, which states that the superset of a non-frequent itemset cannot be frequent. Apriori terminates when there are no new $k+1$-itemsets remaining after pruning. MS-Apriori [13], ARIMA [16], AfRIM [1] and Apriori-Inverse [12] are algorithms that detect rare itemsets. They all use level-wise exploration similar to Apriori, which have candidate generation and pruning steps. The RP-Tree algorithm was proposed by Tsang et al. [18] as a tree based approach. RP-Tree avoids the expensive itemset generation and pruning steps of Apriori and uses a tree structure, based on FP-Tree [7], to find rare patterns. Both the Apriori and tree based approaches find rare patterns in static database environments. Recently the SRP-Tree algorithm [9], an adaptation of the RP-Tree algorithm, was developed to enable the capturing of rare patterns in a stream environment. There is currently no relevant work in rare pattern mining that also considers drifts in data streams.

Sebastiao and Gama [15] present a concise survey on drift detection methods. They point out that methods used fall into four basic categories: Statistical Process Control (SPC), Adaptive Windowing [5], Fixed Cumulative Windowing Schemes [10] and finally other classic statistical drift detection methods such as the Martingale frameworks [8], the Page-Hinkley Test [14], and support vector machines [11]. Gama et al. [5] adapted the SPC approach. They proposed the Drift Detection Method (DDM) based on the fact, that in each iteration an online classifier predicts the decision class of an example. That prediction can be either true or false, thus forming the binary input for the method. More recently Bifet et al. [3] proposed an adaptive windowing scheme called ADWIN that is based on the use of the Hoeffding bound to detect concept change. The ADWIN algorithm was shown to outperform the SPC approach and has the attractive property of providing rigorous guarantees on false positive and false negative rates. ADWIN maintains a window (W) of instances at a given time and compares the mean difference of any two subwindows (W_0 of older instances and W_1 of recent instances) from W. If the mean difference is statistically significant, then ADWIN removes all instances of W_0 considered to represent the old concept and only carries W_1 forward to the next test.

In addition, ADWIN has also been used in the IncTreeNat algorithm [4] that performs mining of frequent closed trees using streaming data which finds frequent closure patterns in closed trees. Currently there is no research that specifically looks at finding changes in rare patterns from data streams.

3 Preliminaries

In this section we present the preliminaries of drift detection in Section 3.1 and formally define rare patterns and itemsets in Section 3.2.

3.1 Drift Detection

Let us frame the problem of drift detection and analysis more formally. Let $S_1 = (x_1, x_2, ..., x_m)$ and $S_2 = (x_{m+1}, ..., x_n)$ with $0 < m < n$ represent two samples of instances from a stream with population means μ_1 and μ_2 respectively. The drift detection problem can be expressed as testing the null hypothesis H_0 that $\mu_1 = \mu_2$, i.e. the two samples are drawn from the same distribution against the alternate hypothesis H_1 that they are drawn from different distributions with $\mu_1 \neq \mu_2$. In practice the underlying data distribution is unknown and a test statistic based on sample means is constructed by the drift detector. A false negative occurs if the null hypothesis is accepted incorrectly when a change has occurred. On the other hand if the drift detector accepts H_1 when no change has occurred in the data distribution then a false positive is said to have occurred. Since the population mean of the underlying distribution is unknown, sample means need to be used to perform the above hypothesis tests. The hypothesis tests can be restated as the following. We accept hypothesis H_1 whenever $Pr(|\hat{\mu_1} - \hat{\mu_2}|) \geq \epsilon) > \delta$, where $\delta \in (0,1)$ and is a parameter that controls the maximum allowable false positive rate, while ϵ is a function of δ and is the test statistic used to model the difference between the sample means.

In all drift detection algorithms, a detection delay is inevitable but should be minimized. A detection delay can be expressed as the distance between $\hat{c}$ and n', where $\hat{c}$ is the true drift point and n' is the instance at which change is actually detected. Thus detection delay is determined by $\left(n' - (\hat{c}+1)\right)$. Detection delay is used as one of our evaluation measures in this research.

3.2   Rare Patterns and Itemsets

Let $\mathcal{I} = \{i_1, i_2, \ldots, i_n\}$ be a set of literals, called items. A set $X = \{i_l, \ldots, i_m\} \subseteq I$ and $l, m \in [1, n]$, is called an itemset, or a k-itemset if it contains k items. A transaction $t = (tid, Y)$ is a tuple where tid is an identifier and Y is a pattern. An association rule is an implication $X \to Y$ such that $X \cup Y \subseteq \mathcal{I}$ and $X \cap Y = \emptyset$. X is the antecedent and Y is the consequent of the rule. The *support* of $X \to Y$ is the proportion of transactions that contain $X \cup Y$. The *confidence* of $X \to Y$ is the proportion of transactions containing X that also contain Y.

We adopt the rare itemsets concept from Tsang et al. [18]. We consider an itemset to be rare when its support is below a threshold, the minimum frequent support (minFreqSupp) threshold. We also define a noise filter threshold to prune out the noise called the minimum rare support (minRareSupp) threshold.

Definition 1. *An itemset X is a* rare itemset *in a window* $\mathcal{W}$ *iff*

$$sup_{\mathcal{W}}(X) \leq minFreqSupp \text{ and } sup_{\mathcal{W}}(X) > minRareSupp$$

However not all rare itemsets that fulfill these properties are interesting so we only consider *rare-item itemsets* in this paper.

Rare item itemsets refer to itemsets which are a combination of only rare items and itemsets that consist of both rare and frequent items.

Definition 2. *An itemset X is a* rare-item itemset *iff X is a* rare itemset ***and***

$$\exists x \in X, sup_{\mathcal{W}}(x) \leq minFreqSupp \text{ and } sup_{\mathcal{W}}(x) > minRareSupp$$

4   Our Algorithm

In this section we discuss our RPDD (Rare Pattern Drift Detector) algorithm for detecting changes in rare patterns. This approach is designed to find rare pattern changes from unlabeled transactional data streams intended for unsupervised association rule mining. Overall the framework has two components. One of the components is the processing of stream data and we detail it in Section 4.1. In Section 4.2 we detail the actual drift detection component of the algorithm. Here we introduce our novel M measure, a consolidated numerical measure that represents the state of item associations, and discuss drift detection techniques used in this paper that monitors the M measure to detect changes.

4.1   Stream Processing and Item Selection

As transactions from the stream are fed into the algorithm, they are processed and maintained using a sliding window W of size $|W|$. A list of item frequencies is recorded while the stream is processed. We use the minFreqSupp and min-RareSupp defined in the preliminaries to identify the rare items. A rare item is an item with a support value between the minFreqSupp and minRareSupp. All the rare items found are forwarded to the drift detection component where individual item tracking is performed.

An example is given Table 1. We set $minFreqSupp = 3$ and $minRareSupp = 1$. Based on the thresholds, items b and c are selected as rare items for tracking from this window of transactions. Although the selection of items for tracking in this example is based on the measures of support, this does not have to be the case. The user can specify other ways of selecting items to be tracked or use a specific list of items that are of interest based on domain knowledge.

4.2   Drift Detection Using M Measure

The selected rare items from the stream processing component will be individually tracked and monitored. Essentially, each selected item will have a separate process that monitors the item associations it has with other items in the stream using our M measure.

For each selected item, a separate local item frequency list is maintained that keeps track of the frequency of occurrence of other items with the selected item. For example, consider the previous example reproduced below:

Table 1. Transactions   **Table 2.** Global Freq List   **Table 3.** Local Freq List
for item c

tid	transaction
1	a b c
2	a b
3	a c
4	a h j

item	freq
a	4
b	2
c	2
h	1
j	1

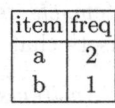

item	freq
a	2
b	1

Since items b and c were selected as tracking items, in this phase, these two items are separately monitored. For each monitored item, a local frequency list is maintained that consists of item associations of the monitored item with other items in the stream. For example, for item c, the local frequency list would consist of the items that occur with item c in the same transaction and their respective co-occurrence frequencies. The local frequency list for item c is shown in Table 3. These local frequency lists for each item will be used by our novel M measure to detect changes in item associations.

The M measure is based on the absolute percentage difference in support of an item. The M measure is a consolidated measure that represents the level of change in the item associations in an item at a given time t.

The M measure for an item x at time t is given by the following formula:

$$M(x)_t = \frac{1}{|X|} \times \sum_{i \in X}^{|X|} \frac{|Supp(i)_t - Supp(i)_{t-1}|}{Supp(i)_{t-1} + c}$$

where X is the set of items that occurs with item x, $Supp(i)$ is the support of i in the local frequency list, and c is the Laplacian coefficient.

The Laplacian coefficient c should be set at a low value such as 0.01 and is introduced to avoid cases of undefined values. In most cases c can be set equal to the minimum rare support threshold specified by the user earlier. The lower bound of the M measure is 0 and is reached when $Supp(i)_t - Supp(i)_{t-1}$ is equal to 0 for all $i \in X$. The upper bound of the M measure is $1/c$ and is reached when $Supp(i)_t = 1$ and $Supp(i)_{t-1} = 0$ for all $i \in X$.

The M measure is calculated for each transaction and for each tracked item individually. Then, we monitor the M measures across time t by applying existing drift detection methods. A fluctuation in the M measure for an item would represent a rare pattern change.

The usage of the M measure is crucial in achieving the goal of detecting changes in item associations and patterns because if users choose to monitor only the support of items, then pattern changes related to change in item associations would not be detected. The M measure is specifically useful for finding rare item association changes because of its property of using percentage difference in support. Since we consider rare items and rare items are characterized by having a lower support value, a small fluctuation in support of these values would result in a much higher percentage difference. For example, if a rare item changes in support from 0.1 to 0.05, the percentage difference is 100%, whereas, if a frequent item goes through the same support change 0.90 to 0.85, the percentage difference is then 6%. The M measure, based on percentage difference, allows us to monitor the changes in rare items more closely and enables the detection of drifts that results from a smaller magnitude of change that would otherwise be missed using pure support based strategies.

In this paper we use two different drift detection techniques to monitor our M measure and find changes in the item associations: the Page-Hinkley Test and the Hoeffding bound with Fixed Size Flushing Window. These two techniques operate on numerical inputs as required by our M measure. We cannot use ADWIN2 as the drift detection technique here because ADWIN2 takes in binary input values and our M measure is not a binary measure.

Page-Hinkley Test. As described by Sebastiao in [15], the Page-Hinkley Test is a sequential analysis technique and designed to detect change in the average of a Gaussian signal. The test monitors a cumulative variable U_T which represents the cumulative difference between observed values and the cumulative mean of the samples. U_T is given by the formula: $U_T = \sum_{t=1}^{T} (x_t - \mu_T - \delta)$ where $\mu_T = \sum_{t=1}^{T} x_t \times \frac{1}{T}$, the cumulative mean at T, and δ is the magnitude threshold.

To detect drifts, it calculates the minimum value of U_t, $m_T = min(U_t, t = 1 \ldots T)$ and the difference between U_T and m_T is monitored at each time t.

A change is signaled when $U_T - m_T > \lambda$ where λ is the detection threshold. The detection threshold, λ, controls the balance between false positives, true positive rates and detection delay of the test. Setting a high λ value will lower false positives but will increase detection delay and might miss drifts, whereas setting a low λ value will increase false positives but will decrease detection delay and increase true positive rates. In practice it is often difficult to find the optimal λ value as the setting of the threshold can vary widely depending on the input.

Hoeffding Bound with Fixed Size Flushing Window. As the Page-Hinkley Test requires the setting of thresholds that can sometimes require an extensive amount of trial and error experimentation, we also use the statistical Hoeffding bound which requires only the setting of one δ parameter while providing statistical guarantees on error bounds. The Hoeffding bound provides an upper bound on the probability that the sum of the random variables deviates from its expected value. The Hoeffding bound states that, with probability $1 - \delta$, the true mean of a random variable r is at least $\bar{r} - \epsilon$ when the mean is estimated over t samples, where R is the range of r.

$$\epsilon = \sqrt{\frac{R^2 \ln(1/\delta)}{2t}}$$

The Hoeffding bound is applied with a Fixed Size Flushing Window (FSFW) in order to perform drift detection. The FSFW consists of a window W with size $|W|$. W is split into two separate blocks of equal size (B_L and B_R). The window is filled as sample values arrive at each time t. When the window is full, the oldest instance is dropped as new instances arrive. At each time t, Hoeffding bound is applied to check for the difference in the mean between the two blocks within the window W. If $|\mu_{B_L} - \mu_{B_R}| > \epsilon$ then a drift is signaled and samples in B_L are flushed and replaced with the samples in B_R. For the purpose of optimization, the FSFW is coded with a circular array to eliminate the cost of shifting items from B_R to B_L. The window size $|W|$ in this case should be set to a reasonably large value to allow a statistically sound comparison of two samples (*e.g.* 1000).

5   Experimental Evaluation

In this section we present the experimental results and evaluations we performed on our RPDD algorithm for finding changes in rare patterns from data streams. Specifically we compare the results of using our M measure in drift detection monitored by the Page-Hinkley Test (PHT) against using M measure in the Hoeffding Bound with Fixed Size Flushing Window. The performance measures used are: True Positives, False Positives, Detection Delay, Execution Time, and Memory. The algorithms are coded in Java and run on an Intel Core i5-2400 CPU @ 3.10 GHz with 8GB of RAM running Windows 7 x64. The execution time reported excludes I/O costs.

In the first experiment we compare the number of false positives obtained by applying the M measure with PHT and Hoeffding bound based drift detection methods. The experiment is set up by generating item transaction streams using a modified IBM Quest Market Basket Data Generator. We control one rare item in the stream by forcing it to undergo a pattern drift at mid-point of the stream (e.g. If R is the rare item then an example of the pattern drift can be: $\{A, B, C, R\}$ to $\{X, Y, Z, R\}$). The controlled item has varying support values (0.05 - 0.2) across the stream and the stream size is 1M. All experiments are run 100 times and the results shown are average values across the 100 runs.

The results for the average number of false positives found are shown in Table 4. For the Page-Hinkley Test experiments we used a constant δ threshold of 0.001 while varying the λ threshold. The Laplacian coefficient c is set at 0.01 for all experiments. The Hoeffding bound is tested using δ values of 0.05, 0.1, and 0.3. The window size $|W|$ is set at 1000 for all experiments.

Overall we observe that PHT produces more false positives than Hoeffding bound. The number of false positives for PHT decreases as λ increases, as expected. The number of false positives for $\lambda = 200$ is extremely high showing that the λ value is too low and that the technique is raising most points as drifts. As we increase the λ value, we see that the number of false positives decreases significantly reaching a more acceptable state. The number of false positives for Hoeffding bound slightly increases as δ increases, also as expected.

Table 4. False positives

	PHT					Hoeffding Bound		
Support	$\lambda = 200$	$\lambda = 400$	$\lambda = 600$	$\lambda = 800$	$\lambda = 1000$	$\delta = 0.05$	$\delta = 0.1$	$\delta = 0.3$
0.05	977.58	36.98	13.13	10.2	8.64	2.29	2.64	3.20
0.1	1884.42	58.38	13.57	10.51	8.95	2.26	2.70	3.38
0.15	2803.16	84.91	13.99	10.74	9.11	2.36	2.75	3.34
0.2	3760.67	116.01	14.15	10.73	9.14	2.34	2.60	3.30

In the second experiment we compare the true positive rates of the techniques. The experiment is set up similar to the false positives experiments except instead of varying the support level of the controlled item, in the true positive experiment we varied the percentage of pattern change. In these experiments the support of the controlled item is set at 0.1. The percentage of pattern change represents the magnitude of pattern change with 0% representing no pattern change and 100% representing a complete pattern change in the controlled item. The percentage of pattern change can be viewed as the change in the items that the controlled item is associated with. It is distinctly different from a change where there is an increase or decrease in the support value of the patterns.

Overall the true positive rates of both techniques at detecting the drift is 1.00 with the exception of Hoeffding bound with the conditions of 25% pattern change and $\delta = 0.05$. This is likely due to the compounded effects of a smaller magnitude of drift produced by the 25% pattern change and a smaller δ value which produces a tighter bound.

In the third experiment we experiment on the detection delay of the two techniques. The experiment is set up exactly the same as the true positive experiments. The detection delay is the distance between $\hat{c}$ and n', where $\hat{c}$ is the

Table 5. True positive rates

% Change	PHT					Hoeffding Bound		
	$\lambda = 200$	$\lambda = 400$	$\lambda = 600$	$\lambda = 800$	$\lambda = 1000$	$\delta = 0.05$	$\delta = 0.1$	$\delta = 0.3$
25%	1.00	1.00	1.00	1.00	1.00	0.99	1.00	1.00
50%	1.00	1.00	1.00	1.00	1.00	1.00	1.00	1.00
75%	1.00	1.00	1.00	1.00	1.00	1.00	1.00	1.00
100%	1.00	1.00	1.00	1.00	1.00	1.00	1.00	1.00

Table 6. PHT: Detection delay

% Change	$\lambda = 200$	$\lambda = 400$	$\lambda = 600$	$\lambda = 800$	$\lambda = 1000$
25%	186.26±(100.86)	626.61±(427.56)	1159.79±(453.03)	1429.05±(508.83)	1693.81±(551.30)
50%	160.83±(89.44)	477.29±(341.77)	949.30±(407.01)	1214.16±(442.61)	1412.18±(506.27)
75%	152.27±(76.47)	398.64±(276.37)	844.03±(315.43)	1109.62±(370.61)	1303.90±(434.68)
100%	122.95±(70.69)	367.04±(292.02)	785.38±(321.16)	1025.64±(377.68)	1294.21±(430.63)

Table 7. Hoeffding Bound: Detection delay

% Change	$\delta = 0.05$	$\delta = 0.1$	$\delta = 0.3$
25%	3338.80±(459.53)	3028.56±(427.60)	2396.52±(346.82)
50%	3083.74±(471.49)	2765.02±(422.96)	2137.10±(338.95)
75%	2897.53±(412.59)	2585.15±(365.43)	1963.74±(282.92)
100%	2863.67±(388.49)	2548.06±(352.91)	1915.35±(279.10)

true drift point and n' is the instance at which change is signaled. Thus, delay is determined by $\left(n' - (\hat{c}+1) \right)$ as described in Section 3.

We observe that as the % change of the pattern is increased, the delay is reduced. This observation meets expectation as a higher % change represents a higher magnitude of change and would result in an earlier detection. PHT generally has a lower detection delay but this is at the cost of a higher false rate.

In the last set of experiments we experiment on the execution time and memory use of tracking a various number of items for drift detection. Since the execution time and memory use is heavily reliant on the number of items selected for tracking and the compounded effects of tracking multiple items, we experimented with a varying number of tracked items. Table 8 shows the results. The execution time is reported in ms and the memory is reported in bytes.

We observe that the execution time and memory use of the algorithm for both PHT and Hoeffding bound increases in a linear fashion as the number of tracked

Table 8. Execution Time and Memory Use

# Items	PHT		Hoeffding Bound	
	Time (ms)	Memory (bytes)	Time (ms)	Memory (bytes)
1	9115±(3758)	586459±(13044)	11618±(5279)	595867±(13014)
2	14780±(6991)	636018±(14317)	21151±(10664)	654649±(14318)
3	20130±(9803)	685151±(15978)	29584±(15102)	712996±(15978)
4	25703±(12312)	733791±(18547)	36870±(18235)	770836±(18548)
5	32209±(14533)	783120±(20315)	47386±(22117)	829373±(20317)
10	61859±(24527)	1026566±(29311)	92722±(37962)	1118854±(29317)
25	130840±(34240)	1751869±(50487)	198476±(51754)	1982639±(50508)
50	250506±(37330)	2936653±(66849)	417208±(64545)	3398090±(66880)

items increase. Overall PHT executes faster and uses less memory due to the extra costs of maintaining the window in the Hoeffding bound technique with Fixed Size Flushing Window. The standard costs of processing and maintaining a sliding window in the stream processing component constitutes the base execution time and memory cost across the two techniques.

6 Conclusion and Future Work

In this paper we proposed a new approach that deals with the problem of detecting rare pattern changes in unlabeled data streams. Our approach uses a novel measure, the M measure, that is a consolidated numerical value which represents the status of item associations in a stream for a given item at a given time. Our experimentation showed that the use of the M measure in conjunction with drift detection techniques enabled the detection of changes in rare patterns that are otherwise undetectable using standard support based detection approaches.

Our future work includes developing a drift detection technique that is optimized with the aim of detecting rare patterns drifts. Even though drift detection techniques such as the Page-Hinkley test works relatively well in this scenario, it requires the setting of the λ parameter and the overall detection scheme is not optimized to the proposed M measure. We also want to adapt the M measure to detect drift in frequent patterns and investigate the possibility of combining and adapting other frequent pattern mining mechanisms such as the CPS-Tree with M measure.

References

1. Adda, M., Wu, L., Feng, Y.: Rare itemset mining. In: Proceedings of the Sixth International Conference on Machine Learning and Applications, ICMLA 2007, pp. 73–80. IEEE Computer Society, Washington, DC (2007)
2. Agrawal, R., Srikant, R.: Fast algorithms for mining association rules in large databases. In: Bocca, J.B., Jarke, M., Zaniolo, C. (eds.) Proceedings of the 20th International Conference on Very Large Data Bases, VLDB, Santiago, Chile, pp. 487–499 (1994)
3. Bifet, A., Gavaldà, R.: Learning from time-changing data with adaptive windowing. In: SIAM International Conference on Data Mining (2007)
4. Bifet, A., Gavaldà, R.: Mining adaptively frequent closed unlabeled rooted trees in data streams. In: Proceedings of the 14th ACM SIGKDD International Conference on Knowledge Discovery and Data Mining, KDD 2008, pp. 34–42. ACM, New York (2008),
   http://doi.acm.org.ezproxy.auckland.ac.nz/10.1145/1401890.1401900
5. Gama, J., Medas, P., Castillo, G., Rodrigues, P.: Learning with drift detection. In: Bazzan, A.L.C., Labidi, S. (eds.) SBIA 2004. LNCS (LNAI), vol. 3171, pp. 286–295. Springer, Heidelberg (2004)
6. Giannella, C., Han, J., Pei, J., Yan, X., Yu, P.S.: Mining frequent patterns in data streams at multiple time granularities. Next Generation Data Mining 212, 191–212 (2003)

7. Han, J., Pei, J., Yin, Y.: Mining frequent patterns without candidate generation. In: Proceedings of the 2000 ACM SIGMOD International Conference on Management of Data, SIGMOD 2000, pp. 1–12. ACM, New York (2000)
8. Ho, S.S., Wechsler, H.: A Martingale framework for detecting changes in data streams by testing exchangeability. IEEE Trans. on Pattern Analysis and Machine Intelligence 32(12), 2113–2127 (2010)
9. Huang, D., Koh, Y.S., Dobbie, G.: Rare pattern mining on data streams. In: Cuzzocrea, A., Dayal, U. (eds.) DaWaK 2012. LNCS, vol. 7448, pp. 303–314. Springer, Heidelberg (2012), http://dx.doi.org/10.1007/978-3-642-32584-7_25
10. Kifer, D., Ben-David, S., Gehrke, J.: Detecting change in data streams. In: Proceedings of the Thirtieth International Conference on VLDB, vol. 30, pp. 180–191. VLDB Endowment (2004)
11. Klinkenberg, R., Joachims, T.: Detecting concept drift with support vector machines. In: Proceedings of the Seventeenth International Conference on Machine Learning (ICML), pp. 487–494 (2000)
12. Koh, Y.S., Rountree, N.: Finding sporadic rules using apriori-inverse. In: Ho, T.-B., Cheung, D., Liu, H. (eds.) PAKDD 2005. LNCS (LNAI), vol. 3518, pp. 97–106. Springer, Heidelberg (2005)
13. Liu, B., Hsu, W., Ma, Y.: Mining association rules with multiple minimum supports. In: Proceedings of the 5th ACM SIGKDD International Conference on Knowledge Discovery and Data Mining, pp. 337–341 (1999)
14. Page, E.S.: Continuous inspection schemes. Biometrika 41(1/2), 100–115 (1954), http://www.jstor.org/stable/2333009
15. Sebastiao, R., Gama, J.: A study on change detection methods. In: 4th Portuguese Conf. on Artificial Intelligence, Lisbon (2009)
16. Szathmary, L., Napoli, A., Valtchev, P.: Towards rare itemset mining. In: Proceedings of the 19th IEEE International Conference on Tools with Artificial Intelligence, ICTAI 2007, vol. 01, pp. 305–312. IEEE Computer Society, Washington, DC (2007)
17. Tanbeer, S.K., Ahmed, C.F., Jeong, B.S., Lee, Y.K.: Efficient frequent pattern mining over data streams. In: Proceedings of the 17th ACM Conference on Information and Knowledge Management, CIKM 2008, pp. 1447–1448. ACM, New York (2008)
18. Tsang, S., Koh, Y.S., Dobbie, G.: RP-Tree: Rare pattern tree mining. In: Cuzzocrea, A., Dayal, U. (eds.) DaWaK 2011. LNCS, vol. 6862, pp. 277–288. Springer, Heidelberg (2011)

Noise-Tolerant Approximate Blocking
for Dynamic Real-Time Entity Resolution*

Huizhi Liang[1], Yanzhe Wang[1], Peter Christen[1], and Ross Gayler[2]

[1] Research School of Computer Science, The Australian National University,
Canberra ACT 0200, Australia
{huizhi.liang,peter.christen}@anu.edu.au, colin788@163.com
[2] Veda, Melbourne VIC 3000, Australia
ross.gayler@veda.com.au

Abstract. Entity resolution is the process of identifying records in one
or multiple data sources that represent the same real-world entity. This
process needs to deal with noisy data that contain for example wrong pronunciation or spelling errors. Many real world applications require rapid
responses for entity queries on dynamic datasets. This brings challenges
to existing approaches which are mainly aimed at the batch matching of
records in static data. Locality sensitive hashing (LSH) is an approximate
blocking approach that hashes objects within a certain distance into the
same block with high probability. How to make approximate blocking approaches scalable to large datasets and effective for entity resolution in
real-time remains an open question. Targeting this problem, we propose
a noise-tolerant approximate blocking approach to index records based
on their distance ranges using LSH and sorting trees within large sized
hash blocks. Experiments conducted on both synthetic and real-world
datasets show the effectiveness of the proposed approach.

Keywords: Entity Resolution, Real-time, Locality Sensitive Hashing,
Indexing.

1 Introduction

The purpose of entity resolution is to find records in one or several databases
that belong to the same real-world entity. Such an entity can for example be a
person (e.g. customer, patient, or student), a consumer product, a business, or
any other object that exists in the real world. Entity resolution is widely used
in various applications such as identity crime detection (e.g. credit card fraud
detection) and estimation of census population statistics [1].

Currently, most available entity resolution techniques conduct the resolution
process in offline or batch mode with static databases. However, in real-world
scenarios, many applications require real-time responses. This requires entity

---

* This research was funded by the Australian Research Council (ARC), Veda Advantage, and Funnelback Pty. Ltd., under Linkage Project LP100200079. Note the first
two authors contributed equally.

V.S. Tseng et al. (Eds.): PAKDD 2014, Part II, LNAI 8444, pp. 449–460, 2014.

resolution on query records that need to be matched within sub-seconds with databases that contain (a possibly large number of) known entities [1]. For example, online entity resolution based on personal identifying details can help a bank to identify fraudulent credit card applications [2], while law enforcement officers need to identify suspect individuals within seconds when they conduct an identity check [1]. Moreover, real-world databases are often dynamic. The requirement of dealing with large-scale dynamic data with quick responses brings challenges to current entity resolution techniques. Only limited research has so far focused on using entity resolution at query time [3,4] or in real-time [5,6].

Typically, pair-wise comparisons between records are used to identify the records that belong to the same entity. The number of record comparisons increases dramatically as the size of a database grows. Indexing techniques such as blocking or canopy formation can help to significantly decrease the number of comparisons [1]. Often phonetic encoding functions, such as Soundex or Double Metaphone, are used to overcome differences in attribute values.

Locality sensitive hashing (LSH) [7] is an approximate blocking approach that uses l length k hash functions to map records within a certain distance range into the same block with a given probability. This approach [8] can filter out records with low similarities, thus decreasing the number of comparisons. However, the tuning of the required parameters k and l is not easy [9]. This is especially true for large-scale dynamic datasets. For some query records, one may need to investigate records with low similarities, while for other query records one only needs to investigate those records with high similarities with the query record. Moreover, entity resolution needs to deal with noise such as pronunciation or spelling errors. Although some LSH approaches such as multi-probe [10] are to decrease the number of hash functions needed, the question of how to make blocking approaches become more noise-tolerant and scalable remains open.

In this paper, we propose a noise-tolerant approximate blocking approach to conduct real-time entity resolution. To deal with noise, an n-gram based approach [1] is employed where attribute values are converted into sets of n-grams (i.e, substring sets of length n). Then, LSH is used to group records into blocks with various distance ranges based on the Jaccard similarity of their n-grams. To be scalable, for blocks that are large (i.e., contains more than a certain number of records), we propose to build dynamic sorting trees inside these blocks and return a small set of nearest neighbor records for a given query record.

2   Related Work

Indexing techniques can help to scale-up the entity resolution process [1]. Commonly used indexing approaches include standard blocking based on inverted indexing and phonetic encoding, n-gram indexing, suffix array based indexing, sorted neighborhood, multi-dimensional mapping, and canopy clustering. Some recent work has proposed automatic blocking mechanisms [11]. Only a small number of approaches have addressed real-time entity resolution. Christen et al. [5] and Ramadan et al. [6] proposed a similarity-aware indexing approach for real-time entity resolution. However, this approach fails to work well for large datasets, as the number of similarity comparisons for new attribute values increases significantly when the size of each encoding block grows.

Approximate blocking techniques such as LSH and tree based indexing [9] are widely used in nearest neighbour similarity search in applications such as recommender systems [12] and entity resolution [8]. In LSH techniques, the collision probability of a LSH family is used as a proxy for a given distance or similarity measure function. Popularly used LSH families include the minHash family for Jaccard distance, the random hyperplane family for Cosine Distance, and the p-stable distribution family for Euclidean Distance [13].

Recently, Gan et al. [14] proposed to use a hash function base with n basic length 1 signatures rather than using fixed l length k signatures or a forest [9] to represent a data point. The data points that are frequently colliding with the query record across all the signatures are selected as the approximate similarity search results. However, as $k=1$ usually leads to large sized blocks, this approach needs to scan all the data points in the blocks to get the frequently colliding records each time. This makes it difficulty to retrieve results quickly for large-scale datasets. Some approaches such as multi-probe [10] have been used to decrease the number of hash functions needed. However, how to explore LSH blocks in a fast way and decrease the retrieval time to facilitate real-time approximate similarity search for large scale datasets still needs to be explored.

3   Problem Definition

To describe the proposed approach, we first define some key concepts.

- **Record Set:** $R = \{r_1, r_2, \ldots, r_{|R|}\}$ contains all existing records in a dataset. Each record corresponds to an entity, such as a person. Let U denote the universe of all possible records, $R \subseteq U$.
- **Element:** An element is an n-gram of an attribute value. Elements may overlap but do not cross attribute boundaries. A record contains $1 \ldots m$ elements, denoted as $r_i = \{v_{i1}, v_{ij}, \ldots, v_{im}\}$.
- **Query Record:** A query record $q_i \in U$ is a record that has the same attribute schema as the records in R. After query processing and matching, q_i will be inserted into R as a new record $r_{|R|+1}$.
- **Query Stream:** $Q = \{q_1, q_2, \ldots, q_{|Q|}\}$ is a set of query records.
- **Entity Set:** $E = \{e_1, e_2, \ldots, e_{|E|}\}$ contains all unique entities in U.

For a given record $r_i \in R$, the decision process of linking r_i with the correspondent entity $e_j \in E$ is denoted as $r_i \rightarrow e_j$, and $e_j = l(r_i)$, where $l(r_i)$ denotes the function of finding the entity of record r_i. The problem of real-time entity resolution is defined as: for each query record q_i in a query stream Q, let e_j be the entity of q_i, $e_j = l(q_i)$, find all the records in R that belong to the same entity as the query record in sub-second time. Let L_{q_i} denote the records in R that belong to e_j, $L_{q_i} = \{r_k \mid r_k \rightarrow l(q_i), r_k \in R\}, L_{q_i} \subseteq R, q_i \in Q$.

Record ID	Entity ID	First Name	Family Name	City	Zip Code
r_1	e_1	Tony	Hua	Sydney	4329
r_2	e_2	Emily	Hu	Perth	1433
r_3	e_3	Yong	Wan	Perth	4320
r_4	e_1	Tonny	Hue	Syndey	4456

(a) An example record dataset

Record ID	2-grams ($n=2$)
r_1	to, on, ny, hu, ua, sy, yd, dn, ne, ey, 43, 32, 29
r_2	em, mi, il, ly, hu, pe, er, rt, th, 14, 43, 33
r_3	yo, on, ng, wa, an, pe, er, rt, th, 43, 22, 20
r_4	to, on, nn, ny, hu, ue, sy, yn, nd, de, ey, 43, 35, 56

(b) The elements(2-grams) for the example records

Fig. 1. An example dataset and the elements(2-grams) of each record

[**Example 1**] Figure 1(a) shows example records r_1, r_2, r_3, and r_4. They belong to three entities e_1, e_2, and e_3. Assume r_4 is a query record, the entity resolution process for r_4 is to find $L_{r_4} = \{r_1\}$ based on the four attribute values.

4 Proposed Approximate Blocking Approach

A blocking schema [15] is an approach to map a set of records into a set of blocks. Blocking schemes generate signatures for records. A blocking scheme can be a LSH function, a canopy clustering function, or a phonetic encoding function [1]. Those records with the same signature will be allocated together in the same block. To make a LSH blocking scheme scalable for large-scale dynamic datasets, we propose to build a dynamic sorting tree to sort the records of large-sized LSH blocks and return a window of nearest neighbors for a query record. Through controlling the window size, we can select nearest neighbors with various approximate similarity ranges for the purpose of being noise-tolerant. Thus, the proposed approach includes two parts: LSH and dynamic sorting tree, which will be discussed in Section 4.1 and 4.2. Then, the discussion of conducting entity resolution based on the proposed blocking approach will be in Section 4.3.

4.1 Locality Sensitive Hashing

LSH can help to find approximate results for a query record for high dimensional data. Let h denote a basic approximate blocking scheme for a given distance measure D, $Pr(i)$ denote the probability of an event i, and p_1 and p_2 are two probability values, $p_1 > p_2$, $0 \leq p_1, p_2 \leq 1$. h is called (d_1, d_2, p_1, p_2)-sensitive for D, if for any records $r_x, r_y \in R$, the following conditions hold:

1. if $D(r_x, r_y) \leq d_1$ then $Pr(h(r_x) = h(r_y)) \geq p_1$
2. if $D(r_x, r_y) > d_2$ then $Pr(h(r_x) = h(r_y)) \leq p_2$

Record ID	h_1	h_2	h_3	h_4
r_1	3	2	1	6
r_2	3	5	4	7
r_3	3	2	4	8
r_4	3	2	1	6

$h_1 \wedge h_2$				$h_3 \wedge h_4$	
3_2	r_1	r_3	r_4	1_6	r_1 r_4
3_5	r_2			4_7	r_1
				4_8	r_3

$H_{2,2} = \{h_1 \wedge h_2, h_3 \wedge h_4\}$ $(k=2, l=2)$

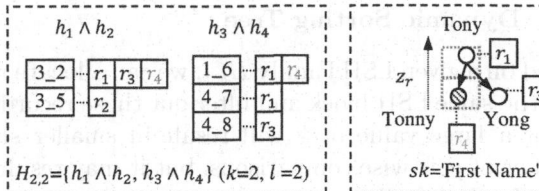

(a) The length 1 minHash values of hash functions h_1, h_2, h_3, h_4.

(b) An example LSH blocking family and generated blocks

(c) Dynamic sorting tree for block 3_2

Fig. 2. Example hash values, LSH family, and generated blocks and a dynamic sorting tree for the example dataset in Figure 1, r_4 is a query record

Minwise hashing (minHash) is a popular LSH approach that estimates the Jaccard similarity [7]. Let $J(r_x, r_y)$ denote the Jaccard similarity for any two records r_x and r_y. This hashing method applies a random permutation Π on the elements of any two records r_x and r_y and utilizes

$$Pr(min(\Pi(r_x)) = min(\Pi(r_y))) = J(r_x, r_y) = \frac{|r_x \bigcap r_y|}{|r_x \bigcup r_y|} \tag{1}$$

to estimate the Jaccard similarity of r_x and r_y, where $min(\Pi(r_x))$ denotes the minimum value of the random permutation of the elements of record r_x. p denotes the hash collision probability $Pr(min(\Pi(r_x)) = min(\Pi(r_y)))$. It represents the ratio of the size of the intersection of the elements of the two records to that of the union of the elements of the two records. A minHash function can generate a basic signature for a given record. The basic signature is called a length 1 signature and the hash function is called a length 1 hash function.

In order to allocate records that have higher similarities with each other into the same block, k length 1 hash functions can be combined to form a length k $(k > 1)$ compound blocking scheme to get the intersection records of the basic length 1 blocking schemes. Let h_c denote a compound LSH blocking scheme that is the AND-construction (conjunction) of k basic LSH blocking schemes h_i, $h_c = \wedge_{i=1}^{k} h_i$. Let p_c denote the collision probability of a length k compound blocking scheme. It can be calculated based on the product of the collision probabilities of its basic length 1 blocking schemes, denoted as $p_c = p^k$. To increase the collision probability, each record is hashed l times to conduct OR-construction (disjunction) and form l hash tables (i.e., l length k signatures), $n = k \cdot l$. Let $H_{k,l}$ denote a LSH family that has l length k hashing blocking schemes, the collision probability of $H_{k,l}$ can be estimated with $p_{k,l} = 1 - (1 - p^k)^l$.

[**Example 2**] (LSH blocking scheme and blocks) Figure 2 (a) shows the example length 1 minHash signatures of the example records in Figure 1. Figure 2 (b) shows an example LSH blocking scheme $H_{2,2}$ and the generated LSH blocks. $H_{2,2} = \{h_1 \wedge h_2, h_3 \wedge h_4\}$, $k = 2$ and $l = 2$. The records are allocated into different blocks based on the given blocking scheme, shown in Figure 2 (b). For example, the length 2 blocking scheme $h_1 \wedge h_2$ generated block signatures 3_2 and 3_5. Records r_1, r_3 and r_4 are allocated in block 3_2 while r_2 is in block 3_5.

4.2   Dynamic Sorting Tree

Based on a given LSH Family $H_{k,l}$, we can allocate records with certain similarity into the same LSH block and filter out those records that have lower similarities. Using a large value of k will result in smaller sized blocks and decrease the number of pair-wise comparisons, but it may result in low recall because a large value for k may filter out some true matched records that have low similarities. A small value for k usually can get higher recall values but may result in large-sized blocks. As scanning the records in large-sized blocks to conduct pair-wise comparisons is usually time-consuming, how to quickly identify a small set of nearest neighbors in a large sized LSH block is very important.

If we sort the records in a LSH block and only return a small set of similar records for each query record, then the exploration of the whole large-sized block can be avoided. $B+$ trees are commonly used to sort the data points for dynamic data in one dimension [9]. We build a $B+$ tree to sort the records inside each large sized LSH block. As forming sub-blocks for those small sized blocks is not necessary, we set up a threshold for building sorting trees in a block. Let γ ($\gamma > 1$)denote this threshold, if the size of a block B_i is greater than γ, then we build a $B+$ tree for B_i, otherwise, no sorting tree will be formed. To build a sorting tree, we firstly discuss how to select a sorting key.

Selecting a Sorting Key Adaptively. Typically, the sorting key can be assigned by a domain expert [16,17]. For example, for the example dataset, we can select 'First Name' as the sorting key. As the LSH blocks are formed by the random permutation of elements, the common elements of the records in block B_i may be different from those in another block B_j. Using a predefined fixed sorting key may result in a whole block being returned as query results, which will fail to return a sub-set of records in a block. On the other hand, we can select the sorting key adaptively for each LSH block B_i individually.

One or several attributes can be selected as sorting key. For a block B_i, a good sorting key should divide the records into small sub-blocks. Thus, we can select those attributes that have a greater number of unique values in γ records. Moreover, if the attribute value occurrences are uniformly distributed, we can get sub-blocks with the same or similar sizes. Thus, for an attribute a_j, we calculate a sorting key selection weight w_j, which consists of the linear combination of two components: attribute cardinality weight and distribution weight:

$$w_j = \alpha \cdot \frac{n_j}{\gamma} + (1 - \alpha) \cdot \sigma_j \qquad (2)$$

Where $0 \leq \alpha \leq 1$, and n_j is the number of unique values of attribute a_j in block B_i. $\frac{n_j}{\gamma}$ measures the attribute cardinality weight, $0 \leq \frac{n_j}{\gamma} \leq 1$. σ_j measures the distribution of occurrences of each unique value of a_j, calculated as the standard deviation of the occurrences of the value of a_j. Let o_{jc} denote the occurrence of attribute value v_{jc} in γ records, m_a denote the maximum occurrence, $m_a = \max_{b \in [1,K]}(o_{jb})$,where K is the number of unique attribute values of attribute j in γ records, m_i denotes the minimum occurrence, $m_i =$

$\min_{b\in[1,K]}$. To get a normalized weight between 0 and 1, we set $n_{jc} = \frac{o_{jc}-m_i}{m_a-m_i}$, thus, $\sigma_j = \frac{1}{K}\sqrt{\Sigma_{c=1}^{K}(n_{jc}-\mu)^2)}$, where μ is the average value of n_{jc} for a_j, $\mu = \frac{\Sigma_{c=1}^{K}o_{jc}}{K}$.

The attribute a_j that has the largest w_j will be selected as the sorting key. Let sk denote a selected sorting key for B_i, then the $B+$ tree is built by sorting the records in B_i on key sk. For text records, we can sort them lexically by the alphabetic order of the sorting key. One advantage of sorting by alphabetic order is that such an ordering is a global unique order for all the attribute values of the sorting key in B_i. The distance of any two attribute values of the sorting key can be measured by the distance of their alphabetic order. Thus, for a query record, a set of nearest neighbors can be obtained through measuring their alphabetic order distance. Each unique sorting key value denoted as v_{sk} is one node of the $B+$ tree. Each node is an inverted index that points to the records that have the same v_{sk}, denoted as $dt_i = (v_{sk}, I(v_{sk}))$, where v_{sk} is an attribute value of sk, and $I(v_{sk})$ is the set of records that have the same v_{sk}. Let $|B_i|$ denote the number of records in a block B_i, the time complexity of searching, insertion and deletion of $B+$ tree is $O(\log|B_i|)$, which is quicker than scanning all the records in a block, $O(\log|B_i|) < O(|B_i|)$, for large blocks.

Selecting Nearest Neighbors. Every record in a block B_i can be selected as a candidate record for query record q_i. However, for those blocks that have a large number of records, we can select a set of nearest neighbors as candidate records to reduce the query time. For a given query record q_i, we firstly insert it into the $B+$ tree of a LSH block based on the alphabetic order of the block's sorting key sk. Then the nearest neighbor nodes of the $B+$ tree will be selected as the candidate records for this query record. Let $v_{i,sk}$ denote the sorting key value of query record q_i, we choose z_r nodes that are greater than $v_{i,sk}$ and z_l nodes that are smaller than $v_{i,sk}$ to form the nearest neighbor nodes of $v_{i,sk}$, $0 \leq z_l + z_r \leq |B_i|$. How to set the z_l and z_r value is important.

If we set z_l and $z_r=0$, then only those records with the exact same value as the query record will be selected. If z_l or $z_r = |B_i|$, then all the records of B_i will be selected. As the distance of two nodes in a sorting $B+$ tree reflects the distance of two records, we set a distance threshold θ of two nodes. Let $v_{i,sk}$ denote the attribute value of sorting key sk for query record q_i. For a node $v_{j,sk}$ of a given $B+$ tree, let $D(v_{i,sk}, v_{j,sk})$ denote the distance of $v_{i,sk}$ and $v_{j,sk}$ for a given distance measure D (e.g., edit distance [1]). If the distance of two nodes is less than θ, then we increase the window size to include node $v_{j,sk}$ inside the window. Thus, we firstly set z_l and $z_r = 0$. The window size expansion is along both directions [16] (i.e., greater than or smaller than the query record node i) of the $B+$ tree. For each direction, we expand the window size (i.e., z_l or z_r) and include neighbor node j, if $D(v_{i,sk}, v_{j,sk}) < \theta$, with $0 \leq \theta \leq 1$.

To further decrease the number of candidate records and select a smaller set of nearest neighbors, we count the collision number of each record of the neighboring nodes inside the window of the sorting tree in all l LSH blocks to rank the records that are attached to the selected neighboring nodes. This is because the co-occurrence of a record r_x that appears together with q_i in the LSH blocks reflects the similarity of the two records [14]. The higher the co-occurrence is, the more similar the two records are. We set a threshold φ to select those records that appear at least φ times with the query record q_i together in LSH blocks. Let g_{ix} denote the co-occurrence of record r_x and query record q_i. Let $N_{v_{i,sk}}$ denote the nearest neighbor record set of query record q_i in block B_i. For each record r_x of the neighbor node j inside the window of the sorting tree DST_i(i.e., $r_x \in I(v_{j,sk})$), we add r_x to $N_{v_{i,sk}}$ if $g_{ix} > \varphi$, $0 \leq \varphi \leq l$.

[**Example 3**] (Dynamic sorting tree) Figure 2(c) shows the dynamic sorting tree $DST_{3.2}$ for block 3.2. $DST_{3.2}$ is sorted by attribute 'First Name'. Through setting window size parameters, we can get a set of nearest neighbor records for a query record. For example, if we set $z_r = 0$, no records will be selected as candidate records, and only the records with node value 'Tonny' will be selected. Assume $\theta = 0.6$, $\varphi = 1$, the Jaccard similarity of the bi-grams of 'Tony' and 'Tonny' is 0.75 and that of the bi-grams of 'Tonny' and 'Yong' is 0.167. Thus, we can set $z_l = 0$ and $z_r = 1$ for $DST_{3.2}$. Record r_1 appears twice together with query record r_4 in blocks 3.2 and 1.6, thus r_1 is selected.

4.3   Real-Time Entity Resolution

For a query record q_i, we can obtain the nearest neighbor records that are being allocated in the same block with q_i as the candidate records. Then, we can conduct pair-wise comparisons for all candidate records with the query record q_i. We use the Jaccard similarity of the n-grams of a candidate record and the query record, or other appropriate approximate distance/similarity measures to rank their similarity [1]. Let C_{q_i} denote the candidate record set, for each candidate record $r_j \in C_{q_i}$ and query record $q_i \in Q$, the similarity can be calculated with $sim(q_i, r_j) = J(q_i, r_j)$. The top N candidate records will be returned as the query results L_{q_i}. The algorithm is shown as Algorithm 1.

5   Experiments and Results

5.1   Data Preparation

To evaluate the proposed approach, we conducted experiments on the following two datasets.

1) **Australian Telephone Directory.** [5] (named OZ dataset). It contains first name, last name, suburb, and postcode, and is sourced from an Australian telephone directory from 2002 (Australia On Disc). This dataset was modified by introducing various typographical and other variations to simulate real 'noisy' data. To allow us to evaluate scalability, we generated three sub-sets of this dataset. The smallest dataset (named OZ-Small) has 34,596 records, the medium sized dataset (OZ-Median) has 345,996 records, and the largest dataset (OZ-Large) has 3,458,758 records. All three datasets have the same features including similarity distribution, duplicate percentages (i.e., 25%) and modification types.

Algorithm 1: Query(q_i, $H_{k,l}$, N)

Input:
- $q_i \in Q$ is a given query record
- $H_{k,l}$ is a given LSH blocking schema
- N is a given number of returned results

Output:
- L_{q_i} is the ranked list of retrieved records

1: $L_{q_i} \leftarrow \{\}$, $C_{q_i} \leftarrow \{\}$ // Initialization, C_{q_i} is the candidate records set
2: $B_{q_i} \leftarrow H_{k,l}(gram(q_i, n))$ // Conduct LSH blocking for the n-grams of q_i
3: **For** each block $B_{bid} \in B_{q_i}$:
4:     If $|B_{bid}| < \gamma$:
5:         $C_{q_i} \leftarrow C_{q_i} \bigcup B_{bid}$ // Get all records in B_{bid} as candidate records
6:     If $|B_{bid}| = \gamma$: // Build sorting tree and select nearest neighbor records. See Section 4.2.
7:         Get sorting key sk
8:         Build dynamic sorting tree DST_{bid}
9:         Get nearest neighbours $N_{v_{i,sk}}$
10:       $C_{q_i} \leftarrow C_{q_i} \bigcup N_{v_{i,sk}}$
11:     If $|B_{bid}| > \gamma$: // Select nearest neighbor records. See Section 4.2.
12:         Insert q_i to DST_{bid}
13:         Get nearest neighbours $N_{v_{i,sk}}$
14:         $C_{q_i} \leftarrow C_{q_i} \bigcup N_{v_{i,sk}}$
15: **For** each candidate record $r_j \in C_{q_i}, r_j \neq q_i$:
16:     Get $sim(q_i, r_j)$ // Conduct pair-wise similarity comparisons
17: $L_{q_i} \leftarrow max\{C_{q_i}, N\}$ // Return top N ranked results

---

2) **North Carolina Voter Registration Dataset.** (i.e., NC dataset). This dataset is a large real-world voter registration database from North Carolina (NC) in the USA [18]. We downloaded this database every two months since October 2011. The attributes used in our experiments are: first name, last name, city, and zip code. The entity identification is the unique voter registration number. This data set contains 2,567,642 records. There are 263,974 individuals (identified by their voter registration numbers) with two records, 15,093 with three records, and 662 with four records.

5.2 Evaluation Approaches

In the experiments, we employ the commonly used Recall, Memory Cost and Query Time to measure the effectiveness and efficiency of real-time top N entity resolution approach [1]. We divided each dataset into a training (i.e., building) and a test (i.e., query) set. Each test dataset contains 50% of the whole dataset. For each test query record, the entity resolution approach will generate a list of ordered result records. The top N records (with the highest rank scores) will be selected as the query results. If a record in the results list has the same entity identification as the test query record, then this record is counted as a hit (i.e., an estimated true match). The Recall value is calculated as the ratio of the total number of hits of all the test queries to the total number of true matches in the test query set. We conducted comparison experiments with three other state-of-the-art methods as described below. All methods were implemented in Python, and the experiments were conducted on a server with 128 GBytes of main memory and two 6-core Intel Xeon CPUs running at 2.4 GHz.

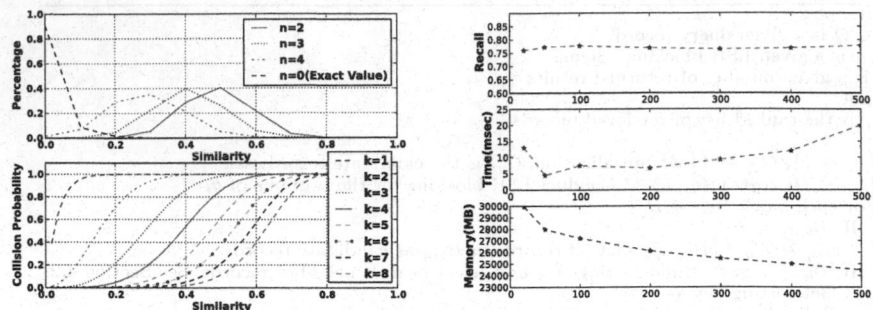

Fig. 3. The similarity distribution of various n values, and the collision probability
of various k values with $l=30$ for OZ-Large

Fig. 4. Top $N=10$ results of NAB with various γ values for the OZ-Large, with the
x-axis showing values for γ

- **NAB**: This is the proposed noise-tolerant approximate blocking approach
  that includes LSH and dynamic sorting trees.
- **DCC**: This is a locality sensitive hashing approach that uses dynamic collision counting [14] approach. It uses a base of length 1 basic hash functions.
- **LSH**: This is the basic LSH approach. It scans the data points in each block
  and conducts pair-wise comparison for all the records inside the blocks.
- **SAI**: This approach pre-calculates the similarity of attribute values to decrease the number of comparisons at query time [5,6].

5.3   Parameter Setting

We firstly discuss the parameter setting for the OZ datasets. To set the parameters k and l, we calculated the Jaccard similarity distribution of the exact values
and n-grams with $n=2$, 3, 4 of the true matched records of the training sets.
The similarity distribution is shown in Figure 3. The Jaccard similarity of 90%
of the exact values of the true matched records is zero. This means that it would
be very difficult to find true matched records if we use the exact value of the
records. Also, the similarity range of the majority (i.e., 95%) of the 2-grams of
the true matched records is between 0.3 and 0.7. Thus, using an n-gram based
approach can help to find those true matched records that contain small variations or errors. n is therefore set to 2 in the experiments. Figure 3 also shows
the collision probability of various k values with $l=30$. We set $k=4$ and $l=30$
for the NAB approach to let most true matched records have a higher collision
probability. To get a similar collision probability, we set $k=4$ and $l=30$ for the
LSH approach, and $k=1$ and $l=20$ for the DCC approach. The settings of the
other parameters of the NAB approach are $\alpha=0.5$, $\theta=2$, $\varphi=0.1$.

Figure 4 shows the top $N=10$ evaluation of NAB for the OZ-Large dataset
with various γ value. With various γ value, recall remains stable while average
query time is increasing, with the increase of γ. This shows that building dynamic

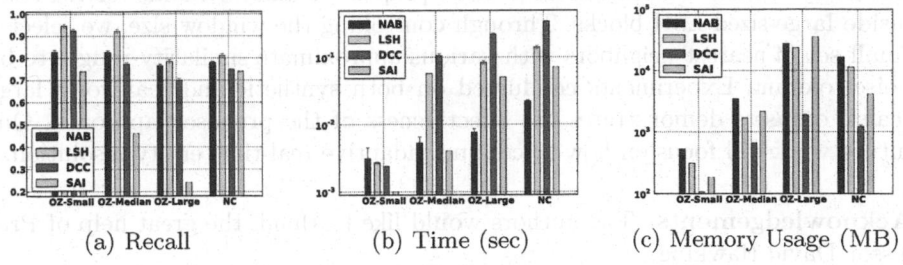

(a) Recall   (b) Time (sec)   (c) Memory Usage (MB)

Fig. 5. The top $N=10$ Recall, Average Query Time and Memory Usage results

sorting trees inside LSH blocks can help to decrease the query time through selecting a small number of nearest neighbor records as candidate records. A small γ value (e.g., $\gamma=20$) will not necessarily decrease the average query time, as the building of trees also takes time and space. When γ is set to a large value (e.g., $\gamma=500$), the average query time increases because scanning and comparing a large number of candidate records is time consuming. We set $\gamma=200$. For the NC dataset, we set $n=3$, $k=2$, $l=20$ for LSH and NAB, and $k=1$, $l=10$ for DCC. The other parameters for NAB are the same as with the OZ datasets.

5.4 Comparison with Baseline Models

The performances of the compared approaches are shown in Figure 5. To eliminate the influence of random permutation, the LSH based approaches LSH, DCC and NAB were run three times. The average query time and memory usage are shown on a logarithmic scale. From Figure 5 we can see that SAI achieved good recall value for OZ-Small and NC data but low recall for OZ-Large. The LSH based approaches (LSH, DCC and NAB) had higher recall than SAI. This can be explained that these approaches can capture the common elements (i.e., n-grams) of the attribute values to deal with the noise of the data. Moreover, through controlling the k and l value, these approaches can filter out the records that have lower similarities with the query record. DCC had very low memory usage but high average query time. NAB had slightly lower recall and higher memory usage than that of LSH, but with the help of dynamic sorting trees, the average query time (e.g., 8 msec for OZ-Large) is much lower than LSH (e.g., 0.1 sec for OZ-Large). Thus, NAB can be effectively and efficiently used for large scale real-time entity resolution, especially for noisy data.

6   Conclusions

We discussed a noise-tolerant approximate blocking approach to facilitate real-time large scaled entity resolution. To deal with noise, we use an LSH approach to group records into blocks with various distance ranges based on the Jaccard similarity of their n-grams. Moreover, we propose to build dynamic sorting trees inside large-sized LSH blocks. Through controlling the window size, we select a small set of nearest neighbors with various approximate similarity ranges to be noise-tolerant. Experiments conducted on both synthetic and real-world large scaled datasets demonstrates the effectiveness of the proposed approach. Our future work will focus on how to conduct adaptive real-time entity resolution.

Acknowledgements. The authors would like to thank the great help of Professor David Hawking.

References

1. Christen, P.: Data Matching. Data-Centric Systems and Appl. Springer (2012)
2. Christen, P., Gayler, R.W.: Adaptive temporal entity resolution on dynamic databases. In: Pei, J., Tseng, V.S., Cao, L., Motoda, H., Xu, G. (eds.) PAKDD 2013, Part II. LNCS, vol. 7819, pp. 558–569. Springer, Heidelberg (2013)
3. Lange, D., Naumann, F.: Cost-aware query planning for similarity search. Information Systems, 455–469 (2012)
4. Bhattacharya, I., Getoor, L., Licamele, L.: Query-time entity resolution. In: SIGKDD, pp. 529–534 (2006)
5. Christen, P., Gayler, R., Hawking, D.: Similarity-aware indexing for real-time entity resolution. In: CIKM, pp. 1565–1568 (2009)
6. Ramadan, B., Christen, P., Liang, H., Gayler, R.W., Hawking, D.: Dynamic similarity-aware inverted indexing for real-time entity resolution. In: Li, J., Cao, L., Wang, C., Tan, K.C., Liu, B., Pei, J., Tseng, V.S. (eds.) PAKDD 2013 Workshops. LNCS (LNAI), vol. 7867, pp. 47–58. Springer, Heidelberg (2013)
7. Gionis, A., Indyk, P., Motwani, R., et al.: Similarity search in high dimensions via hashing. In: VLDB, pp. 518–529 (1999)
8. Kim, H.S., Lee, D.: HARRA: Fast iterative hashed record linkage for large-scale data collections. In: EDBT, pp. 525–536 (2010)
9. Bawa, M., Condie, T., Ganesan, P.: LSH forest: Self-tuning indexes for similarity search. In: WWW, pp. 651–660 (2005)
10. Lv, Q., Josephson, W., Wang, Z., Charikar, M., Li, K.: Multi-probe LSH: Efficient indexing for high-dimensional similarity search. In: VLDB, pp. 950–961 (2007)
11. Das Sarma, A., Jain, A., Machanavajjhala, A., Bohannon, P.: An automatic blocking mechanism for large-scale de-duplication tasks. In: CIKM, pp. 1055–1064 (2012)
12. Li, L., Wang, D., Li, T., Knox, D., Padmanabhan, B.: Scene: A scalable two-stage personalized news recommendation system. In: SIGIR, pp. 125–134 (2011)
13. Anand, R., Ullman, J.D.: Mining of massive datasets. Cambridge University Press (2011)
14. Gan, J., Feng, J., Fang, Q., Ng, W.: Locality-sensitive hashing scheme based on dynamic collision counting. In: SIGMOD, pp. 541–552 (2012)
15. Michelson, M., Knoblock, C.A.: Learning blocking schemes for record linkage. In: AAAI, pp. 440–445 (2006)
16. Yan, S., Lee, D., Kan, M.Y., Giles, L.C.: Adaptive sorted neighborhood methods for efficient record linkage. In: DL, pp. 185–194 (2007)
17. Draisbach, U., Naumann, F., Szott, S., Wonneberg, O.: Adaptive windows for duplicate detection. In: ICDE, pp. 1073–1083 (2012)
18. Christen, P.: Preparation of a real voter data set for record linkage and duplicate detection research. Technical report, Australian National University (2013)

A Relevance Weighted Ensemble Model
for Anomaly Detection in Switching Data Streams

Mahsa Salehi[1], Christopher A. Leckie[1],
Masud Moshtaghi[2], and Tharshan Vaithianathan[3]

[1] National ICT Australia, Department of Computing and Information Systems,
The University of Melbourne, Victoria 3010, Australia
msalehi@student.unimelb.edu.au, caleckie@unimelb.edu.au
[2] Faculty of Information Technology,
Monash University, Victoria 3145, Australia
masud.moshtaghi@monash.edu
[3] National ICT Australia, Department of Electrical and Electronic Engineering,
The University of Melbourne, Victoria 3010, Australia
tharshan@nicta.com.au

Abstract. Anomaly detection in data streams plays a vital role in online data mining applications. A major challenge for anomaly detection is
the dynamically changing nature of many monitoring environments. This
causes a problem for traditional anomaly detection techniques in data
streams, which assume a relatively static monitoring environment. In an
environment that is intermittently changing (known as switching data
streams), static approaches can yield a high error rate in terms of false
positives. To cope with dynamic environments, we require an approach
that can learn from the history of normal behaviour in data streams,
while accounting for the fact that not all time periods in the past are
equally relevant. Consequently, we have proposed a relevance-weighted
ensemble model for learning normal behaviour, which forms the basis of
our anomaly detection scheme. The advantage of this approach is that
it can improve the accuracy of detection by using relevant history, while
remaining computationally efficient. Our solution provides a novel contribution through the use of ensemble techniques for anomaly detection in
switching data streams. Our empirical results on real and synthetic data
streams show that we can achieve substantial improvements compared
to a recent anomaly detection algorithm for data streams.

Keywords: Anomaly detection, Ensemble models, Data streams.

1 Introduction

Anomaly detection (also known as novelty or outlier detection) is an important
component of many data stream mining applications. For example, in network
intrusion detection, anomaly detection is used to detect suspicious behaviour
that deviates from normal network usage. Similarly, in environmental monitoring, anomaly detection is used to detect interesting events in the monitored

V.S. Tseng et al. (Eds.): PAKDD 2014, Part II, LNAI 8444, pp. 461–473, 2014.

environment, as well as to detect faulty sensors that can cause contamination of the collected data. A major challenge for anomaly detection in applications such as these is the presence of nonstationary behaviour in the underlying distribution of normal observations. In addition, the high rate of incoming data in such applications adds another challenge to anomaly detection, as the anomalous data points should be detected in a computationally efficient way with high accuracy. Many anomaly detection techniques are based on an assumption of stationarity. However, such an assumption can result in low detection accuracy when the underlying data stream exhibits nonstationarity. In this paper, we present a novel method for robust anomaly detection for a special class of nonstationarity, known as switching data streams.

Many environments are intermittently changing, and thus exhibit piecewise stationarity. For example, sensors in an office environment may collect observations with different underlying statistical distributions between day-time and night-time. We refer to the data from such environments as switching data streams. In addition to large-scale switching behaviour, there can also be finer-scale variation in the distribution of observations, e.g., variation during day-time observations. To address these challenges, we propose a novel approach to anomaly detection that can selectively learn from previous time periods in order to construct a model of normal behaviour that is relevant to the current time window. This is achieved by maintaining a cluster model of normal behaviour in each previous time period, and then constructing an ensemble model of normal behaviour for the current period, based on the relevance of the cluster models from previous time periods to the current time period. This relevance-weighted ensemble model of normal behaviour can then provide a more robust model of normality in switching data streams.

While ensemble methods have been widely used in supervised learning [1,2], their use for unsupervised anomaly detection in data streams is still an open research challenge. A key advantage of our approach is that it can improve the accuracy of anomaly detection in switching data streams. We have evaluated the effectiveness of our approach on large-scale synthetic and real sensor network data sets, and demonstrated its ability to adapt to the intermittent changes in switching data streams. We show that our approach achieves greater accuracy compared to a state-of-the-art anomaly detection approach for data streams.

2   Related Work

A number of surveys [3,4] have categorized various techniques for anomaly detection. Most focus only on static environments and do not consider dynamic behaviour. In dynamic environments an unknown volume of data (data streams) are produced. Hence the major challenge is how to detect anomalies in such environments in the presence of dynamics in the distribution of the data stream. Based on existing surveys of anomaly detection in data streams [5,6], we categorize these methods into *Model based* and *Distance based* approaches.

Model based approaches build a model over the data set and update it as data evolves with each incoming data point. In this category, some authors learn a probabilistic model. This requires a priori knowledge about the underlying distribution of data, which is not appropriate if the distribution is not known [7,8]. Other approaches use clustering algorithms to build a model for normal behaviour in data streams. However, they are not really designed for anomaly detection [9,10]. An exception is [11], in which a segment based approach is proposed. It uses k-means as a base model and provides guidelines regarding how to use the proposed approach for anomaly detection. As a result, it assumes the number of clusters is known, which may not be the case in changing environments (since the number of clusters may change over time). In addition, while these clustering techniques work for gradually evolving data streams, they are not applicable in switching environments, which is our focus in this paper.

Distance based approaches are the second category of anomaly detection techniques in data streams. We have two types of distance-based outliers: 'global' and 'local'. Distance-based 'global' outliers are first introduced by [12], where a data point x is a distance-based outlier if less than k data points are within a distance R from it. In [13] and [14], a sliding window is used to detect such global outliers. Since parameter R is fixed for all portions of the data, these approaches fail to detect anomalies in non-homogenous densities. In contrast, distance-based 'local' outliers are data points that are outliers with respect to their k nearest neighbours without considering any distance R. Local outliers are first introduced in [15] and a measure of being an outlier - the Local Outlier Factor(LOF) - is assigned to each data point in a static environment. Later, in [16] and [6] two similar approaches are proposed to find local outliers in data streams. While the former made an assumption about the underlying distribution of data, the latter (Incremental LOF) extended [15], and could detect outliers in data streams by assigning the LOF to incoming data points and updating the k nearest neighbors. However, there are still some limitations with this approach in the presence of switching data streams: 1) It has problems with detecting dense drift regions in data streams, which results in false negatives. 2) In switching data streams, the distribution of normal data can change suddenly. Since incremental LOF keeps all of the history of the data points, it could not differentiate between different states, which again results in false negatives (e.g., a data point is an anomaly in one state while it is not an anomaly in another). 3) It is hard to choose the parameter value k in the presence of changing distribution environments. We propose to address these open problems for anomaly detection in switching data streams by using an ensemble approach.

Ensemble approaches have been shown to have benefits over using a single classifier, particularly in dynamic environments. So far, several ensemble learning methods have been proposed for data stream classification [1,2]. In contrast, there is little work done on anomaly detection using ensemble techniques [17]. In a recent comprehensive survey paper [18], a categorization of outlier detection ensemble techniques is based on the constituent components (data/model) in these techniques. Nevertheless, none of these approaches are aimed at handling switching data streams or even streaming data.

Since incremental LOF (iLOF) is the only anomaly detection technique in streaming data which handles different densities and is reported to detect changes in data distributions, we use it as a baseline to evaluate our proposed approach.

3   Our Methodology

In this section, we formally define our problem and proposed algorithm.

3.1   Problem Definition

We begin by describing our notation for switching data streams. We consider the problem of anomaly detection in a switching data stream, where the underlying distribution of observations is only piecewise stationary. That is, the monitored environment switches between a number of "normal" states, such as day vs night in an episodic manner. Let the state of the system be a random variable over the domain of possible states $S = \{s_1, ..., s_D\}$, where the system has D normal states. The distribution of an observation X in state s_d is drawn from a mixture of K_d components. For example, each component of the mixture distribution of state s_d could be a multivariate Gaussian distribution with different means and covariance. Let $\Theta = \{\theta_{d,i}, i = 1, .., K_d\}$ denote the parameters corresponding to the K_d components of state s_d, e.g., $\theta_{d,1} = \{\mu_{d,1}, \Sigma_{d,1}\}$ for a multivariate Gaussian distribution. Further, let $\Phi = \{\phi_{d,i}, i = 1, .., K_d\}$ denote the mixture weights of the K_d components of state s_d, i.e., the prior probability that a random observation in state s_d comes from component i is $\phi_{d,i}$.

We observe a stream of observations, where $X(1 : T)$ denotes the sequence of observation vectors collected over the time period $[t_1, ..., t_T]$. At time t, we receive a vector of observations $X(t) = [x_1(t), .., x_P(t)]^T$ corresponding to P different types of observation variables, e.g., temperature, pressure, humidity, etc, where $X(t) \in \Re^P$. If the environment is in state s_d at time t, then $X(t)$ is sampled from one of the K_d component distributions in Θ_d of state s_d.

The monitored environment is initially in some random state $s_{(1)} \in S$ at time t_1, and remains in that state until a later time t_{c1} when it switches to a different state $s_{(2)} \in S \backslash s_{(1)}$. It then remains in state $s_{(2)}$ until a later time t_{c2} when it switches again to $s_{(3)} \in S \backslash s_{(2)}$, and so on. Our aim is to detect anomalies in this data stream X. In order to detect anomalies, we require a model of "normal" behaviour for the environment, given that the number of possible states, the number and mixture of components in each state, and the parameters of each component in each state are unknown a priori.

In order to learn our model of normal behaviour, we could estimate all of the different component parameters for all states by keeping all data $X(1 : t)$. However, this is impractical due to the memory requirements and it would mix all the states together when building the normal model. Alternatively we can learn only from the window of the w most recent observations $X(t - w + 1 : t)$.

While this is more computationally efficient, it only provides a limited sample of measurements from the current state. If the window size w is small, then this might yield a noisy estimate of the component distribution parameters.

The problem we address is how to find a balance between these two extremes, by maintaining multiple models of normal behaviour based on previous data windows. Our expectation is that this will provide us with a balance between minimising the memory requirements for our anomaly detection scheme while maximising accuracy. However, the key challenge in achieving this balance is that not all previous windows will be relevant to the current window of observations, due to the switching nature of the data stream. Consequently, we require a method to take into consideration the relevance of previous windows, so that we can construct a model of normal behaviour based on the current and previous windows. In particular, we need a way to weight the influence of previous windows on data in the current window, based on the degree of relevance of those previous windows. We consider different kinds of anomalies in the system: either localized in time, i.e., a burst of noise or a drift in the data stream, or uniformly distributed over the data stream.

3.2   Our Ensemble-Based Algorithm

We now describe the main steps of our algorithm for ensemble anomaly detection in switching data streams. Algorithm 1 shows the pseudo code of the ensemble algorithm, comprising three main steps: *Windowing, Weighting* and *Ensemble* formation. We assume the previous data streams have already been clustered based on windows of w observations and the problem is how to detect anomalies in the most recent (current) window of w observations.

1. *Windowing Step*: In general, in applications that generate data steams, the observations arrives sequentially. We consider a data stream that switches between different states, where each state comprises different component distributions. By breaking the whole data stream of observations into windows of size w, we can extract the different underlying distributions of the data streams.

In this step, we cluster the current w observations by using an appropriate clustering algorithm, in order to find out the current underlying component distributions. We chose the HyCARCE clustering algorithm [19], which is a computationally efficient density-based hyperellipsoidal clustering algorithm that automatically detects the number of clusters, and only requires an initial setting for one input parameter, i.e., the grid-cell size. In addition, it has a lower computational complexity in comparison to existing methods. Hence, it is an appropriate clustering algorithm for our data stream analysis problem in which time is a vital issue and the number of clusters is unknown. Interested readers can find a detailed description and pseudo code for the HyCARCE algorithm in [19]. At the end of this step a clustering model is built that comprises a set of cluster boundaries, $C_\kappa = \{c_{\kappa,1}, ..., c_{\kappa,|C_\kappa|}\}$. In this paper we use the boundaries of the ellipsoidal clusters as our decision boundaries.

Input : $W_\kappa = \{X(t - w + 1), ..., X(t)\}$: most recent window of w observations
from the data stream
$\mathcal{E} = \{C_1, C_2, ..., C_{\kappa-1}\}$: set of clusterings corresponding to previous
windows $\{W_1, ..., W_{k-1}\}$, where a clustering $C_j = \{c_{j,1}, ..., c_{j,|C_j|}\}$
Output: $A_\kappa = \{a_1, ..., a_w\}$: set of anomaly scores in most recent window W_κ
1 $C_\kappa \longleftarrow Cluster(W_\kappa)$; // cluster W_κ using the HyCARCE clustering [19]
2 $R \longleftarrow \{\}$; // compute relevance of previous clustering to current one
3 **foreach** *clustering* $C_j \in \mathcal{E}$ **do**
4 | $R \longleftarrow R \cup \{Relevance(C_\kappa, C_j)\}$; // Algorithm 2
5 **end**
6 $R \longleftarrow R \cup \{max(R)\}$; // $R = \{r_1, ..., r_\kappa\}$
7 $\mathcal{E} \longleftarrow \mathcal{E} \cup \{C_\kappa\}$;
   // test each observation in W_κ to check if it is an anomaly
8 **foreach** $X_n \in W_\kappa$ **do**
    // test if X belongs to any cluster in each clustering C_j
    // belongs is defined in Definition 1 using Equation 1 and 2
9   **foreach** $C_j \in \mathcal{E}$ **do**
10   **if** $\exists c_{j,i} \in C_j - belongs\,(X, c_{j,i})$ **then**
11 | $b_j = 0$; // X is normal in clustering C_j
12   **else**
13 | $b_j = 1$; // X is anomalous in clustering C_j
14   **end**
15   **end**
16    $a_n = \dfrac{\sum_{j=1}^{\kappa} b_j r_j}{\sum_{j=1}^{\kappa} r_j}$; [1]
17    $A \longleftarrow A \cup \{a_n\}$;
18 **end**
19 **return** A;

Algorithm 1. Ensemble Anomaly Detection

2. *Weighting Step*: According to our windowing step, all of the previous w sized observation windows have already been clustered. The history is used to better estimate the underlying distribution of the current window. However, not all of the previous clusterings have the same level of importance, and some of them might be more relevant to the current window, as the data stream switches between different states. Our solution, which is called the *Weighting step*, is to assign weights to previous clustering models based on the similarity between the current and previous clustering models.

In this step, the distance between each previous clustering model (C_j) and the current clustering model (C_κ) is computed. As time complexity is a major issue for anomaly detection on data streams, a *greedy* approach is proposed to compute this distance. In this approach, the distance between each pair of cluster boundaries in C_κ and C_j is computed based on their *focal distance* [20]. Focal distance is a measure of the distance between two hyperellipsoids considering their shapes, orientations and locations. According to a recent work [20], this measure works well for computing the similarity between hyperellipsoids.

```
    Input  : C_j = {c_{j,1}, ..., c_{j,|C_j|}}: previous clustering
             C_κ = {c_{κ,1}, ..., c_{κ,|C_κ|}}: current clustering
    Output: r_j: the similarity between C_j and C_κ
    // λ and γ are smallest/equal and largest clustering respectively
 1  if |C_j| ≤ |C_κ| then
 2  |    λ ⟵ j; γ ⟵ κ;
 3  else
 4  |    λ ⟵ κ; γ ⟵ j;
 5  end
 6  D ⟵ {}; // compute distance between every pair of hyperellipsoids
 7  foreach c_{λ,r} ∈ C_λ do
 8  |    foreach c_{γ,c} ∈ C_γ do
 9  |    |   d_{r,c} ⟵ FocalDistance(c_{λ,r}, c_{γ,c}); // Focal Distance[20]
    |    |   // D = {D_1, ..., D_{|C_λ|}} and D_i = {d_{i,1}, ..., d_{i,|C_γ|}}
10  |    end
11  end
12  dist = 0; // compute minimum distance based on a Greedy approach
13  while D ≠ ∅ do
14  |    d_{r,c}^{min} ⟵ find minimum element in D;
15  |    dist ⟵ dist + d_{r,c}^{min}
16  |    foreach D_i ∈ D do
17  |    |   D_i ⟵ D_i - {d_{i,c}}; // removing relevant assigned clusters
18  |    end
19  |    D ⟵ D - {D_r};
20  end
21  r_j = (1/dist); return r_j;
```

Algorithm 2. Relevance Function

After determining the focal distance, the algorithm finds the minimum distance among all computed distances, and assigns the relevant cluster boundaries as a matching pair. It continues this process, until all of the clusters in at least one of the two clustering models are each matched to a corresponding cluster in the other model. The sum of the found minimum distances is used as the distance between two clustering models. Obviously, if the models are less distant, they would be more similar. Therefore, the distances are reversed at the end of the algorithm to show the similarity. The relevant pseudo code is described in Algorithm 2. Finally, in order to use the current clustering model in the ensemble step, the maximum assigned similarity among all previous clusterings is assigned to the current clustering. In this way, we can find a balance between the current window and previous windows for anomaly detection.

3. *Ensemble Step*: In this step, anomaly detection is performed based on the current and previous clusterings. As discussed in the previous subsection, not all historical models are useful, due to the changing environment. Hence, for each current observation, we check if it belongs to a hyperellipsoid in each of the clustering models according to the following definition:

Definition 1. *The observation X **belongs** to hyperellipsoid $c_{j,i}$, if the Mahalanobis distance between them is less than a threshold t^2, where the Mahalanobis distance is:*

$$\mathcal{M}(X, c_{j,i}) = (X - \hat{\mu}_{j,i})^T \hat{\Sigma}_{j,i}^{-1}(X - \hat{\mu}_{j,i}) \tag{1}$$

where $\hat{\mu}_{j,i}$ and $\hat{\Sigma}_{j,i}$ are the mean and covariance of hyperellipsoid $c_{j,i}$ respectively, and the threshold is:

$$t^2 = (\chi_P^2)_\rho^{-1} \tag{2}$$

where t^2 is the inverse of chi-square statistic, ρ is the percentage of data covered by the ellipsoid, and P is the dimensionality of the observations.

Thereafter, the probability of being outlier is computed for the current observation according to the formula in Algorithm 1 (line 16). The formula is based on a relevance voting. The algorithm computes this probability for all data points X in the current window.

3.3   Time Complexity

The total number of observations (data points) N is divided into κ windows of size w. Let l be the average of number of ellipsoids in each C_j. Therefore l is a function of $K_d, d = \{1, .., D\}$ and $K_d \ll N$. The time complexity for the *Windowing Step* is $O(N)$ as the complexity of HyCARCE is near linear with respect to the number of data points. The *Weighting Step* can be computed in $O(\frac{N}{w}l^3)$, considering the 'Relevance' can be computed in $O(l^3)$ in the worst case. In the last step of the algorithm, 'belongs' can be computed in $O(l)$ so the time complexity of the *Ensemble Step* is $O(w\kappa l)$. As a result, the time complexity of Algorithm 1 is $O(N(\frac{l^3}{w} + l))$, which is near linear with respect to the number of data points. Hence, our ensemble approach is computationally efficient. Time complexity of iLOF is also near to linear depending on the parameter k.

4   Evaluation

In this section we aim to compare the accuracy, sensitivity and specificity of our ensemble model with the iLOF algorithm on several dynamic environments.

4.1   Data Sets

We use three different data sets to evaluate our approach.

Synthetic Data Set: In order to generate synthetic data sets, we consider a state machine that can simulate a switching data stream from a changing environment. We assume there are only two different states (S_1, S_2) and changing from one state to the other occurs periodically. The first state (s_1) has three underlying component distributions $\theta_{1,1}$, $\theta_{1,2}$ and $\theta_{1,3}$ with corresponding mixture weights $\phi_{1,1}$, $\phi_{1,2}$ and $\phi_{1,3}$. The second state (s_2) has two underlying component distributions $\theta_{2,1}$ and $\theta_{2,2}$ with mixture weights $\phi_{2,1}$ and $\phi_{2,2}$. Altogether the environment consists of five component distributions. The states are changing periodically based on a constant rate (every $m = 100$ samples). In addition, the distributions are hyperellipsoids and the initial mean and covariance is chosen randomly. As discussed earlier, from one instance of a state to the other, for the same state, we have some perturbations in the mean and covariance of the underlying distribution, i.e.,the parameters of $\theta_{d,i}$ can be slightly perturbed between different occurrences of state s_d. By using this state machine, we have generated 50,000 2-dimensional records. We generated 10 different data sets and averaged the final results. The inserted noise was 2%.

Real Data Sets: In order to evaluate our algorithm over a real data set, we used two public available data sets. These data sets contain periodic measurements over day (state s_1) and night (state s_1). The first is the IBRL (Intel Berkeley Research Lab) data set[1]. A group of 54 sensors were deployed to monitor an office environment, from Feb. 28th until Apr. 5th, 2004. Moreover, by visualising the data collected by all the sensors, we observed that sensor number 45 stopped working in the last two days causing a drift which is labeled as anomalies (4%). The sensors were collecting weather information. We chose two features, humidity and temperature. The measurements were taken every 31 seconds and there are about 50,000 records.

The second real data set is from the TAO (Tropical Atmosphere Ocean) project[2], by the Pacific Marine Environmental Lab of the U.S. National Oceanic and Atmospheric Administration. This monitors the atmosphere in the tropical Pacific ocean. We have used the period of time from Jan. 1st until Sep. 1st, 2006 which is used in [13]. We chose three features, precipitation, relative humidity and sea surface temperature. Among the different monitoring sites, we chose site:(2 °N,165 °E), since this site has all three features available for the mentioned period of time. The measurements were taken every 10 minutes and there are about 37,000 records. This data set has some labels on the quality of measurements and we have used them for our evaluation. After visualization of the low quality data points (2%) and good quality ones, we find that in this data set the noise is a dense separate region from the normal observations.

4.2   Performance Measures

In order to evaluate our algorithm in comparison to iLOF, we have used three performance measures: (1) Area under the ROC curve (AUC). (2) The *accuracy* ($\dfrac{TP + TN}{P + N}$) of the ROC curve's optimal point, where P is the number of positives (anomalies) in the data set, N is the number of negatives (normal observations), TP is the number of true positives (correctly reported anomalies) and TN is the number of true negatives (correctly reported normal observations).

---

[1] http://db.lcs.mit.edu/labdata/labdata.html
[2] www.pmel.noaa.gov/tao

The optimal point on a ROC curve is the point with the maximum distance from diagonal line using the Youden Index ($max_i \dfrac{SE_i + SP_i - 1}{\sqrt{2}}$), where SE_i is the sensitivity for the ith threshold and SP_i is the specificity for the ith threshold. (3) The ratio of correctively detected anomalies $sensitivity$ ($\dfrac{TP}{P}$) and correctly detected normal observations $specificity$ ($\dfrac{TN}{N}$) for the ROC curve's optimal points.

4.3   Results and Discussion

We have compared our ensemble approach with iLOF on the three test data sets. For iLOF we have used the implementation provided in ELKI[21]. All measurements were normalized based on min-max normalization and we set $\rho = 0.99$ in Equation 2 [19]. Moreover, we have studied how the performance of our approach varies over different window sizes. The window size w is initially set to 100 observations, and then it is increased by increments of 500 observations until w is $\frac{1}{3}$ of the whole data set length (because we need at least two window sizes for voting). We also studied the effect of changing the number of nearest neighbours k in the iLOF algorithm, where $k \in \{3, 5, 10, .., 200\}$. Since the computational time of the iLOF algorithm increases with k, we set the upper bound to 200 to obtain a reasonable runtime.

We have computed the ROC curves for different window sizes w and number of neighbours k. Figure 1 depicts only the best ROC curves (thicker graph) and worst ones (thinner graphs) for simplicity for both approaches over three different data sets. The results show that in the two real data sets with dense outliers, our approach is better than iLOF by a large margin for both the best and worst curves (Figure 1a and 1b). In addition in Figure 1b our worst curve is even better than iLOF's best curve. However, in the synthetic data set in which the outliers are uniformly distributed over the data stream, both approaches have approximately the same best curves, and our approach has better AUC in comparison to iLOF in the worst case (Figure 1c). In order to make a more detailed comparison, we computed the optimal points of different ROC curves

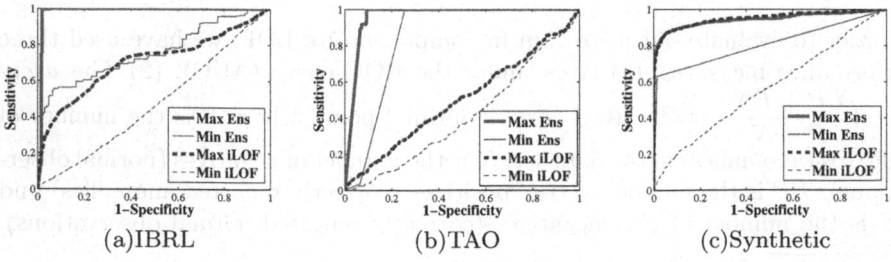

Fig. 1. ROC curves for three data sets: Ensemble (Ens) in red, iLOF in blue

Table 1. Optimal Point Comparison: Our Ensemble Method vs iLOF

Data Set	Algorithm	Accuracy			Specificity			Sensitivity		
		Max	Avg	Min	Max	Avg	Min	Max	Avg	Min
IBRL	Ensemble	98.23	90.62	74.90	98.18	90.22	73.60	100	98.39	55.45
	iLOF	93.68	76.46	07.73	98.06	78.36	03.30	95.81	39.74	03.28
TAO	Ensemble	90.81	86.57	74.62	90.73	86.45	74.40	100	100	100
	iLOF	88.42	84.98	53.38	88.92	76.09	53.37	54.72	39.89	26.06
Synthetic	Ensemble	98.17	97.80	94.42	98.87	98.27	94.56	87.06	72.61	48.66
	iLOF	94.49	90.82	84.74	94.64	91.04	85.51	88.98	79.09	21.82

as we described in Section 4.2. The results of the optimal point's accuracy, sensitivity and specificity are shown in Table 1.

The accuracy and specificity of the optimal points in our approach are higher than iLOF in all data sets. This means that iLOF produces more false negatives. In the IBRL and TAO data sets which are the real data sets, we have dense outliers (either drift or a separate distribution) which yields the false negatives. As can be seen the minimum specificity of iLOF in both real data sets is extremely low. Moreover, since in all three data sets we have switching states our ensemble algorithm can perform better in terms of specificity, whereas iLOF keeps all the history of the data points, and it could not differentiate between different states, which again results in false negatives. The last three columns of the table show the optimal point's sensitivity. The results show that we have much higher sensitivity in comparison to iLOF in the IBRL and TAO data sets and iLOF fails to detect anomalies with a considerably lower rate. However, in the synthetic data set with uniformly distributed anomalies, iLOF performs better in terms of finding anomalies in the average and maximum cases. However, the difference is small, and our ensemble method performs better in terms of overall accuracy. Finally, considering the max, average and min in all measures in Table 1, our method performs more consistently, which shows the results are less dependent on choosing the window size w. Also, for iLOF the range between the max and min cases is larger, indicating that it is sensitive to the choice of the parameter k.

In summary, our approach outperforms iLOF on the two real data sets with dense anomalies. Moreover, it is better than iLOF in accuracy and specificity on the synthetic data set, while iLOF only performs better in terms of sensitivity in the synthetic data set with uniformly distributed anomalies.

5   Conclusion

In this paper we proposed a novel approach to the problem of anomaly detection in data streams where the environment changes intermittently. We introduce an ensemble based approach to construct a robust model of normal behaviour in switching data streams. Although there have been many supervised techniques based on ensemble models for outlier detection in data streams, to the best of our knowledge there is no similar work that tackles the problem of using ensemble techniques for anomaly detection in data streams. The empirical results show the strength of our approach in terms of accuracy. This highlights several interesting directions for future research. First, we can explore alternatives to our greedy approach for comparing clusterings. Second, we will investigate approaches to minimize memory consumption. Third, we will investigate different methods for selecting appropriate window boundaries.

Acknowledgments. The authors would like to thank National ICT Australia (NICTA) for providing funds and support.

References

1. Wang, H., Fan, W., Yu, P.S., Han, J.: Mining concept-drifting data streams using ensemble classifiers. In: SIGKDD, pp. 226–235. ACM (2003)
2. Bifet, A., Holmes, G., Pfahringer, B., Kirkby, R., Gavaldà, R.: New ensemble methods for evolving data streams. In: SIGKDD, pp. 139–148. ACM (2009)
3. Rajasegarar, S., Leckie, C., Palaniswami, M.: Anomaly detection in wireless sensor networks. IEEE Wireless Communications 15(4), 34–40 (2008)
4. Chandola, V., Banerjee, A., Kumar, V.: Anomaly detection: A survey. ACM Computing Surveys 41(3), 1–58 (2009)
5. Gupta, M., Gao, J., Aggarwal, C.C., Han, J.: Outlier detection for temporal data: A survey. Knowledge and Data Eng. 25(1), 1–20 (2013)
6. Pokrajac, D., Lazarevic, A., Latecki, L.J.: Incremental local outlier detection for data streams. In: CIDM, pp. 504–515. IEEE (2007)
7. Yamanishi, K., Takeuchi, J.I., Williams, G., Milne, P.: On-line unsupervised outlier detection using finite mixtures with discounting learning algorithms. In: SIGKDD, pp. 320–324. ACM (2000)
8. Yamanishi, K., Takeuchi, J.I.: A unifying framework for detecting outliers and change points from non-stationary time series data. In: SIGKDD, pp. 676–681. ACM (2002)
9. Aggarwal, C.C., Han, J., Wang, J., Yu, P.S.: A framework for clustering evolving data streams. In: VLDB, pp. 81–92. VLDB Endowment (2003)
10. Cao, F., Ester, M., Qian, W., Zhou, A.: Density-based clustering over an evolving data stream with noise. In: SIAM Conf. on Data Mining, pp. 328–339 (2006)
11. Aggarwal, C.C.: A segment-based framework for modeling and mining data streams. Knowledge and Inf. Sys. 30(1), 1–29 (2012)
12. Knox, E.M., Ng, R.T.: Algorithms for mining distance-based outliers in large datasets. In: VLDB, pp. 392–403. Citeseer (1998)
13. Angiulli, F., Fassetti, F.: Detecting distance-based outliers in streams of data. In: CIKM, pp. 811–820. ACM (2007)
14. Yang, D., Rundensteiner, E.A., Ward, M.O.: Neighbor-based pattern detection for windows over streaming data. In: Advances in DB Tech., pp. 529–540. ACM (2009)
15. Breunig, M.M., Kriegel, H.P., Ng, R.T., Sander, J.: LOF: identifying density-based local outliers. In: ACM SIGMOD, vol. 29, pp. 93–104. ACM (2000)
16. Vu, N.H., Gopalkrishnan, V., Namburi, P.: Online outlier detection based on relative neighbourhood dissimilarity. In: Bailey, J., Maier, D., Schewe, K.-D., Thalheim, B., Wang, X.S. (eds.) WISE 2008. LNCS, vol. 5175, pp. 50–61. Springer, Heidelberg (2008)

17. Lazarevic, A., Kumar, V.: Feature bagging for outlier detection. In: SIGKDD, pp. 157–166. ACM (2005)
18. Aggarwal, C.C.: Outlier ensembles: Position paper. SIGKDD Explorations Newsletter 14(2), 49–58 (2013)
19. Moshtaghi, M., Rajasegarar, S., Leckie, C., Karunasekera, S.: An efficient hyperellipsoidal clustering algorithm for resource-constrained environments. Pattern Recognition 44(9), 2197–2209 (2011)
20. Moshtaghi, M., Havens, T.C., Bezdek, J.C., Park, L., Leckie, C., Rajasegarar, S., Keller, J.M., Palaniswami, M.: Clustering ellipses for anomaly detection. Pattern Recognition 44(1), 55–69 (2011)
21. Achtert, E., Goldhofer, S., Kriegel, H.P., Schubert, E., Zimek, A.: Evaluation of clusterings–metrics and visual support. In: ICDE, pp. 1285–1288. IEEE (2012)

MultiAspectSpotting: Spotting Anomalous Behavior within Count Data Using Tensor

Koji Maruhashi and Nobuhiro Yugami

Fujitsu Laboratories Ltd.
{maruhashi.koji,yugami}@jp.fujitsu.com

Abstract. Methods for finding *anomalous* behaviors are attracting much attention, especially for very large datasets with several attributes with tens of thousands of categorical values. For example, security engineers try to find *anomalous* behaviors, *i.e.*, remarkable attacks which greatly differ from the day's trend of attacks, on the basis of intrusion detection system logs with source IPs, destination IPs, port numbers, and additional information. However, there are large amount of *abnormal* records caused by noise, which can be repeated more *abnormally* than those caused by *anomalous* behaviors, and they are hard to be distinguished from each other. To tackle these difficulties, we propose a two-step anomaly detection. First, we detect *abnormal* records as individual anomalies by using a statistical anomaly detection, which can be improved by *Poisson Tensor Factorization*. Next, we gather the individual anomalies into groups of records with similar attribute values, which can be implemented by *CANDECOMP/PARAFAC (CP) Decomposition*. We conduct experiments using datasets added with synthesized anomalies and prove that our method can spot *anomalous* behaviors effectively. Moreover, our method can spot interesting patterns within some real world datasets such as IDS logs and web-access logs.

Keywords: anomaly detection, tensor decomposition.

1 Introduction

Our work is motivated by anomaly detection in datasets that have several attributes with tens of thousands of categorical values. We want to know the existence of *anomalous* behavior by finding *abnormal* records, *i.e.*, records strangely repeated or strangely less than expected. For example, an intrusion detection system (IDS) monitors network traffic for suspicious activity, and each record in IDS logs has attributes such as *srcIP*, *dstIP*, *port*, and *type* as shown in Table 1. A serious problem for analysts in charge of a company's security system is that IDS logs contain too many records to investigate all of them precisely. Therefore, it is important not only to determine *ordinary* behaviors, *i.e.*, the day's trend of attacks which changes rapidly day-to-day, but also to spot *anomalous* behaviors, *i.e.*, remarkable attacks which greatly differ from *ordinary* behaviors of the day, which are worth investigating. We can build a model of records caused by *ordinary* behavior under the assumption that the majority of records are caused by

V.S. Tseng et al. (Eds.): PAKDD 2014, Part II, LNAI 8444, pp. 474–485, 2014.

Table 1. Example of a dataset. (*a.a.a.a*, *b.b.b.b*, *80*, *type X*) were repeated two times.

source IP (srcIP)	destination IP (dstIP)	port number (port)	possible attack type (type)
a.a.a.a	b.b.b.b	80	type X (DOS attack)
a.a.a.a	b.b.b.b	80	type X (DOS attack)
c.c.c.c	d.d.d.d	12345	type Y (port scan)

ordinary behavior, and distinguish *anomalous* behaviors from them. One possible model is to assume the probability of an *ordinary* record which contains two attribute sets A and B as $P(A)P(B)$, *i.e.*, statistically independent, and to declare a record *anomalous* if their joint appearance $P(A, B)$ is much higher than $P(A)P(B)$. This is an intuition based on *suspicious coincidence* [1]. However, there are large amount of *abnormal* records caused by noise, *e.g.*, false positives in IDS logs [2], and they can be repeated more *abnormally* than those caused by *anomalous* behaviors, and they are hard to be distinguished. We can assume that an *anomalous* behavior can affect a group of records with similar attribute values, and can be distinguished from noise by gathering *abnormal* records into such a group. For example, many *abnormal* records with similar *srcIP*, *dstIP*, *port*, and *type* can be caused by a common remarkable attack, instead of false positives. However, a problem is that it becomes harder to detect *abnormal* records in such a group as the size of the group grow, because they become more likely to be *ordinary* behaviors, *e.g.*, $P(A)P(B)$ gets closer to $P(A, B)$.

To tackle these difficulties, we propose a two-step anomaly detection. In the first step, we detect *abnormal* records as individual anomalies with a statistical anomaly detection that models the distribution of the numbers of records caused by *ordinary* behaviors as Poisson distribution. By making a stronger assumption of the distribution for *ordinary* behaviors, we try to detect *abnormal* records in larger groups more effectively. This step can be improved by using *Poisson Tensor Factorization (PTF)* [3]. In the second step, we gather the individual anomalies into groups of records with similar attribute values. This step can be implemented by using *CANDECOMP/PARAFAC (CP) Decomposition* [4].

Our main contributions are: (1) We propose a novel framework *MultiAspectSpotting* combining statistical anomaly detection with spotting groups of *abnormal* records. (2) By using datasets added with synthesized anomalies, we show our method can spot *anomalous* behaviors effectively. (3) We show our method can spot interesting patterns in real world datasets like IDS logs and web-access logs.

The remainder of this paper is organized as follows. We describe the related literature in Section 2 and introduce our method in Section 3. We describe the accuracy and scalability of our method in Section 4 and the experimental evaluation on real data in Section 5. In Section 6 we summarize our conclusions.

2 Related Work

2.1 Anomaly Detection in Categorical Datasets

Anomaly detection has attracted wide interest in many applications such as security, risk assessment, and fraud analysis [5]. Das et al. [6] proposed an anomaly pattern detection in noisy categorical datasets based on a rule-based anomaly detection [7]. They searched through all possible one or two component rules and detected anomalies whose counts were significantly differed from the expected counts determined by the training dataset. They used the *conditional anomaly detection* [8,6] as a definition of anomalies which is an alternative of *suspicious coincidence* proposed by Barlow [1]. However, they tried to find groups of *abnormal* records which significantly differed from the training dataset, whereas our problem is to spot groups of *abnormal* records which are most remarkable among all records in the dataset.

2.2   Tensor Decomposition

Tensor decomposition is a basic technique that has been widely studied and applied to a wide range of disciplines and scenarios. *CP Decomposition* and *Tucker Decomposition* are two well-known approaches [4], and has been applied to study tensor streams [9]. Non-negative tensor factorizations have been proposed to retain the nonnegative characteristics of the original data [10], as natural expansions of non-negative matrix factorizations[11]. *PTF* is one such technique, that models *sparse count* data by describing the random variation via a Poisson distribution [3]. Our work is also related to the *Boolean Tensor Factorization* [12], which uses Boolean arithmetic, i.e., defining that $1 + 1 = 1$. The problems of *Boolean Tensor Factorization* were proved to be NP-hard, and heuristics for these problems were presented [12]. Some implementations of tensor decomposition algorithms have been made publicly available, such as MATLAB Tensor Toolbox [13]. We combine some of these tensor decompositions effectively to spot *anomalous* behaviors. Moreover, some works detected outliers in a low-dimensional space obtained by tensor decompositions [14], but outliers caused by *anomalous* behaviors were not distinguished from those caused by noise.

3   Proposed Method

3.1   Notation

A *tensor* can be represented as a multi-dimensional array of scalars, and we call each scalar an *entry*. Its *order* is the dimensionality of the array, while each dimension is known as one *mode*. A tensor is *rank one* if it can be written as the outer product of vectors. The *rank* of a tensor is defined as the smallest number of rank-one tensors that can generate the tensor as their sum, and we refer to each rank-one tensor as a *component*. Throughout, scalars are denoted by lowercase letters (a), vectors by boldface lowercase letters ($\mathbf{v}$), matrices by boldface capital letters ($\mathbf{A}$), and higher-order tensors by boldface Euler script letters ($\mathcal{X}$). The jth column of a matrix $\mathbf{A}$ is denoted by $\mathbf{a}_j$, and ith entry of a vector $\mathbf{v}$ is denoted by v_i. We use multi-index notation so that a boldface $\mathbf{i}$ represents the index ($i_1...i_M$) of a tensor of order M. The size of nth mode is denoted as I_n. The notation $\| \cdot \|$ refers to the square root of the sum of the squares of the entries, analogous to the matrix Frobenius norm. The outer product is denoted by $\circ$, and the inner product is denoted by $\langle \cdot, \cdot \rangle$.

3.2   Problem Setting

Our problem can be defined as follows: Given a dataset in which each record i has M categorical attributes and repeated x_i times, how can we detect *abnormal* records repeated strangely more than or less than expected, caused by *anomalous* behaviors as distinguished from those caused by noise?

We make two assumptions. (1) The majority of records are caused by *ordinary* behavior, and we can build a model with minimal harm caused by *anomalous* behaviors and noise. (2) A group of *abnormal* records with similar attribute values is likely to be caused by a common *anomalous* behavior.

3.3   *MultiAspectSpotting* Framework

In this paper, we focus on statistical anomaly detection based on the assumption "*Normal data instances occur in high probability regions of a stochastic model, while anomalies occur in the low probability regions of the stochastic model*" [5]. However, a simple statistical anomaly detection is insufficient to spot interesting anomalies effectively because we cannot distinguish *abnormal* records caused by *anomalous* behaviors from those caused by noise. To tackle this difficulty, we propose a novel framework *MultiAspectSpotting* that can spot *anomalous* behaviors by conducting two-step different tensor decompositions (Fig. 1):

1. Create a tensor $\mathcal{X}$ in which mth mode corresponds to mth attribute of a dataset and entries of $\mathcal{X}$ indicating the numbers of corresponding records. Then calculate anomaly score of each record by conducting *PTF*, and pick up records with larger anomaly scores than a threshold t as individual anomalies. We make a strong assumption that the distribution of the number of records caused by *ordinary* behaviors is a mixture of R Poisson distributions, to detect individual anomalies in larger groups effectively (see Section 3.4).
2. Create a binary tensor $\mathcal{B}$ in which 1s indicate individual anomalies, and spot groups of individual anomalies of the maximum number of S as *anomalous* behaviors by conducting *CP Decomposition* (see Section 3.5).

Deciding threshold t to pick up individual anomalies in the first step is very important. Our strategy is to set the *ratio of noise records* Z, and to decide threshold t so that the ratio Z of distinct records is picked up as individual anomalies. We assume that a specific ratio of records are caused by noise, and that the number of records caused by *anomalous* behavior is relatively small. If no groups are spotted in the second step, we conclude that the dataset is not affected by *anomalous* behaviors.

Now we do not have a clear strategy of the parameter settings of R, S, and Z, and there is a big room for improvement of our framework. However, in Section 4 we show we can achieve better results by using $R > 1$ or $S > 0$ than using $R = 1$ (assuming a single Poisson distribution) or $S = 0$ (without the second step). Moreover, we show the selection of Z does not dramatically affect the results of spotting *anomalous* behaviors.

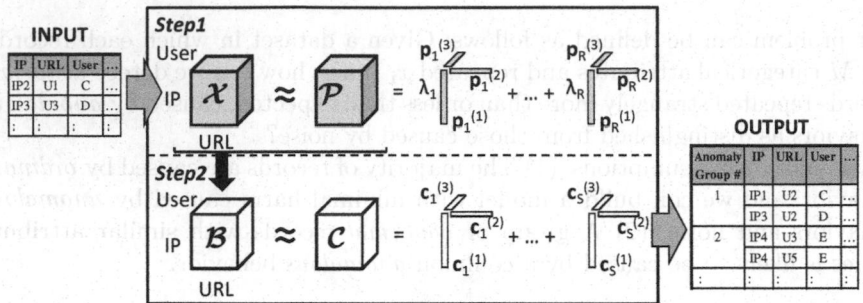

Fig. 1. The *MultiAspectSpotting* framework. Step 1: Conduct *Poisson Tensor Factorization*. Step 2: Create binary tensor indicating individual anomalies and conduct additional tensor decomposition.

3.4 A Statistical Anomaly Detection Approach

We describe details of the first step. The probability of the number of records of **i** to be $x_\mathbf{i}$ in a fixed interval of time can be modeled as the Poisson distribution in which the cumulative probability function is

$$F(x_\mathbf{i}, \mu_\mathbf{i}) = \sum_{k=0}^{x_\mathbf{i}} \frac{\mu_\mathbf{i}^k}{k!} e^{-\mu_\mathbf{i}}, \tag{1}$$

where $\mu_\mathbf{i}$ is the Poisson parameter equal to the expected number of the records of **i** caused by *ordinary* behaviors. Anomaly score is calculated as

$$anomaly\_score(x_\mathbf{i}, \mu_\mathbf{i}) = \begin{cases} (-1) * \log(F(x_\mathbf{i}, \mu_\mathbf{i})) & (x_\mathbf{i} \leq \mu_\mathbf{i}), \\ (-1) * \log(1 - F(x_\mathbf{i}, \mu_\mathbf{i})) & (x_\mathbf{i} > \mu_\mathbf{i}). \end{cases} \tag{2}$$

We consider a distinct record of **i** to be an individual anomaly if the anomaly score is higher than a threshold t. Also, $F(x_\mathbf{i}, \mu_\mathbf{i})$ can be easily computed with the incomplete gamma function. As $\mu_\mathbf{i}$ is the expected number of records of **i**, we can estimate $\mu_\mathbf{i}$ as

$$\mu_\mathbf{i} = \lambda p_{i_1}^{(1)} \cdots p_{i_M}^{(M)}, \tag{3}$$

where λ is total number of records and $p_{i_m}^{(m)}$ is the probability of mth value to be i_m, under the assumption of independence among the attributes. $p_{i_m}^{(m)}$ can be estimated as $p_{i_m}^{(m)} = N_{i_m}^{(m)}/\lambda$, where $N_{i_m}^{(m)}$ is the number of records of which mth value is i_m. Alternatively, we can assume the distribution as the mixture of R Poisson distributions. The Poisson parameters can be estimated as

$$\mu_\mathbf{i} = \sum_{r=1}^{R} \lambda_r p_{ri_1}^{(1)} \cdots p_{ri_M}^{(M)}, \tag{4}$$

under the assumption of independence among the distributions, where λ_r is the expected total number of records emerged from rth distribution and $p_{ri_m}^{(m)}$ is the probability of mth value to be i_m in rth distribution. We can estimate λ_r and $p_{ri_m}^{(m)}$ for all $\mathbf{i}$ and r by conducting *PTF*, which calculates each parameter so as to minimize the generalized Kullback-Leibler divergence, *i.e.*, $\sum_{\mathbf{i}} \mu_{\mathbf{i}} - x_{\mathbf{i}} log \mu_{\mathbf{i}}$ [3]. The details of *PTF* are outside the scope of this paper. Note that *PTF* of $R = 1$ is equivalent to calculating the parameters of equation (3).

3.5 Spotting Anomalous Behaviors by Tensor Decomposition

In the second step, we try to find S product sets $\mathcal{D}_s = \{(i_1, \ldots, i_M)|i_m \in \mathbf{d}_s^{(m)} \forall m = 1, \ldots, M\}$ where $\mathbf{d}_s^{(m)}$ is sth set of values of mth attribute ($s = 1, \ldots, S$), such that each product set contains as many individual anomalies as possible. We propose *DenseSpot*, a tensor decomposition approach (Algorithm 1). *DenseSpot* construct a binary tensor $\mathcal{B}$ of order M in which an entry $b_{\mathbf{i}}$ is

$$b_{\mathbf{i}} = \begin{cases} 1 & (anomaly\_score(x_{\mathbf{i}}, \mu_{\mathbf{i}}) > t), \\ 0 & (otherwise). \end{cases} \tag{5}$$

The aim of *DenseSpot* is to obtain a rank-S tensor

$$\mathcal{C} = \sum_{s=1}^{S} \mathbf{c}_s^{(1)} \circ \ldots \circ \mathbf{c}_s^{(M)} \tag{6}$$

which minimize $\|\mathcal{B}-\mathcal{C}\|$, where $\mathbf{c}_s^{(m)}$ are binary vectors. However, the decision version of this problem is a NP-hard problem similar to *Boolean Tensor Factorization* [12]. Thus, *DenseSpot* first obtains a rank-S tensor $\hat{\mathcal{C}} = \sum_{s=1}^{S} \hat{\mathbf{c}}_s^{(1)} \circ \ldots \circ \hat{\mathbf{c}}_s^{(M)}$ which minimize $\|\mathcal{B} - \hat{\mathcal{C}}\|$, where $\hat{\mathbf{c}}_s^{(m)}$ are real-value vectors. This is a relaxation problem of the above problem, and we can obtain a solution by conducting *CP Decomposition* [4]. After that, *DenseSpot* checks entries in sth component of $\hat{\mathcal{C}}$ corresponding to individual anomalies $(i_1, \ldots, i_M)$ and puts 1 on i_mth element of $\mathbf{c}_s^{(m)}$ if the entries are greater than a threshold h. Finally, *DenseSpot* selects h, which minimizes $\|\mathcal{B} - \mathcal{C}\|$ and returns those $\mathcal{C}$ calculated by the h. We can easily calculate $\|\mathcal{B} - \mathcal{C}\|^2$ as $\|\mathcal{B}\|^2 - 2\langle \mathcal{B}, \mathcal{C} \rangle + \|\mathcal{C}\|^2$. A set $\mathbf{d}_s^{(m)}$ can be created by selecting value 1 entries of $\mathbf{c}_s^{(m)}$.

Also, *Boolean Tensor Factorization* [12] might be a good solution for this. Even though this could improve the efficiency of our method, we explain how our simple heuristics can perform better than baseline methods in Section 4.

4   Evaluation of Accuracy and Scalability

In this section, we present experimental results on the accuracy and scalability of our methods. The running example in this section comes from network traffic logs that consist of packet traces in an enterprise network (LBNL/ICSI Enterprise Tracing Project [1]). We abbreviate them as LBNL logs. Each trace in the logs is a

---

[1] http://www.icir.org/enterprise-tracing/

Algorithm 1. *DenseSpot*

Input: A binary tensor $\mathcal{B}$
Input: Maximum number S of anomalies to spot
Input: A set of thresholds $H = \{h_1, ..., h_d\}$
Output: Rank-S tensor $\mathcal{C} = \sum_{s-1}^{S} \mathbf{c}_s^{(1)} \circ ... \circ \mathbf{c}_s^{(M)}$ where $\mathbf{c}_s^{(m)}$ are binary vectors

1 $\hat{\mathcal{C}} \leftarrow \sum_{s=1}^{S} \hat{\mathbf{c}}_s^{(1)} \circ ... \circ \hat{\mathbf{c}}_s^{(M)}$ s.t. minimize $\|\mathcal{B} - \hat{\mathcal{C}}\|$ ▷ *CP Decomposition*
2 **for** $j = 1$ **to** d **do**
3 | $\mathcal{C}^{(j)} \leftarrow \sum_{s=1}^{S} \mathbf{c}_s^{(1)} \circ ... \circ \mathbf{c}_s^{(M)}$ where $\mathbf{c}_s^{(m)}$ are I_m-length vectors of all 0
4 | **forall the** $(i_1...i_M)$ *of 1 entries in* $\mathcal{B}$ **do**
5 | | **for** $s = 1$ **to** S **do**
6 | | | **if** $\hat{c}_{si_1}^{(1)}...\hat{c}_{si_M}^{(M)} \geq h_j$ **then** $c_{si_1}^{(1)} \leftarrow 1, ..., c_{si_M}^{(M)} \leftarrow 1$
7 | | **end**
8 | **end**
9 **end**
10 $j_{min} \leftarrow \arg\min_{j} \|\mathcal{B} - \mathcal{C}^{(j)}\|$
11 **return** $\mathcal{C}^{(j_{min})}$

---

triplet of {*source IPs (srcIP)*, *destination IPs (dstIP)*, and *port number (port)*}, which can be represented as a 3-mode tensor. First, we evaluate the accuracy of spotting *anomalous* behaviors by using 10 largest LBNL logs added with synthesized anomalies. Then we evaluate the scalability by using many LBNL logs of various numbers of records.

MultiAspectSpotting is implemented in the MATLAB language, and we use implementations of *PTF* (*cp_apr*) and *CP Decomposition* (*cp_als*), publicly available in MATLAB Tensor Toolbox [13]. All the experiments are performed on a 64-bit Windows XP machine with four 2.8GHz cores and 8GB of memory.

4.1   Putting Synthesized Anomalies on Datasets

We create some synthesized anomalies and add into 10 largest LBNL logs, and evaluate how effectively our method can spot these anomalies. These LBNL logs have about $900,000$ to $9,000,000$ records and $15,000$ to $50,000$ distinct records, with $1,400$ to $4,500$ srcIPs, $1,400$ to $4,800$ dstIPs and $5,400$ to $24,000$ ports. Each distinct record is repeated about 50 to 350 times in average, and the standard deviation is about $1,000$ to $22,000$.

Given parameters of *volume V*, *density D* and *maximum number P*, we create N groups of *abnormal* records as follows: (1) For each group, we randomly select three values s,d,p between 0 and 1, and decide the number of srcIPs and dstIPs and ports in accordance with the ratio of three selected values, so that sdp is not lower than V, *e.g.*, the number of srcIPs is $\lceil s(V/(sdp))^{1/3} \rceil$ where $\lceil \cdot \rceil$ is the ceiling function. (2) $\lceil VD \rceil$ distinct records are randomly selected for each group, and (3) the number of each record is decided randomly between 1 and P. We test for $V = 50$, $D = 0.1, 0.3, 0.5, 0.7, 0.9$, $P = 500$, and $N = 10$.

4.2   Methods Compared

We compare the accuracies in spotting synthesized anomalies among the following methods:

MASP-Multi *MultiAspectSpotting* with $R = 10$ and $S = 20$.

MASP-Single *MultiAspectSpotting* with $R = 1$ and $S = 20$, which is equivalent to modeling the probabilities of the numbers of the records caused by *ordinary* behaviors as a single Poisson distribution.

DS-Only Conducting just *DenseSpot* of $S = 20$ by picking up all distinct records as individual anomalies.

SC-DS Using a measure of *suspicious coincidence* proposed by Barlow [1]. For each record, we calculate the ratio $r = P(A, B)/(P(A)P(B))$ where $P(A)$ and $P(B)$ are probabilities of a record having attribute sets A and B (*e.g.*, $\{srcIP\}$, $\{dstIP, port\}$), and $P(A, B)$ is the joint probability. The anomaly score of the record is defined as the minimum value of r among those of all possible combinations of A and B. We pick up individual anomalies and conduct *DenseSpot* as the same as *MultiAspectSpotting*.

Note that we have tried several methods similar to *SC-DS*, such as those using the maximum value of r, or those considering records with lower r as anomalous, or those using the ratio $r = P(A, B, C)/(P(A)P(B)P(C))$ where $P(A), P(B), P(C)$ and $P(A, B, C)$ correspond to attributes A, B and C, but these variations have obtained far worse results than *SC-DS* (not shown).

4.3   Accuracy of Spotting Synthesized Anomalies

We apply the above methods to LBNL logs added with synthesized anomalies and compare a group of records spotted by each method with a group of synthesized anomalies. We conduct chi-square tests of independence, which assess whether these two groups are independent of each other. In short, given these two group, we calculate $\chi^2 = n(a(n-e-g+a)-(e-a)(g-a))^2/(e(n-e)g(n-g))$ where n is the total number of distinct records, a is the number of common distinct records between two groups, e and g are the numbers of distinct records of two groups. If χ^2 is greater than a value of p-value at 0.05 of the chi-squared distribution for 1 degree of freedom, we conclude that the method has successfully spotted the synthesized anomalous group.

Fig. 2 is the number of groups spotted by each method. *MASP-Multi* and *MASP-Single* can spot many more groups than *DS-Only*, which suggests the statistical anomaly detection in the first step works efficiently. However, *SC-DS* is worse than *DS-Only*, which suggests the measure of *suspicious coincidence* is not good at detecting the anomalies we consider in this paper. Moreover, *MASP-Multi* is better than *MASP-Single*, which indicates we can model *ordinary* behaviors better by using a mixture of Poisson distributions. Overall, the more *density* grows, the better *MASP-Multi* and *MASP-Single* can spot than *DS-Only* and *SC-DS*. Moreover, the results of *MASP-Multi* and *MASP-Single* do not dramatically differ between $Z = 0.01$ (Fig. 2 left) and $Z = 0.1$ (Fig. 2 right), especially for higher *density* such as $P = 0.7, 0.9$.

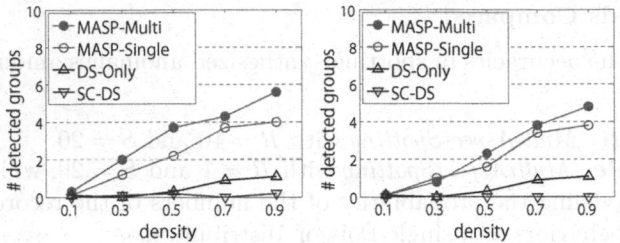

Fig. 2. Vertical axis: average number of spotted groups. Horizontal axis: *density* of groups of synthesized anomalies. (left) $Z = 0.01$. (right) $Z = 0.1$.

4.4 Details of Accuracy

We detail the effectiveness of each step. For the first step, we show the area under the ROC curve (AUC) of each method in detecting distinct synthesized anomalies out of all distinct records at various settings of Z (Fig. 3 left). Overall, AUCs of *MASP-Multi* and *MASP-Single* are better than those of *SC-DS*, and do not become much worse as *density* grow, whereas AUCs of *SC-DS* become much worse. These indicate that this statistical anomaly detection is a better strategy at least in detecting anomalies described here. There are no significant differences between *MASP-Multi* and *MASP-Single* in view of AUCs. For the second step, we select as many records with the highest anomaly scores as the total number of individual anomalies *DenseSpot* has detected, which we call *TopRecords*. We compare the *precision* of detecting synthesized anomalies out of individual anomalies between *DenseSpot* and *TopRecords*. The *precision* of *DenseSpot* and *TopRecords* are calculated as p/k and q/k, where p is the total number of distinct synthesized anomalies that *DenseSpot* has detected, q is the number of those *TopRecords* has selected, and k is the total number of individual anomalies that *DenseSpot* has detected. Fig. 3 (right) shows *precisions* on each method ($Z = 0.01$). The *precisions* of *DenseSpot* are much better than those of *TopRecords* on *MASP-Multi* and *MASP-Single* for higher *density*. This means the synthesized anomalies do not have very high anomaly scores among individual anomalies, whereas *DenseSpot* can pick up these synthesized anomalies, especially for higher density. Additionally, the *precisions* of *DenseSpot* on *MASP-Multi* are better than those on *MASP-Single*, which indicates that the difference in the number of Poisson distributions in the first step strongly affects the second step, even though differences in AUCs are very small. In addition, the *precisions* of *DenseSpot* are worse than those of *TopRecords* on *SC-DS*, possibly due to the poor accuracy of the *suspicious coincidence* in the first step.

4.5 Scalability

We conduct experiments for scalability on 123 different LBNL logs with various numbers of records, from less than 100 to more than $9,000,000$. As shown in Fig. 4, computation time of *PTF* (left) and *DenseSpot* (right) increases linearly

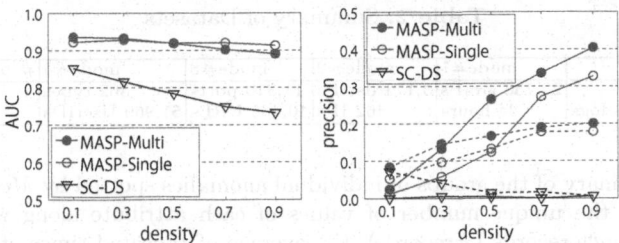

Fig. 3. (left) Average AUC of each method. Vertical axis: average AUC. Horizontal axis: *density* of groups of anomalous records. (right) The *precision* of *DenseSpot* (solid lines) and *TopRecords* (dotted lines) as described in Section 4 ($Z = 0.01$). Vertical axis: average of the *precision*. Horizontal axis: *density* of groups of anomalous records.

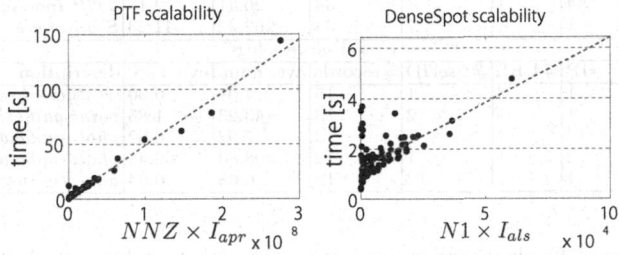

Fig. 4. Computation time have high correlation with $NNZ \times I_{apr}$ on *PTF* of $R = 10$ (left) and $N1 \times I_{als}$ on *DenseSpot* of $S = 20$ (right), where NNZ is the number of non-zeros (NNZ) in $\mathcal{X}$, I_{apr} is the number of inner iterations of *cp_apr*, $N1$ is the number of 1s in $\mathcal{B}$, and I_{als} is number of iterations of *cp_als*. Vertical axis: computation time (s). Horizontal axis: $NNZ \times I_{apr}$ (left), $N1 \times I_{als}$ (right).

along with the number of non-zero entries in the tensor (entries of 1 for *DenseSpot*) multiplied by the number of iterations of *cp_apr* and *cp_als*. These results are consistent with these alternating optimization algorithms implemented for sparse tensor [4,3] and suggest that these two steps of our framework scale linearly along with the number of distinct records in the dataset.

5   Empirical Results on Real Data

We present our experimental results on two sets of real world data: intrusion detection system logs and web-access logs ($R = 20$, $S = 20$, and $Z = 0.1$). We cannot mention the names of the companies from whom we have obtained these datasets because of the business relationship. Table 2 summarizes these datasets.

5.1   Intrusion Detection System Logs

We apply our method to IDS logs of a crowd system of an IT company. We analyze inbound logs on Dec 2011, that is, suspicious packets sent from outside

Table 2. Summary of Datasets

	mode#1	mode#2	mode#3	mode#4	# of records
IDS logs	22,733 srcIPs	3,171 dstIPs	11,310 ports	362 types	228,852
web-access logs	24 hours	462 IPs	40,465 URLs	51,960 UserIDs	462,271

Table 3. Summary of the groups of individual anomalies spotted by *MultiAspectSpotting*. We show the unique number of values of each attribute along with the total number of distinct records (#records), the average of repeated times of each distinct record (ave. num.), the average of Poisson parameters (eve. exp.), and description.

(a) IDS logs

No.	#srcIP	#dstIP	#port	#type	#records	ave. num.	ave. exp.	description
1	1	60	1	1	60	22.52	2.65	*FingerPrint*
2	53	1	1	1	53	5.26	0.00	*FTP Login Fail*
3	1	1	45	1	45	1.00	0.00	*Malicious Javascript*
4	1	34	1	1	34	20.41	2.17	*TCP Invalid flags*
5	5	1	1	7	33	262.42	41.23	*Scanning to Web Server*

(b) web-access logs

No.	#hour	#IP	#URL	#UserID	#records	ave. num.	eve. exp.	description
1	5	1	3	1	15	4.07	0.00	*weather-checking*
2	2	1	5	2	13	43.23	1.85	*point-gathering*
3	1	13	1	1	13	7.92	0.02	*photo-uploading*
4	6	2	1	1	12	408.83	66.48	*photo-uploading*
5	1	1	1	12	12	6.08	0.04	*advertisement-viewing*

the crowd system. Each record represents a report which has attributes of {*source IP (srcIP), destination IP (dstIP), port number (port)*, and *attack type (type)*}.

Table 3(a) summarizes the five largest groups of anomalous records spotted by our method. The descriptions are characteristics of these groups guessed by a specialist knowledgeable about the IDS of this crowd system. These include several kinds of attacks: attacks from many srcIPs including suspicious FTP login trials(#2), attacks on many dstIPs (#1,#4), attacks on many port numbers (#3), and attacks from several srcIPs of various attack types (#5). For example, the group #3 indicates that an outside IP has attacked many port numbers of an inside IP with a specific attack type, and that these attacks are remarkable because they are very rare events. Moreover, it is hard for analysts to notice the existence of this group of attacks because the number of records of this group is almost 0.02% of the total number of records within this dataset.

5.2 Web Access Logs

We also apply our method to web-access logs of a web-service company on Jan 10, 2013. Each record has attributes of {*hour, IP, URL*, and *UserID*} which means an access on the *URL* by the *UserID* from the *IP* at the *hour* of a day. The engineers at this company want to find any strange accesses within web-access logs and surprising or illegal usage of their web pages.

Table 3(b) summarizes the five largest groups of anomalous records spotted by our method, with descriptions of characteristics of these groups guessed by a specialist knowldgeable about the web site. For example, the group #2 is a *point-gathering* group, in which two users have strangely accessed a set of URLs many times from a IP continuously from 16 pm to 17 pm of this day. By accessing these URLs, users can obtain *points* that can be exchanged for some gifts, so the user who accesses just for gathering *points* illegally is suspicious.

6 Conclusion

We proposed a novel framework *MultiAspectSpotting* that can effectively spot *anomalous* behaviors by leveraging a two-step approach of a different kind of tensor decomposition. Experimental results of synthesized anomalies show our method can spot groups of individual anomalies more effectively than some baseline methods and can be improved by using *PTF*. The effectiveness of our method is achieved thanks to the combination of the accuracy of statistical anomaly detection in the first step and the ability of gathering individual anomalies in the second step, even though it might become harder for our method to model *ordinary* behaviors as the number of the attributes grows, *i.e.*, the dataset becomes sparser. Moreover, experimental results on real world data proved that our method could spot interesting patterns within IDS logs and web-access logs.

References

1. Barlow, H.B.: Unsupervised learning. Neural Computation 1, 295–311 (1989)
2. Julisch, K., Dacier, M.: Mining intrusion detection alarms for actionable knowledge. In: KDD, pp. 366–375 (2002)
3. Chi, E.C., Kolda, T.G.: On tensors, sparsity, and nonnegative factorizations. SIAM J. Matrix Analysis Applications 33(4), 1272–1299 (2012)
4. Kolda, T.G., Bader, B.W.: Tensor decompositions and applications. SIAM Review 51(3), 455–500 (2009)
5. Chandola, V., Banerjee, A., Kumar, V.: Anomaly detection: A survey. ACM Comput. Surv. 41(3) (2009)
6. Das, K., Schneider, J.G., Neill, D.B.: Anomaly pattern detection in categorical datasets. In: KDD, pp. 169–176 (2008)
7. Wong, W.K., Moore, A.W., Cooper, G.F., Wagner, M.M.: Rule-based anomaly pattern detection for detecting disease outbreaks. In: AAAI/IAAI, pp. 217–223 (2002)
8. Das, K., Schneider, J.G.: Detecting anomalous records in categorical datasets. In: KDD, pp. 220–229 (2007)
9. Sun, J., Tao, D., Papadimitriou, S., Yu, P.S., Faloutsos, C.: Incremental tensor analysis: Theory and applications. TKDD 2(3) (2008)
10. Shashua, A., Hazan, T.: Non-negative tensor factorization with applications to statistics and computer vision. In: ICML, pp. 792–799 (2005)
11. Lee, D.D., Seung, H.S.: Algorithms for non-negative matrix factorization. In: NIPS, pp. 556–562 (2000)
12. Miettinen, P.: Boolean tensor factorizations. In: ICDM, pp. 447–456 (2011)
13. Bader, B.W., Kolda, T.G., et al.: Matlab tensor toolbox version 2.5. Available online (January 2012), http://www.sandia.gov/~tgkolda/TensorToolbox/
14. Hayashi, K., Takenouchi, T., Shibata, T., Kamiya, Y., Kato, D., Kunieda, K., Yamada, K., Ikeda, K.: Exponential family tensor factorization for missing-values prediction and anomaly detection. In: ICDM, pp. 216–225 (2010)

Outlier Detection Based on Leave-One-Out Density Using Binary Decision Diagrams

Takuro Kutsuna[1,2] and Akihiro Yamamoto[2]

[1] Toyota Central R&D Labs. Inc., Nagakute, Aichi, 480-1192, Japan
kutsuna@mosk.tytlabs.co.jp
[2] Kyoto University, Sakyo-ku, Kyoto, 606-8501, Japan

Abstract. We propose a novel method for detecting outliers based on the leave-one-out density. The leave-one-out density of a datum is defined as a ratio of the number of data inside a region to the volume of the region after the datum is removed from an original data set. We propose an efficient algorithm that evaluates the leave-one-out density of each datum on a set of regions around the datum by using binary decision diagrams. The time complexity of the proposed method is near linear with respect to the size of a data set, while the outlier detection accuracy is still comparable to other methods. Experimental results show the usefulness of the proposed method.

Keywords: Outlier detection, binary decision diagram.

1 Introduction

In this paper, we propose a novel and efficient method for outlier detection, which is an important task in data mining and has been applied to many problems such as fraud detection, intrusion detection, data cleaning and so on [7]. The goal of outlier detection is to find an unusual datum (*outlier*) from a given data set. Although many kinds of notion have been proposed to define an outlier, we consider a datum as an outlier if the *leave-one-out density* is lower than a given threshold for a set of regions around the datum. The leave-one-out density is a ratio of the number of data inside a region to the volume of the region, in which the focused datum is removed from the original data set. Generally, a leave-one-out like method is time consuming because a learning procedure is repeated N-times, where N is the cardinality of a data set. However, the proposed method enables us to evaluate the leave-one-out density efficiently without repeating a learning procedure N-times.

We employ the initial region method proposed in [10], in which a data set is encoded into a Boolean formula and represented as a binary decision diagram. Although a one-class classifier is proposed based on the initial region method in [10], it is not applicable to outlier detection, because the classifier is estimated as an over-approximation of the data set and never classify a datum in the data set as an outlier. We extend the work of [10] to outlier detection by introducing the notion of leave-one-out density and developing an efficient algorithm to evaluate it.

V.S. Tseng et al. (Eds.): PAKDD 2014, Part II, LNAI 8444, pp. 486–497, 2014.

The proposed method is compared to other well-known outlier detection methods, the one-class support vector machine [13] and the local outlier factor [4], with both synthetic data sets and realistic data sets. The experimental results indicate that the computation time of the proposed method is shorter than those of the other methods, keeping the outlier detection accuracy comparable to the other methods.

The outlier detection problem addressed in this paper is formally defined in Sect. 2. We review the initial region method in Sect. 3. In Sect. 4, the outline of the proposed method is first stated, and then, its efficient implementation based on binary decision diagrams is proposed. Section 5 shows experimental results. In Sect. 6, we conclude this paper by discussing limitations and future work.

1.1 Related Work

Kutsuna [10] proposed a one-class classifier that over-approximates a training data set. The approximation is done quite efficiently by manipulating a binary decision diagram that is obtained by encoding the training data set. The situation considered in [10] is that both a training data set and a test data set are given: A classifier is first learned from the training dataset, then the test data set is classified by the classifier. It may seem that we can detect outliers within a data set by using the data set as both the training data set and the test data set simultaneously. However, no datum is detected as an outlier in such a setting, because the classifier is estimated as an over-approximation of the training data set. Therefore, the method in [10] cannot be applied to outlier detection directly.

Schölkopf et al. [13] extended the support vector machine (SVM) to outlier detection, which was originally invented for binary classification. Their method estimates a hyperplane that separates the origin and a data set with maximum margin, in which the hyperplane can be nonlinear by introducing kernel functions. The data that are classified to the origin side are detected as outliers. The SVM has an advantage that various nonlinear hyperplanes are estimated by changing kernel parameters. Some heuristics are proposed to tune kernel parameters, such as [6].

Breunig et al. [4] proposed the local outlier factor (LOF) that is calculated based on the distance to the k-nearest neighbor of each datum and has an advantage that it can detect local outliers, that is, data that are outlying relative to their local neighborhoods. The LOF has been shown to perform very well in realistic problems [12]. An efficient calculation of the k-nearest neighbors is essential in the LOF. Some techniques are proposed to accelerate the k-nearest neighbors calculation, such as [2].

2 Problem Setting

Let D be a data set that includes N data. The i-th datum in D is denoted by $x^{(i)} \in \mathbb{R}^u$ ($i = 1, \ldots, N$). We assume that there is no missing value in D and all the data in D are unlabeled. In this paper, we regard a datum as an outlier if the *leave-one-out density* is lower than a threshold for a set of regions around the datum. The leave-one-out density ρ_{LOO} of the i-th datum is defined as:

$$\rho_{LOO}(i, \mathcal{D}) = \frac{\#\left(\{x^{(j)} \mid x^{(j)} \in \mathcal{D},\ j = 1, \ldots, N,\ j \neq i\}\right)}{vol(\mathcal{D})}$$

where $\mathcal{D}$ is a u-dimensional region such that $x^{(i)} \in \mathcal{D}$, $\#(\cdot)$ is the cardinality of a set and $vol(\cdot)$ is the volume of the region. The *outlier score* S of the i-th datum is defined as:

$$S(i) = \max_{\mathcal{D} \in \tilde{\mathcal{D}}(i)} \rho_{LOO}(i, \mathcal{D})$$

where $\tilde{\mathcal{D}}(i)$ is a set of regions around $x^{(i)}$ defined as:

$$\tilde{\mathcal{D}}(i) = \left\{ u\text{-dimensional region } \mathcal{D} \mid x^{(i)} \in \mathcal{D} \right\}.$$

A datum is detected as an *outlier* if the outlier score of the datum is less than a given threshold. Note that $\tilde{\mathcal{D}}(i)$ is not the set of all possible regions, but rather a fixed family, which is defined in Sect. 4.1. In the following sections, we propose an efficient algorithm that enables us to evaluate the outlier score in near linear time with respect to N.

3 Preliminaries

In this section, we briefly review the initial region method [10] and define notations. Let $\mathcal{H}$ be a u-dimensional hypercube defined as $\mathcal{H} = [0, 2^m)^u$, where m is an arbitrary positive integer. And let σ be an example normalizer such that $\sigma(x^{(i)}) \in \mathcal{H}$ for every $x^{(i)}$ in D. An example of σ is given in [10] as a simple scaling function. The *neighborhood function* ν, which is defined as $\nu(z) := [\lfloor z_1 \rfloor, \lfloor z_1 \rfloor + 1) \times \ldots \times [\lfloor z_u \rfloor, \lfloor z_u \rfloor + 1)$, returns a u-dimensional unit hypercube that subsumes $z = \sigma(x)$, where $\lfloor \cdot \rfloor$ is the floor function. The *initial region* G is a u-dimensional region inside $\mathcal{H}$ that subsumes all the projected data, which is defined as:

$$G = \bigcup_{j=1,\ldots,N} \nu\left(z^{(j)}\right).$$

For example, we consider a data set in $\mathbb{R}^2$. We set $m = 3$, then the data set is projected into $\mathcal{H} = [0, 2^3)^2$ by σ. The projected data z are shown as x-marks and the initial region G is shown as the gray region in Fig. 1(a).

The initial region G is expressed as a Boolean function by using the coding function $\mathtt{CodeZ}$. $\mathtt{CodeZ}$ first truncates each element of z to an integer, and then, code them into a logical formula in the manner of an unsigned-integer-type coding. The set of Boolean variables $\mathbb{B} := \{b_{ij} \mid i = 1, \ldots, u;\ j = 1, \ldots, m\}$ is used to code z, where b_{i1} and b_{im} represent the most and the least significant bit of the i-th element of z, respectively. The *initial Boolean function* F is given as a disjunction of logical formulas such as:

$$F = \bigvee_{i=1,\ldots,N} \mathtt{CodeZ}(z^{(i)}).$$

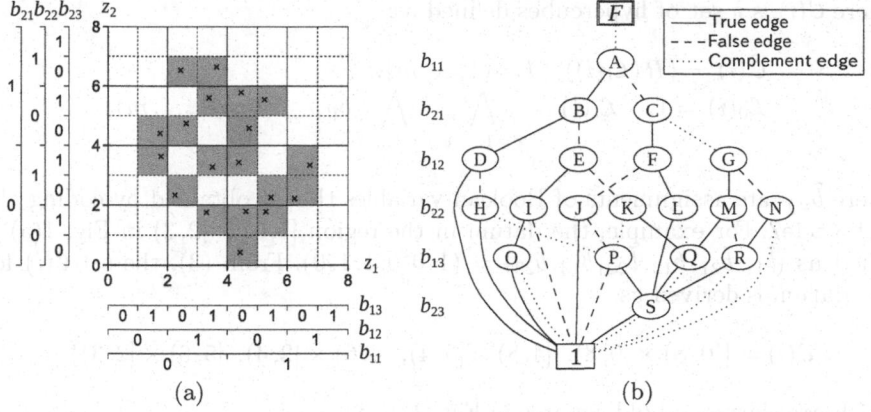

Fig. 1. X-marks mean a normalized data set and the gray region is the initial region G (*left*). A BDD that represents the initial Boolean function F (*right*).

It is shown that the initial Boolean function F is informational equivalent of the initial region G [11]. Let R be a function that decodes a Boolean function defined on $\mathbb{B}$ into the corresponding u-dimensional region. In particular, $R(1) = \mathcal{H}$ and $R(F) = G$. Binary decision diagrams (BDDs) [5] are used to efficiently construct and represent the initial Boolean function F. The order of Boolean variables is set to hold $[b_{11}, \ldots, b_{u1}] \prec \cdots \prec [b_{1m}, \ldots, b_{um}]$, where the variables inside the square brackets can be in arbitrary order, on constructing F as a BDD. For example, Fig. 1(b) shows a BDD that represents the initial Boolean formula F that is obtained from the data set in Fig. 1(a). In Fig. 1(b), square nodes, ellipsoidal nodes and double-squared nodes are referred to as terminal nodes, variable nodes and function nodes, respectively. Boolean variables that variable nodes represent are on the left side. Solid lines, dashed lines and dotted lines are true edges, false edges and complement edges, respectively. A path from a function node to the terminal 1 corresponds to a conjunction of literals. If the path contains an even number of complement edges, the conjunction is included in the function.

4 Proposed Method

4.1 Outline

In the proposed method, we calculate the leave-one-out density based on the normalized data $z^{(i)} = \sigma\left(x^{(i)}\right)$ as follows:

$$\rho_{\text{LOO}}\left(i, \mathcal{C}\right) = \frac{\#\left(\left\{z^{(j)} \mid z^{(j)} \in \mathcal{C}, \ j = 1, \ldots, N, \ j \neq i\right\}\right)}{vol\left(\mathcal{C}\right)} \tag{1}$$

where $\mathcal{C}$ is a region such that $z^{(i)} \in \mathcal{C}$. The outlier score is given as:

$$S\left(i\right) = \max_{\mathcal{C} \in \tilde{\mathcal{C}}(i)} \rho_{\text{LOO}}\left(i, \mathcal{C}\right) \tag{2}$$

where $\tilde{C}(i)$ is a set of hypercubes defined as:

$$\tilde{C}(i) = \{R(\mathcal{L}_l(i)) \mid l = 0, \ldots, m\},$$
$$\mathcal{L}_0(i) = 1, \quad \mathcal{L}_l(i) = \bigwedge_{i'=1,\ldots,u} \bigwedge_{j'=1,\ldots,l} \tilde{b}_{i'j'} \quad (l = 1, \ldots, m) \tag{3}$$

where $\tilde{b}_{i'j'}$ are assignments of Boolean variables that is obtained by coding $z^{(i)}$ with CodeZ. For example, the datum in the region $[5,6) \times [2,3)$ in Fig. 1(a) is coded as $(\tilde{b}_{11}, \tilde{b}_{21}, \tilde{b}_{12}, \tilde{b}_{22}, \tilde{b}_{13}, \tilde{b}_{23}) = (1,0,0,1,1,0)$. From (3), the set $\tilde{C}(i)$ for this datum is derived as

$$\tilde{C}(i) = \{[0,8) \times [0,8), \; [4,8) \times [0,4), \; [4,6) \times [2,4), \; [5,6) \times [2,3)\}$$

which are shown as bold squares in Fig. 2.

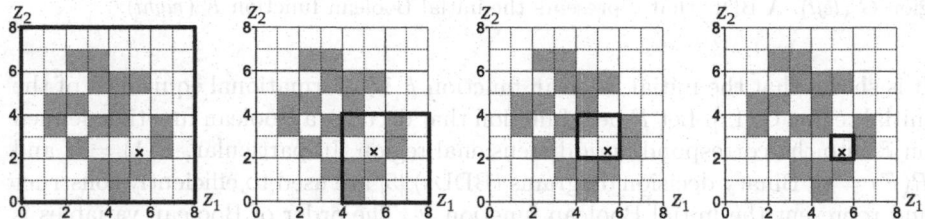

Fig. 2. Bold squares show $\tilde{C}(i)$ for the data shown as the x-mark. The leave-one-out density ρ_{LOO} of the datum are $\frac{18}{64}$, $\frac{6}{16}$, $\frac{1}{4}$ and $\frac{0}{1}$ for each hypercube. The outlier score of the datum is evaluated as $\max\left(\frac{18}{64}, \frac{6}{16}, \frac{1}{4}, \frac{0}{1}\right) \approx 0.38$.

We assume that there is no duplicate in D, that is, at least one of the attributes has a different value for $x^{(i)}$ and $x^{(j)}$ if $i \neq j$. Then, we can construct the initial region G so that each datum in D is allocated in a distinct unit hypercube by setting m large enough and using an appropriate example normalizer. In this case, the number of data inside a region can be calculated as the volume of the region unless the boundary of the region goes across a unit hypercube in which a datum exists. Therefore, the leave-one-out density defined as (1) can be calculated based on the initial region as follows for $C \in \tilde{C}(i)$:

$$\rho_{\text{LOO}}(i, C) = \frac{vol(G_{\text{LOO}}(i) \cap C)}{vol(C)} \tag{4}$$

where G_{LOO} is the leave-one-out region defined as:

$$G_{\text{LOO}}(i) = G \setminus \nu\left(z^{(i)}\right). \tag{5}$$

For example, Fig. 2 illustrates the calculation of the leave-one-out density for the datum shown as the x-mark.

4.2   BDD Based Implementation

The Number of Minterms Calculation. A *minterm* is a conjunction of Boolean variables in which each Boolean variable in the domain appears once. Let $\#_A$ be a function that returns the number of minterms of a Boolean function on the assumption that the domain of the function is A. For example, $\#_a(1) = 2$ and $\#_{\{a,b\}}(1) = 4$ where a and b are Boolean variables.

Lemma 1. For a Boolean formula A that is defined on $\mathbb{B}$, it holds that:

$$vol\left(R\left(A\right)\right) = \#_{\mathbb{B}}\left(A\right).$$

Proof. Since a minterm represents a unit hypercube in the initial region method, the number of minterms equals to the volume of the region that A represents.

Let $\mathcal{N}_\alpha^+$ be a Boolean function that node α in a BDD represents being connected with a non-complement edge. Also, let $\mathcal{N}_\alpha^-$ be a Boolean function being connected with a complement edge. Both of $\mathcal{N}_\alpha^+$ and $\mathcal{N}_\alpha^-$ represent regions inside $\mathcal{H}$. For example, Fig. 3 shows $R\left(\mathcal{N}_E^-\right)$ and $R\left(\mathcal{N}_G^+\right)$ where E and G are nodes in Fig. 1(b). It is possible to efficiently calculate the number of minterms of $\mathcal{N}_\alpha^+$ and $\mathcal{N}_\alpha^-$ for each node α in a BDD in a depth-first manner [15]. For example, Table 1 shows the number of minterms of $\mathcal{N}_\alpha^+$ and $\mathcal{N}_\alpha^-$ for each node in Fig. 1(b). We can see that $\#_{\mathbb{B}}\left(\mathcal{N}_E^-\right)$ and $\#_{\mathbb{B}}\left(\mathcal{N}_G^+\right)$ are equal to the volumes of regions which are shown in Fig. 3.

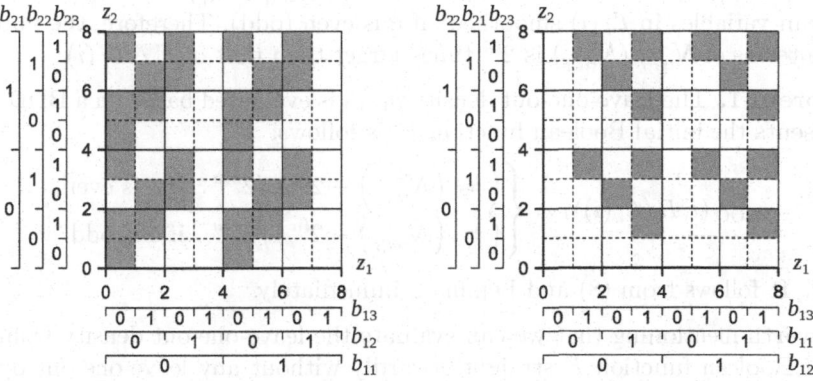

Fig. 3. Examples of regions that BDD nodes represent: $R\left(\mathcal{N}_E^-\right)$ (*left*) and $R\left(\mathcal{N}_G^+\right)$ (*right*) where E and G are nodes in Fig. 1(b)

The Leave-One-Out Density Calculation. Because of the fact that $\nu\left(z^{(i)}\right) \subseteq G$ and $\nu\left(z^{(i)}\right) \subseteq \mathcal{C} \in \tilde{\mathcal{C}}(i)$, it is derived from (4) and (5) that the following equation holds for $\mathcal{C} \in \tilde{\mathcal{C}}(i)$:

$$\rho_{\text{LOO}}\left(i,\mathcal{C}\right) = \frac{vol\left(G \cap \mathcal{C}\right) - 1}{vol\left(\mathcal{C}\right)}. \tag{6}$$

Table 1. The number of minterms of $\mathcal{N}_\alpha^+$ and $\mathcal{N}_\alpha^-$ for each node α in Fig. 1(b)

α	A	B	C	D	E	F	G	H	I	J	K	L	M	N	O	P	Q	R	S	1
$\#_\mathbb{B}\left(\mathcal{N}_\alpha^+\right)$	45	44	46	52	36	44	16	40	48	56	24	32	24	8	32	48	32	16	32	64
$\#_\mathbb{B}\left(\mathcal{N}_\alpha^-\right)$	19	20	18	12	28	20	48	24	16	8	40	32	40	56	32	16	32	48	32	0

By replacing $\mathcal{C}$ in (6) with $R\left(\mathcal{L}_l\left(i\right)\right)$, the following equation is derived.

$$\rho_{\text{LOO}}\left(i, R\left(\mathcal{L}_l\left(i\right)\right)\right) = \frac{vol\left(G \cap R\left(\mathcal{L}_l\left(i\right)\right)\right) - 1}{vol\left(R\left(\mathcal{L}_l\left(i\right)\right)\right)} = \frac{vol\left(R\left(F \wedge \mathcal{L}_l\left(i\right)\right)\right) - 1}{vol\left(R\left(\mathcal{L}_l\left(i\right)\right)\right)} \tag{7}$$

From Lemma 1, (7) is transformed as follows:

$$\rho_{\text{LOO}}\left(i, R\left(\mathcal{L}_l\left(i\right)\right)\right) = \frac{\#_\mathbb{B}\left(F \wedge \mathcal{L}_l\left(i\right)\right) - 1}{\#_\mathbb{B}\left(\mathcal{L}_l\left(i\right)\right)} = \frac{\#_\mathbb{B}\left(F \wedge \mathcal{L}_l\left(i\right)\right) - 1}{2^{(m-l)u}} \tag{8}$$

Lemma 2. In a BDD that represents F, let $\alpha_{i,l}$ be the node that can be reached from the function node through the path defined by $\mathcal{L}_l\left(i\right)$. Let c be the number of complement edges on the path. Then, it holds that:

$$\#_\mathbb{B}\left(F \wedge \mathcal{L}_l\left(i\right)\right) = \begin{cases} \#_\mathbb{B}\left(\mathcal{N}_{\alpha_{i,l}}^+\right)/2^{lu} & \text{if } c \text{ is even,} \\ \#_\mathbb{B}\left(\mathcal{N}_{\alpha_{i,l}}^-\right)/2^{lu} & \text{if } c \text{ is odd.} \end{cases}$$

Proof. $F \wedge \mathcal{L}_l\left(i\right)$ means that $l \times u$ Boolean variables that appear in $\mathcal{L}_l\left(i\right)$ are fixed to specific values in F. On the other hand, $\mathcal{N}_{\alpha_{i,l}}^+$ ($\mathcal{N}_{\alpha_{i,l}}^-$) means F with $l \times u$ Boolean variables in $\mathcal{L}_l\left(i\right)$ smoothed[1] if c is even (odd). Therefore, the number of minterms of $\mathcal{N}_{\alpha_{i,l}}^+$ ($\mathcal{N}_{\alpha_{i,l}}^-$) is 2^{lu} times larger than that of $F \wedge \mathcal{L}_l\left(i\right)$.

Theorem 1. The leave-one-out density ρ_{LOO} is evaluated based on a BDD that represents the initial Boolean function F as follows:

$$\rho_{\text{LOO}}\left(i, R\left(\mathcal{L}_l\left(i\right)\right)\right) = \begin{cases} \left(\#_\mathbb{B}\left(\mathcal{N}_{\alpha_{i,l}}^+\right) - 2^{lu}\right)/2^{mu} & \text{if } c \text{ is even,} \\ \left(\#_\mathbb{B}\left(\mathcal{N}_{\alpha_{i,l}}^-\right) - 2^{lu}\right)/2^{mu} & \text{if } c \text{ is odd.} \end{cases}$$

Proof. It follows from (8) and Lemma 2 immediately.

It is worth mentioning that we can evaluate the leave-one-out density from the initial Boolean function F straightforwardly without any leave-one-out operation by using Theorem 1. For example, we consider a data set in Fig. 1(a) and the datum in $[5, 6) \times [2, 3)$. The path of the datum is $F \bullet A \circ B \circ E \circ K \circ Q \circ S \bullet 1$ in Fig. 1(b), where $\circ$ and $\bullet$ mean non-complement and complement edges, respectively. The leave-one-out density of the datum is calculated from Theorem 1 and Table 1 as follows:

$$\left(\#_\mathbb{B}\left(\mathcal{N}_A^-\right) - 2^0\right)/2^6 = 18/64, \quad \left(\#_\mathbb{B}\left(\mathcal{N}_E^-\right) - 2^2\right)/2^6 = 6/16,$$

$$\left(\#_\mathbb{B}\left(\mathcal{N}_Q^-\right) - 2^4\right)/2^6 = 1/4, \quad \left(\#_\mathbb{B}\left(\mathcal{N}_1^+\right) - 2^6\right)/2^6 = 0.$$

We can see that these values are equal to those in Fig. 2.

---

[1] Smoothing a Boolean function f with respect to x means $(f \wedge x) \vee (f \wedge \bar{x})$.

4.3   The Proposed Algorithm and Computational Complexity

We propose Algorithm 1 that calculates the outlier score of each datum in D. In Algorithm 1, the time complexity of constructing the initial Boolean function F is approximately $O(MN)$, where M is the number of nodes of the created BDD, because logical operations between BDDs are practically almost linear to the size of the BDDs [3]. The size of the created BDD depends on the characteristics of the data set and can be exponentially large in the worst case, but, it is compact for realistic data sets used in our experiments. The time complexity of calculating the number of minterms is $O(M)$ as mentioned in Sect. 4.2. The time complexity of calculating the outlier score is $O(muN)$ because the depth of the BDD is mu. Consequently, the time complexity of Algorithm 1 is $O((M+mu)N)$. Therefore, the proposed method can deal with a large data set efficiently unless the number of Boolean variables and the created BDD are intractably huge.

Algorithm 1: The outlier score calculation.

Input: A data set D.
Output: The outlier score S of each datum in D.
1 Construct the initial Boolean function F as a BDD.;
2 Calculate the number of minterms of each node of the BDD.;
3 **for** $i \leftarrow 1$ **to** N **do**
4 | Search the path that $\mathtt{CodeZ}(z^{(i)})$ represents in the BDD and evaluate $\rho_{\mathrm{LOO}}(i, R(\mathcal{L}_l(i)))$ for $l = 0, \ldots, m$ by using Theorem 1.;
5 | $S(i) \leftarrow \max_{l=0,\ldots,m} \rho_{\mathrm{LOO}}(i, R(\mathcal{L}_l(i)))$;

4.4   Dealing with Categorical Attributes

The proposed method can be extended in order to deal with a data set that consists of both continuous attributes and categorical attributes. Let $y^{(i)}$ be a vector of categorical attributes of the i-th datum. We extend the leave-one-out density defined as (1) as follows:

$$\rho_{\mathrm{LOO}}(i, \mathcal{C}) = \frac{\#\left(\left\{z^{(j)} \mid z^{(j)} \in \mathcal{C},\ y^{(j)} = y^{(i)},\ j = 1, \ldots, N,\ j \neq i\right\}\right)}{vol(\mathcal{C})}$$

Then, the outlier score defined as (2) can be evaluated very efficiently in the same manner as mentioned in the previous sections. The details are skipped because of the page limit.

5   Experimental Results

We compare the proposed method with existing methods, the one-class support vector machine (OCSVM) and the local outlier factor (LOF). The proposed method is referred to as ODBDD. We implemented ODBDD as a C program with the help of CUDD [14]. The `ksvm` function in the `kernlab` package [9] is used for OCSVM and the `lofactor` function in the `DMwR` package [16] is used for LOF. The parameter m that defines the size of the hypercube $\mathcal{H}$ is fixed to $m = 16$ in ODBDD. In OCSVM, the Gaussian kernel is used and the kernel parameter γ is set to one of the 10%, 50% and 90% quantiles of the distance between samples [6], which are referred to as $\gamma_{0.1}$, $\gamma_{0.5}$ and $\gamma_{0.9}$, respectively. The parameter ν is fixed to $\nu = 0.1$ in OCSVM. In LOF, the number of neighbors is set to either $k = 10$ or $k = 50$. In OCSVM and LOF, continuous attributes are scaled and categorical attributes are coded by using dummy variables. The accuracy is evaluated in terms of the *area under an ROC curve (AUC)* [8]. The experiment was performed on a Microsoft Windows 7 machine with an Intel Core i7 CPU (3.20 GHz) and 64 GB RAM.

5.1   Evaluation with Synthetic Data Sets

Ten data set is a synthetic data set that consists of two continuous attributes and no categorical attribute. The 95 % data of Ten data set distributes inside the shape "10" randomly. The remaining 5 % data distributes outside randomly, which are regarded as outliers. The number of data is set to 10^3, 10^4, 10^5 or 10^6. An example of Ten-10^3 data set is shown in the left side of Fig. 4.

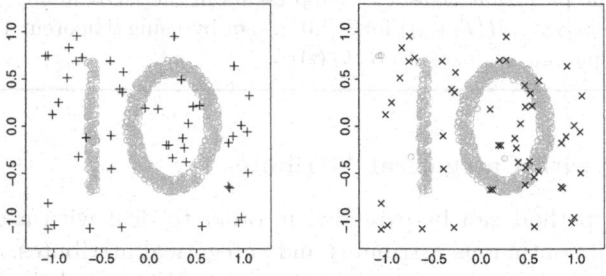

Fig. 4. An example of Ten-10^3 data set in which true outliers are shown as cross-marks (*left*). X-marks mean outliers detected by the proposed method (*right*).

Table 2 shows the mean and the standard deviation of computation time over 10 random trials for Ten data set. We can see that the computation time of the proposed method increases moderately compared to the other methods.

Table 3 shows the mean and the standard deviation of AUC values over 10 random trials for Ten data set. From Table 3, the accuracy of the proposed method is comparable to those of the other methods. For example, the outliers detected by ODBDD is shown in the right side of Fig. 4, in which the top 5 % of the data are detected as outliers based on the outlier score of ODBDD.

Table 2. The mean and the std. dev. of comp. time for Ten data set (sec)

Data set	ODBDD	OCSVM ($\nu = 0.1$)			LOF	
		$\gamma = \gamma_{0.1}$	$\gamma = \gamma_{0.5}$	$\gamma = \gamma_{0.9}$	$k = 10$	$k = 50$
Ten-10^3	0.1 (0.0)	0.0 (0.0)	0.0 (0.0)	0.0 (0.0)	0.4 (0.0)	0.5 (0.0)
Ten-10^4	0.3 (0.0)	3.0 (0.1)	2.7 (0.1)	2.6 (0.1)	22.6 (0.1)	24.8 (0.3)
Ten-10^5	2.3 (0.0)	294.5 (5.7)	267.3 (4.0)	250.6 (4.0)	2942.0 (38.4)	3052.1 (34.3)
Ten-10^6	30.1 (0.9)	timeout	timeout	timeout	timeout	timeout

The timeout limit is 3600 seconds.

Table 3. The mean and the std. dev. of AUC values for Ten data set

Data set	ODBDD	OCSVM ($\nu = 0.1$)			LOF	
		$\gamma = \gamma_{0.1}$	$\gamma = \gamma_{0.5}$	$\gamma = \gamma_{0.9}$	$k = 10$	$k = 50$
Ten-10^3	0.97 (0.02)	0.79 (0.02)	0.80 (0.03)	0.95 (0.01)	0.99 (0.01)	0.97 (0.02)
Ten-10^4	0.99 (0.00)	0.80 (0.02)	0.82 (0.02)	0.97 (0.00)	0.82 (0.02)	1.00 (0.00)
Ten-10^5	0.99 (0.00)	0.80 (0.00)	0.82 (0.00)	0.97 (0.00)	0.60 (0.01)	0.77 (0.01)
Ten-10^6	1.00 (0.00)	timeout	timeout	timeout	timeout	timeout

5.2 Evaluation with Realistic Data Sets

We use seven data sets from UCI machine learning repository [1] as shown in Table 4, where N_a is the size of the original data set. All of these data sets are originally arranged for the classification task. In order to apply these data sets to the evaluation of outlier detection algorithms, we randomly picked out data from each data set to generate a new data set as follows: 1) Pick out all the data whose class are C_m where C_m is the class of the maximum data size. Let N_m be the number of data that belong to class C_m. 2) Pick out $N_o = round(0.01 N_m)$ data randomly from the remaining data set, which are regarded as outliers.

Table 4. An overview of UCI data sets used in the experiment

Data set	N_a	N_m	N_o	# of cate. attr.	# of cont. attr.
abalone	4177	689	7	1	7
adult	32561	24720	248	8	6
bank	45211	39922	400	9	7
ionosphere	351	225	3	0	34
magic	19020	12332	124	0	10
shuttle	43500	34108	342	0	8
yeast	1484	463	5	0	8

Table 5 shows the mean and the standard deviation of computation time over 10 random trials for UCI data sets. The computation time varies drastically depending on the size of the data set in both OCSVM and LOF. On the other hand, ODBDD works quite fast for all of the data sets.

Table 5. The mean and the std. dev. of comp. time for UCI data sets (sec)

Data set	ODBDD	OCSVM ($\nu = 0.1$)			LOF	
		$\gamma = \gamma_{0.1}$	$\gamma = \gamma_{0.5}$	$\gamma = \gamma_{0.9}$	$k = 10$	$k = 50$
abalone	0.2 (0.0)	0.0 (0.0)	0.0 (0.0)	0.0 (0.0)	0.3 (0.0)	0.4 (0.0)
adult	3.1 (0.1)	26.5 (0.2)	24.5 (0.1)	24.2 (0.1)	2688.9 (320.0)	2935.3 (459.5)
bank	6.0 (0.0)	66.2 (0.4)	61.5 (0.5)	60.0 (0.3)	2784.8 (672.6)	2328.6 (155.7)
ionosphere	0.2 (0.0)	0.0 (0.0)	0.0 (0.0)	0.0 (0.0)	0.1 (0.0)	0.1 (0.0)
magic	2.4 (0.0)	4.6 (0.1)	4.1 (0.1)	4.0 (0.1)	52.3 (0.7)	56.6 (0.1)
shuttle	2.9 (0.1)	35.6 (0.4)	32.3 (0.3)	31.0 (0.3)	398.4 (10.9)	424.4 (12.5)
yeast	0.2 (0.0)	0.0 (0.0)	0.0 (0.0)	0.0 (0.0)	0.2 (0.0)	0.2 (0.0)

Table 6 shows the mean and the standard deviation of AUC values over 10 random trials. Although some results of ODBDD are not as good as the best result of the other methods, ODBDD achieves similar accuracy to the others.

Table 6. The mean and the std. dev. of AUC values for UCI data sets

Data set	ODBDD	OCSVM ($\nu = 0.1$)			LOF	
		$\gamma = \gamma_{0.1}$	$\gamma = \gamma_{0.5}$	$\gamma = \gamma_{0.9}$	$k = 10$	$k = 50$
abalone	0.59 (0.13)	0.59 (0.08)	0.57 (0.08)	0.58 (0.08)	0.61 (0.11)	0.66 (0.15)
adult	0.59 (0.02)	0.62 (0.01)	0.62 (0.01)	0.61 (0.01)	0.46 (0.02)	0.52 (0.03)
bank	0.61 (0.02)	0.62 (0.01)	0.62 (0.01)	0.62 (0.01)	0.62 (0.01)	0.69 (0.01)
ionosphere	0.87 (0.12)	0.79 (0.08)	0.87 (0.09)	0.64 (0.13)	0.94 (0.07)	0.95 (0.08)
magic	0.79 (0.02)	0.64 (0.03)	0.67 (0.03)	0.71 (0.03)	0.85 (0.03)	0.83 (0.03)
shuttle	0.93 (0.01)	0.78 (0.01)	0.89 (0.01)	0.93 (0.00)	0.44 (0.02)	0.69 (0.03)
yeast	0.65 (0.15)	0.66 (0.11)	0.64 (0.10)	0.55 (0.06)	0.69 (0.15)	0.73 (0.12)

6   Conclusions

In this work, we proposed a novel approach for outlier detection. A score of being an outlier is defined based on the leave-one-out density, which is evaluated very efficiently by processing a binary decision diagram that represents a data set in a logical formula. The proposed method can deal with a large data set efficiently, because the time complexity is near linear unless the created BDD gets intractably huge.

This work can be extended in several ways. First, the region set $\tilde{\mathcal{D}}(i)$ is enriched by using various normalizers. Then, the accuracy of outlier detection is expected to improve. A simple approach to generate various normalizers is to incorporate a random rotation into a normalizer. Another extension is to employ nonlinear normalizers. If we use nonlinear normalizers, a hypercube in a projected space corresponds to a nonlinear region in the original space, which may lead to more precise outlier detection.

The proposed method may suffer from the curse of dimensionality when a data set has many attributes and the number of data is not enough, because the leave-one-out density is zero in almost every subregion of the whole hypercube in such a situation. A simple solution is to embed some dimension reduction method into a normalizer.

Although we have conducted experiments with several data sets mainly to compare the proposed method to other methods, it is necessary to apply the proposed method to other real world problems in order to examine the practical usefulness of the proposed method and reveal problems to tackle in future.

References

1. Bache, K., Lichman, M.: UCI machine learning repository (2013),
   http://archive.ics.uci.edu/ml
2. Beckmann, N., Kriegel, H., Schneider, R., Seeger, B.: The R*-tree: an efficient and robust access method for points and rectangles. SIGMOD Rec. 19(2), 322–331 (1990)
3. Brace, K., Rudell, R., Bryant, R.: Efficient implementation of a BDD package. In: The 27th ACM/IEEE Design Automation Conference, pp. 40–45 (1990)
4. Breunig, M., Kriegel, H., Ng, R., Sander, J.: LOF: Identifying density-based local outliers. In: SIGMOD Conference, pp. 93–104 (2000)
5. Bryant, R.: Graph-based algorithms for boolean function manipulation. IEEE Trans. Computers 35(8), 677–691 (1986)
6. Caputo, B., Sim, K., Furesjo, F., Smola, A.: Appearance-based object recognition using svms: Which kernel should I use? In: NIPS Workshop on Statistical Methods for Computational Experiments in Visual Processing and Computer Vision (2002)
7. Chandola, V., Banerjee, A., Kumar, V.: Anomaly detection: A survey. ACM Computing Surveys (CSUR) 41(3), 15 (2009)
8. Fawcett, T.: An introduction to ROC analysis. Pattern Recognition Letters 27(8), 861–874 (2006)
9. Karatzoglou, A., Smola, A., Hornik, K., Zeileis, A.: kernlab – an S4 package for kernel methods in R. Journal of Statistical Software 11(9), 1–20 (2004)
10. Kutsuna, T.: A binary decision diagram-based one-class classifier. In: The 10th IEEE International Conference on Data Mining, pp. 284–293 (December 2010)
11. Kutsuna, T., Yamamoto, A.: A parameter-free approach for one-class classification using binary decision diagrams. Intelligent Data Analysis 18(5) (to appear, 2014)
12. Lazarevic, A., Ertoz, L., Kumar, V., Ozgur, A., Srivastava, J.: A comparative study of anomaly detection schemes in network intrusion detection. In: Proceedings of SIAM Conference on Data Mining (2003)
13. Schölkopf, B., Platt, J., Shawe-Taylor, J., Smola, A., Williamson, R.: Estimating the support of a high-dimensional distribution. Neural Computation 13(7), 1443–1471 (2001)
14. Somenzi, F.: CUDD: CU decision diagram package,
    http://vlsi.colorado.edu/~fabio/CUDD/
15. Somenzi, F.: Binary decision diagrams. In: Calculational System Design, vol. 173, pp. 303–366. IOS Press (1999)
16. Torgo, L.: Data Mining with R, learning with case studies (2010)

Rare Category Detection on $O(dN)$ Time Complexity

Zhenguang Liu[1], Hao Huang[2], Qinming He[1], Kevin Chiew[3], and Lianhang Ma[1]

[1] College of Computer Science and Technology, Zhejiang University, Hangzhou, China
{zhenguangliu,hqm,lhma}@zju.edu.cn
[2] School of Computing, National University of Singapore, Singapore
huanghao@comp.nus.edu.sg
[3] Provident Technology Pte. Ltd., Singapore
kev.chiew@gmail.com

Abstract. Rare category detection (RCD) aims at finding out at least one data example of each rare category in an unlabeled data set with the help of a labeling oracle to prove the existence of such a rare category. Various approaches have been proposed for RCD with quadratic or even cubic time complexity. In this paper, by using histogram density estimation and wavelet analysis, we propose FRED algorithm and its prior-free version iFRED algorithm for RCD, both of which achieve linear time complexity w.r.t. either the data set size N or the data dimension d. Theoretical analysis guarantees its effectiveness, and comprehensive experiments on both synthetic and real data sets verify the effectiveness and efficiency of our algorithms.

Keywords: Rare category detection, wavelet analysis, linear time complexity.

1 Introduction

Emerging from anomaly detection, rare category detection (in short as RCD henceforce) [8,9,15] is proposed to figure out which rare categories exist in an unlabeled data set with the help of a labeling oracle. Different from imbalanced clustering or classification [3], RCD verifies the existence of a rare category by finding out at least one data example of this category. This work has a wealth of potential applications such as network intrusion detection [13], financial security [2], and scientific experiments [15].

Generally, RCD is carried out by two phases [11], i.e., (1) analyzing characteristics of data examples in a data set and picking out candidate examples with rare category characteristics such as compactness [4,5,7,11,12,17] and isolation [11,17], followed by (2) querying the category labels of these candidate examples to a labeling oracle (e.g., a human expert). The first phase involves processing a big amount of data, facing an *efficiency challenge* which aims to achieve low time complexity; while the second phase involves limited labeling budget, leading to a *query challenge* which aims to find out at least one data example for each rare category with as less queries as possible.

Most of the existing approaches (e.g., see [4,12,15,17]) focus on query challenge without addressing too much of efficiency challenge with a time complexity not less than $O(dN^{2-\frac{1}{d}})$. To address both query and efficiency challenges simultaneously, we propose FRED (**F**ast **R**are cat**E**gory **D**etection) algorithm and its prior-free version

V.S. Tseng et al. (Eds.): PAKDD 2014, Part II, LNAI 8444, pp. 498–509, 2014.

Table 1. Algorithms' Time Complexity and Used Prior Information

Algorithm	Complexity	Category	Prior Info Used
Interleave	$O(dN^2)$	Prior Dependent	m
NNDM	$O(dN^{2-\frac{1}{d}})$	Prior Dependent	$m, p_1, p_2, ..., p_r$
SEDER	$O(d^2 N^2)$	Prior Free	–
GRADE	$O(dN^3)$	Prior Dependent	$m, p_1, p_2, ..., p_r$
GRADE-LI	$O(dN^3)$	Prior Dependent	p_{max}
RADAR	$O(dN^2)$	Prior Dependent	$m, p_1, p_2, ..., p_r$
CLOVER	$O(dN^{2-\frac{1}{d}})$	Prior Free	–
HMS	$\Omega(dN^2)$	Prior Free	–

iFRED algorithm on $O(dN)$ time complexity which is linear w.r.t. either d or N. This is done by utilizing Histogram Density Estimation (HDE) to estimate local data density and identifying candidate data examples of rare categories through the abrupt density changes via wavelet analysis. On the other hand, the existing RCD approaches [11, 15, 17] are often based on the assumptions of isolation and compactness of rare category examples; in contrast, our algorithms do not require rare categories being isolated from majority categories, and relax the compactness assumption to that every rare category may only be compact on partial dimensions.

2 Related Work

The existing paradigms for RCD can be classified into three groups, namely (1) the mixture model-based [15], (2) the data distribution change-based [4, 5, 7, 8, 11, 12] and (3) the hierarchical clustering-based [17]. A brief review on some representatives of these approaches in terms of time complexity and required prior information about a given data set is shown in Table 1. Note that throughout the paper m stands for the number of all categories, r the number of rare categories, and p_i (where $1 \leqslant i \leqslant r$) the proportion of data examples of rare category R_i out of all data examples in a data set.

Mixture model-based algorithms assume that data examples are generated by a mixture data model and need to iteratively update the model, the computation cost is usually substantial. For example, Interleave algorithm [15] takes $O(dN^2)$ time to update the covariance for each mixture Gaussian.

Data distribution change-based algorithms select data examples with maximal data distribution changes as candidate examples of rare categories. According to the measurements for the data distribution changes, these algorithms can be classified into two sub-groups, namely (1) local density-based, such as SEDER [5], GRADE [7], GRADE-LI [7], and NNDM [4]; and (2) nearest neighborhood-based, such as RADAR [12] and CLOVER [11]. Their time complexities are nearly quadratic or even cubic.

Hierarchical clustering-based algorithms investigate rare category characteristics of clusters on various levels. HMS [17] as a representative uses Mean Shift with increasing bandwidths to create a cluster hierarchy, and adopts *Compactness* and *Isolation* criteria to measure rare category characteristics. Its overall time complexity is $\Omega(dN^2)$.

Besides time complexity, the prior information needed on a given data set leads to the existing algorithms falling into two classes, namely prior-dependent and prior-free.

3   Problem Statement and Assumptions

Adhering to the problem definition by He *et al.* [4, 5, 7] and Huang *et al.* [11, 12], we formally define the problem of rare category detection as follows.

Given: (1) An unlabeled data set $S = \{x_1, x_2, \ldots, x_N\}$ containing m categories; (2) a labeling oracle which is able to give category label for any data example.

Find: At least one data example for each category.

For the data distribution of majority categories, we have the following assumption which is commonly used in the existing work [4, 5, 7, 11, 12, 15] explicitly or implicitly.

Assumption 1. *Data distribution of each majority category is locally smooth on each dimension.*

Gaussian, Poisson, t-, uniform distribution and many other distributions well satisfy this assumption. Thus this assumption can be satisfied by most applications.

For data distribution of rare categories, the exiting work [4–7, 10–12, 15] assumes that each rare category forms a compact cluster in the whole feature space, i.e., data examples from a rare category are similar to each other on every dimension. We relax this assumption to that every rare category may only be compact on partial dimensions.

Assumption 2. *Each rare category forms a compact cluster on partial dimensions or on the whole feature space.*

This assumption is more realistic because in many applications, data examples from a rare category are different from those from a majority category on partial dimensions. For example, panda subspecies are different from giant panda only in fur color and tooth size. According to this assumption, data examples of each rare category should show cohesiveness and form a compact cluster on at least partial dimensions.

Let D_{R_i} be the dimensions such that on each dimension $j \in D_{R_i}$ rare category R_i forms a compact cluster. According to the assumptions, we have following observations.

Observation 1. *In the areas without clusters of rare category examples, data distribution is smooth on each dimension.*

According to Assumption 1, data distribution of each majority category is smooth on each dimension. Due to the additivity of continuous functions, even in the overlapped areas of different majority categories, data distribution is smooth on each dimension. Thus for simplicity, we can assume that there is one majority category R_0 in S.

Observation 2. *Any abrupt change of local data density on each dimension $j \in D_{R_i}$ indicates the presence of rare category R_i.*

According to Assumption 2, data examples of R_i form a compact cluster on dimension $j \in D_{R_i}$, thus the local data density of R_i on dimension j is significant. This significant data density, combining with overlaps of data examples from majority category R_0, brings an abrupt change in local data distribution, which is distinct from the smooth distribution of R_0. Therefore, abrupt changes of local data density on dimension $j \in D_{R_i}$ indicate the presence of rare category R_i.

4 RCD Algorithm via Local Density Change

Based on the observations, we present FRED algorithm for RCD by exploring these abrupt local density changes via three steps. (1) On each dimension of a data set, FRED tabulates data examples into bins of appropriate bandwidth, and estimates the local density of each bin by Histogram Density Estimation (HDE) [16]. (2) By conducting wavelet analysis on estimated density function, FRED locates abrupt changes of local data density and quantitatively evaluates the change rates via our proposed DCR criterion. (3) After summing up each data examples' weighted DCR scores on all dimensions, FRED keeps selecting data examples with maximal DCR scores for labeling until at least one data example is discovered for each category.

4.1 Histogram Density Estimation

To find abrupt changes of local data density, a crucial step is to estimate local data density. We adopt HDE [16] for this goal due to its accuracy and time efficiency.

HDE firstly tabulates the feature space of a single dimension within interval $[s1, s2]$ into w non-overlapped bins $B_1, B_2, \ldots, B_w$, which have the same bandwidth h, and uses the number of data examples in each bin to estimate the local data density.

Let v_k be the number of data examples in the kth bin B_k and $\hat{f}(k)$ be the estimated local data density at bin k. Then we have

$$\hat{f}(k) = \frac{v_k}{N * h}, \quad k = 1, 2, \ldots, w \tag{1}$$

The structure of the histogram is completely determined by two parameters, bandwidth h and bin origin t_0. Well established theories (e.g., [16], [18]) show that bandwidth h has dominant effect and bin origin t_0 is negligible for sufficiently large sample sizes. A very small bandwidth results in a jagged histogram with each distinct observation lying in a separate bin (under-smoothed histogram); and a very large bandwidth results in a histogram with a single bin (over-smoothed histogram) [18]. We propose a criterion on h selection for detecting rare category R_i as

$$|avg(v_k) - C_i| \leqslant \varepsilon, \quad k = 1, 2, \ldots, w \tag{2}$$

where C_i is the number of data examples of R_i and ε a relaxation factor. This criterion guarantees that the average bin count is approximate to C_i, which makes the abrupt density change caused by R_i more significant to be detected.

4.2 Wavelet Analysis

After estimating local density, we perform wavelet analysis on the estimated density function to find abrupt density changes, which is the key to detecting rare categories.

First, we provide a brief review of main concepts on wavelet analysis. We define a mother wavelet as a square integrable function $\psi(x)$ that satisfies (1) $\psi(x)$ has a compact support, i.e., $\psi(x)$ has values in a small range and zeros otherwise; (2) $\psi(x)$ is normalized, i.e., $\int_{-\infty}^{+\infty} \psi(x)\psi^*(x)dx = 1$, where * denotes complex conjugate; and (3) $\psi(x)$ is zero mean, i.e., $\int_{-\infty}^{+\infty} \psi(x)dx = 0$.

A wavelet family can be obtained by translating and scaling the mother wavelet. Mathematically, they are $\psi_{a,b}(x) = \frac{1}{\sqrt{a}}\psi\left(\frac{x-b}{a}\right)$ for $a, b \in \mathbb{R}$ and $a > 0$ where a is the scale, which is inversely proportional to the frequency, and b represents the translation, which indicates the point of location where we concern [14].

Given these, we define the wavelet analysis of a quadratic integrable function $f(x)$ with real-valued wavelet ψ as

$$WT_f(a, b) = \int_{-\infty}^{+\infty} f(x)\psi_{a,b}(x)dx. \tag{3}$$

Note that (1) wavelet analysis maps a 1-D signal to a 2-D domain of scale (frequency) variable a and location variable b, which allow for location-frequency analysis. (2) For a fixed scale a_0 and translation b_0, $\psi_{a_0,b_0}(x)$ is the wavelet chosen, and $WT_f(a_0, b_0)$ is called the wavelet coefficient, which represents the resemblance index of $f(x)$ on neighborhood of b_0 to $\psi_{a_0,b_0}(x)$, where large coefficients correspond to strong resemblance [14]. Once an appropriate wavelet is chosen, $WT_f(a_0, b_0)$ reflects the amplitude of density change at the point of location b_0. As mentioned above, identifying local density changes is the key to detecting rare categories, thus this amplitude can help us fast locate the location of rare categories.

4.3 Data Distribution Change Rate

To quantify local density change rate for bins, we propose a new criterion defined as

Definition 1 (Data distribution change rate (DCR)). *Given bin density function $\hat{f}$, wavelet basis ψ, scale a, the central point b_0 of bin B, DCR of bin B is defined as*

$$DCR(B) = \frac{1}{\sqrt{a}} \int_{-\infty}^{+\infty} \hat{f}(x)\psi(\frac{x - b_0}{a})dx \tag{4}$$

DCR of each bin is calculated by wavelet analysis on $\hat{f}$. In practice, either Mexican hat or Reverse biorthogonal 2.2 (in short as Rbio2.2) wavelet can be chosen as wavelet basis ψ because they are similar in shape as cusps of density function brought by rare categories and have a compact support. Scale a in Eq. (4) is usually set to a positive value smaller than 1, which is the result of balancing the bandwidth of local region and computing cost.

Given DCR definition for bins, DCR of each data example on dimension j can be calculated by four steps. (1) Calculate the optimal bandwidth h by Eq. (2). (2) Divide the feature space of dimension j into bins and calculate bin density function by Eq. (1). (3) Compute DCR score of each bin by Definition 1, negative DCR scores are set to 0 because negative scores indicate drop of local data density and are of no interests to us here. (4) Perform K-means clustering on each bin with $K = v$ as a parameter. Let $x_1, x_2, \ldots, x_K$ be the central data examples of K clusters in bin B, then DCR scores of $x_1, x_2, \ldots, x_K$ are set to the DCR score of B, DCR scores of other data examples in bin B are set to zero.

Algorithm 1. Fast Rare Category Detection Algorithm (FRED)

    Input: $S = \{x_k | 1 \leqslant k \leqslant N\}$, proportions of rare categories $p_1, p_2, \ldots, p_r$, dimension of
        data examples d, number of categories m

    Output: The set Q of queried data examples and the set L of their labels

1   Initialize $Q = \varnothing, L = \varnothing$;

2   **for** $i = 1 : m$ **do**

3   **if** *category i is a majority category* **then**

4          $C_i = N * \max(p_l), 1 \leqslant l \leqslant r$;

5   **else**

6          $C_i = N * p_i$;

7   Set DCR of $\forall x_k \in S$ (denoted as $DCR(x_k)$) to 0;

8   **for** $j = 1 : d$ **do**

9   Calculate DCR of $\forall x_k \in S$ on dimension j (denoted as $DCR_j(x_k)$) by running
             the four steps introduced in Sec. 4.3;

10   calculate DCR of $\forall x_k \in S$, namely $DCR(x_k) = \sum_{j=1}^{d} W_j DCR_j(x_k)$;

11   Set the DCR of $\forall q \in Q$ to $-\infty$;

12   **while** $\max_{x_k \in S}(DCR(x_k)) > 0$ **do**

13   Query $s = \arg\max_{x_k \in S}(DCR(x_k))$ for its category label ℓ;

14          $Q = Q \cup s, L = L \cup \ell$;

15   **if** *s belong to an undiscovered category* **then**

16   break;

17   Set the DCR of s to $-\infty$;

4.4 FRED Algorithm

Algorithm 1 presents FRED algorithm which works as follows. Given proportions of rare categories, data dimension d, and the number of categories m, we first initialize hints set Q and their label set L to empty (line 1). Then for each rare category, (1) we compute the count of data examples C_i (lines 3–6), which will be used in the h selection step of DCR score calculation. (2) Then we calculate DCR score of $\forall x_k \in S$ on each dimension (lines 8–9). (3) For $\forall x_k \in S$, we sum up its weighed DCR score on each dimension as its final DCR score (line 10). It is recommended that $W_1, W_2, \ldots$, and W_k have the same value, whereas users with domain knowledge can modify them. (4) Next, we keep proposing the data example with maximal DCR score to the labeling oracle until a new category is found (lines 12–17). Note that DCR scores of selected data examples are set to $-\infty$ (lines 11 & 17) to prevent them from being chosen twice.

    The time complexity of FRED consists of two parts, (1) DCR score computation and (2) sampling. (1) In DCR computation on each dimension, the most time consuming step is K-means clustering, which takes $O(N)$ time complexity. Note that on each dimension the time complexity of HDE is $O(N)$ and the time complexity of wavelet analysis is $O(w)$, where w is the number of bins and $w < N$. So the overall time complexity of DCR score computation is $O(dN)$. (2) Since one data example will never be selected twice according to Algorithm 1, the time complexity of sampling is $O(N)$. Thus the time complexity of FRED is $O(dN)$ which is linear w.r.t. either d or N.

Algorithm 2.   Prior-free Rare Category Detection Algorithm (iFRED)

Input: $S = \{x_k | 1 \leqslant k \leqslant N\}$, sample size u, parameter β, ϵ
Output: The set Q of queried instances and the set L of their labels
1 Initialize $Q = \varnothing, L = \varnothing$;
2 $scale = 1$;
3 **for** $j = 1 : d$ **do**
4 $\quad$ Find bandwidth h_j on dimension j by Cross Validation;

5 **while** *labeling budget is not exhausted* **and** *scale* $> \epsilon$ **do**
6 $\quad$ Set DCR of $\forall x_k \in S$ to 0;
7 $\quad$ **for** $j = 1 : d$ **do**
8 $\quad\quad$ Calculate DCR of $\forall x_k \in S$ on dimension j (denoted as $DCR_j(x_k)$) with
$\quad\quad$ bandwidth h_j;
9 $\quad$ calculate DCR of $\forall x_k \in S$, namely $DCR(x_k) = \sum_{j=1}^{d} W_j DCR_j(x_k)$;
10 $\quad$ Set the DCR of $\forall q \in Q$ to $-\infty$;
11 $\quad$ **while** $\max_{x_k \in S}(DCR(x_k)) > 0$ *and labeling budget is not exhausted* **do**
12 $\quad\quad$ Query u data examples (denoted as set U) that have the maximum DCR scores
$\quad\quad$ for their category labels L_U;
13 $\quad\quad$ $Q = Q \cup U, L = L \cup L_U$;
14 $\quad\quad$ **if** *U all belong to discovered categories* **then**
15 $\quad\quad\quad$ $h_j = h_j * \beta, 1 \leqslant j \leqslant d$;
16 $\quad\quad\quad$ $scale = scale * \beta$;
17 $\quad\quad\quad$ break;
18 $\quad\quad$ Set the DCR of each data example in U to $-\infty$;

---

5   iFRED Algorithm

We propose iFRED algorithm as a prior-free version of FRED for scenarios where no prior knowledge about the given data set is available.

The difference between iFRED and FRED is twofold. (1) For bandwidth h selection, iFRED algorithm cannot follow the criterion introduced in Eq. (2) because the number of data examples C_i in each rare category is not available. Instead, it uses Cross Validation [16] to find the original bandwidth h. Furthermore, if current h is not efficient in finding rare categories, iFRED reduces h by setting $h = h * \beta, 0 < \beta < 1$. (2) In sampling phase, iFRED does not choose one data example each time for labeling, instead, each time it picks up u ($u \in \mathbb{N}^*$) data examples to measure the efficiency of current h in detecting rare categories. If at least one of the u data examples belongs to a new category, then current h is efficient; otherwise it sets $h = h * \beta, 0 < \beta < 1$.

Algorithm 2 presents iFRED algorithm which works as follows. (1) The initialization phase (lines 1–4) initializes hints set Q and their label set L to empty, *scale* is initialized to 1 and bandwidth on each dimension is initialized by Cross Validation. (2) The computation phase (lines 6–9) calculates DCR of each data example. (3) The sampling phase (lines 10–18) chooses each time u data examples of maximum DCR for labeling. If at least one of the u selected data examples belongs to a new category, we continue the sampling loop; otherwise we break out from the sampling loop, update

h_j $(1 \leqslant j \leqslant d)$ by setting $h_j = h_j * \beta$ and then continue the loop of computation and sampling phase until the labeling budget is exhausted or *scale* is too small (line 5, where ϵ is the threshold).

Time complexity of iFRED consists of two parts, (1) DCR score computation and (2) sampling. (1) In DCR computation, since the time complexity of Cross Validation on each dimension is $O(N)$ and the other three steps of DCR computation on each dimension takes $O(N)$ time complexity as analyzed in Sec. 4.4, the time complexity of DCR score computation is $O(dN)$. (2) Since one data example will never be selected twice according to Algorithm 2, the time complexity of sampling is $O(N)$. Thus the overall time complexity of iFRED is $O(dN)$ which is linear w.r.t. either d or N.

6 Effectiveness Analysis

In this section, we prove that if Assumptions 1 & 2 are fulfilled, our algorithms will sample repeatedly in the region where rare category examples occur with high probability. Without loss of generality, assume that we are searching for rare category $R_i, 1 \leqslant i \leqslant r$. Let B_{R_i} be the bins where data examples of R_i cluster together, D_{R_i} the dimensions that on each dimension $j \in D_{R_i}$ rare category R_i forms a compact cluster.

Claim 1.   *According to the bandwidth selection criterion of FRED and iFRED, a cusp of bin density function will appear in B_{R_i} on each dimension j where $j \in D_{R_i}$.*

Proof. Since $\hat{f}(k) = \frac{v_k}{N*h}$ (see Eq. (1)), a sharp cusp of bin density function is equivalent to a cusp of bin count function. We prove from the following three points that a cusp of bin count function will appear in B_{R_i} on dimension $j \in D_{R_i}$.

(1) On dimension $j \in D_{R_i}$, the compact rare category examples of R_i will cluster together in the same bin or Q adjacent bins where Q is a small integer.

(2) Let γ_1 be the data distribution of majority categories at B_{R_i} and γ_2 be the data distribution of R_i at B_{R_i}. Then the bin count of B_{R_i} should be $\gamma_1 + \gamma_2$; whereas nearby bins without rare category examples have bin count of $\gamma_1 \pm \xi$. By Assumption 2, γ_2 is significant. Note that ξ is very small because data distribution of majority categories changes slowly according to Assumption 1, thus $\gamma_2 \gg \xi$.

(3) Let C_i be the number of data examples of R_i. For FRED, according to the h selection criterion (see Eq. (2)), $avg(v_k) \approx C_i$, thus γ_1 will not be too large than γ_2; for iFRED, the bandwidth h will keep reducing, resulting in smaller and smaller bins where γ_1 will not be too large than γ_2.

Therefore, bins with rare categories will have significantly higher bin counts than nearby bins without rare categories. Claim 1 is proven.   ∎

Claim 2.   *According to DCR criterion, bins with cusp will get significantly high DCR scores while bins without cusp will get low DCR scores approximate to 0.*

Proof. Here we use Mexican hat wavelet (denoted by $\hat{\psi}(x)$) as an example wavelet to prove Claim 2 (wavelet Rbio2.2 can also be used in the same way). The shape of $\hat{\psi}_{a,b}(x) = \frac{1}{\sqrt{a}}\hat{\psi}\left(\frac{x-b}{a}\right)$ is shown in Fig. 1(a). Since the support interval of $\hat{\psi}(x)$ is $[-5, 5]$, thus the support interval of $\hat{\psi}_{a,b}(x)$ is $[-5a + b, 5a + b]$, outside which the

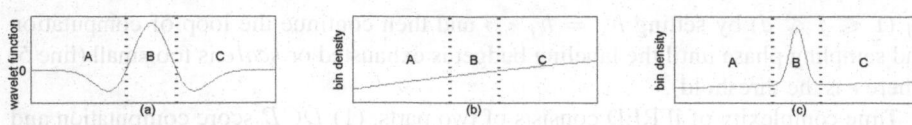

Fig. 1. Wavelet Analysis

value of $\hat{\psi}_{a,b}(x)$ is zero. Thus wavelet analysis of $f(x)$ by $\hat{\psi}_{a,b}(x)$ is $WT_f(a, b) = \int_{-5a+b}^{5a+b} f(x)\hat{\psi}_{a,b}(x)dx$.

Since FRED and iFRED use a positive fixed small scale a to detect local density changes, the integral interval $[-5a + b, 5a + b]$ is very narrow, which means that the data distribution change of majority categories is trivial. Fig. 1(b) shows the bin density function on the interval without cusps, and Fig. 1(c) shows the bin density function on the interval with one cusp. As shown in Fig. 1, wavelet analysis of $f(x)$ by $\hat{\psi}_{a,b}(x)$ is

$$WT_f(a, b) = \int_{A+B+C} f(x)\hat{\psi}_{a,b}(x)dx \qquad (5)$$

For local areas without cusp, the bin density change on this interval is trivial. Thus

$$WT_f(a, b) \approx 0 \qquad (6)$$

For local areas with cusps, Eq. (5) has a significant value because $\int_B f(x)\hat{\psi}_{a,b}(x)dx$ dominates the integration and has a significant value as shown in Fig. 1. Combining this conclusion with Eq. (6), we know that bins with cusps of bin density function will get a significantly high coefficients and bins without cusps will get coefficients approximate to zero. Therefore, Claim 2 is proven. ∎

Claim 3. *In FRED and iFRED, representative data examples of rare category R_i where $1 \leqslant i \leqslant r$ will get significantly high DCR scores, whereas data examples with locally smooth data density will get low DCR scores approximate to 0 .*

Proof. From Claims 1 & 2, we know that B_{R_i} will get significantly high DCR score on each dimension j where $j \in D_{R_i}$. The significantly high dimensional DCR score of B_{R_i} will pass to representative data examples of R_i in the K-means clustering steps of FRED and iFRED. Since the DCR score of each data example is the sum of its weighted DCR scores on each dimension, representative data examples of R_i will have significantly high DCR scores; whereas according to Claim 2, bins with locally smooth data density will get low DCR scores approximate to 0, these low DCR scores will pass to representative data examples of these bins. Claim 3 is proven. ∎

According to Assumption 1, the pdf of majority data examples is locally smooth. Combining this conclusion with Claim 3, we know that representative data examples of rare categories have significantly higher probabilities to be selected for labeling.

7   Experimental Evaluation

In this section, we conduct experiments to verify the efficiency and effectiveness of FRED and iFRED algorithms from two aspects, namely (1) time efficiency and scalability on data size N and dimension d, and (2) number of queries required for rare

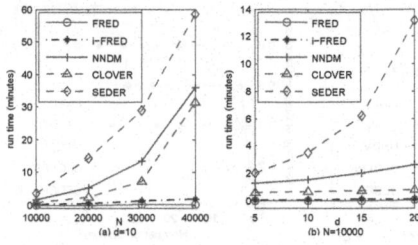

Fig. 2. Time Scalability Over N and d

Table 2. Properties of Real Data Sets

Data set	N	d	m	Largest class (%)	Smallest class (%)
Iris	106	4	3	47.17	5.66
Vertebral	310	6	3	48.39	19.35
Wine Quality	1589	11	5	42.86	1.13
Pen Digits	7143	16	10	10.92	5.15
Letter	19500	16	26	4.17	1.20
Shuttle	43494	9	6	78.42	0.03

category discovery. All algorithms are implemented with MATLAB 7.11 and running on a server computer with Intel Core 4 2.4GHz CPU and 20GB RAM.

7.1 Scalability

In this experiment, we compare our methods with NNDM, SEDER, and CLOVER on a synthetic data set where the pdf of majority categories is Gaussian and the pdf of rare categories is uniform within a small region. The synthetic data set satisfies that (1) the data size N ranges from 10000 to 40000 , (2) rare category R_1 forms a compact cluster in the densest area of the data set and has 395 data examples, (3) rare category R_2 forms a compact cluster in the moderate dense area and has 100 data examples, and (4) rare category R_3 forms a compact cluster in the low dense area and has 157 data examples.

(1) We set data dimension to 10 and vary the data size from 10000 to 40000 with incremental 10000. Fig. 2(a) shows the comparison results which agrees well with the time complexities shown in Table 1; i.e., the SEDER curve raises steeply due to its $O(d^2N^2)$ time complexity, followed by the curves of NNDM and CLOVER with $O(dN^{2-\frac{1}{d}})$ complexity, and lastly the curves of FRED and iFRED with linear complexity w.r.t. N.

(2) We set the size of the data set to 10000 and vary the data dimension from 5 to 20 with incremental 5. Fig. 2(b) shows that (1) iFRED, NNDM and CLOVER consumes much less time than SEDER; (2) time consumption by FRED, iFRED, NNDM and CLOVER grows linearly with data dimension.

(3) Our algorithms are much more efficient than other tested algorithms, e.g., on the data set with data size 40000, the runtime of CLOVER in seconds is 1884, NNDM 2153, SEDER 3499, whereas FRED only needs 5 seconds and iFRED 114 seconds. Experiments on real data sets such as Shuttle and Letter [1] have similar observations.

7.2 Efficiency

In RCD, the efficiency of an algorithm is evaluated by the number of queries needed to discover all categories in a data set. Usually these queries involve expensive human experts' work, thus our goal is to discover each category with minimal number of queries.

This experiment compares our algorithms with other algorithms on six real data sets from the UCI data repository [1], as detailed in Table 2. Specifically, categories with too

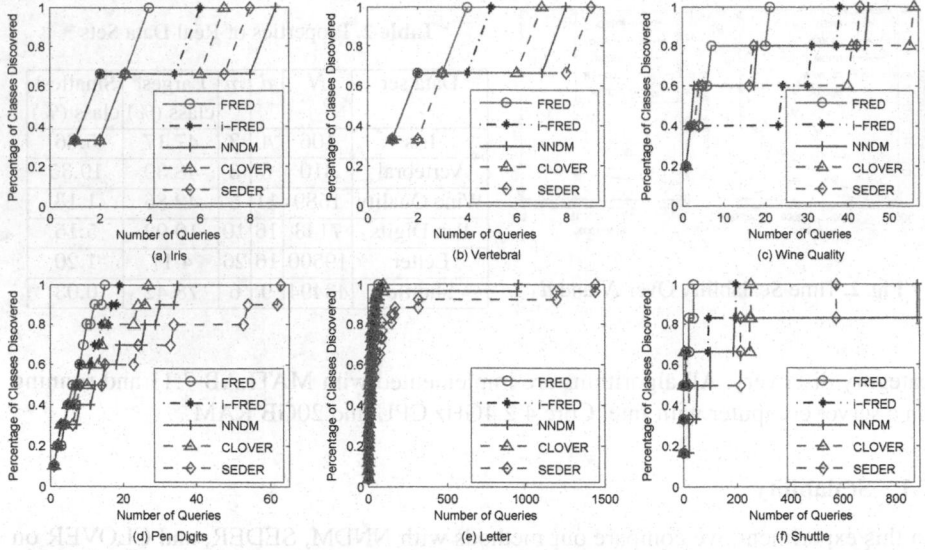

Fig. 3. Performance Comparisons on Real Data Sets

few ($\leqslant 10$) data examples are removed in order to satisfy compactness characteristic of rare categories, and sets Iris, Pen Digits and Letter are sub-sampled to create skewed data sets. Parameter v is set to either 1 or 2 for both FRED and iFRED.

Fig. 3 illustrates the comparison results on six data sets. From the figure, we have the following observations. (1) On all six data sets, FRED requires the least queries to discover all categories among all tested algorithms and performs significantly better than other algorithms. For example, on Shuttle set (Fig. 3(f)), FRED needs 42 labeling queries, iFRED 197, CLOVER 254, SEDER 575, and NNDM 884. (2) On all six data sets, iFRED takes up the second place right after FRED, especially on Vertebral, Pen Digits and Letter sets, its query number is almost as less as that of FRED. (3) A steeper curve means that the algorithm can discover new categories with fewer labeling queries. The curves of FRED and iFRED are much steeper than those of others, meaning that they need much less queries to discover a new category.

8   Conclusion

In this paper, we have proposed FRED and iFRED algorithms for RCD which have achieved $O(dN)$ time complexity and required least labeling queries. After using HDE to estimate local data density, they have been able to effectively identify candidate examples of rare categories through the abrupt density changes via wavelet analysis. Theoretical analysis has proven the effectiveness of our algorithms, and comprehensive experiments have further verified the efficiency and effectiveness.

For the next stage of study, a promising direction is to investigate new methods for the estimation of local data density. Another suggestion for future study is to work on sub-space selection which may bring with breakthroughs on this topic since in many scenarios rare categories are distinct on only a few dimensions.

Acknowledgment. This work is supported by the National Key Technology R&D Program of the Chinese Ministry of Science and Technology under Grant No. 2012BAH94F03.

References

1. Asuncion, A., Newman, D.: UCI Machine Learning Repository (2007)
2. Bay, S., Kumaraswamy, K., Anderle, M., Kumar, R., Steier, D.: Large scale detection of irregularities in accounting data. In: ICDM, Hong Kong, China, pp. 75–86 (2006)
3. Garcia-Pedrajas, N., Garcia-Osorio, C.: Boosting for class-imbalanced datasets using genetically evolved supervised non-linear projections. Progress in AI 2(1), 29–44 (2013)
4. He, J., Carbonell, J.: Nearest-neighbor-based active learning for rare category detection. In: NIPS 2007, Vancouver, British Columbia, Canada, December 3-6, pp. 633–640 (2007)
5. He, J., Carbonell, J.: Prior-free rare category detection. In: SDM 2009, Sparks, Nevada, USA, April 30-May 2, pp. 155–163 (2009)
6. He, J., Carbonell, J.: Coselection of features and instances for unsupervised rare category analysis. Statistical Analysis and Data Mining 3(6), 417–430 (2010)
7. He, J., Liu, Y., Lawrence, R.: Graph-based rare category detection. In: ICDM 2008, Pisa, Italy, December 15-19, pp. 833–838 (2008)
8. He, J., Tong, H., Carbonell, J.: An effective framework for characterizing rare categories. Frontiers of Computer Science 6(2), 154–165 (2012)
9. Hospedales, T.M., Gong, S., Xiang, T.: Finding rare classes: Active learning with generative and discriminative models. TKDE 25(2), 374–386 (2013)
10. Huang, H., Chiew, K., Gao, Y., He, Q., Li, Q.: Rare category exploration. Expert Systems with Applications (2014)
11. Huang, H., He, Q., Chiew, K., Qian, F., Ma, L.: Clover: A faster prior-free approach to rare category detection. Knowledge and Information Systems 35(3), 713–736 (2013)
12. Huang, H., He, Q., He, J., Ma, L.: Radar: Rare category detection via computation of boundary degree. In: Huang, J.Z., Cao, L., Srivastava, J. (eds.) PAKDD 2011, Part II. LNCS (LNAI), vol. 6635, pp. 258–269. Springer, Heidelberg (2011)
13. Khor, K., Ting, C., Phon-Amnuaisuk, S.: A cascaded classifier approach for improving detection rates on rare attack categories in network intrusion detection. Applied Intelligence 36(2), 320–329 (2012)
14. Nenadic, Z., Burdick, J.: Spike detection using the continuous wavelet transform. IEEE Transactions on Biomedical Engineering 52(1), 74–87 (2005)
15. Pelleg, D., Moore, A.: Active learning for anomaly and rare-category detection. In: NIPS 2004, Vancouver, British Columbia, Canada, December 13-18, pp. 1073–1080 (2004)
16. Scott, D.W.: Multivariate Density Estimation: Theory, Practice, and Visualization. Wiley, New York (1992)
17. Vatturi, P., Wong, W.: Category detection using hierarchical mean shift. In: KDD 2009, Paris, France, June 28-July 1, pp. 847–856 (2009)
18. Wand, M.P.: Data-based choice of histogram bin width. The American Statistician 51(1), 59–64 (1997)

Improving iForest with Relative Mass

Sunil Aryal[1], Kai Ming Ting[2], Jonathan R. Wells[1], and Takashi Washio[3]

[1] Monash University, Victoria, Australia
{sunil.aryal,jonathan.wells}@monash.edu
[2] Federation University, Victoria, Australia
kaiming.ting@federation.edu.au
[3] Osaka University, Osaka, Japan
washio@ar.sanken.osaka-u.ac.jp

Abstract. iForest uses a collection of isolation trees to detect anomalies. While it is effective in detecting global anomalies, it fails to detect local anomalies in data sets having multiple clusters of normal instances because the local anomalies are masked by normal clusters of similar density and they become less susceptible to isolation. In this paper, we propose a very simple but effective solution to overcome this limitation by replacing the global ranking measure based on path length with a local ranking measure based on relative mass that takes local data distribution into consideration. We demonstrate the utility of relative mass by improving the task specific performance of iForest in anomaly detection and information retrieval tasks.

Keywords: Relative mass, iForest, ReFeat, anomaly detection.

1 Introduction

Data mining tasks such as Anomaly Detection (AD) and Information Retrieval (IR) require a ranking measure in order to rank data instances. Distance or density based methods are widely used to rank instances in these tasks. The main problem of these methods is that they are computationally expensive in large data sets because of their high time complexities.

Isolation Forest (iForest) [1] is an anomaly detector that does not use distance or density measure. It performs an operation to isolate each instance from the rest of instances in a given data set. Because anomalies have characteristics of being 'few and different', they are more susceptible to isolation in a tree structure than normal instances. Therefore, anomalies have shorter average path lengths than those of normal instances over a collection of isolation trees (iTrees).

Though iForest has been shown to perform well [1], we have identified its weakness in detecting local anomalies in data sets having multiple clusters of normal instances because the local anomalies are masked by normal clusters of similar density; thus they become less susceptible to isolation using iTrees. In other words, iForest can not detect local anomalies because the path length measures the degree of anomaly globally. It does not consider how isolated an instance is from its local neighbourhood.

V.S. Tseng et al. (Eds.): PAKDD 2014, Part II, LNAI 8444, pp. 510–521, 2014.

iForest has its foundation in mass estimation [2]. Ting et al [2] have shown that the path length is a proxy to mass in a tree-based implementation. From this basis, we analyse that iForest's inability to detect local anomalies can be overcome by replacing the global ranking measure based on path length with a local ranking measure based on relative mass using the same iTrees. In general, relative mass of an instance is a ratio of data mass in two regions covering the instance, where one region is a subset of the other. The relative mass measures the degree of anomaly locally by considering the data distribution in the local regions (superset and subset) covering an instance.

In addition to AD, we show the generality of relative mass in IR that overcomes the limitation of a recent IR system called ReFeat [3] that uses iForest as a core ranking model. Even though ReFeat performs well in content-based multimedia information retrieval (CBMIR) [3], the ranking scheme based on path length does not guarantee that two instances having a similar ranking score are in the same local neighbourhood. The new ranking scheme based on relative mass provides such a guarantee.

The contributions of this paper are as follows:

1. Introduce relative mass as a ranking measure.
2. Propose ways to apply relative mass, instead of path length (which is a proxy to mass) to overcome the weaknesses of iForest in AD and IR.
3. Demonstrate the utility of relative mass in AD and IR by improving the task specific performance of iForest and ReFeat using exactly the same implementation of iTrees as employed in iForest.

The rest of the paper is organised as follows. Section 2 introduces the notion of relative mass and proposes ways to apply to AD and IR. Section 3 provides the empirical evaluation followed by conclusions in the last section.

2   Relative Mass: A Mass-Based Local Ranking Measure

Rather than using the global ranking measure based on path length in iForest, an instance can be ranked using a local ranking measure based on relative mass w.r.t its local neighbourhood. In a tree structure, the relative mass of an instance is computed as a ratio of mass in two nodes along the path the instance traverses from the root to a leaf node. The two nodes used in the calculation of relative mass depend on the task specific requirement.

- In AD, we are interested in the relative mass of x w.r.t its local neighbourhood. Hence, the relative mass is computed as the ratio of mass in the immediate parent node and the leaf node where x falls.
- In IR, we are interested in the relative mass of x w.r.t to a query q. Hence, the relative mass is computed as the ratio of mass of the leaf node where q falls and the lowest node where x and q shared along the path q traverses.

We convert iForest [1] and ReFeat [3] using the relative mass, and named the resultant relative mass versions, ReMass-iForest and ReMass-ReFeat, respectively. We describe iForest and ReMass-iForest in AD in Section 2.1; and ReFeat and ReMass-ReFeat in IR in Section 2.2.

2.1   Anomaly Detection: iForest and ReMass-iForest

In this subsection, we first discuss iForest and its weakness in detecting local anomalies and introduce the new anomaly detector, ReMass-iForest, based on the relative mass to overcome the weakness.

iForest

Given a d-variate database of n instances ($D = \{\mathbf{x}^{(1)}, \mathbf{x}^{(2)}, \cdots, \mathbf{x}^{(n)}\}$), iForest [1] constructs t iTrees ($T_1, T_2, \cdots, T_t$). Each T_i is constructed from a small random sub-sample ($\mathcal{D}_i \subset D$, $|\mathcal{D}_i| = \psi < n$) by recursively dividing it into two non-empty nodes through a randomly selected attribute and split point. A branch stops splitting when the height reaches the maximum (H_{max}) or the number of instances in the node is less than $MinPts$. The default values used in iForest are $H_{max} = \log_2(\psi)$ and $MinPts = 1$. The anomaly score is estimated as the average path length over t iTrees as follows:

$$L(\mathbf{x}) = \frac{1}{t} \sum_{i=1}^{t} \ell_i(\mathbf{x}) \qquad (1)$$

where $\ell_i(\mathbf{x})$ is the path length of $\mathbf{x}$ in T_i

As anomalies are likely to be isolated early, they have shorter average path lengths. Once all instances in the given data set have been scored, the instances are sorted in ascending order of their scores. The instances at the top of the list are reported as anomalies.

iForest runs very fast because it does not require distance calculation and each iTree is constructed from a small random sub-sample of data.

iForest is effective in detecting global anomalies (e.g., a_1 and a_2 in Figures 1a and 1b) because they are more susceptible to isolation in iTrees. But it fails to detect local anomalies (e.g., a_1 and a_2 in Figure 1c) as they are less susceptible to isolation in iTrees. This is because the local anomalies and the normal cluster C_3 have about the same density. Some fringe instances in the normal cluster C_3 will have shorter average path lengths than those for a_1 and a_2.

ReMass-iForest

In each iTree T_i, the anomaly score of an instance $\mathbf{x}$ w.r.t its local neighbourhood, $s_i(\mathbf{x})$, can be estimated as the ratio of data mass as follows:

$$s_i(\mathbf{x}) = \frac{m(\check{T}_i(\mathbf{x}))}{m(T_i(\mathbf{x})) \times \psi} \qquad (2)$$

where $T_i(\mathbf{x})$ is the leaf node in T_i in which $\mathbf{x}$ falls, $\check{T}_i(\mathbf{x})$ is the immediate parent of $T_i(\mathbf{x})$, and $m(\cdot)$ is the data mass of a tree node. ψ is a normalisation term which is the training data size used to generate T_i.

$s_i(\cdot)$ is in $(0, 1]$. The higher the score the higher the likelihood of $\mathbf{x}$ being an anomaly. Unlike $\ell_i(\mathbf{x})$ in iForest, $s_i(\mathbf{x})$ measures the degree of anomaly locally.

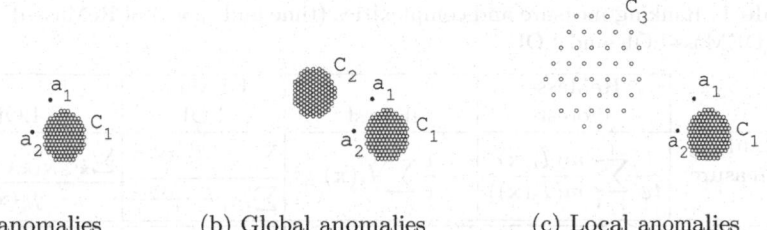

(a) Global anomalies   (b) Global anomalies   (c) Local anomalies

Fig. 1. Global and Local anomalies. Note that both anomalies a_1 and a_2 are exactly the same instances in Figures (a), (b) and (c). In Fig.(a) and Fig.(b), a_1 and a_2 have low density than that in the normal clusters C_1 and C_2. In Fig.(c), a_1, a_2 and the normal cluster C_3 have the same density but a_1 and a_2 are anomalies relative to the normal cluster C_1 with a higher density.

The final anomaly score can be estimated as the average of local anomaly scores over t iTrees as follows:

$$S(\mathbf{x}) = \frac{1}{t} \sum_{i=1}^{t} s_i(\mathbf{x}) \tag{3}$$

Once every instance in the given data set has been scored, instances can be ranked in descending order of their anomaly scores. The instances at the top of the list are reported as anomalies.

Relation to LOF and DEMass-LOF

The idea of relative mass in ReMass-iForest has some relation to the idea of relative density in Local Outlier Factor (LOF) [4]. LOF uses k nearest neighbours to estimate density $\bar{f}_k(\mathbf{x}) = \dfrac{|N(\mathbf{x}, k)|}{n \sum_{\mathbf{x}' \in N(\mathbf{x},k)} distance(\mathbf{x}, \mathbf{x}')}$ where $N(\mathbf{x}, k)$ is the set of k nearest neighbours of $\mathbf{x}$. It estimates its anomaly score as the ratio of the average density of $\mathbf{x}$'s k nearest neighbours to $\bar{f}_k(\mathbf{x})$. In LOF, the local neighbourhood is defined by k nearest neighbours which requires distance calculation. In contrast, in ReMass-iForest, the local neighbourhood is the immediate parent in iTrees. It does not require distance calculation.

DEMass-LOF [5] computes the same anomaly score as LOF from trees, without distance calculation. The idea of relative density of parent and leaf nodes was used in DEMass-LOF. It constructs a forest of t balanced binary trees where the height of each tree is $b \times d$ (b is a parameter that determines the level of division on each attribute and d is the number of attributes). It estimates its anomaly score as the ratio of average density of the parent node to the average density of the leaf node where $\mathbf{x}$ falls. The density of a node is estimated as the ratio of mass to volume. It uses mass to estimate density and ranks instances based on the density ratio. Like iForest, it is fast because no distance calculation is involved. But, it has limitation in dealing problems with even a moderate number of dimensions because each tree has $2^{(b \times d)}$ leaf nodes.

Table 1. Ranking measure and complexities (time and space) of ReMass-iForest, iForest, DEMass-LOF and LOF

	ReMass-iForest	iForest	DEMass-LOF	LOF
Ranking Measure	$\dfrac{1}{t\psi} \sum\limits_{i=1}^{t} \dfrac{m(\breve{T}_i(\mathbf{x}))}{m(T_i(\mathbf{x}))}$	$\dfrac{1}{t} \sum\limits_{i=1}^{t} \ell_i(\mathbf{x})$	$\dfrac{\sum_{i=1}^{t} \frac{m(\breve{T}_i(\mathbf{x}))}{\breve{v}_i}}{\sum_{i=1}^{t} \frac{m(T_i(\mathbf{x}))}{v_i}}$	$\dfrac{\sum_{\mathbf{x}' \in N(\mathbf{x},k)} \frac{\bar{f}_k(\mathbf{x}')}{|N(\mathbf{x},k)|}}{\bar{f}_k(\mathbf{x})}$
Time Complexity	$O(t(n+\psi)\log\psi)$	$O(t(n+\psi)\log\psi)$	$O(t(n+\psi)bd)$	$O(dn^2)$
Space Complexity	$O(t\psi)$	$O(t\psi)$	$O(td\psi)$	$O(dn)$

$\breve{v}_i$ and v_i are the volumes of nodes $\breve{T}_i(\mathbf{x})$ and $T_i(\mathbf{x})$, respectively.

In contrast to LOF and DEMass-LOF, ReMass-iForest does not require density estimation, it uses relative mass directly in order to estimate the local anomaly score from each iTree.

The ranking measure and complexities (time and space) of ReMass-iForest, iForest, DEMass-LOF and LOF are provided in Table 1.

2.2   Information Retrieval: ReFeat and ReMass-ReFeat

In this subsection, we first describe how ReFeat uses iForest in IR and its weakness. Then, we introduce a new IR system, ReMass-ReFeat, based on the relative mass to overcome the weakness.

ReFeat

Given a query instance $\mathbf{q}$, ReFeat [3] assigns a weight $w_i(\mathbf{q}) = \frac{\ell_i(\mathbf{q})}{c} - 1$ (where c is a normalisation constant) to each T_i. The relevance feedback process [6] allows user to refine the retrieved result by providing some 'relevant' and 'irrelevant' examples for the query. Let $\mathcal{Q} = \mathcal{P} \cup \mathcal{N}$ is a set of feedback instances to the query $\mathbf{q}$ where $\mathcal{P}$ and $\mathcal{N}$ are the sets of positive and negative feedbacks, respectively. Note that $\mathcal{P}$ includes $\mathbf{q}$. In a relevance feedback round, ReFeat assigns a weight to T_i using positive and negative feedback instances as: $w_i(\mathcal{Q}) = \dfrac{1}{|\mathcal{P}|} \sum\limits_{\mathbf{y}^+ \in \mathcal{P}} w_i(\mathbf{y}^+) -$

$\gamma \dfrac{1}{|\mathcal{N}|} \sum\limits_{\mathbf{y}^- \in \mathcal{N}} w_i(\mathbf{y}^-)$, where $0 \leq \gamma \leq 1$ is a trade-off parameter for the relative contribution of positive and negative feedbacks. The relevance of $\mathbf{x}$ w.r.t $\mathcal{Q}$ is estimated as the weighted average of its path lengths over t iTrees as follows:

$$R_{ReFeat}(\mathbf{x}|\mathcal{Q}) = \frac{1}{t} \sum_{i=1}^{t} (w_i(\mathcal{Q}) \times \ell_i(\mathbf{x})) \tag{4}$$

Even though ReFeat has been shown to have superior retrieval performance over other existing methods in CBMIR, the ranking scheme does not guarantee that two instances having similar ranking scores are in the same local neighbour-hood. Two instances can have similar score if they have equal path lengths in an iTree even though they lie in two different branches which shares few common nodes. This effect will degrade the performance of ReFeat especially when the tree height (h) is increased. Hence, ReFeat must use a low h (2 or 3) in order to reduce this weakness. The superior performance of ReFeat is mainly due to its large ensemble size $(t = 1000)$. We will discuss the effect of h and t in ReFeat in Section 3.2. In a nutshell, ReFeat does not consider the positions of instances in the feature space as it computes the path length in iTrees.

ReMass-ReFeat

In each iTree T_i, the relevance of $\mathbf{x}$ w.r.t. $\mathbf{q}$, $r_i(\mathbf{x}|\mathbf{q})$, is estimated using relative mass as follows:

$$r_i(\mathbf{x}|\mathbf{q}) = \frac{m(T_i(\mathbf{q}))}{m(T_i(\mathbf{x}, \mathbf{q}))} \qquad (5)$$

where $T_i(\mathbf{x}, \mathbf{q})$ is the smallest region in T_i where $\mathbf{x}$ and $\mathbf{q}$ appear together.

In equation 5, the numerator corresponds with $w_i(\mathbf{q})$ in ReFeat. The denom-inator term measures how relevant $\mathbf{x}$ is to $\mathbf{q}$. In contrast, ReFeat's $\ell_i(\mathbf{x})$ is inde-pendent of $\mathbf{q}$ (it does not examine whether $\mathbf{x}$ and $\mathbf{q}$ are in the same locality [3]); whereas $m(T_i(\mathbf{x}, \mathbf{q}))$ measures how close $\mathbf{x}$ and $\mathbf{q}$ are in the feature space. In each T_i, $r_i(\mathbf{x}|\mathbf{q})$ is in the range of $(0, 1]$. The higher the score the more relevance of $\mathbf{x}$ w.r.t $\mathbf{q}$. If $\mathbf{x}$ and $\mathbf{q}$ lie in the same leaf node in T_i, $r_i(\mathbf{x}|\mathbf{q})$ is 1. This relevance mea-sure gives a high score to an instance which lies deeper in the branch where $\mathbf{q}$ lies.

The final relevance score of $\mathbf{x}$ w.r.t $\mathbf{q}$, $R(\mathbf{x}|\mathbf{q})$, is the average over t iTrees:

$$R(\mathbf{x}|\mathbf{q}) = \frac{1}{t} \sum_{i=1}^{t} r_i(\mathbf{x}|\mathbf{q}) \qquad (6)$$

Once the relevance score of each instance is estimated, the scores can be sorted in descending order. The instances at the top of the list are regarded as the most relevant instances to $\mathbf{q}$.

ReMass-ReFeat estimates the relevance score with relevance feedback as fol-lows:

$$R(\mathbf{x}|\mathcal{Q}) = \frac{1}{|\mathcal{P}|} \sum_{\mathbf{y}^+ \in \mathcal{P}} R(\mathbf{x}|\mathbf{y}^+) - \gamma \frac{1}{|\mathcal{N}|} \sum_{\mathbf{y}^- \in \mathcal{N}} R(\mathbf{x}|\mathbf{y}^-) \qquad (7)$$

Note that equations 5 and 6 do not make use of any distance or similarity measure, and $R(\mathbf{x}|\mathbf{q})$ is not a metric as it does not satisfy all metric axioms. It has the following characteristics. For $\mathbf{x}, \mathbf{y} \in D$,

i. $0 < R(\mathbf{x}|\mathbf{y}) \le 1$ (Non-negativity)
ii. $R(\mathbf{x}|\mathbf{x}) = R(\mathbf{y}|\mathbf{y}) = 1$ (Equal self-similarity; maximal similarity)
iii. $R(\mathbf{x}|\mathbf{y}) \neq R(\mathbf{y}|\mathbf{x})$ (Asymmetric)

Note that ReMass-ReFeat and ReFeat have the same time complexities. If indices of data instances falling in each node are recorded in the modelling stage, the joint mass of $\mathbf{q}$ and every $\mathbf{x} \in D$ can be estimated in one search from

Table 2. Time and space complexities of ReMass-ReFeat and ReFeat

	ReMass-ReFeat	ReFeat
Time Complexity	$O(t(n+\psi)\log\psi)$ (Model building) $O(t(n+\log\psi))$ (On-line query)	$O(t(n+\psi)\log\psi)$ (Model building) $O(t(n+\log\psi))$ (On-line query)
Space Complexity	$O(t(n+\psi))$	$O(n+t\psi)$

the root to $T_i(\mathbf{q})$ in each tree. But, it will increase the space complexity as it requires to store n indices in each iTree. The time and space complexities of ReMass-ReFeat and ReFeat are provided in Table 2.

3 Empirical Evaluation

In this section, we evaluate the utility of relative mass in AD and CBMIR tasks. In AD, we compared ReMass-iForest with iForest [1], DEMass-LOF [5] and LOF [4]. In CBMIR, we compared ReMass-ReFeat with ReFeat [3] and the other existing CBMIR systems: MRBIR [7], InstRank [8] and Qsim [9]. Both the AD and CBMIR experiments were conducted in unsupervised learning settings. The labels of instances were not used in the model building process. They were used as the ground truth in the evaluation stage. The AD results were measured in terms of the area under ROC curve (AUC). In CBMIR, the precision at the top 50 retrieved results (P@50) [3] was used as the performance measure. The presented result was the average over 20 runs for all randomised algorithms. A two-standard-error significance test was conducted to check whether the difference in performance of two methods was significant.

We used the same MATLAB implementation of iForest provided by the authors of ReFeat [3], the JAVA implementation of DEMass-LOF in the WEKA [10] platform, and the JAVA implementation of LOF in the ELKI [11] platform.

We present the empirical evaluation results in the following two subsections.

3.1 Anomaly Detection: ReMass-iForest versus iForest

In the first experiment, we used a synthetic data set to demonstrate the strength of ReMass-iForest over iForest to detect local anomalies. The data set has 263 normal instances in three clusters and 12 anomalies representing global, local and clustered anomalies. The data distribution is shown in Figure 2a. Instances a_1, a_2 and a_3 are global anomalies; four instances in A_4 and two instances in A_5 are clustered anomalies; and a_6, a_7 and a_8 are local anomalies; C_1, C_2 and C_3 are normal instances in three clusters of varying densities.

Figures 2b-2d show the anomaly scores of all data instances obtained from iForest and ReMass-iForest. With iForest, local anomalies a_6, a_7 and a_8 had lower anomaly scores than some normal instances in C_3; and it produced AUC of 0.98. In contrast, ReMass-iForest had ranked local anomalies a_6, a_7, a_8 higher than any instances in normal clusters C_1, C_2 and C_3 along with global anomalies a_1, a_2 and a_3. But, ReMass-iForest with $MinPts = 1$ had some problem in ranking

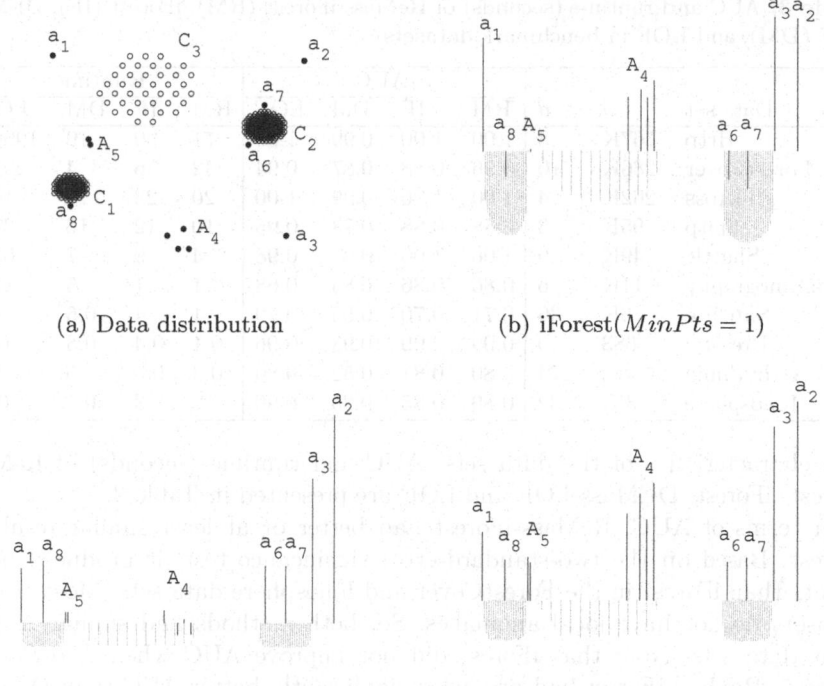

(a) Data distribution

(b) iForest($MinPts = 1$)

(c) ReMass-iForest($MinPts = 1$)   (d) ReMass-iForest($MinPts = 5$)

Fig. 2. Anomaly scores by iForest and ReMass-iForest using $t = 100, \psi = 256$. Note that in anomaly score plots, instances are represented by their values on x_1 dimension. Anomalies are represented by black lines and normal instances are represented by gray lines. The height of lines represents the anomaly scores. In order to differentiate the scores of normal and anomaly instances, the maximum score for normal instances is subtracted from the anomaly scores so that all normal instances have score of zero or less.

clustered anomalies in A_4 and produced AUC of 0.99. One fringe instance in the cluster C_3 was ranked higher than two clustered anomalies in A_4. This is because cluster anomalies have similar mass ratio w.r.t their parents as that for the instances in sparse normal cluster C_3. Clustered anomalies were correctly ranked and AUC of 1.0 was achieved when $MinPts$ was increased to 5. The performance of iForest did not improve when $MinPts$ was increased to any values in the range (2, 3, 4, 5 and 10).

In the second experiment, we used the ten benchmark data sets previously employed by Liu et al (2008) [1]. In ReMass-iForest, iForest and DEMass-LOF, the parameter t was set to 100 as default and the best value for the sub-sample size ψ was searched from 8, 16, 32, 64, 128 to 256. In ReMass-iForest, $MinPts$ was set to 5 as default. iForest uses the default settings as specified in [1], i.e, $MinPts = 1$. The level of subdivision (b) for each attribute in DEMass-LOF was searched from 1, 2, 3, 4, 5, and 6. In LOF, the best k was searched between 5 and 4000 (or to $\frac{n}{4}$ for small data sets), with steps from 5, 10, 20, 40, 60, 80, 150, 250, 300, 500, 1000, 2000, 3000 to 4000. The best results were reported.

Table 3. AUC and runtime (seconds) of ReMass-iForest (RM), iForest (IF), DEMass-LOF (DM), and LOF in benchmark datasets

Data set	n	d	AUC				Runtime			
			RM	IF	DM	LOF	RM	IF	DM	LOF
Http	567K	3	1.00	1.00	0.99	1.00	71	99	19	19965
ForestCover	286K	10	0.96	0.88	0.87	0.94	42	56	4	2918
Mulcross	262K	4	1.00	1.00	0.99	1.00	20	23	16	2169
Smtp	95K	3	0.88	0.88	0.78	0.95	10	12	16	373
Shuttle	49K	9	1.00	1.00	0.95	0.98	4	9	7	656
Mammography	11K	6	0.86	0.86	0.86	0.68	1	1	5	127
Satellite	6K	36	0.71	0.70	0.55	0.79	1	4	0.6	24
Breastw	683	9	0.99	0.99	0.98	0.96	0.1	0.4	0.3	0.4
Arrhythmia	452	274	0.80	0.81	0.52	0.80	0.3	0.5	5	1
Ionosphere	351	32	0.89	0.85	0.85	0.90	2	3	0.5	0.3

The characteristics of the data sets, AUC and runtime (seconds) of ReMass-iForest, iForest, DEMass-LOF and LOF are presented in Table 3.

In terms of AUC, ReMass-iForest had better or at least similar results to iForest. Based on the two-standard-error significance test, it produced better results than iForest in the ForestCover and Ionosphere data sets. Most of these datasets do not have local anomalies. So, both methods had similar AUC in eight data sets. Note that iForest did not improve AUC when $MinPts$ was set to 5. ReMass-iForest had produced significantly better AUC than DEMass-LOF in relatively high dimensional data sets (Arrhythmia - 274, Satellite - 36, Ionosphere - 32, ForestCover - 10, Shuttle - 9). These results show that DEMass-LOF has problem in handling data sets with a moderate number of dimensions (9 or 10). ReMass-iForest was competitive to LOF. It was better than LOF in the Mammography data set, worse in the Smtp and Satellite data sets, and equal performance in the other seven data sets.

As shown in Table 3, the runtime of ReMass-iForest, iForest and DEMass-LOF were of the same order of magnitude whereas LOF was upto three order of magnitude slower in large data sets. Note that we can not conduct a head-to-head comparison of runtime of ReMass-iForest and iForest with DEMass-LOF and LOF because they were implemented in different platforms (MATLAB versus JAVA). The results are included here just to provide an idea about the order of magnitude of runtime. The difference in runtime of ReMass-iForest and iForest was due to the difference in ψ and $MinPts$. $MinPts = 5$ results in smaller size iTrees in ReMass-iForest than those in iForest ($MinPts = 1$). Hence, ReMass-iForest runs faster than iForest even though the same ψ is used.

3.2   CBMIR: ReMass-ReFeat versus ReFeat

The performance of ReMass-ReFeat was evaluated against that of ReFeat in music and image retrieval tasks with GTZAN music data set [12] and COREL image data set [13], respectively. GTZAN is a data set of 1000 songs uniformly distributed in 10 genres. Each song is represented by 230 features. COREL is a data set of 10,000 images uniformly distributed over 100 categories. Each image

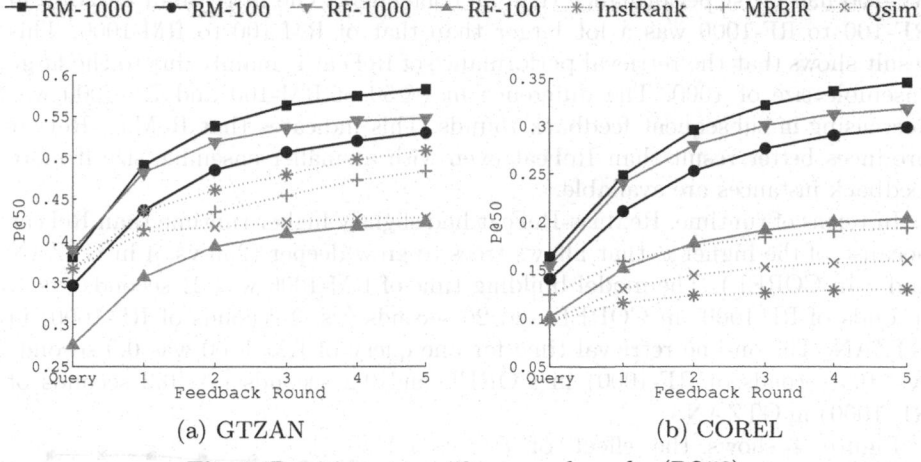

(a) GTZAN　　　　　　　　　　　(b) COREL

Fig. 3. Precision at top 50 returned results (P@50)

is represented by 67 features. These are the same data sets used in [3] to evaluate the performance of ReFeat. The results of the existing CBMIR systems InstRank, Qsim and MRBIR were taken from [3].

We conducted our experiments using the same experimental design as in [3]. Initially five queries were chosen randomly from each class. For each query, instances from the same class were regarded as relevant and the other classes were irrelevant. At each round of feedback, two relevant (instances from the same class) and two irrelevant (instances from the other classes) instances were provided. Upto five rounds of feedback were conducted for each query. The instance was not used in ranking if it was used as a feedback instance. The feedback process was repeated five times with different relevant and irrelevant feedbacks. The above process was repeated 20 times and average P@50 was reported.

In ReMass-ReFeat, the parameters ψ and $MinPts$ were set as default to 256 and 1, respectively. In ReFeat, ψ was set to 4 for GTZAN and 8 for COREL as reported in [3]. Other settings of ψ in ReFeat were found to perform worse than these settings. In order to show how their retrieval performance varies when ensemble size was increased, we used two settings for t: ReMass-ReFeat and ReFeat with (i) $t = 100$ (RM-100 and RF-100) and (ii) $t = 1000$ (RM-1000 and RF-1000). The feedback parameter γ was set as default to 0.5 in ReMass-ReFeat and 0.25 in ReFeat (as used in [3]).

P@50 of ReMass-ReFeat (RM-100 and RM-1000), ReFeat (RF-100 and RF-1000), InstRank, MRBIR and Qsim in the GTZAN and COREL data sets are shown in Figure 3. P@50 curves in both the data sets show that ReMass-ReFeat (RM-1000) has better retrieval performance than all contenders, especially in feedback rounds. In round 1 or no feedback (query only), ReMass-ReFeat (RM-1000) and ReFeat (RF-1000) produced similar retrieval performance but in latter feedback rounds, RM-1000 produced better results than RF-1000.

It is interesting to note that the performance of RF-100 was worse than that of RM-100 in all feedback rounds including query only (no feedback). In GTZAN,

RF-100 had worst performance than all contenders. The increase in P@50 from RF-100 to RF-1000 was a lot larger than that of RM-100 to RM-1000. This result shows that the retrieval performance of ReFeat is mainly due to the large ensemble size of 1000. The difference in P@50 of RM-100 and RF-1000 was decreasing in subsequent feedback rounds. This indicates that ReMass-ReFeat produces better result than ReFeat even with a smaller ensemble size if more feedback instances are available.

In terms of runtime, ReMass-ReFeat had slightly higher runtime than ReFeat because of the higher ψ that allows trees to grow deeper (256 vs. 4 in GTZAN and 8 in COREL). The model building time of RM-1000 was 21 seconds (vs. 4 seconds of RF-1000) in COREL and 20 seconds (vs. 2 seconds of RF-1000) in GTZAN. The on-line retrieval time for one query of RM-1000 was 0.9 seconds (vs. 0.3 seconds of RF-1000) in COREL and 0.2 seconds (vs. 0.2 seconds of RF-1000) in GTZAN.

Figure 4 shows the effect of ψ on the P@50 of ReMass-ReFeat and ReFeat at feedback round 5 (one run) in the GTZAN data set. In ReFeat, when ψ was increased above 4, the retrieval performance degraded. This is due to the increase in the height of iTrees ($h = \log_2(\psi)$) and instances falling in two distinct branches having similar relevance score based on the same path lengths. In contrast, ReMass-ReFeat improved its retrieval performance up to 64 and then remained almost flat beyond that. Similar effect was observed in the COREL data set where the performance of ReFeat degraded when ψ was set above 8.

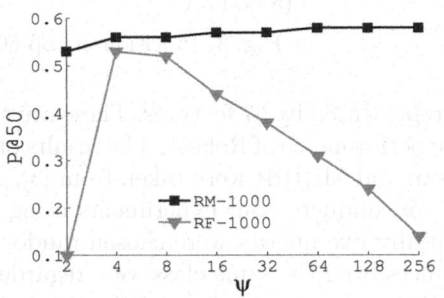

Fig. 4. P@50 at feedback round 5 with varying sample size (ψ) in the GTZAN data set

4   Conclusions

While the relative mass was motivated to overcome the weakness of iForest in detecting local anomalies, we have shown that the idea has a wider application. In information retrieval, we apply it to overcome the weakness of a state-of-the-art system called ReFeat. Our empirical evaluations show that ReMass-iForest and ReMass-ReFeat perform better than iForest and ReFeat, respectively, in terms of task-specific performance. In comparison with other state-of-the-art systems in both tasks, ReMass-iForest and ReMass-ReFeat are found to be either competitive or better.

The idea of relative mass in ReMass-iForest is similar to that of relative density in LOF and our empirical results show that ReMass-iForest and LOF have similar anomaly detection performance. However, ReMass-iForest runs significantly faster than LOF in large data sets because it does not require distance or density calculations.

Acknowledgement. This work is partially supported by the U.S. Air Force Research Laboratory, under agreement#FA2386-13-1-4043. Sunil Aryal is supported by Australian Postgraduate Award (APA), Monash University. The paper on mass-based similarity measure [14] has inspired us in creating the relevance score based on relative mass used in ReMass-ReFeat; though the motivations of the two papers differ.

References

1. Liu, F.T., Ting, K.M., Zhou, Z.H.: Isolation forest. In: Proceedings of the Eighth IEEE International Conference on Data Mining, pp. 413–422 (2008)
2. Ting, K.M., Zhou, G.T., Liu, F.T., Tan, S.C.: Mass estimation. Machine Learning 90(1), 127–160 (2013)
3. Zhou, G.T., Ting, K.M., Liu, F.T., Yin, Y.: Relevance feature mapping for content-based multimedia information retrieval. Pattern Recognition 45(4), 1707–1720 (2012)
4. Breunig, M.M., Kriegel, H.P., Ng, R.T., Sander, J.: LOF: Identifying Density-Based Local Outliers. In: Proceedings of ACM SIGMOD International Conference on Management of Data, pp. 93–104 (2000)
5. Ting, K., Washio, T., Wells, J., Liu, F., Aryal, S.: DEMass: a new density estimator for big data. Knowledge and Information Systems 35(3), 493–524 (2013)
6. Rui, Y., Huang, T., Ortega, M., Mehrotra, S.: Relevance feedback: a power tool for interactive content-based image retrieval. IEEE Transactions on Circuits and Systems for Video Technology 8(5), 644–655 (1998)
7. He, J., Li, M., Zhang, H.J., Tong, H., Zhang, C.: Manifold-ranking based image retrieval. In: Proceedings of the 12th Annual ACM International Conference on Multimedia, pp. 9–16. ACM, New York (2004)
8. Giacinto, G., Roli, F.: Instance-based relevance feedback for image retrieval. In: Advances in Neural Information Processing Systems, vol. 17, pp. 489–496 (2005)
9. Zhou, Z.H., Dai, H.B.: Query-sensitive similarity measure for content-based image retrieval. In: Proceedings of the Sixth International Conference on Data Mining, pp. 1211–1215 (2006)
10. Hall, M., Frank, E., Holmes, G., Pfahringer, B., Reutemann, P., Witten, I.H.: The WEKA Data Mining Software: An Update. SIGKDD Explorations 11(1) (2009)
11. Achtert, E., Hettab, A., Kriegel, H.-P., Schubert, E., Zimek, A.: Spatial outlier detection: Data, algorithms, visualizations. In: Pfoser, D., Tao, Y., Mouratidis, K., Nascimento, M.A., Mokbel, M., Shekhar, S., Huang, Y. (eds.) SSTD 2011. LNCS, vol. 6849, pp. 512–516. Springer, Heidelberg (2011)
12. Tzanetakis, G., Cook, P.: Musical genre classification of audio signals. IEEE Transactions on Speech and Audio Processing 10(5), 293–302 (2002)
13. Zhou, Z.H., Chen, K.J., Dai, H.B.: Enhancing relevance feedback in image retrieval using unlabeled data. ACM Transactions on Information Systems 24(2), 219–244 (2006)
14. Ting, K.M., Fernando, T.L., Webb, G.I.: Mass-based Similarity Measure: An Effective Alternative to Distance-based Similarity Measures. Technical Report 2013/276, Calyton School of IT, Monash University, Australia (2013)

Domain Transfer via Multiple Sources Regularization

Shaofeng Hu[1], Jiangtao Ren[1], Changshui Zhang[2,3,4], and Chaogui Zhang[1]

[1] School of Software, Sun Yat-sen University, Guangzhou, P.R. China
[2] Department of Automation, Tsinghua University, Beijing, P.R. China
[3] State Key Lab. of Intelligent Technologies and Systems, Beijing, P.R. China
[4] Tsinghua National Laboratory for Information Science and Technology (TNList)
{hugoshatzsu,daguizhang}@gmail.com, issrjt@mail.sysu.edu.cn,
zcs@mail.tsinghua.edu.cn

Abstract. The common assumption that training and testing samples share the same distribution is often violated in practice. When this happens, traditional learning models may not generalize well. To solve this problem, domain adaptation and transfer learning try to employ training data from other related source domains. We propose a multiple sources regularization framework for this problem. The framework extends classification model with regularization by adding a special regularization term, which penalizes the target classifier far from the convex combination of source classifiers. Then this framework guarantees the target classifier minimizes the empirical risk in target domain and the distance from the convex combination of source classifier simultaneously. By the way, the weights of the convex combination of source classifiers are embedded into the learning model as parameters, and will be learned through optimization algorithm automatically, which means our framework can identify similar or related domains adaptively. We apply our framework to SVM classification model and develop an optimization algorithm to solve this problem in iterative manner. Empirical study demonstrates the proposed algorithm outperforms some state-of-art related algorithms on real-world datasets, such as text categorization and optical recognition.

Keywords: domain adaptation, multiple sources regularization.

1 Introduction

The common assumption that training and testing samples share the same distribution is often violated in practice. When this happens, traditional learning models may not generalize well even with abundant training samples. *Domain Adaptation* is one of these situations where little labeled data is provided from target domain, but large amount of labeled data from source domains are available. Domain adaptation methods [1,2] learn robust decision function by leveraging labeled data both from target and source domains which usually don't share the same distributions. This problem involves in many real world application such as natural language processing[3], text categorization[4], video concept detection[5], WiFi localization[4], remote sensor network[2], etc.

V.S. Tseng et al. (Eds.): PAKDD 2014, Part II, LNAI 8444, pp. 522–533, 2014.

Most of domain adaptation methods can be classified into two classes according to their strategies of adapting source information: either with sources labeled data or with sources classifier. The former strategy selects source labeled samples that match target distribution to overcome distribution discrepancy. For example, [6] predicts unlabel samples via an ensemble method in local region including labeled samples of sources. [7] iteratively draws sources labeled samples that are in the same cluster with target labeled data in projected subspace. Alternately, the latter strategy try to get the final target classifier by weighted sum of target classifier f_T trained from target domain data and multiple source classifiers $\{f_{S_1}, f_{S_2}, \ldots, f_{S_m}\}$ trained from source domain data. [8] seeks a convex combination of f_T and $\frac{1}{m} \sum_k f_{S_k}$ by cross validation. [9] proposes Adaptive Support Vector Machine (ASVM) to learn f_T by incorporating the weighted sum of source classifiers $\sum_k \lambda_k f_{S_k}$ into the objective function of SVM, where λ_k is evaluated by a meta-learning procedure. [10] obtains the final f_T by maximizing output consensus of source classifiers. [11] modifies f_T and penalizes the output difference between f_T and each f_{S_k} on unlabeled data.

We focus on the strategy of adapting source classifiers in this paer. Based on the related works, it can be summarized one of the simplest methods to adapt source classifiers is treating their weighted sum as a single classifier. However, performance of this strategy is dependent on the weights for target and sources classifiers. It would be appropriate to assign higher weights to sources that are more similar with target domain. To our best of knowledge, although a few works have been addressed on domain weights assignment, little of them try to learn the appropriate weights automatically. [8] weights each source equally. [9] evaluates weights by meta-learning algorithm which is not promising since features of meta-learning are only dependent on the output of source classifiers. [11] determines domain weights by estimating the distribution similarity by MMD.

In this paper, we propose a novel way of adapting source classifiers by considering multiple source classifiers as prior information. Instead of learning the combination weights of target and source classifiers explicitly, we learn the target classifier directly from target domain data while keeping the target classifier approximates a convex combination of source classifiers as closely as possible, and the convex combination weights of source classifiers will be learned jointly with the learning of target classifier through optimization methods.

To illuminate the motivation of our paper, let us consider an example in Figure 1. Because of the rareness of labeled data in target domain, it is hard to learn a good target classifier directly. For example, in Figure 1 (a), only one labeled sample of each class is provided, denoted by □/∗ respectively. There exists a very large classifier space in which every classifier can separate the training samples well with high uncertainty on test samples however. As depicted in Figure 1(a), the horizontal hyperplane (solid line with circle) generalizes best based on the real classes distribution indicated by different colors. But we will get a bad hyperplane (dotted line with triangular) by large margin principle[12]. However, by the introduction of some useful prior information contained in related source domains, we can improve the target classifier performance on test samples.

We can restrict the target classifier approximates the convex combination of source classifiers, because we think the convex combination of the source classifiers is a compact version of the source classifiers. In this way, we can exploit every source classifier with high confidence. Further, when we add the convex combination of the source classifiers as a regularization term to object function, it will shrink the search space of target classifier greatly and provide a good way to optimize it. For example, Figure 1(b) presents two source classifiers (dotted line with diamond), and a gray region which represents the convex combination space of the two source classifiers. It is clear that if the target classifier is in or near to the convex combination space of source classifiers, the target classifier (dotted line with triangular in Figure 1 (b)) will have better generalization performance than the one learned by large margin principle.

Therefore, we propose a multiple sources regularization framework based on the above motivation. The framework extends general classification model with regularization by adding a special regularization term, which penalizes the target classifier far from the convex combination of source classifiers. Then this framework make sure the target classifier minimizes the empirical risk in target domain and the distance from the convex combination of source classifier simultaneously. By the way, the weights of the convex combination of source classifiers are embedded into the learning model as parameters, and will be learned through optimization algorithm automatically, which means our framework can identify similar or related domains adaptively. we propose an iterative algorithms to solve this optimization problem efficiently.

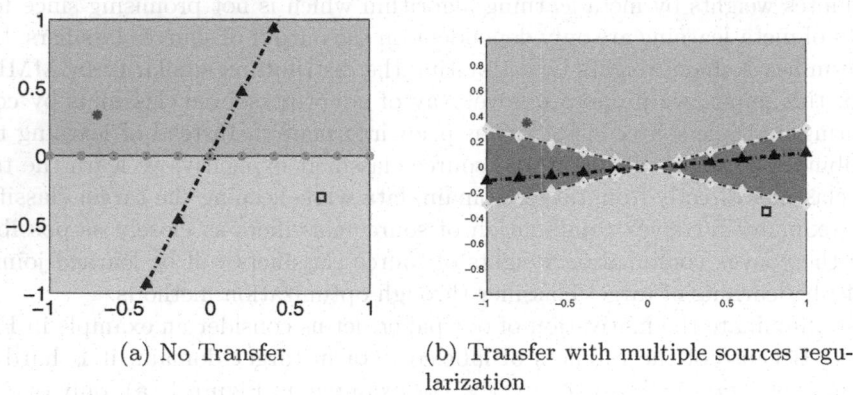

(a) No Transfer   (b) Transfer with multiple sources regularization

Fig. 1. Intuitive example about multiple sources regularization

2   Multiple Sources Regularization Framework

To solve multiple sources domain adaptation problem, we propose a multiple sources regularization framework. Supposed there exist m source domain data sets, denoted by $S = \{S_1, S_2, \ldots, S_m\}$. We assume that all the samples in source data sets are labeled, etc. $S_k = \{X_{s_k}, y_{s_k}\}$ and $|X_{s_k}| = |y_{s_k}|$ for all $k \in \{1, \ldots, m\}$. y_{s_k} is output variable, which can be either continuous or discrete. Correspondingly, target domain is unique and is divided into labeled training set and unlabel testing set, etc. $T = \{(X_L, y_L), X_U\}$. A multiple sources domain adaptation problem can be summarized as: (a) each source domain has a different but similar distribution with target domain, $Pr_{S_k}(X, Y) \neq Pr_T(X, Y)$. (b) scale of training set of target domain is much smaller than that of test set, $|X_L| \ll |X_U|$. (c) source and target domain share the same output variable. (d) the objective of multiple sources domain adaptation is to utilize source data S to improve learning performance of target domain T. We firstly discuss our regularization framework under general linear form. Then the framework is extended to RKHS (Reproducing Kernel Hilbert Space) with SVM hinge loss function for classification problem. Thirdly, an iterative optimization algorithm is proposed to efficiently solve multiple sources regularization SVM.

2.1   Multiple Sources Regularization Framework

In this section, multiple source regularization (MSR) framework is introduced for classification. We start from the linear classification model. Linear model is more intuitive and geometrically interpretable. Denote linear predictive function $f(\mathbf{x}) = \mathbf{w}^T \mathbf{x} + b$, where $\mathbf{w}$ is feature weights and b is bias term of separating hyperplane. Learning algorithms seek to find optimum $\mathbf{w}$ and b that minimize structural risk, such as hinge loss. Generally, structural risk trade off between empirical risk and regularization:

$$\min_{\mathbf{w}, b} \sum_{i=1}^{l} L(x_i, y_i; \mathbf{w}, b) + \lambda \Phi(\mathbf{w}) \tag{1}$$

$L(\cdot)$ is loss function while $\Phi(\cdot)$ is regularization term. $\Phi(\cdot)$ penalizes function complexity to avoid overfitting. When labeled data number l is large enough, Eq(1) is a tight upper bound of expected risk. However, under domain adaptation setting, training samples of T would be scarce. Therefore, structural risk will be too loose to be used as upper bound under supervised learning setting. As we know, loose bound created by Eq(1) will be tightened by introducing unlabeled data which is referred as semi-supervised learning. Alternately, our framework alleviates this problem by including multiple source classifiers trained from source domain labeled data. To do this, we modify Eq(1) by adding an extra regularization term as following:

$$\min_{\mathbf{w}, b, \beta} \sum_{i=1}^{l} L(x_i, y_i; \mathbf{w}, b) + \lambda \Phi(\mathbf{w}) + \rho \|\mathbf{w} - W_s \beta\|_2^2$$

$$s.t. \sum_{k=1}^{m} \beta_k = 1, \beta \geq 0 \tag{2}$$

where $W_s = [w_{s1}, w_{s2}, \ldots, w_{sm}] \in R^{d \times m}$. w_{sk} is a feature weight vector learned from the k-th source domain S_k. Learning model of each source domain should be consistent with that of target domain in order to maintain homogeneity of model coefficients. The last regularization term of Eq(2) penalizes w far away from the convex combination of m source classifiers. $\rho > 0$ trade off between structural risk and multiple source regularization. Moreover, $\beta \in \Delta$ denotes the weight vector that determines convex combination of source classifiers. Δ represents the m dimensional simplex: $\Delta = \{\beta : \sum_{k=1}^{m} \beta_k = 1, \beta_k \geq 0\}$. Our method of determining β also differs from other state-of-art multiple source domain adaptation methods: other than manual setting, meta learning or model selection, our framework embedded the auxiliary domain weights vector β into model Eq(2) as a parameter, then β can be learned by optimization method automatically. When $m = 1$, Eq(2) degenerate to a simple situation. β is fixed to be 1 which make Eq(2) much similar to Eq(1) from optimization aspect. Therefore, we only focus on the situation where $m > 1$ in Eq(2) in our paper.

2.2   Multiple Sources Regularization SVM (MSRSVM)

Loss function L in Eq(2) varies according to different models. It is easy to realize that our framework can be adapted to a wide variety of models including SVM, logistic regression, ridge regression and so on. SVM hinge loss is discussed in detail in the following. We choose SVM for discussion with the following reasons: (a) It is convenient to transform Eq(2) to its dual problem, extending linear model to kernel form.(b) SVM fits for problems with very little training samples which is consistent with the setting of domain adaptation.

Firstly, with hinge loss, Eq(2) can be reformulated as:

$$\min_{w,b,\xi,\beta} \quad \sum_{i=1}^{l} \xi_i + \lambda \Phi(\mathbf{w}) + \rho \|\mathbf{w} - W_s\beta\|_2^2 \tag{3}$$

$$s.t. \quad y_i(\mathbf{w}^T\mathbf{x} + b) \geq 1 - \xi_i, \xi_i \geq 0, i \in L$$

$$\sum_{k=1}^{m} \beta_k = 1, \beta \geq 0$$

where ξ is the slack variable. From the viewpoint of optimization, Eq(3) is a QP problem. While Eq(3) is QP, it can be solved by numeric optimization package directly.

However, we transform Eq(3) to dual form instead of optimizing the prime problem directly for two reasons: a) variable dimension of Eq(3) is $d + n + m + 1$ while dual problem shrinks to $n + m$. Optimizing the dual problem reduces the problem complexity. b) the dual problem can generalize the linear model to nonlinear case in RHKS (Reproduced Hilbert Kernel Space). It is worth to note that we do not transform all variables to dual problem. β remains fixed in prime form. This is because Eq(3) can be transform to an optimization whose structure is very closed to regular SVM dual problem without β. Then the

Lagrange function of Eq(3) can be formulated as:

$$L(\boldsymbol{w}, b, \boldsymbol{\xi}, \boldsymbol{\alpha}, \boldsymbol{\mu}) = C \sum_{i \in L} \xi_i + \frac{1}{2} \|\boldsymbol{w}\|^2 + \|\boldsymbol{w} - W_s \boldsymbol{\beta}\|^2$$
$$- \sum_{i \in L} \alpha_i \left(y_i \boldsymbol{w}^T \phi(\boldsymbol{x_i}) + b - 1 + \xi_i \right)$$
$$+ \sum_{i \in L} \mu_i \xi_i$$

Setting derivative of Lagrange function with (w, b, ξ) to zero and adding other constraints under KKT condition, we obtain the dual problem:

$$\max_{\beta} \min_{\alpha} \ \alpha^T e - \frac{1}{2(1+\rho)} \left(\rho \beta^T \Sigma_s \beta - \alpha^T Y K Y \alpha \right)$$
$$- \rho \beta^T \Omega \alpha$$
$$s.t. \ \alpha^T y = 0$$
$$0 \le \alpha \le C$$
$$\sum_{k=1}^{m} \beta_k = 1, \beta \ge 0 \tag{4}$$

Where $\Sigma_s \in R^{m \times m}$ is a symmetric matrix representing correlation of feature weight among multiple source domains, analogous to covariance matrix of Gaussian distribution. Moreover, $\Omega \alpha$ is a vector related to the correlation of feature weight between target domain and source domain. Σ_s and Ω can be evaluated using definition of its element in kernel space:

$$\Sigma_s = \begin{pmatrix} \alpha_{s_1}^T K_{s_1, s_1} \alpha_{s_1} & \alpha_{s_1}^T K_{s_1, s_2} \alpha_{s_2} & \cdots \\ \alpha_{s_2}^T K_{s_2, s_1} \alpha_{s_1} & \alpha_{s_2}^T K_{s_2, s_2} \alpha_{s_2} & \cdots \\ \vdots & \vdots & \ddots \end{pmatrix}$$
$$\Omega = \begin{pmatrix} \alpha_{s_1}^T K_{s_1, t} \\ \vdots \\ \alpha_{s_m}^T K_{s_m, t} \end{pmatrix} Y \tag{5}$$

Where α and $Y = diag(y)$ denote dual variable and output diagonal matrix respectively. K are the kernel matrix constructed by input patterns from either source or target domains. We need to note that α_{sk} is optimized SVM dual variable training only on S_k.

Eq(4) appears to be more complicated than Eq(3). Actually, Eq(4) is a saddle-point minmax problem, which can be regard as a zero sum game between two players. In section 2.3, we develop a two stage iterative optimization algorithm to solve Eq(4) in a general framework. This problem is similar to the optimization problem referred in DIFFRC[13], SimpleNPKL[14] and SimpleMKL[15].

2.3 Iterative Optimization Algorithm for MSRSVM

The special structure of Eq (4) indicates that MSRSVM needs a customized optimization algorithm. Fortunately, many optimization algorithms have been proposed to solve similar min-max problems. The main idea of such algorithms is to separately optimize part of variables while keep others fixed. It turns out that Eq (4) can be decomposed into two subproblems.

The main steps of our optimization algorithm for MSRSVM is described as following. Denote $J(\alpha, \beta)$ as the objective function of Eq (4) and α^* and β^* as the optimal solution of model variables. The complicated min-max problem of Eq (4) can be decomposed into two simple optimization problems: minimizing $J(\alpha, \tilde{\beta})$ as well as maximizing $J(\tilde{\alpha}, \beta)$ where $\tilde{\alpha}$ and $\tilde{\beta}$ are fixed values. Therefore, after initialization of α and β, the algorithm will loop over two steps until stop condition is met:

- Step 1: solve subproblem $\min_{0 \le \alpha \le C, \alpha^T y = 0} J(\alpha, \tilde{\beta})$ under fixed $\tilde{\beta}$.
- Step 2: solve subproblem $\max_{\sum_k \beta_k = 1, \beta \ge 0} J(\tilde{\alpha}, \beta)$ under fixed $\tilde{\alpha}$.

The optimization problem in Step 1 shares similar problem structure with common SVM. They differ only on the first order term of objective functions. The first order term of common SVM's objective function is all one vector e while Eq (4) has an extra negative term $\rho \beta^T \Omega$. Fast algorithms such as SMO or SVM^{light} could be adapted to solve this subproblem of MSRSVM without many modifications. Thus we optimize subproblem of Step 1 using a modified version of regular SVM algorithms. Without referring any specific implementation of SVM algorithm, we use SVMSolver (K, f, y) to define a general solver for SVM optimization where K, f, y denote kernel matrix, first order term and label vector respectively.

The subproblem of Step 2 is a classical QP problem. Since dimension of β is m and m is not large usually, Newton method is appropriate to optimize β. However, as stated in section 2.1, Eq (4) is a saddle point minmax problem. If both optimization steps are taken to local optimum point, fluctuation happens and progress towards global optimum slows down. Therefore, as an alternately strategy we update β by taking one gradient step at each iteration. Regular gradient update formula can not be used here because the simplex constraint exists, and gradient method is for unconstrained optimization generally. In this paper, reduced gradient method is introduced to handle this simplex constraint optimization problem [15]. This method evaluates ascent gradient firstly, then projects the gradient into simplex using the formula stated below:

$$
D = \begin{cases}
0 & \text{if } \beta_k = 0, \frac{\partial J_{\tilde{\alpha}}}{\partial \beta_k} - \frac{\partial J_{\tilde{\alpha}}}{\partial \beta_\mu} > 0 \\
-\frac{\partial J_{\tilde{\alpha}}}{\partial \beta_k} + \frac{\partial J_{\tilde{\alpha}}}{\partial \beta_\mu} & \text{if } \beta_k = 0, k \neq \mu \\
\sum_{\beta_\nu > 0} \left(\frac{\partial J_{\tilde{\alpha}}}{\partial \beta_\mu} - \frac{\partial J_{\tilde{\alpha}}}{\partial \beta_\nu} \right) & \text{for } k = \mu
\end{cases} \tag{6}
$$

Where $\frac{\partial J_{\tilde{\alpha}}}{\partial \beta_k}$ and $\frac{\partial J_{\tilde{\alpha}}}{\partial \beta_\mu}$ are the gradients of objective function with fixed $\tilde{\alpha}$, k and μ are vector indexes. Then we update β by using: $\beta_{t+1} = \beta_t + \eta D$. η

denotes the step length. Boyed [16] showed that when η is small enough at each iteration, global convergence could be guaranteed. η is choosen to be $O(\frac{1}{t})$. We use **objective gap** as convergence criterion. **Objective gap** represents absolute difference between the objective value after Step 1 and Step 2 within the same iteration. Algorithm 1 summarizes the whole iterative optimization algorithm.

Algorithm 1. Iterative optimization algorithm for MSRSVM

---

1. **Input:** target training data (X_t, y_t), m source data sets $\{(X_{s_1}, y_{s_1}), (X_{s_2}, y_{s_2}), \ldots, (X_{s_m}, y_{s_m})\}$;
2. **Output:** optimized variable α^* and β^*;
3. initialize $\alpha_0 = \mathbf{0}$ and $\beta_0 = \frac{1}{m}\mathbf{e}$;
4. **for** i=1 to m **do**
5.   construct kernel matrix K_{s_i} using X_{s_i}, $K_{s_i,t}$ using X_{s_i} and X_t;
6.   optimize corresponding dual variable $\alpha_{s_i} = $ SVMSolver (K_{s_i}, e, y_{s_i});
7.   **for** j=1 to m **do**
8.   construct kernel matrix K_{s_i,s_j} using X_{s_i} and X_{s_j};
9.   **end for**
10. **end for**
11. calculate Σ_s and Ω by Eq (5) with K_{s_i,s_j} and $K_{s_i,t}$ for $i,j \in \{1, \ldots, m\}$;
12. **while** convergence criterion is not met **do**
13.   solve modified SVM subproblem, $\alpha_t = $ SVMslover $(K_t, (1+\rho)(e - \rho\beta_t^T \Omega), y_t)$;
14.   calculate D using Eq (6), then β_t and update β: $\beta_{t+1} = \beta_t + \eta D$, and $\eta = \frac{1}{t}$;
15. **end while**

---

3 Experiment

To demonstrate the effectiveness of our proposed framework MSRSVM, we perform experiments on multiple transfer learning data sets. They are real world data sets that frequently used in the context of transfer learning or multitask learning. Performance of MSRSVM are compared with some other state-of-art algorithms that can handle multiple source domains.

3.1 Data Sets and Experiment Setup

Three data collections are used in our experiment study, they are Reuters-21578[17], 20-Newsgroups[18] and Letters. Among them, Reuters-21578 and 20-Newsgroups are benchmark of text categorization for transfer learning. Letters is optical recognition dataset that is preprocessed for multitask learning.

Data Sets. All data sets that have been used in our experiment study are binary classification tasks. Reuters-21578 and 20-Newsgroups are both text categorization data collections with hierarchical class structure. For each dataset, we need to construct both target and source domain dataset. Target and source domain datasets are sampled from different subcategories of the same top categories. For example, for dataset "comp vs rec", its source task dataset is sampled from subcategories "comp.windows.x" and "rec.autos", while target task dataset is sampled from subcategories "comp.graphics" and "rec.motocycles". Therefore, source and target domain datasets share the same feature space but different words distribution. But in our multiple source adaptation setting, we need more than one source domain datasets for one target domain prediction task. To solve this problem, all the source domain datasets are grouped and shared as source domain datasets. For example, in 20-Newsgroups task, the source domain datasets of "comp vs sci", "rec vs talk", "rec vs sci", "sci vs talk", "comp vs rec" and "comp vs talk" constitute of the multiple source domains. While keeping the 6 source domains fixed, we can construct different multiple source adaptation problems with different target domain datasets.

For Letters dataset without hierarchical class structure, we build different learning tasks by randomly sampling from two different handwritten digit letters that are difficult to be distinguished. For example, "c/e" denotes a prediction task that "c" is the positive class while "e" is the negative class. Each task is treated as target task and all the other tasks as source tasks. For example, if "c/e" is target task, then task "g/y", "m/n", "a/g", "i/j", "a/o", "f/t" and "h/n" form the 7 source domain tasks.

Baseline. We compare the performance of MSRSVM with other SVM based learning algorithms which can cope with multiple sources adaptation problems. They are ASVM[9], FR[19], MCCSVM[8]and regular SVM without any transfer. ASVM can be obtained online, which is based on LibSVM and programmed in C++. Others including MSRSVM are implemented in matlab basing on SMO. ASVM combines source classifiers with weights by an independent meta learning algorithm. SVM classification parameter C is fixed to 10. Other related parameter are set to default values. Moreover, RBF kernel $k(x, y) = e^{-\sigma \|x-y\|^2}$ is chosen as kernel function, where σ is set to 0.0001 for text data and 0.01 for optical recognition. For MSRSVM, model parameter ρ is set to 1.

3.2 Performance Study

We adopt classification accuracy as evaluation metric to compare MSRSVM with other four state-of-art methods. All of the accuracy results in this paper are the average results of 10 experiments. The accuracy comparison results are summarized in Table 1 and Table 2 for text and Letters dataset respectively. Training ratio are fixed to 20% for text datasets, and 30 points are randomly selected as training set for Letters dataset. Note that the best results are highlighted in bold in the Table 1 and Table 2. On Reuters-21578 dataset, MSRSVM performs better than all of the baseline algorithms on all of the 3 tasks. For example, MSRSVM get the accuracy of 60.81% on **Pe vs Pl** dataset, while ASVM get the accuracy of 59.27%, which is the best one of baseline methods. On 20-Newsgroup dataset, MSRSVM improves the accuracies significantly in most of time, comparing with the baseline methods. MSRSVM performs at least 3% better than regular SVM on 4 of 6 data sets. Meanwhile, MSRSVM outperforms other methods on all data sets except **Comp Vs Talk** where MCCSVM achieves highest accuracy, slightly

Table 1. Accuracy comparison on text data sets(%)

Method	Reuters			20-Newsgroup					
	O vs Pe	O vs Pl	Pe vs Pl	C vs S	T vs R	R vs S	S vs T	C vs R	C vs T
SVM	61.25	57.25	58.91	83.22	86.19	93.76	88.01	90.37	86.94
MCCSVM	60.73	55.63	58.46	84.37	87.80	89.27	89.66	88.06	**90.76**
FR	53.17	55.28	48.26	65.87	75.99	52.80	57.54	53.07	66.91
ASVM	57.28	56.68	59.27	83.00	64.80	76.43	50.23	51.12	65.26
MSRSVM	**62.50**	**59.40**	**60.81**	**85.92**	**91.40**	**94.76**	**93.02**	**93.38**	89.27

better than MSRSVM (less than 2%). Moreover, ASVM performs surprisingly poor on some tasks of 20-Newsgroup such as **Sci vs Talk** and **Comp vs Rec**, while MSRSVM behaves stable on all of the text datasets. Similar conclusions can be reached according to Letters dataset. MSRSVM achieves the highest accuracy on 5 of 8 datasets, and MCCSVM achieves on 3 of 8. And the performance of MSRSVM is still more stable than the others. Thus on the whole, MSRSVM significantly improves the accuracy most of the time.

Table 2. Accuracy comparison on Letters dataset(%)

Method	c vs e	g vs y	m vs n	a vs g	i vs j	a vs o	f vs t	h vs n
SVM	84.89	67.98	81.02	83.59	67.15	82.49	80.79	81.87
MCCSVM	84.07	68.99	87.58	**89.79**	**94.49**	**88.25**	78.14	90.14
FR	50.00	50.68	50.26	82.50	47.18	53.91	54.42	46.17
ASVM	78.11	50.00	50.38	60.64	37.60	50.32	52.21	62.71
MSRSVM	**89.48**	**71.52**	**87.59**	87.73	90.07	86.86	**83.20**	**91.10**

Performance of classifier may be dependent on the number of training data. When we refer training data here, it means training data of target domain. As mention before, samples of sources domains are fixed and all labeled. Figure 2 depicts the performance of MSRSVM, regular SVM and MCCSVM, with respect to different ratio or number of training data in target domain. Training data of target domain is assumed to be sparse in domain adaptation problem. Thus the ratio of training data varies from 0.05 to 0.3 for text datasets, and number of training data for Letters datasets varies from 18 to 38(1~2% of the whole sample set) in the experiments. MSRSVM is compared with regular SVM and MCCSVM because they are more sensitive to the size of training data. Two important conclusions can be reached based on Figure 2. Firstly, performance of the three algorithms improve with the increase of the size of training data most of time. This is because the target classifier can get more information about target domain with more and more labeled data coming from target domain. Secondly, MSRSVM outperforms regular SVM and MCCSVM steadily most of time, especially when the size of training data are small. For example, MSRSVM wins on nearly all 20-Newsgroup data sets with only 5% of training data except

532   S. Hu et al.

for **Rec vs Sci**. Similar phenomena happens for Letters datasets. The accuracy gap between MSRSVM and MCCSVM is maximum when the number of training data is about 18-22. The curves also demonstrate MSRSVM can utilize the information of source domains more effectively than MCCSVM.

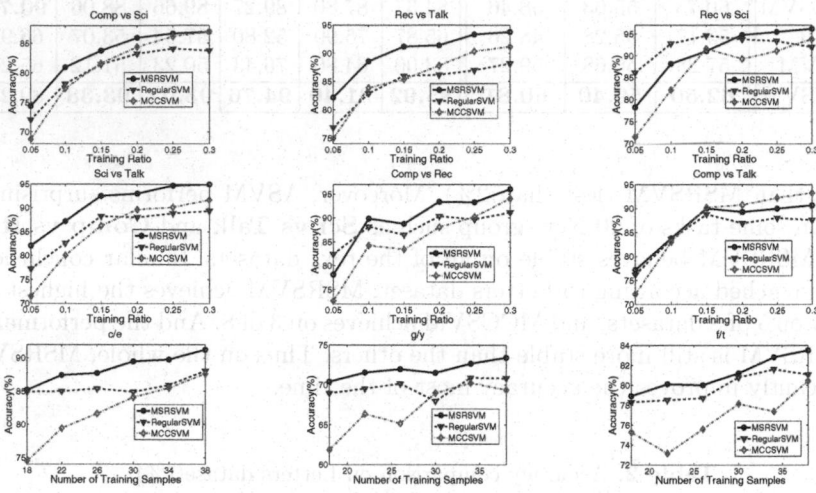

Fig. 2. Accuracy wrt ration or number of training data

4   Conclusion

We address multiple source domain adaptation problem in this paper. There exist more than one similar or related source domains whose distributions are not identical with the target domain. To adaptively utilize the information of sources domains and improve the performance of target classifier, we propose a simple framework named Multiple Source Regularization framework. This framework regularizes target classifier and make it approximate the convex combination of sources' classifier, while the combination weights will be learned adaptively. Our idea is that the sources information in regularization function acts as a prior to target domain. By substituting SVM's loss function into MSR framework, we propose a Multiple Source Regularization SVM (MSRSVM) model, and develop an optimization algorithm to solve this model in iterative manner. Experiments on both text and optical recognition datasets verify that MSRSVM outperforms many other state-of-art domain adaptation algorithms.

Acknowledgements. This work is supported by the Fundamental Research Funds for the Central Universities under grant 12lgpy40, Guangdong Natural Science Foundation under grant S2012010010390 and Beijing Municipal Education Commission Science and Technology Development Plan key project under grant KZ201210005007.

References

1. Crammer, K., Kearns, M., Wortman, J.: Learning from multiple sources. Journal of Machine Learning Research 9, 1757–1774 (2008)
2. Bruzzone, L., Marconcini, M.: Domain adaptation problems: A dasvm classification technique and a circular validation strategy. IEEE Trans. Pattern Anal. Mach. Intell. 32(5), 770–787 (2010)
3. Jiang, J., Zhai, C.: Instance weighting for domain adaptation in nlp. In: ACL. The Association for Computer Linguistics (2007)
4. Pan, S.J., Tsang, I.W., Kwok, J.T., Yang, Q.: Domain adaptation via transfer component analysis. In: IJCAI, pp. 1187–1192 (2009)
5. Duan, L., Tsang, I.W.-H., Xu, D., Maybank, S.J.: Domain transfer svm for video concept detection. In: CVPR, pp. 1375–1381. IEEE (2009)
6. Gao, J., Fan, W., Jiang, J., Han, J.: Knowledge transfer via multiple model local structure mapping. In: KDD, pp. 283–291 (2008)
7. Zhong, E., Fan, W., Peng, J., Zhang, K., Ren, J., Turaga, D.S., Verscheure, O.: Cross domain distribution adaptation via kernel mapping. In: KDD, pp. 1027–1036 (2009)
8. Schweikert, G., Widmer, C., Schölkopf, B., Rätsch, G.: An empirical analysis of domain adaptation algorithms for genomic sequence analysis. In: NIPS, pp. 1433–1440 (2008)
9. Yang, J., Yan, R., Hauptmann, A.G.: Cross-domain video concept detection using adaptive svms. ACM Multimedia, 188–197 (2007)
10. Luo, P., Zhuang, F., Xiong, H., Xiong, Y., He, Q.: Transfer learning from multiple source domains via consensus regularization. In: CIKM, pp. 103–112 (2008)
11. Duan, L., Tsang, I.W., Xu, D., Chua, T.-S.: Domain adaptation from multiple sources via auxiliary classifiers. In: ICML, p. 37 (2009)
12. Vapnik, V.: Statistical Learning Theory. JohnWiley, NewYork (1998)
13. Bach, F., Harchaoui, Z.: Diffrac: a discriminative and flexible framework for clustering. In: NIPS (2007)
14. Zhuang, J., Tsang, I.W., Hoi, S.C.H.: Simplenpkl: simple non-parametric kernel learning. In: ICML, p. 160 (2009)
15. Szafranski, M., Grandvalet, Y., Rakotomamonjy, A.: Composite kernel learning. Machine Learning 79(1-2), 73–103 (2010)
16. Boyd, S., Xiao, L.: Least-squaures covariance matrix adjustment. SIAM Journal of Matrix Anal. Appl. 27, C532–C546 (2005)
17. Asuncion, A., Newman, D.J.: UCI machine learning repository (2007), http://www.ics.uci.edu/mlearn/ML-Repository.html
18. Davidov, D., Gabrilovich, E., Markovitch, S.: Parameterized generation of labeled datasets for text categorization based on a hierarchical directory. In: SIGIR, pp. 250–257 (2004)
19. Daumé III, H.: Frustratingly easy domain adaptation. In: Conference of the Association for Computational Linguistics (ACL), Prague, Czech Republic (2007)

Mining Diversified Shared Decision Tree Sets for Discovering Cross Domain Similarities

Guozhu Dong and Qian Han

Knoesis Center, and Department of Computer Science and Engineering,
Wright State University, Dayton, Ohio, USA
{guozhu.dong,han.6}@wright.edu

Abstract. This paper studies the problem of mining diversified sets of shared decision trees (SDTs). Given two datasets representing two application domains, an SDT is a decision tree that can perform classification on both datasets and it captures class-based population-structure similarity between the two datasets. Previous studies considered mining just one SDT. The present paper considers mining a small diversified set of SDTs having two properties: (1) each SDT in the set has high quality with regard to "shared" accuracy and population-structure similarity and (2) different SDTs in the set are very different from each other. A diversified set of SDTs can serve as a concise representative of the huge space of possible cross-domain similarities, thus offering an effective way for users to examine/select informative SDTs from that huge space. The diversity of an SDT set is measured in terms of the difference of the attribute usage among the SDTs. The paper provides effective algorithms to mine diversified sets of SDTs. Experimental results show that the algorithms are effective and can find diversified sets of high quality SDTs.

Keywords: Knowledge transfer oriented data mining, research by analogy, shared decision trees, cross dataset similarity, shared accuracy similarity, matching data distribution similarity, tree set diversity.

1 Introduction

Shared knowledge structures across multiple domains play an essential role in assisting users to transfer understanding between applications and to perform analogy based reasoning and creative thinking [3,6,9,8,10], in supporting users to perform research by analogy [4], and in assessing similarities between datasets in order to avoid negative learning transfer [16]. Motivated by the above, Dong and Han [5] studied the problem of mining knowledge structures shared by two datasets, with a focus on mining a single shared decision tree. However, providing only one shared decision tree may present only a limited view of shared knowledge structures that exist across multiple domains and does not offer users a concise representative of the space of possible shared knowledge structures. Moreover, computing all possible shared decision trees is infeasible. The purpose of this paper is to overcome the above limitations by studying the problem of mining diversified sets of shared decision trees across two application domains.

V.S. Tseng et al. (Eds.): PAKDD 2014, Part II, LNAI 8444, pp. 534–547, 2014.

In a diversified set of shared decision trees, each individual tree is a high quality shared decision tree in the sense that (a) the tree has high accuracy in classifying data for each dataset and the tree has high cross-domain class-distribution similarity at all tree nodes, and (b) different trees are structurally highly different from each other in the sense that they use very different sets of attributes. The requirements in (a) will ensure that each shared decision tree captures high quality shared knowledge structure between the two given datasets, providing the benefit that each root-to-leaf path in the tree corresponds to a similar rule (having similar support and confidence) for the two datasets and the benefit that the tree nodes describe similar data populations in the two datasets connected by similar multi-node population relationships.

Presenting too many shared decision trees to human users will imply that users will need to spend a lot of time to understand those trees in order to select the ones most appropriate for their application. Efficient algorithms solving the problem of mining diversified sets of shared decision trees meeting the requirements in (b) can offer a *small representative* set of high quality shared decision trees that can be understood without spending a lot of time, hence allowing users to more effectively select the tree most appropriate for their situation. Figure 1 illustrates the points given above, with the six stars as the diversified representatives of all shared decision trees.

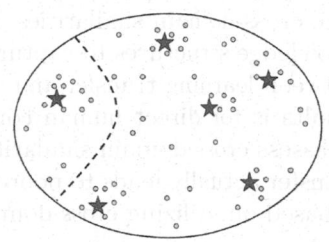

Fig. 1. Diversified Representatives of SDT Space

The main contributions of this paper include the following: (1) The paper motivates and formulates the diversified shared decision tree set problem. (2) It presents two effective algorithms to construct diversified high quality shared decision tree sets. (3) It reports an extensive experimental evaluation on the diversified shared decision tree set mining algorithms. (4) The shared decision trees reported in the experiments are mined from high dimensional microarray datasets for cancers, which can be useful to medical researchers.

The rest of the paper is organized as follows. Section 2 summarizes related work. Section 3 formally introduces the diversified shared decision tree set problem and associated concepts. Sections 4 presents our algorithms for mining diversified shared decision tree sets. Section 5 reports our experimental evaluation. Section 6 summarizes the results and discusses several future research problems.

2   Related Work

Limited by space, we focus on previous studies in four highly related areas.

Importance of Similarity/Analogy from Psychology and Cognitive Science: Psychology/cognitive science studies indicate that analogy plays a vital role in human thinking and reasoning, including *creative thinking*. For example, Fauconnier [6] states that *"Our conceptual networks are intricately structured by* *analogical and metaphorical mappings, which play a key role in the synchronic construction of meaning ...*". Gentner and Colhoun [9] state that "*Much of humankind's remarkable mental aptitude can be attributed to analogical ability.*" Gentner and Markman [10] suggest "that both similarity and analogy involve a process of structural alignment and mapping." Christie and Gentner [3] suggest, based on psychological experiments, that "structural alignment processes are crucial in developing new relational abstractions" and *forming new hypothesis*.

Learning Transfer: In learning transfer [16], it is typical to use available structure/knowledge of an auxiliary application domain to help build better classifiers/clusters for a target domain where there is a lack of data with class label or other domain knowledge. The constructed classifiers are not intended to capture cross-domain similarities. In contrast, our work focuses on mining (shared) knowledge structures to capture cross domain similarity; this is a key difference between learning transfer and our work. One of the intended uses of our mining results is for direct human consumption. Our mining results can also be used to assess cross domain similarity, to help avoid negative transfer where learning transfer actually leads to poorer results, since learning transfer is a process that is based on utilizing cross domain similarities that exist.

Shared Knowledge Structure Mining: Reference [4] defined the *cross domain similarity mining* (CDSM) problem, and motivated CDSM with several potential applications. CDSM has big potential in (1) supporting understanding transfer and (2) supporting research by analogy, since similarity is vital to understanding/meaning and to identifying analogy, and since analogy is a fundamental approach frequently used in hypothesis generation and in research. CDSM also has big potential in (3) advancing learning transfer since cross domain similarities can shed light on how to best adapt classifiers/clusterings across given domains and how to avoid negative transfer. CDSM can also be useful for (4) solving the schema/ontology matching problem. Reference [5] motivated and studied the shared decision tree mining problem, but that paper focused on mining just one shared decision tree. Several concepts of this paper were borrowed from [5], including shared accuracy, data distribution similarity, weight vector pool (for two factors), and information gain for two datasets. We enhance that paper by considering mining diversified sets of shared decision trees.

Ensemble Diversity: Much has been done on using ensemble diversity among member classifiers [14] to improve ensemble accuracy, including data based diversity approaches such as Bagging [2] and Boosting [7]. However, most previous studies in this area focused on classification behavior diversity, in the sense that different classifiers make highly different classification predictions. Several studies used attribute usage diversity to optimize ensemble accuracy, in a less systematic manner, including the random subspace method [13] and the distinct tree root method [15]. Our work focuses on attribute usage diversity aimed at providing diversified set of shared knowledge structures between datasets. The concept of attribute usage diversity was previously used in [12], to improve classifier ensemble diversity and classification accuracy for just one dataset.

3   Problem Definition

In this section, we introduce the problem of mining diversified sets of shared decision trees. To that end, we also need to define a quality measure on diversified sets of shared decision trees, which is based on the quality of individual shared decision trees and on the diversity measure for sets of shared decision trees.

As mentioned earlier, a shared decision tree (SDT) for a given dataset pair $(D_1 : D_2)$ is a decision tree that can classify both data in D_1 and data in D_2.

We assume that D_1 and D_2 are datasets with (1) an identical set of class names and (2) an identical set[1] of attributes. Our aim is to mine a small diversified set of high quality SDTs with these properties: (a) each tree in the set (1) is highly accurate in each D_i and (2) has highly similar data distribution in D_1 and D_2, and (b) different trees in the set are highly different from each other.

3.1   Shared Accuracy and Data Distribution Similarity of SDT Set

The shared accuracy and data distribution similarity measures for shared decision tree (SDT) sets are based on similar measures defined for individual SDTs [5], which we will review below. Let T be an SDT for a dataset pair $(D_1 : D_2)$, and let $Acc_{D_i}(T)$ denote T's accuracy[2] on D_i.

Definition 1. The *shared accuracy* of T (denoted by $\mathsf{SA}(T)$) is defined as the minimum of T's accuracies on D_1 and D_2: $\mathsf{SA}(T) = \min(Acc_{D_1}(T), Acc_{D_2}(T))$.

The *data distribution similarity* of T reflects population-structure (or class distribution) similarity between the two datasets across the nodes of T. The *class distribution vector* of D_i at a tree node V is defined by

$$CDV_i(V) = (Cnt(C_1, SD(D_i, V)),\ Cnt(C_2, SD(D_i, V))),$$

where $Cnt(C_j, SD(D_i, V)) = |\{t \in SD(D_i, V) \mid t\text{'s class is } C_j\}|$, and $SD(D_i, V)$ is the subset of D_i for V (satisfying the conditions on the root-to-V path). The *distribution similarity* (DSN) at V is defined as $DSN(V) = \frac{CDV_1(V) \cdot CDV_2(V)}{\|CDV_1(V)\| \cdot \|CDV_2(V)\|}$.

Definition 2. The *data distribution similarity* of an SDT T over $(D_1 : D_2)$ is defined as $\mathsf{DS}(T) = avg_V DSN(V)$, where V ranges over nodes of T.

We can now define SA and DS for shared decision tree sets.

Definition 3. Let TS be a set of SDTs over dataset pair $(D_1 : D_2)$. The *shared classification accuracy* of TS is defined as $\mathsf{SA}(TS) = avg_{T \in TS}\mathsf{SA}(T)$, and the *data distribution similarity* of TS is defined as $\mathsf{DS}(TS) = avg_{T \in TS}\mathsf{DS}(T)$.

---

[1] If D_1 and D_2 do not have identical classes and attributes, one will need to identify an 1-to-1 mapping between the classes of the two datasets, and an 1-to-1 mapping between the attributes of the two datasets. The 1-to-1 mappings can be real or hypothetical (for "what-if" analysis) equivalence relations on the classes/attributes.

[2] When D_i is small, one may estimate $Acc_{D_i}(T)$ directly using D_i. Holdout testing can be used when the datasets are large.

3.2 Diversity of SDT Set

To define the *diversity of SDT sets*, we need to define tree-pair difference, and we need a way to combine the tree-pair differences for all possible SDT pairs.[3]

We measure the difference between two SDTs in terms of their attribute usage summary (AUS). Let $A_1, A_2, \ldots, A_n$ be a fixed list enumerating all shared attributes of D_1 and D_2.

Definition 4. The *level-normalized-count* AUS ($\mathsf{AUS_{LNC}}$) of an SDT T over $(D_1 : D_2)$ is defined as

$$\mathsf{AUS_{LNC}}(T) = \left(\frac{Cnt_T(A_1)}{avgLvl_T(A_1)}, \; \ldots, \; \frac{Cnt_T(A_n)}{avgLvl_T(A_n)} \right),$$

where $Cnt_T(A_i)$ denotes the number of occurrences of attribute A_i in T, and $avgLvl_T(A_i)$ denotes the average level of $A_i's$ occurrences in T.

The root is at level 1, the children of the root are at level 2, and so on. In the $\mathsf{AUS_{LNC}}$ measure, nodes near the root have high impact since those nodes have small level number and attributes used at those nodes often have small $avgLvl_T$.

One can also use the *level-listed count* ($\mathsf{AUS_{LLC}}$) approach for AUS. Here we use a matrix in which each row represents the attribute usage in one tree level: Given an SDT T with L levels, for all attributes A_i and integers l satisfying $1 \leq l \leq L$, the (l, i) component of $\mathsf{AUS_{LLC}}$ has the value $Cnt_T(l, A_i)$ (the occurrence frequency count for attribute A_i in the lth level of T).

Remark: $\mathsf{AUS_{LNC}}$ pays more attention to nodes near the root, while $\mathsf{AUS_{LLC}}$ gives more emphasis to levels near the leaves (there are many nodes at those levels).

Definition 5. Given an attribute usage summary measure AUS_μ, the *tree pair difference* (TPD) for two SDTs T_1 and T_2 is defined as

$$\mathsf{TPD}_\mu(T_1, T_2) = 1 - \frac{\mathsf{AUS}_\mu(T_1) \cdot \mathsf{AUS}_\mu(T_2)}{\|\mathsf{AUS}_\mu(T_1)\| \cdot \|\mathsf{AUS}_\mu(T_2)\|}.$$

We can now define the SDT set diversity concept.

Definition 6. Given an SDT set TS and an AUS measure AUS_μ, the *diversity* of TS is defined as $\mathsf{TD}_\mu(TS) = avg\{\mathsf{TPD}_\mu(T_i, T_j) \mid T_i, T_j \in TS, and \; i \neq j\}$.

3.3 Diversified Shared Decision Tree Set Mining Problem

To mine desirable diversified SDT sets, we need an objective function. This section defines our objective function, which combines the quality of the SDTs and the diversity among the SDTs.

Definition 7. Given an attribute usage summary method AUS_μ, the quality score of an SDT set TS is defined as:

$$\mathsf{SDTSQ}_\mu(TS) = min(\mathsf{SA}(TS), \mathsf{DS}(TS), \mathsf{TD}_\mu(TS)) * avg(\mathsf{SA}(TS), \mathsf{DS}(TS), \mathsf{TD}_\mu(TS)).$$

---

[3] We borrow the diversity concepts from [12], which considered mining diversified decision tree ensembles for one dataset.

We also considered other definitions using e.g. average, weighted average, and the harmonic mean of the three factors. They were not selected, since they give smaller separation of quality scores or require parameters from users. The above formula is chosen since it allows each of SA, DS, and TD to play a role, it is simple to compute, and it does not require any parameters from users.

We now turn to defining the diversified SDT set mining problem.

Definition 8 (Diversified Shared Decision Tree Set Mining Problem). Given a dataset pair $(D_1{:}D_2)$ and a positive integer k, the *diversified shared decision tree set mining problem* (KSDT) is to mine a diversified set of k SDTs with high SDTSQ from the dataset pair.

An example SDT set mined from two cancer datasets will be given in § 5.

4   KSDT-Miner

This section presents two KSDT mining algorithms, for mining diversified high quality SDT sets. One is the parallel KSDT-Miner (PKSDT-Miner), which builds a tree set concurrently and splits one-node-per-tree in a round-robin fashion; the other one is the sequential KSDT-Miner (SKSDT-Miner), which mines a tree set by building one complete tree after another.

In comparison, PKSDT-Miner gives *all SDTs* almost equal opportunity[4] in selecting desirable attributes for use in high impact nodes near the roots of SDTs, whereas SKSDT-Miner gives SDTs built earlier more possibilities in selecting desirable attributes (even for use at low impact nodes near the leaves), which deprives the chance of later SDTs in using those attributes at high impact nodes.

Limited by space and due to the similarity in most ideas except the node split order, we present PKSDT-Miner and omit the details of SKSDT-Miner.

4.1   Overview of PKSDT-Miner

PKSDT-Miner builds a set of SDTs in parallel, in a node-based round-robin manner. In each of the round-robin loop, the trees are processed in an ordered manner; for each tree, one node is selected and spilt. Figure 2 illustrates with two consecutive states in such a loop: 2(a) gives three (partial) trees (blank rectangles are nodes to be split), and 2(b) gives those trees after splitting node V_2 of T_2. Here, PKSDT-Miner splits node V_2 in T_2 even though V_1 in T_1 can be split. PKSDT-Miner will select a node in T_3 to split next.

4.2   Aggregate Tree Difference

To build highly diversified tree sets, the *aggregate tree difference* (ATD) is used to measure the differences between a new/modified tree T and the set of other trees TS. One promising approach is to define ATD as the average of ℓ smallest

---
[4] An attribute may be highly desirable in more than one tree.

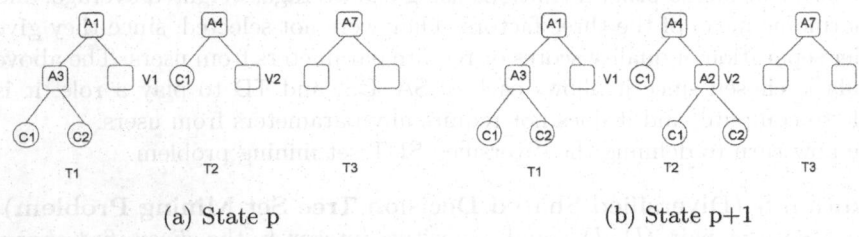

(a) State p   (b) State p+1

Fig. 2. PKSDT-Miner Builds Trees in Round-robin Manner

$\mathsf{TPD}_\mu(T, T')$ values ($T' \in TS$). We define a new aggregation function called $avgmin_\ell$, where $avgmin_\ell(S)$ is the average of the ℓ smallest values in a set S of numbers. (In experiments our best choice for ℓ was 3.) Then the μ-ℓ-minimal aggregate tree difference ($\mathsf{ATD}_{\mu,avgmin\ell}$) is defined as the average TPD_μ between T and the ℓ most similar trees in TS:

$$\mathsf{ATD}_{\mu,avgmin\ell}(T, TS) = avgmin_\ell(\{\mathsf{TPD}_\mu(T, T')|T' \in TS\}).$$

PKSDT-Miner selects an attribute and a value to split a tree node by maximizing an objective function IDT that combines *information gain* (to be denoted by IG and defined below) and *data distribution similarity* (DS) on two datasets, and *aggregate tree difference* (ATD) between the current tree and the other trees. To tradeoff the three factors, they are combined using a weighted sum based on a weight vector $w = (w_{IG}, w_{DS}, w_{ATD})$ whose three weights are required to satisfy $0 < w_{IG}, w_{DS}, w_{ATD} < 1$ and $w_{IG} + w_{DS} + w_{ATD} = 1$.

Given an AUS measure μ, an aggregation method $\alpha \in \{avg, min, avgmin_\ell\}$, a tree T and a tree set TS, the μ-α aggregate tree difference is defined as

$$\mathsf{ATD}_{\mu,\alpha}(T, TS) = \alpha(\{\mathsf{TPD}_\mu(T, T') \mid T' \in TS\}).$$

For example, $\mathsf{ATD}_{\mu,min}(T, TS) = \min_{T' \in TS}(\mathsf{TPD}_\mu(T, T'))$ when α is *min*. Each variant of ATD can be used in our two SDT set mining algorithms, resulting a number of variant algorithms. For instance, the standard version of the PKSDT-Miner algorithm can be written as PKSDT-Miner(LNC,$avgmin\ell$) and we can replace LNC by LLC to get PKSDT-Miner(LLC,$avgmin\ell$).

4.3 IG for Two Datasets

This paper uses the union-based definition of IG for two datasets of [5]. ([5] discussed other choices and the union-based way was shown to be the best by experiments.) For each attribute A and split value a, and dataset pair ($D'_1 : D'_2$) (associated with a given tree node), the *union-based information gain* is defined as $\mathsf{IG}(A, a, D'_1, D'_2) = \mathsf{IG}(A, a, D'_1 \cup D'_2)$. (The IG function is overloaded: IG in the LHS is 4-ary while IG in the RHS is 3-ary.) $\mathsf{IG}(A, a, D')$ is defined in terms of entropy, as used for decision tree node splitting.

4.4   The Algorithm

PKSDT-Miner has six inputs: Two datasets D_1 and D_2, a set *AttrSet* of candidate attributes that can be used in shared trees, a dataset size threshold *MinSize* for node-splitting termination, a weight vector w (w_{IG} on information gain, w_{DS} on data distribution similarity, w_{ATD} on aggregate tree difference), and an integer k for the desired number of trees. PKSDT-Miner calls PKSDT-SplitNode (Function 1) to split nodes for each tree.

Algorithm 1. PKSDT-Miner

Input: ($D_1 : D_2$): Two datasets
      AttrSet: Set of candidate attributes that can be used
      MinSize: Dataset size threshold for splitting termination
      $w = (w_{IG}, w_{DS}, w_{ATD})$: A weight vector on IG, DS, and ATD
      k: Desired number of trees
Output: A diversified shared decision tree set TS for ($D_1 : D_2$).
Method:
  1.   Create root node V_p for each tree T_p ($1 \le p \le k$);
  2.   Repeat
  3.   For $p = 1$ to k do
  4.   let node V be the next node[a] of tree T_p to split;
  5.   Call PKSDT-SplitNode($T, V, D_1, D_2, AttrSet, TS, MinSize, w$);
  6.   Until there are no more trees with nodes that can be split[b];
  7.   Output the diversified shared decision tree set TS.

---

[a] The next node of a tree to split is determined by a tree traversal method, which can be depth first, breadth first, and so on. We use depth first here.

[b] No more node to split means that all candidate split nodes satisfy the termination conditions defined in Function ShouldTerminate.

PKSDT-SplitNode splits the data of a node V of a tree T by picking the split attribute and split value that optimize the IDT score. Let T be the tree that we wish to split, and let TS be the other trees that we have built. Let V be T's current node to split, and A and a_V be resp. a candidate splitting attribute/value. Let $T(A, a_V)$ be the tree obtained by splitting V using A and a_V. Then the IDT scoring function is defined by:

$$IDT(T(A, a_V), TS) = w_{IG} * IG(A, a_V) +$$
$$w_{DSN} * DSN(A, a_V) + w_{ATD} * ATD(T(A, a_V), TS),$$

where $IG(A, a_V)$ is the information gain for V when split by A and a_V, $DSN(A, a_V)$ is the average DSN value of the two children nodes of V.

Function ShouldTerminate determines if nodes splitting should terminate. (Our algorithms aim to build simple trees and avoid "overfitting".) It uses two techniques. (1) When many attribtues are available, we restrict the candidate attributes to those whose IG is ranked high in both datasets, so avoiding non-discriminative attributes that are locally discriminative at a given node. (2) We stop splitting for a given tree node when at least one dataset is small or pure.

Function 1. PKSDT-SplitNode($T, V, D'_1, D'_2, AttrSet, TS, MinSize, w$)

1. If ShouldTerminate($V, D'_1, D'_2, MinSize, AttrSet$) then assign
   the majority class in D'_1 and D'_2 as class label of V and return;
2. Select the attribute B and value b_V that maximize IDT, that is
   $IDT(T(B, b_V), TS)$=max$\{IDT(T(A, a_V), TS) \mid A \in AttrSet$, and a_V
   is a common candidate split value for A at $V\}$;
3. Create left child node V_l of V, with "$B \leq b_V$" as the corresponding
   edge's label, and let $D'_{il} = \{t \in D'_i \mid t$ satisfies "$B \leq b_V$"$\}$ for $i = 1, 2$;
4. Create right child node V_r of V, with "$B > b_V$" as the corresponding
   edge's label, and let $D'_{ir} = \{t \in D'_i \mid t$ satisfies "$B > b_V$"$\}$ for $i = 1, 2$.

4.5 Weight Vector Pools

Different dataset pairs have different characteristics concerning IG, DS and ATD. To mine the best SDT set, we need to treat different characteristics using appropriate focus/bias. (It is open if one can determine the characteristics of a dataset pair without performing SDT set mining.) We solve the problem by using a pool of weight vectors to help mine (near) optimal SDTs efficiently. Such a pool is a small representative set of all possible weight vectors.

We consider two possible weight vector pools: WVP_1 contains 36 weight vectors, defined by $WVP_1 = \{x \mid x$ is a multiple of 0.1 and $0 < x < 1\}$. WVP_2={(0.1,0.1,0.8),(0.1,0.3,0.6),(0.1,0.5,0.4),(0.1,0.7,0.2),(0.3,0.1,0.6), (0.3, 0.4, 0.3),(0.3,0.5,0.2),(0.5,0.1,0.4),(0.5,0.3,0.2),(0.7,0.2,0.1)}. So WVP_2 contains 10 representative vectors selected from WVP_1. For each of the three factors, each pool contains some vectors where the given factor plays the dominant role.

5   Experimental Evaluation

This section uses experiments to evaluate KSDT-Miner, using real-world and also (pseudo) synthetic datasets. It reports that (1) PKSDT-Miner tends to build more diversified high quality SDT sets on average, which confirms the advantages of PKSDT-Miner analyzed in Section 4, and (2) KSDT-Miner is scalable w.r.t. number of tuples/attributes/trees. It discusses (3) how KSDT-Miner performs concerning the use of weight vectors. It also examines (4) how KSDT-Miner performs when it uses different AUS$_\mu$ and ATD$_{\mu,\alpha}$ measures. Finally, it reports that KSDT-Miner outperforms SDT-Miner regarding mining one single SDT.

In the experiments, we set $\ell = 3$, $MinSize = 0.02 * min(|D_1|, |D_2|)$ and $AttrSet = \{A \mid rank_1(A) + rank_2(A)$ is among the smallest 20% of all shared attributes, where $rank_i(A)$ is the position of A when D_i's attributes are listed in decreasing IG order$\}$. Experiments were conducted on a 2.20 GHz AMD Athlon with 2 GB memory running Windows XP, with codes implemented in Matlab. To save space, the tables may list results on subsets of the 15 dataset pairs on the 6 microarray datasets, although the listed averages are for all 15 pairs.

5.1   Datasets and Their Preprocessing

Our experiments used six real-world microarray gene expression datasets for cancers.[5] ArrayTrack [20] was used to identify shared (equivalent) attributes. Two genes are shared if they represent the same gene in different gene name systems. Table 1 lists the number of shared attributes for the 15 dataset pairs.

Table 1. Number of Shared Attributes (NSA) between Dataset Pairs

Dataset Pair	NSA
(BC:CN), (BC:DH), (BC:LM)	5114
(CN:DH), (CN:LM), (DH:LM)	7129
(CN:LB), (DH:LB), (LB:LM)	5313
(CN:PC), (DH:PC), (LM:PC)	5317
(BC:LB)	8123
(BC:PC)	8124
(LB:PC)	9030

In some dataset pairs the two datasets have very different class ratios. The class ratio of a dataset is likely an artifact of the data collection process and may not have practical implications. However, class ratio difference can make it hard to compare quality values for results mined from different dataset pairs. To address this, we use sampling with replacement method to replicate tuples so that class ratios for the two datasets are nearly the same.

5.2   Example Diversified SDT Set Mined from (DH:LM)

We now give[6] an example diversified set of two shared decision trees mined from real (cancer) dataset pair (DH:LM) in Figures 3 and 4. For each tree, data in two datasets have very similar distributions at tree nodes (the average DSN for each tree is about 0.977) and the leaf nodes are very pure with average shared classification accuracy of leaf nodes being about 0.963. For the diversified tree set $\{T_1, T_2\}$, tree diversity is 1 since these two trees don't share any splitting attributes, and the SDTSQ is about 0.944.

5.3   KSDT-Miners Mine Diversified High Quality SDT Sets

Experiments show that KSDT-Miners are able to mine diversified high quality SDT sets. Table 2 lists the statistics of best SDT sets mined by either PKSDT-Miner or SKSDT-Miner from each of the 15 dataset pairs in Table 1 (using all weight vectors). For the 15 dataset pairs, PKSDT-Miner got the best SDT sets in 9 pairs, and SKSDT-Miner got the best in 6 pairs. We include the result for (BC: LM) to indicate that it is not possible to always have high quality SDTs (as expected).

---

[5] References for the datasets: *BC* (breast cancer) [21], *CN* (Central Nervous System) [17], *DH* (DLBCL-Harvard*) [18], *LB* (Lung Cancer-BAWH) [11], *LM* (Lung Cancer-Michigan*) [1], *PC* (Prostate Cancer) [19].

[6] We draw shared decision tree figures as follows: For each node V , we show $CDV_1(V)$ for D_1 at $V's$ left, show $CDV_2(V)$ for D_2 at $V's$ right, and show $DSN(V)$ below V.

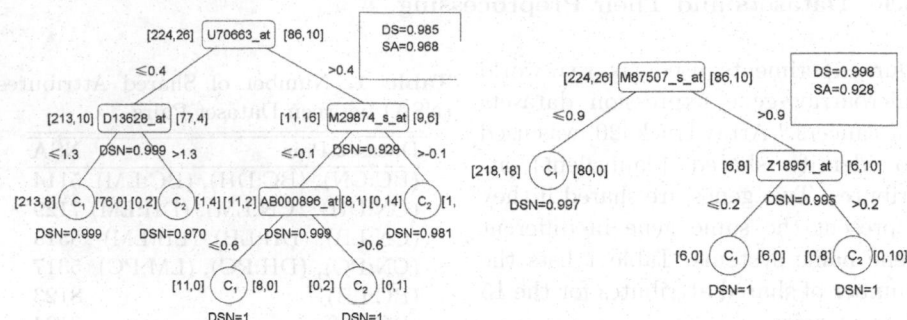

Fig. 3. Shared Decision Tree T_1 **Fig. 4.** Shared Decision Tree T_2

Table 2. Stats of Best SDT Sets

Dataset Pair	DS	SA	TD	SDTSQ
(BC: CN)	0.98	0.98	0.99	0.96
(BC: DH)	0.98	0.97	1	0.95
(BC: LB)	0.97	0.97	1	0.95
(BC: LM)	0.94	0.74	1	0.67
(BC: PC)	0.97	0.97	1	0.95
(CN: PC)	0.98	0.98	0.99	0.96
Average*	0.96	0.95	0.99	0.92

Table 3. PKSDT(P) vs SKSDT(S)

Dataset Pair	TD P	TD S	SDTSQ P	SDTSQ S
(BC: CN)	0.97	0.97	0.93	0.93
(BC: LB)	0.94	0.91	0.88*	0.85
(BC: LM)	0.92	0.92	0.61	0.63*
(BC: PC)	0.95	0.94	0.89	0.88
(CN: LB)	0.98	0.97	0.93	0.92
(DH: LB)	0.96	0.95	0.91*	0.88
Average*	0.94	0.93	0.86	0.85

5.4 Comparison between PKSDT-Miner and SKSDT-Miner

Experiments show that PKSDT-Miner is better than SKSDT-Miner. Indeed, PKSDT-Miner gets SDT sets of higher quality values on average, albeit slightly, and it never gets SDT sets of lower quality values (see Table 3, which gives the average TD and SDTSQ values for best diversified tree sets mined by PKSDT-Miner and SKSDT-Miner respectively when using all weight vectors). As noted for Table 2, PKSDT-Miner got the best SDT sets in 9, whereas SKSDT-Miner got the best in only 6, out of the 15 dataset pairs. Below we only consider PKSDT-Miner.

5.5 Comparison of AUS and ATD Variants

Experimental results demonstrate that (1) AUS_{LNC} produces better results than AUS_{LLC}, and (2) $ATD_{\mu,avgmin\ell}$ outperforms $ATD_{\mu,min}$ (reason: when it is used a highly similar outlier may give too much influence) and $ATD_{\mu,avg}$ (reason: when it is used the highly dissimilar cases may give big influence). The details are omitted to save space.

5.6 Weight Vector Issues

We examined the "best" and "worst" $(w_{IG}, w_{DS}, w_{ATD})$ weight vectors, which produce the SDT sets with the highest and lowest SDTSQ mined by PKSDT-Miner(LNC, $avgmin\ell$). (1) We observed that the average relative improvement of the "best" over the "worst" is an impressive 4.8% and the largest is 20.5%. This indicates that the choice of weight vector has significant impact on the tree set quality mined by KSDT-Miner. (2) We also saw that no single weight vector is the best weight vector for all dataset pairs. This reflects the fact that different dataset pairs have different characteristics regarding which of IG, DS and ATD is most important.

Regarding which weight vectors may be better suited for which kinds of dataset pairs, we observed that there are three cases. (A) For some dataset pairs (e.g. (LM:PC)), weight vectors with high IG weight (and low DS weight, low ATD weight) tend to yield SDT sets with high SDTSQ. (B) For some dataset pairs (e.g. (BC:DH)), weight vectors with high DS weight tend to yield SDT sets with high SDTSQ. (C) For some dataset pairs (e.g. (BC:CN)), weight vectors with high ATD weight tend to yield SDT sets with high SDTSQ.

Experiment showed that using multiple weight vectors leads to much better performance than using a single weight vector. Moreover, SDTSQ scores of best SDT sets obtained using WVP_2 are almost identical to those obtained using WVP_1. Since WVP_2 is smaller (having 10 weight vectors) than WVP_1 (having 36 weight vectors), WVP_2 is preferred since it requires less computation time.

5.7 KSDT-Miner **Outperforms** SDT-Miner **on SDTQ**

Both KSDT-Miner and SDT-Miner can be used to mine a single high quality SDT, by having KSDT-Miner return the best tree in the SDT set it constructs. Experiments show that KSDT-Miner gives better performance than SDT-Miner. Indeed, the average relative SDTQ improvement by KSDT-Miner over SDT-Miner for all dataset pairs is 13.8%. For some dataset pairs, the relative improvement is about 45.3%. Through more detailed comparison, the average relative improvement on DS by KSDT-Miner over SDT-Miner for all dataset pairs is 3.2%, and on SA is 5.4%. Clearly, better single SDT can be mined when tree set diversity is considered.

5.8 **Scalability of** KSDT-Miner

We experimented to see how KSDT-Miner's execution time changes when the number of tuples/attributes/trees increases. Experiments show that execution time increases roughly linearly. (The figure is omitted to save space.) The experiments used synthetic datasets obtained by replicating tuples with added random noises up to a bound given by $P\%$ of the maximum attribute value magnitude (in order to get a desired number N of tuples), and by attribute elimination.

5.9 **Using Fewer Attributes Leads to Poor SDT Sets**

Incidently, we compared the SDTSQ of SDT sets mined from real dataset pairs using all available attributes against those obtained from projected data using fewer attributes (e.g. 100). On average, SDTSQ using all attributes is about 34.2% better than SDTSQ using only the first 100 attributes.

6 Concluding Remarks and Future Directions

In this paper we motivated the diversified shared decision tree set mining problem, presented two algorithms of KSDT-Miner, and evaluated the algorithms using real microarray gene expression data for cancers and using synthetic datasets. Experimental results show that KSDT-Miner can efficiently mine high quality shared decision trees. Future research directions include mining other types of shared knowledge structures (including those capturing alignable differences, defined as shared knowledge structures that capture cross-domain similarities and cross-domain differences within the context of the similarities given elsewhere in the shared knowledge structures) and utilizing such mined results to solve various research and development problems in challenging domains.

References

1. Beer, D.G., et al.: Gene-expression profiles predict survival of patients with lung adenocarcinoma. Nature Medicine 8, 816–824 (2002)
2. Breiman, L.: Bagging predictors. Machine Learning 24(2), 123–140 (1996)
3. Christie, S., Gentner, D.: Where hypotheses come from: Learning new relations by structural alignment. Journal of Cognition and Development 11(3), 356–373 (2010)
4. Dong, G.: Cross domain similarity mining: Research issues and potential applications including supporting research by analogy. ACM SIGKDD Explorations (June 2012)
5. Dong, G., Han, Q.: Mining accurate shared decision trees from microarray gene expression data for different cancers. In: International Conference on Bioinformatics and Computational Biology, BIOCOMP 2013 (2013)
6. Fauconnier, G.: Mappings in Thought and Language. Cambridge University Press (1997)
7. Freund, Y., Schapire, R.E.: Experiments with a new boosting algorithm. In: ICML, pp. 148–156 (1996)
8. Gentner, D.: Structure mapping: A theoretical framework for analogy. Cognitive Science 7, 155–170 (1983)
9. Gentner, D., Colhoun, J.: Analogical processes in human thinking and learning. In: Glatzeder, B., Goel, V., von Müller, A. (eds.) Towards a Theory of Thinking. On Thinking, vol. 2. Springer, Heidelberg (2010)
10. Gentner, D., Markman, A.B.: Structure mapping in analogy and similarity. American Psychologist 52(1), 45–56 (1997)
11. Gordon, G.J., et al.: Translation of microarray data into clinically relevant cancer diagnostic tests using gene expression ratios in lung cancer and mesothelioma. Cancer Research 62, 4963–4967 (2002)
12. Han, Q., Dong, G.: Using attribute behavior diversity to build accurate decision tree committees for microarray data. J. Bioinformatics and Computational Biology 10(4) (2012)
13. Ho, T.K.: The random subspace method for constructing decision forests. IEEE Transactions on Pattern Analysis and Machine Intelligence 20(8), 832–844 (1998)
14. Kuncheva, L.I., Whitaker, C.J.: Measures of diversity in classifier ensembles and their relationship with the ensemble accuracy. Machine Learning 51(2), 181–207 (2003)

15. Li, J., Liu, H.: Ensembles of cascading trees. In: ICDM, pp. 585–588 (2003)
16. Pan, S.J., Yang, Q.: A survey on transfer learning. IEEE Transactions on Knowledge and Data Engineering 22(10), 1345–1359 (2010)
17. Pomeroy, S.L., et al.: Prediction of central nervous system embryonal tumour outcome based on gene expression. Nature 415, 436–442 (2002)
18. Shipp, M.A., et al.: Diffuse large b-cell lymphoma outcome prediction by gene-expression profiling and supervised machine learning. Nature Medicine 8, 68–74 (2002)
19. Singh, D., et al.: Gene expression correlates of clinical prostate cancer behavior. Cancer Cell 1(2), 203–209 (2002)
20. Tong, W., et al.: ArrayTrack-Supporting toxicogenomic research at the FDA's National Center for Toxicological Research (NCTR). EHP Toxicogenomics 111(15), 1819–1826 (2003)
21. Van't Veer, L.J., et al.: Gene expression profiling predicts clinical outcome of breast cancer. Nature 415, 530–536 (2002)

Semi-supervised Clustering
on Heterogeneous Information Networks

Chen Luo[1], Wei Pang[2], and Zhe Wang[1,*]

[1] College of Computer Science and Technology, Jilin University,
Changchun 130012, China
rackingroll@163.com, wz2000@jlu.edu.com
[2] School of Natural and Computing Sciences, University of Aberdeen,
Aberdeen, AB24 3UE, UK
pang.wei@abdn.ac.uk

Abstract. Semi-supervised clustering on information networks combines both the labeled and unlabeled data sets with an aim to improve the clustering performance. However, the existing semi-supervised clustering methods are all designed for homogeneous networks and do not deal with heterogeneous ones. In this work, we propose a semi-supervised clustering approach to analyze heterogeneous information networks, which include multi-typed objects and links and may contain more useful semantic information. The major challenge in the clustering task here is how to handle multi-relations and diverse semantic meanings in heterogeneous networks. In order to deal with this challenge, we introduce the concept of *relation-path* to measure the similarity between two data objects of the same type. Thereafter, we make use of the labeled information to extract different weights for all *relation-paths*. Finally, we propose SemiRPClus, a complete framework for semi-supervised learning in heterogeneous networks. Experimental results demonstrate the distinct advantages in effectiveness and efficiency of our framework in comparison with the baseline and some state-of-the-art approaches.

Keywords: Heterogeneous information network, Semi-supervised clustering.

1 Introduction

The real world is interconnected: objects and inter-connections between these objects constitute various information networks. Clustering methods in information networks [1] become more and more popular in recent years. One can discover much interesting knowledge from the information networks by using appropriate clustering methods, and the clustering result can also be used in many fields such as information retrieval [2] and recommendation systems [3]. In particular, the real world information networks are often heterogeneous [4], which means in these networks objects and links between these objects may belong to different

---

* Corresponding Author.

V.S. Tseng et al. (Eds.): PAKDD 2014, Part II, LNAI 8444, pp. 548–559, 2014.
© Springer International Publishing Switzerland 2014

types. In order to handle the multi-relational data in heterogeneous networks, semi-supervised learning methods [5] can be an appropriate tool. In this paper, we focus on the semi-supervised clustering task in heterogeneous information networks.

Up till now many semi-supervised clustering algorithms have been proposed for information networks [6, 7, 8]. Some of these algorithms consider the labeled information as constraints for clustering tasks [6]. These constraints can guide the clustering process to achieve better results. Others focus on semi-supervised learning on graphs [7], which uses a small portion of labeled objects to label all the other objects in the same network by propagating the labeled information. The semi-supervised algorithm, proposed in [8], integrates both the constraint-based learning and distance-function learning methods. All the above-mentioned link-based clustering methods are specifically designed for homogeneous information networks, in which all the links in the network are assumed to be of the same type [1]. However, most of the real-world networks are heterogeneous ones [4]. In KDD-2012, Sun *et.al* proposed PathSelClus [9], a user guided clustering method in heterogeneous information networks. PathSelClus integrates both the meta-path selection and clustering processes. The experimental results produced by PathSelClus also showed that more meaningful results could be obtained by considering the clustering task on the heterogeneous information networks instead of the homogeneous ones. However, in PathSelCLus, the number of clusters needs to be pre-specified at the beginning of the algorithm, which is not realistic in many real-world problems.

In this paper, we will investigate semi-supervised clustering [5] in heterogeneous information networks, and intend to develop a clustering algorithm that does not need to pre-specify the number of clusters. In a heterogeneous information network, two objects may be connected via different relation paths or sequences of relations [9]. These different relation paths have different semantic meanings. For example, in the academic community network, two authors can be connected via either the co-author relationship or the co-institution relationship, but these two relations have very different meanings. Sun *et.al* proposes the concept of 'meta-path' [10] to indicate the relation sequence. In this research we propose a similar definition —'*relation-path*', which is specifically for our clustering task. Correspondingly, we also propose a topological measure for our relation-path, which is different from existing path topological measures [10, 11]. By using a logistic regression approach, we evaluate each weight of the relation-path. Finally, SemiRPClus, a novel framework for clustering in heterogeneous information network, is presented. Experiments on DBLP showed the distinct advantages in effectiveness and efficiency of SemiRPClus in comparison with some clustering methods on information networks.

The rest of this paper is organized as follows: in Section 2 some important definitions used in this paper are introduced. The proposed framework, named SemiRPClus, is described in Section 3. In Section 4, we present a series of experiments on DBLP, which demonstrated the effectiveness and efficiency of SemiRPClus. Finally, we conclude our work in Section 5.

2   Problem Definition

As in [4], we use $G = \langle V, E, W \rangle$ to represent a heterogeneous network, where $V = \bigcup_{i=1}^{m} X_i$, and $X_1 = \{x_{11}, ..., x_{1n_1}\}, ..., X_m = \{x_{m1}, ..., x_{mn_m}\}$ denote the m different types of nodes. E is the set of links between any two data objects of V, and W is the set of weight values on the links. $T_G = \langle A, R \rangle$ denotes the network schema [10], which is a directed graph defined over object types A, with edges as relations from R. For more details about heterogeneous networks, please refer [4, 10]. First, the semi-supervised clustering in heterogeneous network is given below:

Definition 1 (Semi-supervised clustering in heterogeneous informa-tion network). *In a heterogeneous information network $G = \langle V, E, W \rangle$ fol-lowing a network schema $T_G = \langle A, R \rangle$, suppose V' is a subset of V and $V' \subseteq V \in X_i$, where X_i is the target type for clustering, and each data object O in V' is labeled with a value γ indicating which cluster O should be in. Given a set of relation-path (see Definition 2), the learning task is to predict the labels for all the unlabeled objects $V - V'$.*

Second, we give the definition of *relation-path*, which can be considered as a special case of meta-path [10]:

Definition 2 (Relation-path). *Given a network schema $T_G = \langle A, R \rangle$, a relation-path RP is in the form of $A_t \xrightarrow{R_1} A_2 \xrightarrow{R_2} ...A_{l-1} \xrightarrow{R_{l-1}} A_t$, which defines a composite relation $RP = R_1 \circ R_2 \circ ... \circ R_{l-1}$ between two objects in the same target type A_t, and $\circ$ is the composition operator on relations.*

Different from the definition of meta-path [10], in *relation-path* the starting ob-ject and the end object of the relation-path must belong to the same target type. From Definition 2 we can see that a *relation-path* is always a meta-path, but not vice versa. More importantly, as the relation-path is defined for objects of the same target type, it will be more suitable for our clustering task. Third, we define a transform of our *relation-path* named '*inverse relation-path*' as follow:

Definition 3 (Inverse Relation-path). *Given a relation-path RP: $A_t \xrightarrow{R_1} A_2 \xrightarrow{R_2} ...A_{l-1} \xrightarrow{R_{l-1}} A_t$, RP^{-1} is the Inverse Relation-path of RP, if RP^{-1} is $A_t \xrightarrow{R_{l-1}^{-1}} A_{l-1} \xrightarrow{R_{l-2}^{-1}} ...A_1 \xrightarrow{R_l} A_t$, where R^{-1} is the inverse relation of R.*

After defining all the above concepts, we introduce a typical heterogeneous in-formation network used in the experiments of our research: the DBLP network, which has been used as test cases in a number of papers [12, 4, 9].

Example 1 (The DBLP bibliographic network). DBLP, computer science bib-liography database, is a representative of heterogeneous information networks. The DBLP schema is shown in Figure 1. There are four types of objects in the schema: Paper, Author, Term, and Conference. Links between Author and Pa-per are defined by the relation of "write" and "written by", denoted as "write^{-1}".

Relation between Term and Paper is "mention" and "mentioned by", denoted as "mention⁻¹". Relation between Paper and Conference is "publish" and "published by", denoted as "publish⁻¹". The "cite" relation exists between the papers in the schema. In this research we extract the "cite" relation from the Microsoft Academic Search API.

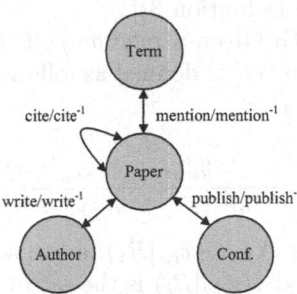

Fig. 1. DBLP schema

3 The SemiRPClus Framework

As mentioned in [4], there are two constraints which determine the clustering results on heterogeneous information networks: first, the clustering result should be consistent with the network structure; second, the clustering results should be consistent with the labeled information pre-assigned for some data objects. Our semi-supervised clustering process will follow these two constraints.

In this section, we first introduce in detail the proposed framework, SemiRP-Clus, which includes two components: (1) the linear regression based topological measure and (2) the relation extraction model. Then we present the overall clustering framework.

3.1 Linear Regression Based Topological Measure

Topological features, also called structural features, are connectivity properties extracted from a network for some pairs of objects. Many topological features have been proposed for the homogeneous networks, and see more details in [13]. There are also some topological features proposed for heterogeneous networks, and we redefine them based on *relation-path* as below:

- **Path Count Measure [12].** Given a *relation-path*, denoted as RP, the Path Count can be calculated as the number of path instances of RP between two objects, say $x_{t,i}$ and $x_{t,j}$, denoted as $S_{RP}^{PC}(x_{t,i}, x_{t,j})$, where $x_{t,i}, x_{t,j} \in X_t$ and X_t is the target type.
- **Random Walk Measure [12].** Random walk measure following a *relation-path* RP is defined as $S_{RP}^{RW}(x_{t,i}, x_{t,j}) = \frac{S_{RP}^{PC}(x_{t,i}, x_{t,j})}{S_{RP}^{PC}(x_{t,i}, :)}$. Here, $S_{RP}^{PC}(x_{t,i}, :)$ denotes the path count value following RP starting from $x_{(t,i)}$.

- **PathSim Measure [10].** Given a *relation-path* RP, PathSim between two objects $x_{t,i}, x_{t,j}$ is defined as :

$$S_{RP}^{PS}(x_{t,i}, x_{t,j}) = \frac{2*S_{RP}^{PC}(x_{t,i}, x_{t,j})}{S_{RP}^{PC}(x_{t,i}, x_{t,i}) + S_{RP-1}^{PC}(x_{t,j}, x_{t,j})}$$

Here, S_{RP}^{PC} is a path count measure. In the above RP^{-1} denotes the *inverse path-relation* of RP (see Definition 3);

- **HeteSim Measure [11]:** Given a *relation-path* $RP = R_1 \circ R_2 \circ ... \circ R_l$ (as in Definition 2), HeteSim [11] is defined as follows:

$$S_{RP(R_1 \circ R_2 \circ ... \circ R_l)}^{HS}(x_{t,i}, x_{t,j}) =$$

$$\frac{\sum_{p=1}^{|O(x_{t,i}|R_1)|} \sum_{q=1}^{|I(x_{t,j}|R_l)|} S_{RP(R_2 \circ R_3 \circ ... \circ R_{l-1})}^{HS}(O_p(x_{t,i}|R_1), I_q(x_{t,j}|R_l))}{S_{RP}^{PC}(x_{t,i},:) + S_{RP-1}^{PC}(:, x_{t,j})}$$

In the above, $x_{t,i}, x_{t,j} \in X_t$, $O(x_{t,i}|R_1)$ is the set of out-neighbors of $x_{t,i}$ based on relation R_1, and $I(x_{t,j}|R_l)$ is the set of in-neighbors of $x_{t,j}$ based on relation R_l.

All the above-described topological measures only focus on the topological structure of the networks. However, in the semi-supervised clustering process, different labeled information will lead to different similarity measures and different clustering results [9]. As a result, we propose a linear regression based measure which considers the small amount of labeled information. We use the labeled information as guidance, and propose a linearly combined measure, which is defined as follows:

$$S_{RP}^{LS}(x_{t,i}, x_{t,j}) = \sum_{d=1}^{m} \alpha_d s_d(x_{t,i}, x_{t,j}) \tag{1}$$

where S_{RP}^{LS} is the linear regression based measure of $x_{t,i}, x_{t,j} \in X_t$, and $s(x_{t,i}, x_{t,j}) = [s_1(x_{t,i}, x_{t,j}), s_2(x_{t,i}, x_{t,j}), ..., s_m(x_{t,i}, x_{t,j})]^T$ is the topological features following the given *relation-path* RP, and each feature is calculated using one of the formulae introduced at the beginning of this section, and m is the number of measures used in the framework. For example, $s_1(x_{t,i}, x_{t,j}) = S_{RP}^{PC}(x_{t,i}, x_{t,j})$, $s_2 = S_{RP}^{RW}(x_{t,i}, x_{t,j})$, $s_3 = S_{RP}^{PS}(x_{t,i}, x_{t,j})$, $s_4 = S_{RP}^{HS}(x_{t,i}, x_{t,j})$. Here, $\alpha = [\alpha_1, \alpha_2, ..., \alpha_m]^T$ denotes the weights for all measures. An optimization algorithm can be used to solve the following approximation problem:

$$\alpha^{opt} = \arg \min_{\alpha} \left\| \sum_{i=1}^{n} \sum_{j=1}^{n} (\mathbb{R}(x_{t,i}, x_{t,j}) - \sum_{d=1}^{m} \alpha_d s_d(x_{t,i}, x_{t,j})) \right\| \tag{2}$$

In the above, n denotes the number of labeled objects. Matrix $\mathbb{R}$ is obtained from the pre-labeled information as follow:

$$\mathbb{R}(x_{t,i}, x_{t,j}) = \begin{cases} 1 & x_{t,i}, x_{t,j} \ are \ labeled \ as \ the \ same \ label \\ 0 & otherwise \end{cases} \tag{3}$$

where $x_{t,i}, x_{t,j} \in X_t$ and X_t is the target type for clustering. Eq. (2) is actually a linear regression problem, which can be efficiently solved by many existing algorithms [14]. In this paper, we use the gradient descent method [14] to solve this problem.

3.2 Relationship Extraction Model

In our clustering task, given a set of *relation-paths*, each object within the target type may be connected via these *relation-paths*. By using these *relation-paths*, a heterogeneous information network can be reduced into a multi-relational network [15], in which the objects correspond to those of the target type in the original heterogeneous network and the different relations correspond to the given different relation-paths.

Similar to the model proposed in [15], the basic motivation of our relationship extraction model is as follows: different *relation-paths* correspond to different relation graphs[1], which can provide different clustering results. By combining these different clustering results through different weights of corresponding *relation-paths*, the final clustering result may be improved [15]. In this paper, logistic regression method [16] is used to handle the relation extraction problem.

The set of pre-labeled objects of the target type are regarded as the training set. Each two objects in the labeled set is regarded as a training pair, denoted as $x_{t,i}, x_{t,j} \in X_t$, and X_t is the target type for clustering. We first extract the topological features for these objects, and then build an extraction model to learn the weight values associated with these features. For each training pair $x_{t,i}, x_{t,j} \in X_t$, all the features are calculated by the linear regression based measure as in Eq. (1), denoted as $\boldsymbol{F} = [f_1, f_2, ..., f_d]$, where d is the number of *relation-paths* between the two objects. Here we denote the training set as $\boldsymbol{X}^{Train} = [X^1, X^2, ..., X^n]$, where X^i denotes the $i-th$ training pair, and n is the number of pairs in the training set. We define y^i as a label indicating whether these two objects are in the same cluster: if these two objects are in the same cluster, $y^i = 1$, and 0 otherwise, which is denoted as $\boldsymbol{Y}^{Train} = [y^1, y^2, ..., y^n]$, where $y^i \in \{0, 1\}$. The set of weights for all *relation-paths* is denoted as $\Lambda = [\lambda_1, \lambda_2, ..., \lambda_d]$. We use the Entropy maximization [17] method to calculate Λ: first, the conditional probability of the two objects in X^i belong to the same cluster can be modeled as:

$$p_\Lambda(y = y^i | X^i) = \frac{1}{Z(X^i)} exp(\sum_{i=1}^{d} f_i(X^i) * \lambda_i) \tag{4}$$

---

[1] A **relation graph** is a homogeneous network reduced by the heterogeneous network using a typical relation-path.

where Z is the normalization term calculated as $Z(X^i) = 1 + exp(\sum_{i=1}^{d} f_i(X^i) * \lambda_i)$. Second, we use the MLE (Maximum Likelihood Estimation) approach to derive Λ by maximizing the likelihood of all the training pairs:

$$L(\Lambda) = \prod_{i}^{n} [p_\Lambda(y = y^i | X^i)]^{y^i} [p_\Lambda(y = y^i | X^i)]^{1-y^i} \qquad (5)$$

Third, $\Lambda = [\lambda_1, \lambda_2, ..., \lambda_d]$ can be obtained as follows:

$$\Lambda^* = \arg max_{\Lambda = [\lambda_1, \lambda_2, ..., \lambda_d]} L(\Lambda) \qquad (6)$$

In this research we use the gradient descent [14] method to calculate Λ^*. Finally the combined affinity matrix $W^{combine}$ is defined by the following equation:

$$W^{combine} = \sum_{i=0}^{d} \lambda_i * W_i \qquad (7)$$

Here W_i is the similarity matrix of $i - th$ homogeneous network reduced by $i - th$ relation-path. It is noted that the similarity of each two objects in W_i is calculated by the measure introduced in section 3.1. We can see from the Eq. (7) that the combined affinity matrix $W^{combine}$ is a linear combination of different relation graphs. After obtaining $W^{combine}$, we then perform clustering on $W^{combine}$ to obtain the finally result.

3.3 The Detailed Steps of SemiRPClus

After presenting the calculation method for each relevant variable, the detailed steps of SemiRPClus is given as follows:

Step 1 Given a heterogeneous information network $G = \langle V, E, W \rangle$, a set of relation-paths, the target type X_t for clustering and labeled information Y.

Step 2 Use Eq. (3) to calculate the relation matrix $\mathbb{R}$.

Step 3 Calculation of the linear based similarity measure

    Step 3-a Calculate each kind of relation-path measure using the measure methods introduced in Section 3.1.

    Step 3-b Use Eq. (2) to calculate the weight $\alpha = [\alpha_1, \alpha_2, ..., \alpha_m]^T$ for each measure.

    Step 3-c Calculate the linear based similarity S_{RP}^{LS} using Eq. (1).

Step 4 Use Eq. (6) to obtain the weight of each relation-path: $\Lambda = [\lambda_1, \lambda_2, ..., \lambda_d]$, then use Eq. (7) to calculate the combine affinity matrix $W^{combine}$.

Step 5 Cluster the relation matrix $W^{combine}$ to obtain the final clustering result.

It is pointed out that many useful clustering methods [15, 18] can be used in Step 5, no matter whether the cluster number is pre-assigned or not.

3.4 Complexity Analysis

In this section, we consider the time complexity of the SemiRPClus. At the beginning of our framework, all the traditional topological features are calculated, and the time complexity is $O(k_{path}n^2)$. Here k_{path} is the number of *relation-paths* selected for the framework, and n is the number of target objects. For the linear based measure calculation process, the time complexity is $O(t_1 \sum_m |T_m|)$. Here t_1 is the number of iterations, and m is the dimension of the training dataset. For the Relationship Selection Model, the time complexity is $O(t_2 \sum_m |T_m|)$. Here t_2 is the number of iterations of this model. Assuming that the time complexity of the clustering used in Step 5 is $O_{cluster}$, the overall time complexity of our framework is $O(k_{path}n^2) + O(t_1 \sum_m |T_m|) + O(t_2 \sum_m |T_m|) + O_{cluster}$.

4 Experimental Results

In this section, we use the DBLP dataset[2] as a test bed to evaluate both the effectiveness and efficiency of our approach compared with some existing methods.

4.1 Datasets

The DBLP dataset is used for the performance test. Following [9], we extract a sub network of DBLP, "four-area dataset", which contains 20 major conferences in four areas: Data Mining, Database, Information Retrieval and Machine Learning. Each area contains five top conferences. In this dataset, the term is extracted from the paper titles, and the paper citation relationship is obtained by the Microsoft academic search API[3]. We use the following three datasets for our experiments.

DataSet-1 top 100 authors in the DBLP within the 20 major conferences, and the corresponding papers published by these authors after 2007.
DataSet-2 top 500 authors in the DBLP within the 20 major conferences, and the corresponding papers.
DataSet-3 top 2000 authors in the DBLP within the 20 major conferences, and the corresponding papers published by them after 2007.

The ground truth used in our experiment is obtained from the "four-area dataset [9]". It is pointed out that the labeled information and the true clustering result are all obtained from the ground truth.

As in [12], we choose 10 types of *relation-paths* as bellow: *author − paper − author, author → paper → paper − author, author → paper ← paper → author, author → paper → conference ← paper ← author, author → paper ← author, author → paper → author, author → paper → term ← paper ← author, author → paper → paper → paper ← author, author → paper → paper ←*

---

[2] http://www.informatik.uni-trier.de/~ley/db/
[3] http://academic.research.microsoft.com/

paper ← *author, author* → *paper* ← *paper* ← *paper* ← *author, author* →
paper ← *paper* → *paper* ← *author*. For example, *author* − *paper* − *author*
denotes the co-author relationship, and *author* → *paper* ← *paper* → *paper* ←
author denotes two authors' papers are cited in the same paper.

4.2   Case Study on Effectiveness

In this section, we study the effectiveness of our algorithm by comparing it with
several existing methods on the three datasets given in Section 4.1. For the ease
of comparison, we choose the hierarchical-cluster algorithm [19] to cluster the
similarity matrix, and the cluster number is pre-assigned as the input of the
cluster algorithm.

Three clustering methods are used in our experiment for comparison: Path-
SelClus [9], GNetMine [20] and LP [7]. The first two algorithms are proposed for
heterogeneous networks, and they are regarded as the state-of-the-art clustering
algorithms. LP is proposed for homogeneous network, thus we use two homogeneous networks reduced by two corresponding *relation-paths*. The two selected
relation-paths have the highest weight in SemiRPClus.

Two evaluation methods are used for testing the clustering result: Accuracy
[9], which is calculated as the percentage of target objects clustered into the
correct clusters; and Normalized Mutual Information (NMI) [9], which is one
of the most popular evaluation methods to evaluate the quality of clustering
results.

The clustering results are presented in Table 1. In the table, performance
under different percentage of labeled information (5%, 10% and 20%) in each
cluster is tested. All the results are averaged for 10 times. In Table 1, results in
bold indicate the best performance among all algorithms.

From Table 1 we see that the performance of SemiRPClus is comparable with
PathSelClus and GNetMine, and better in some cases. From the result we can
also see that SemiRPClus can have a better result evaluated by NMI in some
cases. NMI considers not only the accuracy, but also the distribution of the
objects within each cluster. From this perspective, SemiRPClus is more effective
than the other three algorithms. The LP algorithm always performs worse than
all other three heterogeneous clustering algorithms. This demonstrates that by
considering the heterogeneous information better results can be obtained. On
the other hand, we can see that mining heterogeneous networks can gain more
useful information than homogeneous ones.

4.3   Case Study on Efficiency

In this section, we study the efficiency of SemiRPClus. We use the same hard-
ware configuration to run SemiRPClus and the other three algorithms. Every
algorithm is run 10 trials and the average performance is calculated. The CPU
execution time for each algorithm is showed in Fig. 2. It is pointed out that we
use LP-RP1 to represent LP algorithm in this section. In Fig. 2, we can see that

Table 1. Cluster Accuracy and NMI for Three Dataset

Labeled	Evaluation	SemiRPClus	LP-RP1[4]	LP-RP5[5]	PathSelClus	GNetMine
Dataset-1						
5%	NMI	.048±.015	.078±.012	.020±.002	**.457±.095**	.387±.089
5%	Accuracy	.380±.021	.350±.036	.280±.016	**.570±.040**	.520±.073
10%	NMI	.318±.032	.056±.021	.031±.015	**.523±.026**	.408±.127
10%	Accuracy	.510±.024	.320±.041	.390±.012	**.710±.096**	.550±.048
20%	NMI	**.696±.032**	.069±.009	.036±.009	.541±.081	.488±.057
20%	Accuracy	.680±.042	.320±.023	.320±.084	**.730±.070**	.620±.058
Dataset-2						
5%	NMI	.621±.103	.014±.007	.004±.008	.609±.045	**.677±.042**
5%	Accuracy	.720±.087	.306±.085	.272±.052	.786±.116	**.884±.034**
10%	NMI	**.698±.038**	.023±.010	.006±.005	.646±.073	.664±.117
10%	Accuracy	.736±.085	.316±.034	.284±.094	**.830±.042**	.664±.042
20%	NMI	**.774±.034**	.026±.006	.013±.024	.718±.049	.702±.039
20%	Accuracy	.862±.046	.356±.088	.282±.044	.854±.0350	**.900±.028**
Dataset-3						
5%	NMI	**.798±.046**	.001±.006	.007±.002	.652±.089	.621±.045
5%	Accuracy	.750±.048	.254±.012	.207±.034	**.872±.015**	.862±.065
10%	NMI	**.759±.095**	.003±.014	.004±.002	.664±.015	.632±.015
10%	Accuracy	.784±.014	.254±.074	.271±.045	**.880±.034**	.868±.034
20%	NMI	**.868±.015**	.002±.001	.004±.002	.697±.095	.676±.024
20%	Accuracy	.800±.031	.261±.049	.275±.041	**.897±.012**	.889±.025

SemiRPClus is more efficient than PathSelClus and GNetMine: it is three to four orders of magnitude faster than PathSelClus in all experiments.

4.4   Case Study on Relation Path Weight

In this section, we study the learned weights for different *relation-paths* obtained by SemiRPClus compared with PathSelClus. As the ranking of the *relation-paths* showed in Table 2, the ranking learned by SemiRPClus is fundamentally the same as the ranking learned by the PathSelClus.

From the result we can see the *relation-path author − paper − author* always has a more trusted weight than other relation paths. This is consistent with human intuition: two authors having the co-author relationship means that they are very likely to have a very similar research interest. On the other hand, the *relation-path author → paper → term ← paper ← author* has the lowest weight in both algorithms. This is consistent with real-world scenarios: it is not rare that two papers from different areas can have the same term. For example, the words "optimization" and "method" appear in many papers from different areas.

---

[4] *author − paper − author.*

[5] *author → paper ← author → paper ← author.*

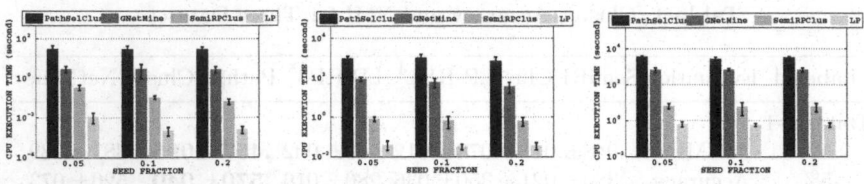

(a) Efficiency in DataSet-1 (b) Efficiency in DataSet-2 (c) Efficiency in DataSet-3

Fig. 2. Running time of SemiRPClus compared with the other three algorithms

Table 2. *Relation-Paths* Weight Comparison

Rank	PathSelClus	PathSelClus
1	$A - P - A^6$	$A - P - A$
2	$A \rightarrow P \leftarrow A \rightarrow P \leftarrow A$	$A \rightarrow P \leftarrow A \rightarrow P \leftarrow A$
3	$A \rightarrow P \leftarrow P \leftarrow P \leftarrow A$	$A \rightarrow P \leftarrow P \rightarrow P \leftarrow A$
4	$A \rightarrow P \rightarrow P - A$	$A \rightarrow P \leftarrow P \rightarrow A$
5	$A \rightarrow P \leftarrow P \rightarrow A$	$A \rightarrow P \rightarrow T \leftarrow P \leftarrow A$
6	$A \rightarrow P \rightarrow P \rightarrow P \leftarrow A$	$A \rightarrow P \rightarrow P - A$
7	$A \rightarrow P \rightarrow T \leftarrow P \leftarrow A$	$A \rightarrow P \rightarrow P \rightarrow P \leftarrow A$
8	$A \rightarrow P \leftarrow P \rightarrow P \leftarrow A$	$A \rightarrow P \rightarrow C \leftarrow P \leftarrow A$
9	$A \rightarrow P \rightarrow C \leftarrow P \leftarrow A$	$A \rightarrow P \leftarrow P \leftarrow P \leftarrow A$
10	$A \rightarrow P \rightarrow T \leftarrow P \leftarrow A$	$A \rightarrow P \rightarrow T \leftarrow P \leftarrow A$

5   Conclusions and Future Work

In this work, we explore the semi-supervised clustering analysis in heterogeneous information networks. Firstly, a similarity measure, which is more suitable for the semi-supervised clustering task, is proposed for measuring the similarity between objects in heterogeneous information networks. Secondly, a logistic regression model is used for extracting the relations. At last, an overall computational framework is proposed to perform semi-supervised clustering in heterogeneous information networks. Experimental results on the DBLP dataset demonstrate the effectiveness and efficiency of SemiRPClus.

In the future, we intend to apply SemiRPClus to more real-world clustering problems. In addition, another direction of our future research is to explore the potential of SemiRPClus on big data problems, such as massive social media and bioinformatics problems.

Acknowledgements. WP is supported by the partnership fund from dot.rural, RCUK Digital Economy research. This work is supported by the National Natural Science Foundation of China (NSFC) under Grant No. 61373051; the National Science and Technology Pillar Program (Grant No. 2013BAH07F05), Jilin Province Science and Technology Development Program (Grant No. 20111020);

---

[6] A: author, P: paper, C: conference, T: term.

Project of Science and Technology Innovation Platform of Computing and Software Science (985 engineering), and the Key Laboratory for Symbolic Computation and Knowledge Engineering, Ministry of Education, China.

References

[1] Fortunato, S.: Community detection in graphs. Physics Reports 486(3), 75–174 (2010)

[2] Lipka, N., Stein, B., Anderka, M.: Cluster-based one-class ensemble for classification problems in information retrieval. In: SIGIR 2012, pp. 1041–1042. ACM (2012)

[3] Pham, M.C., Cao, Y., et al.: A clustering approach for collaborative filtering recommendation using social network analysis. J. UCS 17(4), 583–604 (2011)

[4] Sun, Y., Han, J., Zhao, P., et al.: Rankclus: integrating clustering with ranking for heterogeneous information network analysis. In: ICDT 2009, pp. 565–576. ACM (2009)

[5] Zhu, X.: Semi-supervised learning literature survey. Computer Science, University of Wisconsin-Madison 2, 3 (2006)

[6] Basu, S., Banerjee, A., Mooney, R.J.: Semi-supervised clustering by seeding. In: ICML, vol. 2, pp. 27–34 (2002)

[7] Zhou, D., Bousquet, O., Lal, T.N., et al.: Learning with local and global consistency. Advances in Neural Information Processing Systems 16(16), 321–328 (2004)

[8] Bilenko, M., Basu, S., Mooney, R.J.: Integrating constraints and metric learning in semi-supervised clustering. In: ICML, p. 11. ACM (2004)

[9] Sun, Y.E.: Integrating meta-path selection with user-guided object clustering in heterogeneous information networks. In: KDD 2012, pp. 1348–1356. ACM (2012)

[10] Sun, Y., Han, J., Yan, X., Yu, P.S., Wu, T.: Pathsim: Meta path-based top k similarity search in heterogeneous information networks. In: VLDB 2011 (2011)

[11] Shi, C., Kong, X., Yu, P.S., Xie, S., Wu, B.: Relevance search in heterogeneous networks. In: ICDT 2012, pp. 180–191. ACM (2012)

[12] Sun, Y., Barber, R., Gupta, M., et al.: Co-author relationship prediction in heterogeneous bibliographic networks. In: ASONAM 2011, pp. 121–128. IEEE (2011)

[13] Lü, L., Zhou, T.: Link prediction in complex networks: A survey. Physica A: Statistical Mechanics and its Applications 390(6), 1150–1170 (2011)

[14] Montgomery, D.C., Peck, E.A., Vining, G.G.: Introduction to linear regression analysis, vol. 821. Wiley (2012)

[15] Cai, D., Shao, Z., He, X., Yan, X., Han, J.: Mining hidden community in heterogeneous social networks. In: LinkKDD, pp. 58–65. ACM (2005)

[16] Hosmer Jr., D.W., Lemeshow, S., Sturdivant, R.X.: Applied logistic regression. Wiley. com (2013)

[17] Berger, A.L., Pietra, V.J.D., Pietra, S.A.D.: A maximum entropy approach to natural language processing. Computational Linguistics 22(1), 39–71 (1996)

[18] Frey, B.J., Dueck, D.: Clustering by passing messages between data points. Science 315(5814), 972–976 (2007)

[19] Murtagh, F.: A survey of recent advances in hierarchical clustering algorithms. The Computer Journal 26(4), 354–359 (1983)

[20] Ji, M., Sun, Y., Danilevsky, M., Han, J., Gao, J.: Graph regularized transductive classification on heterogeneous information networks. In: Balcázar, J.L., Bonchi, F., Gionis, A., Sebag, M. (eds.) ECML PKDD 2010, Part I. LNCS (LNAI), vol. 6321, pp. 570–586. Springer, Heidelberg (2010)

A Content-Based Matrix Factorization Model
for Recipe Recommendation

Chia-Jen Lin, Tsung-Ting Kuo, and Shou-De Lin

Department of Computer Science and Information Engineering,
National Taiwan University, Taipei, Taiwan
heartherlin@gmail.com, {d97944007,sdlin}@csie.ntu.edu.tw

Abstract. This paper aims at bringing recommendation to the culinary domain in recipe recommendation. Recipe recommendation possesses certain unique characteristics unlike conventional item recommendation, as a recipe provides detailed heterogeneous information about ingredients and cooking procedure. Thus, we propose to treat recipes as an aggregation of features, which are extracted from ingredients, categories, preparation directions, and nutrition facts. We then propose a content-driven matrix factorization approach to model the latent dimension of recipes, users, and features. We also propose novel bias terms to incorporate time-dependent features. The recipe dataset is available at http://mslab.csie.ntu.edu.tw/~tim/recipe.zip

Keywords: Recipe recommendation, content-based recommendation, matrix factorization.

1 Introduction

With the prevalence of the Internet, people share huge amounts of recipes online, be a family recipe passed down through generations or one bright idea put into action in one afternoon. Currently there are over 10,000 cooking websites [1] providing various forms of information (e.g., texts, dish photos, cooking videos), as well as useful functions for searching and filtering by certain criteria. Conceivably, discovering appropriate recipes from such overwhelming database can be time-consuming. A recommendation system for recipes offers a desirable solution.

The task of recommending recipes does present several unique challenges. First, each recipe can be considered as a combination of several ingredients together with some contextual information such as cooking process and nutrition facts, or even certain meta-information such as its order in a course meal, type of cuisine, etc. As a result, a suitable recommendation system should take such profound and heterogeneous information into consideration. Second, there is no limit on the number of ingredients that can be used in a recipe, and generally recipes are not rated by as many viewers as movie or music does, we are facing a serious sparse rating and cold start problem. As shown in Table 1, the density of a recipe rating matrix is much lower than that of a movie rating. Such challenges can bring serious problems for traditional collaborative filtering models as these models rely heavily on the correlation among ratings to identify the latent connection between users and items.

V.S. Tseng et al. (Eds.): PAKDD 2014, Part II, LNAI 8444, pp. 560–571, 2014.

Table 1. Statistics of Netflix and FOOD.COM

Data	Netflix	FOOD.COM
User	480189	24741
Item	17770 (movies)	226025 (recipes)
Rating	100480507	956826
Sparsity	1.18%	0.02%
Average rating/per user	5654.50	4.23
Average rating/per item	209.25	38.67

Taking advantage of content information can be a solution to address the data sparsity and cold start problems. Unfortunately, such approach also has its own limitation in recipe recommendation since it fails to model the relationship among different features (e.g., different ingredients). An example is that the opinion of a user for an ingredient can be dramatically different depending on the type of dish to be prepared.

For instance, raw fish is a signature Japanese cuisine called Sashimi, but does not fit well with fries in traditional British fish and chips recipe. Therefore we cannot simply determine the usefulness of an ingredient without considering its correlation with other ingredients or preparation methods. This imposes a serious challenge for a content-based recommender.

In this paper, we propose a collaborative filtering approach called *content-driven temporal-regularized matrix factorization* (CTRMF), which aims at integrating heterogeneous content information into a Matrix Factorization (MF) model for a recipe recommendation system. The reason to choose an MF-based model is two-fold. First, MF-based models have been proven empirically as one of the most effective approaches for recommendation systems [2] [3]. Second, MF-based models allow us to exploit the latent correlation among objects, which is critical for recipes which include set of ingredients, preparation methods, and other meta-information. To incorporate the heterogeneous information of a recipe into an MF model, we propose to work on the *feature-matrix* instead of the original user-rating matrix. Feature matrix encodes the latent information about ingredients, categories, preparation directions, nutrition facts, and authors. We introduce several temporal biases into our model, including a novel idea to exploit the concept of Recency-Frequency-Monetary in different context.

1. We propose a content-driven MF-based model that incorporates the heterogeneous information of a recipe, including ingredients, dietary facts, preparation methods, serving order, cuisine type, and occasion. To our knowledge, this is the first proposal on using heterogeneous content information to perform recipe recommendation. Our experiments demonstrate decent improvement over the state-of-the-art models.

2. We propose a set of novel bias terms using the concept of Recency-Frequency-Monetary in different context. Such bias terms can potentially be applied to design recommendation systems in other domains.

3. Several works have been proposed on recipe recommendation. However, no benchmark test has been conducted to compare the performance of the proposed model with that of other competitors. This paper extracts real-life data from FOOD.COM to compare our model with two competitors to establish the performance benchmark on recipe recommendation.

2   Related Work

Personalized recommendation is important in consumer industry with huge variety of applications. Two common set of approaches are exploited for recommendation. (1) *Content-based ltering* is a paradigm that has been used mainly in the context of recommending items, for which informative content descriptors exist. Standard machine learning methods (e.g., SVM) have been used in this context. (2) *Collaborative ltering* exploits correlations between ratings across a population of users by finding users most similar to the active user and forming a weighted vote over these neighbors to predict unobserved ratings [11].

Recipe recommendation tasks have only been tackled by a small amount of researchers. Svensson et al. [4] propose a recipe recommendation system based on a user's explicit and implicit feedbacks through social interactions. Sobecki et al. [5] present a hybrid recommendation system, using fuzzy reasoning to recommend recipes. The above methods treat a recipe as a whole item, and require the social network between users for recommendation. In contrast, we break a recipe down into individual features, and need only the ratings but not social information to make recommendations.

There are also some recipe recommendation systems using content based techniques. Zhang et al. [6] construct a learning model using knowledge sources (e.g., WordNet) and a classifier (kNN) to make recommendations by finding similar recipes. Wang et al. [7] utilize NLP technique to parse preparation directions of recipes, and represent the recipes as cooking graphs consist of ingredients and cooking directions. They demonstrate that graph representations can be used to characterize Chinese dishes, by modeling the flow of cooking steps and the sequence of added ingredients. However, their work models the occurrence of ingredients and cooking methods but fails to take into account the relationships between ingredients. Neither do they consider users' preferences on specific recipes or ingredients. The main drawback of such language-dependent methods lies in the limited generality to non-Chinese recipes.

Freyne et al [8] proposes an Intelligent Food Planning (IFP) system, which breaks a recipe into core ingredients and gives each ingredient a weight. Then, IFP uses the weights of the ingredients to predict the rating of a new recipe. However, IFP does not take other information such as cooking style into account.

Forbes et al [9] propose content-boosted matrix factorization (CBMF), which is an extension of the matrix factorization model, to model hidden factors between users and ingredients.

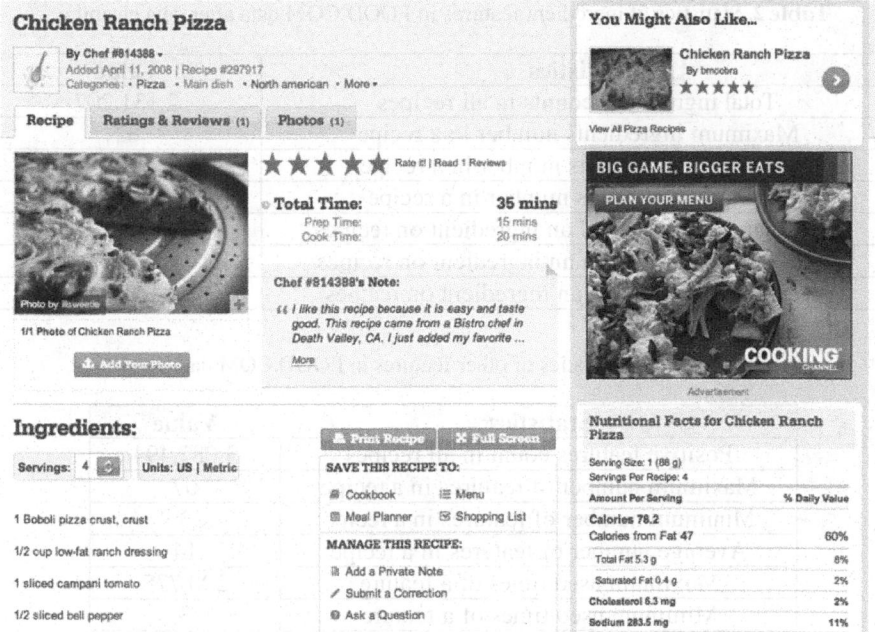

Fig. 1. A recipe from FOOD.COM

Although CBMF incorporates content information using linear constraints and proves the potential usefulness in the experiment, it only considers ingredients and does not use other information. Unlike our model, CBMF does not use temporal-regularized bias to improve accuracy. Experiment shows that our proposed model outperforms the CBMF model significantly. In this paper, we implemented IFP and CBMF as benchmark to compare with our proposed model.

3   Dataset and Features

3.1   Data Source

We collect data from 2000/2/25 to 2012/3/9 from FOOD.COM (www.food.com), one of the largest online recipe sharing communities. Figure 1 shows a sample recipe on FOOD.COM, which includes detail information such as ingredients, preparation directions, categories added by users, and the nutrition facts of this recipe. Our goal is to construct a recommendation engine that takes into consideration the profound types of information available.

We first filter out recipes that are rated no more than 3 times, as well as the users who rate no more than 5 times. Table 1 compares the statistics of FOOD.COM data with the Netflix data. We have found that this data is much sparser than the Netflix dataset. The average ratings per users/items are also much smaller. Such sparsity limits the effectiveness of a conventional collaborative filtering model and justifies the needs of adding content or meta-information into the recommendation system.

Table 2. Statistics of ingredient features in FOOD.COM data after data cleaning

Statistics	Value
Total ingredients counts in all recipes	2,131,207
Maximum ingredients number in a recipe	82
Minimum ingredients number in a recipe	1
Average ingredients number in a recipe	9
Maximum appearance of an ingredient on recipes	91,560
Minimum appearance of an ingredient on recipes	3
Average appearance of an ingredient on recipes	419

Table 3. Statistics of other features in FOOD.COM data

Statistics	Value
Positive features count in all recipes	3,087,494
Maximum number of features in a recipe	67
Minimum number of features in a recipe	3
Average number of features in a recipe	14
Maximum used times of a feature	220,775
Minimum used times of a feature	3
Average frequency for a feature being used	6,366

3.2   Features

We try to extract diverse features for each recipe. Originally, the dataset consists of 576,292 distinct ingredients, which requires certain level of data cleaning. We first correct some typos, and then merge ingredients of similar constituent, usually with different modifiers. For instance, "big red potato" and "small white potato" are both changed to "potato". We then remove ingredients used no more than 3 times to obtain 5,365 binary ingredients features. Those features cover about 99.8% of all the ingredients used in the recipes. Table 2 shows the statistics of ingredient features.

Besides ingredients, we extract features from categories, preparation directions, and nutrition facts to create the profile of a recipe. We group these features into 6 groups:

- *Main Ingredient*: Ingredient with maximum weight in recipe, excluding water/stock/bouillon.
- *Dietary*: Based on the FDA reference daily intake (RDI) [100], healthy terms such as low-fat (i.e., Recipes only contains 2% of fat), high fiber (i.e., 20% or more for fiber) are defined as binary features.
- *Preparation*: Describe the preparation process of a recipe, such as ways of cooking (stir-fry, oven bake, etc.). Note that we only choose terms with sufficiently high TFIDF values as binary features.
- *Courses*: describe the order of the dish being served in a coursed meal. For instance, appetizers, main dish, or desserts.
- *Cuisines*: describe style of food in terms of countries, such as Italian, Asian, etc.
- *Occasion*: describe the situation of food being served (e.g., brunch, dinner party)

Table 4. Top 10 features in six groups

Main Ingredients	Courses	Preparation	Cuisines	Occasion	Dietary
Meat	Main dish	Time to make	North U.S.	Taste/mood	Low fat
Vegetables	Dessert	Easy	U.S.	Dinner party	Low sodium
Fruit	Side dishes	Equipment	European	Holiday/event	Healthy
Eggs/dairy	Lunch/snacks	< 60 minutes	Asian	Comfort food	Low carb
Pasta, rice & grains	Appetizers	Number of servings	Italian	Seasonal	Low cholesterol
Poultry	One dish meal	< 30 minutes	Southern U.S.	To go	Low calorie
Chicken	Salads	< 4 hours	Mexican	Weeknight	Vegetarian
Beef	Breads	< 15 minutes	Canadian	Brunch	Low protein
Cheese	Breakfast	3 steps or less	South west pacific	Potluck	Low sat. fat
Seafood	Cookies and brownies	5 ingredients or less	Southwestern U.S.	Summer	Kid friendly

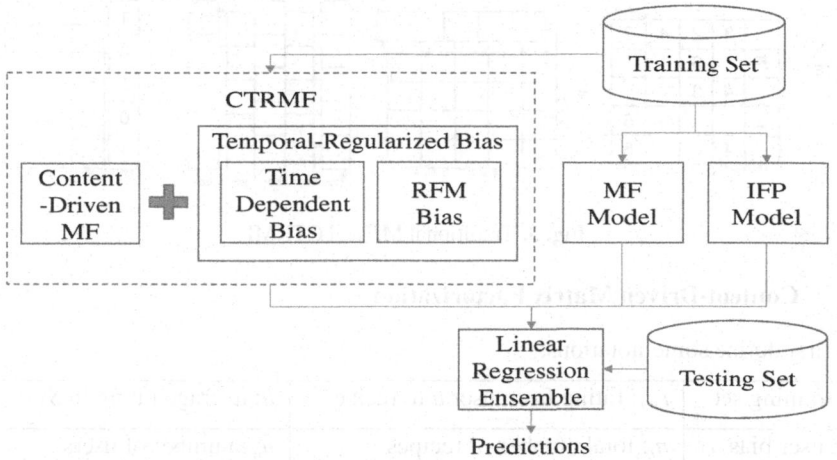

Fig. 2. Flowchart of our methodology

Here we obtain 485 additional features (not counting the original 5,635 ingredients) from FOOD.COM. Finally we merge highly similar features and remove extremely frequent, indiscriminative features such as salt and sugar. Finally we choose 5,538 features, 5,073 ingredients and 465 additional features. The statistics of those features are shown in Table 3. We list top 10 most frequent features in each group in Table 4.

4   Methodology

Figure 2 shows the flow chart of our proposed framework for recommendation. The heart of this system is the CTRMF engine, which will be described in section 4.1 and 4.2. As have been suggested by several researchers [2] [3] that the ensemble of models usually leads to the better results, we then linearly combine results from CTRMF with two diverse models, MF and IFP, to show that CTRMF can further improve the performance.

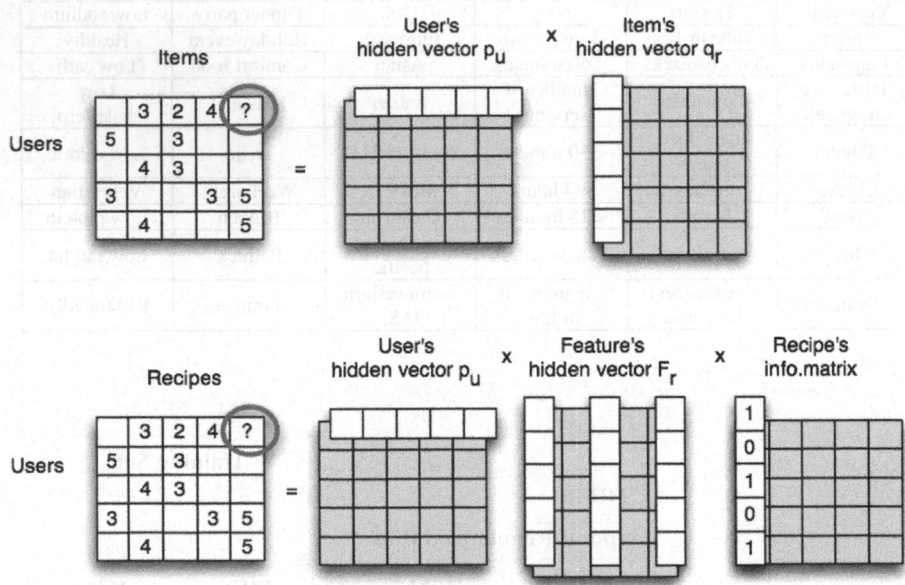

Fig. 3. Traditional MF and CTRMF

4.1   Content-Driven Matrix Factorization

We first define some notations:

S: training set	$r_{u,r}$: rating from user u to recipe r	μ: average rating in S
b_u: user bias	n_r: total number of recipes	n_u: number of users
b_r: recipe bias	n_f: total number of features	n_h: size of the hidden vector
P: a user-hidden matrix of dimension $n_u * n_h$, where each column represents a hidden vector of eash user		
F: a hidden-feature matrix of dimension $n_h * n_f$, where each row represents a hidden vector of each feature		
R: a recipe-feature matrix of dimension $n_r * n_f$, where each element is 1 if the recipe contains the corresponding feature, 0 otherwise		

Traditional MF tries to model hidden factors by decomposing the original user-item matrix into two low-dimensional matrices as below:

$$\hat{r}_{(u,r)} = p_u^T q_r$$

It models the interaction between latent user feature vector and item feature vector. That is, if a user likes a specific latent factor and an item has that factor, we conjecture that the user likes the item.

However, such model does not consider other useful information. Here we assume that recipes of common latent features are favored by certain group of users having similar latent features. Therefore our model predicts the rating using the following equation:

$$\hat{r}_{(u,r)} = p_u^T q_r^T R$$

Here p is a user-latent matrix, q represents latent-feature information, and R is the feature-recipe mapping. Note that p and q are learned from data and R is a matrix that encodes the heterogeneous information of each recipe. Figure 3 compares MF and CTRMF. Different from CBMF which does not include bias terms, here we add user bias and item bias; both are proven to be effective in our experiments. The objective function can then be defined as follows:

$$\min_{p*,b*,S} \sum_{(u,i)} (r_{u,r} - \mu - b_u - b_r - p_u^T F^T R_r)^2 + \lambda (\|p_u\|^2 + \|F\|^2 + b_u^2 - b_r^2)$$

The update function used in training can be derived as the follows:

$$e_{ur} \stackrel{def}{=} (r_{u,r} - \mu - b_u - b_r - p_u^T F^T R_r) \qquad p_u \leftarrow p_u + \eta (\sum_{r \in S_u} e_{ur} F^T R_r - \lambda p_u)$$

$$F \leftarrow F + \eta (\sum_{r \in S_u} e_{ur} R_r p_u^T - \lambda F) \qquad b_u \leftarrow b_u + \eta (e_{ur} - \lambda b_u)$$

$$b_r \leftarrow b_r + \eta (e_{ur} - \lambda b_r$$

4.2   Temporal-Regularized Bias

Bell et al. [2] have discovered from the Netflix data that there generally are some temporal patterns among ratings that can be exploited for better prediction accuracy. We also find similar patterns among the most active users in the FOOD.COM dataset. As shown in Figure 4, during the early days of the website, more than 30% of the ratings are relatively low (1, 2 and 3 in a five-star rating system). As the website becomes more mature, the percentage of low rates decreases to about 10%. Based on such observation, we add a time-aware bias to both users and recipes. We further propose to use the idea of Recency-Frequency-Monetary (RFM) Bias into our model. RFM is a concept proposed for analyzing customer behavior in customer relationship management (CRM). It is commonly used in database marketing and has received high attention in retail domain. The three main components of RFM are:

1. *Recency*:   whether the customer purchased something recently?
2. *Frequency*: whether the customer purchased something frequently?
3. *Monetary*: whether the customer spends lots of money on something?

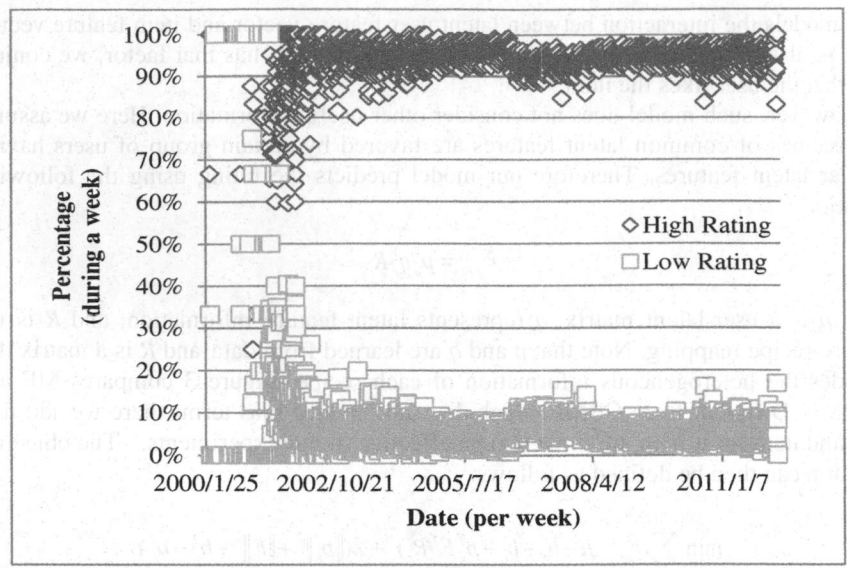

Fig. 4. Rating percentage distribution by week

We adopted the concept of RFM to incorporate more temporal biases into our model. For a certain user, R, F, and M become three binary variables indicating a user's Recent, Frequent, and Monetary rating behaviors. These three binary values then categorize the users into 8 different groups, and we assign each group a bias value to be learned. Similarly, items and authors are also divided into 8 groups, each correspond to a bias term. For each group, we try to learn a different bias value. Below we define the meaning of each group for users, recipes and authors.

4.2.1   User
First, from users' perspective, RFM of a user u can be defined as:

1. *Recency*: whether u rates a recipe more recently than u's average rating recency in the past?
2. *Frequency*: whether u rates a recipe more frequently than u's average rating frequency in the past?
3. *Monetary*: whether the most recent rating u provided rates higher than u's average rating?

Figure 5 is an example showing that u had provided a rating of 3 on May 1st, 3 on May 8th, 4 on May 15, and 5 on May 19. In this example, the current Recency value is 21-19=2, lower than the average past Recency ((8–1) + (15–8) + (19–15)) / 3=6. Similarly, the current Frequency 5/21 (rated 5 times in 21 days) is higher than the average of user u's past Frequency, 4/19 (rated 4 times in 19 days). For Monetary term, the last rating provided, a score of 5, is higher than user u's past average rating, (3+3+4) / 3=3.3.

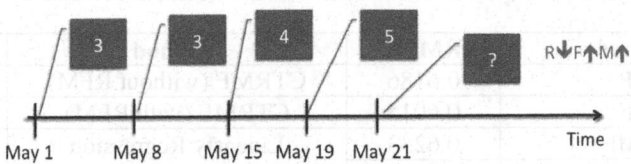

Fig. 5. Example of RFM in user side

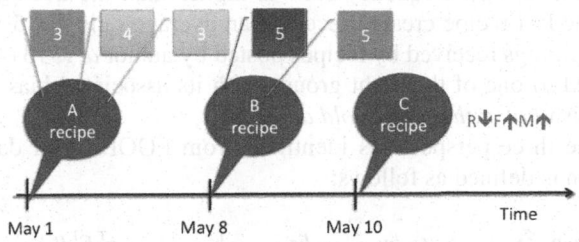

Fig. 6. Example of RFM in author side

Therefore, this user is assigned to group {R=0, F=1, M=1} and the corresponding bias terms is imposed. Such grouping allows us distinguish *hot users* from *cold users*.

4.2.2   Recipe
Similarly, from recipes' perspective, the RFM of a recipe r can be defined as:

1. *Recency*: whether r is rated more recently than its average recency of rating?
2. *Frequency*: whether r is rated more frequently than average frequency of rating?
3. *Monetary*: whether the most recent rating of r is higher than its average rating?

Similar to users, the recipes can now be divided into eight groups and each group is assigned a bias value to be learned. Such bias helps us distinguish *hot recipes* from *cold recipes*.

4.2.3   Author
From authors' perspective, RFM of an author a can be defined as:

1. *Recency*: does a create new recipe more recently than a's average recency?
2. *Frequency*: does a create new recipe more frequently than a's average frequency?
3. *Monetary*: does a's last recipe received higher rating than a's average rating received?

Note that the definition of Monetary here is slightly different from those of users and recipes. Figure 6 is an example showing that author a created the first recipe A on May 1st, second recipe B on May 8th, and recipe C on May 10th.

Table 5. RMSE results of baseline, our method, and ensemble

Method	RMSE	Method	RMSE
IFP	0.6186	CTRMF (without RFM)	0.5931
MF	0.6015	CTRMF (with RFM)	0.5901
CBMF	0.6233	Linearly Regression Ensemble	**0.5813**
Content-Driven MF	0.6013		

In this example, the current recency is 2, lower than the average past recency, 7 / 1=7. Similarly, the current frequency 3/10 is higher than the average frequency of author a, 2/8. The last recipe created received an average rating of 5 which is higher than the average ratings received by recipes posted by author a, (3+3+4) / 3=3.3. Each author is assigned to one of the eight groups with its associated bias term. This bias helps us distinguish *hot authors* and *cold authors*.

Combining the three perspectives identified from FOOD.COM dataset, our final objective function is defined as follows:

$$\min_{p_*, b_*, F} \sum_{(u,i)\in k} (r_{u,r} - \mu - bu_{Time(t)} - br_{Time(t)} - b_{rfm(u,r,a,t)} - p_u^T F^T R_r)^2$$

$$+ \lambda(\|p_u\|^2 + \|F\|^2 + (bu_{Time(t)})^2 + (br_{Time(t)})^2 + b_{rfm(u,r,a,t)}^2)$$

Note that b_{rfm} term is the multiplication of three terms, user, recipe, and author biases, defined above. And, the update functions are as follows:

$$e_{ur} \stackrel{def}{=} (r_{u,r} - \mu - bu_{Time(t)} - br_{Time(t)} - b_{rfm(u,r,a,t)} - p_u^T F^T R) \qquad p_u \leftarrow p_u + \eta(\sum_{r\in S_u} e_{ur}F^T R_r - \lambda p_u)$$

$$F \leftarrow F + \eta(\sum_{r\in S_u} e_{ur}R_r p_u^T - \lambda F) \qquad\qquad bu_{Time(t)} \leftarrow bu_{Time(t)} + \eta(e_{ur} - \lambda bu_{Time(t)})$$

$$br_{Time(t)} \leftarrow br_{Time(t)} + \eta(e_{ur} - \lambda br_{Time(t)}) \qquad\qquad b_{rfm(u,r,a,t)} \leftarrow b_{rfm(u,r,a,t)} + \eta(e_{ur} - \lambda b_{rfm(u,r,a,t)})$$

Here we train our model using stochastic gradient decent (SGD). We set λ to 0.01, η to 0.001, and the number of hidden factors to 100.

5   Experiments

We randomly select 4/5 of data from the users' ratings as training data, and use the rest as testing data. We compare our model (CTRMF) with IFP, standard MF, and CBMF models. The results showing in Table 5 reveal that the content-driven MF (introduced in Section 4.1) is better than CBMF, proving that the bias terms are useful. CTRMF has significant improvement over the existing methods with better RMSE. Also, adding RFM bias terms can improve CTRMF. Then we use linear regression to create an ensemble of IFP, MF, and our method. We divide the training data into training and validation to learn the parameters (i.e., the testing data remains unseen during ensemble). The ensemble RMSE can be further boosted to 0.5813.

6 Conclusion

This paper, to our knowledge, is the first ever attempt that incorporates 6 different types of content information, main ingredient, dietary, preparation, course order, cuisine type, and occasion, with user ratings for recipe recommendation. Such data will be released and become the only benchmark data so far for recipe recommendation. We also proposed the CTRMF model which is the first recommendation model that adopts the concept of RFM-based bias for recommendation, which can be potentially applied to domains other than recipe recommendation. Finally, this paper is the first to provide empirical comparison on different state-of-the-art models. For the future, we intent to extend the recommendation into a set of courses, such as appetizer, main dish, soup, dessert, and so on.

References

1. Alexa, http://www.alexa.com/topsites/category/Top/Home
2. Bell, R.M., Koren, Y., Volinsky, C.: The BellKor solution to the Netflix Prize. Technical Report, AT&T Labs Research (2007)
3. Koren, Y., Bell, R.M., Volinsky, C.: Matrix factorization techniques for recommender systems. Computer (2009)
4. Svensson, M., Laaksolahti, J., Höök, K., Waern, A.: A recipe based on-line food store. In: IUI 2000: Proceedings of the 5th International Conference on Intelligent User Interfaces, pp. 260–263 (2000)
5. Sobecki, J., Babiak, E., Słanina, M.: Application of hybrid recommendation in web-based cooking assistant. In: Gabrys, B., Howlett, R.J., Jain, L.C. (eds.) KES 2006, Part III. LNCS (LNAI), vol. 4253, pp. 797–804. Springer, Heidelberg (2006)
6. Zhang, Q., Hu, R., Namee, B., Delany, S.: Back to the future: Knowledge light case base cookery. Technical report, Technical report, Dublin Institute of Technology (2008)
7. Wang, L., Li, Q., Li, N., Dong, G., Yang, Y.: Substructure similarity measurement in Chinese recipes. In: Proceeding of the 17th International Conference on World Wide Web, WWW 2008, Beijing, China, April 21-25, pp. 979–988. ACM, New York (2008)
8. Freyne, J., Berkovsky, S.: Intelligent food planning: personalized recipe recommendation. In: Proceeding of the 14th International Conference on Intelligent User Interfaces, IUI 2010, pp. 321–324. ACM, New York (2010)
9. Peter, F., Zhu, M.: Content-boosted matrix factorization for recommender systems: experiments with recipe recommendation. In: Proceedings of the Fifth ACM Conference on Recommender Systems. ACM (2011)
10. http://en.wikipedia.org/wiki/Reference_Daily_Intake,
    http://en.wikipedia.org/wiki/Dietary_Reference_Intake,
    http://www.fda.gov/downloads/Food/GuidanceRegulation/UCM2654
    46.pdf
11. Basilico, J., Hofmann, T.: Unifying collaborative and content-based filtering. In: ICML (2004)

Unsupervised Analysis of Web Page Semantic Structures by Hierarchical Bayesian Modeling

Minoru Yoshida[1], Kazuyuki Matsumoto[1], Kenji Kita[1], and Hiroshi Nakagawa[2]

[1] Institute of Technology and Science, University of Tokushima
2-1, Minami-josanjima, Tokushima, 770-8506, Japan
{mino,matumoto,kita}@is.tokushima-u.ac.jp
[2] Information Technology Center, University of Tokyo
7-3-1, Hongo, Bunkyo-ku, Tokyo 113-0033, Japan
nakagawa@dl.itc.u-tokyo.ac.jp

Abstract. We propose a Bayesian probabilistic modeling of the semantic structures of HTML documents. We assume that HTML documents have logically hierarchical structures and model them as links between blocks. These links or dependency structures are estimated by sampling methods. We use hierarchical Bayesian modeling where each block is given labels such as "heading" or "contents", and words and layout features (i.e., symbols and HTML tags) are generated simultaneously, based on these labels.

Keywords: Hierarchical Bayesian modeling, Web document analysis, Gibbs sampling.

1 Introduction

In this study, we propose a new model for HTML documents that can extract *document structures* from them. *Document structures* are hierarchical structures of documents that decompose documents into smaller parts recursively. For example, scientific papers typically consist of several *sections*, each of which can be decomposed into *subsections*. In addition, *titles*, *abstracts*, and so on, are included in the document structure.

Web document analysis is a challenge to extract such document structures from *HTML documents*. Web documents can be decomposed into *subdocuments*, typically with their *headings* representing titles of each subdocuments. Figure 1 shows an example of subdocuments found in the web page shown in Figure 2, where each subdocument is presented as a heading. For example, in this document, "Age: 25" is a subdocument with the heading "Age" and content "25". Our purpose is to extract such lists of subdocuments such that those in the same list have parallel relations (such as "TEL:..." and "FAX:..." in Figure 1.) Note that there are "nested" lists – the element starting with "Contact:" contains another list "TEL:..." and "FAX:...".

We assume generative models for documents. The most basic way to collect parallel subdocuments is to use clustering algorithms such as K-means.

V.S. Tseng et al. (Eds.): PAKDD 2014, Part II, LNAI 8444, pp. 572–583, 2014.

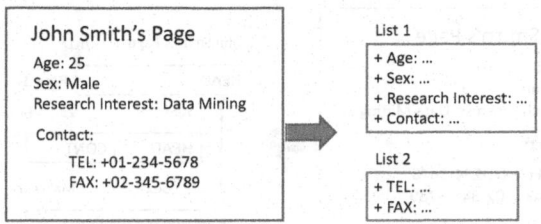

Fig. 1. Example Document and Its Repeated Tuples

The model we propose in this study is somewhat similar; however, it uses a hierarchical Bayesian framework to not only model similarities of visual effects, but also model

- hierarchical structures by using dependency trees as constraints to prior probabilities, and,
- simultaneously model local tag similarity and general visual effect usage, by using the hierarchical Bayesian model.

2 Related Work

Research on extracting repeated patterns from Web documents has a long history. The most popular approach is to make DOM trees and find frequency patterns on them[1, 2]. The problem with this approach is that it does not work for repeated patterns indicated by non-DOM patterns including patterns with symbols as shown in Figure 2.

Several studies have addressed the problem of extracting logical structures from general HTML pages *without labeled training examples*. One of these studies used domain-specific knowledge to extract information used to organize logical structures [3]. However, the approach in these studies cannot be applied to domains without any knowledge. Another study employed algorithms to detect *repeated patterns* in a list of HTML tags and texts [4, 5] or more structured forms [6–8] such as DOM trees. This approach might be useful for certain types of Web documents, particularly those with highly regular formats such as www.yahoo.com and www.amazon.com. However, there are also many cases in which HTML tag usage does not have significant regularity, or the HTML tag patterns do not reflect semantic structures (whereas symbol patterns do.) Therefore, this type of algorithm may be inadequate for the task of heading extraction from arbitrary Web documents. Nguyen [9] proposed a method for web document analysis using supervised machine learning. However, our proposal is to use probabilistic modeling for Web documents to obtain their structures in an unsupervised manner.

Some studies on extracting titles or headlines have been reported in [10, 11]. Our task differs from these, in that their methods focus only on titles

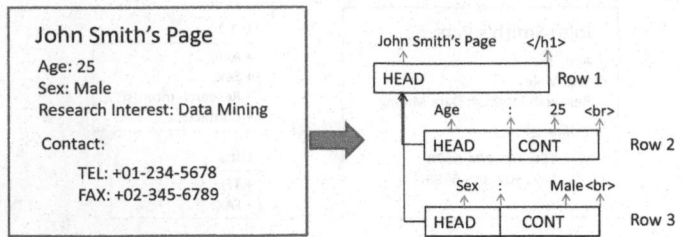

Fig. 2. Conversion from HTML Document to Dependency Structure. HEAD represents "heading" and CONT represents "contents".

(and headlines) and ignore the other parts of Web documents, while our algorithm handles *all parts* of the Web documents and provides a tree structure of the entire document; this algorithm enables the system to extract various types of heading other than titles and headlines, such as attributes. In particular, our approach has the advantage that it can handle symbols as well as HTML tags, making the system applicable to many private (less formal) Web documents.

Cai et al. [12] proposed VIPS that can extract content structure without DOM trees. Their algorithm depends on several heuristic rules for visual representation of each block in Web documents. However, we propose unsupervised analysis based on a Bayesian probabilistic model for Web document structures. This has several advantages, including easy adaptation to some specific document layouts, easiness of tuning its parameters (because we only have to change hyperparameters), and ability of obtaining probabilities for words and symbols that may be used for other type of documents such as text files.

Weninger et al.[13] compared several list extraction algorithm. One of the contributions of this study is that we propose a model for *nested structures* of lists, which has not been tried in most previous studies.

3 Preliminaries

3.1 Problem Setting and Terms Definition

We model an HTML document as a list of *rows*. Each row r_i is a list of *blocks*, i.e, $r_i =< b_{i_1}, ..., b_{i_{|r_i|}} >$. Here $|r_i|$ is the *size* of row r_i, which is the number of blocks in the row. A block b_j is a pair $(w_j, \mathbf{s}_j)$ of the *representative word* and *symbol list* $\mathbf{s} =< s_{j_1}, ..., s_{j_{|s_j|}} >$.

Because our experiments currently use Japanese documents, which do not contain word breaking symbols, it is not a trivial task to extract the representative word for each document. In our current system, we extract the predefined length of suffix from each block[1] and call them *representative words*. We did

---

[1] Length changes according to character types such as "trigrams for alphabets and numbers", "unigrams for Kanji characters", etc.

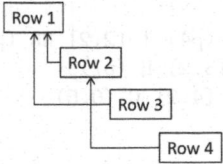

Fig. 3. Prohibited Dependency Relations

not use word breaker tools, partly because they are not for short strings such as those that frequently appear in Web documents, partly because we do not need human-comprehensible features (because our purpose is not information extraction but document structure recognition), and partly because simple suffix extraction rules contribute to stability of extraction results.

We assume several hidden variables associated with the above observed variables. First, each block b_j is labeled with *label* l_j, which can have one of two values $\{H, C\}$. Here, H means *heading* and C means *contents*. Headings are titles of subdocuments, and we assume that the heading is presented in the first few blocks of the subdocument, followed by other blocks that we call content blocks. (See Figure 2.) Blocks labeled with H are called *heading blocks*, and blocks labeled with C are called *content blocks*. *Heading rows* are the rows that contain one or more heading blocks, while *content rows* are the rows that contain no heading blocks. In addition a hidden variable p_k is associated with each symbol s_k. It indicates whether the symbol is a linefeed-related one, used in the Gibbs sampling step described later.

Next, we assume the dependency structures in documents. Here a *dependency relation* between two rows means that one row is modified by another. In Figure 2, the row "Age: 25" (*depending row*) depends on the row "John Smith" (*depended row*). We assume that a pair of hidden variables $(dep_i, bound_i)$ for the i-th row. (We also write $dep(r_i) = j$ if $dep_i = j$.) Here, dep_i is the row id that the i-th row depends on. Note that the structure is augmented with an additional variable $bound_i$, which denotes the position of the *boundary* between heading blocks and content blocks in the i-th row. If $bound_i = 0$, it means that there is no heading block in the row (and $dep_i = -1$ in this case), and if $bound_i = |r_i|$, it means that there is no content block in the line. The *sisters* of row r_i denote the list of all the rows $r_{i1}, r_{i2}, \ldots$ depending on the same row as r_i (i.e., $dep(r_{i1}) = dep(r_{i2}) = \ldots = dep(i)$).

Dependency Structures. Our definition of *dependency structures* in this study is slightly different from that used in natural language processing communities. One main difference is that it allows *isolated rows* that do not depend on any other rows. We consider that isolated rows link to a special row called the *null row*, indicated by the id number -1. We consider that two dependency structures are different if at least one row has different links. Note that we prohibit crossings in the dependency structures, and their probability is set to 0 (See Figure 3).

```
[1] R_nolink -> CL (1.0)
[2] [2.1] R_haslink -> HL CL (p4) | [2.2] HL (p5)
[3] [3.1] HL -> H HL (p6) | [3.2] H (p7)
[4] [4.1] CL -> C CL (0.5) | [4.2] C (0.5)
[5] H -> word1 | word2 | ...
[6] C -> word1 | word2 | ...
```

Fig. 4. PCFG Rules

Dependency structures with no crossing links are called *possible dependency structures*.

4   Probability Model and Estimation

We define the probability of generating the whole document as $P(d, T) = P_{prior}(T) \cdot P_{block}(d|T)$, where T is the assignment of (*dep, bound*) for all rows in document d.

Our idea is to divide the process of generating blocks into *vertical* and *horizontal generation* processes. The former generates rows under the constraints of row dependency structures. Currently, the probability of row dependency structures is defined as uniform distribution among all possible dependency structures. After each row is generated, all blocks in the row are generated horizontally with probabilities induced by CFG rules. Dividing the generation process in this way reduces the size of the search space. One of the merits of using CFG for prior probability calculation is that it can naturally model the ratio of headings and contents in each row, regardless of how many blocks are in the line. For example, if we directly model the probability of the value of $bound_i$, different lengths of rows require different models, which makes the model complicated. Instead, in our model, the ratio can be modeled by generation probabilities of a few of rules.

Figure 4 shows our PCFG rules used in our model. H means *headings* and C means *contents*. HL and CL mean *heading list* and *contents list*, which generates a list of heading blocks and content blocks, respectively. R means *rows*, which consist of one heading list, optionally followed by a content list. Here R_nolink is a nonterminal that indicates content (isolated) rows, and R_haslink is a nonterminal for heading rows. Note that this model prohibits headings from starting in the middle of each row.

Then, probabilities are calculated on the basis of the resulting CFG tree structures, using the PCFG rules shown in Figure 4.

Probability for Heading Rows. A heading row needs a probability of p4 or p5 before generating its heading and content lists. However, content rows do not need such probability (because they generate content lists with a probability of 1 by rule [1].)

Probability by Heading Blocks. As rule [3] shows, for each heading row, the last heading block needs a probability of p7, and other heading blocks need a probability of p6, to be generated. We define *heading probability* of the row as $P_h(r) = p6^{n_h(r)-1} \cdot p7$, where $n_h(r)$ is the number of heading blocks in row r.

Probability by Content Blocks. From rule [4], it is easily shown that each content block needs a probability of 0.5 to be generated. We define *content probability* of the row as $P_c = 0.5^{n_c}$, where n_c is the number of content blocks in the document.

4.1 Block Probability

The remaining part of the probability is the probability for blocks $P_{block}(D|T)$. First, note that each block is labeled H or C, according to the CFG tree in the horizontal generation. Each word in the block is generated from a distribution selected according to this label. We assign one of the labels $\{B, N, E, L\}$ to each symbol in the document using the following rules. Intuitively, B denotes *boundary* between heading and contents, N denotes *non-boundary*, E denotes *end of subdocuments*, and L denotes *line symbols* that are used in most of the linefeeds, which are not likely to have any semantic meaning, such as heading-associated tags such as <h1>.

- If the separator s_{i_k} is in the last block of row r_i, and the value of p_{i_k} is 1, then it is labeled L.
- If the separator s_{i_k} is in the last block of row r_i, the value of p_{i_k} is 0, and $bound_i \neq i_k$, then it is labeled E.
- Otherwise, if $bound_i = i_k$, separators in block b_{i_k} are labeled B.
- Otherwise, separators are labeled N.

Based on these labels, $P_{block}(d|T)$ is defined as a multinomial distribution of the bag of words: $P_{block}(d, T) = P(\mathbf{w}_H) \cdot P(\mathbf{w}_C) \cdot P(\mathbf{s}_B) \cdot P(\mathbf{s}_N) \cdot P(\mathbf{s}_E) \cdot P(\mathbf{s}_L)$ where $\mathbf{w}_H$ is a list of words labeled H, $\mathbf{w}_C$ is a list of words labeled C, $\mathbf{s}_B$ is a list of symbols labeled B, $\mathbf{s}_N$ is a list of symbols labeled N, $\mathbf{s}_L$ is a list of symbols labeled E, $\mathbf{s}_L$ is a list of symbols labeled L, in document d.

Our word/symbol generation model is a hierarchical Bayesian model. We assume the following generative process for words assigned nonterminal H.

1. Each word w in a document labeled H (i.e, $w \in \mathbf{w}_H$) is drawn from the distribution H_d: $w \sim H_d$.
2. H_d, the heading distribution for document d, is drawn from the Dirichlet distribution with base distribution H_{base} and concentration parameter α_H: $H_d \sim Dir(\alpha_H H_{base})$.
3. The base distribution H_{base} is drawn from the Dirichlet distribution with measure B: $H_{base} \sim Dir(B)$.

Words labeled C, and separators labeled N, E, or L are distributed in the same manner. Base distributions and concentration parameters for them are denoted by C_{base}, N_{base}, E_{base}, and L_{base}, and α_C, α_N, α_E, α_L, respectively.

Sampling from B is slightly different, because distributions are generated not for each document, but for each *sister*. Here sisters are a group of rows that depends on the same row.

1. Each separator s in the class labeled B (i.e., $s \in s_B$) is drawn from B_i: $s \sim B_i$.
2. B_i, the bound distribution for sisters class i, is drawn from the Dirichlet distribution with base distribution B_{base} and concentration parameter α_B: $B_i \sim Dir(\alpha_B B_{base})$.
3. The base distribution B_{base} is drawn from the Dirichlet distribution with measure B_B: $B_{base} \sim Dir(B_B)$.

Parallel subdocuments tend to have similar layouts, and such similarity typically appears in the boundary between headings and contents. We intend to model similar layouts by modeling boundary separators in the same list as that drawn from the same distribution.

We collapse the distribution for each document drawn from base distributions. For example, assume that $w_1, w_2, ..., and w_{n-1}$ have been drawn from H_d. Then, distribution for w_n is obtained by integrating out the multinomial distribution H_d, which results in the following equation.

$$P(w) = \frac{n_w}{\alpha_H + n.} + \frac{\alpha}{\alpha_H + n.} H_{base}(w) \tag{1}$$

where n_w is the number of times w occurs, and $n.$ is the number of all word occurrences, in the list $w_1, ..., w_{n-1}$. This equation can be obtained using the one for Pitman-Yor process [14] by assigning the discount parameter to be zero. By using this backoff-smoothing style equation, we can model the locality of the usage of words/separators by the first term, which corresponds to the number of occurrences of w in the same context, and global usage by the second term.

4.2   Sampling of Dependency Relations

Gibbs sampling is executed by looking at each row r_i and sampling the pair $(dep_i, bound_i)$ for the row according to the probability $P((dep_i, bound_i)|d_{-i})$. Here d_{-i} means the document without current row r_i. We calculate the relative probability of the document for all possible values for $(dep_i, bound_i)$ by taking all possible values for $(dep_i, bound_i)$, calculating the probability $P(d, T)$ for each dependency value, and normalizing them by the sum of all calculated values.

4.3   Sampling of Base Distributions

Another important part of our Gibbs sampling is sampling distributions given the document structures, which model the *general tendency of usage* of words and symbols. We use the fast sampling scheme described in [14] which omit time-consuming "bookkeeping" operations for sampling base distributions. First, the parameter $\mathbf{m}$, which indicates "how many times each word w was drawn *from the second term* of the equation 1, is drawn from the following distribution. $p(m_w = m | \mathbf{z}, \mathbf{m}^{-w}, \beta) = \frac{\gamma(\alpha\beta_w)}{\gamma(\alpha\beta_w + n_w)} s(n_w, m)(\alpha\beta_w)^m$ where $s(n, m)$ are unsigned Stirling numbers of the first kind. (Note that the factor k in Teh's representation corresponds to word w in our representation.) After drawing $\mathbf{m}$, the base parameter β is drawn from the following Dirichlet distribution: $(\beta_1, ..., \beta_K) \sim Dir(m_1 + \alpha'\gamma_1, ..., m_K + \alpha'\gamma_K)$ where α and γ are the strength parameter and the base distribution for the distribution for drawing H, respectively.

The following base distributions are sampled by using this scheme: H_{base} is sampled from $\mathbf{w}_H$, C_{base} is sampled from $\mathbf{w}_C$, N_{base} is sampled from $\mathbf{s}_N$, B_{base} is sampled from $\mathbf{s}_N$, and L_{base} is sampled from $\mathbf{s}_L$.

5   Implementation Issues

5.1   Sentence Row Finder

In HTML layout structure detection, *sentence blocks* are critical performance bottlenecks. For example, it is relatively easy to detect the suffixes of the rows that indicate sentences. However, it is difficult to decide whether the row *starts* with headings, especially when the sentences are decorated with HTML tags or symbols. (e.g., "Hobby: I like to hear music!")

Our idea is to use *prefixes* to decide whether the row contains headings. We assume that *rows starting with sentences contain no headers*, and the algorithm finds sentences by using the *ratio of obvious sentences* in all rows, starting with the prefix. The obvious sentences are detected by using simple heuristics that "if symbols in the row are only commas and periods, then the row surely consists of only sentences." Currently, if the ratio exceeds the threshold value 0.3, the row is determined as a sentence. Note that the sentence row finder is also applied to the baseline algorithm described later.

5.2   Hyperparameter Settings

We assume a Dirichlet prior for each rule of probability where Dirichlet hyperparameters are set $\alpha_1 = \alpha 2 = 5.0$ for rules [2.1] and [2.2], and $(\alpha_1, \alpha_2) = (10.0, 90.0)$ for rules [3.1] and [3.2] heuristically. The latter parameters suggest that our observation that the length of heading lists is not so large, giving high-probability values to short heading lists. This sampling of parameters helps to stabilize sampled dependency relations.

5.3   Parallelization

Parallelization of the above Gibbs sampling is straightforward because each sampling of tuples $(dep, rel, bound)$ only uses the state of other tuples in the same document, along with the base distributions such as H_{base} and B_{base}, which are not changed in tuple sampling. The task of sampling tuples is therefore divided into several groups as each group consists of one or more whole documents, and the sampling of tuples for each group is executed in parallel. (Sampling base distributions is not easily parallelized.)

5.4 Dependency Structure Estimation

Gibbs sampling can be used as a scheme for *sampling* the latent variables; however, it is not obvious how to extract *highly probable* states using this sampling scheme. Plausible base distributions can be obtained by taking several samples and averaging them. However, dependency structures are so complicated that it is almost impossible to see the same sample of structure two times or more. We thus use the following heuristic steps to obtain highly probable structures.

- Run the Gibbs sampling for some burn-in period.
- Take several samples for base distributions, and average them as an estimation for the base distribution and PCFG rule probabilities (these parameters are fixed thereafter.)
- Initialize the latent categorical variables.
- Run the Gibbs sampling again, but only for categorical variables for some burn-in period and calculate the marginal likelihood of the selected structures in each time.
- Take the structures with the maximum marginal likelihood so far.
- Greedy finalization: for each line, fix the state to the one with the highest probability. This step is executed over all rows sequentially, and repeated several times.

6 Experiments

Our corpus consists of 1,012 personal web pages found in the Japanese web site @nifty. We randomly selected 50 Web documents from them. We excluded 10 documents that contain <table> tags because table structures need special treatment for proper analysis and including them into the corpus harms the reliability of the evaluation. We extracted all repeated subdocuments in the remaining 40 documents manually. Among them, 14 documents contained no repeated subdocuments. For each algorithm, we extracted each set of sisters from dependency structures and regarded them as resulting sets of lists. We used purity, inverse purity, and their f-measure for evaluation, which is a popular measure for clustering evaluation.

6.1 Evaluation Measure

To evaluate the quality of extracted lists, we use purity and inverse purity measures[15], which are popular for cluster evaluation. We regard each extracted list as a cluster of subdocuments and represent it with the pair (i, j), where i is the start position and the j is the end position of the subdocuments.

The end position is set just before the start position of the next subdocument in the list.[2] Subdocument extraction is evaluated by comparing this cluster to manually constructed subdocument clusters.

Assume that C_i is a cluster in the algorithm results, and L_i is a cluster in the manual annotation. Purity is computed by $P = \sum_i \frac{|C_i|}{N} \max_j Precision(C_i, L_j)$ where $Precision(C_i, L_j) = \frac{|C_i \cap L_j|}{|C_i|}$ and $N = \sum_i |C_i|$. Inverse purity is defined as $IP = \sum_i \frac{|L_i|}{N} \max_j Precision(L_i, C_j)$ where $N = \sum_i |L_i|$.

Quality of the output lists is evaluated by the F-measure, which is the harmonic mean of purity and inverse purity: $F = \frac{1}{1/P + 1/IP}$.

We did not used B-cubed evaluation measures[16] because B-cubed is an element-wise definition, which calculates correctness of all rows in the corpus, indicating that we would have to consider rows that have no headings, for which no clusters are generated. B-cubed measures are developed as a metric that works for soft-clustering, whereas our task can be regarded as hard clustering, in which P-IP measures work well.

We used *micro-averaged* and *macro-averaged f-measures* for cluster evaluation. Macro-averaged f-measure compute f-value for each document that has any repeated patterns (i.e., 26 documents in the test set) and average all the f-values. However, micro-averaged f-measures regard all 40 documents in the test set as one document, and calculate P, IP, and F on this one large document. Thus, we can evaluate how each method *does not* extract unnecessary lists from documents with no repeated lists by using a micro-averaged f-measure.

6.2   Baseline

We use the baseline algorithm that uses some heuristic rules to extract subdocuments. We test several configurations (e.g., what header tags are used for extraction, whether rows with $|r| = 1$ are extracted as headings, etc.) and select the one that performed the best on the test set. This baseline algorithm selects heading rows among all rows (except the ones discarded by the sentence row finder) using following heuristic rules.

First, it uses "header tag heuristics". For example, if the row is in an <h2> tag, we assume that the row is a heading that modifies the following blocks until the next <h2> or larger headers (<h1> in this case) appear. Header tags <h1>, <h2>, <h3>, and <h4> are used in this heuristics.[3]

Second, it uses the *block number heuristics* which showed good performance in our preliminary experiments. Assume that $|r|$ is the number of blocks in the row $|r|$. If $|r| \geq 2$, the algorithms regard r as heading row (we assume that this row is bracketed by <h8>, which is smaller than all other h tags.) If $|r| = 1$ and r is not a sentence row, we assume r is bracketed by <h7>, which indicates that it will be the heading of the next rows (if the next row has more than one blocks.)

---

[2] It is set to the end position of the document for the last subdocument in the list.
[3] We also used <h7> and <h8> generated by the block number heuristics described below.

Table 1. Averaged F-measure (%) for Each Method

Method	w/o no-repeat (26 docs.)	w/ no-repeat (40 docs.)
micro-averaged		
Proposed	50.23	47.68
Baseline	46.93	42.42
macro-averaged		
Proposed	49.63	—
Baseline	42.05	—

Note that this simple heuristics can extract many sub-documents in Figure 2 including "Age:25" and "TEL:+01-234-5678".

6.3   Results

We run our Gibbs sampling with 1000 initial iterations and 500 final iterations. Values of parameters $(\alpha_B, \alpha_E, \alpha_L, \alpha_N)$ were set to $(10, 100, 100, 1000)$ heuristically. We use the uniform distribution for each base distribution. Results were obtained by running Gibbs sampling 5 times and averaging all the averaged f-measure values.

Table 1 shows the results. Our algorithm outperformed the baseline algorithm by about 3.3 – 7.6 points. The performance gain of our algorithm in micro-averaged f-measure increased from 3.3 to 5.3 by using 14 "no-repeat" documents. This result suggests that our method works well in detecting "no-repeat" documents to avoid incorrect repeated lists.

Performance gain was mainly obtained by detection of heading blocks that could not be found by the baseline algorithm and detection of content blocks that could not be found by sentence row finder heuristics. However, the performance of our algorithm for documents with heading blocks that were easily detected by the baseline algorithm tended to be lower. We need an algorithm that takes the strength of both our method and the baseline method for better performance.

7   Conclusion

In this study, we proposed a probabilistic model for document structures for HTML documents that uses Bayesian hierarchical modeling. Our model can simultaneously manage both local coherence and global tendencies of layout usage, thanks to hierarchical modeling and cache effects obtained by integrating out of distributions. Experimental results showed that document structures obtained by our model were better than those obtained by the heuristic baseline method. For future study, we are keen to improve the performance of our method by, for example, using larger data sets to obtain more reliable knowledge about layout usage, or using more sophisticated methods to obtain maximum-likelihood states for our model.

References

1. Miao, G., Tatemura, J., Hsiung, W.P., Sawires, A., Moser, L.E.: Extracting data records from the web using tag path clustering. In: Proceedings of WWW 2009, pp. 981–990 (2009)
2. Liu, B., Grossman, R.L., Zhai, Y.: Mining data records in web pages. In: Proceedings of KDD 2003, pp. 601–606 (2003)
3. Chung, C.Y., Gertz, M., Sundaresan, N.: Reverse engineering for web data: From visual to semantic structures. In: ICDE (2002)
4. Yang, Y., Zhang, H.: HTML page analysis based on visual cues. In: Proceedings of the Sixth International Conference on Document Analysis and Recognition, ICDAR 2001 (2001)
5. Nanno, T., Saito, S., Okumura, M.: Structuring web pages based on repetition of elements. In: Proceedings of the Second International Workshop on Web Document Analysis, WDA 2003 (2003)
6. Mukherjee, S., Yang, G., Tan, W., Ramakrishnan, I.: Automatic discovery of semantic structures in HTML documents. In: Proceedings of the Seventh International Conference on Document Analysis and Recognition, ICDAR 2003 (2003)
7. Crescenzi, V., Mecca, G., Merialdo, P.: ROADRUNNER: Towards automatic data extraction from large web sites. In: Proceedings of the 27th International Conference on Very Large Data Bases (VLDB 2001), pp. 109–118 (2001)
8. Chang, C.H., Lui, S.C.: IEPAD: Information extraction based on pattern discovery. In: Proceedings of the 10th International WWW Conference (WWW 2001), pp. 681–688 (2001)
9. Nguyen, C.K., Likforman-Sulem, L., Moissinac, J.C., Faure, C., Lardon, J.: Web document analysis based on visual segmentation and page rendering. In: Proceedings of International Workshop on Document Analysis Systems (DAS 2012), pp. 354–358. IEEE Computer Society (2012)
10. Hu, Y., Xin, G., Song, R., Hu, G., Shi, S., Cao, Y., Li, H.: Title extraction from bodies of HTML documents and its application to web page retrieval. In: Proceedings of the 28th Annual International ACM SIGIR Conference (SIGIR 2005), pp. 250–257 (2005)
11. Tatsumi, Y., Asahi, T.: Analyzing web page headings considering various presentation. In: Proceedings of the 14th International Conference on World Wide Web Special Interest Tracks and Posters, pp. 956–957 (2005)
12. Cai, D., Yu, S., Wen, J.R., Ma, W.Y.: Extracting content structure for web pages based on visual representation. In: Zhou, X., Zhang, Y., Orlowska, M.E. (eds.) APWeb 2003. LNCS, vol. 2642, pp. 406–417. Springer, Heidelberg (2003)
13. Weninger, T., Fumarola, F., Barber, R., Han, J., Malerba, D.: Unexpected results in automatic list extraction on the web. ACM SIGKDD Explorations Newsletter 12(2), 26–30 (2010)
14. Teh, Y.W., Jordan, M.I., Beal, M.J., Blei, D.M.: Hierarchical dirichlet processes. Journal of the American Statistical Association 101(476), 1566–1581 (2006)
15. Artiles, J., Gonzalo, J., Sekine, S.: The semeval-2007 weps evaluation: Establishing a benchmark for the web people search task. In: Proceedings of the Workshop on Semantic Evaluation (SemEval 2007) at ACL 2007, pp. 64–69 (2007)
16. Artiles, J., Gonzalo, J., Sekine, S.: Weps 2 evaluation campaign: overview of the web people search clustering task. In: Proceedinsg of the 2nd Web People Search Evaluation Workshop (WePS 2009), 18th WWW Conference (2009)

For User-Driven Software Evolution: Requirements Elicitation Derived from Mining Online Reviews

Wei Jiang[1], Haibin Ruan[2], Li Zhang[1], Philip Lew[1], and Jing Jiang[1]

[1] School of Computer Science and Engineering, Beihang University, Beijing, China
jiangwei@cse.buaa.edu.cn, {lily,jiangjing}@buaa.edu.cn,
philiplew@gmail.com
[2] Investment Department, Central University of Finance and Economics, Beijing, China
nivadacufe@gmail.com

Abstract. Online reviews that manifest user feedback have become an available resource for eliciting requirements to design future releases. However, due to complex and diverse opinion expressions, it is challenging to utilize automated analysis for deriving constructive feedback from these reviews. What's more, determining important changes in requirements based on user feedback is also challenging. To address these two problems, this paper proposes a systematic approach for transforming online reviews to evolutionary requirements. According to the characteristics of reviews, we first adapt opinion mining techniques to automatically extract opinion expressions about common software features. To provide meaningful feedback, we then present an optimized method of clustering opinion expressions in terms of a macro network topology. Based on this feedback, we finally combine user satisfaction analysis with the inherent economic attributes associated with the software's revenue to determine evolutionary requirements. Experimental results show that our approach achieves good performance for obtaining constructive feedback even with large amounts of review data, and furthermore discovers the evolutionary requirements that tend to be ignored by developers from a technology perspective.

Keywords: Software evolution, requirements elicitation, online reviews, opinion mining, user satisfaction analysis.

1 Introduction

Successful software systems are always able to evolve as stakeholder requirements and environments where the deployed systems operate [1]. In today's competitive market, meeting changing user demands[1] is a critical driving factor of software evolution. The software system should adapt to the social environment where users form opinions based on their experience with it [3]. Therefore, systematically and effectively eliciting evolutionary requirements is critical for the software to adapt and improve.

---

[1] In economics, demand is an economic principle that describes a consumer's desire, willingness and ability to pay a price for a specific good or service.

V.S. Tseng et al. (Eds.): PAKDD 2014, Part II, LNAI 8444, pp. 584–595, 2014.

User feedback provides useful information that can help to improve software quality and identify missing features [4]. However, with software delivered via the Internet, the scopes and types of users are uncertain before delivering software systems, since there are differences of space and time between users and developers. Fortunately, user generated content is becoming mainstream in web platforms, and consumers are willing to publish online reviews to express their opinions about software systems. These reviews that manifest user demands in real contexts of use have become an available feedback resource for eliciting requirements to design future software releases. Moreover, the reviews that come from large quantities of disparate users contain abundant data and its expressions [5]. For example, the Apple App Store has more than five hundred million active registered users and billions of online reviews.

Current software engineering research and practice favor requirements elicitation derived from online reviews. Several approaches have been developed, regarding the techniques and processes to consolidate, analyze and determine requirements in accordance with online feedback [6-7]. However, these approaches mostly rely on manual content analysis and as such, are not efficient for dealing with large amounts of online reviews in order to shorten time-to-market. Obviously, automated techniques, such as text mining, information retrieval, and machine learning can be effective tools for identifying software features and associated opinions mentioned in user comments [8-9]. However, due to complex and diverse opinion expressions, it is challenging to utilize automated analysis for accurately deriving constructive feedback from the reviews of software systems. What's more, assisting developers with determining evolutionary requirements based on user feedback is also challenging.

In this paper, we present a systematic approach for the transformation of online reviews to evolutionary requirements. For the first problem of automated text analysis we analyze the characteristics of online software reviews and then adapt the syntactic relation-based propagation approach (SRPA) to automatically identify opinion expressions about common software features. In order to provide meaningful feedback, we present a method S-GN for clustering opinion expressions from a network perspective, which uses the Girvan-Newman algorithm with the Explicit Semantic Analysis similarity. To address the second problem of assisting developers, we consider an economic impact to analyze user satisfaction based on this feedback and then determine important changes for requirements.

The contributions of our research are as follows:

- SRPA+, an expanded version of SRPA, is more suitable for mining complete opinion expressions from software reviews.
- The proposed method S-GN optimizes the clusters of opinion expressions by considering a macro network topology.
- We combine user satisfaction analysis with the inherent economic attributes associated with the software revenue to determine the evolutionary requirements.
- We show experimentally that our approach achieves good performance for deriving constructive feedback even with large amounts of review data, and furthermore discovers the evolutionary requirements that tend to be ignored by developers.

The remainder of this paper is organized as follows. Section 2 elaborates our approach. Section 3 describes the experiment and analyzes the results. Section 4 introduces related work and Section 5 discusses conclusions and future work.

2 Systematic Approach of Requirements Elicitation

In our approach, we first extract opinions about software features from large amounts of online reviews and determine whether opinions are positive or negative. Then, we categorize opinions and select corresponding review sentences to represent the feedback on software features. Finally, we generate a document of evolutionary requirements from an economic perspective.

2.1 Extracting Targets and Sentiment Words

A user opinion mentioned in a review is defined as a target-sentiment pair. The target is a topic on the software feature that is a prominent or distinctive user-visible aspect, quality, or characteristic of the system [10]. The sentiment is the user evaluation of its target. Sentiment words and targets are often related syntactically and their relations can be modeled using the dependency grammar [11]. There have been many methods used to model sentiment words, targets and their dependency relationships [21-23]. We adapt the syntactic relation-based propagation approach (SRPA) due to its modeling naturally for the opinion mining task. SRPA extracts targets and sentiment words iteratively using known and extracted words through the identification of syntactic relations [12]. The bootstrapping process starts with a seed sentiment lexicon.

The key of SRPA is the propagation rules based on syntactic relations. As targets and sentiment words are usually nouns/noun phrases and adjectives respectively [13], SRPA only defines the relations between nouns and adjectives, between adjectives, and between nouns. However, software systems have the dynamic features of computing and processing. Users may comment on the software's behavior and impact on the application environment using opinion expressions that depend on the relations between verbs and adverbs. In addition, sentiment verbs such as *love* can also express opinions. Accordingly, we define new propagation rules based on Stanford POS tagger[2] and syntactic parser[3], as shown in Table 1.

Table 1. New propagation rules

ID	Description	Output	Example
$R1_1$	$S{\rightarrow}S\text{-}Dep{\rightarrow}T$ s.t. $S{\in}\{S\}$, $POS(T){\in}\{VB\}$, $S\text{-}Dep{\in}\{MR\}$	$t{=}T$	The software updates quickly.
$R1_2$	$S{\rightarrow}S\text{-}Dep{\rightarrow}T$ s.t. $T{\in}\{T\}$, $POS(S){\in}\{RB\}$, $S\text{-}Dep{\in}\{MR\}$	$s{=}S$	This software operates well with Firefox.
R2	$T_{i(j)}{\rightarrow}T_{i(j)}\text{-}Dep{\rightarrow}T_{j(i)}$ s.t. $T_{j(i)}{\in}\{T\}$, $POS(T_{i(j)}){\in}\{VB\}$, $T_{i(j)}\text{-}Dep{\in}\{CONJ\}$	$t{=}T_{i(j)}$	It is easy to download and update the software.
R3	$S_{i(j)}{\rightarrow}S_{i(j)}\text{-}Dep{\rightarrow}S_{j(i)}$ s.t. $S_{j(i)}{\in}\{S\}$, $POS(S_{i(j)}){\in}\{RB\}$, $S_{i(j)}\text{-}Dep{\in}\{CONJ\}$	$s{=}S_{i(j)}$	Norton runs smoothly and quietly.
R4	$S{\rightarrow}S\text{-}Dep{\rightarrow}T$ s.t. $S{\in}\{S\}$, $POS(T){\in}\{NN\}$, $S\text{-}Dep{\in}\{OBJ\}$	$t{=}T$	Installation destabilizes the correct operation.
R5	$S{\rightarrow}S\text{-}Dep{\rightarrow}T$ s.t. $S{\in}\{S\}$, $POS(T){\in}\{VB\}$, $S\text{-}Dep{\in}\{COMP\}$	$t{=}T$	I love to use Kaspersky.

*S (or T) is the sentiment word (or target). $\{S\}$ (or $\{T\}$) is the set of known sentiment words (or the set of targets). $POS(S)$ (or $POS(T)$) represents the POS tag of the sentiment word (or target). $\{RB\}$: RB, RBR, RBS; $\{VB\}$: VB, VBD, VBG, VBN, VBP, VBZ; $\{NN\}$: NN, NNS, NNP. S-Dep (or T-Dep) represents the dependency relation of the word S (or T). $\{MR\}$: advmod; $\{CONJ\}$: conj; $\{OBJ\}$: dobj, pobj; $\{COMP\}$: xcomp.

2.2 Identifying Complete Expressions of Opinions

Since the targets extracted in the previous step are individual words, we need to identify the complete target expressions. Generally, a sentence clause contains only one target-sentiment pair unless there are conjunctions [12]. However, in the sentence

---

[2] http://nlp.stanford.edu/software/tagger.shtml
[3] http://nlp.stanford.edu/software/lex-parser.shtml

Kaspersky didn't update to a correct version quickly, there are two target-sentiment expressions, namely *Kaspersky update quickly* and *a correct version*. Therefore, we further merge the opinion expressions on the same software feature, and remove the noisy ones caused by unconstrained propagation rules.

The target words consist of nouns and verbs. For noun target expressions, we exploit the phrase-structure trees to identify the noun phrases that contain the extracted target words. There are two cases of verb target expressions. If the target words are verbs with direct objects or prepositional objects, the verb phrases in the phrase-structure trees serve as the target expressions. Note that the parser doesn't distinguish prepositional objects and adverb prepositional phrases. We check the compound phrases of verbs and prepositions through the online dictionary[4]. If the target words are verbs without objects, the verbs and their subjects compose the target expressions.

Based on the identified target expressions, we employ the following two rules to merge the opinion expressions in the same sentence. **Rule 1**: if the opinion expressions share some words, they must describe the same software feature. **Rule 2**: if the words in opinion expressions have direct dependency relations, they are likely to represent the same software feature.

We prune noisy opinion expressions according to word frequency. The sentiment words, negations, and stop words are removed from the merged opinion expressions and then the scores of processed expressions are calculated as follows:

$$O-Score(e_i) = \alpha * \frac{W_{e_i}}{W_E} + \beta * \frac{N_{e_i}}{N_E} + \gamma * \frac{F_{e_i}}{F_E} \qquad (1)$$

where e_i is a processed expression; E is the set of processed expressions; W_x is the number of words in x; N_{ei} is the number of processed expressions that contain the words in e_i; N_E is the number of processed expressions in E; F_x is the number of frequent words in x; α, β, and γ are weights. For a sentence, we only choose the opinion expession with the highest score unless there are conjunctions.

2.3   Assigning Sentiment Polarities to Opinion Expressions

We propose a two-stage method for assigning polarities to opinions. In the first stage, the polarities for the newly extracted sentiment words are inferred through the rules in [12]. The extracted words are assigned with the same polarities as the known words in the current review. The polarity changes when there are an odd number of negations or contrary words between the extracted word and the known word. The polarity value of the sentiment word that has either no or multiple polarities is computed as the sum of polarity values of known sentiment words in the current review.

In the second stage, the polarity score of a complete opinion expression is estimated by the following ordered weighted averaging (OWA) aggregation [14] of the contained sentiment words.

$$p = \sum_{i=1}^{m} \frac{w_{s_i}}{N_{s_i}} s_i \qquad (2)$$

---

[4] http://thesaurus.com/

where s_i is the polarity of a sentiment word in the opinion expression; N_{si} is the number of sentiment words that have the same type with s_i; w_{si} is the OWA weight of s_i. The types of sentiment words consis of verbs (v), adjectives (adj), and adverbs (adv). As verbs are the core of sentences whereas adjectives and adverbs are modifiers, the types of sentiment words are ranked as (v, adv, adj) from big to small according to their importance. The value of w_{si} is concerned with the position of the type of s_i. The OWA operators aggregate both the polarities of sentiment words and the importance of their types. If the polarity score of an opinion expression is greater than 0, its polarity is positive, and otherwise negative. The polarity changes if there are an odd number of negations associated with sentiment words in the opinion expression.

2.4   Organizing Opinions into Structured Feedback

In order to provide meaningful feedback, we first group similar opinion expressions about software features. Each category represents a unique overall, functional or quality requirement. Then we produced the structured feedback classified by software features, including several corresponding sentences.

The opinion expressions and the semantic associations between them construct an undirected graph $G = (V, E)$, where V is the set of vertices for the opinion expressions and E is the set of edges for the semantic links between any two vertices. There is a semantic link between two opinion expressions if their semantic similarity is greater than a certain threshold λ. We rely on the Explicit Semantic Analysis (ESA) algorithm to compute the semantic similarity between opinion expressions. ESA first builds an inverted index for all Wikipedia concepts, then represents any text as a weighted vector of Wikipedia concepts by retrieving its terms from the inverted index and finally assesses the relatedness in concept space using conventional metrics [15].

Based on the graph G, we adopt the Girvan-Newman (GN) algorithm to cluster the opinion expressions. The GN algorithm is a typical method for detecting network communities. Without the prior information for determining the centers of communities, the GN algorithm constructs communities by progressively removing edges from the original graph [16].   Each community represents a feature category.

For describing the feedback of each feature category, we select 3-5 sentences in which the opinion expressions are nearest to the center of the category. Finally, we manually label the names of feature categories in accordance with the corresponding sentences and then merge the feature categories with the same name.

2.5   Generating Document of Evolutionary Requirements

The feedback for each feature category implies an overall, function or quality requirement of the software system. However, it is critical to assess the priority and importance of feedback for software evolution. Because user satisfaction is the best indicator of a company's future profits [17], it is the direct driving factor of software evolution. From an economic perspective, our decision to determine evolutionary requirements depends on the user satisfaction indexes evaluated by the feedback for software features. The satisfaction score of a software feature is defined as follows:

$$S - Score(f) = \frac{1}{(n_p + n_n)} \left(\sum_{i=1}^{n_p} P_i(f) + E_d * \sum_{i=1}^{n_n} N_i(f) \right) \qquad (3)$$

where n_p and n_n are the numbers of positive and negative opinion expressions about the software feature f; $P_i(f)$ and $N_i(f)$ are the positive and negative polarity value of an opinion expression about the software feature f; E_d is the price elasticity of demand, which is a measure used in economics to show the responsiveness, or elasticity, of the quantity demanded of a good or service to a change in its price[5]. Formulated as [18], E_d indicates the substitutability and importance of the software in customer purchases. Introducing E_d to user satisfaction evaluation emphasizes user acceptance of the technological level in the market environment. If E_d is greater, even a few negative opinions may result in so massive loss of users to reduce the software revenue.

According to Equation (3), if the user satisfaction score of a software feature is greater than 0, the corresponding requirement is reusable or added, and changed otherwise. We compute the user satisfaction score for high frequent features mentioned in the reviews and then manually generate the document of evolutionary requirements in the light of those review sentences for each software features. Those evolutionary requirements that drive more economic gain are prioritized systematically.

3   Experiment

To demonstrate the practicality of our approach in eliciting evolutionary requirements even with large amounts of online reviews, we first introduce the data sets and settings. Then, we evaluate the opinion mining techniques including the identification and classification of opinions. Finally, we analyze the usefulness of the evolutionary requirements document for developers.

3.1   Data Sets and Settings

As can be seen in Table 2, we used two data sets of online reviews: the packaged software of Kaspersky Internet Security 2011 3-Users (KIS 2011) from Amazon.com and the mobile application of TuneIn Radio Pro V3.6 (TuneIn 3.6) from the Apple App Store. For each testing data set, we manually labeled the potential software features, opinions and their polarities mentioned in the reviews, and then classified the review sentences according to the semantics of related opinions.

Table 2. Statistics of the data sets of online reviews

Data set	#Reviews	#Sentences	#Words	#Sentence per review	#Words per sentence
KIS 2011	380	3392	52682	8.9	15.5
TuneIn 3.6	461	1211	12711	2.6	10.5

The Stanford POS Tagger and parser are respectively used to tag and parse the data sets. The seed sentiment lexicon is provided by Liu's sentiment words[6]. The Wikipedia

---

[5] http://en.wikipedia.org/wiki/Price_elasticity_of_demand
[6] http://www.cs.uic.edu/~liub/FBS/sentiment-analysis.html

version on Sep. 1, 2011 is adopted for computing the ESA semantic similarity. α=0.5, β=0.3, γ=0.2, and λ=0.4 are experimentally set for our approach.

As the exact quantity demanded for determining E_d is not available, we modified the evaluation model of user satisfaction in Section 2.5 by replacing E_d with the price elasticity of sales rank E_r. The sales rank implies the demand for a software product/ application relative to other competitors. Since it is observed that the association between sales rank and demand approximately conforms to a Pareto distribution, the formula of E_r is similar with $-E_d$.

3.2   Evaluation of Opinion Identification

As shown in Table 3, our method for opinion identification, SRPA+, outperforms SRPA in all conditions. Significantly higher recall, especially for KIS 2011 indicates that the new propagation rules work well. The improvement in precision implies that merging opinion expressions avoids one-sided identification. Note that the results of TuneIn 3.6 produced by SRPA+ are basically lower than those of KIS 2011. This is because the decreased performance of natural language processing techniques for dealing with more phrases and incomplete sentences in the reviews of TuneIn 3.6.

Table 3. The comparison results of opinion identification

Data set	Recall		Precision		F-score	
	SRPA	SRPA+	SRPA	SRPA+	SRPA	SRPA+
KIS 2011	0.67	0.83	0.74	0.78	0.70	0.80
TuneIn 3.6	0.69	0.81	0.71	0.73	0.70	0.77

Table 4 illustrates the results of polarity assignment using SRPA+. Clearly, the good recall reveals that the propagation performs well in discovering new sentiment words. We can observe that the precision of opinion expressions is significantly higher than that of new sentiment words. There are two main reasons for relatively worse performance of new sentiment words in precision. First, the review data sets often have errors of spelling and grammatical structure or non-standard sentences so that automatic tagging and parsing don't always work correctly. Second, the propagation rules have only the constraints of POS tags so that more ordinary words are introduced with the increase of the review data sets. In spite of this, our methods of merging and pruning opinion expressions reduce the effect of noisy sentiment words.

Table 4. The results of polarity assignment

Data set	Recall			Precision			F-score		
	New words	Words	Expressions	New words	Words	Expressions	New words	Words	Expressions
KIS 2011	0.71	0.73	0.76	0.62	0.67	0.72	0.66	0.70	0.73
TuneIn 3.6	0.68	0.71	0.72	0.63	0.65	0.70	0.65	0.68	0.71

3.3   Evaluation of Opinion Classification

The proposed method for opinion classification is called S-GN. The baseline methods include J-Kmeans and S-Kmeans. Both are k-means clustering algorithms based on the Jaccard coefficient and the ESA similarity respectively. S-GN produces k clusters automatically. The inputs of J-Kmeans and S-Kmeans are set as the same number of clusters produced by S-GN.

Table 5 shows the results of opinion classification in recall, precision and F-score. S-GN and S-Kmeans significantly outperform J-Kmeans in all conditions. S-GN has better results than S-Kmeans especially in precision. Such results imply that the ESA similarity is better than the Jaccard coefficient. The Jaccard coefficient is essentially a keyword-based similarity measurement. As ESA enhances the semantic representations of texts using expanded Wikipedia concepts, it alleviates the clusters of duplicated categories. In addition, the k-means algorithm requires a priori number of clusters. It is difficult to optimize k seeds for avoiding poor clusters. The GN algorithm produces the optimized number of clusters considering the global network topology so that it reduces the clusters containing mixed categories.

Table 5. The comparison results of opinion classification

Data set	Recall			Precision			F-score		
	J-Kmeans	S-Kmeans	S-GN	J-Kmeans	S-Kmeans	S-GN	J-Kmeans	S-Kmeans	S-GN
KIS 2011	0.58	0.68	0.71	0.63	0.72	0.76	0.60	0.70	0.73
TuneIn 3.6	0.55	0.66	0.67	0.60	0.71	0.74	0.57	0.68	0.70

3.4   Evaluation of Generated Evolutionary Requirements Document

We organized a human subjective study to evaluate the usefulness of the generated evolutionary requirements document. 50 participants that have over three years experience in software development were required to report the evolutionary requirements for TuneIn 3.6. We compared the generated document with participants' reports to validate that our approach can discover requirements that were ignored by developers.

In the first stage, the participants make decisions about requirements evolution based on their experience with TuneIn 3.6. Table 6 indicates the common results designed by developers, with which more than 30% of the participants agreed. Table 7 shows the results generated by our approach. For functional requirements, developers paid more attention to key, special and value-added features while users were more concerned with the features relative to user habits. As TuneIn 3.6 is one of the best sellers in the *music* category, its main functional features are implicitly desirable by the mass market. Such features should be improved or changed only when there are significant issues and bugs, for example "*pause*". However, developers have difficulty predicting user preferences for these features in the real world, such as "*favorites*", "*schedule*" and "*streaming*". In terms of quality requirements, developers assessed the objective features for Internet Radio applications as "*connection*", "*reliability*", and "*efficiency*" whereas users revealed subjective features like "*operability*". Consequently, user feedback can assist in determining evolutionary requirements that developers sometimes overlook.

Table 6. Evolutionary requirements of TuneIn 3.6 designed by developers

ID	Type	Requirements	Feature	#Participants
1	functionality	Fix issues that are causing playback or stream errors	record	15
2		Fix issue that is causing skipping to the current radio after pausing	pause	21
3		Make connection smoother and more stable	connection	32
4	quality	Fix issues that are causing crashing, freezing, and restarting	reliability	48
5		Improve the speed for connecting stations and favorites in personal account	efficiency	26

Table 7. Evolutionary requirements of TuneIn 3.6 generated by our approach

ID	Type	Requirements	Feature	Frequency	E-score
1		Improve reliability during pausing and the quality after pausing	pause	17	-0.263
2	functionality	Keep the schedule of stations	schedule	16	-0.789
3		Keep the local favorites	favorites	19	-0.413
4		Provide high quality streaming and allow users to choose streaming quality	streaming	39	-0.193
5		Work in the background	operability	16	-0.678
6	quality	Make connection smoother and more stable	connection	18	-0.392
7		Reduce the crash, freezing, reboot and skipping during listening to radios	reliability	35	-0.432

In the second stage, we provided the structured feedback derived from mining the reviews of TuneIn 3.6 for the participants to revise their presented evolutionary requirements. The most common strategy used by 76% participants is to decide the priority of feedback on each software feature by the frequency of its occurrence and its user satisfaction. The features revised by the participants contains "*user interface*" in addition to those in Table 7. The ordinary satisfaction score and economic satisfaction score of the feature are -0.0714 and 0.0489. Intuitively, it is difficult to determine evolving "*user interface*" because its ordinary satisfaction is only slightly negative. However, the economic satisfaction indicates the acceptance of a feature compared with the overall technical level in the market. Although the negative ordinary satisfaction score implies that "*user interface*" implemented in TuneIn 3.6 is lower than common user expectation with it, the positive economic satisfaction score suggests that the same is true for other competitors in the market. Even if users are not satisfied with TuneIn 3.6, they have no other better alternatives. In other words, improving "*user interface*" cannot significantly have positive impact on the revenue of TuneIn 3.6. Thereby, our approach argues even though a feature of the software product or application has poor ordinary satisfaction, it does not have to be changed until its economic satisfaction is negative. Determining evolutionary requirements based on economic satisfaction manifests that software evolution does not blindly pursue user interests, but rather balances the interests of users and developers.

In addition, "*user interface*" is a feature relative to specific contexts of use. As TuneIn 3.6 is an application for the public, the requirements for "*user interface*" are diverse in terms of specific user habits. To reduce the potential risk, this type of features should only be improved when it has significantly low user satisfaction.

4   Related Work

In other research, there are techniques developed for eliciting requirements from online user feedback. Gebauer et al. use content analysis of user reviews to identify functional and non-functional requirements of mobile devices through finding the factors that are significantly related to overall user evaluation [6]. Lee et al. elicit customer requirements using their opinions gathered from social network services [7]. Such works capture changing requirements without limited range of users and insufficient expressions. However, these approaches mostly rely on manual content analysis.

Cleland-Huang et al. adopt a classification algorithm to detect non-functional requirements (NFRs) from freeform documents including stakeholder comments [19]. One problem is that limited documents hinder identifying changing NFRs in a timely manner. Hao et al. utilize machine learning techniques to extract the aspects of service quality from Web reviews for conducting automatic service quality evaluation [8]. Carreño et al. adapt information retrieval techniques including topic modeling for exploring the rich user comments of mobile applications to extract new/changed requirements for future releases [9]. Li et al. compare the changes in user satisfaction before and after software evolution to provide instructive information for designing future systems [20]. Although these approaches initially access the validity of using automated techniques to discover software requirements, they lack the deep analysis about how user feedback influences changes in requirements.

Our research is inspired by opinion mining techniques that make it possible to automatically elicit requirements from huge volumes of user feedback data. The mainstream approaches are divided into two categories. One is to identify opinions based on word co-occurrence and grammatical structures [21-23]. Such approaches have good performance for extracting fine-grained features as well as opinions. However, the integrity of extraction rules/templates and domain knowledge have an obvious impact on their accuracy. The other one is to identify and group opinion pairs using topic modeling [24-26]. While it is not hard for topic models to find those very general and frequent features from a large document collection, it is not easy to find those locally frequent but globally not so frequent features [2].

5 Conclusions

This paper presented a novel approach for eliciting evolutionary requirements through analysis of online review data by integrating various techniques of SRPA+, S-GN and user satisfaction analysis including economic factors. To conduct our research, we first accessed a broad spectrum of review data with complex and diverse opinion expressions and then evaluated the performance of automated techniques for consolidating, analyzing and structuring feedback information. Furthermore, the proposed method of user satisfaction analysis assisted developers with finding a set of evolutionary requirements associated with the software revenue. We reported a human subjective study with fifty developers, evaluating the usefulness of the evolutionary requirements document generated by our approach. As a result, the generated document could help developers understand why and what to evolve for future software releases. In particular, they were led to focus on the improvements in specific functions and quality in use that they had previously ignored.

Future work will refine our opinion mining method to improve the performance of automated requirements elicitation in the big data era. In addition, we will further evaluate our approach using a broader data set from different domains.

Acknowledgments. The work in this paper was partially fund by National Natural Science Foundation of China under Grant No. 61170087 and State Key Laboratory of Software Development Environment of China under Grant No. SKLSDE-2012ZX-13.

References

1. Nuseibeh, B., Easterbrook, S.M.: Requirements Engineering: A Roadmap. In: 2000 Conf. on The Future of Software Engeering, pp. 35–46. ACM, New York (2000)

2. Liu, B.: Sentiment Analysis and Opinion Mining. Morgan & Claypool Publishers (2012)
3. Godfrey, M.W., German, D.M.: The Past, Present, and Future of Software Evolution. In: 2008 Frontiers of Software Maintenance, FoSM 2008, pp. 129–138 (2008)
4. Pagano, D., Brügge, B.: User Involvement in Software Evolution Practice: A Case Study. In: 2013 Int'l Conf. on Software Engineering, pp. 953–962. IEEE Press, New York (2013)
5. Vasa, R., Hoon, L., Mouzakis, K., Noguchi, A.: A Preliminary Analysis of Mobile App User Reviews. In: 24th Conf. on Australian Computer-Human Interaction, pp. 241–244. ACM, New York (2012)
6. Gebauer, J., Tang, Y., Baimai, C.: User Requirements of Mobile Technology: Results from A Content Analysis of User Reviews. Inf. Syst. E-Bus. Manage 6, 361–384 (2008)
7. Lee, Y., Kim, N., Kim, D., Lee, D., In, H.P.: Customer Requirements Elicitation based on Social Network Service. KSII Trans. on Internet and Information Systems 5(10), 1733–1750 (2011)
8. Hao, J., Li, S., Chen, Z.: Extracting Service Aspects from Web Reviews. In: Wang, F.L., Gong, Z., Luo, X., Lei, J. (eds.) WISM 2010. LNCS, vol. 6318, pp. 320–327. Springer, Heidelberg (2010)
9. Galvis, L.V., Winbladh, K.: Analysis of User Comments: An Approach for Software Requirements Evolution. In: 35th International Conference on Software Engeering, pp. 582–591. IEEE Press, New York (2013)
10. Kang, K.C., Cohen, S.G., Hess, J.A., Novak, W.E., Peterson, A.S.: Feature-Oriented Domain Analysis (FODA) Feasibility Study. Carnegie-Mellon University Software Engeering Institute (1990)
11. Tesniere, L.: Élements de Syntaxe Structurale: Préf. de Jean Fourquet. C. Klincksieck (1959)
12. Qiu, G., Liu, B., Bu, J., Chen, C.: Sentiment Word Expansion and Target Extraction through Double Propagation. Comput. Linguist. 37, 9–27 (2011)
13. Hu, M., Liu, B.: Mining and Summarizing Customer Reviews. In: Tenth ACM SIGKDD Int'l Conf. on Knowledge Discovery and Data Mining, pp. 168–177. ACM, New York (2004)
14. Yager, R.R.: On Ordered Weighted Averaging Aggregation Operators in Multicriteria Decision Making. IEEE Trans. on Systems, Man and Cybernetics 18(1), 183–190 (1988)
15. Gabrilovich, E., Markovitch, S.: Computing Semantic Relatedness Using Wikipedia-based Explicit Semantic Analysis. In: 20th Int'l Joint Conf. on Artificial intelligence, pp. 1606–1611. Morgan Kaufmann Publishers Inc. (2007)
16. Girvan, M., Newman, M.E.: Community Structure in Social and Biological Networks. The National Academy of Sciences 99(12), 7821–7826 (2002)
17. Kotler, P.: Marketing Management: Analysis, Planning, Implementation, and Control. Prentice Hall College Div. (1999)
18. Michael, P., Melanie, P., Kent, M.: Economics. Addison-Wesley, Harlow (2002)
19. Cleland-Huang, J., Settimi, R., Xuchang, Z., Solc, P.: The Detection and Classification of Non-functional Requirements with Application to Early Aspects. In: 14th IEEE Int'l Requirements Engeering Conf., pp. 39–48. IEEE CS (2006)
20. Li, H., Zhang, L., Zhang, L., Shen, J.: A User Satisfaction Analysis Approach for Software Evolution. In: Int'l. Conf. on Progress in Informatics and Computing, pp. 1093–1097. IEEE Press, New York (2010)
21. Popescu, A., Etzioni, O.: Extracting Product Features and Opinions from Reviews. In: Conf. on Human Language Tech. and Empirical Methods in Natural Language Processing, pp. 339–346. ACL (2005)

22. Wu, Y., Zhang, Q., Huang, X., Wu, L.: Phrase Dependency Parsing for Opinion Mining. In: Conf. on Empirical Methods in Natural Language Processing, pp. 1533–1541. ACL (2009)
23. Ding, X., Liu, B., Yu, P.S.: A Holistic Lexicon-based Approach to Opinion Mining. In: Int'l Conf. on Web Search and Web Data Mining, pp. 231–240. ACM (2008)
24. Zhao, W., Jiang, J., Yan, H., Li, X.: Jointly Modeling Aspects and Opinions with A Max-Ent-LDA Hybrid. In: Conf. on Empirical Methods in Natural Language Processing, pp. 56–65. ACL (2010)
25. Jo, Y., Oh, A.H.: Aspect and Sentiment Unification Model for Online Review Analysis. In: Fourth ACM Int'l Conf. on Web Search and Data Mining, pp. 815–824. ACM (2011)
26. Mei, Q., Ling, X., Wondra, M., Su, H., Zhai, C.: Topic Sentiment Mixture: Modeling Facets and Opinions in Weblogs. In: 16th Int'l Conf. on World Wide Web, pp. 171–180. ACM, New York (2007)

Automatic Fake Followers Detection
in Chinese Micro-blogging System

Yi Shen[1,2,*], Jianjun Yu[2], Kejun Dong[2], and Kai Nan[2]

[1] University of Chinese Academy of Sciences
[2] Computer Network Information Center, Chinese Academy of Sciences
{shenyi,yujj,kevin,nankai}@cnic.ac.cn

Abstract. Micro-blogging, which has greatly influenced people's life, is experiencing fantastic success in the worldwide. However, during its rapid development, it has encountered the problem of content pollution. Various pollution in the micro-blogging platforms has hurt the credibility of micro-blogging and caused significantly negative effect. In this paper, we mainly focus on detecting fake followers which may lead to a problematic situation on social media networks. By extracting major features of fake followers in Sina Weibo, we propose a binary classifier to distinguish fake followers from the legitimate users. The experiments show that all the proposed features are important and our method greatly outperforms to detect fake followers. We also present an elaborate analysis on the phenomenon of fake followers, infer the supported algorithms and principles behind them, and finally provide several suggestions for micro-blogging systems and ordinary users to deal with the fake followers.

Keywords: Micro-blogging, Fake followers, Classification, Feature extraction.

1 Introduction

Due to its simplicity and rapid velocity, micro-blogging is experiencing tremendous success. However, the micro-blogging services have also encountered several serious troubles during their booming development, one of which is the fake followers problem. The phenomenon of fake followers emerges soon after the birth of micro-blogging systems and now has flooded in the mainstream micro-blogging services such as Twitter[twitter.com] and Sina Weibo[weibo.com]. According to Yahoo reports[1], a considerable part of the followers of celebrities on Twitter are fake, and the proportion may be as high as over 50%. Despite both Twitter and Sina Weibo have made much effort on struggling with the fake accounts, nevertheless, the effect is not very significant.

Generally speaking, people purchase fake followers mainly for two motivates according to the investigation from Theweek[2]. The first one is that people purchase followers just to achieve fame and feed their vanity. They misunderstand that, the more

---

* This research was supported by NSFC Grant No.61202408 and CAS 125 Informatization Project XXH12503.

[1] http://news.yahoo.com/blogs/prot-minded/10-people-won-t-believe-fake-followers-twitter-215539518.html

[2] http://theweek.com/article/index/243357/how-celebrities-buy-twitter-followings-mdash-and-how-you-can-too

V.S. Tseng et al. (Eds.): PAKDD 2014, Part II, LNAI 8444, pp. 596–607, 2014.
© Springer International Publishing Switzerland 2014

the followers, the greater their influence. The second, some merchants or brands seek a huge number of followers so as to push Ads. The follower merchants take advantage of algorithms and softwares to produce a mass of fake accounts automatically. These accounts act as real followers to meet the needs of their customers. There are already some websites for trading fake twitter followers in public, such as [intertwitter.com] and [fakefollowerstwitter.com]. The prices for 1K fake followers provided by these online merchants are around 5-20$. A recent report from NYTimes pointed out that the fake Twitter followers have become a multimillion-dollar business.

The fake followers are so epidemic in the micro-blogging systems, which have caused plenty of hazards, such as making noise on personalized recommendation and user influence analysis, risking privacy to unknown people, receiving valueless Ads. Therefore, it is very significant to present an effective method to distinguish the fake followers from the legitimate users in micro-blogging systems. Researchers have developed effective tools[3] based on the inactive characteristic and spam-related features to detect fake followers of Twitter. However, the fake followers are distinguished to traditional spammers. The spammers mainly utilize the freedom and rapid nature of micro-blogging platforms to push unsolicited advertisements or malicious information[1], while fake followers aim to follow the users who are urgent to be popular users. Therefore, in many cases, fake followers do not send spam to others, and pretend to be legitimate users. At the same time, fake followers in Sina Weibo seem to be more sophisticated than Twitter's. For they not only appear as active as the legitimate users, but also almost have no spam in their posts. We have checked the tweets of the followers purchased from the four markets (1000 followers each) manually. An account will be labeled as a spammer once there is any spam appear in the most recent 20 tweets of its timeline. Finally, we found about 62% of these fake Twitter followers are belong to spammers. We define the accounts which have less than 10 posts or 10 followers as inactive accounts. In general, 57% of the fake followers for Twitter are inactive. However, according our statistic on fake followers for Sina Weibo, the spammer-ratio and inactive-ratio are just 8% and 6% respectively. As a result, we aim to address the issue of detecting the fake followers for Sina Weibo in this paper. Considering that there are no related public datasets for validating the effects of the proposed method, we just purchased a considerable number of fake followers from different merchants as our datasets.

We frame our contributions as follows:

(1) We examine a number of properties of fake followers in Sina Weibo, and present an effective strategy for automatic detection of fake followers.

(2) We give an in-depth analysis of the profile-evolution of fake followers, and try to understand the principles and algorithms for fake followers' generation.

(3) We provide several suggestions for both the micro-blogging systems and legitimate users on how to deal with fake followers.

The rest of this paper is organized as follows. After a brief review of the related work in Section 2, we introduce datasets used in this paper in Section 3. Next, we analyze the features of fake followers and propose a voting classier for detecting the fake followers

---

[3] http://www.socialbakers.com/twitter/fakefollowercheck/
   methodology/

automatically in Section 4. Then, we present our experiments in Section 5. Section 6 gives the analysis on the results and the discussion. Finally, we conclude the paper in Section 7 with future work.

2   Related Work

Due to its dual role of social network and news media[2], micro-blogging has become an important platform for people to access information. However, with its rapid development, micro-blogging is plagued by various credibility problems for a long time. The information on micro-blogging systems has been polluted heavily by the rumors [3][4] and the spams [5][6]. As fake followers usually accompany with illusive following-actions, they will also pollute the social connections in the micro-blogging systems.

There are two main methods for the follower merchants to provide fake followers to their customers. One way is to utilize some third-party applications to defraud the authorizations from some legitimate users, then manipulate these compromised users to follow the customers so as to promote their follower-count[7] [8]. Another method is to utilize specific softwares to batch produce a mass of fake accounts that disguised as real users.

The compromised users have a fatal weakness, that is their loyalty to the customers is very low[9]. Since the compromised users are real users, once they discover strangers appeared in their followee-list or receive valueless posts from the customers, they may initiatively remove the customers from their followee-list. In contrast, the fake accounts act as followers are very stable. Moreover, the fake followers have been very tricky, especially in Sina Weibo. As a result, we mainly focus on how to automatically detect the fake followers created by the follower merchants in this paper.

As the spammers in social networks are also created by certain bots [10]. Therefore, the detection of spammers in social networks is related to our work. Thomas et al. [11] analyzed the profiles of suspended spammers list provided by Twitter, then got some significant key points about the techniques of the spammers. [12] presented graph-based methods to analyze the network structure of the spammers. In [13] [14] [15], the researchers proposed feature-based systems to address the spammer-detection problem. All of these works have provided meaningful investigation on the characteristics of the forged accounts in social networks.

Our work is quite different from the works mentioned above, and presented as the first effort on detecting fake followers as far as we know. First, according to our real dataset, only a small part of fake followers in Sina Weibo belong to spammers, and thus the traditional methods for detecting the spammers in social networks are proved not appropriate to identify fake followers. Secondly, previous approaches did not take account of the evolution of fake accounts. We extract several evolutionary features for our detection task, which are proved significantly important to promote detection performance.

3   Data Collection

We notice that fake followers have two kinds of following-behaviors: following the customers who have purchased their service, and randomly following other users so as to disguise themselves. To investigate these two behaviors respectively, we collect two different datasets. We buy 20,000 fake followers in Sina Weibo from 4 different follower merchants, and use these followers on 4 new-created accounts[4].

We keep monitoring these fake accounts for about 4 months(02/03/2013-30/06/2013), and only 17 of them are suspended by Sina Weibo during this period. It means these fake followers are very deceptive and seem to be good at disguising themselves. The remaining fake followers are made as our dataset(DATASET1) for the proposed method, which contains 19,146 records after removing the duplicated ones.

20 new Sina Weibo accounts are registered and used as baits to attract randomly following from fake followers. We keep these baits no actions (no posts and no following behaviors) for a long period(10/03/2013-15/06/2013), and recognize their followers as fake follows since legitimate users only follow the users they are really interested in. Totally, 724 accounts are captured by our baits and used as DATASET2.

In order to compare with fake followers, a dataset of legitimate users is also necessary. 114 volunteers are invited to identify the accounts of their acquaintances from their respective followee-list on Sina Weibo. In this way, 14,873 different accounts of real users are obtained. We also crawl the accounts of 6,472 celebrities whose identity have been officially verified by Sina Weibo("Big V"). After merging these two parts of datasets, a legitimate dataset(DATASET3) contains 20,211 users. Finally, we utilize Sina Weibo API[5] to access the profiles of all users in DATASET1, DATASET2 and DATASET3 for further research.

4 Proposed Method for Detecting Fake Followers

In this section, we mainly present our approach of detecting fake followers in Sina Weibo. The fake followers detection issue can be considered as a binary classification task. Through comprehensive analysis of the datasets, we extract numerous features with great discrimination between legitimate followers and fake followers to build the classifier. These features can be divided into three types: the post-related features, the user relationship features and the evolutionary features. We randomly sample 2,000 items from DATASET1 and DATASET3 respectively, depict their differences on various features, and provide a voting-SVM classifier.

4.1 The User Relationship Features

The Ratio of Followee Count and Follower Count (RFF). RFF of fake followers is surprisingly high due to their large number of followees and very few followers. According to our statistics, for a typical legitimate user, this ratio is usually within a range of [0.5,3]. Since some celebrities have a huge number of followers, their RFF are often close to zero. RFF is shrunk with a logarithmic function as follows:

---

[4] All these four merchants do not have public website. We get in touch with them via searching in Sina Weibo with the keyword "add followers". The prices of them are all around ¥ 60-80 per 1K followers.

[5] http://open.weibo.com

$$RFF(U) = \lg \frac{FolloweeCount + 1}{FollowerCount + 1} \qquad (1)$$

Figure 1(a) shows the cumulative distribution function(CDF) of RFF for fake followers and legitimate users. It is very clear that RFF is able to discriminate the two types of users distinctly.

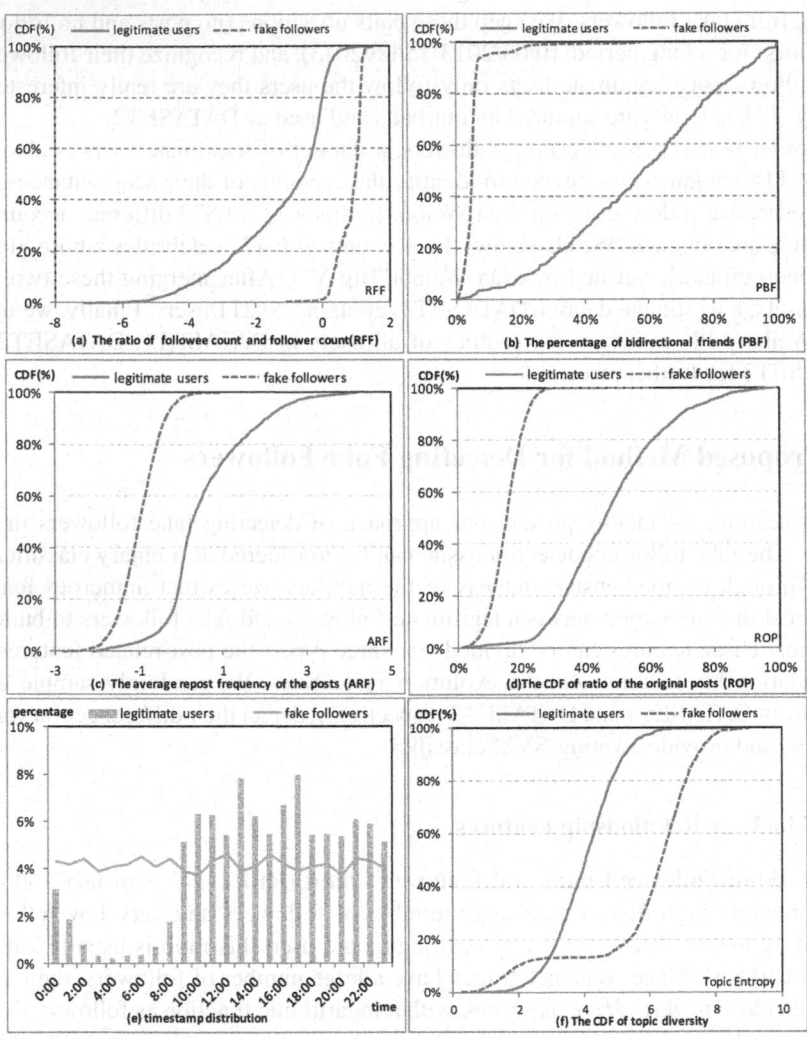

Fig. 1. The comparison between legitimate users and fake followers

The Percentage of Bidirectional Friends (PBF). PBF is calculated with Equation (2). In real scenarios, people have their own social communities, and often the legitimate users belong to the same community always follow each other in micro-blogging systems. Certainly, fake followers have no friends, no classmates and no colleagues, therefore, their PBF should be lower than legitimate users'. The CDF curves of PBF in Figure 1(b) have proven this assumption.

$$PBF\,(U) = \frac{CountOf(Followee \cap Follower) + 1}{FolloweeCount + 1} \tag{2}$$

4.2 The Post-related Features

Average Repost Frequency of the Posts (ARF). ARF can reflect the influence of the user[16]. The ARF of a fake follower is usually very low for two reasons. One reason is that a fake follower has hardly any high quality followers. The other is that its post behavior is manipulated by softwares, as a result, their posts are meaningless for others in most cases. According to our datasets, the average ARF of normal users is much higher than fake accounts. The CDF in Figure 1(c) shows that the ARF of most normal users is over 0, which means the posts from normal users will be reposted at least once in average. In contrast, it is obvious that over 80% of the fake followers have a low ARF between [-2,-1].

$$ARF\,(U) = \lg \frac{\sum_{P \in Posts(U)} RepostCount(P)}{TotalPostsCount(U) + 1} \tag{3}$$

Ratio of the Original Posts (ROP). We find the vast majority of posts from fake followers are belong to repost. In Figure 1(d), we find that ROP for almost all fake followers are less than 20%, while ROP for legitimate users are in the range of(20%-60%).

$$ROP\,(U) = \frac{OriginalPostsCount + 1}{TotalPostsCount + 1} \tag{4}$$

Proportion of Nighttime Posts (PNP). We make a comparison of the post-timestamp distributions of users in DATASET1 and DATASET3. We notice that legitimate users rarely publish posts during the nighttime. However, the timestamp distribution of fake followers obeys an approximative uniform distribution, which is contrary to the common sense since we all know that people need to sleep and with a very low probability to publish posts in the middle of the night. Consequently, we suspect that the post-creation behavior of fake followers is controlled by a periodic creation algorithm. As the distributions in Figure 1(e) have shown a remarkable discrimination between fake followers and legitimate users, we are confident to employ the proportion of nighttime (1:00am-7:00am) posts as an indicator for classification.

$$PNP\,(U) = \frac{NightPostsCount + 1}{TotalPostsCount + 1} \tag{5}$$

Topic Diversity. We apply author topic model[17] to mine the themes of the accounts in DATASET1 and DATASET3. Normally, a legitimate user usually has countable interests of topics, while the fake followers usually present variable interests since they repost from others randomly. As a result, we assume that the topics of fake followers should be more diverse than legitimate users. We use topic entropy [18] to measure the topic diversity as Equation 6. Correspondingly, fake followers present higher topic entropy than legitimate users'.

$$H(u) = -\sum_{i=1}^{K} P(z_i|u) \log_2 P(z_i|u) \tag{6}$$

Where z_i denotes the topics generated by author topic model. K is the count of topics.

We notice that there is a step appears in the beginning of the CDF curve of fake followers as shown in Figure 1(f). This is because a small part of fake followers publish spams frequently, which causes their lower topic entropy.

4.3   The Evolutionary Features

As we observed, the fake followers always evolve when time changes. To better capture the pattern of their evolution, we keep tracking the accounts in DATASET1 for a period of 60 days (07/04/2013-06/06/2013), and several important parameters(i.e. the count of posts, followees, followers) have been recorded every day. We then display the evolution model of typical fake accounts in Figure 2 comparing with legitimate users[6].

We can find that the evolution of post-count of fake followers is consistent with gradient trend every day, which means they share the same post-frequency. This also implies that they are manipulated by the same algorithm. The evolutionary curves of post-count from legitimate users in Figure 2(b) are flatter than fake followers'. This is because most of legitimate users publish fewer posts than fake followers and are with low probability of mutation.

Because people rarely remove excessive amount of users from their followee-lists during a short time[19], their followee count will tend to increase in general, whereas the followee-count of five fake followers fluctuate violently for many times as shown in Figure 2(c). According to this surprising investigation, we are more confident to confirm that the fake followers are manipulated by software. In Sina Weibo, an account can only follows 2,000 users at most, the software has to make the fake followers to remove some followees who do not purchase their service in order to release the room for their new customers.

The Figure 2(e) and Figure 2(f) demonstrate that follower-count of both fake followers and legitimate users maintain a relatively stable trend. However, the fake followers seem to decrease more frequently.

Base on the analysis above, we extract 6 features to model the evolutionary characteristics:

---

[6] Due to the space limitation, we just plot 15-day evolution(23/05/2013-06/06/2013) in Figure 2.

(1) the standard deviation of post-count(σ_{post}).
(2) the general slope of post-count(g_{post}).
(3) the standard deviation of followee-count($\sigma_{followee}$).
(4) the decrease frequency of followee-count($DF_{followee}$).
(5) the standard deviation of follower-count ($\sigma_{follower}$).
(6) the decrease frequency of followee-count($DF_{follower}$).

The following Equations give the calculation of these features.

$$\sigma = \sqrt{\frac{1}{N}\sum_{i=1}^{N}(X_i - \mu)^2} \tag{7}$$

$$DF = \frac{\sum_{i=2}^{N} sigmoid(X_i - X_{i-1})}{N} \tag{8}$$

$$g_{post} = \frac{(X_N - X_1)}{N} \tag{9}$$

Among these features, the three standard deviations are used to model the degree of fluctuation. The decrease frequency can capture the exceptions of followee-count and follower-count. In addition, g_{post} is able to reflect the rate of increase of posts during a period of time. We believe that these evolutionary features are also significant for the classification task. For each account, we calculate these features above based on the profiles in the latest 30 days, and integrate them into our classifier.

4.4   Classifier for Detecting Fake Followers

Due to its prominent fame on solving multidimensional classification problem, we utilize Support Vector Machine(SVM) as our basic classifier. We exploit LibSVM[20] to implement a SVM classifier with RBF (Radial Basis Function) kernel function. Since we do not know the actual proportion of fake followers in Sina Weibo, the classifier directly trained from the original dataset may be biased. To this end, we adopt "bagging"[21] strategy to overcome this problem. DATASET1 and DATASET3 are merged as the final dataset(DATASET4), therefore, it contains 13,873 fake followers and 20,211 legitimate users. 3,400 accounts in DATASET4 are stochastically selected as test dataset, and the left is used as training dataset. We randomly sample 2,000 accounts from DATASET4 for 15 times, then use 15 subsets to train a SVM classifier respectively. Finally, the category for an account in test dataset will be determined by the voting results of these 15 SVM classifiers. A 10-fold cross validation on DATASET4 shows that our voting classifier works well. The average accuracy is about 98.7%, and the average false positive rate is very close to 0%.

5   Experiments

In this section, we first make a comparison between our voting-SVM classifier to several baselines. Next, we utilize our proposed classifier to evaluate the accounts in DATASET2

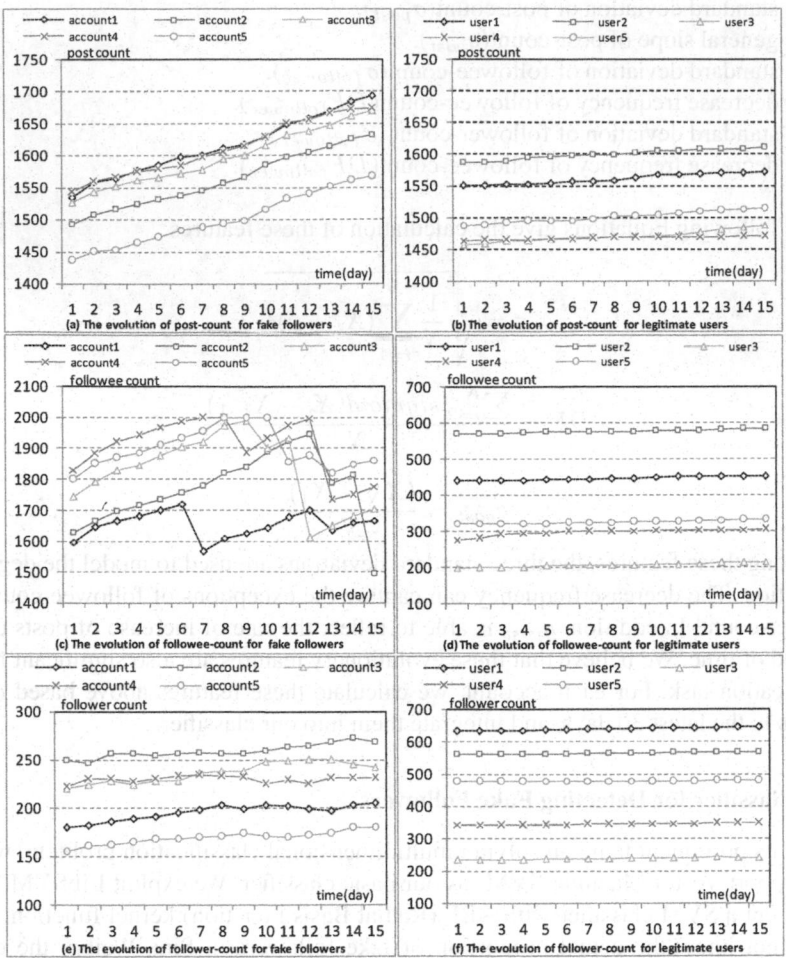

Fig. 2. The evolution of typical fake followers and legitimate users

which are attracted by our baits. Finally, we adopt our voting-SVM classifier to detect fake followers in the wild.

To examine the importance of evolutionary features, we implement a version of voting-SVM classifier with no evolutionary features. A normal SVM classifier is employed as a baseline to validate the improvement of the "bagging" strategy. We also implement the spammer classifier introduced in [13] as another baseline. Our voting-SVM outperforms others with all the metrics as shown in Table 1. The evolutionary features and the "bagging" strategy both play an important role in the final results. The accuracy of spammer classifier is much lower than others'. It is because most fake followers in Sina Weibo do not have obvious spam-related characteristics.

As we observed, the suspicious accounts attracted by our baits in DATASET2 are very similar to fake followers in DATASET1 in many aspects(i.e. the features

Table 1. The percentage of fake followers for different groups

Classifier	Accuracy	False Positive Rate	F1
Spammer Classifier	63.4%	0.7%	0.712
SVM(no evolution)	91.3%	3.8%	0.834
SVM	95.1%	1.2%	0.933
Voting-SVM(no evolution)	93.9%	3.5%	0.917
Voting-SVM	**98.7%**	**0.4%**	**0.964**

discussed above), we guess that they have a great probability to be fake followers. To prove our conjecture, we apply our trained voting-SVM classifier to classify the accounts in DATASET2. Up to 95.6% account in DATASET2 are judged as fake followers, which indicates that the accounts in DATASET1 and DATASET2 are extremely homogeneous. Therefore, we confirm that most followers of our baits are indeed caused by randomly following from fake followers.

In order to understand the global state in the whole Sina Weibo system, we pick up 100 legitimate users. Among the sampled accounts, 50 of them are registered by ordinary people with different occupations. Another half are those celebrities in different fields. We exploit our classifier to analyze the quality of their followers, and the average percentage of fake followers is about 37.2%. As listed in Table 2, the celebrity accounts contain more fake followers than ordinary users. We guess that most celebrities have the suspects on purchasing fake followers.

Table 2. The percentage of fake followers for different groups

Ordinary Users	Celebrity Users	Fake Followers	Baits	Average
14.4%	42.3%	94.1%	95.6%	37.2%

6 Discussion

6.1 The Principle Behind Fake Followers

Based on all analysis mentioned above, we give an investigation on the behaviors and characteristics of fake followers, which helps to find several key points of principal control algorithms behind fake followers in Sina Weibo:

(1) The fake follower merchants exploit certain register tools to produce massive accounts in Sina Weibo.

(2) Then they make fresh fake accounts to follow each other, as a result, both the followee-count and the follower-count will increase rapidly within a short time.

(3) Next, the fake accounts will send random following-action to many legitimate users. On the one hand, this operation may attract some back-following from legitimate users, which can improve their level of camouflage; for the other hand, the fake accounts would get the source of the reposts.

(4) Since it is very complicated to automatically generate diverse posts with high quality, so reposting the posts from legitimate users is a conservative method, especially

for those Chinese posts. The control algorithm manages fake accounts to repost many posts every day, which not only ensures the quality of post-content, but also makes fake followers appear to be much more realistic.

(5) When someone has purchased the fake-follow service, the merchants will manipulate fake accounts to follow this customer, so as to make he/she appears to have many followers.

(6) Once the followee-count of a fake follower is close to the limit(2,000 in Sina Weibo), the control software will remove some legitimate users or other fake accounts from the followee-list of fake followers.

6.2   How to Struggle with Fake Followers

It is a comprehensively difficult work for micro-blogging systems to identify fake followers. The micro-blogging services should make crucial measure to deal with fake followers in a fair and just manner. We believe that our work provides quite good suggestions for micro-blogging systems. As normal users, we need to be cautious to follow others in micro-blogging systems, or we may receive many valueless reposts from fake followers. We discover that fake followers prefer to choose to follow some novice users. Due to their less experience, many novice users tend to follow back when they are followed by fake followers. Also it is easy for a user to utilize the features above to judge whether a follower is fake manually through PNP, RFF and OPR, since these features of an account are public.

In fact, the count of followers alone means very little about the influence of a user[16]. As a result, it's unnecessary for us to purchase followers. We should share this fact with the customers of fake followers, so as to fundamentally undermine the follower market.

7   Conclusion

The fake followers have severely hurt the credibility of micro-blogging systems. In this paper, we mainly focus on automatic detection of fake followers in Sina Weibo. We extract many discriminative features especially several evolutionary features to build a classifier detecting fake followers. The proposed classifier performs satisfactorily on the standard metrics for classification. We also summarize several principles behind fake followers. Finally, we give suggestions for combating with fake followers.

Still there is much work to do in the future. We believe that the camouflage algorithm of fake followers has huge space for improvement. The manipulator can adjust the characteristics of fake followers to evade our detection algorithm. The race between the detection algorithms and the camouflage strategies will exist for a long time. As a result, it is necessary for us to keep tracking the evolution of fake followers and constantly sum up new features to deal with them.

References

1. Stringhini, G., Kruegel, C., Vigna, G.: Detecting spammers on social networks. In: ACSAC 2010 Conference Proceedings, pp. 1–9 (2010)

2. Kwak, H., Lee, C., Park, H., Moon, S.: What is twitter, a social network or a news media? In: WWW 2010 Conference Proceedings, pp. 591–600 (2010)
3. Qazvinian, V., Rosengren, E., Radev, D.R., Mei, Q.: Rumor has it: identifying misinformation in microblogs. In: EMNLP 2011 Conference Proceedings, pp. 1589–1599 (2011)
4. Mendoza, M., Poblete, B., Castillo, C.: Twitter under crisis: Can we trust what we rt? In: Proceedings of the First Workshop on Social Media Analytics, pp. 71–79 (2010)
5. Grier, C., Thomas, K., Paxson, V., Zhang, M.: @spam: the underground on 140 characters or less. In: CCS 2010 Conference Proceedings, pp. 27–37 (2010)
6. McCord, M., Chuah, M.: Spam detection on twitter using traditional classifiers. In: Calero, J.M.A., Yang, L.T., Mármol, F.G., García Villalba, L.J., Li, A.X., Wang, Y. (eds.) ATC 2011. LNCS, vol. 6906, pp. 175–186. Springer, Heidelberg (2011)
7. Egele, M., Stringhini, G., Kruegel, C., Vigna, G.: Compa: Detecting compromised accounts on social networks. In: Symposium on Network and Distributed System Security, NDSS (2013)
8. Gao, H., Hu, J., Wilson, C., Li, Z., Chen, Y., Zhao, B.Y.: Detecting and characterizing social spam campaigns. In: IMC 2010 Conference Proceedings, pp. 35–47 (2010)
9. Stringhini, G., Egele, M., Kruegel, C., Vigna, G.: Poultry markets: On the underground economy of twitter followers. In: WOSN 2012 Conference Proceedings, pp. 1–6 (2012)
10. Ghosh, S., Viswanath, B., Kooti, F., Sharma, N.K., Korlam, G., Benevenuto, F., Ganguly, N., Gummadi, K.P.: Understanding and combating link farming in the twitter social network. In: WWW 2012 Conference Proceedings, pp. 61–70 (2012)
11. Thomas, K., Grier, C., Song, D., Paxson, V.: Suspended accounts in retrospect: an analysis of twitter spam. In: IMC 2011 Conference Proceedings, pp. 243–258 (2011)
12. Yang, C., Harkreader, R., Zhang, J., Shin, S., Gu, G.: Analyzing spammers' social networks for fun and profit: a case study of cyber criminal ecosystem on twitter. In: WWW 2012 Conference Proceedings, pp. 71–80 (2012)
13. Benevenuto, F., Magno, G., Rodrigues, T., Almeida, V.: Detecting spammers on twitter. In: Collaboration, Electronic Messaging, Anti-abuse and Spam Conference (CEAS), vol. 6 (2010)
14. Lee, K., Caverlee, J., Webb, S.: Uncovering social spammers: social honeypots+ machine learning. In: SIGIR 2010 Conference Proceedings, pp. 435–442 (2010)
15. Zhu, Y., Wang, X., Zhong, E., Liu, N.N., Li, H., Yang, Q.: Discovering spammers in social networks. In: AAAI 2012 Conference Proceedings, pp. 171–177 (2012)
16. Cha, M., Haddadi, H., Benevenuto, F., Gummadi, P.K.: Measuring user influence in twitter: The million follower fallacy. ICWSM 10, 10–17 (2010)
17. Rosen-Zvi, M., Griffiths, T., Steyvers, M., Smyth, P.: The author-topic model for authors and documents. In: UAI 2004 Conference Proceedings, pp. 487–494 (2004)
18. Chang, J., Gerrish, S., Wang, C., Boyd-graber, J.L., Blei, D.M.: Reading tea leaves: How humans interpret topic models. In: Advances in Neural Information Processing Systems, pp. 288–296 (2009)
19. Kwak, H., Chun, H., Moon, S.: Fragile online relationship: a first look at unfollow dynamics in twitter. In: CHI 2011 Conference Proceedings, pp. 1091–1100 (2011)
20. Chang, C.C., Lin, C.J.: Libsvm: a library for support vector machines. ACM Transactions on Intelligent Systems and Technology (TIST) 2, 1–27 (2011)
21. Quinlan, J.R.: Bagging, boosting, and c4. 5. In: AAAI/IAAI, vol. 1, pp. 725–730 (1996)

Constrained-hLDA for Topic Discovery
in Chinese Microblogs

Wei Wang, Hua Xu, Weiwei Yang, and Xiaoqiu Huang

State Key Laboratory of Intelligent Technology and Systems,
Tsinghua National Laboratory for Information Science and Technology,
Department of Computer Science and Technology, Tsinghua University,
Beijing 100084, China
{ww880412,alexalexhxqhxq}@gmail.com,
xuhua@tsinghua.edu.cn, ywwbill@163.com

Abstract. Since microblog service became information provider on web scale, research on microblog has begun to focus more on its content mining. Most research on microblog context is often based on topic models, such as: Latent Dirichlet Allocation(LDA) and its variations. However,there are some challenges in previous research. On one hand, the number of topics is fixed as a priori, but in real world, it is input by the users. On the other hand, it ignores the hierarchical information of topics and cannot grow structurally as more data are observed. In this paper, we propose a semi-supervised hierarchical topic model, which aims to explore more reasonable topics in the data space by incorporating some constraints into the modeling process that are extracted automatically. The new method is denoted as constrained hierarchical Latent Dirichlet Allocation (constrained-hLDA). We conduct experiments on Sina microblog, and evaluate the performance in terms of clustering and empirical likelihood. The experimental results show that constrained-hLDA has a significant improvement on the interpretability, and its predictive ability is also better than that of hLDA.

Keywords: Hierarchical Topic Model, Constrained-hLDA, Topic Discovery.

1 Introduction

In the information explosion era, social network not only contains relationships, but also much unstructured information such as context. Furthermore, how to effectively dig out latent topics and internal semantic structures from social network is an important research issue. Early work on microblogs mainly focused on user relationship and community structure. [1] studied the topological and geographical properties of Twitter. Others work such as [2] studied user behaviors and geographic growth patterns of Twitter. Only little research on content analysis of microblog was proposed recently. [3] was mainly based on traditional text mining algorithms. [4] proposed MB-LDA by overall considering contactor relevance relation and document relevance relation of microblogs. In this paper, we propose a novel probabilistic generative model based on hLDA, called constrained-hLDA, which focuses on both text content and topic hierarchy.

V.S. Tseng et al. (Eds.): PAKDD 2014, Part II, LNAI 8444, pp. 608–619, 2014.

Previous work on microblog text was mainly based on LDA. To our best knowledge, there was little research on the topic hierarchy on microblog text. However, hierarchical topic modeling is able to obtain the relations between topics. [5] proposed an unsupervised hierarchical topic model, called hierarchical Latent Dirichlet Allocation (hLDA), to detect automatically new topics in the data space after fixing the level. Based on the stick-breaking process, [6] proposed the fully nonparametric hLDA without fixing the level. After that, some modifications of hLDA were proposed [7–9]. Given a parameter L indicating the depth of the hierarchy, hLDA makes use of nested Chinese Restaurant Process(nCRP) to automatically find useful sets of topics and learn to organize the topics according to a hierarchy in which more abstract topics are near the root of the hierarchy and more concrete topics are near the leaves. However, the traditional hLDA is an unsupervised learning which does not incorporate any prior knowledge. In this paper, we attempt to extract some prior knowledge and incorporate them to the sampling process.

The rest of the paper is organized as follows. Section 2 introduces the previous work related to this paper. Section 3 describes the hLDA briefly. Section 4 introduces the novel model constrained-hLDA. The experiment is introduced in Section 5, which is followed by the conclusion in Section 6.

2   Related Work

There have been many variations of probabilistic topic models, which was first introduced by [10]. The probabilistic topic model is based on the idea that documents are generated by mixtures of topics which is a multinomial distribution over words. One limitation of Hofmann's model is that it is not clear how the mixing proportions for topics in a document are generated. To overcome this limitation, [11] propose Latent Dirichlet Allocation(LDA). In LDA, the topic proportion of every document is a K-dimensional hidden variable randomly drawn from the same Dirichlet distribution, where K is the number of topics. Thus, generative semantics of LDA are complete, and LDA is regarded as the most popular approach for building topic models in recent years[12–16].

LDA is a useful algorithm for topic modeling, but it fails to draw the relationship between one topic and another and fails to indicate the level of abstract for a topic. To address this problem, many models have been proposed to build the relations, such as hierarchical LDA(hLDA) [5, 6], Hierarchical Dirichlet processes(HDP) [17], Pachinko Allocation Model(PAM) [18] and Hierarchical PAM(HPAM) [19] etc. These models extend the "flat" topic models into hierarchical versions for extracting hierarchies of topics from text collections. [6] proposed the most up-to-date hLDA model, which is a fully nonparametric model. It simultaneously learns the structure of a topic hierarchy and the topics that are contained within that hierarchy. Furthermore, it can also learn the most appropriate levels and hyper-parameters although it is time-consuming. In recent years, some modifications of hLDA has also been proposed. [7] proposed a supervised hierarchical topic model, called hierarchical Labeled Latent Dirichlet Allocation(hLLDA), which uses hierarchical labels to automatically build corresponding topic for each label. [8] propose an unsupervised hierarchical topic model, called

Semi-Supervised Hierarchical Latent Dirichlet Allocation (SSHLDA), which can not only make use of the information from the hierarchy of observed labels, but also can explore new latent topics in the data space. Although our work has some slight resemblance with their work, there still exist several important differences:

1. Our constrained-hLDA mainly focuses on the text of microblogs or reviews without observed labels.
2. The prior knowledge is extracted automatically from the corpus instead of first-hand observation.
3. The constraints are alterable by different parameters.

3 Preliminaries

The nested Chinese restaurant process (nCRP) is a distribution over hierarchical partitions[5, 6]. It generalizes the Chinese restaurant process (CRP), which is a single parameter distribution over partitions of integers. It has been used to represent the uncertainty over the number of components in a mixture model. The generative process is as follow:

1. There are N customers entering the restaurant in sequence, which is labeled with the integers $\{1, ..., N\}$.
2. First customer sits at the first table.
3. The nth customer sit at:
   (a) Table i with probability $\frac{n_i}{\gamma+n-1}$, where n_i is the number of customers currently sitting at table i, which has been occupied.
   (b) A new table with probability $\frac{\gamma}{\gamma+n-1}$.
4. After N customers have sat down, their seating plan describes a partition of N items.

In the nested CRP, suppose there are an infinite number of infinite-table Chinese restaurants in a city. One restaurant is identified as the root restaurant and its every table has a card with the name that refers to another restaurant. This structure repeats infinitely many times, thus, the restaurants in the city are organized into an infinitely branched, infinitely-deep tree. When a tourist arrives at the city, he selects a table, which is associated with a restaurant at next level, using the CRP distribution at each level. After M tourists have visited in this city, the path collection, which they selected, describes a random subtree of the infinite tree.

Based on identifying documents with the paths generated by the nCRP, the hierarchical topic model, which consists of an infinite tree, is defined. Each node in the tree is associated with a topic, which is a probability distribution across words. Each document is assumed to be generated by a mixture of topics on a path from the root to a leaf. For each token in the document, one picks a topic randomly according to the distribution, and draws a word from the multinomial distribution of that topic. To infer the topic hierarchy, the per-document paths c_d and the per-word level allocation to topics in those paths $z_{d,n}$ must be sampled. Then we will introduce the process briefly.

For the path sampling, the path associated with each document conditioned on all other paths and the observed words need to be sampled. Assume the depth is finite and let T denotes it, the posterior distribution of path $\mathbf{c}_d$ is as denote:

$$p(\mathbf{c}_d|\mathbf{w}, \mathbf{c}_{-d}, \mathbf{z}, \eta, \gamma) \propto p(\mathbf{c}_d|\mathbf{c}_{-d}, \gamma)p(\mathbf{w}_d|\mathbf{c}, \mathbf{w}_{-d}, \mathbf{z}, \eta) \tag{1}$$

In Equation 1, two factors influence the probability that a document belongs to a path. The first factor is the prior on paths implied by the nested CRP. The second factor is the probability of observing the words in the document given a particular choice of path with equation organized as follows:

$$p(\mathbf{w}_d|\mathbf{c}, \mathbf{w}_{-d}, \mathbf{z}, \eta) = \prod_{t=1}^{T} \frac{\Gamma(n_{c_d,t,-d}^{(\cdot)} + V\eta)}{\prod_w \Gamma(n_{c_d,t,-d}^{(w)} + \eta))} \frac{\prod_w \Gamma(n_{c_d,t,-d}^{(w)} + n_{c_d,t,d}^{(w)} + \eta))}{\Gamma(n_{c_d,t,-d}^{(\cdot)} + n_{c_d,t,d}^{(\cdot)} + V\eta)} \tag{2}$$

where $n_{c_d,t,-d}^{(w)}$ is the number of word w that have been allocated to the topic indexed by $c_{d,t}$, not including those in the current document, V denotes the total vocabulary size, and $\Gamma(\cdot)$ is the standard gamma function. When $\mathbf{c}$ contains a previously unvisited restaurant, $n_{c_d,t,-d}^{(w)}$ is zero.

After selecting the current path assignments, the level allocation variable $z_{d,n}$ for word n in document d conditioned on the current values of all other variables need to be sampled as:

$$p(z_{d,n}|\mathbf{z}_{-(d,n)}, \mathbf{c}, \mathbf{w}, m, \pi, \eta) \propto p(w_{d,n}|\mathbf{z}, \mathbf{c}, \mathbf{w}_{-(d,n)}, \eta)p(z_{d,n}|\mathbf{z}_{d,-n}, m, \pi) \tag{3}$$

where $\mathbf{z}_{-(d,n)}$ and $\mathbf{w}_{-(d,n)}$ are the vectors of level allocations and observed words leaving out $z_{d,n}$ and $w_{d,n}$, $\mathbf{z}_{d,-n}$ denotes the level allocations in document d , leaving out $z_{d,n}$.

4   Constrained-hLDA

In this section, we will introduce a constrained hierarchical topic model, i.e., the constrained herarchical Latent Dirichlet Allocation(constrained-hLDA). As we have known, similar to LDA, the original hLDA is a purely unsupervised model without considering any pre-existing knowledge. However, in semi-supervised clustering framework, the prior knowledge can help clustering algorithm produce more meaningful clusters. In our algorithm, the extracted prior knowledge can help to pre-establish a part of the infinite tree structure. In this section, we will give an introduction to the constraint extraction and the proposed constrained-hLDA which can use pre-existing knowledge expressed as constraints.

4.1   Path Constraints Extraction

To construct constrained hierarchical topic model, we adopt hLDA and incorporate the constraints from the pre-existing knowledge. Compared with hLDA, constrained-hLDA has one more input for improving path sampling. The input is a set of constrained indicators, which is in the form of $\{\{w_{1,1}, w_{1,2}, \ldots\}, \ldots, \{w_{N,1}, w_{N,2}, \ldots\}\}$. Each subset

$\{w_{i,1}, w_{i,2}, \ldots\}$, which corresponds to a node in constrained-hLDA, consists of several high correlation words. In our work, these words, which can indicate the correlation of a path and a document, are called **constrained indicators**. These corresponding nodes, which are pre-allocated several constrained indicators, are called **constrained nodes**.

The intuition of above idea is very simple and easy to follow. In this paper, we just attempt to solve it based on a correlation approach, more novel and efficient method will be further explored in the future. Algorithm 1 summarizes the main steps of constraints extraction. First, the FP-tree algorithm is adopted to extract the one-dimension frequent items according to the minimum support and maximum support(Line 1). The maximum support is used to filter some common words in order to make sure that the occurrences of each candidate are close, therefore, there will not be hierarchical relationship of these frequent items. Next, for each fis_i, it is added to an empty collection CS_i first, and then the correlation of fis_i with other items is computed. If the correlation of fis_i and fis_j is greater than the given threshold, it is assumed that fis_i and fis_j should constitute a must-link and fis_j is appended to CS_i (Line 2 - Line 9). In this work, the correlation is calculated by overlap as follows:

$$overlap(A, B) = \frac{P_{A\&B}}{min(P_A, P_B)} \tag{4}$$

where $P_{A\&B}$ is the co-occurrence of word A and word B, P_A is the occurrence of word A, and P_B is the occurrence of word B. The range of equation 4 is between $[0, 1.0]$, so the threshold can be easily given for different corpora. In the end, we delete the same set only retaining one from CS (Line 10 - Line 14). Based on Algorithm 1, the prior set CS, each of which contains several high correlation indicators, can be acquired. In this paper, the threshold of overlap is set as 0.4, the maximum support is set as five times as minimum support, all these parameters are estimated number. Additionally, we attempt to utilize different minimum supports to obtain different set so that different experiment results can be made for sure.

Algorithm 1. Constraints extraction

---

1. Frequent Item Set $FIS \leftarrow$ FP-tree (D, min_sup, max_sup)
2. **for** each fis_i in FIS **do**
3. $CS_i \leftarrow fis_i$
4.   **for** each $fis_{j|j!=i}$ in FIS **do**
5.   **if** $correlation(fis_i, fis_j) > threshold$ **then**
6. $CS_i \leftarrow fis_j$
7.   **end if**
8.   **end for**
9. **end for**
10. **for** each CS_i in Constraint Set CS **do**
11.   **if** there exists $CS_j == CS_i$ **then**
12.   delete CS_i from CS
13.   **end if**
14. **end for**

4.2 Path Constraints Incorporation

To integrate constrained indicators into hLDA, we extend the nCRP to a more realistic situation. Suppose the root restaurant has infinite tables, some tables have a menu containing some special dishes. Suppose N tourists arrive at the city, some of them have a list of special dishes that they want to taste. When a tourist enters into the root restaurant, if he has a list, he will select a table whose menu contains the special dishes of his list. Otherwise, according to his willingness to taste the special dishes, he will use CRP equation to select a table among those tables without menus. To keep it simple in this paper, we assume only the root restaurant has menus.

In constrained-hLDA model, each constrained set CS_i corresponds to a menu and each constrained indicator corresponds to a special dish. Then the documents in a corpus are assumed drawn from the following generative process:

1. For each table $k \in T$ in the infinite tree
   (a) Draw a topic $\beta_k \sim Dir(\eta)$
2. For each document, $d \in \{1, 2, \ldots, D\}$
   (a) Let c_1 be the root node.
   (b) For level $l = 2$:
      i. If d contains the constrained indicators $\{i_{d,1}, i_{d,2}, \ldots\}$, select a table c_2 with probability $\frac{n_{i_{d,i}} + \gamma}{\sum(n_{i_{d,i}} + \gamma)}$, where $n_{i_{d,i}}$ is the number of table which contains $i_{d,i}$.
      ii. Otherwise, draw a table c_2 among $C_{nm,2}$ from restaurant using CRP, where $C_{nm,2}$ is the set of tables which have no menus on root restaurant.
   (c) For each level $l \in \{3, \ldots, L\}$
      i. Draw a table c_l from restaurant using CRP.
   (d) Draw a distribution over levels in the tree, $\theta_d | \{m, \pi\} \sim GEM(m, \pi)$
   (e) For each word $n \in \{1, 2, \ldots, N\}$
      i. Choose level $z | \theta_d \sim Discrete(\theta_d)$.
      ii. Choose word $w | \{z, \mathbf{c}, \beta\} \sim Discrete(\beta_{c_z})$, which is parameterized by the topic associated with restaurant c_z.

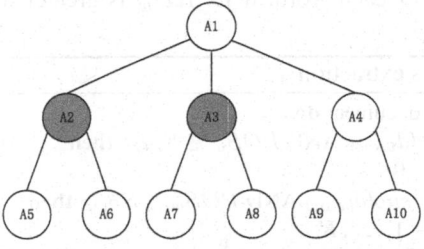

Fig. 1. One illustration of constrained-hLDA. The tree has 3 levels. The shaded nodes are constrained topics, which is pre-defined. The circled nodes are latent topics. After learning, each node in this tree is a topic, which is a corresponding probability distribution over words.

As the example shown in Figure 1, we assume that the height of the desired tree is $L = 3$, and the constrained-topics extracted are $\{A2, A3\}$. The constrained topics amount to the tables containing menu, each of which is pre-defined as the constrained indicators coming from a CS_i, and the constrained indicators amount to special dishes. In our work, because microblogs are mainly short texts, the maximum level is truncated to 3. Furthermore, it is notable that the constraints can be extended to the deeper level. For example, the constrained set can be extracted again from the documents which pass by the node $A2$, and then the constrained indicators set corresponding to $A2$ can be drawn from these documents.

In constrained-hLDA, the idea of incorporating prior knowledge derives from [20], and the most important process is incorporating the constraints to the path sampling process according to the probabilities calculated using Equation 5:

$$p(\mathbf{c}_d|\mathbf{w}, \mathbf{c}_{-d}, \mathbf{z}, \eta, \gamma) \propto (\eta'\delta(\mathbf{w}_d, \mathbf{c}_d) + 1 - \eta')p(\mathbf{c}_d|\mathbf{c}_d, \gamma)p(\mathbf{w}_d|\mathbf{c}, \mathbf{w}_{-d}, \mathbf{z}, \eta) \quad (5)$$

where $\delta(\mathbf{w}_d, \mathbf{c}_d)$ is an indicator function, which indicates whether the nodes from $\mathbf{c}_d$ contain the same constrained indicator with that of $\mathbf{w}_d$: If $\mathbf{w}_d$ contains such node, $\delta(\mathbf{w}_d, \mathbf{c}_d) = 1$, otherwise, $\delta(\mathbf{w}_d, \mathbf{c}_d) = 0$. The hard constraint indicator can be relaxed by η', Let $0 \leq \eta' \leq 1$ be the strength of our constraint, where $\eta' = 1$ recovers a hard constraint, $\eta' = 0$ recovers unconstrained sampling and $0 < \eta' < 1$ recovers a soft constraint sampling.

4.3   Level Constraints Extraction and Incorporation

After revising the path sampling process and selecting a particular path, some prior knowledge can also integrate into level sampling process. As we have known, hLDA can discover the function words in root topic, furthermore, these words have no effect on the document interpretability and often appear in many documents. Therefore, we hope to improve level sampling process by pre-discriminating some function words and non-function words. In our work, the function words are discriminated according to the Part-Of-Speech(POS) and the term frequency in each document. Algorithm 2 describes our purpose, where RD_w denotes the ratio of the documents containing the word w in the current corpus. For each word, if its RD_w is greater than the given threshold

---

Algorithm 2 Constraints extraction

---

1. **for** each w_i in current document **do**
2.   **if** $RD_w > threshold_{upper}$ AND $POS_w \notin S_{POS}$ **then**
3. $samleLevel_w \leftarrow 0$
4.   **else if** $RD_w < threshold_{below}$ AND $POS_w \in S_{POS}$ **then**
5. $samleLevel_w \leftarrow 1, \ldots, K$
6.   **else**
7. $samleLevel_w \leftarrow 0, \ldots, K$
8.   **end if**
9.   sample the level according $samleLevel_w$
10. **end for**

$threshold_{upper}$ and it does not belong to the pre-defined POS set S_{POS}, it would be likely to be a function word that is allocated to root node directly(Line 2 - Line 3). If its RD_w is less than the given threshold $threshold_{below}$ and it belongs to the pre-defined POS set S_{POS} at the same time, it would be likely to be a non-function word without being allocated to root node(Line 4 - Line 5). Finally, we sample the level according to these prior knowledge (Line 9). In this paper, $threshold_{upper}$ and $threshold_{below}$ are set to 0.02 and 0.005, and the pre-defined POS set S_{POS} is set as noun, adjective and verb.

5 Experiment

5.1 Data Sets

Due to the lack of standard data set for this kind of research yet, we collected the experiment data from sina microblog[1] by ourselves. It is generally known that Ya'an Earthquake[2] on 20th, April, 2013 was a catastrophe shocking everyone, which is exactly an ideal hot issue for research. We crawled 19811 microblog users all coming from Ya'an, and also crawled their posted microblogs from 8am 20th April 2013 to 8am 25th April 2013. There are 58476 original microblogs released by these users, each of which contains several sentences. As time passed by, people's concern level on this issue would decline gradually, therefore, we use the data on a daily level for further analysis. Table 1 depicts the data sets for evaluation. The designed experiments and sampling results can also be referred in [6]. For hLDA and constrained-hLDA, there is a restriction that documents can only follow a single path in the tree. In order to make each sentence of a document can follow different paths, we split texts into sentences, such a change can get a remarkable improvement for hLDA and constrained-hLDA in the corpus of microblogs. In our experiment, hLDA algorithm is completed with Java codes by ourselves according to [6]. In our constrained-hLDA, the stick-breaking procedures are truncated at three levels to facilitate visualization of results. The topic Dirichlet hyper-parameters are fixed at $\eta = \{1.0, 1.0, 1.0\}$, The nested CRP parameter γ is fixed at 0.5, the GEM parameters are fixed at $\pi = 100$ and $m = 0.25$.

Table 1. Experiment data

Time	Token	Number of microblogs	Number of sentences
20-21	T1	19709	50631
21-22	T2	11678	32311
22-23	T3	9779	28215
23-24	T4	9308	27213
24-25	T5	8002	23159

---

[1] http://weibo.com/

[2] http://en.wikipedia.org/wiki/2013_Lushan_earthquake

5.2 Hierarchy Topic Discovery

Figure 2 depicts the hierarchical structure of cluster results. It is natural to conclude that the constrained-hLDA can well discover the underlying hierarchical structure of the content of micorblogs, and each topic and its child node mainly relate to pre-allocated the constrained indicator, which is the underlined word. For example, there are three sub-topic of *Ya' an*, the first sub-topic relates to *blessing*, the second talks about the *situations* of *Ya' an*, the third talks about *relief* of *Ya' an*. Furthermore, as we can find, the latent topic of second level is a meaningless topic, which is hard to summarize the interpretability of these topics. This phenomenon illustrates that the irrelevant information in microblog context that can be filtered well by our algorithm.

5.3 Comparison with hLDA

In this section, we compare the experimental results with hLDA, and the per-document distribution over levels is truncated at three levels. In order to evaluate our model, we use predictive held-out likelihood as a measure of performance to compare the two approaches quantitatively. The procedure is to divide the corpus into D_1 observed documents and D_2 held-out documents, and approximate the conditional probability of the held-out set given the training set:

$$p(\mathbf{w}_1^{held-out}, \ldots, \mathbf{w}_{D_2}^{held-out} | \mathbf{w}_1^{obs}, \ldots, \mathbf{w}_{D_1}^{obs}) \tag{6}$$

For this evaluation method, more details can be found in [6].

Figure 3 depicts the performance of constrained-HLDA on several data sets by different minimum support. Table 2 depicts the best performance of different constraints on several data sets. According to these experimental results, we can conclude that: (1) Both path sampling constraints and level sampling constraints can improve hLDA. (2) The smaller minimum support can obtain more constrained indicators so that it can achieve better log likelihood. (3) The likelihood of constrained-hLDA is better than the likelihood of hLDA, but for different corpus, the degree of improvement is different. When the topic of corpus is more concentrated, the improvement seems to be better.

Table 2. The Best Results of Different Prior Constraints(800 samplers)

Data set token	hLDA	hLDA + level constrains	hLDA + path constraints	constrained-hLDA
T1	-233776.355	-226566.45	-225814.798	-218583.987
T2	-169646.036	-164147.048	-162871.358	-158950.632
T3	-137735.633	-133976.967	-134931.846	-130895.533
T4	-130254.675	-127269.889	-128493.374	-124552.455
T5	-106223.172	-104269.033	-104670.728	-100796.114

In order to avoid interference from the values of hyperparameters, as with [6]'work, we also interleave Metropolis-Hastings (MH) steps between iterations of the Gibbs sampler to obtain new values of m, π, γ and η. Table 3 present the results by sampling the hyperparameters in the same case, from which we can see that constrained-hLDA still performs better than hLDA.

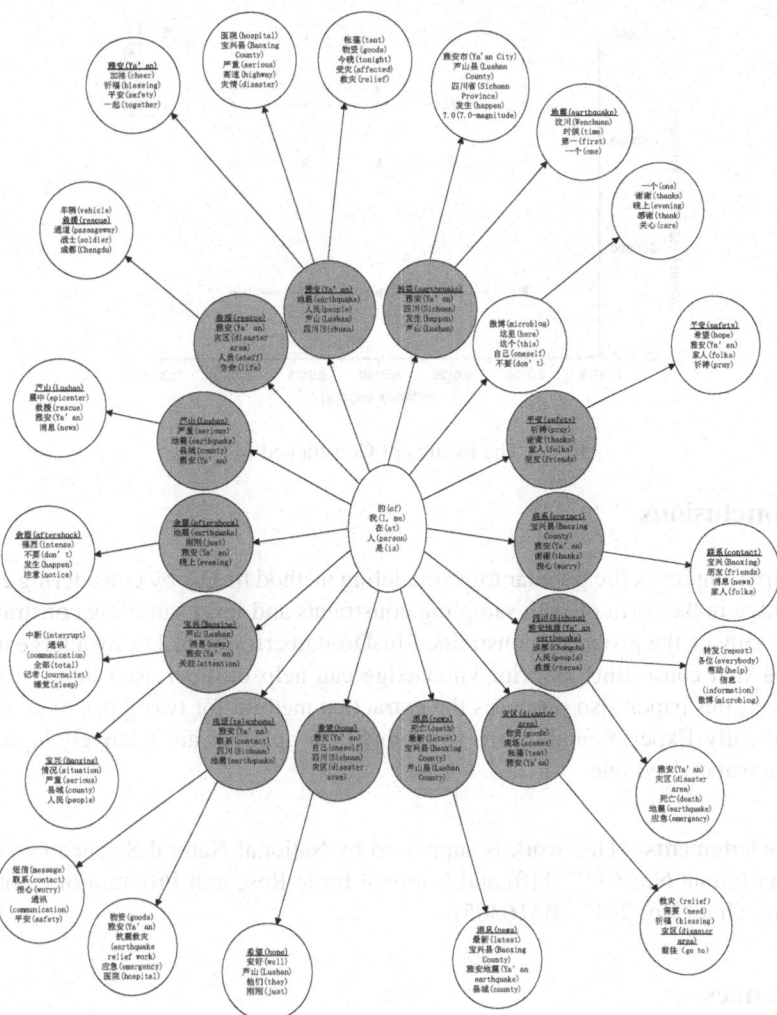

Fig. 2. A portion of the hierarchy learned from T1 Data. The shaded nodes are constrained topics, the bold and underlining words are the constrained indicators extracted by Algorithm 1.

Table 3. The Results by sampling the hyperparameters (800 samplers)

Data set token	hLDA	constrained-hLDA
T1	-210794.652	-195034.999
T2	-151989.286	-140812.547
T3	-123816.852	-115215.844
T4	-117760.271	-110516.351
T5	-94291.395	-90791.294

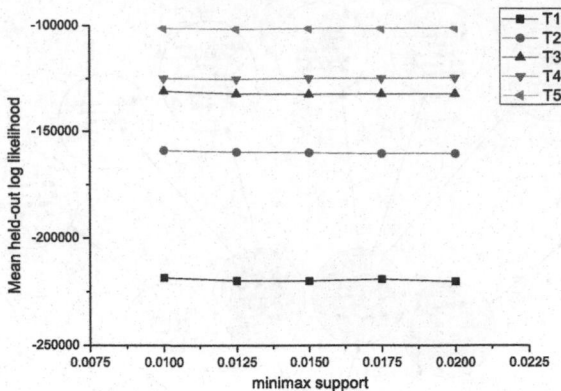

Fig. 3. The Results of Constrained-hLDA

6 Conclusions

This paper improves the popular topic modeling method hLDA by considering existing knowledge in the form of path sampling constraints and level sampling constraints. In the experiment, the proposed constrained-hLDA outperforms hLDA by a large margin, showing that constraints as prior knowledge can help unsupervised topic modeling. Moreover, this paper also proposes the extraction method for two types of constraints automatically. Experimental results show that their qualities are relatively higher than that of unsupervised one.

Acknowledgments. This work is supported by National Natural Science Foundation of China (Grant No: 61175110) and National Basic Research Program of China (973 Program, Grant No: 2012CB316305).

References

1. Java, A., Song, X., Finin, T., Tseng, B.: Why we twitter: understanding microblogging usage and communities. In: Proceedings of the 9th WebKDD and 1st SNA-KDD 2007 Workshop on Web Mining and Social Network Analysis, pp. 56–65. ACM (2007)
2. Krishnamurthy, B., Gill, P., Arlitt, M.: A few chirps about twitter. In: Proceedings of the First Workshop on Online Social Networks, pp. 19–24. ACM (2008)
3. Ramage, D., Dumais, S., Liebling, D.: Characterizing microblogs with topic models. In: International AAAI Conference on Weblogs and Social Media, vol. 5, pp. 130–137 (2010)
4. Zhang, C., Sun, J.: Large scale microblog mining using distributed mb-lda. In: Proceedings of the 21st International Conference Companion on World Wide Web, pp. 1035–1042. ACM (2012)
5. Blei, D.M., Griffiths, T.L., Jordan, M.I., Tenenbaum, J.B.: Hierarchical topic models and the nested chinese restaurant process. In: NIPS (2003)
6. Blei, D.M., Griffiths, T.L., Jordan, M.I.: The nested chinese restaurant process and bayesian nonparametric inference of topic hierarchies. Journal of the ACM 57(2), 7 (2010)

7. Petinot, Y., McKeown, K., Thadani, K.: A hierarchical model of web summaries. In: Proceedings of the 49th Annual Meeting of the Association for Computational Linguistics: Human Language Technologies, pp. 670–675 (2011)

8. Mao, X.L., Ming, Z.Y., Chua, T.S., Li, S., Yan, H., Li, X.: Sshlda: a semi-supervised hierarchical topic model. In: Proceedings of the 2012 Joint Conference on Empirical Methods in Natural Language Processing and Computational Natural Language Learning, pp. 800–809. Association for Computational Linguistics (2012)

9. Mao, X.L., He, J., Yan, H., Li, X.: Hierarchical topic integration through semi-supervised hierarchical topic modeling. In: Proceedings of the 21st ACM International Conference on Information and Knowledge Management, pp. 1612–1616. ACM (2012)

10. Hofmann, T.: Probabilistic latent semantic analysis. In: Proceedings of the Fifteenth Conference on Uncertainty in Artificial Intelligence, pp. 289–296. Morgan Kaufmann Publishers Inc. (1999)

11. Blei, D.M., Ng, A.Y., Jordan, M.I.: Latent dirichlet allocation. The Journal of Machine Learning Research 3, 993–1022 (2003)

12. Chemudugunta, C., Steyvers, P.S.M.: Modeling general and specific aspects of documents with a probabilistic topic model. In: Advances in Neural Information Processing Systems 19: Proceedings of the 2006 Conference, vol. 19, p. 241. The MIT Press (2007)

13. Griffiths, T.L., Steyvers, M.: Finding scientific topics. Proceedings of the National Academy of Sciences of the United States of America 101(suppl. 1), 5228–5235 (2004)

14. Boyd-Graber, J., Blei, D., Zhu, X.: A topic model for word sense disambiguation. In: Proceedings of the 2007 Joint Conference on Empirical Methods in Natural Language Processing and Computational Natural Language Learning (EMNLP-CoNLL), pp. 1024–1033 (2007)

15. Rosen-Zvi, M., Griffiths, T., Steyvers, M., Smyth, P.: The author-topic model for authors and documents. In: Proceedings of the 20th Conference on Uncertainty in Artificial Intelligence, pp. 487–494. AUAI Press (2004)

16. Mei, Q., Ling, X., Wondra, M., Su, H., Zhai, C.: Topic sentiment mixture: modeling facets and opinions in weblogs. In: Proceedings of the 16th International Conference on World Wide Web, pp. 171–180. ACM (2007)

17. Teh, Y.W., Jordan, M.I., Beal, M.J., Blei, D.M.: Hierarchical dirichlet processes. Journal of the American Statistical Association 101(476) (2006)

18. Li, W., McCallum, A.: Pachinko allocation: Dag-structured mixture models of topic correlations. In: Proceedings of the 23rd International Conference on Machine Learning, pp. 577–584. ACM (2006)

19. Mimno, D., Li, W., McCallum, A.: Mixtures of hierarchical topics with pachinko allocation. In: Proceedings of the 24th International Conference on Machine Learning, pp. 633–640. ACM (2007)

20. Andrzejewski, D., Zhu, X.: Latent dirichlet allocation with topic-in-set knowledge. In: Proceedings of the NAACL HLT 2009 Workshop on Semi-Supervised Learning for Natural Language Processing, pp. 43–48. Association for Computational Linguistics (2009)

Author Index